CALCULUS

FOR MANAGEMENT, LIFE, AND SOCIAL SCIENCES FOURTH EDITION

CALCULUS

FOR MANAGEMENT, LIFE, AND SOCIAL SCIENCES FOURTH EDITION

RAYMOND A. BARNETT
Merritt College

MICHAEL R. ZIEGLER
Marquette University

DELLEN PUBLISHING COMPANY
San Francisco, California

COLLIER MACMILLAN PUBLISHERS
London

Divisions of Macmillan, Inc.

On the cover: "Upright Vee-Angle Slabs," by Ronald Davis, 1986; vinyl-acrylic copolymer and dry pigments on canvas, 32 by 36⅛ inches. Ronald Davis is involved in expressing the illusion of a three-dimensional object rendered on a flat surface. His work can be seen in leading museums throughout the world, including the San Francisco Museum of Modern Art, the Los Angeles County Museum of Art, and the Whitney Museum of American Art. In Los Angeles, Davis is represented by Asher/Faure.

Printed in the United States of America

Permissions: Dellen Publishing Company
400 Pacific Avenue
San Francisco, California 94133

Orders: Dellen Publishing Company
c/o Macmillan Publishing Company
Front and Brown Streets
Riverside, New Jersey 08075

Collier Macmillan Canada, Inc.

Library of Congress Cataloging-in-Publication Data

Barnett, Raymond A.
Calculus for management, life, and social sciences.

Includes indexes.
1. Calculus. 2. Social sciences—Mathematics.
3. Biomathematics. I. Ziegler, Michael R. II. Title.
QA303.B283 1987 515 86-24211
ISBN 0-02-306211-8

Printing: 1 2 3 4 5 6 7 8 9 Year: 6 7 8 9 0

ISBN 0-02-306211-8

Contents

Chapter Dependencies

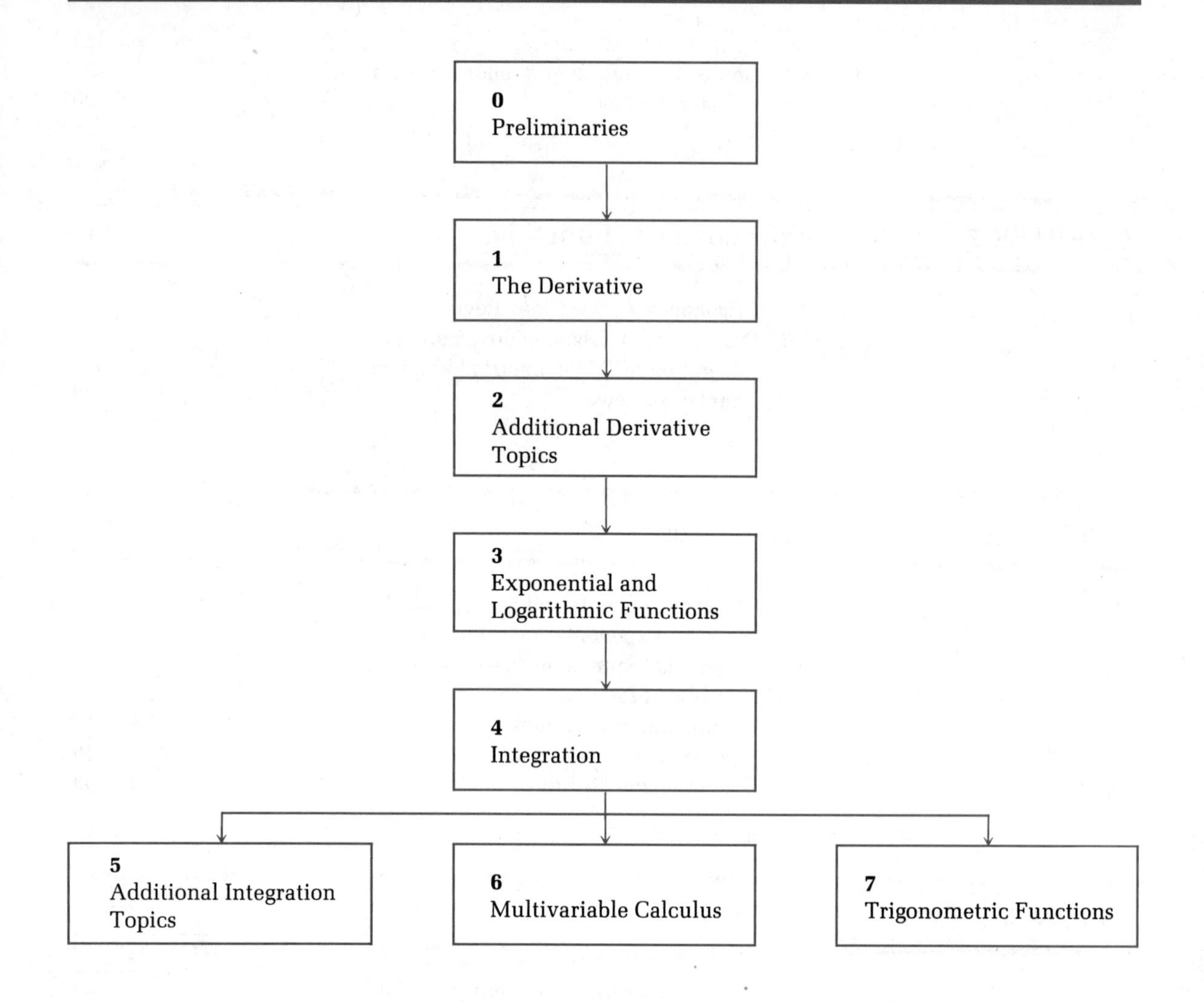

Preface

Many colleges and universities now offer mathematics courses that emphasize topics that are most useful to students in business and economics, life sciences, and social sciences. Because of this trend, the authors have surveyed instructors, course outlines, and college catalogs from a large number of colleges and universities, and on the basis of these surveys, selected the topics, applications, and emphasis found in this text.

The material in this book is suitable for a one-semester course on the calculus for functions of one variable, including the exponential and logarithmic functions, followed by an introduction to multivariable calculus. The choice and organization of topics make the book readily adaptable to a variety of courses. (See the diagram on page viii for chapter dependencies.)

The book is designed for students who have had $1\frac{1}{2}$–2 years of high school algebra or the equivalent. However, because much of this material is forgotten due to lack of use, Chapter 0 contains a review of the basic topics in intermediate algebra that are most pertinent to the course, and portions of Appendix A review some more fundamental concepts. Any of this material can be studied systematically at the beginning of the course or referred to as needed. In addition, certain key topics are reviewed immediately before their use (see Section 3-3), while others are discussed in Appendix A.

■ Major Changes from the Third Edition

The fourth edition of *Calculus for Management, Life, and Social Sciences* reflects the experiences and recommendations of a large number of the users of the first three editions. Much of the material has been reorganized to provide a more efficient presentation of the topics. In addition, certain topics of marginal importance have been deleted. The result is a shorter book that permits a broader coverage of the topics most relevant to its intended audience. Many examples and exercises have been changed to provide clear illustrations of mathematical concepts without undue algebraic complexity. Special attention has been paid to increasing the quantity and quality of applications throughout the book. It is impossible to use actual real-world models in many applied problems, but it is possible to provide simplified versions of such models that reflect the important features of the application and involve mathematical operations appropriate for the level of this book. Such simplified, yet realistic, applications have been included in every chapter of the book.

The intermediate algebra material has been reorganized. Chapter 0 now contains the most frequently used algebra topics, including quadratic equations and logarithmic and exponential functions, and accurately reflects the level of algebraic ability necessary for the material in the remainder of the book. Portions of Appendix A review more fundamental algebra concepts for students whose algebraic skills have become rusty due to lack of use.

The material on differentiation (Chapters 1–3) has been reorganized, the treatment of limits and asymptotes has been simplified, the chain rule has been moved to the chapter on exponential and logarithmic functions (Chapter 3), and the sections on implicit differentiation, related rates, higher-order derivatives, and elasticity of demand have been eliminated.

In Chapters 4 and 5, integration by substitution is now covered earlier (Section 4-2) and in a more fundamental manner, the section on integral tables has been deleted, and the Riemann sum approach to integration has been moved to Chapter 5. The discussion of consumers' and producers' surplus has been expanded and placed in a new section (Section 5-2), along with a discussion of continuous income streams.

A section on the method of least squares has been added to the chapter on multivariable calculus (Chapter 6).

▪ General Comments

Chapters 1–3 present the differential calculus for functions of one variable, including the exponential and logarithmic functions. Limits and continuity are presented in an intuitive fashion, utilizing numerical approximations and graphical concepts. The basic rules of differentiation for algebraic functions are covered in Chapter 1. The relationship between derivatives and graphs is discussed in Sections 2-1 and 2-2, and then applied to optimization problems in Section 2-3 and to curve sketching in Section 2-4. A short section on differentials is included to facilitate the use of differentials in integration by substitution (Chapter 4) and integration by parts (Chapter 5). The chain rule is covered in the chapter on exponential and logarithmic functions (Chapter 3), the first place where its use is really required.

Chapters 4 and 5 deal with integral calculus. Differential equations and exponential growth and decay are included as an application of antidifferentiation. In Chapter 4, the definite integral is intuitively introduced in terms of an area function and then, in Chapter 5, it is formally defined as the limit of a Riemann sum. Integration by parts, some additional applications of integration, and improper integrals are also covered in Chapter 5.

Chapter 6 introduces multivariable calculus, including partial derivatives, optimization, Lagrange multipliers, least squares, and double integrals. If desired, this chapter can be covered immediately after Chapter 4. (See the diagram on page viii.)

Finally, Chapter 7 reviews the basic properties of the trigonometric functions and then develops differentiation and integration formulas for these functions.

Important Features

Emphasis

Emphasis is on computational skills, ideas, and problem solving rather than mathematical theory. Most derivations and proofs are omitted except where their inclusion adds significant insight into a particular concept. General concepts and results are usually presented only after particular cases have been discussed.

Examples and Matched Problems

Over 230 completely worked examples are included. Each example is followed by a similar problem for the student to work while reading the material. This actively involves the student in the learning process. The answers to these matched problems are included at the end of each section for easy reference.

Exercise Sets

This book contains over 2,500 exercises. Each exercise set is designed so that an average or below-average student will experience success and a very capable student will be challenged. Exercise sets are mostly divided into A (routine, easy mechanics), B (more difficult mechanics), and C (difficult mechanics and some theory) levels.

Applications

Enough applications are included to convince even the most skeptical student that mathematics is really useful. The majority of the applications are included at the end of exercise sets and are generally divided into business and economics, life science, and social science groupings. An instructor with students from all three disciplines can let them choose applications from their own field of interest; if most students are from one of the three areas, then special emphasis can be placed there. Most of the applications are simplified versions of actual real-world problems taken from professional journals and professional books. No specialized experience is required to solve any of the applications in this book.

Student and Instructor Aids

Student Aids

Dashed **"think boxes"** are used to enclose steps that are usually performed mentally (see Section 0-2).

Examples and developments are often **annotated** to help students through critical stages (see Section 0-2).

A **second color** is used to indicate key steps (see Section 0-2).

Boldface type is used to introduce new terms and highlight important comments.

Answers to odd-numbered problems are included in the back of the book.

Chapter review sections include a review of all important terms and symbols and a comprehensive review exercise. Answers to all review exercises are included in the back of the book.

A **student's solutions manual** is available at a nominal cost through a book store. The manual includes detailed solutions to all odd-numbered problems and all review exercises.

A **computer applications supplement** is available at a nominal cost through a book store. The supplement contains examples, computer program listings, and exercises that demonstrate the use of a computer to solve a variety of problems in calculus. No previous computing experience is necessary to use this supplement.

Instructor Aids

A unique **computer-generated random test system** is available to instructors without cost. The system, utilizing an IBM PC computer and a number of commonly used dot matrix printers, will generate an almost unlimited number of chapter tests and final examinations, each different from the other, quickly and easily. At the same time, the system produces an answer key and a student worksheet with an answer column that exactly matches the answer column on the answer key. Graphing grids are included on the answer key and on the student worksheet for problems requiring graphs.

A **printed and bound test battery** is also available to instructors without cost. The battery contains several chapter tests for each chapter, answer keys, and student worksheets with answer columns that exactly match the answer columns on the answer keys. Graphing grids are included on the answer key and on the student worksheet for problems requiring graphs.

An **instructor's answer manual** containing answers to the even-numbered problems not included in the text is available to instructors without charge.

A **solutions manual** (see Student Aids) is available to instructors without charge from the publisher.

A **computer applications supplement** (see Student Aids) is available to instructors without charge from the publisher. The programs in this supplement are also available on diskettes for APPLE II and IBM PC computers. The publisher will supply one of these diskettes without charge to institutions using this book.

■ Related Books in the Series

This book is one in a series of three books by the same authors. The other two are:

Finite Mathematics for Management, Life, and Social Sciences, Fourth Edition: A companion text designed for a one-quarter or one-semester course in finite mathematics and employing the same style, emphasis, and features as *Calculus.*

College Mathematics for Management, Life, and Social Sciences, Fourth Edition: A combined version of *Finite Mathematics* and *Calculus* designed for a two-quarter or two-semester course in finite mathematics and calculus.

■ Acknowledgments

In addition to the authors, many others are involved in the successful publication of a book. We wish to thank personally:

Ronald Barnes, University of Houston–Downtown; Carl Bedell, Philadelphia College of Textiles and Science; Paul Boonstra, Calvin College; Miriam Connellan, Marquette University; Edward Connors, University of Massachusetts at Amherst; William Conway, University of Arizona; John Daly, Saint Louis University; Ryness Doherty, Metropolitan State College; Garry Etgen, University of Houston; James Flynn, Cleveland State University; Gerald Goff, Oklahoma State University; Roy Luke, Pierce College; Carolyn Meitler, Marquette University; Donald Minassian, Butler University; Robert Moreland, Texas Tech University; Frank Shirley, University of Texas; Martha Sklar, Los Angeles City College; Louis Talman, Metropolitan State College; Vance Underhill, East Texas State University; T. D. Worosz, Metropolitan State College; and Robert Zahn, San Jose State University.

We also wish to thank:

John Williams for a strong and effective cover design.

John Drooyan for the many sensitive and beautiful photographs seen throughout the book.

Phillip Bender and Stephen Merrill for carefully checking all examples and problems (a tedious but extremely important job).

Phyllis Niklas and Janet Bollow for another outstanding book design and for guiding the book smoothly through all production details.

Don Dellen, the publisher, who continues to provide all the support services and encouragement an author could hope for.

Producing this new edition with the help of all these extremely competent people has been a most satisfying experience.

R. A. Barnett
M. R. Ziegler

Preliminaries

0

CHAPTER 0 Contents

This chapter and Appendix A are provided for those of you whose prerequisite skills are a little rusty. Depending on the degree of your rust, you can either refer to selected sections briefly as needed, you can review certain sections in depth at the start of the course, or you can study them in depth at appropriate points of the course.

0-1 Sets

- Set Properties and Set Notation
- Set Operations
- Application

In this section we will review a few key ideas from set theory. Set concepts and notation not only help us talk about certain mathematical ideas with greater clarity and precision, but are indispensable to a clear understanding of probability.

Set Properties and Set Notation

We can think of a **set** as any collection of objects specified in such a way that we can tell whether any given object is or is not in the collection. Capital letters, such as A, B, and C, are often used to designate particular sets. Each object in a set is called a **member** or **element** of the set. Symbolically,

$a \in A$	means	"a is an element of set A"
$a \notin A$	means	"a is not an element of set A"

A set without any elements is called the **empty** or **null set.** For example, the set of all people over 10 feet tall is an empty set. Symbolically,

$\varnothing$ represents "the empty or null set"

A set is usually described either by listing all its elements between braces { } or by enclosing a rule within braces that determines the elements of the set. Thus, if $P(x)$ is a statement about x, then

$S = \{x|P(x)\}$ means "S is the set of all x such that $P(x)$ is true"

Recall that the vertical bar in the symbolic form is read "such that." The following example illustrates the rule and listing methods of representing sets.

Example 1

Rule	*Listing*
$\{x \mid x \text{ is a weekend day}\}$	$= \{\text{Saturday, Sunday}\}$
$\{x \mid x^2 = 4\}$	$= \{-2, 2\}$
$\{x \mid x \text{ is an odd positive counting number}\}$	$= \{1, 3, 5, \ldots\}$

The three dots . . . in the last set in Example 1 indicate that the pattern established by the first three entries continues indefinitely. The first two sets in Example 1 are **finite sets** (we intuitively know that the elements can be counted); the last set is an **infinite set** (we intuitively know that there is no end in counting the elements). When listing the elements in a set, we do not list an element more than once.

Problem 1 Let G be the set of all numbers such that $x^2 = 9$.*

(A) Denote G by the rule method.
(B) Denote G by the listing method.
(C) Indicate whether the following are true or false: $3 \in G$, $9 \notin G$.

If each element of a set *A* is also an element of set *B*, we say that *A* is a **subset** of *B*. For example, the set of all women students in a class is a subset of the whole class. Note that the definition allows a set to be a subset of itself. If set *A* and set *B* have exactly the same elements, then the two sets are said to be **equal.** Symbolically,

* Answers to matched problems are found near the end of each section just before the exercise set.

$A \subset B$	means	"A is a subset of B"
$A = B$	means	"A and B have exactly the same elements"
$A \not\subset B$	means	"A is not a subset of B"
$A \neq B$	means	"A and B do not have exactly the same elements"

It can be proved that $\varnothing$ **is a subset of every set.**

Example 2 If $A = \{-3, -1, 1, 3\}$, $B = \{3, -3, 1, -1\}$, and $C = \{-3, -2, -1, 0, 1, 2, 3\}$, then each of the following statements is true:

$A = B$	$A \subset C$	$A \subset B$
$C \neq A$	$C \not\subset A$	$B \subset A$
$\varnothing \subset A$	$\varnothing \subset C$	$\varnothing \notin A$

Problem 2 Given $A = \{0, 2, 4, 6\}$, $B = \{0, 1, 2, 3, 4, 5, 6\}$, and $C = \{2, 6, 0, 4\}$, indicate whether the following relationships are true (T) or false (F):

(A) $A \subset B$ (B) $A \subset C$ (C) $A = C$
(D) $C \subset B$ (E) $B \not\subset A$ (F) $\varnothing \subset B$

Example 3 List all the subsets of the set $\{a, b, c\}$.

Solution $\{a, b, c\}$, $\{a, b\}$, $\{a, c\}$, $\{b, c\}$, $\{a\}$, $\{b\}$, $\{c\}$, $\varnothing$

Problem 3 List all the subsets of the set $\{1, 2\}$.

Set Operations

The **union** of sets A and B, denoted by $A \cup B$, is the set of all elements formed by combining all the elements of A and all the elements of B into one set. Symbolically,

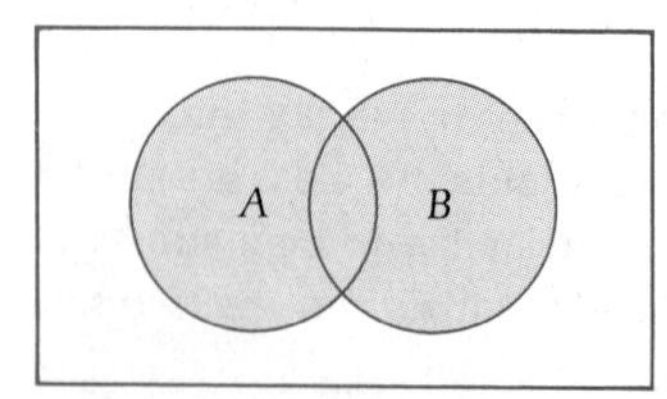

Figure 1 $A \cup B$ is the shaded region.

Union

$$A \cup B = \{x | x \in A \textbf{ or } x \in B\}$$

Here we use the word *or* in the way it is always used in mathematics; that is, x may be an element of set A or set B or both.

Venn diagrams are useful in visualizing set relationships. The union of two sets can be illustrated as shown in Figure 1. Note that

$A \subset A \cup B$ and $B \subset A \cup B$

The **intersection** of sets A and B, denoted by $A \cap B$, is the set of elements in set A that are also in set B. Symbolically,

Intersection

$$A \cap B = \{x | x \in A \quad \textbf{and} \quad x \in B\}$$

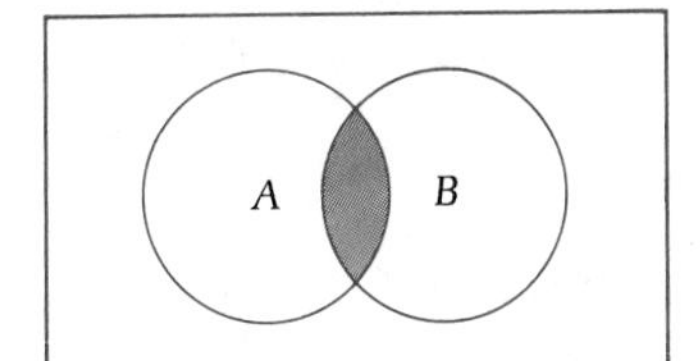

Figure 2 $A \cap B$ is the shaded region.

This relationship is easily visualized in the Venn diagram shown in Figure 2. Note that

$A \cap B \subset A$ and $A \cap B \subset B$

If $A \cap B = \varnothing$, then the sets A and B are said to be **disjoint;** this is illustrated in Figure 3.

The set of all elements under consideration is called the **universal set** U. Once the universal set is determined for a particular discussion, all other sets in that discussion must be subsets of U.

We now define one more operation on sets, called the *complement*. The **complement** of A (relative to U), denoted by A', is the set of elements in U that are not in A (see Fig. 4). Symbolically,

Complement

$$A' = \{x \in U | x \notin A\}$$

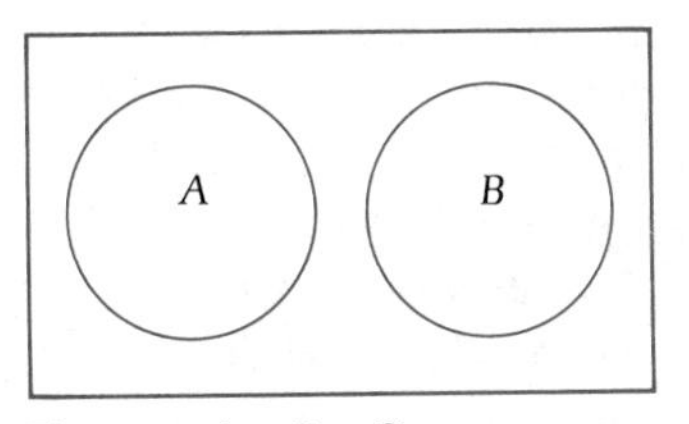

Figure 3 $A \cap B = \varnothing$

U

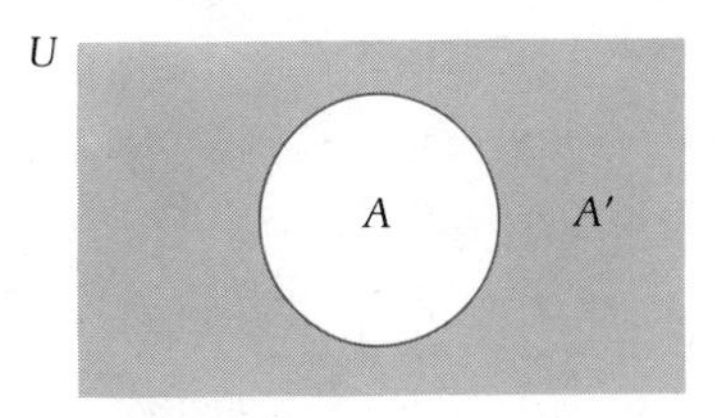

Figure 4 The complement of A is A'.

Example 4 If $A = \{3, 6, 9\}$, $B = \{3, 4, 5, 6, 7\}$, $C = \{4, 5, 7\}$, and $U = \{1, 2, 3, 4, 5, 6, 7, 8, 9\}$, then

$$A \cup B = \{3, 4, 5, 6, 7, 9\}$$
$$A \cap B = \{3, 6\}$$
$$A \cap C = \varnothing \qquad A \text{ and } C \text{ are disjoint}$$
$$B' = \{1, 2, 8, 9\}$$

Problem 4 If $R = \{1, 2, 3, 4\}$, $S = \{1, 3, 5, 7\}$, $T = \{2, 4\}$, and $U = \{1, 2, 3, 4, 5, 6, 7, 8, 9\}$, find:

(A) $R \cup S$ (B) $R \cap S$ (C) $S \cap T$ (D) S'

Application

Example 5 From a survey of 100 college students, a marketing research company found that 75 students owned stereos, 45 owned cars, and 35 owned cars and stereos.

(A) How many students owned either a car or a stereo?
(B) How many students did not own either a car or a stereo?

Solutions Venn diagrams are very useful for this type of problem. If we let

U = Set of students in sample (100)
S = Set of students who own stereos (75)
C = Set of students who own cars (45)
$S \cap C$ = Set of students who own cars and stereos (35)

then

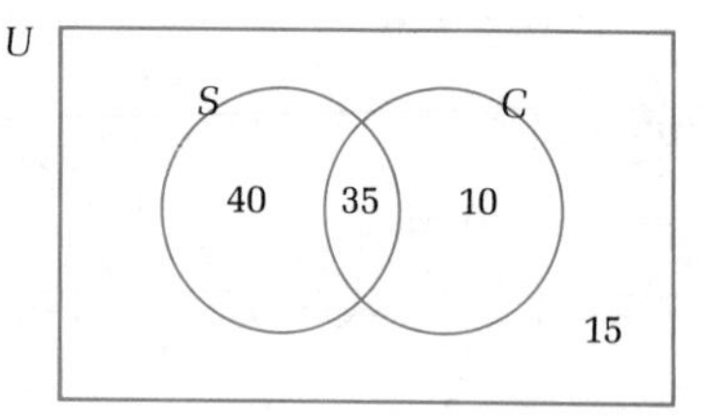

Place the number in the intersection first, then work outward:

$$40 = 75 - 35$$
$$10 = 45 - 35$$
$$15 = 100 - (40 + 35 + 10)$$

(A) The number of students who own either a car or a stereo is the number of students in the set $S \cup C$. You might be tempted to say that this is just the number of students in S plus the number of students in C, $75 + 45 = 120$, but this sum is larger than the sample we started with! What is wrong? We have actually counted the number in the intersection (35) twice. The correct answer, as seen in the Venn diagram, is

$$40 + 35 + 10 = 85$$

(B) The number of students who do not own either a car or a stereo is the number of students in the set $(S \cup C)'$; that is, 15.

Problem 5 Referring to Example 5:

(A) How many students owned a car but not a stereo?
(B) How many students did not own both a car and a stereo?

Note in Example 5 and Problem 5 that the word *and* is associated with intersection and the word *or* is associated with union.

Answers to Matched Problems

1. (A) $\{x|x^2 = 9\}$ (B) $\{-3, 3\}$ (C) True, True
2. All are true
3. $\{1, 2\}, \{1\}, \{2\}, \varnothing$
4. (A) $\{1, 2, 3, 4, 5, 7\}$ (B) $\{1, 3\}$ (C) $\varnothing$ (D) $\{2, 4, 6, 8, 9\}$
5. (A) 10 [the number in $S' \cap C$] (B) 65 [the number in $(S \cap C)'$]

Exercise 0-1

A *Indicate true (T) or false (F).*

1. $4 \in \{2, 3, 4\}$
2. $6 \notin \{2, 3, 4\}$
3. $\{2, 3\} \subset \{2, 3, 4\}$
4. $\{3, 2, 4\} = \{2, 3, 4\}$
5. $\{3, 2, 4\} \subset \{2, 3, 4\}$
6. $\{3, 2, 4\} \in \{2, 3, 4\}$
7. $\varnothing \subset \{2, 3, 4\}$
8. $\varnothing = \{0\}$

In Problems 9–14 write the resulting set using the listing method.

9. $\{1, 3, 5\} \cup \{2, 3, 4\}$
10. $\{3, 4, 6, 7\} \cup \{3, 4, 5\}$
11. $\{1, 3, 4\} \cap \{2, 3, 4\}$
12. $\{3, 4, 6, 7\} \cap \{3, 4, 5\}$
13. $\{1, 5, 9\} \cap \{3, 4, 6, 8\}$
14. $\{6, 8, 9, 11\} \cap \{3, 4, 5, 7\}$

B *In Problems 15–20 write the resulting set using the listing method.*

15. $\{x|x - 2 = 0\}$
16. $\{x|x + 7 = 0\}$
17. $\{x|x^2 = 49\}$
18. $\{x|x^2 = 100\}$
19. $\{x|x$ is an odd number between 1 and 9, inclusive$\}$
20. $\{x|x$ is a month starting with M$\}$
21. For $U = \{1, 2, 3, 4, 5\}$ and $A = \{2, 3, 4\}$, find A'.
22. For $U = \{7, 8, 9, 10, 11\}$ and $A = \{7, 11\}$, find A'.

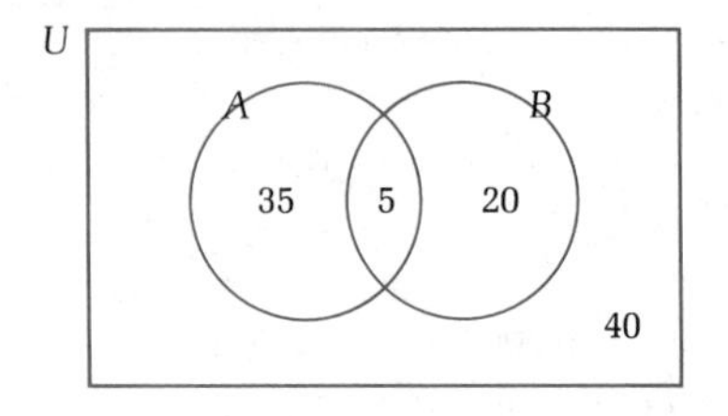

Problems 23–34 refer to the Venn diagram in the margin. How many elements are in each of the indicated sets?

23. A	24. U	25. A'	26. B'
27. $A \cup B$	28. $A \cap B$	29. $A' \cap B$	30. $A \cap B'$
31. $(A \cap B)'$	32. $(A \cup B)'$	33. $A' \cap B'$	34. U'

35. If $R = \{1, 2, 3, 4\}$ and $T = \{2, 4, 6\}$, find:

    (A) $\{x \mid x \in R \text{ or } x \in T\}$ (B) $R \cup T$

36. If $R = \{1, 3, 4\}$ and $T = \{2, 4, 6\}$, find:

    (A) $\{x \mid x \in R \text{ and } x \in T\}$ (B) $R \cap T$

37. For $P = \{1, 2, 3, 4\}$, $Q = \{2, 4, 6\}$, and $R = \{3, 4, 5, 6\}$, find $P \cup (Q \cap R)$.
38. For P, Q, and R in Problem 37, find $P \cap (Q \cup R)$.

C *Venn diagrams may be of help in Problems 39–44.*

39. If $A \cup B = B$, can we always conclude that $A \subset B$?
40. If $A \cap B = B$, can we always conclude that $B \subset A$?
41. If A and B are arbitrary sets, can we always conclude that $A \cap B \subset B$?
42. If $A \cap B = \varnothing$, can we always conclude that $B = \varnothing$?
43. If $A \subset B$ and $x \in A$, can we always conclude that $x \in B$?
44. If $A \subset B$ and $x \in B$, can we always conclude that $x \in A$?
45. How many subsets does each of the following sets have? Also, try to discover a formula in terms of n for a set with n elements.

    (A) $\{a\}$ (B) $\{a, b\}$ (C) $\{a, b, c\}$

46. How do the sets $\varnothing$, $\{\varnothing\}$, and $\{0\}$ differ from each other?

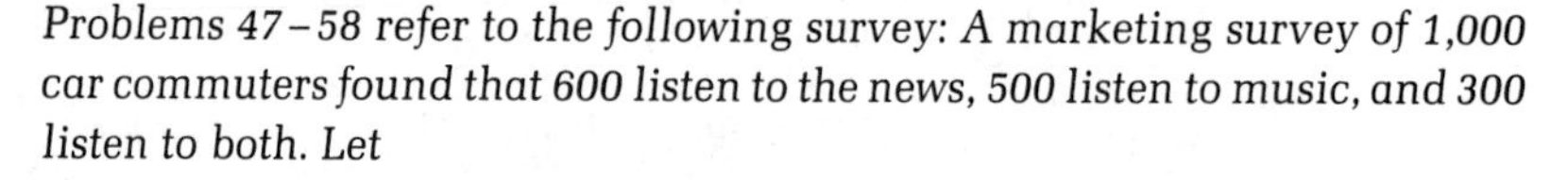

Applications

Business & Economics

Problems 47–58 refer to the following survey: A marketing survey of 1,000 car commuters found that 600 listen to the news, 500 listen to music, and 300 listen to both. Let

N = Set of commuters in the sample who listen to news

M = Set of commuters in the sample who listen to music

Following the procedures in Example 5, find the number of commuters in each set described below.

47. $N \cup M$	48. $N \cap M$	49. $(N \cup M)'$
50. $(N \cap M)'$	51. $N' \cap M$	52. $N \cap M'$

53. Set of commuters who listen to either news or music
54. Set of commuters who listen to both news and music
55. Set of commuters who do not listen to either news or music

56. Set of commuters who do not listen to both news and music
57. Set of commuters who listen to music but not news
58. Set of commuters who listen to news but not music
59. The management of a company, a president and three vice-presidents, denoted by the set $\{P, V_1, V_2, V_3\}$, wish to select a committee of two people from among themselves. How many ways can this committee be formed; that is, how many two-person subsets can be formed from a set of four people?
60. The management of the company in Problem 59 decides for or against certain measures as follows: The president has two votes and each vice-president has one vote. Three favorable votes are needed to pass a measure. List all minimal winning coalitions; that is, list all subsets of $\{P, V_1, V_2, V_3\}$ that represent exactly three votes.

Life Sciences

Blood types. When receiving a blood transfusion, a recipient must have all the antigens of the donor. A person may have one or more of the three antigens A, B, and Rh, or none at all. Eight blood types are possible, as indicated in the following Venn diagram, where U is the set of all people under consideration:

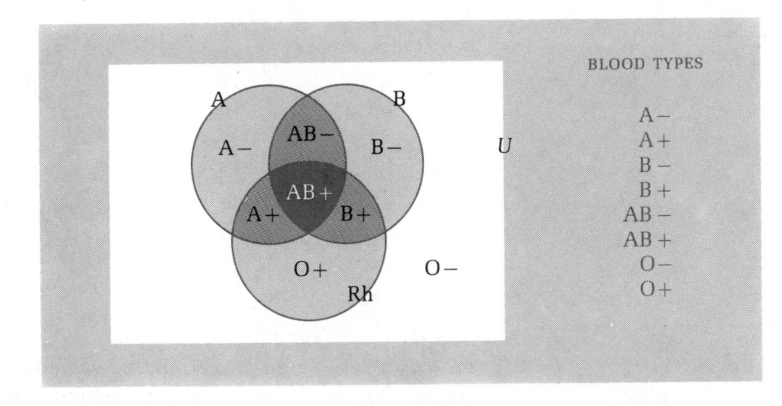

An A− person has A antigens but no B or Rh; an O+ person has Rh but neither A nor B; an AB− person has A and B antigens but no Rh; and so on.

Using the Venn diagram, indicate which of the eight blood types are included in each set.

61. $A \cap Rh$
62. $A \cap B$
63. $A \cup Rh$
64. $A \cup B$
65. $(A \cup B)'$
66. $(A \cup B \cup Rh)'$
67. $A' \cap B$
68. $Rh' \cap A$

Social Sciences

Group structures. R. D. Luce and A. D. Perry, in a study on group structure (*Psychometrika*, 1949, 14: 95–116), used the idea of sets to formally define the notion of a clique within a group. Let G be the set of all persons in the group and let $C \subset G$. Then C is a clique provided that:

1. C contains at least three elements.
2. For every $a, b \in C$, $a \mathrel{R} b$ and $b \mathrel{R} a$.
3. For every $a \notin C$, there is at least one $b \in C$ such that $a \not\mathrel{R} b$ or $b \not\mathrel{R} a$ or both.

[*Note:* Interpret "$a \mathrel{R} b$" to mean "a relates to b," "a likes b" "a is as wealthy as b," and so on. Of course, "$a \not\mathrel{R} b$" means "a does not relate to b," and so on.]

69. Translate statement 2 into ordinary English.
70. Translate statement 3 into ordinary English.

0-2 Linear Equations and Inequalities in One Variable

- Linear Equations
- The Real Number Line
- Linear Inequalities
- Applications

The equation

$$3 - 2(x + 3) = \frac{x}{3} - 5$$

and the inequality

$$\frac{x}{2} + 2(3x - 1) \geqslant 5$$

are both first degree (linear) in one variable.* A **solution** of an equation (or inequality) involving a single variable is a number that when substituted for the variable makes the equation (or inequality) true. The set of all solutions is called the **solution set.** When we say that we **solve an equation** (or inequality), we mean that we find its solution set.

Knowing what is meant by the solution set is one thing; finding it is another. We start by recalling the idea of equivalent equations and equivalent inequalities. If we perform an operation on an equation (or inequality) that produces another equation (or inequality) with the same solution set, then the two equations (or inequalities) are said to be **equivalent.** The basic idea in solving equations and inequalities is to perform operations on these

* An equation (or inequality) is **first degree (linear)** in one variable if it can be transformed into an equation (or inequality) where the left side is of the form $ax + b$, $a \neq 0$, and the right side is zero.

forms that produce simpler equivalent forms, and to continue the process until we obtain an equation or inequality with an obvious solution.

Linear Equations

The following properties of equality produce equivalent equations when applied:

Equality Properties

For a, b, and c real numbers:

1. If $a = b$, then $a + c = b + c$. Addition property
2. If $a = b$, then $a - c = b - c$. Subtraction property
3. If $a = b$, then $ca = cb$, $c \neq 0$. Multiplication property
4. If $a = b$, then $\frac{a}{c} = \frac{b}{c}$, $c \neq 0$. Division property

Several examples should remind you of the process of solving.

Example 6 Solve $8x - 3(x - 4) = 3(x - 4) + 6$.

Solution

$$\begin{aligned} 8x - 3(x - 4) &= 3(x - 4) + 6 \\ 8x - 3x + 12 &= 3x - 12 + 6 \\ 5x + 12 &= 3x - 6 \\ 2x &= -18 \\ x &= -9 \end{aligned}$$

Problem 6 Solve $3x - 2(2x - 5) = 2(x + 3) - 8$.

Example 7 What operations can we perform on

$$\frac{x+2}{2} - \frac{x}{3} = 5$$

to eliminate the denominators? If we can find a number that is exactly divisible by each denominator, then we can use the multiplication property of equality to clear the denominators. The LCD (least common denominator)* of the fractions, 6, is exactly what we are looking for! Actually, any common denominator will do, but the LCD results in a simpler equivalent equation. Thus, we multiply both sides of the equation by 6:

* Recall that the **least common denominator** (LCD) of two or more natural number denominators is the smallest natural number exactly divisible by each denominator.

$$6\left(\frac{x+2}{2} - \frac{x}{3}\right) = 6 \cdot 5$$

$$\overset{3}{\cancel{6}} \cdot \frac{(x+2)}{\underset{1}{\cancel{2}}} - \overset{2}{\cancel{6}} \cdot \frac{x}{\underset{1}{\cancel{3}}} = 30$$ *

$$3(x+2) - 2x = 30$$

$$3x + 6 - 2x = 30$$

$$x = 24$$

Problem 7 Solve $\frac{x+1}{3} - \frac{x}{4} = \frac{1}{2}$.

In many applications of algebra, formulas or equations must be changed to alternate equivalent forms. The following examples are typical.

Example 8 Solve the amount formula for simple interest, $A = P + Prt$, for:

(A) r in terms of the other variables
(B) P in terms of the other variables

Solutions (A) $A = P + Prt$

$P + Prt = A$	Reverse equation
$Prt = A - P$	Now isolate r on the left side
$r = \frac{A - P}{Pt}$	Divide both members by Pt

(B) $A = P + Prt$

$P + Prt = A$	Reverse equation
$P(1 + rt) = A$	Factor out P (note the use of the distributive property)
$P = \frac{A}{1 + rt}$	Divide by $(1 + rt)$

Problem 8 Solve $M = Nt + Nr$ for: (A) t (B) N

▪ The Real Number Line

Figure 5 breaks down the **set of real numbers** into its important subsets

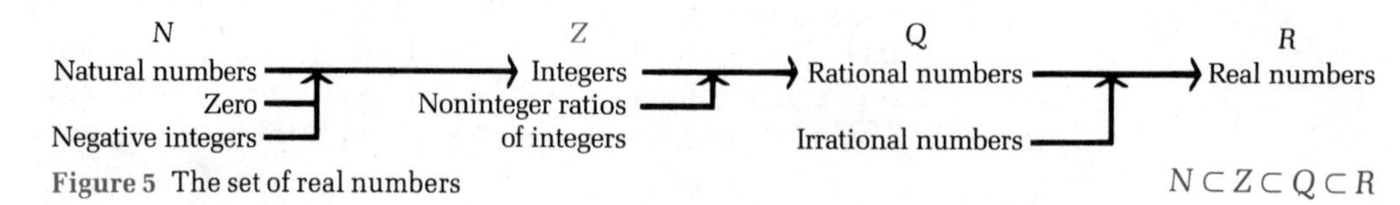

Figure 5 The set of real numbers

* The dashed boxes indicate steps that are usually done mentally.

A one-to-one correspondence exists between the set of real numbers and the set of points on a line; that is, each real number corresponds to exactly one point, and each point to exactly one real number. A line with a real number associated with each point, and vice versa, as in Figure 6, is called a **real number line,** or simply a **real line.** Each number associated with a point is called the **coordinate** of that point. The point with coordinate zero is called the **origin.** The arrow indicates a positive direction; the coordinates of all points to the right of the origin are called **positive real numbers,** and those to the left of the origin are called **negative real numbers.**

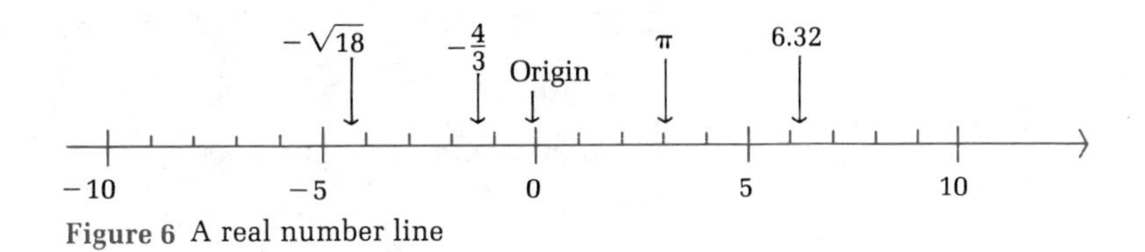

Figure 6 A real number line

▪ Linear Inequalities

Before we start solving linear inequalities, let us recall what we mean by $<$ (less than) and $>$ (greater than). If a and b are real numbers, then we write

$$a < b$$

if there exists a positive number p such that $a + p = b$. Certainly, we would expect that if a positive number was added to any real number, the sum would be larger than the original. That is essentially what the definition states. We write

$$b > a$$

if $a < b$.

Example 9

(A) $3 < 5$ Since $3 + 2 = 5$
(B) $-6 < -2$ Since $-6 + 4 = -2$
(C) $0 > -10$ Since $-10 < 0$

Problem 9 Replace each question mark with either $<$ or $>$.

(A) $2 ? 8$ (B) $-20 ? 0$ (C) $-3 ? -30$

The inequality symbols have a very clear geometric interpretation on the real number line. If $a < b$, then a is to the left of b on the number line; if $c > d$, then c is to the right of d (Fig. 7).

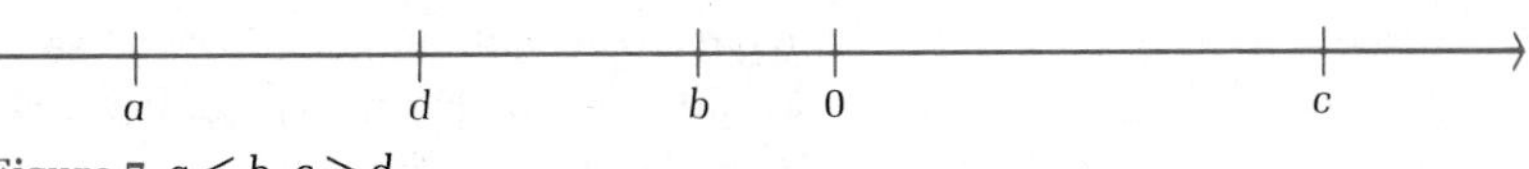

Figure 7 $a < b$, $c > d$

Now let us turn to the problem of solving linear inequalities in one variable. Recall that a solution of an inequality involving one variable is a number that, when substituted for the variable, makes the inequality true. The set of all solutions is called the solution set. When we say that we solve an inequality, we mean that we find its solution set. The procedures used to solve linear inequalities in one variable are almost the same as those used to solve linear equations in one variable but with two important exceptions (as will be noted below). The following properties of inequalities produce equivalent inequalities when applied:

Inequality Properties

For a, b, and c real numbers:

1. If $a > b$, then $a + c > b + c$.
2. If $a > b$, then $a - c > b - c$.
3. If $a > b$ and c is positive, then $ca > cb$. } Note difference
4. If $a > b$ and c is negative, then $ca < cb$. }
5. If $a > b$ and c is positive, then $\frac{a}{c} > \frac{b}{c}$. } Note difference
6. If $a > b$ and c is negative, then $\frac{a}{c} < \frac{b}{c}$. }

Similar properties hold if each inequality sign is reversed or if $>$ is replaced with $\geqslant$ (greater than or equal to) and $<$ is replaced with $\leqslant$ (less than or equal to). Thus, we can perform essentially the same operations on inequalities that we perform on equations with the exception that *the sense of the inequality reverses if we multiply or divide both sides by a negative number*. Otherwise, the sense of the inequality does not change. For example, if we start with the true statement

$$-3 > -7$$

and multiply both sides by 2, we obtain

$$-6 > -14$$

and the sense of the inequality stays the same. But if we multiply both sides of $-3 > -7$ by -2, then the left side becomes 6 and the right side becomes 14, so we must write

$$6 < 14$$

to have a true statement. Thus, the sense of the inequality reverses.

Recall that the double inequality $a \leqslant x \leqslant b$ means that $a \leqslant x$ **and** $x \leqslant b$.

Other variations, as well as a useful interval notation, are indicated in Table 1. Note that an end point on a line graph has a square bracket through it if it is included in the inequality and a parenthesis through it if it is not.

Table 1

Interval Notation	Inequality Notation	Line Graph
$[a, b]$	$a \leq x \leq b$	
$[a, b)$	$a \leq x < b$	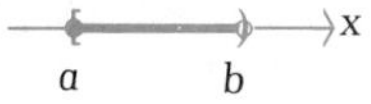
$(a, b]$	$a < x \leq b$	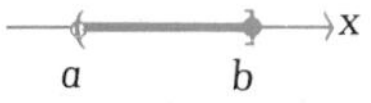
(a, b)	$a < x < b$	
$(-\infty, a]$	$x \leq a$	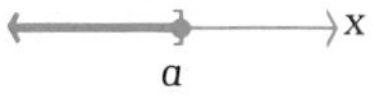
$(-\infty, a)$	$x < a$	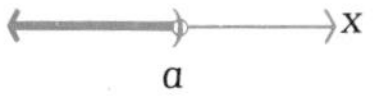
$[b, \infty)$*	$x \geq b$	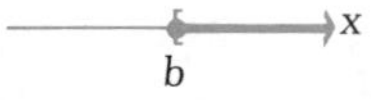
(b, ∞)	$x > b$	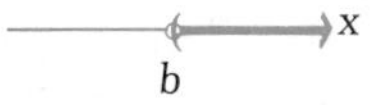

* The symbol ∞ (read "infinity") is not a number. When we write $[b, \infty)$, we are simply referring to the interval starting at b and continuing indefinitely to the right. We would never write $[b, \infty]$.

Example 10 Solve and graph $2(2x + 3) < 6(x - 2) + 10$.

Solution

$$2(2x + 3) < 6(x - 2) + 10$$
$$4x + 6 < 6x - 12 + 10$$
$$4x + 6 < 6x - 2$$
$$-2x + 6 < -2$$
$$-2x < -8$$
$$x > 4 \quad \text{or} \quad (4, \infty)$$

Notice that the sense of the inequality reverses when we divide both sides by -2

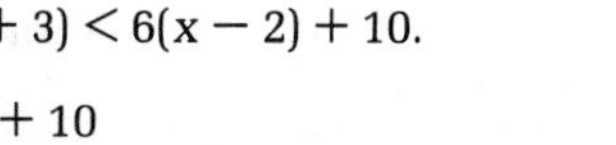

Notice that in the graph of $x > 4$, we use a parenthesis through 4, since the point 4 is not included in the graph.

Problem 10 Solve and graph $3(x - 1) \leq 5(x + 2) - 5$.

Example 11 Solve and graph $-3 < 2x + 3 \leq 9$.

Solution We are looking for all numbers x such that $2x + 3$ is between -3 and 9, including 9 but not -3. We proceed as above except that we try to isolate x in the middle:

$$-3 < 2x + 3 \leq 9$$

$$-3 - 3 < 2x + 3 - 3 \leq 9 - 3$$

$$-6 < 2x \leq 6$$

$$\frac{-6}{2} < \frac{2x}{2} \leq \frac{6}{2}$$

$$-3 < x \leq 3 \quad \text{or} \quad (-3, 3]$$

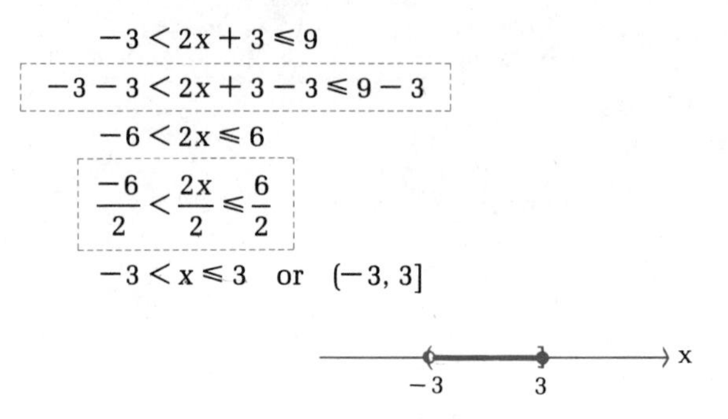

Problem 11 Solve and graph $-8 \leq 3x - 5 < 7$.

Note that a linear equation usually has exactly one solution, while a linear inequality usually has infinitely many solutions.

▪ Applications

To realize the full potential of algebra, we must be able to translate real-world problems into mathematical forms. In short, we must be able to do *word problems.*

Example 12
Break-even Analysis

It costs a record company \$9,000 to prepare a record album—recording costs, album design costs, etc. These costs represent a one-time **fixed cost.** Manufacturing, marketing, and royalty costs (all **variable costs**) are \$3.50 per album. If the album is sold to record shops for \$5 each, how many must be sold for the company to **break even?**

Solution Let x = Number of records sold
C = Cost for producing x records
R = Revenue (return) on sales of x records

The company breaks even if $R = C$, with

$$C = \text{Fixed costs} + \text{Variable costs}$$
$$= \$9{,}000 + \$3.50x$$
$$R = \$5x$$

Find x such that $R = C$; that is, such that

$$5x = 9{,}000 + 3.5x$$
$$1.5x = 9{,}000$$
$$x = 6{,}000$$

Check For $x = 6{,}000$,

$$\begin{aligned} C &= 9{,}000 + 3.5x \\ &= 9{,}000 + 3.5(6{,}000) \\ &= \$30{,}000 \end{aligned} \qquad \text{and} \qquad \begin{aligned} R &= 5x \\ &= 5(6{,}000) \\ &= \$30{,}000 \end{aligned}$$

Thus, the company must sell 6,000 records to break even; any sales over 6,000 will produce a profit; any sales under 6,000 will result in a loss.

Problem 12 What is the break-even point in Example 12 if fixed costs are \$9,900, variable costs are \$3.70 per record, and the records are sold for \$5.50 each?

Algebra has many different types of applications—so many, in fact, that no single approach applies to all. However, the following suggestions may help you get started:

Suggestions for Solving Word Problems

1. Read the problem very carefully.
2. Write down important facts and relationships.
3. Identify unknown quantities in terms of a single letter, if possible.
4. Write an equation or inequality relating the unknown quantities and the facts in the problem.
5. Solve the equation (or inequality).
6. Write all solutions asked for in the original problem.
7. Check the solution(s) in the original problem.

Example 13 The consumer price index for several years is given in Table 2.

Consumer Price Index (CPI)

Table 2 CPI (1967 = 100)

Year	Index
1950	72
1955	80
1960	89
1965	95
1970	116
1975	161
1980	247

What net monthly salary in 1980 would have the same purchasing power as a net monthly salary of \$900 in 1950? Compute the answer to the nearest dollar.

Solution To have the same purchasing power, the ratio of a salary in 1980 to a salary in 1950 would have to be the same as the ratio of the CPI in 1980 to the CPI in 1950. Thus, if x is the net monthly salary in 1980, we solve the equation.

$$\frac{x}{900} = \frac{247}{72}$$

$$x = 900 \cdot \frac{247}{72}$$

$$= \$3{,}088 \text{ per month}$$

Problem 13 What net monthly salary in 1960 would have the same purchasing power as a net monthly salary of $2,000 in 1975? Compute the answer to the nearest dollar.

Answers to Matched Problems

6. $x = 4$
7. $x = 2$
8. (A) $t = \frac{M - Nr}{N}$ (B) $N = \frac{M}{t + r}$
9. (A) $<$ (B) $<$ (C) $>$
10. $x \geq -4$ or $[-4, \infty)$
11. $-1 \leq x < 4$ or $[-1, 4)$
12. 5,500
13. $1,106

Exercise 0-2

A *Solve.*

1. $2m + 9 = 5m - 6$
2. $3y - 4 = 6y - 19$
3. $x + 5 < -4$
4. $x - 3 > -2$
5. $-3x \geq -12$
6. $-4x \leq 8$

Solve and graph.

7. $-4x - 7 > 5$
8. $-2x + 8 < 4$
9. $2 \leq x + 3 \leq 5$
10. $-3 < y - 5 < 8$

Solve.

11. $\frac{y}{7} - 1 = \frac{1}{7}$
12. $\frac{m}{5} - 2 = \frac{3}{5}$
13. $\frac{x}{3} > -2$
14. $\frac{y}{-2} \leq -1$
15. $\frac{y}{3} = 4 - \frac{y}{6}$
16. $\frac{x}{4} = 9 - \frac{x}{2}$

B

17. $10x + 25(x - 3) = 275$
18. $-3(4 - x) = 5 - (x + 1)$
19. $3 - y \leqslant 4(y - 3)$
20. $x - 2 \geqslant 2(x - 5)$
21. $\frac{x}{5} - \frac{x}{6} = \frac{6}{5}$
22. $\frac{y}{4} - \frac{y}{3} = \frac{1}{2}$
23. $\frac{m}{5} - 3 < \frac{3}{5} - m$
24. $u - \frac{2}{3} > \frac{u}{3} + 2$
25. $0.1(x - 7) + 0.05x = 0.8$
26. $0.4(u + 5) - 0.3u = 17$

Solve and graph.

27. $2 \leqslant 3x - 7 < 14$
28. $-4 \leqslant 5x + 6 < 21$
29. $-4 \leqslant \frac{9}{5}C + 32 \leqslant 68$
30. $-1 \leqslant \frac{2}{3}t + 5 \leqslant 11$

C

Solve for the indicated variable.

31. $3x - 4y = 12$, for y
32. $y = -\frac{2}{3}x + 8$, for x
33. $Ax + By = C$, for y $(B \neq 0)$
34. $y = mx + b$, for m
35. $F = \frac{9}{5}C + 32$, for C
36. $C = \frac{5}{9}(F - 32)$, for F
37. $A = Bm - Bn$, for B
38. $U = 3C - 2CD$, for C

Solve and graph.

39. $-3 \leqslant 4 - 7x < 18$
40. $-1 < 9 - 2u \leqslant 5$

Applications

Business & Economics

41. A jazz concert brought in \$60,000 on the sale of 8,000 tickets. If the tickets sold for \$6 and \$10 each, how many of each type of ticket were sold?

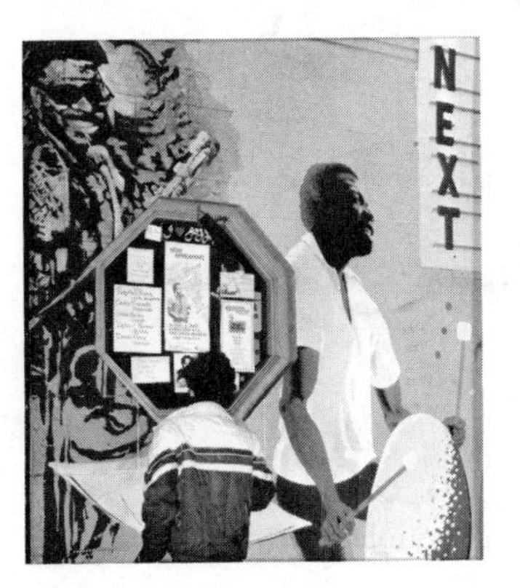

42. An all-day parking meter takes only dimes and quarters. If it contains 100 coins with a total value of \$14.50, how many of each type of coin are in the meter?
43. You have \$12,000 to invest. If part is invested at 10% and the rest at 15%, how much should be invested at each rate to yield 12% on the total amount?
44. An investor has \$20,000 to invest. If part is invested at 8% and the rest at 12%, how much should be invested at each rate to yield 11% on the total amount?
45. *Inflation.* If the price change of cars parallels the change in the CPI (see Table 2 in Example 13), what would a car sell for in 1980 if a comparable model sold for \$3,000 in 1965?
46. *Break-even analysis.* For a business to realize a profit, it is clear that revenue R must be greater than costs C; that is, a profit will result only if $R > C$ (the company breaks even when $R = C$). A record manufacturer has a weekly cost equation $C = 300 + 1.5x$ and a revenue equation $R = 2x$, where x is the number of records produced and sold in a week. How many records must be sold for the company to make a profit?

Life Sciences

47. *Wildlife management.* A naturalist for a fish and game department estimated the total number of rainbow trout in a certain lake using the popular capture–mark–recapture technique. He netted, marked, and released 200 rainbow trout. A week later, allowing for thorough mixing, he again netted 200 trout and found 8 marked ones among them. Assuming that the proportion of marked fish in the second sample was the same as the proportion of all marked fish in the total population, estimate the number of rainbow trout in the lake.
48. *Ecology.* If the temperature for a 24 hour period at an Antarctic station ranged between $-49°$F and $14°$F (that is, $-49 \leq F \leq 14$), what was the range in degrees Celsius? [*Note:* $F = \frac{9}{5}C + 32$.]

Social Sciences

49. *Psychology.* The IQ (intelligence quotient) is found by dividing the mental age (MA), as indicated on standard tests, by the chronological age (CA) and multiplying by 100. For example, if a child has a mental age of 12 and a chronological age of 8, the calculated IQ is 150. If a 9-year-old girl has an IQ of 140, compute her mental age.
50. *Anthropology.* In their study of genetic groupings, anthropologists use a ratio called the *cephalic index*. This is the ratio of the breadth of the head to its length (looking down from above) expressed as a percentage. Symbolically,

$$C = \frac{100B}{L}$$

where C is the cephalic index, B is the breadth, and L is the length. If an Indian tribe in Baja California (Mexico) had an average cephalic index of 66 and the average breadth of their heads was 6.6 inches, what was the average length of their heads?

0-3 Quadratic Equations

- Solution by Square Root
- Solution by Factoring
- Quadratic Formula

A **quadratic equation** in one variable is any equation that can be written in the form

$$ax^2 + bx + c = 0 \qquad a \neq 0$$

where x is a variable and a, b, and c are constants. We will refer to this form as the **standard form.** The equations

$$5x^2 - 3x + 7 = 0 \qquad \text{and} \qquad 18 = 32t^2 - 12t$$

are both quadratic equations since they are either in the standard form or can be transformed into this form.

We will restrict our review to finding real solutions to quadratic equations. Square root radicals are reviewed in Appendix A.

■ Solution by Square Root

The easiest type of quadratic equation to solve is the special form where the first-degree term is missing:

$$ax^2 + c = 0 \qquad a \neq 0$$

The method makes use of the definition of square root given in Appendix A.

Example 14 Solve by the square root method.

(A) $x^2 - 7 = 0$ (B) $2x^2 - 10 = 0$ (C) $3x^2 + 27 = 0$

Solutions (A) $x^2 - 7 = 0$

$x^2 = 7$ What real number squared is 7?

$x = \pm\sqrt{7}$ Short for $\sqrt{7}$ and $-\sqrt{7}$

(B) $2x^2 - 10 = 0$

$2x^2 = 10$

$x^2 = 5$ What real number squared is 5?

$x = \pm\sqrt{5}$

(C) $3x^2 + 27 = 0$

$3x^2 = -27$

$x^2 = -9$ What real number squared is -9?

No real solution. (Why?)

Problem 14 Solve by the square root method.

(A) $x^2 - 6 = 0$ (B) $3x^2 - 12 = 0$ (C) $x^2 + 4 = 0$

■ Solution by Factoring

If the left side of a quadratic equation when written in standard form can be factored, then the equation can be solved very quickly. The method of solution by factoring rests on the following important property of real numbers: *If a and b are real numbers, then $ab = 0$ if and only if $a = 0$ or $b = 0$ (or both).*

Example 15 Solve by factoring, if possible.

(A) $3x^2 - 6x - 24 = 0$ (B) $3y^2 = 2y$ (C) $x^2 - 2x - 1 = 0$

Solutions (A) $3x^2 - 6x - 24 = 0$ — Divide both sides by 3, since 3 is a factor of each coefficient

$$x^2 - 2x - 8 = 0$$ Factor the left side, if possible

$$(x - 4)(x + 2) = 0$$

$$x - 4 = 0 \quad \text{or} \quad x + 2 = 0$$

$$x = 4 \quad \text{or} \quad x = -2$$

(B) $3y^2 = 2y$ — We lose the solution $y = 0$ if both sides are divided by y ($3y^2 = 2y$ and $3y = 2$ are not equivalent)

$$3y^2 - 2y = 0$$

$$y(3y - 2) = 0$$

$$y = 0 \quad \text{or} \quad 3y - 2 = 0$$

$$3y = 2$$

$$y = \tfrac{2}{3}$$

(C) $x^2 - 2x - 1 = 0$

This equation cannot be factored using integer coefficients. We will solve this type of equation by another method, considered below.

Problem 15 Solve by factoring, if possible.

(A) $2x^2 + 4x - 30 = 0$ (B) $2x^2 = 3x$ (C) $2x^2 - 8x + 3 = 0$

The factoring and square root methods are fast and easy to use when they apply. However, there are quadratic equations that look simple but cannot be solved by either method. For example, as was noted in Example 15C, the polynomial in

$$x^2 - 2x - 1 = 0$$

cannot be factored using integer coefficients. This brings us to the well-known and widely used quadratic formula.

■ Quadratic Formula

There is a method called *completing the square* that will work for all quadratic equations. After briefly reviewing this method, we will then use it to develop the famous quadratic formula—a formula that will enable us to solve any quadratic equation quite mechanically.

The method of **completing the square** is based on the process of transforming a quadratic equation in standard form,

$$ax^2 + bx + c = 0$$

into the form

$$(x + A)^2 = B$$

where A and B are constants. Then, this last equation can easily be solved (if it has a real solution) by the square root method discussed above.

Consider the equation

$$x^2 - 2x - 1 = 0 \tag{1}$$

Since the left side does not factor using integer coefficients, we add 1 to each side to remove the constant term from the left side:

$$x^2 - 2x = 1 \tag{2}$$

Now we try to find a number that we can add to each side to make the left side a square of a first-degree polynomial. Note the following two squares:

$$(x + m)^2 = x^2 + 2mx + m^2 \qquad (x - m)^2 = x^2 - 2mx + m^2$$

We see that the third term on the right is the square of one-half the coefficient of x in the second term on the right. To complete the square in equation (2), we add the square of one-half the coefficient of x, $(-\frac{2}{2})^2 = 1$, to each side. (This rule works only when the coefficient of x^2 is 1, that is, $a = 1$.) Thus,

$$x^2 - 2x + 1 = 1 + 1$$

The left side is the square of $x - 1$, and we write

$$(x - 1)^2 = 2$$

What number squared is 2?

$$x - 1 = \pm\sqrt{2}$$
$$x = 1 \pm \sqrt{2}$$

And equation (1) is solved!

Let us try the method on the general quadratic equation

$$ax^2 + bx + c = 0 \qquad a \neq 0 \tag{3}$$

and solve it once and for all for x in terms of the coefficients a, b, and c. We start by multiplying both sides of (3) by $1/a$ to obtain

$$x^2 + \frac{b}{a}x + \frac{c}{a} = 0$$

Add $-c/a$ to both members:

$$x^2 + \frac{b}{a}x = -\frac{c}{a}$$

Now we complete the square on the left side by adding the square of one-half the coefficient of x, that is, $(b/2a)^2 = b^2/4a^2$, to each side:

$$x^2 + \frac{b}{a}x + \frac{b^2}{4a^2} = \frac{b^2}{4a^2} - \frac{c}{a}$$

Writing the left member as a square and combining the right side into a single fraction, we obtain

$$\left(x + \frac{b}{2a}\right)^2 = \frac{b^2 - 4ac}{4a^2}$$

Now we solve by the square root method:

$$x + \frac{b}{2a} = \pm\sqrt{\frac{b^2 - 4ac}{4a^2}}$$

$$x = -\frac{b}{2a} \pm \frac{\sqrt{b^2 - 4ac}}{2a}$$

When this is written as a single fraction, it becomes the quadratic formula:

Quadratic Formula

If $ax^2 + bx + c = 0$, $a \neq 0$, then

$$x = \frac{-b \pm \sqrt{b^2 - 4ac}}{2a}$$

Table 3

$b^2 - 4ac$	$ax^2 + bx + c = 0$
Positive	Two real solutions
Zero	One real solution
Negative	No real solutions

This formula is generally used to solve quadratic equations when the square root or factoring methods do not work. The quantity $b^2 - 4ac$ under the radical is called the **discriminant,** and it gives us the useful information about solutions listed in Table 3.

Example 16 Solve $x^2 - 2x - 1 = 0$ using the quadratic formula.

Solution $x^2 - 2x - 1 = 0$

$$x = \frac{-b \pm \sqrt{b^2 - 4ac}}{2a} \qquad a = 1, \quad b = -2, \quad c = -1$$

$$= \frac{-(-2) \pm \sqrt{(-2)^2 - 4(1)(-1)}}{2(1)}$$

$$= \frac{2 \pm \sqrt{8}}{2} = \frac{2 \pm 2\sqrt{2}}{2} = 1 \pm \sqrt{2}$$

Check $x^2 - 2x - 1 = 0$

When $x = 1 + \sqrt{2}$,

$$(1 + \sqrt{2})^2 - 2(1 + \sqrt{2}) - 1 = 1 + 2\sqrt{2} + 2 - 2 - 2\sqrt{2} - 1 = 0$$

When $x = 1 - \sqrt{2}$,

$$(1 - \sqrt{2})^2 - 2(1 - \sqrt{2}) - 1 = 1 - 2\sqrt{2} + 2 - 2 + 2\sqrt{2} - 1 = 0$$

Problem 16 Solve $2x^2 - 4x - 3 = 0$ using the quadratic formula.

If we try to solve $x^2 - 6x + 11 = 0$ using the quadratic formula, we obtain

$$x = \frac{6 \pm \sqrt{-8}}{2}$$

which is not a real number. (Why?)

Answers to Matched Problems

14. (A) $\pm\sqrt{6}$ (B) ± 2 (C) No real solution
15. (A) $-5, 3$ (B) $0, \frac{3}{2}$
    (C) Cannot be factored using integer coefficients
16. $(2 \pm \sqrt{10})/2$

Exercise 0-3

Find only real solutions in the problems below. If there are no real solutions, say so.

A *Solve by the square root method.*

1. $x^2 - 4 = 0$
2. $x^2 - 9 = 0$
3. $2x^2 - 22 = 0$
4. $3m^2 - 21 = 0$

Solve by factoring.

5. $2u^2 - 8u - 24 = 0$
6. $3x^2 - 18x + 15 = 0$
7. $x^2 = 2x$
8. $n^2 = 3n$

Solve by using the quadratic formula.

9. $x^2 - 6x - 3 = 0$
10. $m^2 + 8m + 3 = 0$
11. $3u^2 + 12u + 6 = 0$
12. $2x^2 - 20x - 6 = 0$

B *Solve, using any method.*

13. $2x^2 = 4x$
14. $2x^2 = -3x$
15. $4u^2 - 9 = 0$
16. $9y^2 - 25 = 0$
17. $8x^2 + 20x = 12$
18. $9x^2 - 6 = 15x$
19. $x^2 = 1 - x$
20. $m^2 = 1 - 3m$
21. $2x^2 = 6x - 3$
22. $2x^2 = 4x - 1$
23. $y^2 - 4y = -8$
24. $x^2 - 2x = -3$
25. $(x + 4)^2 = 11$
26. $(y - 5)^2 = 7$

C

27. Solve $A = P(1 + r)^2$ for r in terms of A and P; that is, isolate r on the left side of the equation (with coefficient 1) and end up with an algebraic expression on the right side involving A and P but not r. Write the answer using positive square roots only.
28. Solve $x^2 + mx + n = 0$ for x in terms of m and n.

Applications

Business & Economics

29. *Supply and demand.* The demand equation for a certain brand of popular records is $d = 3{,}000/p$. Notice that as the price (p) goes up, the number of records people are willing to buy (d) goes down, and vice versa. The supply equation is given by $s = 1{,}000p - 500$. Notice again, as the price (p) goes up, the number of records a supplier is willing to sell (s) goes up. At what price will supply equal demand; that is, at what price will $d = s$? In economic theory the price at which supply equals demand is called the **equilibrium point**—the point where the price ceases to change.

30. If P dollars is invested at $100r$ percent compounded annually, at the end of 2 years it will grow to $A = P(1 + r)^2$. At what interest rate will $100 grow to $144 in 2 years? [*Note:* If $A = 144$ and $P = 100$, find r.]

Life Sciences

31. *Ecology.* An important element in the erosive force of moving water is its velocity. To measure the velocity v (in feet per second) of a stream we have only to find a hollow L-shaped tube, place one end under the water pointing upstream and the other end pointing straight up a couple of feet out of the water. The water will then be pushed up the tube a certain distance h (in feet) above the surface of the stream. Physicists have shown that $v^2 = 64h$. Approximately how fast is a stream flowing if $h = 1$ foot? If $h = 0.5$ foot?

Social Sciences

32. *Safety research.* It is of considerable importance to know the least number of feet d in which a car can be stopped, including reaction time of the driver, at various speeds v (in miles/hour). Safety research has produced the formula $d = 0.044v^2 + 1.1v$. If it took a car 550 feet to stop, estimate the car's speed at the moment the stopping process was started. You might find a hand calculator of help in this problem.

0-4 Cartesian Coordinate System and Straight Lines

- Cartesian Coordinate System
- Graphing Linear Equations in Two Variables
- Slope
- Equations of Lines—Special Forms
- Application

Cartesian Coordinate System

Recall that a **Cartesian (rectangular) coordinate system** in a plane is formed by taking two mutually perpendicular real number lines intersecting at their origins **(coordinate axes)**, one horizontal and one vertical, and then assigning unique **ordered pairs** of numbers **(coordinates)** to each point P in the plane (Fig. 8). The first coordinate **(abscissa)** is the distance of P from the vertical axis, and the second coordinate **(ordinate)** is the distance of P from the horizontal axis. In Figure 8, the coordinates of point P are (a, b). By reversing the process, each ordered pair of real numbers can be associated with a unique point in the plane. The coordinate axes divide the plane into four parts **(quadrants)**, numbered I to IV in a counterclockwise direction.

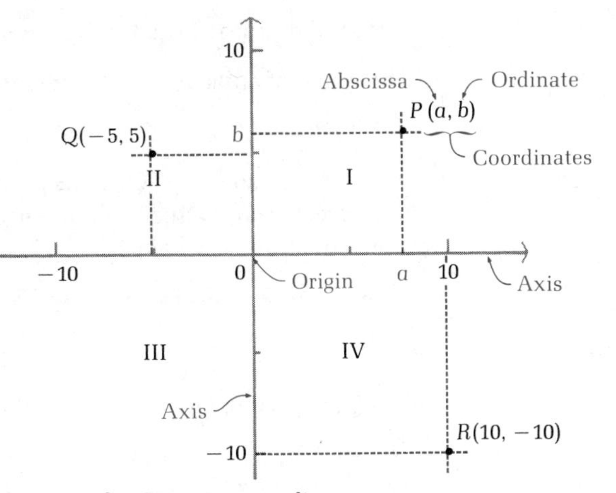

Figure 8 The Cartesian coordinate system

■ Graphing Linear Equations in Two Variables

A linear equation in two variables is an equation that can be written in the form

$$Ax + By = C \qquad \text{Standard form}$$

with A and B not both zero. For example,

$$2x - 3y = 5 \qquad x = 7 \qquad y = \tfrac{1}{2}x - 3 \qquad y = -3$$

can all be considered linear equations in two variables. The first is in standard form, while the other three can be written in standard form as follows:

	Standard form
$x = 7$	$x + 0y = 7$
$y = \frac{1}{2}x - 3$	$-\frac{1}{2}x + y = -3$ or $x - 2y = 6$
$y = -3$	$0x + y = -3$

A **solution** of an equation in two variables is an ordered pair of real numbers that satisfy the equation. For example, $(0, -3)$ is a solution of $3x - 4y = 12$. The **solution set** of an equation in two variables is the set of all solutions of the equation. When we say that we **graph an equation** in two variables, we mean that we graph its solution set on a rectangular coordinate system.

We state the following important theorem without proof:

Theorem 1

Graph of a Linear Equation in Two Variables

The graph of any equation of the form

$$Ax + By = C \qquad \text{Standard form} \tag{1}$$

where A, B, and C are constants (A and B not both zero), is a straight line. Every straight line in a Cartesian coordinate system is the graph of an equation of this type.

Also, the graph of any equation of the form

$$y = mx + b \tag{2}$$

where m and b are constants, is a straight line. Form (2) is simply a special case of (1) for $B \neq 0$. To graph either (1) or (2), we plot any two points of their solution set and use a straightedge to draw the line through these two points. The points where the line crosses the axes—called the **intercepts**—are often the easiest to find when dealing with form (1). To find the ***y* intercept,** we let $x = 0$ and solve for y; to find the ***x* intercept,** we let $y = 0$ and solve for x. It is sometimes wise to find a third point as a check.

Example 17 (A) The graph of $3x - 4y = 12$ is

x	y
0	−3
4	0
8	3

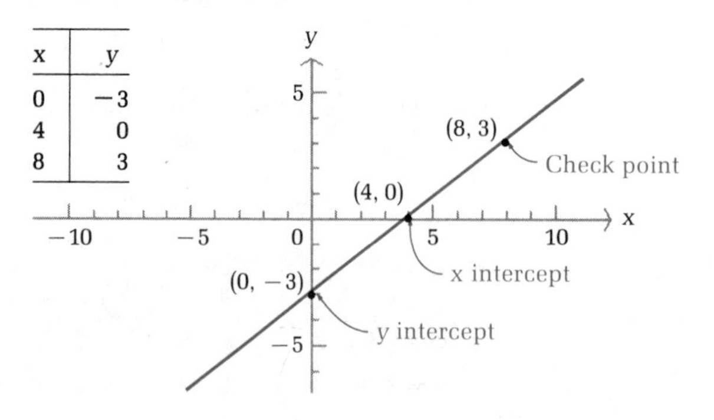

(B) The graph of $y = 2x - 1$ is

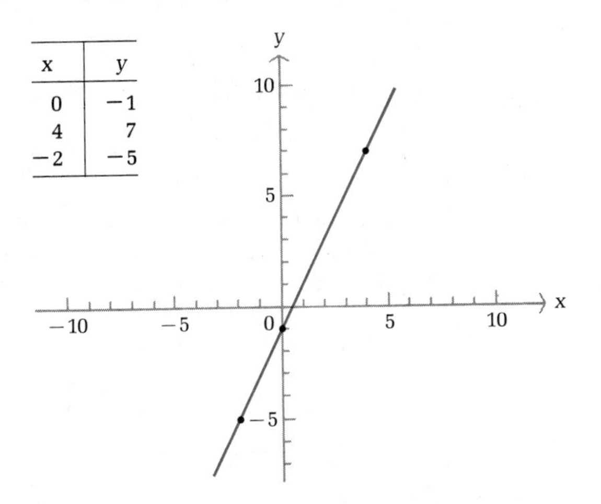

x	y
0	−1
4	7
−2	−5

Problem 17 Graph: (A) $4x - 3y = 12$ (B) $y = \frac{x}{2} + 2$

■ Slope

It is very useful to have a numerical measure of the "steepness" of a line. The concept of slope is widely used for this purpose. The **slope** of a line through the two points (x_1, y_1) and (x_2, y_2) is given by the following formula:

Slope

$$m = \frac{y_2 - y_1}{x_2 - x_1} \qquad x_1 \neq x_2$$

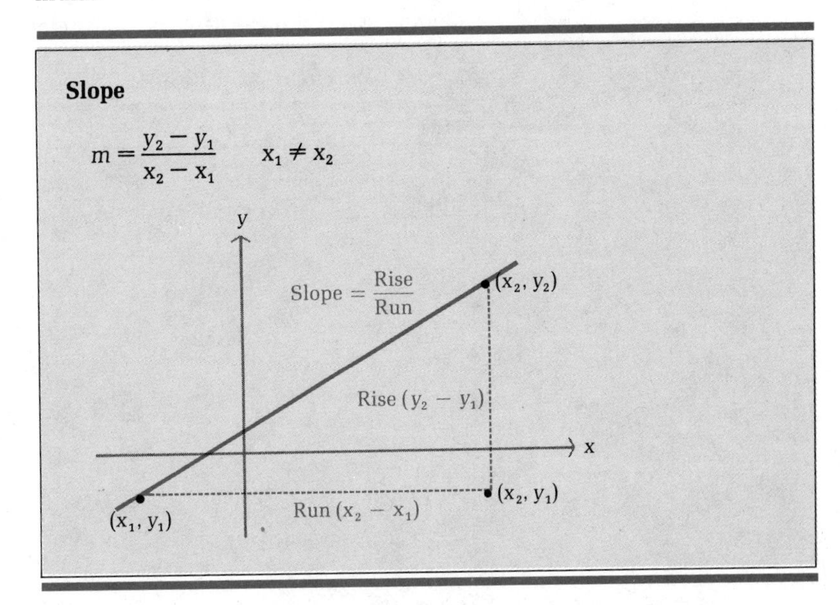

The slope of a vertical line is not defined. (Why? See Example 18B.)

Example 18 Find the slope of the line through each pair of points:

(A) $(-2, 5), (4, -7)$ (B) $(-3, -1), (-3, 5)$

Solutions (A) Let $(x_1, y_1) = (-2, 5)$ and $(x_2, y_2) = (4, -7)$. Then

$$m = \frac{y_2 - y_1}{x_2 - x_1} = \frac{-7 - 5}{4 - (-2)} = \frac{-12}{6} = -2$$

Note that we also could have let $(x_1, y_1) = (4, -7)$ and $(x_2, y_2) = (-2, 5)$, since this simply reverses the sign in both the numerator and the denominator and the slope does not change:

$$m = \frac{5 - (-7)}{-2 - 4} = \frac{12}{-6} = -2$$

(B) Let $(x_1, y_1) = (-3, -1)$ and $(x_2, y_2) = (-3, 5)$. Then

$$m = \frac{y_2 - y_1}{x_2 - x_1} = \frac{5 - (-1)}{-3 - (-3)} = \frac{6}{0} \qquad \text{Not defined!}$$

Notice that $x_1 = x_2$. This is always true for a vertical line, since the abscissa (first coordinate) of every point on a vertical line is the same. Thus, the slope of a vertical line is not defined (that is, the slope does not exist).

Problem 18 Find the slope of the line through each pair of points:

(A) $(3, -6), (-2, 4)$ (B) $(-7, 5), (3, 5)$

In general, the slope of a line may be positive, negative, zero, or not defined. Each of these cases is interpreted geometrically in Table 4.

Table 4 Going from Left to Right

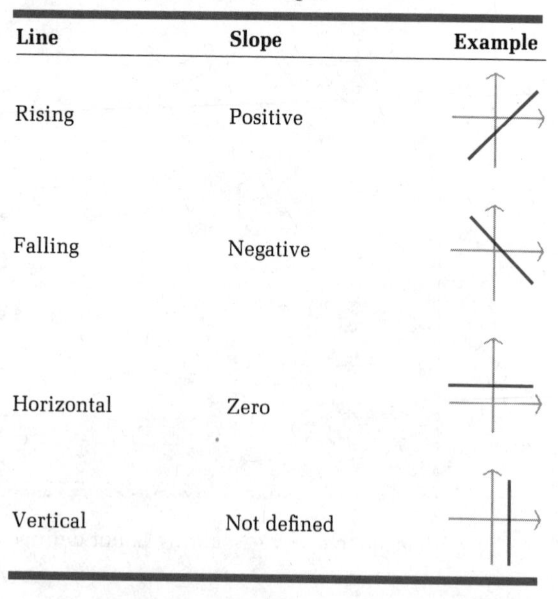

Line	Slope	Example
Rising	Positive	
Falling	Negative	
Horizontal	Zero	
Vertical	Not defined	

Equations of Lines—Special Forms

The constants m and b in the equation

$$y = mx + b \tag{3}$$

have special geometric significance.

If we let $x = 0$, then $y = b$, and we observe that the graph of (3) crosses the y axis at $(0, b)$. The constant b is the *y intercept*. For example, the y intercept of the graph of $y = -4x - 1$ is -1.

To determine the geometric significance of m, we proceed as follows: If $y = mx + b$, then by setting $x = 0$ and $x = 1$, we conclude that $(0, b)$ and $(1, m + b)$ lie on its graph (a line). Hence, the slope of this graph (line) is given by:

$$\text{Slope} = \frac{y_2 - y_1}{x_2 - x_1} = \frac{(m + b) - b}{1 - 0} = m$$

Thus, m is the slope of the line given by $y = mx + b$.

Slope-Intercept Form

The equation

$$y = mx + b \qquad \begin{matrix} m = \text{Slope} \\ b = y \text{ intercept} \end{matrix} \tag{4}$$

is called the **slope-intercept form** of an equation of a line.

Example 19 (A) Find the slope and y intercept, and graph $y = -\frac{2}{3}x - 3$.
(B) Write the equation of the line with slope $\frac{2}{3}$ and y intercept -2.

Solutions (A) Slope $= m = -\frac{2}{3}$
y intercept $= b = -3$

(B) $m = \frac{2}{3}$ and $b = -2$;
thus, $y = \frac{2}{3}x - 2$

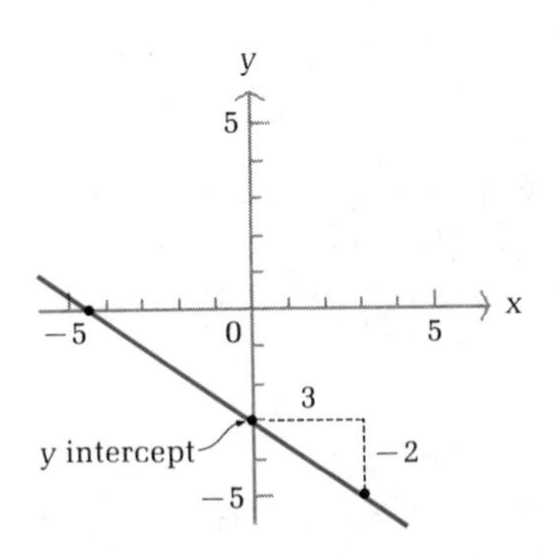

Problem 19 Write the equation of the line with slope $\frac{1}{2}$ and y intercept -1. Graph.

Suppose a line has slope m and passes through a fixed point (x_1, y_1). If the point (x, y) is any other point on the line (Fig. 9), then

$$\frac{y - y_1}{x - x_1} = m$$

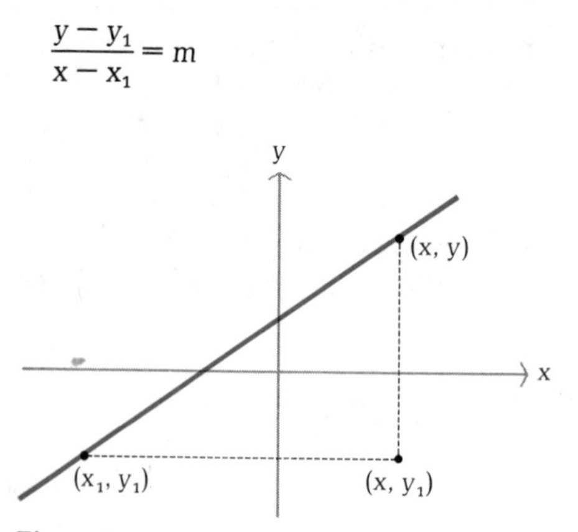

Figure 9

that is,

$$y - y_1 = m(x - x_1)$$

We now observe that (x_1, y_1) also satisfies this equation and conclude that this is an equation of a line with slope m that passes through (x_1, y_1).

Point–Slope Form

An equation of a line with slope m that passes through (x_1, y_1) is

$$y - y_1 = m(x - x_1) \qquad (5)$$

which is called the **point–slope form** of an equation of a line.

The point–slope form is extremely useful, since it enables us to find an equation for a line if we know its slope and the coordinates of a point on the line or if we know the coordinates of two points on the line.

Example 20

(A) Find an equation for the line that has slope $\frac{1}{2}$ and passes through $(-4, 3)$. Write the final answer in the form $Ax + By = C$.

(B) Find an equation for the line that passes through the two points $(-3, 2)$ and $(-4, 5)$. Write the resulting equation in the form $y = mx + b$.

Solutions

(A) $y - y_1 = m(x - x_1)$

Let $m = \frac{1}{2}$ and $(x_1, y_1) = (-4, 3)$. Then

$$y - 3 = \tfrac{1}{2}[x - (-4)]$$

$$y - 3 = \tfrac{1}{2}(x + 4) \qquad \text{Multiply by 2}$$

$$2y - 6 = x + 4$$

$$-x + 2y = 10 \quad \text{or} \quad x - 2y = -10$$

(B) First, find the slope of the line by using the slope formula:

$$m = \frac{y_2 - y_1}{x_2 - x_1} = \frac{5 - 2}{-4 - (-3)} = \frac{3}{-1} = -3$$

Now use

$$y - y_1 = m(x - x_1)$$

with $m = -3$ and $(x_1, y_1) = (-3, 2)$:

$$y - 2 = -3[x - (-3)]$$

$$y - 2 = -3(x + 3)$$

$$y - 2 = -3x - 9$$

$$y = -3x - 7$$

Problem 20

(A) Find an equation for the line that has slope $\frac{2}{3}$ and passes through $(6, -2)$. Write the resulting equation in the form $Ax + By = C$, $A > 0$.

(B) Find an equation for the line that passes through $(2, -3)$ and $(4, 3)$. Write the resulting equation in the form $y = mx + b$.

The simplest equations of a line are those for horizontal and vertical lines. A **horizontal line** has slope 0; thus its equation is of the form

$$y = 0x + c \qquad \text{Slope} = 0, \quad y \text{ intercept } c$$

or simply

$$y = c$$

Figure 10 illustrates the graph of $y = 3$ and $y = -2$.

If a line is vertical, then its slope is not defined. All x values (abscissas) of points on a vertical line are equal, while y can take on any value (Fig. 11).

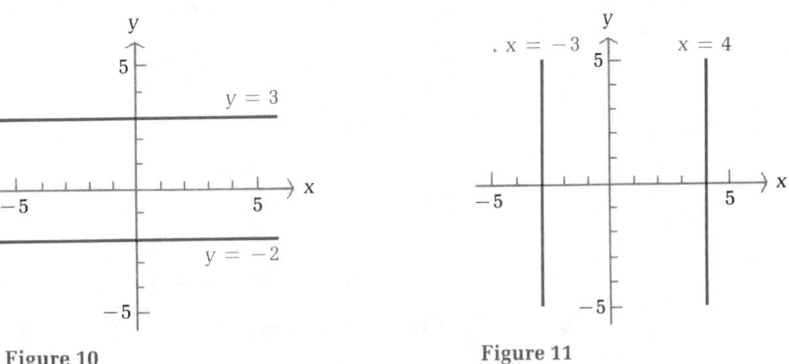

Figure 10

Figure 11

Thus, a **vertical line** has an equation of the form

$x + 0y = c$ x intercept c

or simply

$x = c$

Figure 11 illustrates the graph of $x = -3$ and $x = 4$.

Equations of Horizontal and Vertical Lines

Horizontal line with y intercept c: $y = c$

Vertical line with x intercept c: $x = c$

Example 21 The equation of a horizontal line through $(-2, 3)$ is $y = 3$, and the equation of a vertical line through the same point is $x = -2$.

Problem 21 Find the equations of the horizontal and vertical lines through $(4, -5)$.

It can be shown that if two nonvertical lines are parallel, then they have the same slope. And if two lines have the same slope, they are parallel. It can also be shown that if two nonvertical lines are perpendicular, then their slopes are the negative reciprocals of each other (that is, $m_2 = -1/m_1$, or, equivalently, $m_1 m_2 = -1$). And if the slopes of two lines are the negative reciprocals of each other, the lines are perpendicular. Symbolically:

Parallel and Perpendicular Lines

Given nonvertical lines L_1 and L_2 with slopes m_1 and m_2, respectively, then

$L_1 \parallel L_2$ if and only if $m_1 = m_2$

$L_1 \perp L_2$ if and only if $m_1 m_2 = -1$ or $m_2 = -\dfrac{1}{m_1}$

[*Note:* $\parallel$ means "is parallel to" and $\perp$ means "is perpendicular to."]

Example 22 Given the line $x - 2y = 4$, find the equation of a line that passes through $(2, -3)$ and is:

(A) Parallel to the given line (B) Perpendicular to the given line

Write final equations in the form $y = mx + b$.

Solution First find the slope of the given line by writing $x - 2y = 4$ in the form $y = mx + b$:

$$x - 2y = 4$$
$$y = \tfrac{1}{2}x - 2$$

The slope of the given line is $\tfrac{1}{2}$.

(A) The slope of a line parallel to the given line is also $\tfrac{1}{2}$. We have only to find the equation of a line through $(2, -3)$ with slope $\tfrac{1}{2}$ to solve part A:

$$y - y_1 = m(x - x_1) \qquad m = \tfrac{1}{2} \text{ and } (x_1, y_1) = (2, -3)$$
$$y - (-3) = \tfrac{1}{2}(x - 2)$$
$$y + 3 = \tfrac{1}{2}x - 1$$
$$y = \tfrac{1}{2}x - 4$$

(B) The slope of the line perpendicular to the given line is the negative reciprocal of $\tfrac{1}{2}$; that is, -2. We have only to find the equation of a line through $(2, -3)$ with slope -2 to solve part B:

$$y - y_1 = m(x - x_1) \qquad m = -2 \text{ and } (x_1, y_1) = (2, -3)$$
$$y - (-3) = -2(x - 2)$$
$$y + 3 = -2x + 4$$
$$y = -2x + 1$$

Problem 22 Given the line $2x = 6 - 3y$, find the equation of a line that passes through $(-3, 9)$ and is:

(A) Parallel to the given line (B) Perpendicular to the given line

Write final equations in the form $y = mx + b$.

■ Application

We will now see how equations of lines occur in certain applications.

Example 23 The management of a company that manufactures roller skates has fixed costs (costs at zero output) of $300 per day and total costs of $4,300 per day at an output of 100 pairs of skates per day. Assume that cost C is linearly related to output x.

(A) Find the slope of the line joining the points associated with outputs of 0 and 100; that is, the line passing through (0, 300) and (100, 4,300).

(B) Find an equation of the line relating output to cost. Write the final answer in the form $C = mx + b$.

(C) Graph the cost equation from part B for $0 \leq x \leq 200$.

Solutions

(A) $m = \frac{y_2 - y_1}{x_2 - x_1} = \frac{4{,}300 - 300}{100 - 0} = \frac{4{,}000}{100} = 40$

(B) We must find an equation of the line that passes through (0, 300) with slope 40. We use the slope–intercept form:

$$C = mx + b$$
$$C = 40x + 300$$

(C)

x	C
0	300
100	4,300
200	8,300

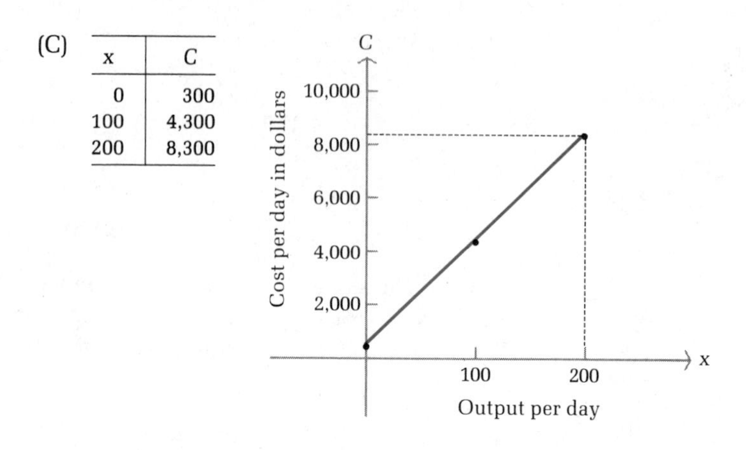

Problem 23 Answer parts A and B in Example 23 for fixed costs of \$250 per day and total costs of \$3,450 per day at an output of 80 pairs of skates per day.

Answers to Matched Problems

17. (A)

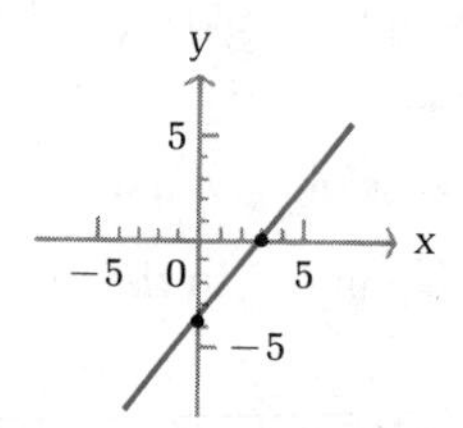

(B)

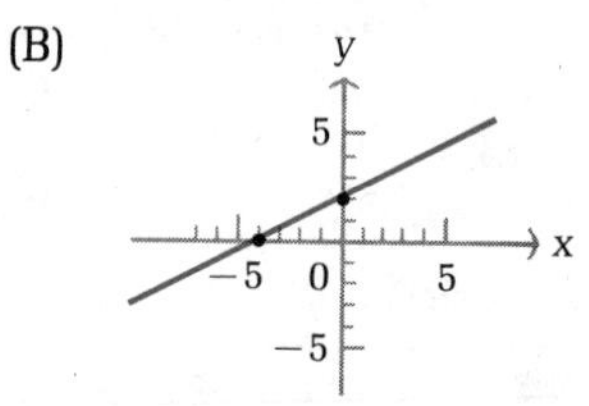

18. (A) -2 (B) 0 (Zero is a number—it exists! It is the slope of a horizontal line.)

19. $y = \frac{1}{2}x - 1$

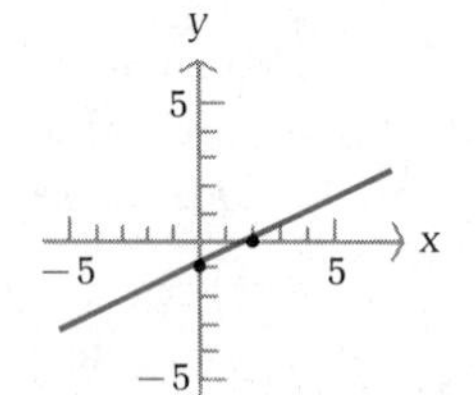

20. (A) $2x - 3y = 18$
(B) $y = 3x - 9$

21. $y = -5$, $x = 4$

22. (A) $y = -\frac{2}{3}x + 7$
(B) $y = \frac{3}{2}x + \frac{27}{2}$

23. (A) $m = 40$ (B) $C = 40x + 250$

Exercise 0-4

A *Graph in a rectangular coordinate system.*

1. $y = 2x - 3$
2. $y = \frac{x}{2} + 1$
3. $2x + 3y = 12$
4. $8x - 3y = 24$

Find the slope and y intercept of the graph of each equation.

5. $y = 2x - 3$
6. $y = \frac{x}{2} + 1$
7. $y = -\frac{2}{3}x + 2$
8. $y = \frac{3}{4}x - 2$

Write an equation of the line with the indicated slope and y intercept.

9. Slope $= -2$
   y intercept $= 4$
10. Slope $= -\frac{2}{3}$
    y intercept $= -2$
11. Slope $= -\frac{3}{5}$
    y intercept $= 3$
12. Slope $= 1$
    y intercept $= -2$

B *Graph in a rectangular coordinate system.*

13. $y = -\frac{2}{3}x - 2$
14. $y = -\frac{3}{2}x + 1$
15. $3x - 2y = 10$
16. $5x - 6y = 15$
17. $x = 3$ and $y = -2$
18. $x = -3$ and $y = 2$

Find the slope of the graph of each equation. (First write the equation in the form $y = mx + b$.)

19. $3x + y = 5$
20. $2x - y = -3$
21. $2x + 3y = 12$
22. $3x - 2y = 10$

Write an equation of the line through each indicated point with the indicated slope. Transform the equation into the form $y = mx + b$.

23. $m = -3$, $(4, -1)$
24. $m = -2$, $(-3, 2)$
25. $m = \frac{2}{3}$, $(-6, -5)$
26. $m = \frac{1}{2}$, $(-4, 3)$

Find the slope of the line that passes through the given points.

27. $(1, 3)$ and $(7, 5)$
28. $(2, 1)$ and $(10, 5)$
29. $(-5, -2)$ and $(5, -4)$
30. $(3, 7)$ and $(-6, 4)$

Write an equation of the line through each indicated pair of points. Write the final answer in the form $Ax + By = C$, $A > 0$.

31. $(1, 3)$ and $(7, 5)$
32. $(2, 1)$ and $(10, 5)$
33. $(-5, -2)$ and $(5, -4)$
34. $(3, 7)$ and $(-6, 4)$

Write equations of the vertical and horizontal lines through each point.

35. $(3, -5)$ 36. $(-2, 7)$ 37. $(-1, -3)$ 38. $(6, -4)$

Find an equation of the line, given the information in each problem. Write the final answer in the form $y = mx + b$.

39. Line passes through $(-2, 5)$ with slope $-\frac{1}{2}$.
40. Line passes through $(3, -1)$ with slope $-\frac{2}{3}$.
41. Line passes through $(-2, 2)$ and is
(A) Parallel (B) Perpendicular to $y = -\frac{1}{2}x + 5$.
42. Line passes through $(-4, -3)$ and is
(A) Parallel (B) Perpendicular to $y = 2x - 3$.
43. Line passes through $(-2, -1)$ and is
(A) Parallel (B) Perpendicular to $x - 2y = 4$.
44. Line passes through $(-3, 2)$ and is
(A) Parallel (B) Perpendicular to $2x + 3y = -6$.

C

45. Graph $y = mx - 2$ for $m = 2$, $m = \frac{1}{2}$, $m = 0$, $m = -\frac{1}{2}$, and $m = -2$, all on the same coordinate system.
46. Graph $y = -\frac{1}{2}x + b$ for $b = -4$, $b = 0$, and $b = 4$, all on the same coordinate system.

Write an equation of the line through the indicated points. Be careful!

47. $(2, 7)$ and $(2, -3)$
48. $(-2, 3)$ and $(-2, -1)$
49. $(2, 3)$ and $(-5, 3)$
50. $(-3, -3)$ and $(0, -3)$

Applications

Business & Economics

51. *Simple interest.* If \$P (the principal) is invested at an interest rate of r, then the amount A that is due after t years is given by

$$A = Prt + P$$

If \$100 is invested at 6% ($r = 0.06$), then $A = 6t + 100$, $t \geq 0$.

(A) What will \$100 amount to after 5 years? After 20 years?
(B) Graph the equation for $0 \leq t \leq 20$.
(C) What is the slope of the graph? (The slope indicates the increase in the amount A for each additional year of investment.)

52. *Cost equation.* The management of a company manufacturing surfboards has fixed costs (zero output) of \$200 per day and total costs of \$1,400 per day at a daily output of twenty boards.

(A) Assuming the total cost per day (C) is linearly related to the total output per day (x), write an equation relating these two quantities. [*Hint:* Find an equation of the line that passes through (0, 200) and (20, 1,400).]

(B) What are the total costs for an output of twelve boards per day?

(C) Graph the equation for $0 \leq x \leq 20$.

[*Note:* The slope of the line found in part A is the increase in total cost for each additional unit produced and is called the *marginal cost.* More will be said about the concept of marginal cost later.]

53. *Demand equation.* A manufacturing company is interested in introducing a new power mower. Its market research department gave the management the demand-price forecast listed in the table.

Price	Estimated Demand
$ 70	7,800
$120	4,800
$160	2,400
$200	0

(A) Plot these points, letting d represent the number of mowers people are willing to buy (demand) at a price of $\$p$ each.

(B) Note that the points in part A lie along a straight line. Find an equation of that line.

[*Note:* The slope of the line found in part B indicates the decrease in demand for each $1 increase in price.]

54. *Depreciation.* Office equipment was purchased for $20,000 and is assumed to have a scrap value of $2,000 after 10 years. If its value is depreciated linearly (for tax purposes) from $20,000 to $2,000:

(A) Find the linear equation that relates value (V) in dollars to time (t) in years.

(B) What would be the value of the equipment after 6 years?

(C) Graph the equation for $0 \leq t \leq 10$.

[*Note:* The slope found in part A indicates the decrease in value per year.]

Life Sciences

55. *Nutrition.* In a nutrition experiment, a biologist wants to prepare a special diet for the experimental animals. Two food mixes, A and B, are available. If mix A contains 20% protein and mix B contains 10% protein, what combination of each mix will provide exactly 20 grams of protein? Let x be the amount of A used and let y be the amount of B used. Then write a linear equation relating x, y, and 20. Graph this equation for $x \geq 0$ and $y \geq 0$.

56. *Ecology.* As one descends into the ocean, pressure increases linearly. The pressure is 15 pounds per square inch on the surface and 30 pounds per square inch 33 feet below the surface.

(A) If p is the pressure in pounds and d is the depth below the surface in feet, write an equation that expresses p in terms of d. [*Hint:* Find an equation of the line that passes through (0, 15) and (33, 30).]
(B) What is the pressure at 12,540 feet (the average depth of the ocean)?
(C) Graph the equation for $0 \leq d \leq 12{,}540$.

[*Note:* The slope found in part A indicates the change in pressure for each additional foot of depth.]

Social Sciences

57. *Psychology.* In an experiment on motivation, J. S. Brown trained a group of rats to run down a narrow passage in a cage to obtain food in a goal box. Using a harness, he then connected the rats to an overhead wire that was attached to a spring scale. A rat was placed at different distances d (in centimeters) from the goal box, and the pull p (in grams) of the rat toward the food was measured. Brown found that the relationship between these two variables was very close to being linear and could be approximated by the equation

$$p = -\tfrac{1}{5}d + 70 \qquad 30 \leq d \leq 175$$

(See J. S. Brown, *Journal of Comparative and Physiological Psychology*, 1948, 41:450–465.)

(A) What was the pull when $d = 30$? When $d = 175$?
(B) Graph the equation.
(C) What is the slope of the line?

0-5 Functions and Graphs

- Definition of a Function
- Functions Specified by Equations
- Function Notation
- Linear Functions and Their Graphs
- Quadratic Functions and Their Graphs
- Application: Market Research

The function concept is one of the most important concepts in mathematics. The idea of correspondence plays a central role in its formulation. You have already had experiences with correspondences in everyday life. For example:

To each person there corresponds an annual income.

To each item in a supermarket there corresponds a price.

To each day there corresponds a maximum temperature.

For the manufacture of x items there corresponds a cost.

For the sale of x items there corresponds a revenue.

To each square there corresponds an area.

To each number there corresponds its cube.

One of the most important aspects of any science (managerial, life, social, physical, etc.) is the establishment of correspondences among various types of phenomena. Once a correspondence is known, predictions can be made. A cost analyst would like to predict costs for various levels of output in a manufacturing process; a medical researcher would like to know the correspondence between heart disease and obesity; a psychologist would like to predict the level of performance after a subject has repeated a task a given number of times; and so on.

Definition of a Function

What do all of the preceding examples have in common? Each deals with the matching of elements from one set with the elements in a second set. Consider the following three tables of the cube, square, and square root.

Table 5

Domain (Number)	Range (Cube)
0 →	0
1 →	1
2 →	8

Table 6

Domain (Number)	Range (Square)
−2	4
−1	1
0	0
1	
2	

(Arrows: −2 → 4, −1 → 1, 0 → 0, 1 → 1, 2 → 4)

Table 7

Domain (Number)	Range (Square Root)
0 →	0
1 →	1
1 →	−1
4 →	2
4 →	−2
9 →	3
9 →	−3

Tables 5 and 6 specify functions, but Table 7 does not. The very important term *function* is now defined.

> **Definition of a Function**
>
> A **function** is a rule (process or method) that produces a correspondence between one set of elements, called the **domain,** and a second set of elements, called the **range,** such that to each element in the domain there corresponds *one and only one* element in the range.

Tables 5 and 6 are functions, since to each domain value there corresponds exactly one range value (for example, the square of -2 is 4 and no other number). On the other hand, Table 7 is not a function, since to at least one domain value there corresponds more than one range value (for example, to the domain value 9 there corresponds -3 and 3, both square roots of 9).

Since in a function elements in the range are paired with elements in the domain by some rule or process, this correspondence (pairing) can be illustrated by using ordered pairs of elements where the first component represents a domain element and the second component a corresponding range element. Thus, we can write functions 1 and 2 (Tables 5 and 6) as follows:

Function 1 = $\{(0, 0), (1, 1), (2, 8)\}$

Function 2 = $\{(-2, 4), (-1, 1), (0, 0), (1, 1), (2, 4)\}$

As a consequence of these definitions, we find that a function can be specified in many different ways: by an equation, by a table, by a set of ordered pairs of elements, and by a graph, to name a few of the more common ways (see Table 8). All that matters is that we are given a set of elements called the domain and a rule (method or process) of obtaining unique corresponding range values for each domain value. (Incidentally, the **graph of a function** specified by an equation in two variables is the graph of the set of all ordered pairs of real numbers that satisfies the equation.)

Table 8 Common Ways of Specifying Functions

Method	Illustration	Example
Equation	$y = x^2 + x \quad x \in R^*$	If $x = 2$, then $y = 6$.
Table	p | C 2 | 14 4 | 18 6 | 22	If $p = 4$, then $C = 18$.
Set of ordered pairs of elements	$\{(2, 14), (4, 18), (6, 22)\}$	6 corresponds to 22.
Graph	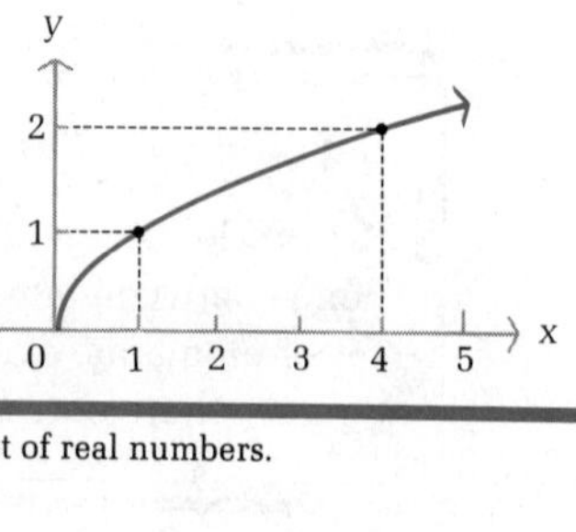 y; x; 0 1 2 3 4 5; 1 2	If $x = 4$, $y = 2$.

* Recall that R is the set of real numbers.

Functions Specified by Equations

Frequently, domains and ranges of functions are sets of numbers, and the rules associating range values with domain values are equations in two variables. Consider the equation

$$y = x^2 - x \qquad x \in R$$

For each **input** x we obtain one **output** y. For example,

If $x = 3$, then $y = 3^2 - 3 = 6$.

If $x = -\frac{1}{2}$, then $y = (-\frac{1}{2})^2 - (-\frac{1}{2}) = \frac{1}{4} + \frac{1}{2} = \frac{3}{4}$.

The input values are domain values and the output values are range values. The equation (a rule) assigns each domain value x a range value y. The variable x is called an *independent variable* (since values are "independently" assigned to x from the domain), and y is called a *dependent variable* (since the value of y "depends" on the value assigned to x). In general, any variable used as a placeholder for domain values is called an **independent variable;** any variable that is used as a placeholder for range values is called a **dependent variable.**

When does an equation specify a function?

Equations and Functions

In an equation in two variables, if there corresponds exactly one value of the dependent variable (output) to each value of the independent variable (input), then the equation specifies a function. If there is more than one output for at least one input, then the equation does not specify a function.

Unless stated to the contrary, we shall adhere to the following convention regarding domains and ranges for functions specified by equations.

Agreement on Domains and Ranges

If a function is specified by an equation and the domain is not indicated, then we shall assume that the domain is the set of all real number replacements of the independent variable (inputs) that produce real values for the dependent variable (outputs). The range is the set of all outputs corresponding to input values.

▪ Function Notation

We have just seen that a function involves two sets of elements, a domain and a range, and a rule of correspondence that enables one to assign to each element in the domain exactly one element in the range. We use different letters to denote names for numbers; in essentially the same way, we will now use different letters to denote names for functions. For example, f and g may be used to name the two functions

$$f: \quad y = 2x + 1$$
$$g: \quad y = x^2 + 2x - 3$$

If x represents an element in the domain of a function f, then we will often use the symbol

$$f(x)$$

in place of y to designate the number in the range of the function f to which x is paired (Fig. 12).

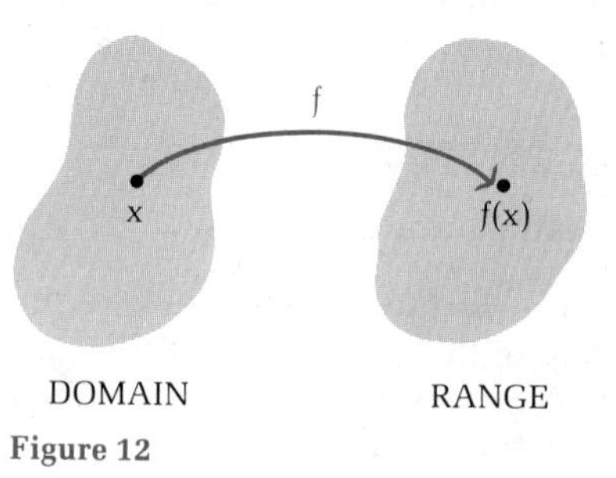

Figure 12

It is important not to think of $f(x)$ as the product of f and x. The symbol $f(x)$ is read "f of x" or "the value of f at x." The variable x is an independent variable; both y and $f(x)$ are dependent variables.

This function notation is extremely important, and its use should be mastered as quickly as possible. For example, in place of the more formal representation of the functions f and g above, we can now write

$$f(x) = 2x + 1 \qquad \text{and} \qquad g(x) = x^2 + 2x - 3$$

The function symbols $f(x)$ and g(x) have certain advantages over the variable y in certain situations. For example, if we write $f(3)$ and g(5), then each symbol indicates in a concise way that these are range values of particular functions associated with particular domain values. Let us find $f(3)$ and g(5).

To find $f(3)$, we replace x by 3 wherever x occurs in

$$f(x) = 2x + 1$$

and evaluate the right side:

$$f(3) = 2 \cdot 3 + 1$$
$$= 6 + 1$$
$$= 7$$

Thus,

$f(3) = 7$ The function f assigns the range value 7 to the domain value 3; the ordered pair (3, 7) belongs to f

To find g(5), we replace x by 5 whenever x occurs in

$$g(x) = x^2 + 2x - 3$$

and evaluate the right side:

$$g(5) = 5^2 + 2 \cdot 5 - 3$$
$$= 25 + 10 - 3$$
$$= 32$$

Thus,

$g(5) = 32$ The function g assigns the range value 32 to the domain value 5; the ordered pair (5, 32) belongs to g

It is very important to understand and remember the definition of $f(x)$:

The $f(x)$ Symbol

For any element x in the domain of the function f, the symbol $\boldsymbol{f(x)}$ represents the element in the range of f corresponding to x in the domain of f. If x is an input value, then $f(x)$ is the corresponding output value; or, symbolically, $f: x \to f(x)$. The ordered pair $(x, f(x))$ belongs to the function f.

Figure 13 on the next page, which illustrates a "function machine," may give you additional insight into the nature of functions and the symbol $f(x)$. We can think of a function machine as a device that produces exactly one output (range) value for each input (domain) value on the basis of a set of instructions such as those found in an equation, graph, or table. (If more than one output value was produced for an input value, then the machine would not be a function machine.)

For the function $f(x) = 2x + 1$, the machine takes each domain value (input), multiplies it by 2, then adds 1 to the result to produce the range

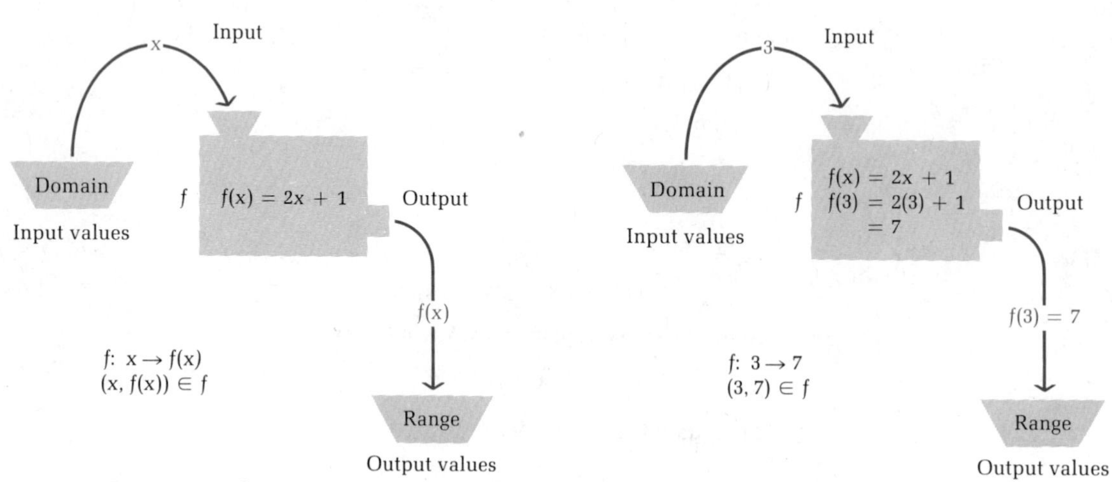

Figure 13 Function machine—exactly one output for each input

value (output). Different rules inside the machine result in different functions.

Example 24 If

$$f(x) = \frac{12}{x-2} \qquad g(x) = 1 - x^2 \qquad h(x) = \sqrt{x-1}$$

then:

(A) $f(6) = \dfrac{12}{6-2} = \dfrac{12}{4} = 3$

(B) $g(-2) = 1 - (-2)^2 = 1 - 4 = -3$

(C) $f(0) + g(1) - h(10) = \dfrac{12}{0-2} + (1 - 1^2) - \sqrt{10-1}$

$$= \frac{12}{-2} + 0 - \sqrt{9}$$

$$= -6 - 3 = -9$$

Problem 24 Use the functions f, g, and h in Example 24 to find:

(A) $f(-2)$ (B) $g(-1)$ (C) $f(3)/h(5)$

Example 25 Find the domains of f, g, and h in Example 24.

Domain f $12/(x-2)$ represents a real number for all replacements of x by real numbers except for $x = 2$ (division by 0 is not defined). Thus, the domain of f is the set of all real numbers except 2. We would often indicate this by writing

$$f(x) = \frac{12}{x-2} \qquad x \neq 2$$

Domain g The domain is all real numbers R, since $1 - x^2$ represents a real number for all replacements of x by real numbers.

Domain h The domain is $[1, \infty)$, since $\sqrt{x-1}$ represents a real number for all real x such that $x - 1$ is not negative; that is, such that

$$x - 1 \geqslant 0$$
$$x \geqslant 1$$

Problem 25 Find the domains of F, G, and H defined by

$$F(x) = x^2 - 3x + 1 \qquad G(x) = \frac{5}{x+3} \qquad H(x) = \sqrt{2-x}$$

Example 26 For $f(x) = 2x - 3$, find:

(A) $f(a)$ (B) $f(a + h)$ (C) $\dfrac{f(a+h) - f(a)}{h}$

Solutions

(A) $f(a) = 2a - 3$

(B) $f(a + h) = 2(a + h) - 3 = 2a + 2h - 3$

(C) $$\frac{f(a+h) - f(a)}{h} = \frac{[2(a+h) - 3] - (2a - 3)}{h}$$
$$= \frac{2a + 2h - 3 - 2a + 3}{h} = \frac{2h}{h} = 2$$

Problem 26 Repeat Example 26 for $f(x) = 3x - 2$.

▪ Linear Functions and Their Graphs

We aleady know how to graph **linear functions**—that is, functions specified by equations of the form

$$\boldsymbol{f(x) = mx + b}$$

This is equivalent to graphing the equation

$$y = mx + b \qquad \text{Slope} = m, \quad y \text{ intercept} = b$$

which we studied in detail in Section 0-4.

Graph of $f(x) = mx + b$

The graph of a linear function f is a nonvertical straight line with slope m and y intercept b.

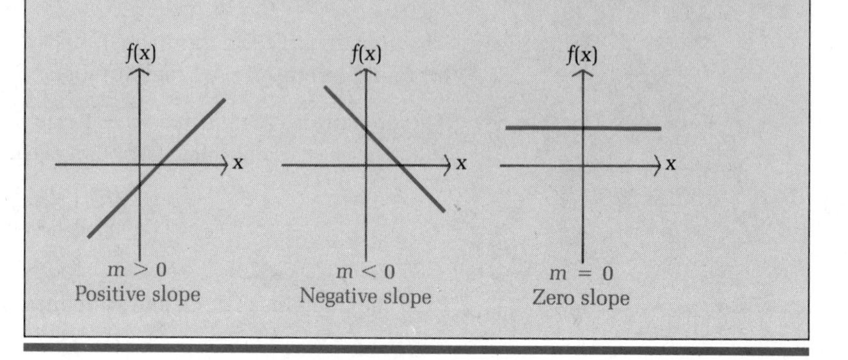

Example 27 Graph the linear function defined by

$$f(x) = -\frac{x}{2} + 3$$

and indicate its slope and y intercept.

Solution

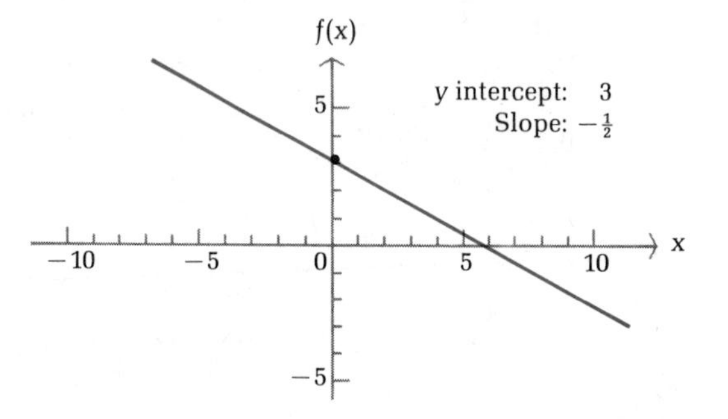

Problem 27 Graph the linear function defined by

$$f(x) = \frac{x}{3} + 1$$

and indicate its slope and y intercept.

▪ Quadratic Functions and Their Graphs

Any function defined by an equation of the form

$$f(x) = ax^2 + bx + c \qquad a \neq 0$$

where a, b, and c are constants and x is a variable, is called a **quadratic function.**

Let us start by graphing the simple quadratic function:

$$f(x) = x^2$$

We evaluate this function for integer values from its domain, find corresponding range values, then plot the resulting ordered pairs and join these points with a smooth curve. The first two steps are usually done mentally or on scratch paper.

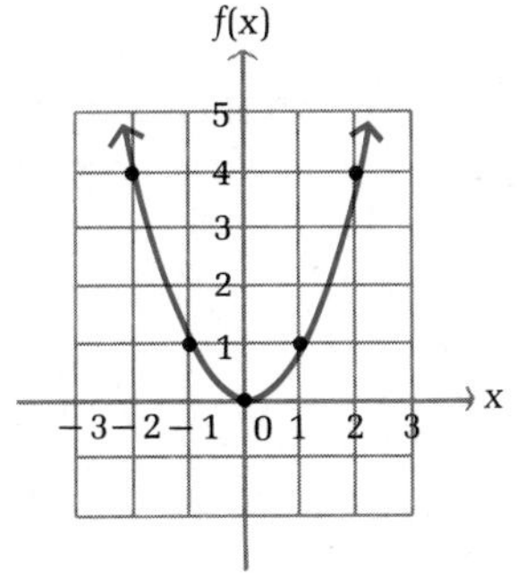

Figure 14

Graphing $f(x) = x^2$

Domain Values	**Range Values**	**Elements of f**
x	$y = f(x)$	$(x, f(x))$
-2	$y = f(-2) = (-2)^2 = 4$	$(-2, 4)$
-1	$y = f(-1) = (-1)^2 = 1$	$(-1, 1)$
0	$y = f(0) = 0^2 = 0$	$(0, 0)$
1	$y = f(1) = 1^2 = 1$	$(1, 1)$
2	$y = f(2) = 2^2 = 4$	$(2, 4)$

The curve shown in Figure 14 is called a **parabola.** It is shown in a course in analytic geometry that the graph of any quadratic function is also a parabola. In general:

Graph of $f(x) = ax^2 + bx + c$, $a \neq 0$

The graph of a quadratic function f is a parabola that has its **axis** (line of symmetry) parallel to the vertical axis. It opens upward if $a > 0$ and downward if $a < 0$. The intersection point of the axis and parabola is called the **vertex.**

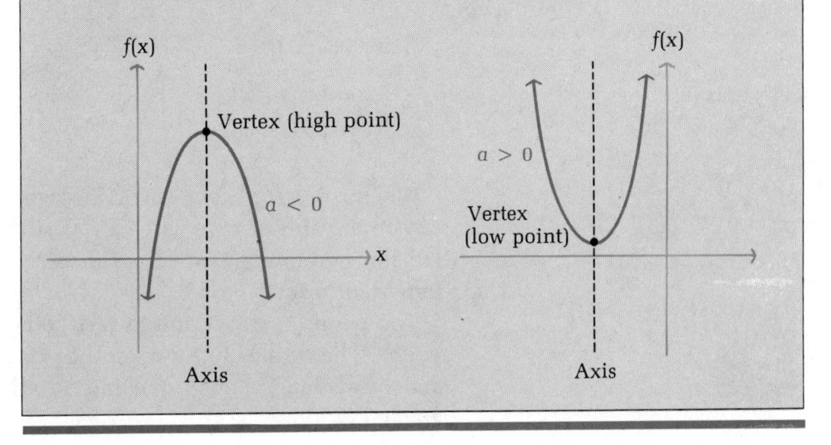

In addition to the point-by-point method of graphing quadratic functions described above, let us consider another approach that will give us added insight into these functions. (A brief review of completing the square, which is discussed in Section 0-3, may prove useful first.) We illustrate the method through an example, and then generalize the results.

Consider the quadratic function given by

$$f(x) = 2x^2 - 8x + 5$$

If we can find the vertex of the graph, then the rest of the graph can be sketched with relatively few points. In addition, we will then have found the maximum or minimum value of the function. We start by transforming the equation into the form

$$f(x) = a(x - h)^2 + k \qquad a, h, k \text{ constants}$$

by completing the square:

$f(x) = 2x^2 - 8x + 5$	Factor the coefficient of x^2 out of the first two terms
$f(x) = 2(x^2 - 4x) + 5$	
$= 2(x^2 - 4x + ?) + 5$	Complete the square within parentheses
$= 2(x^2 - 4x + 4) + 5 - 8$	We added 4 to complete the square inside the parentheses; but because of the 2 on the outside we have actually added 8, so we must subtract 8
$= 2(x - 2)^2 - 3$	The transformation is complete

Thus,

$$f(x) = \underbrace{2(x - 2)^2}_{\substack{\text{Never negative} \\ \text{(Why?)}}} - 3$$

When $x = 2$, the first term on the right vanishes, and we add 0 to -3. For *any* other value of x we will add a positive number to -3, thus making $f(x)$ larger. Therefore, $f(2) = -3$ is the minimum value of $f(x)$ for *all* x. A very important result!

The point $(2, -3)$ is the lowest point on the parabola and is also the vertex. The vertical line $x = 2$ is the axis of the parabola. We plot the vertex and the axis and a couple of points on either side of the axis to complete the graph (Fig. 15).

x	$f(x)$
2	−3
1	−1
3	−1
0	5
4	5

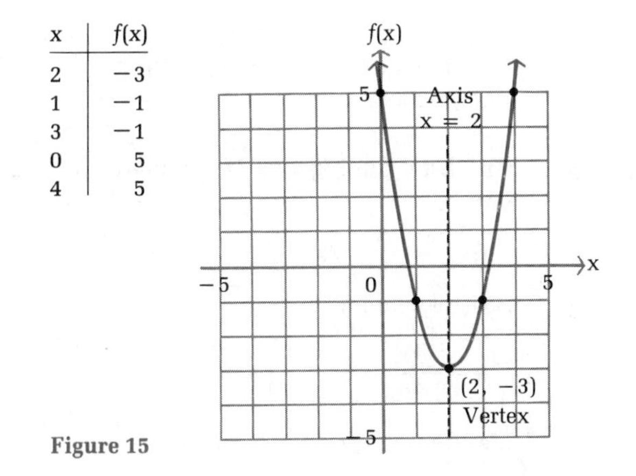

Figure 15

Note the important results we have obtained with this approach. We have found:

1. Axis of the parabola
2. Vertex of the parabola
3. Minimum value of $f(x)$
4. Graph of $y = f(x)$

By proceeding in essentially the same way with the general quadratic function given by

$$f(x) = ax^2 + bx + c \qquad a \neq 0$$

we can obtain the following general results:

Quadratic Function $f(x) = ax^2 + bx + c, \quad a \neq 0$

1. Axis (of symmetry) of the parabola:

$$x = -\frac{b}{2a}$$

2. Maximum or minimum value of $f(x)$:

$$f\left(-\frac{b}{2a}\right) \qquad \begin{array}{l}\text{Minimum if } a > 0\\ \text{Maximum if } a < 0\end{array}$$

3. Vertex of the parabola:

$$\left(-\frac{b}{2a}, f\left(-\frac{b}{2a}\right)\right)$$

To graph a quadratic function using the method of completing the square, we can either actually complete the square as in the earlier example or use the properties listed in the box—some people can more readily remember a formula, others a process. We will use the boxed properties in the next example.

Example 28 Graph by finding axis of symmetry, maximum or minimum of $f(x)$, and vertex:

$$f(x) = 12x - 2x^2 \qquad 0 \leq x \leq 6$$

Solution Axis of symmetry:

$$x = -\frac{b}{2a} = -\frac{12}{2(-2)} = 3$$

Maximum value of $f(x)$ (since $a = -2 < 0$):

$$f(3) = 12(3) - 2(3)^2 = 18$$

Vertex: (3, 18)

x	$f(x)$
3	18
2	16
4	16
1	10
5	10
0	0
6	0

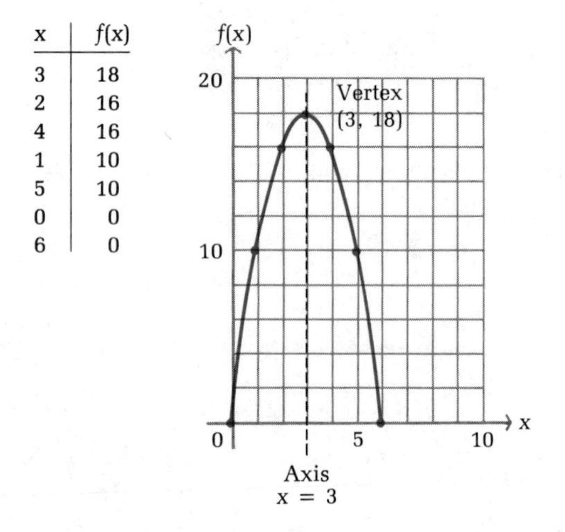

Problem 28 Graph as in Example 28.

$$f(x) = x^2 - 2x - 3 \qquad -1 \leq x \leq 4$$

■ Application: Market Research

The market research department of a company recommended to management that the company manufacture and market a promising new product. After extensive surveys, the research department backed up the recommendation with the **demand equation**

$$x = f(p) = 6{,}000 - 30p \tag{1}$$

where x is the number of units that retailers are likely to buy per month at

$p per unit. Notice that as the price goes up, the number of units goes down. From the financial department, the following **cost equation** was obtained:

$$C = g(x) = 72{,}000 + 60x \tag{2}$$

where $72,000 is the fixed cost (tooling and overhead) and $60 is the variable cost per unit (materials, labor, marketing, transportation, storage, etc.). The **revenue equation** (the amount of money, R, received by the company for selling x units at $p per unit) is

$$R = xp \tag{3}$$

And, finally, the **profit equation** is

$$P = R - C \tag{4}$$

where P is profit, R is revenue, and C is cost.

We notice that the cost equation (2) expresses C as a function of x and the demand equation (1) expresses x as a function of p. Substituting (1) into (2), we obtain cost C as a linear function of price p:

$$\begin{aligned} C &= 72{,}000 + 60(6{,}000 - 3p) \\ &= 432{,}000 - 1{,}800p \end{aligned} \qquad \text{Linear function} \tag{5}$$

Similarly, substituting (1) into (3), we obtain revenue R as a quadratic function of price p:

$$\begin{aligned} R &= (6{,}000 - 30p)p \\ &= 6{,}000p - 30p^2 \end{aligned} \qquad \text{Quadratic function} \tag{6}$$

Now let us graph equations (5) and (6) in the same coordinate system. We obtain Figure 16 on the next page. Notice how much information is contained in this graph. Let us compute the **break-even points;** that is, the prices at which cost equals revenue (the points of intersection of the two graphs in Figure 16). Find p so that

$$\begin{aligned} C &= R \\ 432{,}000 - 1{,}800p &= 6{,}000p - 30p^2 \\ 30p^2 - 7{,}800p + 432{,}000 &= 0 \\ p^2 - 260p + 14{,}400 &= 0 \\ p &= \frac{260 \pm \sqrt{260^2 - 4(14{,}400)}}{2} \\ &= \frac{260 \pm 100}{2} \\ &= \$80, \quad \$180 \end{aligned}$$

Solve using the quadratic formula (Section 0-3)

Thus, at a price of $80 or $180 per unit the company will break even. Between these two prices it is predicted that the company will make a profit.

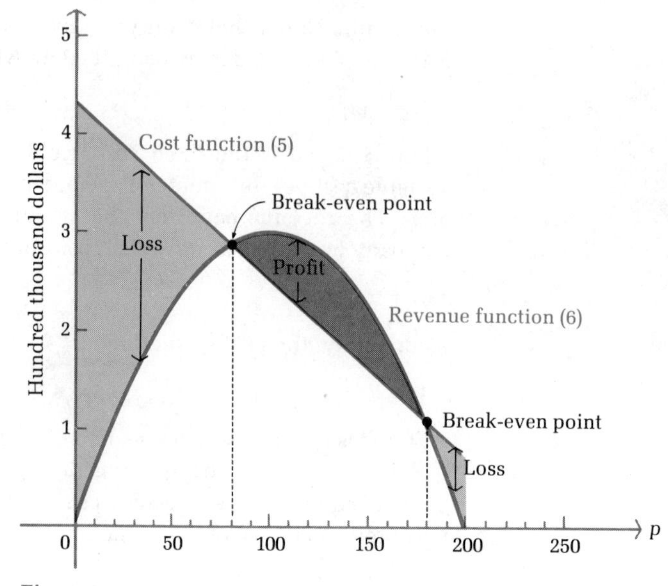

Figure 16

At what price will a **maximum profit** occur? To find out, we write

$$P = R - C$$
$$= (6{,}000p - 30p^2) - (432{,}000 - 1{,}800p)$$
$$= -30p^2 + 7{,}800p - 432{,}000$$

Since this is a quadratic function, the maximum profit occurs at

$$p = -\frac{b}{2a} = -\frac{7{,}800}{2(-30)} = \$130$$

Note that this is not the price at which the maximum revenue occurs. The latter occurs at $p = \$100$, as shown in Figure 16.

Answers to Matched Problems

24. (A) -3 (B) 0 (C) 6
25. Domain of F: R
    Domain of G: All R except -3
    Domain of H: $x \leq 2$ Inequality notation
    $(-\infty, 2]$ Interval notation
26. (A) $3a - 2$ (B) $3a + 3h - 2$ (C) 3
27.

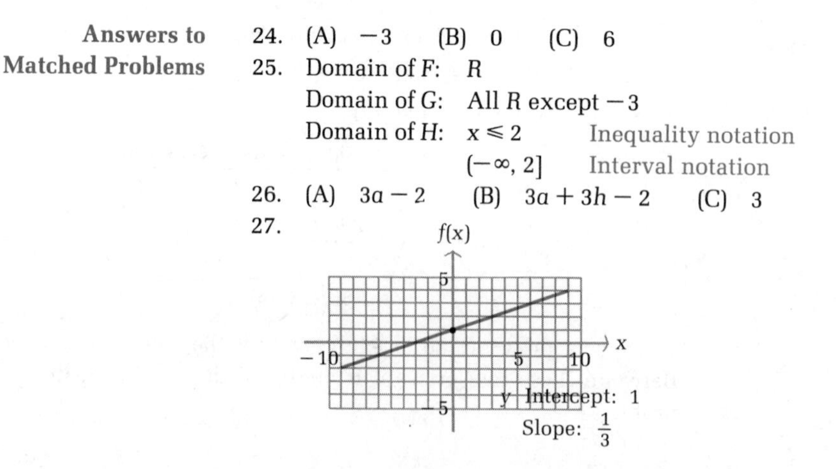

28. Minimum: $f(1) = -4$

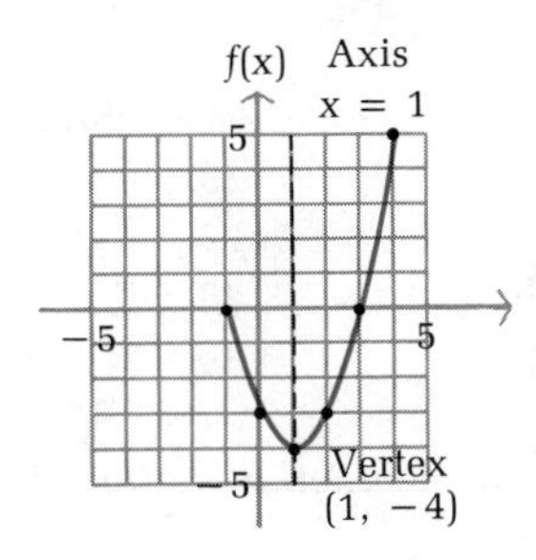

Exercise 0-5

A *Indicate whether each table specifies a function.*

1.

Domain	Range
3 →	0
5 →	1
7 →	2

2.

Domain	Range
−1 →	5
−2 →	7
−3 →	9

3.

Domain	Range
3 →	5
	6
4 →	7
5 →	8

4.

Domain	Range
8 →	0
9 →	1
	2
10 →	3

5.

Domain	Range
3	5
6	
9	6
12	

6.

Domain	Range
−2	
−1	6
0	
1	

Indicate whether each graph specifies a function.

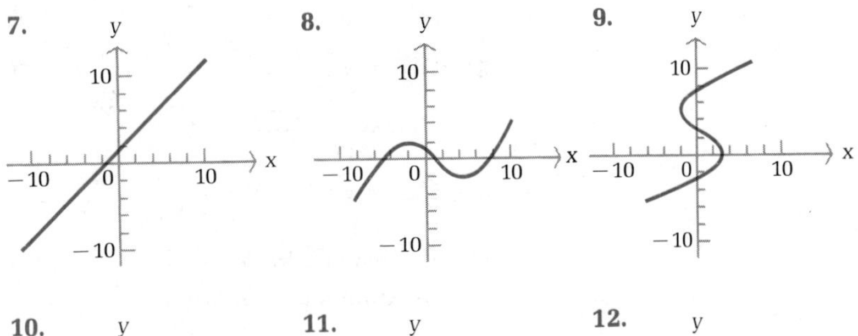

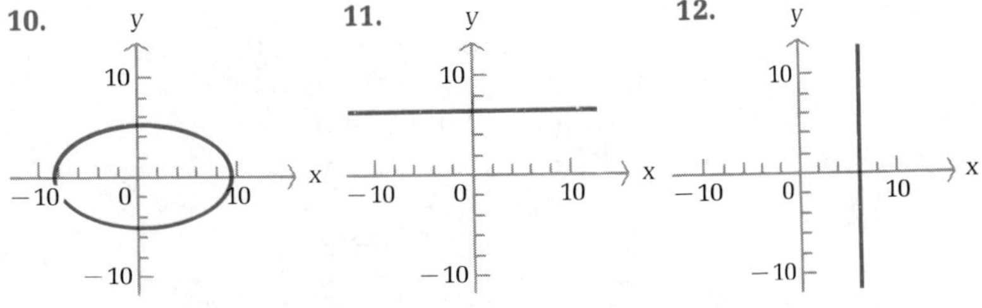

If $f(x) = 3x - 2$ and $g(x) = x - x^2$, find each of the following:

13. $f(2)$
14. $f(1)$
15. $f(-1)$
16. $f(-2)$
17. $g(3)$
18. $g(1)$
19. $f(0)$
20. $f(\frac{1}{3})$
21. $g(-3)$
22. $g(-2)$
23. $f(1) + g(2)$
24. $g(1) + f(2)$
25. $g(2) - f(2)$
26. $f(3) - g(3)$
27. $g(3) \cdot f(0)$
28. $g(0) \cdot f(-2)$
29. $g(-2)/f(-2)$
30. $g(-3)/f(2)$

Graph each linear function, and indicate its slope and y intercept.

31. $f(x) = 2x - 4$
32. $g(x) = \dfrac{x}{2}$
33. $h(x) = 4 - 2x$
34. $f(x) = -\dfrac{x}{2} + 3$
35. $g(x) = -\frac{2}{3}x + 4$
36. $f(x) = 3$

B *If $f(x) = 2x + 1$, $g(x) = x^2 - x$, and $k(x) = \sqrt{x}$, find each of the following:*

37. $f(3) + g(-2)$
38. $g(-1) - f(1)$
39. $k(9) - g(-2)$
40. $g(-2) - k(4)$
41. $f[k(4)]$
42. $k[f(4)]$
43. $k[g(2)]$
44. $g[k(9)]$
45. $g(e)$
46. $f(a)$
47. $k(u)$
48. $g(t)$
49. $g(2 + h)$
50. $f(2 + h)$
51. $f(a + h)$
52. $g(a + h)$
53. $\dfrac{f(2 + h) - f(2)}{h}$
54. $\dfrac{f(a + h) - f(a)}{h}$
55. $\dfrac{g(2 + h) - g(2)}{h}$
56. $\dfrac{g(a + h) - g(a)}{h}$

Find the domain of each function in Problems 57–62.

57. $f(x) = \sqrt{x}$
58. $f(x) = 1/\sqrt{x}$
59. $f(x) = \dfrac{x - 3}{(x - 5)(x + 3)}$
60. $f(x) = \dfrac{x + 1}{x - 2}$
61. $f(x) = \sqrt{x + 5}$
62. $f(x) = \sqrt{7 - x}$

Graph each quadratic function, and include the axis of symmetry, vertex, and maximum or minimum value.

63. $f(x) = x^2 + 8x + 16$
64. $h(x) = x^2 - 2x - 3$
65. $f(u) = u^2 - 2u + 4$
66. $f(x) = x^2 - 10x + 25$
67. $h(x) = 2 + 4x - x^2$
68. $g(x) = -x^2 - 6x - 4$
69. $f(x) = 6x - x^2$
70. $G(x) = 16x - 2x^2$
71. $F(s) = s^2 - 4$
72. $g(t) = t^2 + 4$
73. $F(x) = 4 - x^2$
74. $G(x) = 9 - x^2$

C

75. If

$$f(x) = \begin{cases} x^2 & \text{when } x < 1 \\ 2x & \text{when } x \geq 1 \end{cases}$$

find: (A) $f(-1)$ (B) $f(0)$ (C) $f(1)$ (D) $f(3)$

76. If

$$f(x) = \begin{cases} -x & \text{when } x \leq 0 \\ x & \text{when } x > 0 \end{cases}$$

find: (A) $f(-3)$ (B) $f(-1)$ (C) $f(0)$ (D) $f(5)$

Graph each quadratic function, and include the axis of symmetry, vertex, and maximum or minimum value.

77. $f(x) = x^2 - 7x + 10$

78. $g(t) = t^2 - 5t + 2$

79. $g(t) = 4 + 3t - t^2$

80. $h(x) = 2 - 5x - x^2$

Applications

Business & Economics

81. *Cost equation.* The cost equation (in dollars) for a particular company that produces stereos is found to be

$$C = g(n) = 96{,}000 + 80n$$

where \$96,000 represents fixed costs (tooling and overhead) and \$80 is the variable cost per unit (material, labor, etc.). Graph this function for $0 \leq n \leq 1{,}000$.

82. *Demand equation.* After extensive surveys, the research department of a stereo manufacturing company produced the demand equation

$$n = f(p) = 8{,}000 - 40p \qquad 100 \leq p \leq 200$$

where n is the number of units that retailers are likely to purchase per week at a price of \$$p$ per unit. Graph the function for the indicated domain.

83. *Packaging.* A candy box is to be made out of a piece of cardboard that measures 8 by 12 inches. Equal-sized squares x inches on a side will be cut out of each corner, and then the ends and sides will be folded up to form a rectangular box.

(A) Express the volume of the box $V(x)$ in terms of x.

(B) What is the domain of the function V (determined by the physical restrictions)?

(C) Complete the table:

x	V(x)
1	
2	
3	

Notice how the volume changes with different choices of x

84. *Packaging.* A parcel delivery service will only deliver packages with length plus girth (distance around) not exceeding 108 inches. A rectangular shipping box with square ends, x inches on a side, is to be used.

(A) If the full 108 inches is to be used, express the volume of the box $V(x)$ in terms of x.

(B) What is the domain of the function V (determined by the physical restrictions)?

(C) Complete the table:

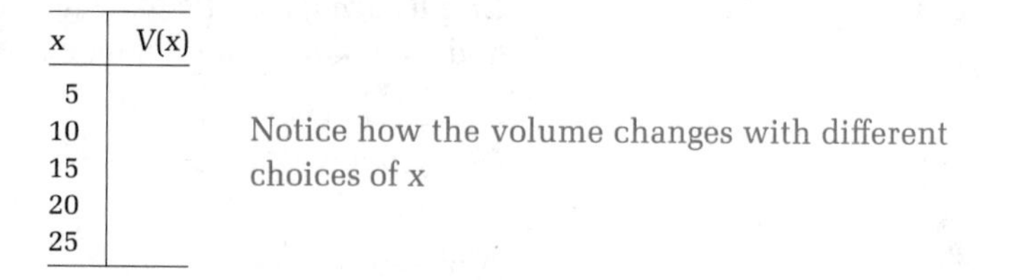

x	$V(x)$
5	
10	
15	
20	
25	

Notice how the volume changes with different choices of x

85. Suppose that in the market research example in this section the demand equation (1) is changed to $x = 9{,}000 - 30p$ and the cost equation (2) is changed to $C = 90{,}000 + 30x$.

(A) Express cost C as a linear function of price p.

(B) Express revenue R as a quadratic function of price p.

(C) Graph the cost and revenue functions found in parts A and B in the same coordinate system, and identify the regions of profit and loss on your graph.

(D) Find the break-even points; that is, find the prices to the nearest dollar at which $R = C$. (A hand calculator might prove useful here.)

(E) Find the price that produces the maximum revenue.

Life Sciences

86. *Air pollution.* On an average summer day in a large city, the pollution index at 8:00 AM is 20 parts per million and it increases linearly by 15 parts per million each hour until 3:00 PM. Let $P(x)$ be the amount of pollutants in the air x hours after 8:00 AM.

(A) Express $P(x)$ as a linear function of x.

(B) What is the air pollution index at 1:00 PM?

(C) Graph the function P for $0 \leq x \leq 7$.

(D) What is the slope of the graph? (The slope is the amount of increase in pollution for each additional hour of time.)

Social Sciences

87. *Psychology—sensory perception.* One of the oldest studies in psychology concerns the following question: Given a certain level of stimulation (light, sound, weight lifting, electric shock, and so on), how much should the stimulation be increased for a person to notice the difference? In the middle of the nineteenth century, E.H. Weber (a German physiologist) formulated a law that still carries his name: If Δs is the change in stimulus that will just be noticeable at a stimulus level s, then the ratio of Δs to s is a constant:

$$\frac{\Delta s}{s} = k$$

Hence, the amount of change that will be noticed is a linear function of the stimulus level, and we note that the greater the stimulus, the more it takes to notice a difference. In an experiment on weight lifting, the constant k for a given individual was found to be $\frac{1}{30}$.

(A) Find Δs (the difference that is just noticeable) at the 30-pound level; at the 90-pound level.
(B) Graph $\Delta s = s/30$ for $0 \leqslant s \leqslant 120$.
(C) What is the slope of the graph?

0-6 Exponential Functions

- Exponential Functions and Their Graphs
- Base e
- Basic Exponential Properties

Exponential Functions and Their Graphs

Many students, if asked to graph equations such as $y = 2^x$ or $y = 2^{-x}$, would not hesitate at all. [*Note:* $2^{-x} = 1/2^x = (\frac{1}{2})^x$.] They would likely make up tables by assigning integers to x, plot the resulting points, and then join these points with a smooth curve as in Figure 17. The only catch is that 2^x

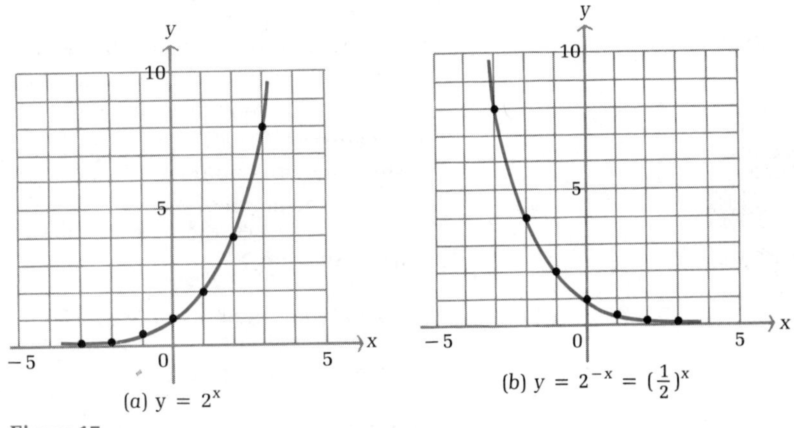

(a) $y = 2^x$ (b) $y = 2^{-x} = (\frac{1}{2})^x$

Figure 17

has not been defined at this point for all real numbers. From Appendix A we know what 2^5, 2^{-3}, $2^{2/3}$, $2^{-3/5}$, $2^{1.4}$, and $2^{-3.15}$ all mean (that is, 2^p, where p is a rational number), but what does

$$2^{\sqrt{2}}$$

mean? The question is not easy to answer at this time. In fact, a precise definition of $2^{\sqrt{2}}$ must wait for more advanced courses, where we can show that

$$2^x$$

names a real number for x any real number, and that the graph of $y = 2^x$ is as indicated in Figure 17A. We can also show that for x irrational, 2^x can be approximated as closely as we like by using rational number approximations for x. Since $\sqrt{2} = 1.414213\ldots$, for example, the sequence

$$2^{1.4}, 2^{1.41}, 2^{1.414}, \ldots$$

approximates $2^{\sqrt{2}}$, and as we move to the right the approximation improves.

Both $y = 2^x$ and $y = 2^{-x}$ are examples of *exponential functions.* (Do not confuse $y = 2^x$ with $y = x^2$; the latter is a quadratic function.) In general, an **exponential function** is a function defined by an equation of the form:

Exponential Function

$$f(x) = b^x \qquad b > 0, \quad b \neq 1$$

where b is a constant, called the base, and the exponent x is a variable. The replacement set for the exponent, the **domain of *f*,** is the set of real numbers R (with the assumption that we "know" what b^x means for x irrational). The **range of *f*** is the set of positive real numbers. We require b to be positive to avoid nonreal quantities such as $(-2)^{1/2}$ and 0^0. It is useful to note that the graph of

$$f(x) = b^x \qquad b > 1$$

will look very much like Figure 17A, and the graph of

$$f(x) = b^x \qquad 0 < b < 1$$

will look very much like Figure 17B. [*Note:* In both cases, the graphs approach, but never touch, the horizontal axis.]

Example 29 Graph $y = \left(\frac{1}{2}\right) 4^x$ for $-3 \leq x \leq 3$.

Solution

x	y
−3	0.01
−2	0.03
−1	0.13
0	0.50
1	2.00
2	8.00
3	32.00

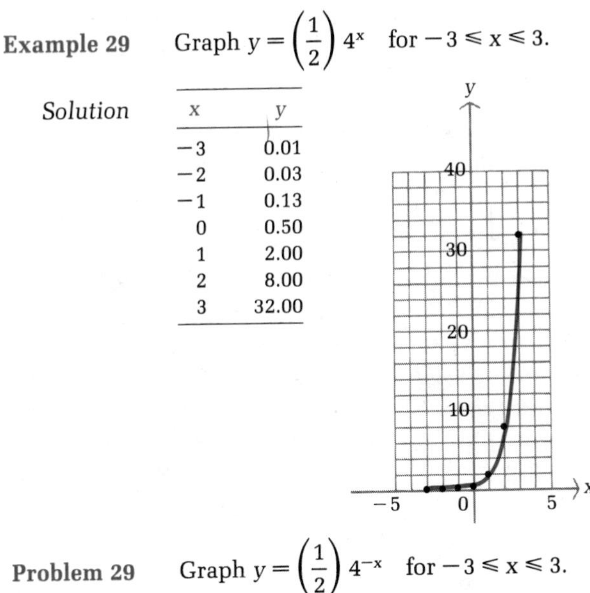

Problem 29 Graph $y = \left(\frac{1}{2}\right) 4^{-x}$ for $-3 \leq x \leq 3$.

A great variety of growth phenomena can be described by exponential functions, which is the reason such functions are often referred to as **growth functions.** They are used to describe the growth of money at compound interest; population growth of people, animals, and bacteria; radioactive decay (negative growth); and the growth of learning a skill such as typing or swimming relative to practice.

■ Base e

For introductory purposes, the bases 2 and $\frac{1}{2}$ were convenient choices; however, a certain irrational number, denoted by e, is by far the most frequently used exponential base for both theoretical and practical purposes. In fact,

$$f(x) = e^x$$

is often referred to as *the* exponential function because of its widespread use. The reasons for the preference for e as a base is made clear in the study of calculus. And at that time, it is shown that e is approximated by $(1 + 1/n)^n$ to any decimal accuracy desired by making n (an integer) sufficiently large. The irrational number e to eight decimal places is

$$e \approx 2.718\ 281\ 83$$

Since, for large n,

$$\left(1 + \frac{1}{n}\right)^n \approx e$$

we can raise each side to the xth power to obtain

$$\left(1+\frac{1}{n}\right)^{nx} \approx e^x$$

Thus, for any x, e^x can be approximated as close as we like by making n (an integer) sufficiently large in

$$\left(1+\frac{1}{n}\right)^{nx}$$

Because of the importance of e^x and e^{-x}, tables for their evaluation are readily available. In fact, all scientific and financial calculators can evaluate these functions directly.* A short table for evaluating e^x and e^{-x} is provided in Table I of Appendix B. The important constant e, along with two other important constants—$\sqrt{2}$ and π—are shown on the number line in Figure 18A. The graph of $y = e^x$ is shown in Figure 18B.

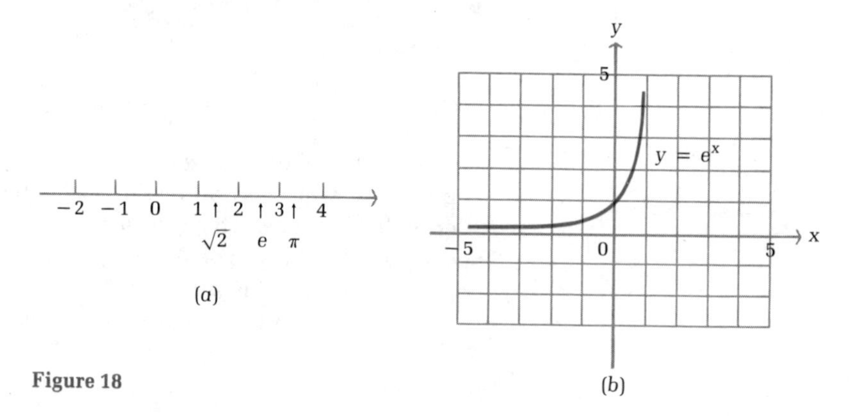

Figure 18

Example 30 If \$$P$ is invested at 100r% compounded continuously, then the amount A in the account at the end of t years is given by (from the mathematics of finance and calculus):

$$A = Pe^{rt}$$

If \$100 is invested at 12% compounded continuously, graph the amount in the account relative to time for a period of 10 years. [*Note:* Many financial institutions use continuous compounding; look at rates in any financial publication to find some examples.]

Solution We wish to graph

$$A = 100e^{0.12t} \qquad 0 \leq t \leq 10$$

We make up a table of values using a calculator, graph the points from the table, and then join the points with a smooth curve.

* Consult the manual for your calculator to determine the procedure for evaluating e^x. Some calculators do not have a key labeled e^x. Instead they use a combination of two keys, such as INV and ln x, to evaluate e^x.

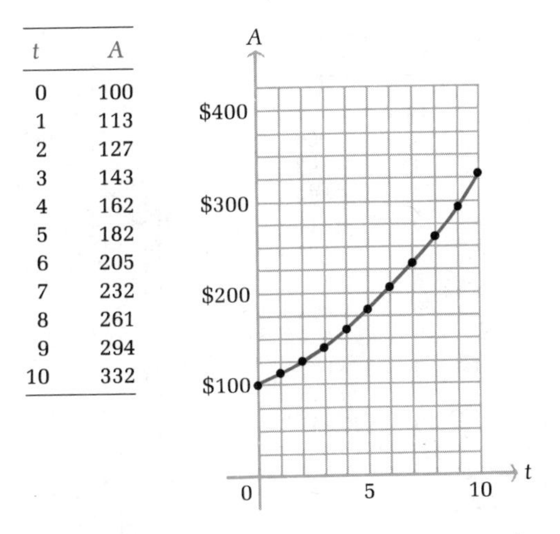

t	A
0	100
1	113
2	127
3	143
4	162
5	182
6	205
7	232
8	261
9	294
10	332

Problem 30 Repeat Example 30 with $5,000 invested at 20% compounded continuously.

▪ Basic Exponential Properties

In Appendix A we discussed five laws for integer and rational exponents. It can be shown that these laws also hold for irrational exponents. Thus, we now assume that all five laws of exponents hold for *any* real exponents as long as the involved bases are positive. In addition,

$$b^m = b^n \quad \text{if and only if} \quad m = n, \quad b > 0, \quad b \neq 1$$

Thus, if $2^{15} = 2^{3x}$, then $3x = 15$ and $x = 5$.

Answers to Matched Problems

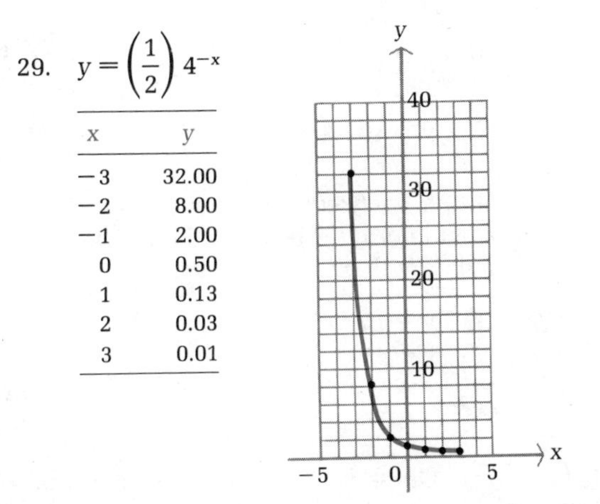

29. $y = \left(\frac{1}{2}\right)4^{-x}$

x	y
−3	32.00
−2	8.00
−1	2.00
0	0.50
1	0.13
2	0.03
3	0.01

30. $A = 5{,}000e^{0.2t}$

t	A
0	5,000
1	6,107
2	7,459
3	9,111
4	11,128
5	13,591
6	16,601
7	20,276
8	24,765
9	30,248
10	36,945

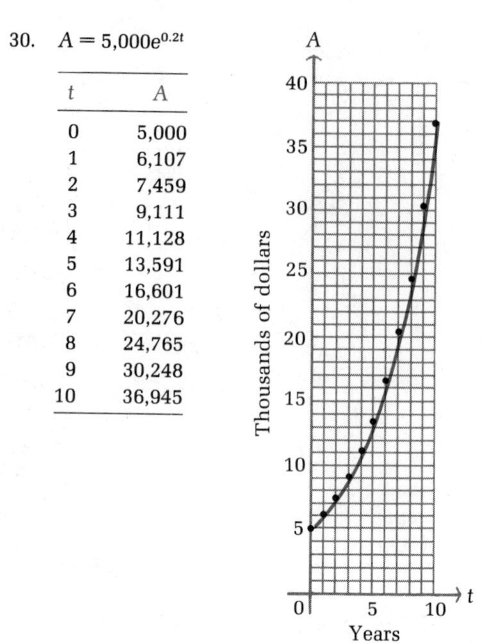

Exercise 0-6

A *Graph each equation for $-3 \leq x \leq 3$. Plot points using integers for x, and then join the points with a smooth curve.*

1. $y = 3^x$
2. $y = 10 \cdot 2^x$ [*Note:* $10 \cdot 2^x \neq 20^x$]
3. $y = \left(\frac{1}{3}\right)^x = 3^{-x}$
4. $y = 10 \cdot \left(\frac{1}{2}\right)^x = 10 \cdot 2^{-x}$
5. $y = 10 \cdot 3^x$
6. $y = 10 \cdot \left(\frac{1}{3}\right)^x = 10 \cdot 3^{-x}$

B *Graph each equation for $-3 \leq x \leq 3$. Use Table I of Appendix B or a calculator if the base is e. Plot points using integers for x, and then join the points with a smooth curve.*

7. $y = 10 \cdot 2^{2x}$
8. $y = 10 \cdot 2^{-3x}$
9. $y = e^x$
10. $y = e^{-x}$
11. $y = 10e^{0.2x}$
12. $y = 100e^{0.1x}$
13. $y = 100e^{-0.1x}$
14. $y = 10e^{-0.2x}$

C 15. Graph $y = e^{-x^2}$ for $x = -1.5, -1.0, -0.5, 0, 0.5, 1.0, 1.5$, and then join these points with a smooth curve. Use Table I of Appendix B or a calculator. (This is a very important curve in probability and statistics.)

16. Graph $y = y_0 2^x$, where y_0 is the value of y when $x = 0$. (Express the vertical scale in terms of y_0.)
17. Graph $y = 2^x$ and $x = 2^y$ on the same coordinate system.
18. Graph $y = 10^x$ and $x = 10^y$ on the same coordinate system.

Applications

Business & Economics

19. *Exponential growth.* If we start with 2¢ and double the amount each day, we would have 2^n¢ after n days. Graph $f(n) = 2^n$ for $1 \leq n \leq 10$. (Label the vertical scale so that the graph will not go off the paper.)
20. *Compound interest.* If a certain amount of money P (the principal) is invested at $100r\%$ interest compounded annually, the amount of money (A) after t years is given by

$$A = P(1 + r)^t$$

Graph this equation for $P = \$100$, $r = 0.10$, and $0 \leq t \leq 6$. How much money would a person have after 10 years if no interest were withdrawn? (Technically t should be restricted to nonnegative integers. However, as a visual aid in observing compound growth, join the points with a smooth curve.)

Life Sciences

21. *Bacteria growth.* A single cholera bacterium divides every $\frac{1}{2}$ hour to produce two complete cholera bacteria. If we start with 100 bacteria, in t hours (assuming adequate food supply) we will have

$$A = 100 \cdot 2^{2t}$$

bacteria. Graph this equation for $0 \leq t \leq 5$.

22. *Ecology.* The atmospheric pressure (P, in pounds per square inch) may be calculated approximately from the formula

$$P = 14.7e^{-0.21h}$$

where h is the altitude above sea level in miles. Graph this equation for $0 \leq h \leq 12$.

Social Sciences

23. *Learning curves.* The performance record of a particular person learning to type is given approximately by

$$N = 100(1 - e^{-0.1t})$$

where N is the number of words typed per minute and t is the number of weeks of instruction. Graph this equation for $0 \leq t \leq 40$. What does N approach as t approaches ∞?

24. *Small group analysis.* After a lengthy investigation, sociologists Stephan and Mischler found that if the members of a discussion group of ten were ranked according to the number of times each participated,

then the number of times, $N(k)$, the kth-ranked person participated was given approximately by

$$N(k) = N_1 e^{-0.11(k-1)} \qquad 1 \leq k \leq 10$$

where N_1 was the number of times the top-ranked person participated in the discussion. Graph the equation assuming $N_1 = 100$. [For a general discussion of this phenomenon, see J. S. Coleman, *Introduction to Mathematical Sociology* (London: The Free Press of Glencoe, 1964), pp. 28–31.]

0-7 Logarithmic Functions

- Definition of Logarithmic Functions
- From Logarithmic to Exponential Form and Vice Versa
- Properties of Logarithmic Functions
- Calculator Evaluation of Common and Natural Logarithms
- Application

Now we are ready to consider logarithmic functions, which are closely related to exponential functions.

Definition of Logarithmic Functions

If we start with an exponential function f defined by

$$y = 2^x \tag{1}$$

and interchange the variables, we obtain an equation that defines a new relation g defined by

$$x = 2^y \tag{2}$$

Any ordered pair of numbers that belongs to f will belong to g if we interchange the order of the components. For example, (3, 8) satisfies equation (1) and (8, 3) satisfies equation (2). Thus, the domain of f becomes the range of g and the range of f becomes the domain of g. Graphing f and g on the same coordinate system (Figure 19), we see that g is also a function. We call this new function the **logarithmic function with base 2,** and write

$$y = \log_2 x \qquad \text{if and only if} \qquad x = 2^y$$

Note that if we fold the paper along the dashed line $y = x$ in Figure 19, the two graphs match exactly.

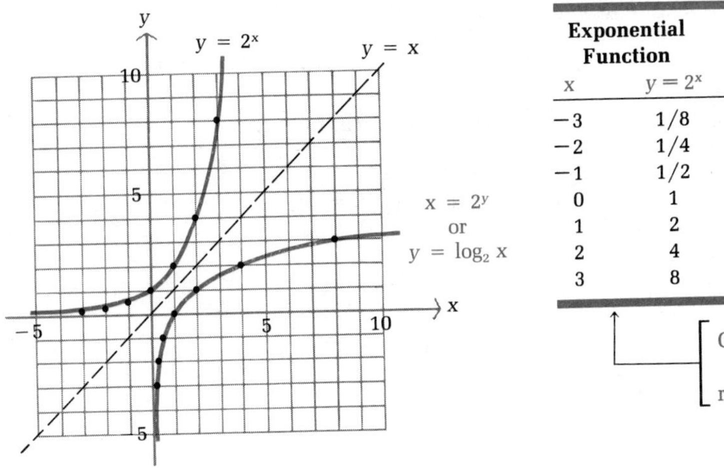

Exponential Function		Logarithmic Function	
x	$y = 2^x$	$x = 2^y$	y
−3	1/8	1/8	−3
−2	1/4	1/4	−2
−1	1/2	1/2	−1
0	1	1	0
1	2	2	1
2	4	4	2
3	8	8	3

Ordered pairs reversed

Figure 19

In general, we define the logarithmic functions with base b as follows:

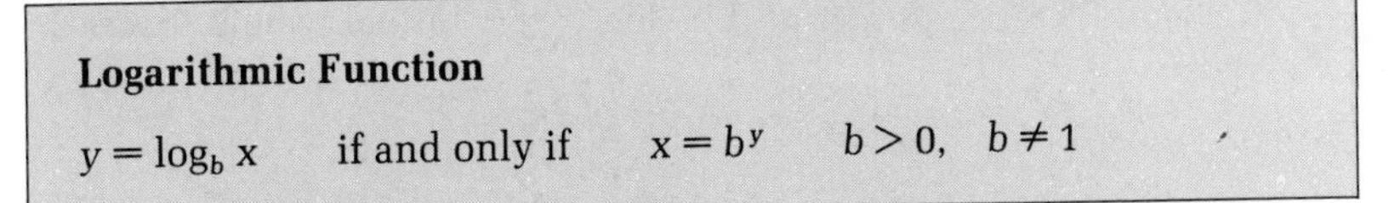

Logarithmic Function

$y = \log_b x$ if and only if $x = b^y$ $b > 0,\ b \neq 1$

In words, **the logarithm of a number x to a base b ($b > 0$, $b \neq 1$) is the exponent to which b must be raised to equal x.** It is important to remember that $y = \log_b x$ and $x = b^y$ describe the same function, while $y = b^x$ is the related exponential function. Look at Figure 19 again.

Since the domain of an exponential function includes all real numbers and its range is the set of positive real numbers, the **domain** of a logarithmic function is the set of all positive real numbers and its **range** is the set of all real numbers. Remember that **the logarithm of 0 or a negative number is not defined.**

■ From Logarithmic to Exponential Form and Vice Versa

We now consider the matter of converting logarithmic forms to equivalent exponential forms and vice versa.

Example 31 Change from logarithmic form to exponential form.

(A) $\log_5 25 = 2$ is equivalent to $25 = 5^2$
(B) $\log_9 3 = 1/2$ is equivalent to $3 = 9^{1/2}$
(C) $\log_2 (1/4) = -2$ is equivalent to $1/4 = 2^{-2}$

Problem 31 Change to an equivalent exponential form.

(A) $\log_3 9 = 2$ (B) $\log_4 2 = 1/2$ (C) $\log_3 (1/9) = -2$

Example 32 Change from exponential form to logarithmic form.

(A) $64 = 4^3$ is equivalent to $\log_4 64 = 3$
(B) $6 = \sqrt{36}$ is equivalent to $\log_{36} 6 = 1/2$
(C) $1/8 = 2^{-3}$ is equivalent to $\log_2 (1/8) = -3$

Problem 32 Change to an equivalent logarithmic form.

(A) $49 = 7^2$ (B) $3 = \sqrt{9}$ (C) $1/3 = 3^{-1}$

Example 33 Find y, b, or x.

(A) $y = \log_4 16$ (B) $\log_2 x = -3$
(C) $y = \log_8 4$ (D) $\log_b 100 = 2$

Solutions (A) $y = \log_4 16$ is equivalent to $16 = 4^y$. Thus,

$$y = 2$$

(B) $\log_2 x = -3$ is equivalent to $x = 2^{-3}$. Thus,

$$x = \frac{1}{2^3} = \frac{1}{8}$$

(C) $y = \log_8 4$ is equivalent to

$$4 = 8^y \quad \text{or} \quad 2^2 = 2^{3y}$$

Thus,

$$3y = 2$$

$$y = \frac{2}{3}$$

(D) $\log_b 100 = 2$ is equivalent to $100 = b^2$. Thus,

$$b = 10$$ Recall that b cannot be negative.

Problem 33 Find y, b, or x.

(A) $y = \log_9 27$ (B) $\log_3 x = -1$ (C) $\log_b 1{,}000 = 3$

Example 34 Graph $y = \log_2(x + 1)$ by converting to an equivalent exponential form first.

Solution Changing $y = \log_2(x + 1)$ to an equivalent exponential form, we have

$$x + 1 = 2^y \quad \text{or} \quad x = 2^y - 1$$

Even though x is the independent variable and y is the dependent variable, it is easier to assign y values and solve for x.

x	y
$-\frac{7}{8}$	-3
$-\frac{3}{4}$	-2
$-\frac{1}{2}$	-1
0	0
1	1
3	2
7	3

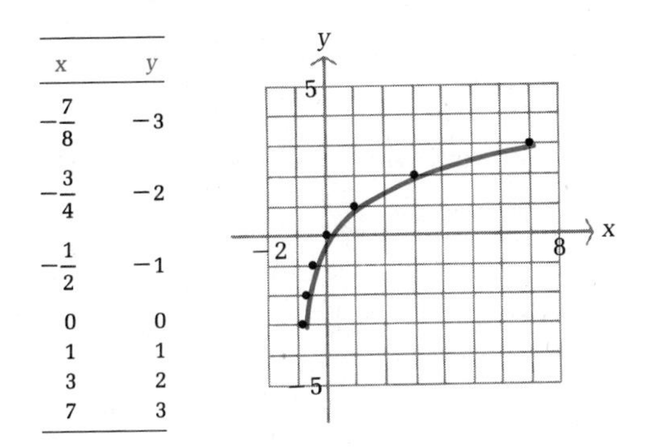

Problem 34 Graph $y = \log_3 (x - 1)$ by converting to an equivalent exponential form first.

▪ Properties of Logarithmic Functions

Logarithmic functions have several very useful properties that follow directly from their definitions. These properties will enable us to convert multiplication problems into addition problems, division problems into subtraction problems, and power and root problems into multiplication problems. We will also be able to solve exponential equations such as $2 = 1.06^n$.

Logarithmic Properties

$(b > 0, \quad b \neq 1, \quad M > 0, \quad N > 0)$

1. $\log_b b^x = x$
2. $\log_b MN = \log_b M + \log_b N$
3. $\log_b \dfrac{M}{N} = \log_b M - \log_b N$
4. $\log_b M^p = p \log_b M$
5. $\log_b M = \log_b N$ if and only if $M = N$
6. $\log_b 1 = 0$

The first property follows directly from the definition of a logarithmic function. Here, we will sketch a proof for property 2. The other properties are established in a similar way. Let

$$u = \log_b M \qquad \text{and} \qquad v = \log_b N$$

Or, in equivalent exponential form,

$$M = b^u \qquad \text{and} \qquad N = b^v$$

Now, see if you can provide reasons for each of the following steps:

$$\log_b MN = \log_b b^u b^v = \log_b b^{u+v} = u + v = \log_b M + \log_b N$$

Example 35 (A)

$$\log_b \frac{wx}{yz} = \log_b wx - \log_b yz$$
$$= \log_b w + \log_b x - (\log_b y + \log_b z)$$
$$= \log_b w + \log_b x - \log_b y - \log_b z$$

(B)

$$\log_b (wx)^{3/5} = \frac{3}{5} \log_b wx$$
$$= \frac{3}{5} (\log_b w + \log_b x)$$

Problem 35 Write in simpler logarithmic forms, as in Example 35.

(A) $\log_b \frac{R}{ST}$ (B) $\log_b \left(\frac{R}{S}\right)^{2/3}$

The following examples and problems, though somewhat artificial, will give you additional practice in using basic logarithmic properties.

Example 36 Find x so that

$$\frac{3}{2} \log_b 4 - \frac{2}{3} \log_b 8 + \log_b 2 = \log_b x$$

Solution

$$\frac{3}{2} \log_b 4 - \frac{2}{3} \log_b 8 + \log_b 2 = \log_b x$$

$$\log_b 4^{3/2} - \log_b 8^{2/3} + \log_b 2 = \log_b x \qquad \text{Property 4}$$

$$\log_b 8 - \log_b 4 + \log_b 2 = \log_b x$$

$$\log_b \frac{8 \cdot 2}{4} = \log_b x \qquad \text{Properties 2 and 3}$$

$$\log_b 4 = \log_b x$$

$$x = 4 \qquad \text{Property 5}$$

Problem 36 Find x so that

$$3 \log_b 2 + \frac{1}{2} \log_b 25 - \log_b 20 = \log_b x$$

Example 37 Solve $\log_{10} x + \log_{10}(x + 1) = \log_{10} 6$.

Solution

$$\log_{10} x + \log_{10}(x + 1) = \log_{10} 6$$

$$\log_{10} x(x + 1) = \log_{10} 6 \qquad \text{Property 2}$$

$$x(x + 1) = 6 \qquad \text{Property 5}$$

$$x^2 + x - 6 = 0 \qquad \text{Solve by factoring.}$$

$$(x + 3)(x - 2) = 0$$

$$x = -3, 2$$

We must exclude $x = -3$, since negative numbers are not in the domains of logarithmic functions; hence,

$$x = 2$$

is the only solution.

Problem 37 Solve $\log_3 x + \log_3(x - 3) = \log_3 10$.

Calculator Evaluation of Common and Natural Logarithms

Of all possible logarithmic bases, the base e and the base 10 are used almost exclusively. Before we can use logarithms in certain practical problems, we need to be able to approximate the logarithm of any number either to base 10 or to base e. And conversely, if we are given the logarithm of a number to base 10 or base e, we need to be able to approximate the number. Historically, tables such as Tables II and III of Appendix B were used for this purpose, but now with inexpensive scientific hand calculators readily available, most people will use a calculator, since it is faster and far more accurate.

Common logarithms (also called **Briggsian logarithms**) are logarithms with base 10. **Natural logarithms** (also called **Napierian logarithms**) are logarithms with base e. Most scientific calculators have a button labeled "log" (or "LOG") and a button labeled "ln" (or "LN"). The former represents a common (base 10) logarithm and the latter a natural (base e) logarithm. In fact, "log" and "ln" are both used extensively in mathematical literature, and whenever you see either used in this book without a base indicated they will be interpreted as follows:

Logarithmic Notation

$$\log x = \log_{10} x$$

$$\ln x = \log_e x$$

Finding the common or natural logarithm using a scientific calculator is very easy: you simply enter a number from the domain of the function and push the log or ln button.

Example 38 Use a scientific calculator to find each to six decimal places:

(A) $\log 3{,}184$ (B) $\ln 0.000\ 349$ (C) $\log(-3.24)$

Solutions

	Enter	*Press*	*Display*
(A)	3184	log	3.502973
(B)	0.000 349	ln	−7.960439
(C)	−3.24	log	Error

An error is indicated in part C because −3.24 is not in the domain of the log function.

Problem 38 Use a scientific calculator to find each to six decimal places:

(A) $\log 0.013\ 529$ (B) $\ln 28.693\ 28$ (C) $\ln(-0.438)$

We now turn to the second problem to be discussed in this section: Given the logarithm of a number, find the number. We make direct use of the logarithmic–exponential relationships, which follow directly from the definition of logarithmic functions at the beginning of this section.

Logarithmic–Exponential Relationships

$\log x = y$ is equivalent to $x = 10^y$

$\ln x = y$ is equivalent to $x = e^y$

Example 39 Find x to three significant digits, given the indicated logarithms:

(A) $\log x = -9.315$ (B) $\ln x = 2.386$

Solutions (A) $\log x = -9.315$ Change to equivalent exponential form.

$x = 10^{-9.315}$

$x = 4.84 \times 10^{-10}$ The answer is displayed in scientific notation in the calculator.

(B) $\ln x = 2.386$ Change to equivalent exponential form.

$x = e^{2.386}$

$x = 10.9$

Problem 39 Find x to four significant digits, given the indicated logarithms.

(A) $\ln x = -5.062$ (B) $\log x = 12.082\ 1$

■ Application

If P dollars are invested at $100i\%$ interest per period for n periods, and interest is paid to the account at the end of each period, then the amount of money in the account at the end of period n is given by

$$A = P(1 + i)^n \qquad \text{Compound interest formula}$$

The fact that interest paid to the account at the end of each period earns interest during the following periods is the reason this is called a **compound interest** formula.

Example 40
Doubling Time

How long (to the next whole year) will it take money to double if it is invested at 20% interest compounded annually?

Solution Find n for $A = 2P$ and $i = 0.2$.

$$A = P(1 + i)^n$$

$$2P = P(1 + 0.2)^n$$

$$1.2^n = 2 \qquad \text{Solve for } n \text{ by taking the natural or common log of both sides.}$$

$$\ln 1.2^n = \ln 2$$

$$n \ln 1.2 = \ln 2 \qquad \text{Property 4}$$

$$n = \frac{\ln 2}{\ln 1.2} \qquad \text{Use a calculator or a table.}$$

$$= 3.8 \text{ years} \qquad \left[\textit{Note: } \frac{\ln 2}{\ln 1.2} \neq \ln 2 - \ln 1.2\right]$$

$$\approx 4 \text{ years} \qquad \text{To the next whole year}$$

When interest is paid at the end of 3 years, the money will not be doubled; when paid at the end of 4 years, the money will be slightly more than doubled.

Problem 40 How long (to the next whole year) will it take money to double if it is invested at 13% interest compounded annually?

It is interesting and instructive to graph the doubling times for various rates compounded annually. We proceed as follows:

$$A = P(1 + i)^n$$

$$2P = P(1 + i)^n$$

$$2 = (1 + i)^n$$

$$(1+i)^n = 2$$
$$\ln(1+i)^n = \ln 2$$
$$n \ln(1+i) = \ln 2$$
$$n = \frac{\ln 2}{\ln(1+i)}$$

Figure 20 shows the graph of this equation (doubling times in years) for interest rates compounded annually from 1% to 70%. Note the dramatic changes in doubling times from 1% to 20%.

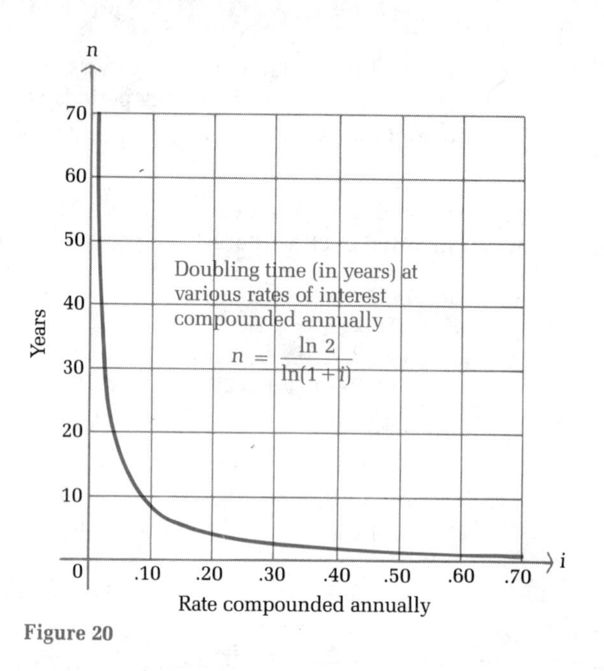

Figure 20

Answers to Matched Problems

31. (A) $9 = 3^2$ (B) $2 = 4^{1/2}$ (C) $1/9 = 3^{-2}$
32. (A) $\log_7 49 = 2$ (B) $\log_9 3 = 1/2$ (C) $\log_3 (1/3) = -1$
33. (A) $y = 3/2$ (B) $x = 1/3$ (C) $b = 10$
34. $y = \log_3(x-1)$ is equivalent to $x = 3^y + 1$

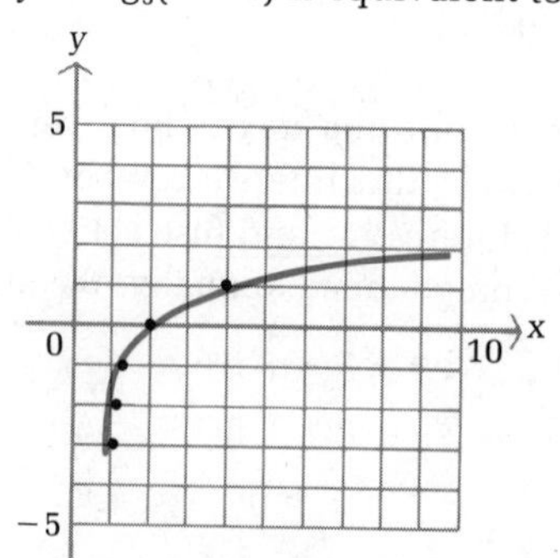

35. (A) $\log_b R - \log_b S - \log_b T$ (B) $(2/3)(\log_b R - \log_b S)$
36. $x = 2$
37. $x = 5$
38. (A) $-1.868\ 734$ (B) $3.356\ 663$ (C) Not defined
39. (A) 6.333×10^{-3} (B) 1.208×10^{12}
40. 6 years

Exercise 0-7

A *Rewrite in exponential form.*

1. $\log_3 27 = 3$
2. $\log_2 32 = 5$
3. $\log_{10} 1 = 0$
4. $\log_e 1 = 0$
5. $\log_4 8 = \frac{3}{2}$
6. $\log_9 27 = \frac{3}{2}$

Rewrite in logarithmic form.

7. $49 = 7^2$
8. $36 = 6^2$
9. $8 = 4^{3/2}$
10. $9 = 27^{2/3}$
11. $A = b^u$
12. $M = b^x$

Find each of the following:

13. $\log_{10} 10^3$
14. $\log_{10} 10^{-5}$
15. $\log_2 2^{-3}$
16. $\log_3 3^5$
17. $\log_{10} 1{,}000$
18. $\log_6 36$

Write in terms of simpler logarithmic forms as in Example 35.

19. $\log_b \frac{P}{Q}$
20. $\log_b FG$
21. $\log_b L^5$
22. $\log_b w^{15}$
23. $\log_b \frac{p}{qrs}$
24. $\log_b PQR$

B *Find x, y, or b.*

25. $\log_3 x = 2$
26. $\log_2 x = 2$
27. $\log_7 49 = y$
28. $\log_3 27 = y$
29. $\log_b 10^{-4} = -4$
30. $\log_b e^{-2} = -2$
31. $\log_4 x = \frac{1}{2}$
32. $\log_{25} x = \frac{1}{2}$
33. $\log_{1/3} 9 = y$
34. $\log_{49} \frac{1}{7} = y$
35. $\log_b 1{,}000 = \frac{3}{2}$
36. $\log_b 4 = \frac{2}{3}$

Write in terms of simpler logarithmic forms going as far as you can with logarithmic properties (see Example 35).

37. $\log_b \frac{x^5}{y^3}$

38. $\log_b x^2y^3$

39. $\log_b \sqrt[3]{N}$

40. $\log_b \sqrt[5]{Q}$

41. $\log_b x^2 \sqrt[3]{y}$

42. $\log_b \sqrt[3]{\frac{x^2}{y}}$

43. $\log_b(50 \cdot 2^{-0.2t})$

44. $\log_b(100 \cdot 1.06^t)$

45. $\log_b P(1 + r)^t$

46. $\log_e Ae^{-0.3t}$

47. $\log_e 100e^{-0.01t}$

48. $\log_{10} (67 \cdot 10^{-0.12x})$

Find x.

49. $\log_b x = \frac{2}{3} \log_b 8 + \frac{1}{2} \log_b 9 - \log_b 6$

50. $\log_b x = \frac{2}{3} \log_b 27 + 2 \log_b 2 - \log_b 3$

51. $\log_b x = \frac{3}{2} \log_b 4 - \frac{2}{3} \log_b 8 + 2 \log_b 2$

52. $\log_b x = 3 \log_b 2 + \frac{1}{2} \log_b 25 - \log_b 20$

53. $\log_b x + \log_b(x - 4) = \log_b 21$

54. $\log_b(x + 2) + \log_b x = \log_b 24$

55. $\log_{10}(x - 1) - \log_{10}(x + 1) = 1$

56. $\log_{10}(x + 6) - \log_{10}(x - 3) = 1$

Graph by converting to exponential form first.

57. $y = \log_2(x - 2)$

58. $y = \log_3(x + 2)$

In Problems 59 and 60, evaluate to five decimal places using a scientific calculator.

59. (A) log 3,527.2 (B) log 0.006 913 2
(C) ln 277.63 (D) ln 0.040 883

60. (A) log 72.604 (B) log 0.033 041
(C) ln 40,257 (D) ln 0.005 926 3

In Problems 61 and 62, find x to four significant digits.

61. (A) $\log x = 3.128\ 5$ (B) $\log x = -2.049\ 7$
(C) $\ln x = 8.776\ 3$ (D) $\ln x = -5.887\ 9$

62. (A) $\log x = 5.083\ 2$ (B) $\log x = -3.157\ 7$
(C) $\ln x = 10.133\ 6$ (D) $\ln x = -4.328\ 1$

C

63. Find the logarithm of 1 for any permissible base.
64. Why is 1 not a suitable logarithmic base? [*Hint:* Try to find $\log_1 8$.]
65. Write $\log_{10} y - \log_{10} c = 0.8x$ in an exponential form that is free of logarithms.
66. Write $\log_e x - \log_e 25 = 0.2t$ in an exponential form that is free of logarithms.

Applications

Business & Economics

67. *Doubling time.* How long (to the next whole year) will it take money to double if it is invested at 6% interest compounded annually?
68. *Doubling time.* How long (to the next whole year) will it take money to double if it is invested at 3% interest compounded annually?
69. *Tripling time.* Write a formula similar to the doubling time formula in Figure 20 for the tripling time of money invested at $100i\%$ interest compounded annually.
70. *Tripling time.* How long (to the next whole year) will it take money to triple if invested at 15% interest compounded annually?

Life Sciences

71. *Sound intensity—decibels.* Because of the extraordinary range of sensitivity of the human ear (a range of over 1,000 million millions to 1), it is helpful to use a logarithmic scale, rather than an absolute scale, to measure sound intensity over this range. The unit of measure is called the *decibel,* after the inventor of the telephone, Alexander Graham Bell. If we let N be the number of decibels, I the power of the sound in question (in watts per square centimeter), and I_0 the power of sound just below the threshold of hearing (approximately 10^{-16} watt per square centimeter), then

$$I = I_0 10^{N/10}$$

Show that this formula can be written in the form

$$N = 10 \log \frac{I}{I_0}$$

72. *Sound intensity—decibels.* Use the formula in Problem 71 (with $I_0 = 10^{-16}$ watt/cm^2) to find the decibel ratings of the following sounds:

(A) Whisper: 10^{-13} watt/cm^2
(B) Normal conversation: 3.16×10^{-10} watt/cm^2
(C) Heavy traffic: 10^{-8} watt/cm^2
(D) Jet plane with afterburner: 10^{-1} watt/cm^2

74

Social Sciences

73. *World population.* If the world population is now 4 billion (4×10^9) people and if it continues to grow at 2% per year compounded an-

nually, how long will it be before there is only 1 square yard of land per person? (The earth contains approximately 1.68×10^{14} square yards of land.)

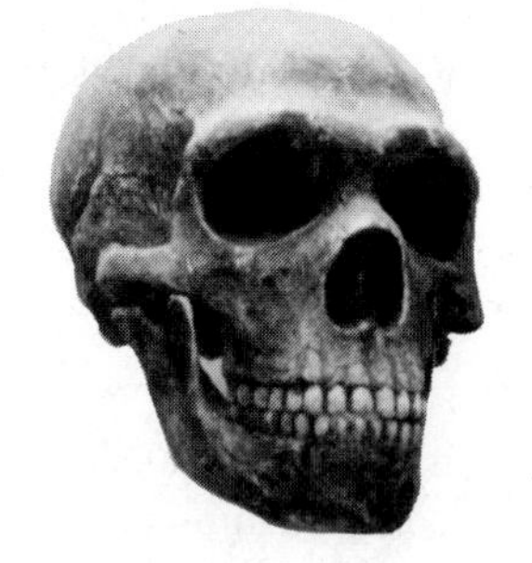

74. *Archaeology—carbon-14 dating.* Cosmic-ray bombardment of the atmosphere produces neutrons, which in turn react with nitrogen to produce radioactive carbon-14. Radioactive carbon-14 enters all living tissues through carbon dioxide which is first absorbed by plants. As long as a plant or animal is alive, carbon-14 is maintained at a constant level in its tissues. Once dead, however, it ceases taking in carbon and the carbon-14 diminishes by radioactive decay according to the equation

$$A = A_0 e^{-0.000124t}$$

where t is time in years. Estimate the age of a skull uncovered in an archaeological site if 10% of the original amount of carbon-14 is still present. [*Hint:* Find t such that $A = 0.1A_0$.]

0-8 Chapter Review

Important Terms and Symbols

0-1 *Sets.* Member, element, empty set, null set, finite set, infinite set, subset, equal sets, union, intersection, disjoint, universal set, complement, $\in$, $\notin$, $\varnothing$, $\subset$, $\not\subset$, $\cup$, $\cap$, A'

0-2 *Linear equations and inequalities in one variable.* Linear equation, linear inequality, solution, solution set, solution of an equation, equivalent equations, real numbers, real number line, equivalent inequalities, equality properties, least common denominator, inequality properties, interval notation, $a < b$, $a > b$, $a \leq x \leq b$, $[a, b]$, $[a, b)$, $(a, b]$, (a, b), $(-\infty, a]$, $(-\infty, a)$, $[b, \infty)$, (b, ∞)

0-3 *Quadratic equations.* Quadratic equation, standard form, completing the square, quadratic formula, discriminant,

$$ax^2 + bx + c = 0, \quad a \neq 0, \quad x = \frac{-b \pm \sqrt{b^2 - 4ac}}{2a}$$

0-4 *Cartesian coordinate system and straight lines.* Cartesian coordinate system, rectangular coordinate system, coordinate axes, ordered pair, coordinates, abscissa, ordinate, quadrants, solution of an equation in two variables, solution set, graph of an equation, x intercept, y intercept, slope, slope–intercept form, point–slope form, horizontal line, vertical line, parallel lines, perpendicular lines, $y = mx + b$, slope $= (y_2 - y_1)/(x_2 - x_1)$, $y - y_1 = m(x - x_1)$, $y = c$, $x = c$

0-5 *Functions and graphs.* Function, domain, range, input, output, independent variable, dependent variable, function notation, linear function, quadratic function, parabola, axis of a parabola, vertex of a parabola, maximum, minimum, demand equation, cost equation, revenue equation, profit equation, break-even point, $f(x)$, $f(x) = ax + b$, $f(x) = ax^2 + bx + c$, $a \neq 0$

0-6 *Exponential functions.* Exponential function, graphs of exponential functions, base e, exponential properties, b^x, e^x

0-7 *Logarithmic functions.* Logarithmic function, logarithmic properties, common logarithms, natural logarithms, calculator evaluation, $\log_b x$, $\log x$, $\ln x$

Exercise 0-8 Chapter Review

Work through all the problems in this chapter review and check your answers in the back of the book. (Answers to all review problems are there.) Where weaknesses show up, review appropriate sections in the text.

A

1. True (T) or false (F)?

   (A) $7 \notin \{4, 6, 8\}$ (B) $\{8\} \subset \{4, 6, 8\}$
   (C) $\varnothing \in \{4, 6, 8\}$ (D) $\varnothing \subset \{4, 6, 8\}$

2. Solve $\frac{u}{5} = \frac{u}{6} + \frac{6}{5}$.
3. Solve and graph on a real number line: $2(x + 4) > 5x - 4$
4. Solve $x^2 = 5x$.
5. Graph the equation below in a rectangular coordinate system. Indicate the slope and the y intercept.

   $$y = \frac{x}{2} - 2$$

6. Write the equation of a line that passes through (4, 3) with slope $\frac{1}{2}$. Write the final answer in the form $y = mx + b$.
7. Graph $x - y = 2$ in a rectangular coordinate system. Indicate the slope.
8. For $f(x) = 2x - 1$ and $g(x) = x^2 - 2x$, find $f(-2) + g(-1)$.
9. Graph the linear function f given by the equation

   $$f(x) = \tfrac{2}{3}x - 1$$

   Indicate the slope of the graph.
10. Write $\log_{10} y = x$ in exponential form.
11. Write $\log_b \frac{wx}{y}$ in terms of simpler logarithms.

B

12. If $A = \{1, 2, 3\}$ and $B = \{2, 3, 4\}$, find

    (A) $A \cup B$ (B) $\{x | x \in A \text{ and } x \in B\}$

13. If $U = \{2, 4, 5, 6, 8\}$, $M = \{2, 4, 5\}$, and $N = \{5, 6\}$, find

    (A) $M \cup N$ (B) $M \cap N$ (C) $(M \cup N)'$ (D) $M \cap N'$

14. Indicate true (T) or false (F) for U, M, and N in Problem 13.

    (A) $N \subset M$ (B) $\varnothing \subset U$ (C) $6 \notin M$ (D) $5 \in N$

15. Given the Venn diagram shown with the number of elements indicated in each part, determine how many elements are in each of the following sets:

    (A) $M \cup N$ (B) $M \cap N$ (C) $(M \cup N)'$ (D) $M \cap N'$

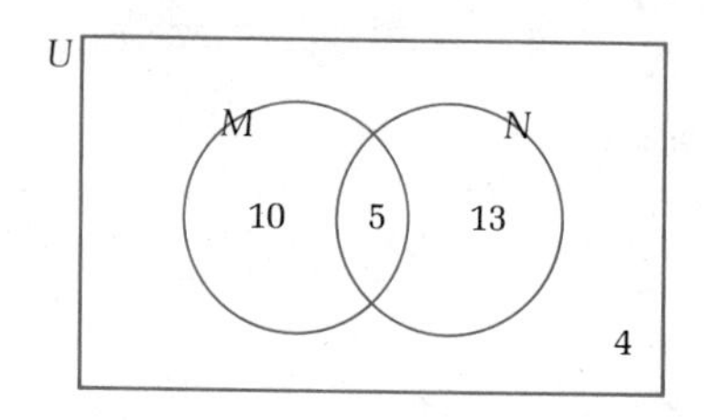

16. In a freshman class of 100 students, 70 are taking English, 45 are taking math, and 25 are taking English and math.

    (A) How many students are taking either English or math?
    (B) How many students are taking English and not math?

17. Solve $\frac{x}{12} - \frac{x-3}{3} = \frac{1}{2}$.

Solve and graph on a real number line.

18. $1 - \frac{x-3}{3} \leq \frac{1}{2}$

19. $-2 \leq \frac{x}{2} - 3 < 3$

Solve for y in terms of x.

20. $2x - 3y = 6$

21. $xy - y = 3$

Solve problems 22–24.

22. $3x^2 - 21 = 0$

23. $x^2 - x - 20 = 0$

24. $2x^2 = 3x + 1$

25. Graph $3x + 6y = 18$ in a rectangular coordinate system. Indicate the slope, x intercept, and y intercept.
26. Find an equation of the line that passes through $(-2, 3)$ and $(6, -1)$. Write the answer in the form $Ax + By = C$, $A > 0$. What is the slope of the line?

27. Write the equations of the vertical line and the horizontal line that pass through $(-5, 2)$. Graph both equations on the same coordinate system.
28. Find an equation of the line that passes through $(-2, 5)$ and $(2, -1)$. Write the answer in the form $y = mx + b$.
29. For $f(x) = 10x - 7$, $g(t) = 6 - 2t$, $F(u) = 3u^2$, and $G(v) = v - v^2$, find

    (A) $2g(-1) - 3G(-1)$ (B) $4G(-2) - g(-3)$

    (C) $\dfrac{f(2) \cdot g(-4)}{G(-1)}$ (D) $\dfrac{F(-1) \cdot G(2)}{g(-1)}$

30. For $f(x) = \sqrt{x}$ and $g(x) = x^2 + 2x$, find

    (A) $f[g(2)]$ (B) $g[f(a)]$
31. Find the domains of the functions f and g if

$$f(x) = 2x - x^2 \qquad g(x) = \frac{1}{x - 2}$$

32. Graph $g(x) = 8x - 2x^2$, $x \geq 0$, in a rectangular coordinate system. Write the coordinates of the vertex and the equation of the axis. What is the maximum or minimum value of $g(x)$?
33. Graph $y = 10 \cdot 2^{3x}$ and $y = 10 \cdot 2^{-3x}$, $-2 \leq x \leq 2$, on the same coordinate system.
34. Graph $y = 100e^{-0.1x}$, $0 \leq x \leq 10$, using a calculator or a table.
35. (A) Find b: $\log_b 9 = 2$ (B) Find x: $\log_4 x = -3$
36. Write in terms of simpler logarithmic forms:

$$\log_b(100 \cdot 1.06^t)$$

37. Find x: $\log_b x = 3 \log_b 2 - \frac{3}{2} \log_b 4 - \frac{1}{2} \log_b 36$.
38. Evaluate to five decimal places using a scientific calculator.

    (A) $\log 0.009\,108\,5$ (B) $\ln 9{,}843.3$

39. Find x to four significant digits.

    (A) $\log x = -3.805\,5$ (B) $\ln x = 12.814\,3$

40. Solve for x: $\ln x + \ln(x - 3) = \ln 28$.

C

41. If $A \cap B = A$, then is it always true that $A \subset B$?
42. Solve $x^2 + jx + k = 0$ for x in terms of j and k.
43. Write an equation of the line that passes through the points $(4, -3)$ and $(4, 5)$.
44. Write an equation of the line that passes through $(2, -3)$ and is (A) parallel (B) perpendicular to $2x - 4y = 5$. Write the final answers in the form $Ax + By = C$, $A > 0$.
45. Find the domain of the function f specified by each equation.

    (A) $f(x) = \dfrac{5}{x - 3}$ (B) $f(x) = \sqrt{x - 1}$

46. For $f(x) = 2x - 1$, find $\frac{f(3+h) - f(3)}{h}$.

47. Find t: $240 = 80e^{0.12t}$.

48. Write $\ln y - \ln c = -0.2x$ in an exponential form free of logarithms.

Applications

Business & Economics

49. *Marketing survey.* A survey company sampled 1,000 students at a university. Out of the sample, 550 students smoked cigarettes, 820 drank alcoholic beverages, and 470 did both.

    (A) How many smoked or drank?
    (B) How many drank but did not smoke?

50. *Investment.* An investor has $60,000 to invest. If part is invested at 8% and the rest at 14%, how much should be invested at each rate to yield 12% on the total amount?

51. *Inflation.* If the CPI was 89 in 1960 and 247 in 1980, how much would a net salary of $800 in 1960 have to be in 1980 in order to keep up with inflation? Set up an equation and solve.

52. *Finance.* If P dollars is invested at 100r% compounded annually, at the end of 2 years it will grow to $A = P(1 + r)^2$. At what interest rate will $1,000 grow to $1,210 in 2 years?

53. *Linear depreciation.* A word-processing system was purchased by a company for $12,000 and is assumed to have a salvage value of $2,000 after 8 years (for tax purposes). If its value is depreciated linearly from $12,000 to $2,000:

    (A) Find the linear equation that relates value V in dollars to time t in years.
    (B) What would be the value of the system after 5 years?

54. *Pricing.* A sporting goods store sells a tennis racket that cost $30 for $48 and a pair of jogging shoes that cost $20 for $32.

    (A) If the markup policy of the store for items that cost over $10 is assumed to be linear and is reflected in the pricing of these two items, write an equation that relates retail price R to cost C.
    (B) What should be the retail price of a pair of skis that cost $105?

55. *Construction.* A Wyoming rancher has 20 miles of fencing to enclose a rectangular piece of grazing land along a river.

    (A) If no fence is required along the river and x is the width of the rectangle (at right angles to the river), express the area A(x) of the rectangle in terms of x.
    (B) What is the domain of the function A (due to physical restrictions)?

(C) Complete the table:

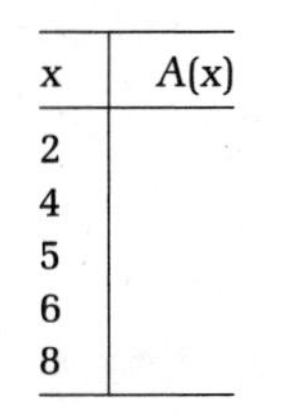

x	A(x)
2	
4	
5	
6	
8	

56. *Finance.* Find the tripling time (to the next higher year if not exact) for money invested at 15% compounded annually $[A = P(1 + r)^t]$.

57. *Finance.* Find the doubling time (to two decimal places) for money invested at 10% compounded continuously $[A = Pe^{rt}]$.

The Derivative

1

CHAPTER 1 Contents

How do algebra and calculus differ? The two words *static* and *dynamic* probably come as close as any in expressing the difference between the two disciplines. In algebra, we solve equations for a particular value of a variable—a static notion. In calculus, we are interested in how a change in one variable affects another variable—a dynamic notion.

Figure 1 illustrates three basic problems in calculus. It may surprise you to learn that all three problems—as different as they appear—are mathematically related. The solutions to these problems and the discovery of their relationship required the creation of a new kind of mathematics. Isaac Newton (1642–1727) of England and Gottfried Wilhelm von Leibniz (1646–1716) of Germany simultaneously and independently developed this new mathematics, called **the calculus**—it was an idea whose time had come.

In addition to solving the problems described in Figure 1, calculus will enable us to solve many important problems. Until fairly recently, calculus was used primarily in the physical sciences, but now people in many other disciplines are finding it a useful tool.

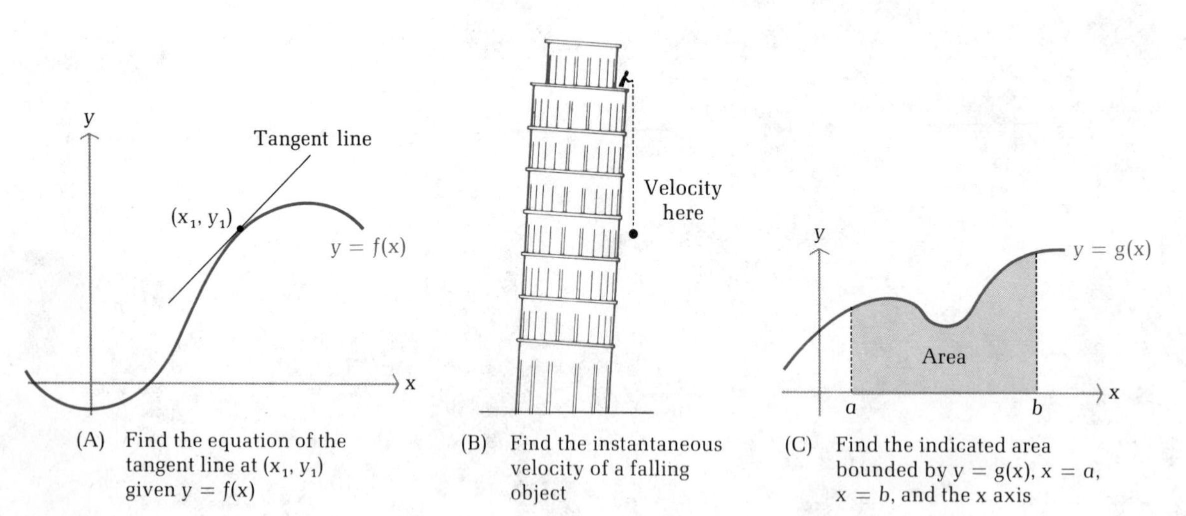

(A) Find the equation of the tangent line at (x_1, y_1) given $y = f(x)$

(B) Find the instantaneous velocity of a falling object

(C) Find the indicated area bounded by $y = g(x)$, $x = a$, $x = b$, and the x axis

Figure 1

1-1 Limits

- Introduction
- Definition of Limit
- Properties of Limits
- Limits at Infinity

Basic to the study of calculus is the concept of *limit*. This concept helps us to describe, in a precise way, the behavior of $f(x)$ when x is close to but not equal to a particular value c. And as we will soon see, it is fundamental to the two main topics of calculus—*the derivative* and the *integral*. In our discussion, we will concentrate on concept development and understanding rather than on the formal details.

Introduction

We introduce the concept of limit through a problem that goes back to early Grecian times. The problem concerns estimating the circumference of a circle using perimeters of regular polygons inscribed in the circle. Figure 2 illustrates three-sided, six-sided, and twelve-sided regular polygons inscribed in a circle. It appears that if we continue to double the number of sides of an inscribed regular polygon, we can make the perimeter as close to the circumference of the circle as we like. We say that the circumference C of the circle is the *limit* of the perimeter of the inscribed regular polygon as the number of sides increases without bound. Archimedes, a Greek mathematician and inventor (287–212 B.C.), approximated the value of π as the limit of perimeters of inscribed regular polygons in a circle with diameter $D = 1$. (Recall that $C = \pi D$. If $D = 1$, then $\pi = C$.)

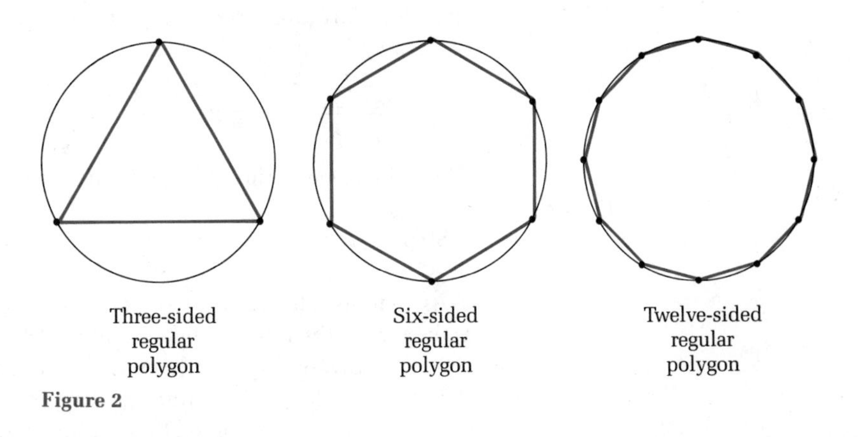

Figure 2

We now turn to another geometric example related to circles. A line that intersects a circle in one and only one point is called a **tangent line** (see Fig. 3A). One of the basic problems of calculus is the generalization of the concept of a tangent line to the graph of an arbitrary function, as illustrated in Figure 1A. The key to this generalization is the relationship between the slope of the tangent line and the slopes of secant lines. Recall that a **secant line** is a line that intersects a circle in two points. If P is a fixed point on the circle and Q is any other point on the circle, then as Q moves closer to P, the secant lines seem to approach the tangent line (see Fig. 3B). It seems reasonable to assume that the slopes of the secant lines will be approaching the slope of the tangent line.

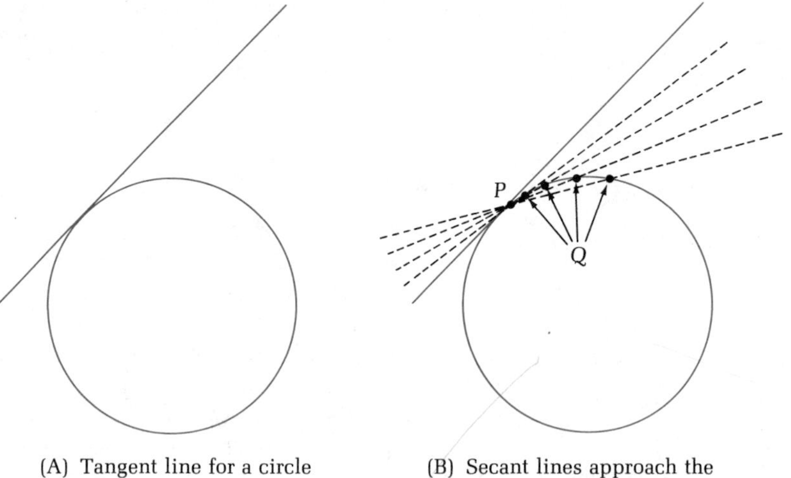

(A) Tangent line for a circle

(B) Secant lines approach the tangent line as Q approaches P

Figure 3

Now consider the graph of $f(x) = x^2$, a parabola, and the line through the (fixed) point $(2, 4)$ and another arbitrary point (x, x^2) on the graph (see Fig. 4A). As with circles, a line through two points on a graph is called a *secant line*. The formula for the slope of the line passing through (x_1, y_1) and (x_2, y_2) is

$$m = \frac{y_2 - y_1}{x_2 - x_1} \qquad x_1 \neq x_2$$

See Section 0-4.

Thus, the slope of the secant line through $(2, 4)$ and (x, x^2) is given by

$$\text{Slope of secant line} = m_s = \frac{x^2 - 4}{x - 2}$$

As x approaches 2 (from either side of 2), the secant lines seem to be approaching the line graphed in Figure 4B, which we will call the *tangent line*. For now, we assume that m_s, the slope of the secant line, approaches the slope of the tangent line as x approaches 2. (Precise definitions of tangent line and slope of the tangent line will be given in Section 1-3.) How

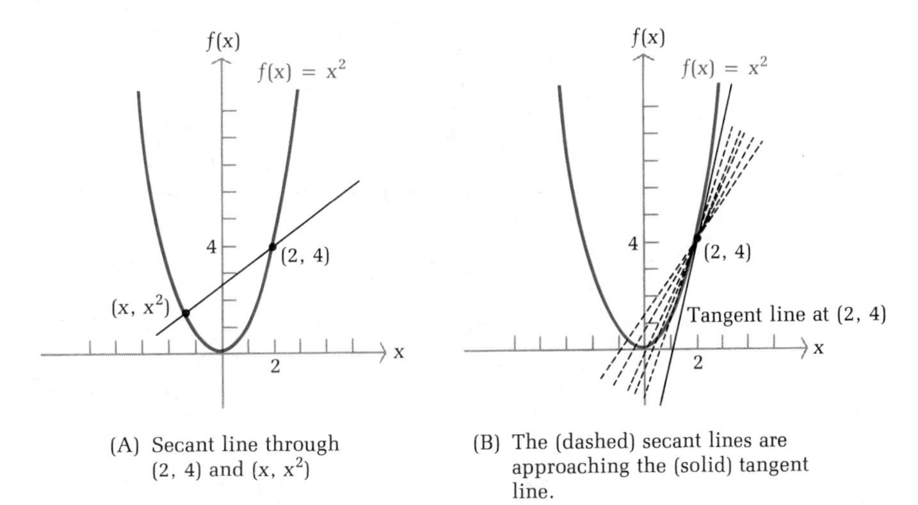

(A) Secant line through (2, 4) and (x, x^2)

(B) The (dashed) secant lines are approaching the (solid) tangent line.

Figure 4

can we find the slope of this tangent line? If we substitute $x = 2$ in the formula for m_s, we get

$$m_s = \frac{2^2 - 4}{2 - 2} = \frac{0}{0} = ?$$

Since division by 0 is not permitted, the expression $\frac{0}{0}$ is meaningless and m_s is not defined at $x = 2$. Thus, we cannot find the slope of the tangent line by evaluating the formula for m_s at $x = 2$. But this formula is defined for all other values of x, and in particular for values of x that are very close to 2.

What happens to m_s when x approaches 2 from either side of 2? Let us investigate this question using a calculator experiment. Table 1 shows the secant line slopes m_s for values of x approaching 2 from the left and for values of x approaching 2 from the right. It appears that m_s approaches 4 ($m_s \to 4$) as x approaches 2 ($x \to 2$) from either side of 2, and the closer x is to 2, the closer m_s will be to 4. We say that 4 is the "limit of m_s as x approaches 2" and write

$$\lim_{x \to 2} \frac{x^2 - 4}{x - 2} = 4$$

As x approaches 2, $(x^2 - 4)/(x - 2)$ approaches 4, and it is this number 4 that we call the *limit*, even though $(x^2 - 4)/(x - 2)$ is not defined at $x = 2$.

In Figure 4B we associate 4 with the slope of the tangent line to the graph at (2, 4).

Table 1

x approaches 2 from the left → 2 ← x approaches 2 from the right

x	1.5	1.8	1.9	1.99	1.999	→ 2 ←	2.001	2.01	2.1	2.2	2.5
m_s	3.5	3.8	3.9	3.99	3.999	→ ? ←	4.001	4.01	4.1	4.2	4.5

Definition of Limit

Finding slopes of tangent lines is not the only problem involving limits we will encounter. In general, we will be concerned with finding **the limit of a function f as x approaches a number c.** We now state an informal definition of this concept. A precise definition will not be needed for our discussion, but one is given in the footnote.*

Limit (Informal Definition)

We write

$$\lim_{x \to c} f(x) = L$$

if the functional value $f(x)$ is close to the single real number L whenever x is close to but not equal to c (on either side of c).

Example 1 Graph $f(x) = x^2 + 2$ and use the graph to find $\lim_{x \to 1} f(x)$.

Solution The graph of f is a parabola opening upwards with vertex (0, 2), as shown in the figure. For each number x, the functional value $f(x)$ represents the

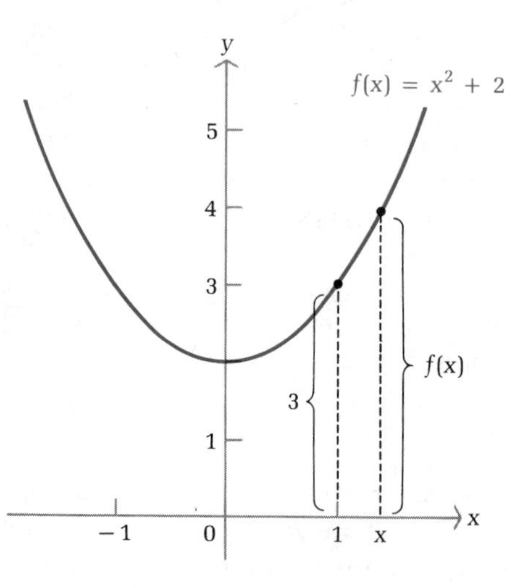

* To make the informal definition of limit precise, the use of the word *close* must be made more precise. This is done as follows: We write $\lim_{x \to c} f(x) = L$ if for each $e > 0$, there exists a $d > 0$ such that $|f(x) - L| < e$ whenever $0 < |x - c| < d$. This definition is used to establish particular limits and to prove many useful properties of limits that will be helpful to us in finding particular limits. [Even though intuitive notions of limit existed for a long time, it was not until the nineteenth century that a precise definition was given by the German mathematician, Karl Weierstrass (1815–1897).]

vertical distance from the x axis to the point $(x, f(x))$ on the graph of f. If x is close to 1, the graph shows that this vertical distance is close to the vertical distance from the x axis to the point (1, 3). Thus, we conclude that $\lim_{x\to 1} f(x) = 3$.

Problem 1 Graph $f(x) = 2x + 1$ and use the graph to find $\lim_{x\to 2} f(x)$.

Example 2 Graph $f(x) = |x|/x$ and use the graph to find:

(A) $\lim_{x\to 2} f(x)$ (B) $\lim_{x\to 0} f(x)$

Solutions We start by sketching a graph of f:

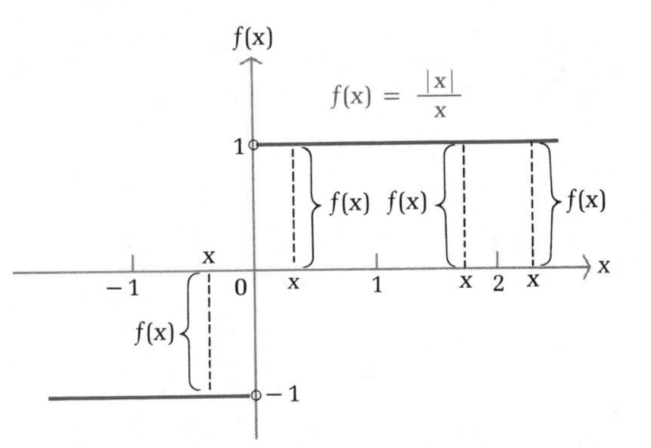

For $x > 0$, $f(x) = |x|/x = x/x = 1$.
For $x < 0$, $f(x) = |x|/x = -x/x = -1$.
For $x = 0$, $f(x)$ is undefined.

(A) If x is close to 2 (on either side of 2), then $f(x) = 1$. Thus, $\lim_{x\to 2} f(x) = 1$.
(B) If x is close to 0 and greater than 0, then $f(x) = 1$, but if x is close to 0 and less than 0, then $f(x) = -1$. Since the functional values $f(x)$ are not close to a single number L for x close to 0 on both sides of 0, we conclude that $\lim_{x\to 0} f(x)$ does not exist.

Problem 2 Graph $f(x) = x + |x|/x$ and use the graph to find:

(A) $\lim_{x\to 2} f(x)$ (B) $\lim_{x\to 0} f(x)$

Example 3 Let $f(x) = 1/(x + 1)$. Use a table of values to find $\lim_{x\to -1} f(x)$.

Solution Notice that f is not defined at $x = -1$ (division by 0 is not permitted). We construct a table of values of $f(x)$ for x near -1 on both sides of -1:

Table 2

	x approaches -1 from the left				$\to -1 \leftarrow$	x approaches -1 from the right			
x	-1.1	-1.01	-1.001	-1.0001	$\to -1 \leftarrow$	-0.9999	-0.999	-0.99	-0.9
$f(x)$	-10	-100	$-1{,}000$	$-10{,}000$	$\to\ ?\ \leftarrow$	10,000	1,000	100	10

As x approaches -1 from either side, the values of $f(x)$ are becoming larger and larger in absolute value. Since the functional values are not approaching a number L, we conclude that $\lim_{x\to -1} f(x)$ does not exist. This behavior is illustrated graphically by drawing a dashed vertical line through $x = -1$. This line is called a *vertical asymptote* for the graph of function f (see the figure).

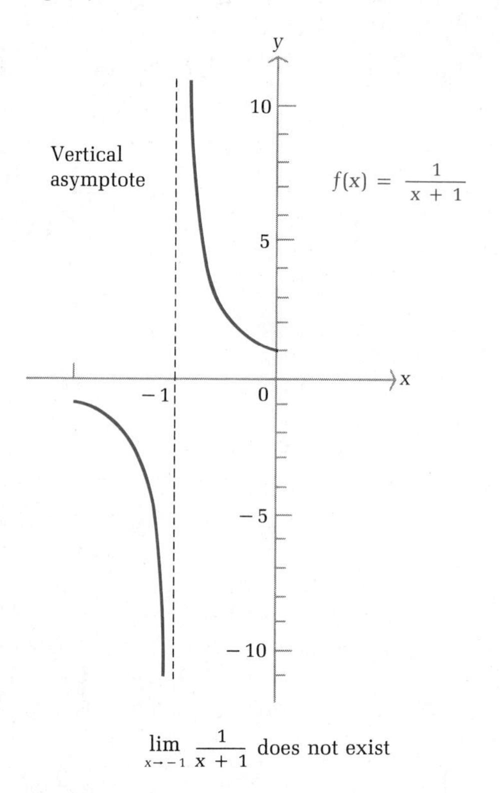

$$\lim_{x\to -1} \frac{1}{x + 1} \text{ does not exist}$$

Problem 3 Let $f(x) = 1/(2 - x)$. Use a table of values to find $\lim_{x\to 2} f(x)$. (Do not graph f.)

▪ Properties of Limits

We now turn to some basic properties of limits that will enable us to evaluate limits of a rather large class of functions without resorting to tables, figures, or graphs. We state some important properties without proof in Theorem 1.

Theorem 1

Properties of Limits

If k and c are constants, n is a positive integer, and

$$\lim_{x \to c} f(x) = A \qquad \lim_{x \to c} g(x) = B$$

then:

1. $\lim_{x \to c} k = k$
2. $\lim_{x \to c} kf(x) = k \lim_{x \to c} f(x) = kA$
3. $\lim_{x \to c} [f(x) \pm g(x)] = \lim_{x \to c} f(x) \pm \lim_{x \to c} g(x) = A \pm B$
4. If P is a polynomial function,* then $\lim_{x \to c} P(x) = P(c)$
5. $\lim_{x \to c} [f(x) \cdot g(x)] = [\lim_{x \to c} f(x)][\lim_{x \to c} g(x)] = AB$
6. $\lim_{x \to c} \frac{f(x)}{g(x)} = \frac{\lim_{x \to c} f(x)}{\lim_{x \to c} g(x)} = \frac{A}{B} \qquad B \neq 0$
7. $\lim_{x \to c} \sqrt[n]{f(x)} = \sqrt[n]{\lim_{x \to c} f(x)} = \sqrt[n]{A}$

   (x is restricted to avoid even roots of negative numbers)

Example 4 Use the properties of limits to evaluate each limit.

(A) $\lim_{x \to 2} (3x^5 - 2x^2 + 3x - 1)$

$= 3(2)^5 - 2(2)^2 + 3(2) - 1$

$= 93$

Use property 4, since $P(x) = 3x^5 - 2x^2 + 3x - 1$ is a polynomial function.

(B) $\lim_{x \to 4} \sqrt{x^2 - 4} = \sqrt{\lim_{x \to 4} (x^2 - 4)}$

$= \sqrt{4^2 - 4}$

$= \sqrt{12} = 2\sqrt{3}$

Use properties 7 and 4. Note that $x^2 - 4$ is positive for x near 4.

* A **polynomial function** is any function that can be specified by an equation of the form

$$P(x) = a_n x^n + a_{n-1} x^{n-1} + \cdots + a_1 x + a_0$$

where the coefficients $a_0, a_1, \ldots, a_n$ are real numbers and n is a nonnegative integer. [*Examples:* $P(x) = 2x - 3$, $P(x) = 5$, $P(x) = 4x^5 - 2x^3 + 3x^2 - x + 1$.]

(C) $$\lim_{x\to-1}\frac{x^2+1}{x+4}=\frac{\lim_{x\to-1}(x^2+1)}{\lim_{x\to-1}(x+4)}$$ Use properties 6 and 4. Note that $\lim_{x\to-1}(x+4)=3\neq 0$.

$$=\frac{(-1)^2+1}{(-1)+4}=\frac{2}{3}$$

Problem 4 Use properties of limits to evaluate each limit.

(A) $\lim_{x\to-2}(2x^4-3x^3+5)$ (B) $\lim_{x\to3}\sqrt{x^2-1}$ (C) $\lim_{x\to0}\frac{x^2+5}{x+3}$

Example 5 Use properties of limits and algebraic manipulations to find each limit, if it exists.

(A) $\lim_{x\to3}\frac{x^2-x-6}{x-3}$ (B) $\lim_{x\to-1}\frac{x-1}{x^2-1}$ (C) $\lim_{x\to9}\frac{\sqrt{x}-3}{x-9}$

Solutions (A) We cannot use property 6, since $\lim_{x\to3}(x-3)=0$. We factor the numerator to see if we can simplify the function:

$$\frac{x^2-x-6}{x-3}=\frac{\overset{1}{\cancel{(x-3)}}(x+2)}{\underset{1}{\cancel{(x-3)}}}=x+2 \qquad x\neq 3 \tag{1}$$

The left and right sides of (1) are equal for all values of x except $x=3$. Since the limit process involves functional values for x near 3 but not equal to 3, we can write

$$\lim_{x\to3}\frac{x^2-x-6}{x-3}=\lim_{x\to3}(x+2)$$ Use property 4.

$$=5$$

In practice, the algebraic manipulations and the limit evaluation are often combined. The solution to this problem can be written more compactly as

$$\lim_{x\to3}\frac{x^2-x-6}{x-3}=\lim_{x\to3}\frac{\overset{1}{\cancel{(x-3)}}(x+2)}{\underset{1}{\cancel{(x-3)}}}=\lim_{x\to3}(x+2)=5 \tag{2}$$

Notice that in (2) we do not have to explicitly state that $x\neq 3$. It is understood that the limit operation is concerned with the functional values for x near 3 but not equal to 3. On the other hand, the restriction $x\neq 3$ is essential in (1). It is incorrect to write

$$\frac{x^2-x-6}{x-3}=x+2$$

without stating that $x\neq 3$.

(B) Since property 6 does not apply, we proceed as in part A.

$$\lim_{x\to -1} \frac{x-1}{x^2-1} = \lim_{x\to -1} \frac{\cancel{(x-1)}^{\,1}}{\cancel{(x-1)}_{\,1}(x+1)}$$

$$= \lim_{x\to -1} \frac{1}{x+1}$$

Does not exist

The values of $1/(x+1)$ become arbitrarily large in absolute value as x approaches -1. See Example 3.

(C) $$\lim_{x\to 9} \frac{\sqrt{x}-3}{x-9} = \lim_{x\to 9} \frac{(\sqrt{x}-3)}{(x-9)} \cdot \frac{(\sqrt{x}+3)}{(\sqrt{x}+3)}$$

Property 6 does not apply. We rationalize the numerator, then use properties 6, 1, and 7.

$$= \lim_{x\to 9} \frac{\cancel{(x-9)}^{\,1}}{\cancel{(x-9)}_{\,1}(\sqrt{x}+3)}$$

$$= \lim_{x\to 9} \frac{1}{\sqrt{x}+3}$$

$$= \frac{1}{\sqrt{9}+3} = \frac{1}{6}$$

Problem 5 Use properties of limits and algebraic manipulations to find each limit, if it exists.

(A) $\lim_{x\to 3} \frac{x^2-2x-3}{x-3}$ (B) $\lim_{x\to 2} \frac{2+x}{4-x^2}$ (C) $\lim_{x\to 1} \frac{\sqrt{x}-1}{x-1}$

▪ Limits at Infinity

It is also of interest to determine what happens to the functional values $f(x)$ as x assumes large positive values and large negative values. We begin by considering an example.

Example 6 Consider the function $f(x) = 2x^2/(1+x^2)$, which is defined for all real numbers. What happens to the functional value $f(x)$ as x assumes larger and larger positive values?

Solution Let us investigate this question using a calculator. Table 3 shows the values of $f(x)$ for increasingly large values of x.

Table 3

	x assumes larger and larger positive values							
x	0	5	10	20	50	100	500	1,000
$f(x)$	0	1.92	1.98	1.995	1.9992	1.9998	1.999992	1.999998

The calculations shown here suggest that as the values of x continue to increase, the functional value $f(x)$ approaches the number 2. It seems that we can make the functional value $f(x)$ come as close to 2 as we like by taking a sufficiently large value of x. If we use the symbol "$x \to \infty$" to indicate that x is increasing with no upper limit on its size, then we can write

$$\lim_{x \to \infty} \frac{2x^2}{1 + x^2} = 2$$

It is important to understand that the symbol "∞" does not represent an actual number that x is approaching, but is used to indicate only that the value of x is increasing with no upper limit on its size. In particular, the statement "$x = \infty$" is meaningless since ∞ is not a symbol for a real number. We will also use the statement "$x \to -\infty$" to indicate that the value of x is decreasing with no lower limit on its size. Since for the function in this example $f(-x) = f(x)$, the values in Table 3 also suggest that

$$\lim_{x \to -\infty} \frac{2x^2}{1 + x^2} = 2$$

Examining the graph of this function (see Figure 5) provides us with a geometric interpretation of these two limit statements. The graph of f is approaching the horizontal line with equation $y = 2$ as $x \to \infty$ and as $x \to -\infty$. The line $y = 2$ is called a **horizontal asymptote** for the graph of f.

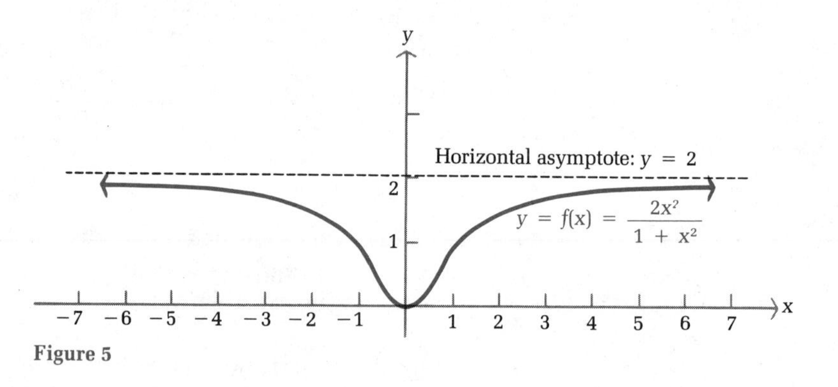

Figure 5

Problem 6 Construct a table such as Table 3 in order to estimate (do not graph):

$$\lim_{x \to \infty} \frac{2}{1 + x^2}$$

We now state an informal definition of **the limit of a function f as x approaches ∞ or $-\infty$.**

Limits at Infinity

We write

$$\lim_{x \to \infty} f(x) = L$$

if the functional value $f(x)$ is close to the single real number L whenever x is a very large positive number. We write

$$\lim_{x \to -\infty} f(x) = M$$

if the functional value $f(x)$ is close to the single real number M whenever x is a negative number with very large absolute value.

How can we evaluate limits of the form $\lim_{x \to \infty} f(x)$ and $\lim_{x \to -\infty} f(x)$ without drawing figures or performing calculator experiments? Fortunately, all the limit properties listed in Theorem 1, except property 4,* are valid if we replace the statement $x \to c$ with the statement $x \to \infty$ (or $x \to -\infty$). These properties, together with Theorem 2, will enable us to evaluate limits at infinity for many functions.

Theorem 2

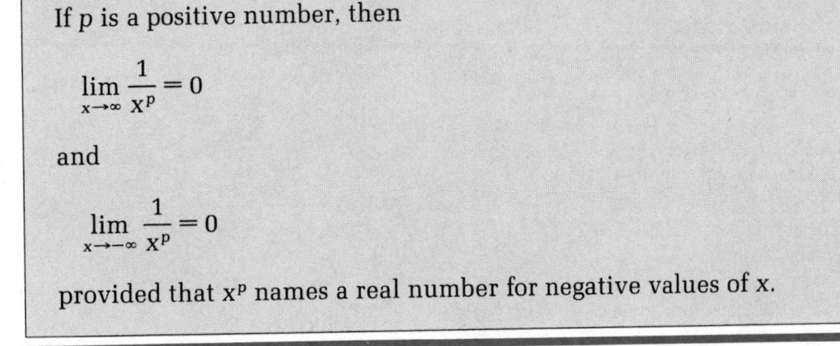

If p is a positive number, then

$$\lim_{x \to \infty} \frac{1}{x^p} = 0$$

and

$$\lim_{x \to -\infty} \frac{1}{x^p} = 0$$

provided that x^p names a real number for negative values of x.

Figure 6 (at the top of the next page) illustrates the theorem for several values of p.

* Limit property 4, $\lim_{x \to c} P(x) = P(c)$, does not make sense if c is replaced with ∞. Remember, ∞ is not a number and it is impossible to evaluate a function at ∞.

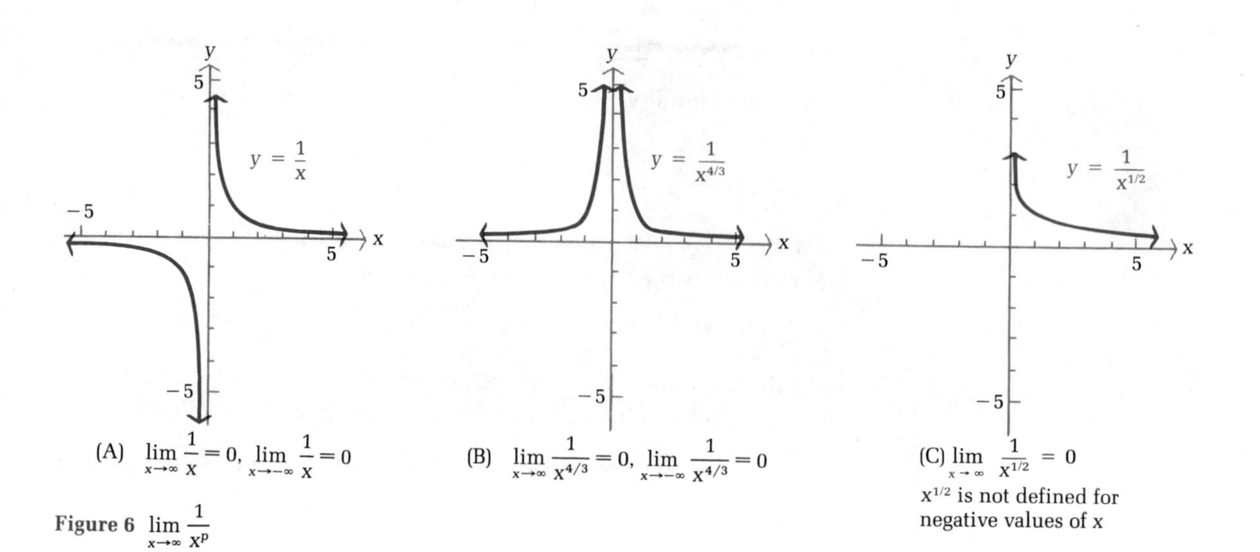

Figure 6 $\lim_{x\to\infty} \frac{1}{x^p}$

Example 7 Find each limit.

(A) $\lim_{x\to\infty} \frac{5x+4}{2x+3}$ (B) $\lim_{x\to-\infty} \frac{4x^2+3x+2}{2x^3+5}$ (C) $\lim_{x\to\infty} \frac{3x^4+6x}{x^2+4}$

Solutions (A) $\lim_{x\to\infty} \frac{5x+4}{2x+3} \neq \frac{\lim_{x\to\infty} 5x+4}{\lim_{x\to\infty} 2x+3}$ since $\lim_{x\to\infty} (5x+4)$ and $\lim_{x\to\infty} (2x+3)$ do not exist

$$= \lim_{x\to\infty} \frac{\frac{5x+4}{x}}{\frac{2x+3}{x}}$$

We divide numerator and denominator by x in order to express the fraction in a form where the limits of the numerator and denominator do exist.

$$= \lim_{x\to\infty} \frac{5+\frac{4}{x}}{2+\frac{3}{x}} = \frac{5+0}{2+0} = \frac{5}{2}$$

Use limit properties 6 and 3 and Theorem 2.

(B) $\lim_{x\to-\infty} \frac{4x^2+3x+2}{2x^3+5}$

Divide numerator and denominator by x^3, the highest power of x that occurs in the numerator or the denominator, simplify, and proceed as before.

$$= \lim_{x\to-\infty} \frac{\frac{4}{x}+\frac{3}{x^2}+\frac{2}{x^3}}{2+\frac{5}{x^3}}$$

Use limit properties 6 and 3 and Theorem 2.

$$= \frac{0+0+0}{2+0} = \frac{0}{2} = 0$$

(C) $\lim_{x\to\infty} \frac{3x^4 + 6x}{x^2 + 4}$

This time, divide numerator and denominator by x^4.

$$= \lim_{x\to\infty} \frac{3 + \frac{6}{x^3}}{\frac{1}{x^2} + \frac{4}{x^4}}$$

Since the numerator of this fraction approaches 3 and the denominator approaches 0, the fraction can be made as large as you like; hence, the limit does not exist.

Does not exist

Problem 7 Find each limit.

(A) $\lim_{x\to\infty} \frac{2x - 4}{3x + 5}$ (B) $\lim_{x\to\infty} \frac{3x^3 + 4}{2x^2 + 6}$ (C) $\lim_{x\to\infty} \frac{2x + 1}{x^2 + 4}$

The methods used to evaluate the limits in Example 7 can be applied to any rational function.* The results are summarized in Theorem 3.

Theorem 3

Limits at Infinity for Rational Functions

If

$$f(x) = \frac{p(x)}{q(x)}$$

where

$$p(x) = a_n x^n + a_{n-1} x^{n-1} + \cdots + a_0 \qquad a_n \neq 0$$

and

$$q(x) = b_m x^m + b_{m-1} x^{m-1} + \cdots + b_0 \qquad b_m \neq 0$$

then

$$\lim_{x\to\pm\infty} f(x) = \begin{cases} 0 & \text{if } n < m & (3) \\ \frac{a_n}{b_m} & \text{if } n = m & (4) \\ \text{Does not exist} & \text{if } n > m & (5) \end{cases}$$

[*Note:* If $n > m$, then the limit fails to exist because $f(x)$ is unbounded as $x \to \infty$ and as $x \to -\infty$.]

* A **rational function** is any function that can be specified by an equation of the form

$$R(x) = \frac{P(x)}{Q(x)}$$

where $P(x)$ and $Q(x)$ are polynomials.

Examples: $R(x) = \frac{x}{x - 3}$, $R(x) = \frac{x^2 - 3x + 1}{2x^3 - 3x^2 + 5}$, $R(x) = \frac{5}{x^3 - 9}$

Thus, we see that the limit at infinity of a rational function can be determined by simply comparing the degree of the numerator and the degree of the denominator. Notice that the value of the limit is the same, whether $x \to \infty$ or $x \to -\infty$. This implies that a rational function can have at most one horizontal asymptote.

Example 8 Use equations (3), (4), and (5) in Theorem 3 to find each limit.

(A) $\lim_{x \to \pm\infty} \frac{5x^2 + 3}{3x^2 + 4}$

(B) $\lim_{x \to \pm\infty} \frac{2x^5 + 7}{6x^3 + 4}$

(C) $\lim_{x \to \pm\infty} \frac{3x^4 + 9}{8x^6 + 5}$

Solutions (A) $\lim_{x \to \pm\infty} \frac{5x^2 + 3}{3x^2 + 4} = \frac{5}{3}$

Use (4): $n = m = 2, \quad a_n = 5, \quad b_m = 3$

(B) $\lim_{x \to \pm\infty} \frac{2x^5 + 7}{6x^3 + 4}$ Does not exist

Use (5): $n = 5, \quad m = 3, \quad n > m$

(C) $\lim_{x \to \pm\infty} \frac{3x^4 + 9}{8x^6 + 5} = 0$

Use (3): $n = 4, \quad m = 6, \quad n < m$

Problem 8 Use equations (3), (4), and (5) in Theorem 3 to find each limit.

(A) $\lim_{x \to \pm\infty} \frac{4x^3 + 5}{2x^4 + 4}$

(B) $\lim_{x \to \pm\infty} \frac{x^6 + 2}{x^5 + 4}$

(C) $\lim_{x \to \pm\infty} \frac{4x^2 + 7}{9x^2 + 11}$

Answers to Matched Problems

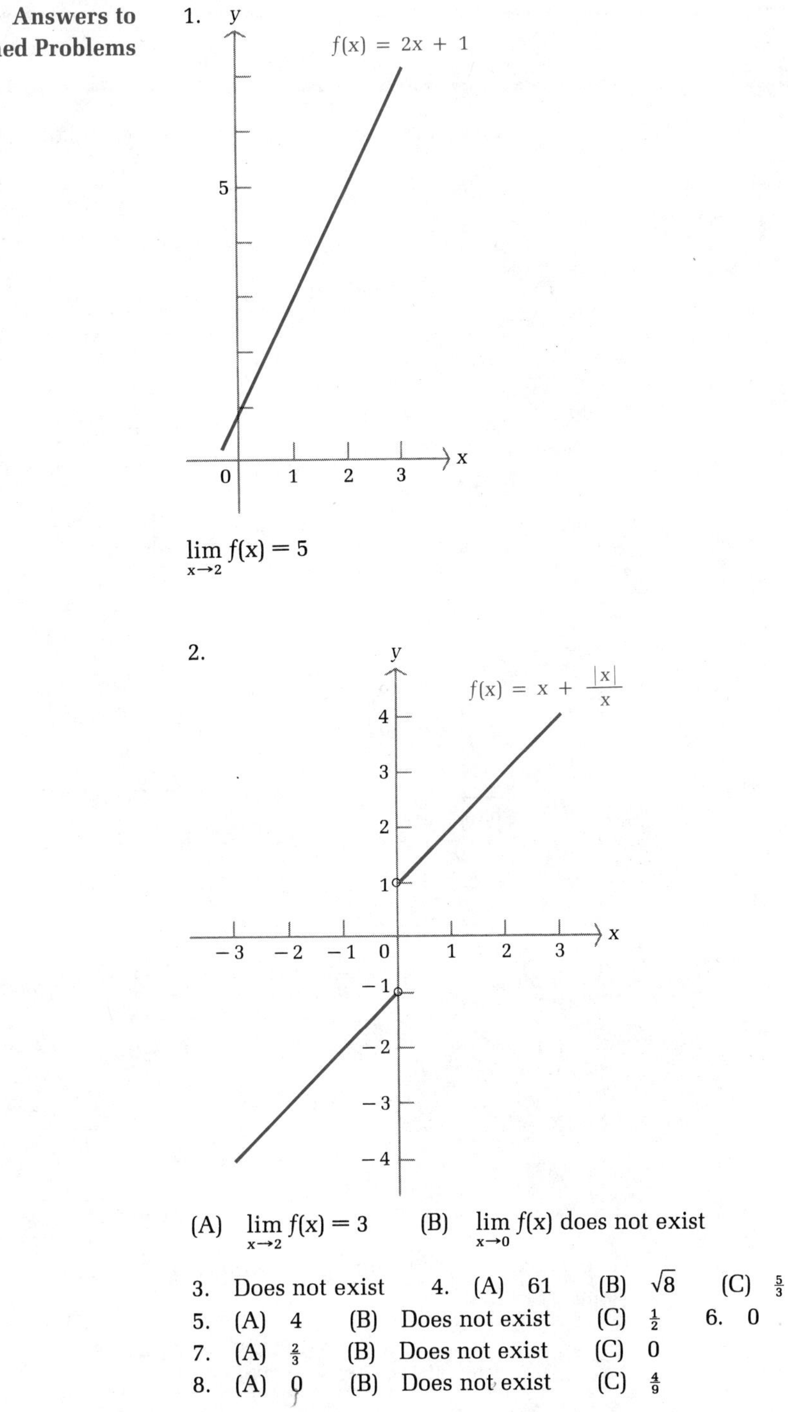

1. $\lim_{x \to 2} f(x) = 5$

2. (A) $\lim_{x \to 2} f(x) = 3$ (B) $\lim_{x \to 0} f(x)$ does not exist

3. Does not exist
4. (A) 61 (B) $\sqrt{8}$ (C) $\frac{5}{3}$
5. (A) 4 (B) Does not exist (C) $\frac{1}{2}$
6. 0
7. (A) $\frac{2}{3}$ (B) Does not exist (C) 0
8. (A) 0 (B) Does not exist (C) $\frac{4}{9}$

Exercise 1-1

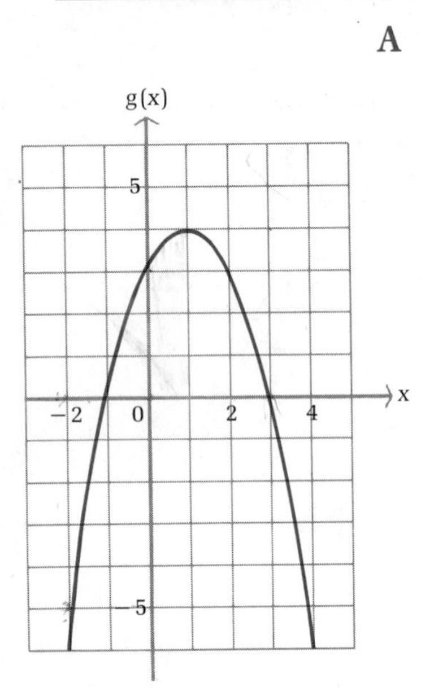

A

1. Use the graph of function g to estimate each limit, if it exists.

   (A) $\lim_{x \to -2} g(x)$ (B) $\lim_{x \to 0} g(x)$ (C) $\lim_{x \to 1} g(x)$ (D) $\lim_{x \to 3} g(x)$

2. Use the graph of function f to estimate each limit, if it exists.

   (A) $\lim_{x \to 1} f(x)$ (B) $\lim_{x \to 2} f(x)$ (C) $\lim_{x \to 4} f(x)$ (D) $\lim_{x \to 5} f(x)$

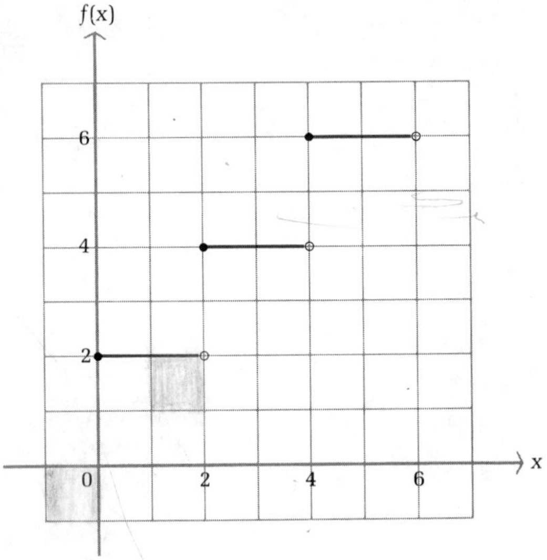

Find each limit.

3. $\lim_{x \to 5} (2x^2 - 3)$

4. $\lim_{x \to 2} (x^2 - 8x + 2)$

5. $\lim_{x \to 4} (x^2 - 5x)$

6. $\lim_{x \to -2} (3x^3 - 9)$

7. $\lim_{x \to 2} \dfrac{5x}{2 + x^2}$

8. $\lim_{x \to 10} \dfrac{2x + 5}{3x - 5}$

9. $\lim_{x \to 2} (x + 1)^3(2x - 1)^2$

10. $\lim_{x \to 3} (x + 2)^2(2x - 4)$

Use a calculator to complete the following table for each function in Problems 11–14:

x	0.9	0.99	0.999	$\rightarrow 1 \leftarrow$	1.001	1.01	1.1
$f(x)$				$\rightarrow ? \leftarrow$			

Use the completed table to estimate $\lim_{x \to 1} f(x)$, *if it exists.*

11. $f(x) = \dfrac{x^2 - 1}{(x - 1)^2}$

12. $f(x) = \dfrac{x^2 - 1}{x - 1}$

13. $f(x) = \dfrac{|x - 1|}{x - 1}$

14. $f(x) = \dfrac{x^2 - 1}{|x - 1|}$

Use a calculator to complete the following table for each function in Problems 15–18:

x	10	100	1,000	10,000	$\to \infty$
f(x)					$\to ?$

Use the completed table to estimate $\lim_{x \to \infty} f(x)$, *if it exists.*

15. $f(x) = \dfrac{1}{x + 1}$

16. $f(x) = \dfrac{x}{x + 1}$

17. $f(x) = \dfrac{x^2}{x + 1}$

18. $f(x) = \dfrac{x + 1}{x^2}$

B **19.** Use the graph of function f to estimate each limit, if it exists.

(A) $\lim_{x \to -\infty} f(x)$ (B) $\lim_{x \to -1} f(x)$

(C) $\lim_{x \to 1} f(x)$ (D) $\lim_{x \to \infty} f(x)$

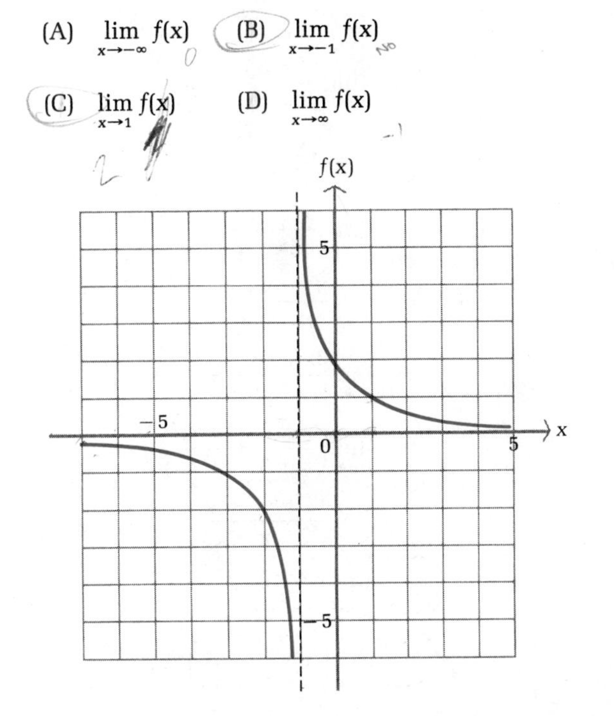

20. Use the graph of function g to estimate each limit, if it exists.

(A) $\lim_{x \to -\infty} g(x)$ (B) $\lim_{x \to -1} g(x)$ (C) $\lim_{x \to 1} g(x)$ (D) $\lim_{x \to \infty} g(x)$

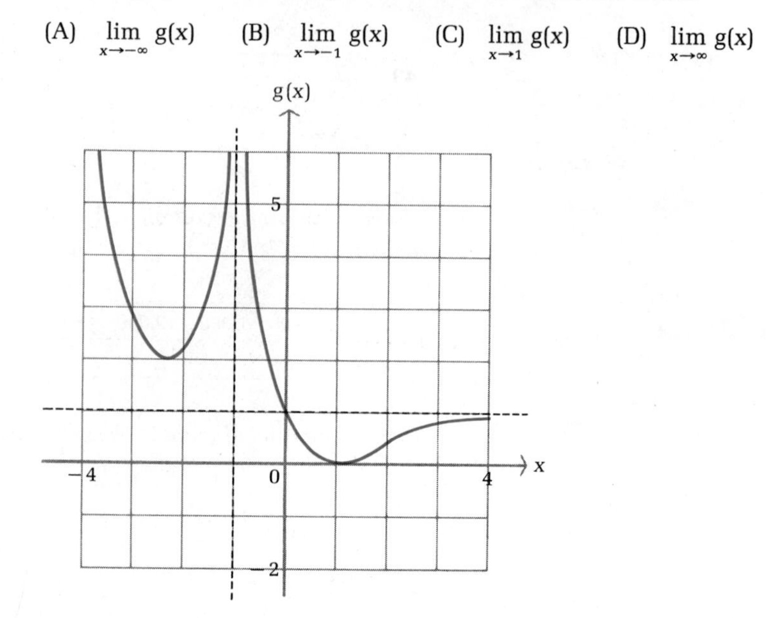

Find each limit, if it exists. (Use algebraic manipulations where necessary.)

21. $\lim_{x \to 0} \frac{x^2 - 3x}{x}$

22. $\lim_{x \to 0} \frac{2x^2 + 5x}{x}$

23. $\lim_{x \to 5} \frac{x^2 - 25}{x - 5}$

24. $\lim_{x \to 4} \frac{x^2 - 16}{x - 4}$

25. $\lim_{x \to -5} \frac{x^2 - 25}{x - 5}$

26. $\lim_{x \to -4} \frac{x^2 - 16}{x - 4}$

27. $\lim_{x \to -2} \frac{x^2 - x - 6}{x + 2}$

28. $\lim_{x \to -4} \frac{2x^2 + 7x - 4}{x + 4}$

29. $\lim_{x \to 3} \frac{x^2 - x - 6}{x^2 - 9}$

30. $\lim_{x \to 2} \frac{x^2 + 2x - 8}{x^2 - 2x}$

31. $\lim_{x \to 1} \frac{2x^3 - 3x + 2}{x^2 + x}$

32. $\lim_{x \to 2} \frac{x^3 - x^2 + 1}{x - x^2}$

33. $\lim_{x \to 3} \left(\frac{x}{x + 3} + \frac{x - 3}{x^2 - 9} \right)$

34. $\lim_{x \to 2} \left(\frac{1}{x + 2} + \frac{x - 2}{x^2 - 4} \right)$

35. $\lim_{x \to 0} \frac{(2 + x)^2 - 4}{x}$

36. $\lim_{x \to 0} \frac{(3 + x)^2 - 9}{x}$

37. $\lim_{x \to \infty} \frac{2x + 4}{x}$

38. $\lim_{x \to \infty} \frac{3x^2 + 5}{x^2}$

39. $\lim_{x \to \infty} \left(4 + \frac{2}{x} - \frac{3}{x^2} \right)$

40. $\lim_{x \to -\infty} \left(3 - \frac{5}{x^3} + \frac{2}{x^4} \right)$

41. $\lim_{x\to\infty} \dfrac{2x^3}{3x^3+5}$

42. $\lim_{x\to\infty} \dfrac{4x^4}{9x^4+10}$

43. $\lim_{x\to-\infty} \dfrac{2x^3}{4x^4+7}$

44. $\lim_{x\to-\infty} \dfrac{3x^2}{x+2}$

45. $\lim_{x\to\infty} \dfrac{3x^3+5}{4x^2+2}$

46. $\lim_{x\to\infty} \dfrac{7x^2}{x^5+7}$

In Problems 47–50, graph the function f and use the graph to estimate each limit, if it exists.

47. $f(x) = \begin{cases} 1 & -1 \leqslant x < 1 \\ 2 & 1 \leqslant x < 3 \\ 3 & 3 \leqslant x \leqslant 5 \end{cases}$

(A) $\lim_{x\to1} f(x)$ (B) $\lim_{x\to2} f(x)$ (C) $\lim_{x\to3} f(x)$

48. $f(x) = \begin{cases} 1 & -1 \leqslant x < 1 \\ x & 1 \leqslant x < 3 \\ 3 & 3 \leqslant x \leqslant 5 \end{cases}$

(A) $\lim_{x\to1} f(x)$ (B) $\lim_{x\to2} f(x)$ (C) $\lim_{x\to3} f(x)$

49. $f(x) = \begin{cases} 2 & x \leqslant -1 \\ x^2 & -1 < x < 1 \\ 1 & x \geqslant 1 \end{cases}$

(A) $\lim_{x\to-1} f(x)$ (B) $\lim_{x\to0} f(x)$ (C) $\lim_{x\to1} f(x)$

50. $f(x) = \begin{cases} -x-1 & x \leqslant -1 \\ 1-x^2 & -1 < x < 1 \\ x-1 & x \geqslant 1 \end{cases}$

(A) $\lim_{x\to-1} f(x)$ (B) $\lim_{x\to0} f(x)$ (C) $\lim_{x\to1} f(x)$

C

Find each limit. [Note: $a^3 - b^3 = (a-b)(a^2+ab+b^2)$ and $a^3 + b^3 = (a+b)(a^2-ab+b^2)$.]

51. $\lim_{x\to4} \sqrt[3]{x^2-3x}$

52. $\lim_{x\to2} \sqrt{x^2+2x}$

53. $\lim_{x\to4} \dfrac{\sqrt{x}-2}{x^2-4}$

54. $\lim_{x\to25} \dfrac{5-\sqrt{x}}{x+25}$

55. $\lim_{x\to4} \dfrac{\sqrt{x}-2}{x-4}$

56. $\lim_{x\to0} \dfrac{\sqrt{x+4}-2}{x}$

57. $\lim_{x\to 9} \frac{x+3}{\sqrt{x}-3}$

58. $\lim_{x\to 25} \frac{x+5}{\sqrt{x}-5}$

59. $\lim_{x\to 16} \frac{x-16}{\sqrt{x}-4}$

60. $\lim_{x\to 36} \frac{x-36}{\sqrt{x}-6}$

61. $\lim_{x\to 2} \frac{x^3-8}{x-2}$

62. $\lim_{x\to 1} \frac{x^2-1}{x^3-1}$

63. $\lim_{x\to -2} \frac{x+2}{x^3+8}$

64. $\lim_{x\to -1} \frac{x^3+1}{(x-1)^2}$

65. $\lim_{x\to 0} \frac{(a+x)^2-a^2}{x}$, a any real number

66. $\lim_{x\to 0} \frac{(a+x)^3-a^3}{x}$, a any real number

67. $\lim_{x\to 0} \frac{\sqrt{a+x}-\sqrt{a}}{x}$, $a>0$

68. $\lim_{x\to 0} \frac{\frac{1}{a+x}-\frac{1}{a}}{x}$, $a\neq 0$

Applications

Business & Economics

69. *Average cost.* The cost equation for manufacturing a particular compact disk album is

$$C(x) = 20{,}000 + 3x$$

where x is the number of disks produced. The average cost per disk, denoted by $\overline{C}(x)$, is found by dividing $C(x)$ by x:

$$\overline{C}(x) = \frac{C(x)}{x} = \frac{20{,}000 + 3x}{x}$$

If only ten disks were manufactured, for example, the average cost per disk would be \$2,003. Find:

(A) $\overline{C}(1{,}000)$

(B) $\overline{C}(100{,}000)$

(C) $\lim_{x\to 10{,}000} \overline{C}(x)$

(D) $\lim_{x\to\infty} \overline{C}(x)$

70. *Employee training.* A company producing computer components has established that on the average, a new employee can assemble $N(t)$ components per day after t days of on-the-job training, as given by

$$N(t) = \frac{100t}{t + 9}$$

Find:

(A) $N(1)$ (B) $N(11)$ (C) $\lim_{t \to 11} N(t)$ (D) $\lim_{t \to \infty} N(t)$

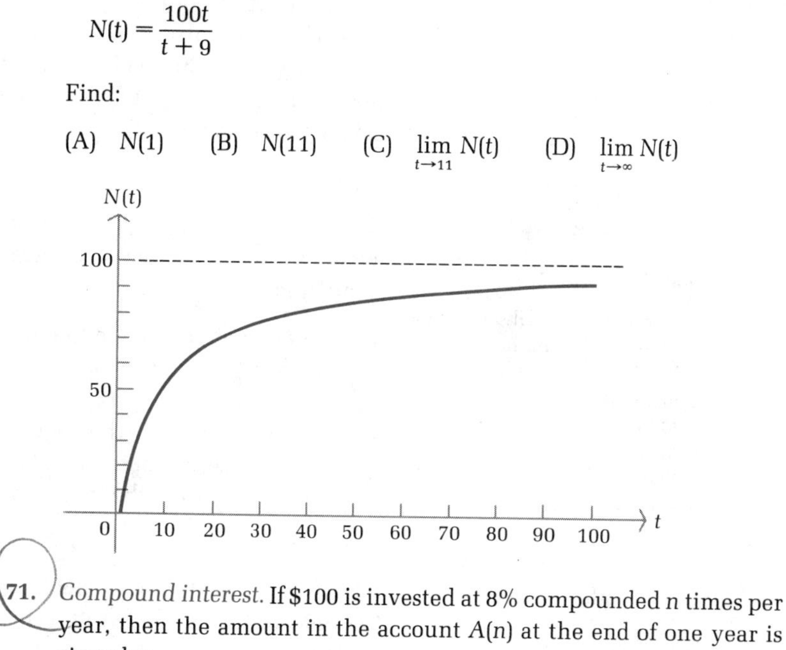

71. *Compound interest.* If \$100 is invested at 8% compounded n times per year, then the amount in the account $A(n)$ at the end of one year is given by

$$A(n) = 100\left(1 + \frac{0.08}{n}\right)^n$$

(A) Use a calculator with a y^x button to complete the following table (each entry to the nearest cent):

Compounded	n	$A(n)$
Annually	1	\$108.00
Semiannually	2	\$108.16
Quarterly	4	\$108.24
Monthly	12	
Weekly	52	
Daily	365	
Hourly	8,760	

(B) Using the results of part A, guess the following limit:

$$\lim_{n \to \infty} A(n) = ?$$

(This example leads to the important concept of compounding continuously, which will be discussed in detail in Section 3-1.)

72. *Pollution.* In Silicon Valley, California, a number of computer related manufacturing firms were found to be contaminating underground water supplies with toxic chemicals stored in leaking underground containers. A water quality control agency ordered the companies to take immediate corrective action and to contribute to a monetary pool for testing and cleanup of the underground contamination. Suppose the required monetary pool (in millions of dollars) for the testing and cleanup is estimated by

$$P(x) = \frac{2x}{1 - x}$$

where x is the percentage (expressed as a decimal fraction) of the total contaminant removed.

Percentage Removed	Pool Required
0.50 (50%)	$2 million
0.60 (60%)	$3 million
0.70 (70%)	
0.80 (80%)	
0.90 (90%)	
0.95 (95%)	
0.99 (99%)	

(A) Complete the table in the margin.

(B) Find $\lim_{x \to 0.80} P(x)$.

(C) What happens to the required monetary pool as the desired percentage of contaminant removed approaches 100% (x approaches 1 from the left)?

Life Sciences

73. *Medicine.* A drug is injected into the bloodstream of a patient through her right arm. The concentration of the drug in the bloodstream of the left arm t hours after the injection is given by

$$C(t) = \frac{0.14t}{t^2 + 1}$$

Find:

(A) $C(0.5)$ (B) $C(1)$ (C) $\lim_{t \to 1} C(t)$ (D) $\lim_{t \to \infty} C(t)$

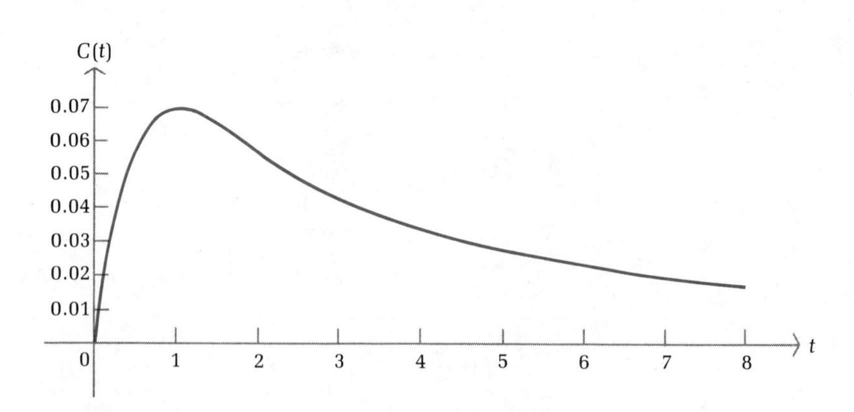

74. *Physiology.* In a study on the speed of muscle contraction in frogs under various loads, researchers W. O. Fems and J. Marsh found that the speed of contraction decreases with increasing loads. More precisely, they found that the relationship between speed of contraction S (in centimeters/second) and load w (in grams) is given approximately by

$$S(w) = \frac{26 + 0.06w}{w} \qquad w > 5$$

Find:

(A) $S(10)$ (B) $S(50)$ (C) $\lim_{w \to 50} S(w)$ (D) $\lim_{w \to \infty} S(w)$

Social Sciences

75. *Psychology—Learning theory.* In 1917, L. L. Thurstone, a pioneer in quantitative learning theory, proposed the function

$$f(x) = \frac{a(x + c)}{(x + c) + b}$$

to describe the number of successful acts per unit time that a person could accomplish after x practice sessions. Suppose for a particular person enrolling in a typing school

$$f(x) = \frac{60(x + 1)}{x + 5}$$

where $f(x)$ is the number of words per minute that the person is able to type after x weeks of lessons. Find:

(A) $f(3)$ (B) $f(10)$ (C) $\lim_{x \to 10} f(x)$ (D) $\lim_{x \to \infty} f(x)$

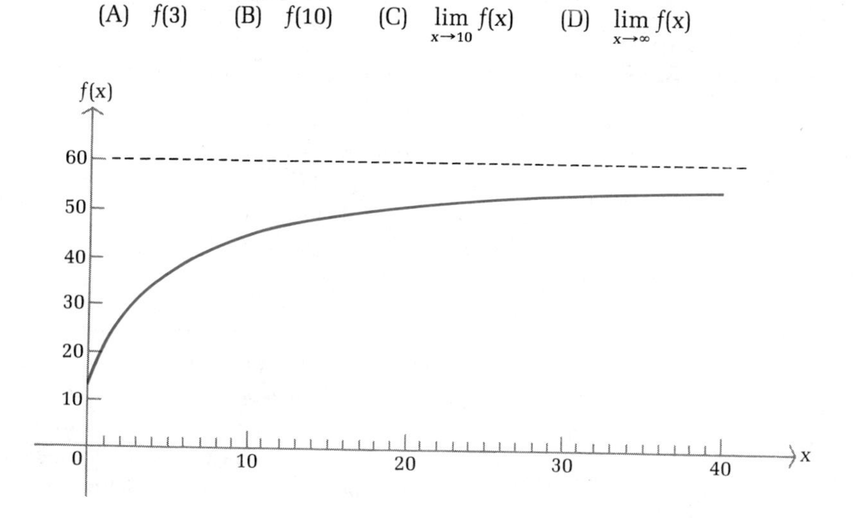

76. *Psychology—Retention.* An experiment on retention is conducted in a psychology class. Each student in the class is given one day to memorize the same list of thirty special characters. The lists are turned in at the end of the day, and for each succeeding day for thirty days each student is asked to turn in a list of as many of the symbols as can be recalled. Averages are taken and it is found that

$$N(t) = \frac{5t + 20}{t} \qquad t \geq 1$$

provides a good approximation of the average number of symbols, $N(t)$, retained after t days. Find:

(A) $N(2)$ (B) $N(10)$ (C) $\lim_{t \to 10} N(t)$ (D) $\lim_{t \to \infty} N(t)$

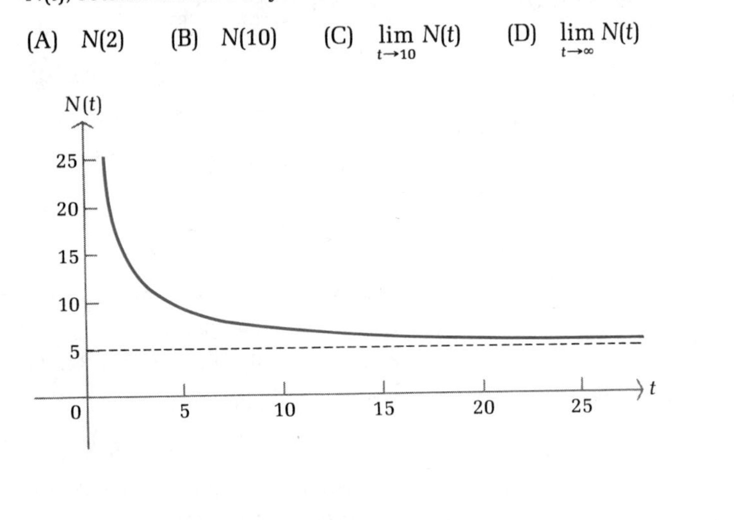

1-2 Continuity

- Introduction
- Definition of Continuity
- Properties of Continuity
- Solving Nonlinear Inequalities Using Continuity Properties

▪ Introduction

In this section we will use the limit concept to describe an important property possessed by many functions. We begin by discussing an example.

Most daily newspapers include hourly temperatures in the weather report. Table 4 lists hourly temperatures from midnight to noon. We can represent the temperature at any time during this 12 hour period by plotting these points and connecting them with a smooth curve, as illustrated in Figure 7.

Table 4

Time	Temperature (°F)
12 midnight	30
1 AM	29
2 AM	25
3 AM	25
4 AM	27
5 AM	22
6 AM	20
7 AM	28
8 AM	30
9 AM	32
10 AM	38
11 AM	40
12 noon	48

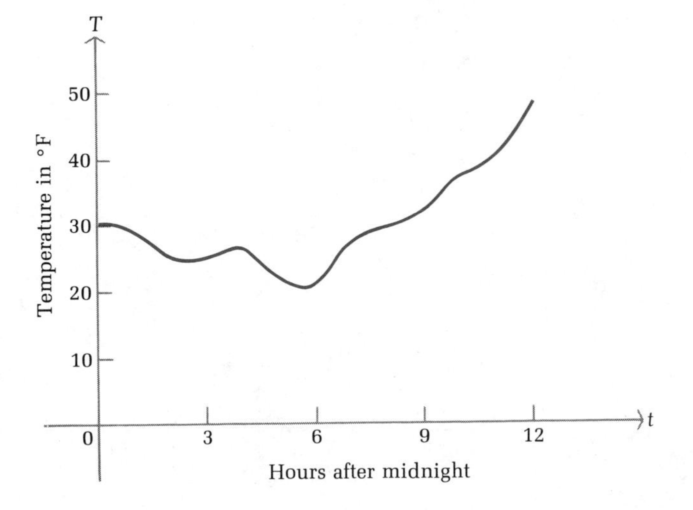

Figure 7

Notice that this curve can be drawn without lifting a pen off the paper. Informally, we say that this is a *continuous* curve. We drew the curve in this manner because our intuition tells us that temperature varies continuously with time. That is, if the temperature is 29°F at 1 AM and 25°F at 2 AM, then the temperature must have gradually and continuously changed from 29°F to 25°F during that hour.

Most graphs of natural phenomena (temperature, growth, decay, and so on) vary continuously with time, whereas many graphs in business and economics do not. The next example illustrates a graph that is not continuous.

A national delivery service uses weight of a package to determine the charge for delivery. The charge is \$11.50 for the first pound (or any fraction thereof) and \$1 for each additional pound (or fraction thereof). If $C(x)$ is the charge for delivering a package weighing x pounds, then

$$C(x) = \begin{cases} 11.50 & \text{for } 0 < x \leqslant 1 \\ 12.50 & \text{for } 1 < x \leqslant 2 \\ 13.50 & \text{for } 2 < x \leqslant 3 \\ \text{and so on} & \end{cases}$$

The function C is graphed for $0 < x \leqslant 3$ in Figure 8. Notice that it is not possible to draw the graph of C without lifting a pen off the paper. There are breaks in the graph at $x = 1$ and at $x = 2$. We say that the graph of C is *discontinuous* at $x = 1$ and $x = 2$. The graph of C is continuous on the intervals $(0, 1)$, $(1, 2)$, and $(2, 3)$.

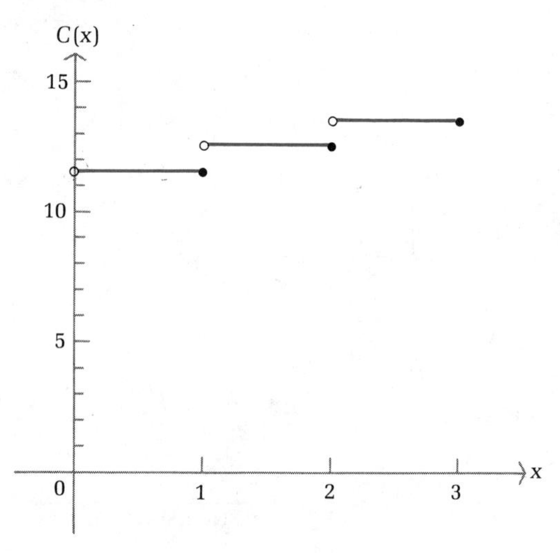

Figure 8

■ Definition of Continuity

If we have a graph of a function, then it is usually easy to identify points of discontinuity. If a function is defined by an equation, how can we identify points of discontinuity without looking at its graph? Figure 9 and Table 5 suggest some procedures as well as a formal definition of continuity in terms of limits. Study the figure and table carefully before proceeding further.

The function shown in Figure 9 is not the type of function that you are likely to encounter with great frequency. It was designed to illustrate most of the kinds of points of discontinuity exhibited by various types of func-

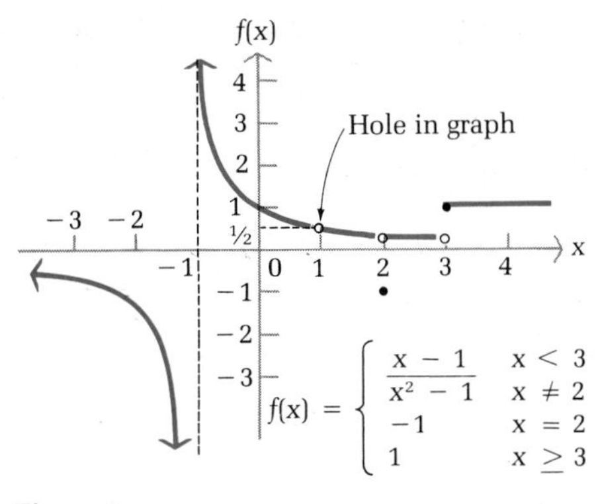

Figure 9

Table 5

c	$\lim_{x\to c} f(x)$	$f(c)$	**Graph**
−2	−1	−1	No break in graph
−1	Does not exist	Does not exist	Break in graph
0	1	1	No break in graph
1	1/2	Does not exist	Break in graph
2	1/3	−1	Break in graph
3	Does not exist	1	Break in graph
4	1	1	No break in graph

tions. Looking at Table 5, we are led to the following precise definition of continuity:

Continuity

A function f is **continuous at the point $x = c$** if

1. $\lim_{x\to c} f(x)$ exists.
2. $f(c)$ exists.
3. $\lim_{x\to c} f(x) = f(c)$.

A function is **continuous on the open interval (a, b)** if it is continuous at each point on the interval.

If one or more of the three conditions in the definition fails, then a function is **discontinuous** at $x = c$. Note that at least one of the three conditions fails at $x = -1, 1, 2$, and 3 in Figure 9 (determine which for each number).

Example 9 Using the definition of continuity, discuss the continuity of each function at the indicated value of c.

(A) $f(x) = \dfrac{x^2 - 4}{x - 2}$ at $c = 1$ and $c = 2$

(B) $f(x) = \begin{cases} x^2 & x < 0 \\ 1 & x \geq 0 \end{cases}$ at $c = 0$

(C) $f(x) = \begin{cases} 4 - x^2 & x \neq 1 \\ 2 & x = 1 \end{cases}$ at $c = 1$

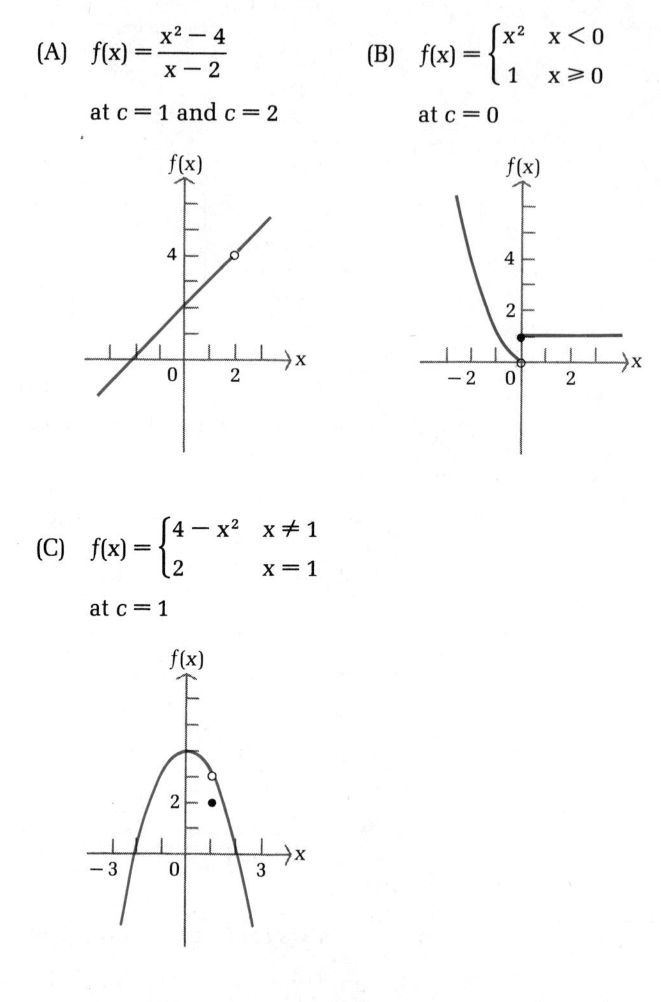

Solutions (A) $\lim_{x\to 1} f(x) = 3 = f(1)$; thus, all three conditions in the definition are satisfied and f is continuous at $x = 1$. $\lim_{x\to 2} f(x) = 4$, but $f(2)$ is not defined. Condition 2 in the definition is not satisfied and f is not continuous at $x = 2$.

(B) $f(0) = 1$ [thus, $f(0)$ is defined], but $\lim_{x\to 0} f(x)$ does not exist. Condition 1 in the definition is not satisfied and f is not continuous at $x = 0$.

(C) $\lim_{x\to 1} f(x) = 3$ and $f(1) = 2$. But $\lim_{x\to 1} f(x) \neq f(1)$. Condition 3 in the definition is not satisfied and f is not continuous at $x = 1$.

Problem 9 Using the definition of continuity, discuss the continuity of each function at the indicated value of c.

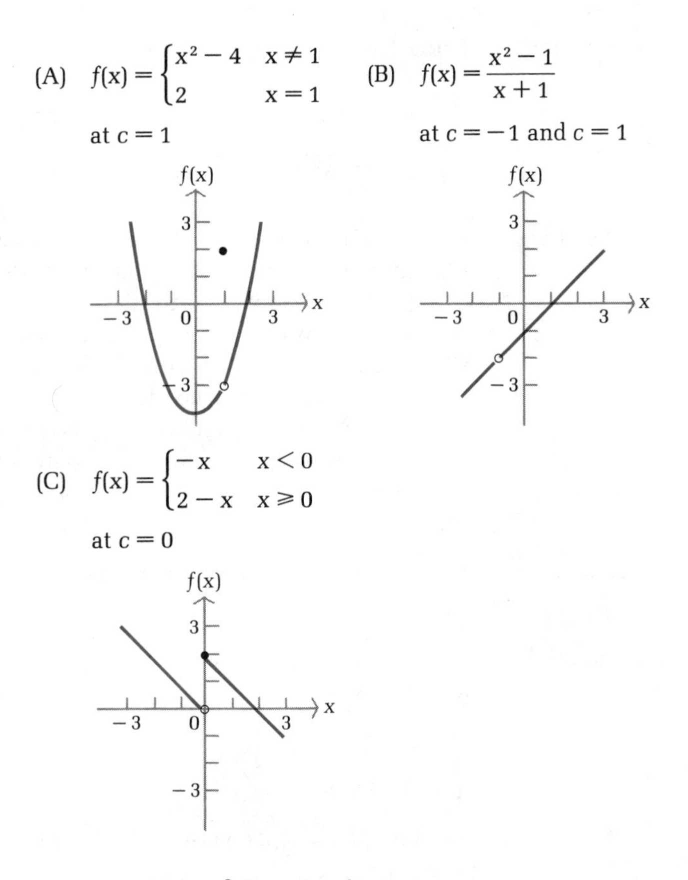

▪ Properties of Continuity

Functions have continuity properties similar to limit properties. For example, **the sum, difference, product, and quotient of two continuous functions are continuous, except for values of x that make a denominator 0.** These properties, along with Theorem 4, enable us to determine intervals of continuity for some important classes of functions without having to look at their graphs or use the three conditions in the definition.

Theorem 4

(A) Polynomial functions are continuous for all values of x.
(B) Rational functions are continuous for all values of x except those that make a denominator 0.
(C) If n is an odd positive integer greater than 1, then $\sqrt[n]{f(x)}$ is continuous wherever $f(x)$ is continuous.
(D) If n is an even positive integer, then $\sqrt[n]{f(x)}$ is continuous wherever $f(x)$ is continuous and positive.

Example 10 Using Theorem 4, determine where each function is continuous.

(A) $f(x) = x^2 - 2x + 1$ (B) $f(x) = \dfrac{x}{(x+2)(x-3)}$

(C) $f(x) = \sqrt[3]{x^2 - 4}$ (D) $f(x) = \sqrt{x - 2}$

Solutions

(A) Since f is a polynomial function, f is continuous for all x.

(B) Since f is a rational function, f is continuous for all x except -2 and 3 (values of x that make the denominator 0). Using interval notation, f is continuous on $(-\infty, -2)$, $(-2, 3)$, and $(3, \infty)$.

(C) The polynomial function $x^2 - 4$ is continuous for all x. Since $n = 3$ is odd, f is continuous for all x.

(D) The polynomial function $x - 2$ is continuous for all x and positive for $x > 2$. Since $n = 2$ is even, f is continuous for $x > 2$ or on the interval $(2, \infty)$.

Problem 10 Using Theorem 4, determine where each function is continuous.

(A) $f(x) = x^4 + 2x^2 + 1$ (B) $f(x) = \dfrac{x^2}{(x+1)(x-4)}$

(C) $f(x) = \sqrt{4 - x}$ (D) $f(x) = \sqrt[3]{x^3 + 1}$

■ Solving Nonlinear Inequalities Using Continuity Properties

We will soon see that it is useful to be able to solve nonlinear inequalities of the form

$$(x+1)(x-2) > 0 \quad \text{and} \quad \frac{(x^2-1)}{(x-3)} < 0$$

Theorem 5 states a property of continuous functions that can be of great use in solving such inequalities.

Theorem 5

> If f is continuous on (a, b) and $f(x) \neq 0$ for any x in (a, b), then either $f(x) > 0$ for all x in (a, b) or $f(x) < 0$ for all x in (a, b).

In other words, if f is continuous and $f(x) \neq 0$ on (a, b), then $f(x)$ cannot change sign on (a, b). To see why this is the case, suppose that there exist numbers x_1 and x_2 in (a, b) such that $f(x_1)$ and $f(x_2)$ do not have the same sign. For purposes of illustration, we assume that $a < x_1 < x_2 < b$, $f(x_1) < 0$, $f(x_2) > 0$ and graph this information in Figure 10A.

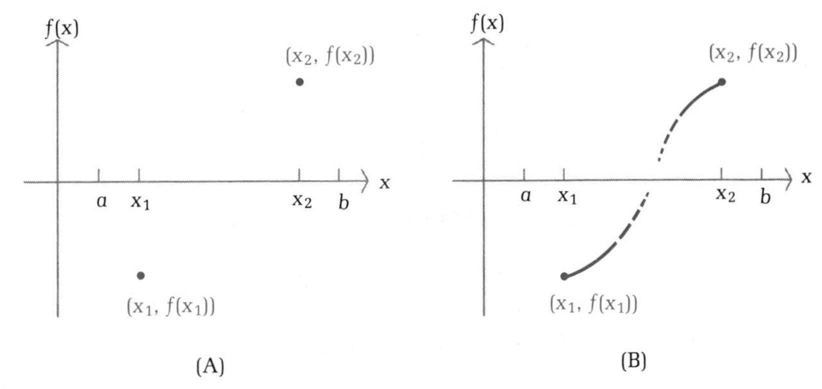

Figure 10

Since we are given that $f(x)$ is not zero on (a, b), its graph must not cross the x axis. Try to draw a continuous curve from $(x_1, f(x_1))$ to $(x_2, f(x_2))$ without crossing the x axis (see Fig. 10B). It can't be done! Thus, if we know that f is continuous and $f(x) \neq 0$ on (a, b), then it is impossible to find numbers x_1 and x_2 in (a, b) such that $f(x_1)$ and $f(x_2)$ have opposite signs; that is, $f(x)$ must either be positive for all x in (a, b) or negative for all x in (a, b).

The next example illustrates the application of Theorem 5 to the solution of nonlinear inequalities.

Example 11 Solve $\dfrac{x+1}{x-2} > 0$.

Solution We start by using the left side of the inequality to form the function f:

$$f(x) = \frac{x+1}{x-2}$$

The function f is discontinuous at $x = 2$, and $f(x) = 0$ for $x = -1$ (a fraction is 0 when the numerator is 0 and the denominator is not 0). We plot $x = 2$ and $x = -1$ on a real number line:

−2 −1 0 1 2 3 x

[*Note:* The dot at 2 is open; the function is not defined at $x = 2$.] The numbers 2 and -1 divide the real line into three intervals: $(-\infty, -1)$, $(-1, 2)$, and $(2, \infty)$. The function f is continuous and nonzero on each of these intervals. From Theorem 5 we know that $f(x)$ does not change sign on any of these intervals. Thus, we can find the sign of $f(x)$ on each of these intervals by selecting a **test number** in each interval and evaluating $f(x)$ at that number. Since any number in each subinterval will do, choose test numbers that are easy to evaluate. We choose -2, 0, and 3:

x	-2	0	3
$f(x)$	$\frac{1}{4}$	$-\frac{1}{2}$	4
	$+$	$-$	$+$

Test numbers

The sign of $f(x)$ at each test number determines the sign of $f(x)$ over the interval containing that test number. Using this information, we construct a **sign chart** for $f(x)$:

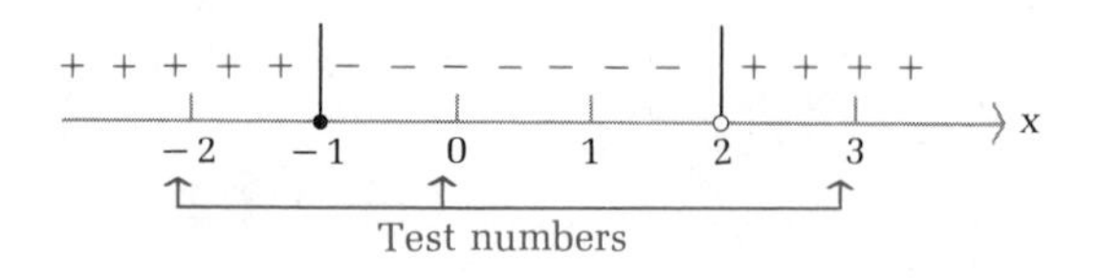

Using this sign chart, we can easily write the solution for the nonlinear inequality at the beginning of this discussion: $f(x) > 0$ for

$x < -1$ or $x > 2$ Inequality notation
$(-\infty, -1) \cup (2, \infty)$ Interval notation

Most of the inequalities we will encounter will involve strict inequalities ($>$ or $<$). If it is necessary to solve inequalities of the $\geq$ or $\leq$ form, simply include the end point of any interval if it is a zero of f [that is, if it is a value of x such that $f(x) = 0$]. Referring to the sign chart above, the solution of the inequality

$\frac{x+1}{x-2} \geq 0$ is $x \leq -1$ or $x > 2$ Inequality notation
$(-\infty, -1] \cup (2, \infty)$ Interval notation

We summarize the procedure for constructing sign charts in the box below.

Constructing Sign Charts

Given a function f:

Step 1. Find all numbers where f is discontinuous.
Step 2. Find all numbers where $f(x) = 0$.
Step 3. Plot the numbers found in steps 1 and 2 on a number line, dividing the number line into intervals.
Step 4. Select a test number in each interval found in step 3 and evaluate $f(x)$ at each number.
Step 5. Construct a sign chart using the real number line in step 3 and the results of step 4. This will show the sign of $f(x)$ on each interval.

Problem 11 Solve $\frac{x^2 - 1}{x - 3} < 0$.

Example 12
Profit Analysis

The marketing research and financial departments of a company estimate that at a price of p dollars per unit, the weekly cost C and revenue R (in thousands of dollars) will be given by the equations

$C = 14 - p$ Cost equation

$R = 8p - p^2$ Revenue equation

Find the prices for which the company has a profit.

Solution A profit will result if revenue is greater than cost; that is, if

$$R > C$$
$$8p - p^2 > 14 - p$$
$$-p^2 + 9p - 14 > 0$$
$$p^2 - 9p + 14 < 0$$
$$(p - 2)(p - 7) < 0$$

Multiply both sides by -1 (inequality sign reverses). Factor left side.

We form a function f using the left side of the inequality,

$$f(p) = (p - 2)(p - 7)$$

and construct a sign chart for $f(p)$.

Step 1. Find all numbers where f is discontinuous. There are no points of discontinuity (f is a polynomial function that is continuous for all p).

Step 2. Find all numbers where $f(p) = 0$:

$$f(p) = 0$$
$$(p - 2)(p - 7) = 0$$
$$p = 2, 7$$

Step 3. Plot the numbers found in steps 1 and 2 on a number line:

0 1 2 3 4 5 6 7 8 p

Step 4. Evaluate $f(p)$ for a test number in each interval:

p	0	5	8
$f(p)$	14 +	-6 $-$	6 +

Test numbers

Step 5. Form a sign chart for f using the results of steps 3 and 4:

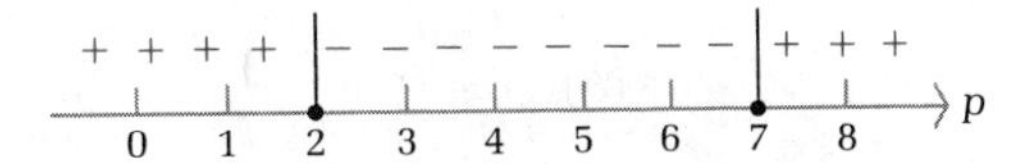

Thus, a profit will result for $f(p) < 0$; that is, for any price between \$2 and \$7. (The break-even prices are \$2 and \$7; losses will occur for prices less than \$2 or greater than \$7. Later we will show how to find a price that will result in the maximum profit.)

Problem 12 Repeat Example 12 for

$C = 8 - p$ Cost equation
$R = 5p - p^2$ Revenue equation

Answers to Matched Problems

9. (A) f is not continuous at $x = 1$; condition 3 is not satisfied.
   (B) f is not continuous at $x = -1$; condition 2 is not satisfied. f is continuous at $x = 1$.
   (C) f is not continuous at $x = 0$; condition 1 is not satisfied.
10. (A) For all x (B) $(-\infty, -1)$, $(-1, 4)$, and $(4, \infty)$ (C) $(-\infty, 4)$
    (D) For all x
11. $-\infty < x < -1$ or $1 < x < 3$, $(-\infty, -1) \cup (1, 3)$
12. The company makes a profit for $\$2 < p < \$4$.

Exercise 1-2

A *Problems 1–6 refer to the function f in the following graph. Use the graph to estimate limits.*

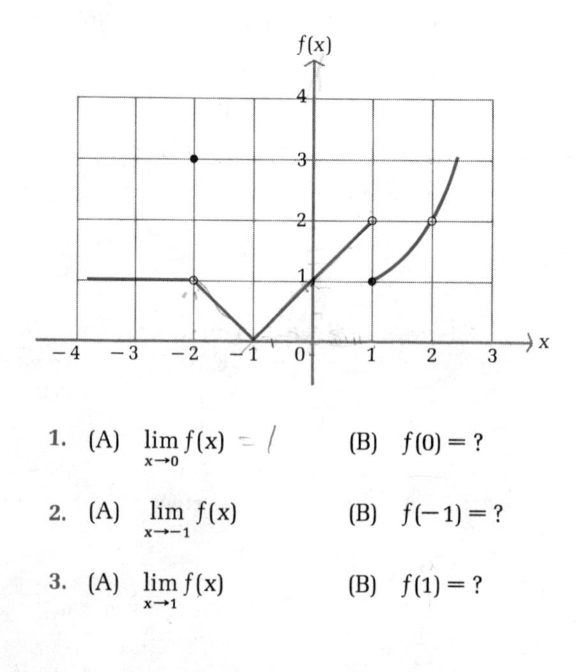

1. (A) $\lim_{x \to 0} f(x)$ (B) $f(0) = ?$ (C) Is f continuous at $x = 0$?
2. (A) $\lim_{x \to -1} f(x)$ (B) $f(-1) = ?$ (C) Is f continuous at $x = -1$?
3. (A) $\lim_{x \to 1} f(x)$ (B) $f(1) = ?$ (C) Is f continuous at $x = 1$?

4. (A) $\lim_{x \to 2} f(x)$ (B) $f(2) = ?$ (C) Is f continuous at $x = 2$?

5. (A) $\lim_{x \to -2} f(x)$ (B) $f(-2) = ?$ (C) Is f continuous at $x = -2$?

6. (A) $\lim_{x \to 0.5} f(x)$ (B) $f(0.5) = ?$ (C) Is f continuous at $x = 0.5$?

Use Theorem 4 to determine where each function is continuous. Express the answer in interval notation.

7. $f(x) = 2x - 3$

8. $g(x) = 3 - 5x$

9. $h(x) = \dfrac{2}{x - 5}$

10. $k(x) = \dfrac{x}{x + 3}$

11. $g(x) = \dfrac{x - 5}{(x - 3)(x + 2)}$

12. $F(x) = \dfrac{1}{x(x + 7)}$

Solve each inequality. Express the answer in inequality notation.

13. $x^2 - x - 12 < 0$

14. $x^2 - 2x - 8 < 0$

15. $x^2 + 21 > 10x$

16. $x^2 + 7x + 10 > 0$

17. $x^2 \leqslant 8x$

18. $x^2 + 6x \geqslant 0$

B *Problems 19–24 refer to the function f in the following graph. Use the definition of continuity to discuss the continuity of f at the indicated value of c.*

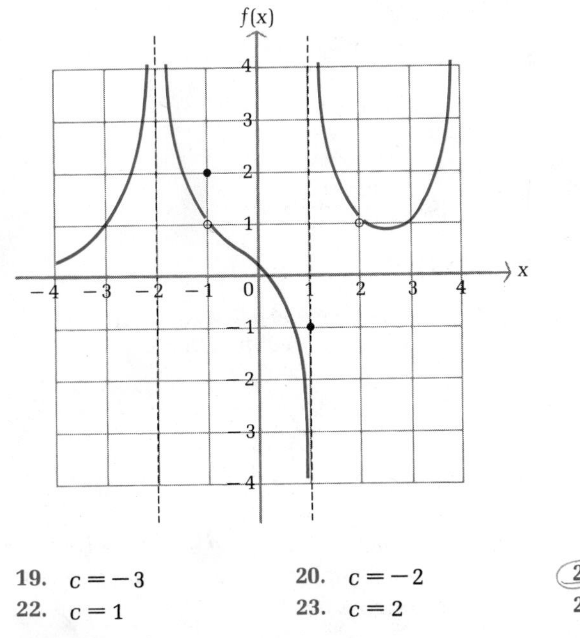

19. $c = -3$

20. $c = -2$

21. $c = -1$

22. $c = 1$

23. $c = 2$

24. $c = 3$

In Problems 25–30, graph f and locate all points of discontinuity.

25. $f(x) = \begin{cases} 1+x & x \leq 1 \\ 5-x & x > 1 \end{cases}$

26. $f(x) = \begin{cases} x^2 & x \leq 1 \\ 2x & x > 1 \end{cases}$

27. $f(x) = \begin{cases} 1+x & x \leq 2 \\ 5-x & x > 2 \end{cases}$

28. $f(x) = \begin{cases} x^2 & x \leq 2 \\ 2x & x > 2 \end{cases}$

29. $f(x) = \begin{cases} -x & x < 0 \\ 1 & x = 0 \\ x & x > 0 \end{cases}$

30. $f(x) = \begin{cases} 1 & x < 0 \\ 0 & x = 0 \\ 1+x & x > 0 \end{cases}$

Use Theorem 4 to determine where each function is continuous. Express the answer in interval notation.

31. $F(x) = 2x^8 - 3x^4 + 5$

32. $h(x) = \dfrac{x^4 - 3x + 5}{x^2 + 2x}$

33. $g(x) = \sqrt{x-5}$

34. $f(x) = \sqrt{3-x}$

35. $K(x) = \sqrt[3]{x-5}$

36. $H(x) = \sqrt[3]{3-x}$

37. $f(x) = \dfrac{x^2 - 1}{x^2 - 3x + 2}$

38. $k(x) = \dfrac{x^2 - 4}{x^2 + x - 2}$

Solve each inequality. Express the answer in inequality notation.

39. $(x-1)(x-3)(x-5) > 0$

40. $(x-2)(x-4)(x-6) < 0$

41. $(x-4)^3(x+5)^2 < 0$

42. $(x+2)^2(x-3)^3 > 0$

43. $\dfrac{x-2}{x+4} \leq 0$

44. $\dfrac{x+3}{x-1} \geq 0$

45. $\dfrac{x^2+5x}{x-3} \geq 0$

46. $\dfrac{x-4}{x^2+2x} \leq 0$

47. $\dfrac{(x+4)^3}{(1-x)^2} < 0$

48. $\dfrac{(3-x)^2}{(x+5)^3} > 0$

49. $\dfrac{x^3(x+3)}{(2x-3)^2} \geq 0$

50. $\dfrac{x(x-4)^2}{(2x+5)^3} \leq 0$

C *Use Theorem 4 to determine where each function is continuous. Express the answer in interval notation.*

51. $f(x) = \sqrt{4 - x^2}$

52. $g(x) = \sqrt{x^2 - 9}$

53. $h(x) = \sqrt{x^3 + x^2 - 6x}$

54. $k(x) = \sqrt{8x - 2x^2 - x^3}$

55. $F(x) = \sqrt{\dfrac{1+x}{1-x}}$

56. $G(x) = \sqrt{\dfrac{2-x}{x+3}}$

In Problems 57–60, the function f is not defined at the indicated value of c. Graph f and determine whether f(c) can be assigned a value that will make f continuous at c.

57. $f(x) = \dfrac{x}{|x|}, \quad c = 0$

58. $f(x) = \dfrac{x^2}{|x|}, \quad c = 0$

59. $f(x) = \dfrac{x^2 - 1}{x - 1}, \quad c = 1$

60. $f(x) = \dfrac{x(x - 1)}{|x - 1|}, \quad c = 1$

Applications

Business & Economics

61. *Postal rates.* First-class postage in 1986 was \$0.22 for the first ounce (or any fraction thereof) and \$0.17 for each additional ounce (or fraction thereof) up to 12 ounces. If $P(x)$ is the amount of postage for a letter weighing x ounces, then we can write

$$P(x) = \begin{cases} \$0.22 & \text{for } 0 < x \leq 1 \\ \$0.39 & \text{for } 1 < x \leq 2 \\ \$0.56 & \text{for } 2 < x \leq 3 \\ \text{and so on} & \end{cases}$$

(A) Graph P for $0 < x \leq 5$.
(B) Find $\lim_{x\to 4.5} P(x)$ and $P(4.5)$.
(C) Find $\lim_{x\to 4} P(x)$ and $P(4)$.
(D) Is P continuous at $x = 4.5$? At $x = 4$?

62. *Telephone rates.* A person placing a station-to-station call on Saturday from San Francisco to New York is charged \$0.30 for the first minute (or any fraction thereof) and \$0.20 for each additional minute (or fraction thereof). If the length of a call is x minutes, then the long-distance charge $R(x)$ is

$$R(x) = \begin{cases} \$0.30 & \text{for } 0 < x \leq 1 \\ \$0.50 & \text{for } 1 < x \leq 2 \\ \$0.70 & \text{for } 2 < x \leq 3 \\ \text{and so on} & \end{cases}$$

(A) Graph R for $0 < x \leq 6$.
(B) Find $\lim_{x\to 2.5} R(x)$ and $R(2.5)$.
(C) Find $\lim_{x\to 2} R(x)$ and $R(2)$.
(D) Is R continuous at $x = 2.5$? At $x = 2$?

63. *Income.* A personal computer salesperson receives a base salary of \$1,000 per month and a commission of 5% of all sales over \$10,000 during the month. If the monthly sales are \$20,000 or more, the salesperson is given an additional \$500 bonus. Let $E(s)$ represent the person's earnings during the month as a function of the monthly sales s.

(A) Graph $E(s)$ for $0 \leq s \leq 30{,}000$.

(B) Find $\lim_{s \to 10{,}000} E(s)$ and $E(10{,}000)$.
(C) Find $\lim_{s \to 20{,}000} E(s)$ and $E(20{,}000)$.
(D) Is E continuous at $s = 10{,}000$? At $s = 20{,}000$?

64. *Equipment rental.* An office equipment rental and leasing company rents electric typewriters for $10 per day (and any fraction thereof) or for $50 per 7 day week. Let $C(x)$ be the cost of renting a typewriter for x days.

(A) Graph $C(x)$ for $0 \leq x \leq 10$.
(B) Find $\lim_{x \to 4.5} C(x)$ and $C(4.5)$.
(C) Find $\lim_{x \to 8} C(x)$ and $C(8)$.
(D) Is C continuous at $x = 4.5$? At $x = 8$?

65. *Profit/loss analysis.* At a price of p dollars per unit the financial department in a company estimates that the weekly cost C and revenue R (in thousands of dollars) will be given by the equations

$C = 28 - 2p$ Cost equation

$R = 9p - p^2$ Revenue equation

Find the prices for which the company has a loss. A profit.

66. *Profit/loss analysis.* At a price of p dollars per unit the financial department in a company estimates that the weekly cost C and revenue R (in thousands of dollars) will be given by the equations

$C = 27 - 2p$ Cost equation

$R = 10p - p^2$ Revenue equation

Find the prices for which the company has a loss. A profit.

Life Sciences

67. *Animal supply.* A medical laboratory raises its own rabbits. The number of rabbits $N(t)$ available at any time t depends on the number of births and deaths. When a birth or death occurs, the function N generally has a discontinuity, as shown in the figure.

(A) Where is the function N discontinuous?
(B) $\lim_{t \to t_5} N(t) = ?$, $N(t_5) = ?$
(C) $\lim_{t \to t_3} N(t) = ?$, $N(t_3) = ?$

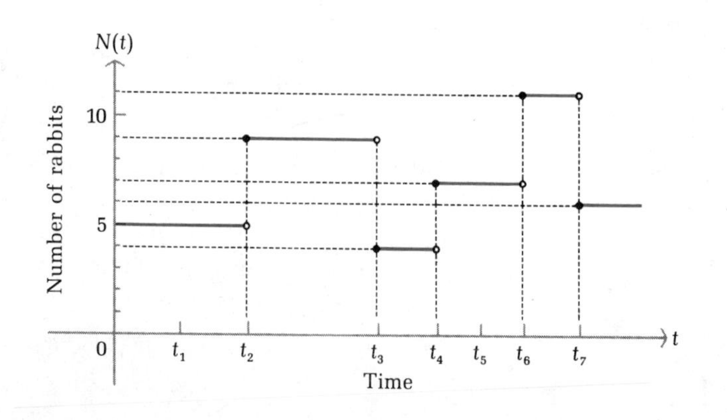

Social Sciences

68. *Learning.* The graph shown here might represent the history of a particular person learning the material on limits and continuity in this book. At time t_2, the student's mind goes blank during a quiz. At time t_4, the instructor explains a concept particularly well, and suddenly, a big jump in understanding takes place.

(A) Where is the function p discontinuous?
(B) $\lim_{t \to t_1} p(t) = ?$, $p(t_1) = ?$
(C) $\lim_{t \to t_2} p(t) = ?$, $p(t_2) = ?$
(D) $\lim_{t \to t_4} p(t) = ?$, $p(t_4) = ?$

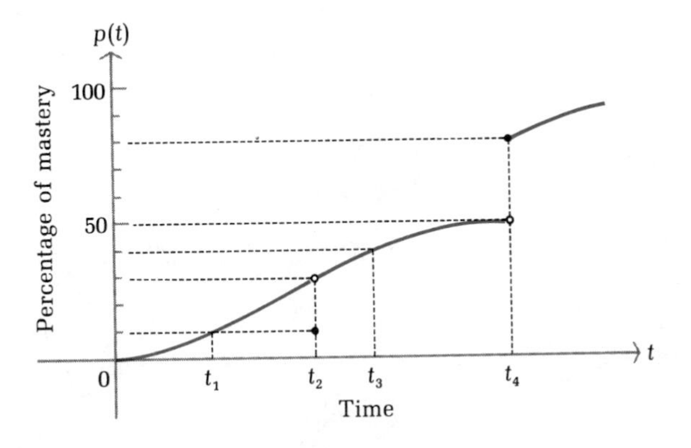

1-3 Increments, Tangent Lines, and Rates of Change

- Increments
- Slope and Tangent Line
- Average and Instantaneous Rates of Change

We will now use the concept of limit to solve two of the three basic problems of calculus stated at the beginning of this chapter. The parts of Figure 1 that we will concentrate on are repeated in Figure 11 (on the next page).

Increments

Before pursuing these problems, we digress for a moment to introduce *increment* notation. If we are given a function defined by $y = f(x)$ and the independent variable x changes from x_1 to x_2, then the dependent variable y will change from $y_1 = f(x_1)$ to $y_2 = f(x_2)$ (see Figure 12 on the next page). Mathematically, the change in x and the corresponding change in y, called **increments in x and y,** respectively, are denoted by Δx and Δy (read "delta x" and "delta y").

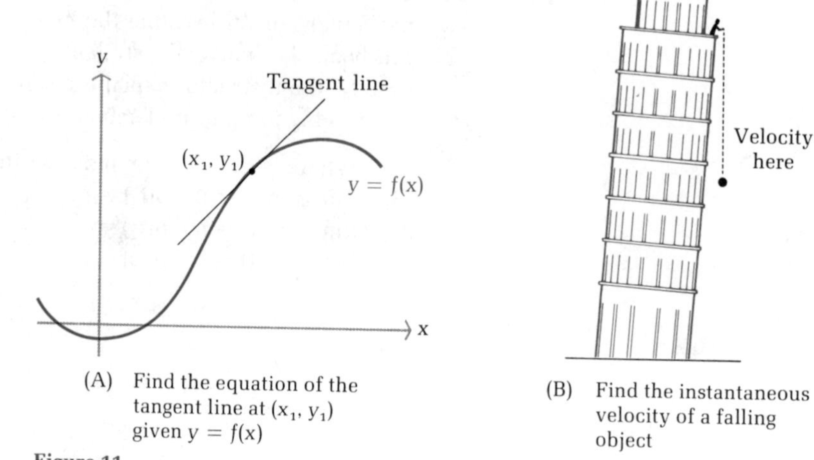

(A) Find the equation of the tangent line at (x_1, y_1) given $y = f(x)$

(B) Find the instantaneous velocity of a falling object

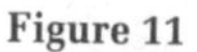

Figure 11

Increments

For $y = f(x)$ (see Figure 12)

$$\begin{aligned}\Delta x &= x_2 - x_1\\ x_2 &= x_1 + \Delta x\\ \Delta y &= y_2 - y_1\\ &= f(x_2) - f(x_1)\\ &= f(x_1 + \Delta x) - f(x_1)\end{aligned}$$

Δy represents the change in y corresponding to a Δx change in x.

[*Note:* Δy depends on the function f, the input x, and the increment Δx.]

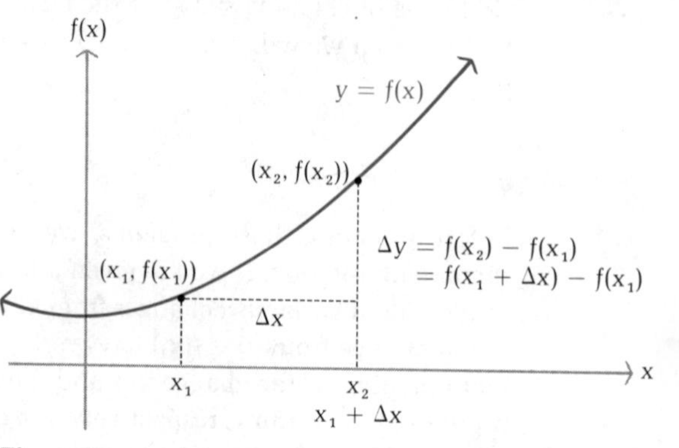

Figure 12

Example 13 Given the function

$$f(x) = \frac{x^2}{2}$$

(A) Find Δx, Δy, and $\Delta y/\Delta x$ for $x_1 = 1$ and $x_2 = 2$.
(B) Find

$$\frac{f(x_1 + \Delta x) - f(x_1)}{\Delta x}$$

for $x_1 = 1$ and $\Delta x = 2$.

Solutions (A) $\Delta x = x_2 - x_1 = 2 - 1 = 1$

$$\Delta y = f(x_2) - f(x_1)$$

$$= f(2) - f(1) = \frac{4}{2} - \frac{1}{2} = \frac{3}{2}$$

$$\frac{\Delta y}{\Delta x} = \frac{f(x_2) - f(x_1)}{x_2 - x_1} = \frac{\frac{3}{2}}{1} = \frac{3}{2}$$

(B)
$$\frac{f(x_1 + \Delta x) - f(x_1)}{\Delta x} = \frac{f(1 + 2) - f(1)}{2}$$

$$= \frac{f(3) - f(1)}{2} = \frac{\frac{9}{2} - \frac{1}{2}}{2} = \frac{4}{2} = 2$$

Problem 13 Given the function $f(x) = x^2 + 1$:

(A) Find Δx, Δy, and $\Delta y/\Delta x$ for $x_1 = 2$ and $x_2 = 3$.
(B) Find

$$\frac{f(x_1 + \Delta x) - f(x_1)}{\Delta x}$$

for $x_1 = 1$ and $\Delta x = 2$.

▪ Slope and Tangent Line

From plane geometry, we know that a tangent to a circle is a line that passes through one and only one point on the circle, but how do we define and find a tangent line to a graph of a function at a point? The concept of the slope of a straight line (see Section 0-4) will play a central role in the process. If we pass a straight line through two points on the graph of $y = f(x)$, as in Figure 13 on the next page, we obtain a secant line. Given the coordinates of the two points, we can find the slope of the secant line using the point–slope formula from Section 0-4. (This is exactly what we did in Figure 4 in Section 1-1 to motivate the concept of limit.)

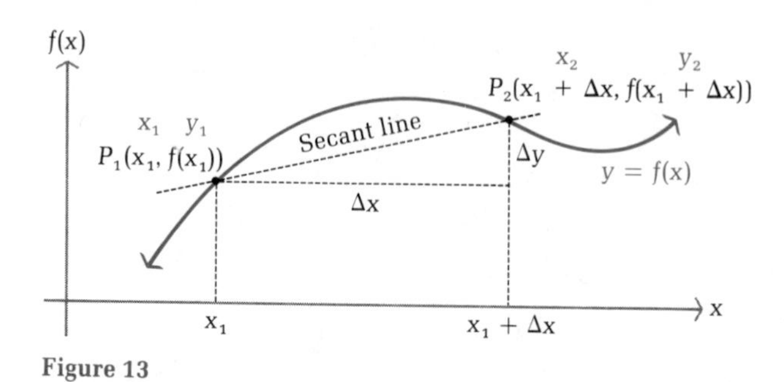

Figure 13

$$\textbf{Secant line slope} = \frac{y_2 - y_1}{x_2 - x_1} = \frac{f(x_1 + \Delta x) - f(x_1)}{\Delta x} = \frac{\Delta y}{\Delta x}$$

As we let Δx tend to 0, P_2 will approach P_1, and it appears that the secant lines will approach a limiting position and the secant slopes will approach a limiting value (see Figure 14). If they do, then we will call the line that the secant lines approach the *tangent line to the graph* at $(x_1, f(x_1))$, and the limiting slope will be the slope of the tangent line. This leads to the following definition of a tangent line:

Tangent Line

Given the graph of $y = f(x)$, then the **tangent line** at $(x_1, f(x_1))$ is the line that passes through this point with slope

$$\textbf{Tangent line slope} = \lim_{\Delta x \to 0} \frac{f(x_1 + \Delta x) - f(x_1)}{\Delta x} \tag{1}$$

if the limit exists. The slope of the tangent line is also referred to as the **slope of the graph** at $(x_1, f(x_1))$. [Actually, in much of the work that follows, our main interest will be in the *slope* of the graph of $y = f(x)$ at $(x_1, f(x_1))$ rather than in the tangent line itself.]

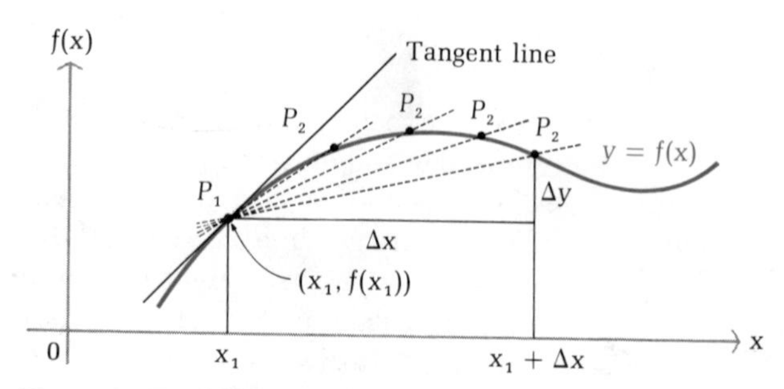

Figure 14 Dotted lines are secant lines for smaller and smaller Δx.

Example 14 Given $f(x) = x^2$, find the slope and equation of the tangent line at $x = 1$. Sketch the graph of f, the tangent line at $(1, f(1))$, and the secant line passing through $(1, f(1))$ and $(2, f(2))$.

Solution First, we find the slope of the tangent line using equation (1):

$$\frac{f(1+\Delta x) - f(1)}{\Delta x} = \frac{(1+\Delta x)^2 - 1^2}{\Delta x}$$

We are computing the slope of a secant line passing through $(1, f(1))$ and $(1 + \Delta x, f(1 + \Delta x))$—see Figure 13.

$$= \frac{1 + 2\Delta x + (\Delta x)^2 - 1}{\Delta x}$$

$$= \frac{2\Delta x + (\Delta x)^2}{\Delta x}$$

$$= \frac{\Delta x(2 + \Delta x)}{\Delta x} = 2 + \Delta x \qquad \Delta x \neq 0$$

$$\textbf{Tangent line slope} = \lim_{\Delta x \to 0} \frac{f(1+\Delta x) - f(1)}{\Delta x}$$

$$= \lim_{\Delta x \to 0} (2 + \Delta x) = 2$$

This is also the slope of the graph of $f(x) = x^2$ at $(1, f(1))$.

Next, we find the point on the graph of f corresponding to $x = 1$:

$f(1) = 1^2 = 1$

$(1, f(1)) = (1, 1)$ Point on the graph of f

Now we use the point–slope formula for the equation of a line to find the **equation of the tangent line:**

$(x_1, y_1) = (1, 1)$ Point

$m = 2$ Slope

$y - y_1 = m(x - x_1)$ Point–slope formula

$y - 1 = 2(x - 1)$

$y - 1 = 2x - 2$ or $y = 2x - 1$ Tangent line equation

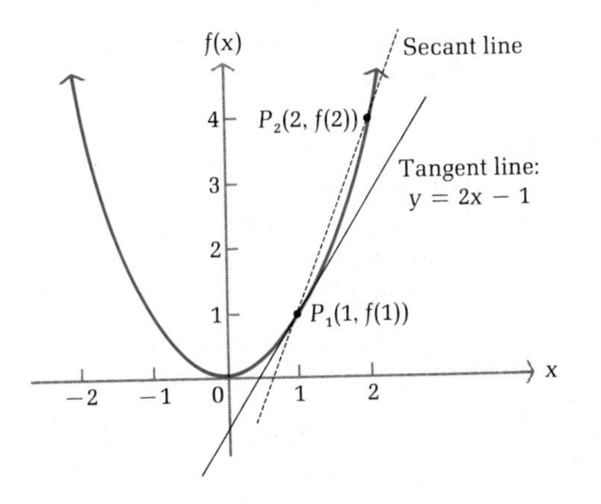

Problem 14 Find the equation of the tangent line for the graph of $f(x) = x^2$ at $x = 2$. Write the answer in the form $y = mx + b$.

▪ Average and Instantaneous Rates of Change

We now show how increments and limits can be used to analyze rate problems. In the process, we will solve the second basic calculus problem we stated at the beginning of the chapter.

Example 15 Velocity

A small steel ball dropped from a tower will fall a distance of y feet in x seconds, as given approximately by the formula (from physics) $y = f(x) = 16x^2$. Let us determine the ball's position on a coordinate line at various times (see Figure 15). Our ultimate objective is to find the ball's *velocity* at a given instant, say, at the end of 2 seconds.

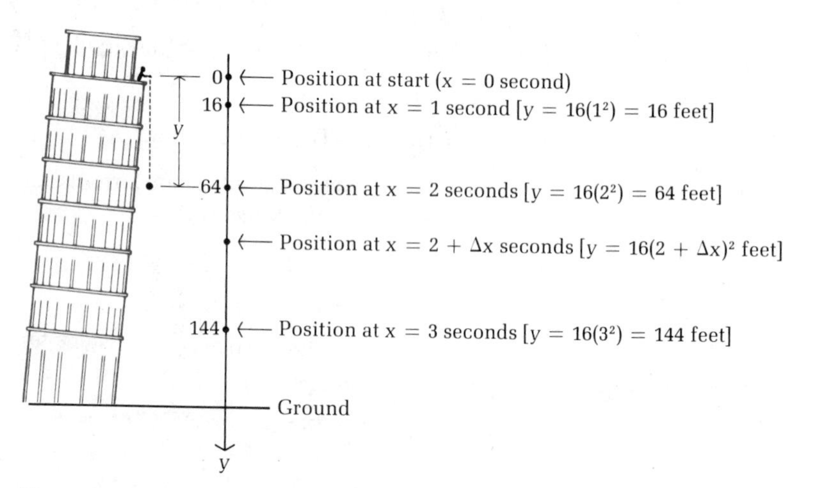

Figure 15 *Note:* Positive y direction is down.

(A) Find x_2 and Δy for $x_1 = 2$ and $\Delta x = 1$.
(B) Find the average velocity for the time change in part A.
(C) Find an expression for the average velocity from $x = 2$ to $x = 2 + \Delta x$, where Δx represents a small but arbitrary change in time and $\Delta x \neq 0$ (see Figure 15).
(D) Find $\lim_{\Delta x \to 0} (\Delta y/\Delta x)$ using $\Delta y/\Delta x$ from part C.

Solutions (A) $x_2 = x_1 + \Delta x = 2 + 1 = 3$

$$\Delta y = f(x_1 + \Delta x) - f(x_1)$$
$$= f(3) - f(2)$$
$$= 16(3^2) - 16(2^2)$$
$$= 144 - 64 = 80 \text{ feet}$$

Distance fallen from end of 2 seconds to end of 3 seconds (see Figure 15)

(B) Recall the formula $d = rt$, which can be written in the form

$$r = \frac{d}{t} = \frac{\text{Total distance}}{\text{Elapsed time}} = \text{Average rate}$$

For example, if a person drives from San Francisco to Los Angeles—a distance of about 420 miles—in 10 hours, then the average rate is

$$r = \frac{d}{t} = \frac{420}{10} = 42 \text{ miles per hour}$$

Sometimes the person will be traveling faster and sometimes slower, but the *average rate* is 42 miles per hour. In our present problem, it is clear from Figure 15 that the ball is *accelerating* (falling faster and faster), but we can compute an average rate, or average velocity, just as we did for the trip from San Francisco to Los Angeles:

$$\textbf{Average velocity} = \frac{\text{Total distance}}{\text{Elapsed time}}$$
$$= \frac{\Delta y}{\Delta x} = \frac{f(3) - f(2)}{1} = \frac{80}{1} = 80 \text{ feet per second}$$

Thus, the average velocity from the end of 2 seconds to the end of 3 seconds is 80 feet per second.

(C) $$\text{Average velocity} = \frac{\Delta y}{\Delta x} = \frac{f(2 + \Delta x) - f(2)}{\Delta x} \qquad \Delta x \neq 0$$
$$= \frac{16(2 + \Delta x)^2 - 16(2^2)}{\Delta x}$$
$$= \frac{16[4 + 4\Delta x + (\Delta x)^2] - 64}{\Delta x}$$
$$= \frac{64 + 64\Delta x + 16(\Delta x)^2 - 64}{\Delta x}$$
$$= \frac{64\Delta x + 16(\Delta x)^2}{\Delta x} = \frac{\Delta x(64 + 16\Delta x)}{\Delta x}$$
$$= 64 + 16\Delta x \qquad \Delta x \neq 0$$

Note that if $\Delta x = 1$, the average velocity is 80 feet per second; if $\Delta x = 0.5$, then the average velocity is 72 feet per second; if $\Delta x = 0.01$, then the average velocity is 64.16 feet per second; and so on. The smaller Δx gets, the closer the average velocity gets to 64 feet per second.

(D) $$\lim_{x \to 0} \frac{\Delta y}{\Delta x} = \lim_{\Delta x \to 0} \frac{f(2 + \Delta x) - f(2)}{\Delta x}$$
$$= \lim_{\Delta x \to 0} (64 + 16\Delta x)$$
$$= 64 \text{ feet per second}$$

We call 64 feet per second the **instantaneous velocity** at $x = 2$ seconds, and we have solved the second basic problem stated at the beginning of this chapter!

The discussion in Example 15 leads to the following general definitions of average rate and instantaneous rate:

Average and Instantaneous Rates

For $y = f(x)$

$$\textbf{Average rate} = \frac{\Delta y}{\Delta x} = \frac{f(x_2) - f(x_1)}{x_2 - x_1} = \frac{f(x_1 + \Delta x) - f(x_1)}{\Delta x}$$

$$\textbf{Instantaneous rate} = \lim_{\Delta x \to 0} \frac{\Delta y}{\Delta x} = \lim_{\Delta x \to 0} \frac{f(x_1 + \Delta x) - f(x_1)}{\Delta x}$$

if the limit exists

The ratio $[f(x_1 + \Delta x) - f(x_1)]/\Delta x$ is also called the **difference quotient.**

Problem 15 For the falling steel ball in Example 15, find:

(A) The average velocity from $x = 1$ to $x = 2$.
(B) The average velocity from $x = 1$ to $x = 1 + \Delta x$.
(C) The instantaneous velocity at $x = 1$.

Now we consider a slightly different type of rate problem, but we will use the same approach as in Example 15.

Example 16 Advertising

An advertising agency for a chain of pizzerias has determined that the relationship between the number of pizzas N sold each day and the num-

ber of television ads x shown each day is

$$N(x) = 100 + 100x - x^2$$

Thus, showing 20 ads would result in sales of $N(20) = 1{,}700$ pizzas and showing 30 ads would result in sales of $N(30) = 2{,}200$ pizzas.

(A) What is the average rate of change in the number of pizzas sold from 20 daily ads to 30 daily ads?
(B) What is the average rate of change in the number of pizzas sold from 20 daily ads to $20 + \Delta x$ daily ads?
(C) What value does $\Delta N/\Delta x$ in part B approach as Δx tends to 0?

Solutions

(A) $$\frac{\Delta N}{\Delta x} = \frac{N(30) - N(20)}{30 - 20} = \frac{2{,}200 - 1{,}700}{10}$$
$$= 50 \text{ pizzas per ad}$$

(B) $$\frac{\Delta N}{\Delta x} = \frac{N(20 + \Delta x) - N(20)}{\Delta x}$$
$$= \frac{[100 + 100(20 + \Delta x) - (20 + \Delta x)^2] - 1{,}700}{\Delta x}$$
$$= \frac{100 + 2{,}000 + 100\Delta x - 400 - 40\Delta x - (\Delta x)^2 - 1{,}700}{\Delta x}$$
$$= \frac{60\Delta x - (\Delta x)^2}{\Delta x} = \frac{\Delta x(60 - \Delta x)}{\Delta x} = 60 - \Delta x \qquad \Delta x \neq 0$$

(C) $$\lim_{\Delta x \to 0} \frac{\Delta N}{\Delta x} = \lim_{\Delta x \to 0} (60 - \Delta x) = 60 \text{ pizzas per ad}$$

Thus, the instantaneous rate of change of N with respect to x at $x = 20$ is 60 pizzas per ad.*

Problem 16 For Example 16, find:

(A) The average rate of change in number of pizzas from 30 ads to 40 ads.
(B) The average rate of change in number of pizzas from 30 ads to $30 + \Delta x$ ads.
(C) The limiting value of $\Delta N/\Delta x$ in part B as Δx tends to 0.

* Technically, the function N(x) is defined only for *n* a nonnegative integer (it is not possible to show 3.7 or $\sqrt{23}$ ads per day). However, to find the instantaneous rate of change of N with respect to x at $x = 20$, we must assume that N is defined for an interval of real numbers containing the number 20. When applying calculus techniques to real-world problems, it is common practice to automatically extend the domain of a function from a discrete set of integers to a set of real numbers. Thus, we assume that the domain of N(x) is the set of all real numbers $x \geq 0$ and not just $x = 0$, 1, 2, 3, ...

Answers to Matched Problems

13. (A) $\Delta x = 1$, $\Delta y = 5$, $\Delta y/\Delta x = 5$ (B) 4
14. $y = 4x - 4$
15. (A) 48 feet per second (B) $32 + 16\Delta x$
    (C) 32 feet per second
16. (A) 30 pizzas per ad (B) $40 - \Delta x$
    (C) 40 pizzas per ad

Exercise 1-3

In Problems 1–14, find the indicated quantities for $y = f(x) = 3x^2$.

A

1. Δx, Δy, and $\Delta y/\Delta x$, given $x_1 = 1$ and $x_2 = 4$
2. Δx, Δy, and $\Delta y/\Delta x$, given $x_1 = 2$ and $x_2 = 5$
3. $\dfrac{f(x_1 + \Delta x) - f(x_1)}{\Delta x}$, given $x_1 = 1$ and $\Delta x = 2$
4. $\dfrac{f(x_1 + \Delta x) - f(x_1)}{\Delta x}$, given $x_1 = 2$ and $\Delta x = 1$
5. $\dfrac{y_2 - y_1}{x_2 - x_1}$, given $x_1 = 1$ and $x_2 = 3$
6. $\dfrac{y_2 - y_1}{x_2 - x_1}$, given $x_1 = 2$ and $x_2 = 3$
7. $\dfrac{\Delta y}{\Delta x}$, given $x_1 = 1$ and $x_2 = 3$
8. $\dfrac{\Delta y}{\Delta x}$, given $x_1 = 2$ and $x_2 = 3$

B

9. The average rate of change of y, for x changing from 1 to 4
10. The average rate of change of y, for x changing from 2 to 5
11. (A) $\dfrac{f(2 + \Delta x) - f(2)}{\Delta x}$ (simplify)
    (B) What does the ratio in part A approach as Δx approaches 0?
12. (A) $\dfrac{f(3 + \Delta x) - f(3)}{\Delta x}$ (simplify)
    (B) What does the ratio in part A approach as Δx approaches 0?
13. (A) $\dfrac{f(4 + \Delta x) - f(4)}{\Delta x}$ (simplify)
    (B) What does the ratio in part A approach as Δx approaches 0?
14. (A) $\dfrac{f(5 + \Delta x) - f(5)}{\Delta x}$ (simplify)
    (B) What does the ratio in part A approach as Δx approaches 0?

Suppose an object moves along the y axis so that its location is $y = f(x) = x^2 + x$ at time x (y is in meters and x is in seconds). Find:

15. (A) The average velocity (the average rate of change of y) for x changing from 1 to 3 seconds
(B) The average velocity for x changing from 1 to $(1 + \Delta x)$ seconds
(C) The instantaneous velocity at $x = 1$

16. (A) The average velocity (the average rate of change of y) for x changing from 2 to 4 seconds
(B) The average velocity for x changing from 2 to $(2 + \Delta x)$ seconds
(C) The instantaneous velocity at $x = 2$

In Problems 17 and 18, find each of the following for the graph of $y = f(x) = x^2 + x$:

17. (A) The slope of the secant line joining $(1, f(1))$ and $(3, f(3))$
(B) The slope of the secant line joining $(1, f(1))$ and $(1 + \Delta x, f(1 + \Delta x))$
(C) The slope of the tangent line at $(1, f(1))$
(D) The equation of the tangent line at $(1, f(1))$

18. (A) The slope of the secant line joining $(2, f(2))$ and $(4, f(4))$
(B) The slope of the secant line joining $(2, f(2))$ and $(2 + \Delta x, f(2 + \Delta x))$
(C) The slope of the tangent line at $(2, f(2))$
(D) The equation of the tangent line at $(2, f(2))$

C

19. If an object moves on the x axis so that it is at $x = f(t) = t^2 - t$ at time t (t measured in seconds and x measured in meters), find the instantaneous velocity of the object at $t = 2$.

20. Find the equation of the tangent line for the graph of $y = x^2 - x$ at $x = 2$.

Applications

Business & Economics

21. *Income.* The per capita income in the United States from 1969 to 1973 is given approximately in the table. Find the average rate of change of per capita income for a time change from:

(A) 1969 to 1971 (B) 1971 to 1973

Year	1969	1970	1971	1972	1973
Income	\$3,700	\$3,900	\$4,100	\$4,500	\$5,000

22. *Demand function.* Suppose in a given grocery store people are willing to buy $D(x)$ pounds of chocolate candy per day at $\$x$ per pound, as given by the demand function

$$D(x) = 100 - x^2 \qquad \$1 \leq x \leq \$10$$

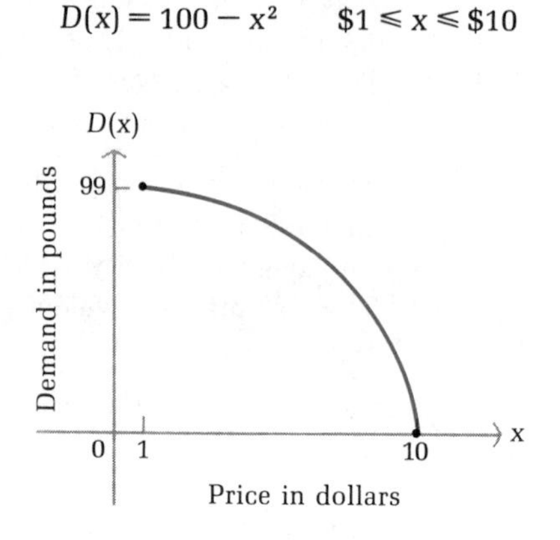

Note that as price goes up, demand goes down (see the figure).

(A) Find the average rate of change in demand for a price change from \$2 to \$5; that is, find $\Delta y/\Delta x$ for $x_1 = 2$ and $x_2 = 5$.

(B) Simplify:

$$\frac{D(2 + \Delta x) - D(2)}{\Delta x}$$

(C) What does the ratio in part B approach as Δx approaches 0? [This is called the instantaneous rate of change of $D(x)$ with respect to x at $x = 2$.]

23. *Advertising.* A discount appliance store uses television ads to promote weekend specials on clothes dryers. Suppose that the store can sell $N(x)$ dryers by showing x ads where

$$N(x) = 5 + 40x - x^2$$

(A) Find the average rate of change in the number of dryers sold from 5 ads to 10 ads.

(B) Find the average rate of change in the number of dryers sold from 5 ads to $5 + \Delta x$ ads.

(C) Find the limiting value of $\Delta N/\Delta x$ in part B as Δx tends to 0. (This is the instantaneous rate of change of N with respect to x at $x = 5$.)

24. *Depreciation.* Office equipment was purchased for \$20,000 and is assumed to have a scrap value of \$2,000 after 10 years. If its value is depreciated linearly (for tax purposes) from \$20,000 to \$2,000, then the value $V(t)$ after t years is given by

$$V(t) = 20{,}000 - 1{,}800t \qquad 0 \leq t \leq 10$$

(A) Find the average rate of change in the value of the equipment from 2 years to 6 years.
(B) Find the average rate of change in the value of the equipment from 3 years to 5 years.
(C) Find the average rate of change in the value of the equipment from t_1 years to t_2 years.

Life Sciences

25. *Medicine.* The area of a small (healing) wound in square millimeters, where time is measured in days, is given in the table.

Area	400	360	180	120	90	72	60
Days	0	1	2	3	4	5	6

Find the average rate of change of area for the time change from:

(A) 0 to 2 days (B) 4 to 6 days

26. *Weight–height.* A formula relating the approximate weight of an average person and his or her height is

$$W(h) = 0.0005h^3$$

where $W(h)$ is in pounds and h is in inches.

(A) Find the average rate of change of weight for a height change from 60 to 70 inches.
(B) Simplify:

$$\frac{W(60 + \Delta h) - W(60)}{\Delta h}$$

(C) What does the ratio in part B approach as Δh approaches 0? [This is called the instantaneous rate of change of $W(h)$ with respect to h at $h = 60$.]

Social Sciences

27. *Illegitimate births.* The approximate numbers of illegitimate births per 1,000 live births in the United States from 1940 to 1970 are given in the table. Find the average rate of change of illegitimate births per 1,000 live births for the time change from:

(A) 1940 to 1945 (B) 1965 to 1970

Year	1940	1945	1950	1955	1960	1965	1970
Illegitimate Births per 1,000 Live Births	38	41	40	47	54	80	120

28. *Learning.* A certain person learning to type has an achievement record given approximately by the function

$$N(t) = 60\left(1 - \frac{2}{t}\right) \qquad 3 \leq t \leq 10$$

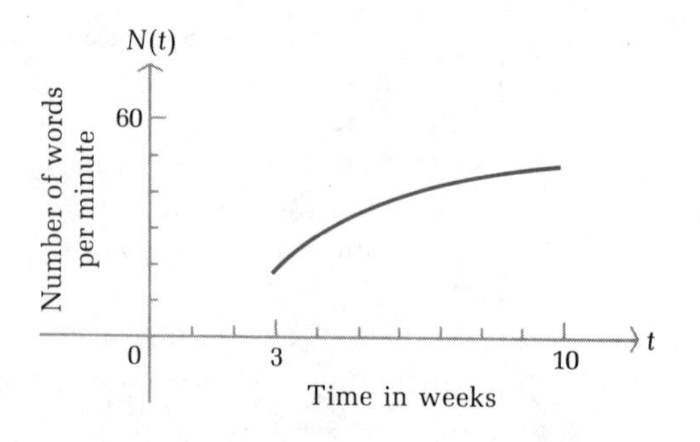

where $N(t)$ is in number of words per minute and t is in weeks. Find the average rate of change of the number of words per minute for the change in time from:

(A) 4 to 6 weeks (B) 8 to 10 weeks

1-4 The Derivative

- The Derivative
- Tangent Lines
- Nonexistence of the Derivative
- Instantaneous Rates of Change
- Marginal Cost
- Summary

The Derivative

In the last section we found that the special limit

$$\lim_{\Delta x \to 0} \frac{f(x_1 + \Delta x) - f(x_1)}{\Delta x} \tag{1}$$

if it exists, gives us the slope of the tangent line to the graph of $y = f(x)$ at $(x_1, f(x_1))$. It also gives us the instantaneous rate of change of y per unit change in x at $x = x_1$. Formula (1) is of such basic importance to calculus and to the applications of calculus that we will give it a name and study it in detail. To keep formula (1) simple and general, we will drop the subscript on x_1 and think of the ratio

$$\frac{f(x + \Delta x) - f(x)}{\Delta x}$$

as a function of Δx, with x held fixed as we let Δx tend to 0. We are now ready to define one of the basic concepts in calculus, the *derivative*:

Derivative

For $y = f(x)$ we define the **derivative of f at x,** denoted by $f'(x)$, to be

$$f'(x) = \lim_{\Delta x \to 0} \frac{f(x + \Delta x) - f(x)}{\Delta x} \qquad \text{if the limit exists}$$

If $f'(x)$ exists, then f is said to be a **differentiable function** at x.

The process of finding the derivative of a function is called **differentiation.** That is, the derivative of a function is obtained by **differentiating** the function. Differentiating a function f creates a new function f' that gives, among other things, the instantaneous rate of change of $y = f(x)$ and the slope of the tangent line to the graph of $y = f(x)$ for each x. The domain of f' is a subset of the domain of f, which will become clearer as we progress through this section.

Example 17 Find $f'(x)$, the derivative of f at x, for $f(x) = 4x - x^2$.

Solution To find $f'(x)$, we find

$$\lim_{\Delta x \to 0} \frac{f(x + \Delta x) - f(x)}{\Delta x}$$

To make the computation easier, we introduce a two-step process:

Step 1. Find $[f(x + \Delta x) - f(x)]/\Delta x$ and simplify.

$$\begin{aligned}
\frac{f(x + \Delta x) - f(x)}{\Delta x} &= \frac{[4(x + \Delta x) - (x + \Delta x)^2] - (4x - x^2)}{\Delta x} \\
&= \frac{4x + 4\Delta x - x^2 - 2x\Delta x - (\Delta x)^2 - 4x + x^2}{\Delta x} \\
&= \frac{4\Delta x - 2x\Delta x - (\Delta x)^2}{\Delta x} \\
&= \frac{\Delta x}{\Delta x}(4 - 2x - \Delta x) \\
&= 4 - 2x - \Delta x \qquad \Delta x \neq 0
\end{aligned}$$

Step 2. Find the limit of the result of step 1.

$$\begin{aligned}
f'(x) = \lim_{\Delta x \to 0} \frac{f(x + \Delta x) - f(x)}{\Delta x} &= \lim_{\Delta x \to 0} (4 - 2x - \Delta x) \\
&= 4 - 2x
\end{aligned}$$

Thus, if $f(x) = 4x - x^2$, then $f'(x) = 4 - 2x$.

Problem 17 Find $f'(x)$, the derivative of f at x, for $f(x) = 8x - 2x^2$.

Now that we are performing more complicated algebraic operations involving the symbol Δx, it is important to remember that Δx is a single symbol representing the change in x, not the product of Δ and x. In particular,

$$(\Delta x)x \neq \Delta x^2 \qquad \text{and} \qquad (\Delta x)^2 \neq \Delta^2 x^2$$

Example 18 Find $f'(x)$, the derivative of f at x, for $f(x) = \sqrt{x} + 2$.

Solution To find $f'(x)$, we find

$$\lim_{\Delta x \to 0} \frac{f(x + \Delta x) - f(x)}{\Delta x}$$

We use the two-step process presented in Example 17.

Step 1. Find $[f(x + \Delta x) - f(x)]/\Delta x$ and simplify.

$$\begin{aligned} \frac{f(x + \Delta x) - f(x)}{\Delta x} &= \frac{(\sqrt{x + \Delta x} + 2) - (\sqrt{x} + 2)}{\Delta x} \\ &= \frac{\sqrt{x + \Delta x} + 2 - \sqrt{x} - 2}{\Delta x} \\ &= \frac{\sqrt{x + \Delta x} - \sqrt{x}}{\Delta x} \end{aligned}$$

Trying to apply the quotient property of limits, we find that $\lim_{\Delta x \to 0} \Delta x = 0$; hence, we cannot use it. We try rationalizing the numerator:

$$\begin{aligned} \frac{\sqrt{x + \Delta x} - \sqrt{x}}{\Delta x} \cdot \frac{\sqrt{x + \Delta x} + \sqrt{x}}{\sqrt{x + \Delta x} + \sqrt{x}} &= \frac{x + \Delta x - x}{\Delta x(\sqrt{x + \Delta x} + \sqrt{x})} \\ &= \frac{\Delta x}{\Delta x(\sqrt{x + \Delta x} + \sqrt{x})} \\ &= \frac{1}{\sqrt{x + \Delta x} + \sqrt{x}} \qquad \Delta x \neq 0 \end{aligned}$$

Step 2. Find the limit of the result of step 1.

$$\begin{aligned} f'(x) &= \lim_{\Delta x \to 0} \frac{\sqrt{x + \Delta x} - \sqrt{x}}{\Delta x} \\ &= \lim_{\Delta x \to 0} \frac{1}{\sqrt{x + \Delta x} + \sqrt{x}} \\ &= \frac{1}{\sqrt{x} + \sqrt{x}} = \frac{1}{2\sqrt{x}} \end{aligned}$$

Thus, if $f(x) = \sqrt{x} + 2$, then $f'(x) = 1/(2\sqrt{x})$. [*Note:* The domain of $f(x) = \sqrt{x} + 2$ is $[0, \infty)$. Since $f'(0)$ is undefined, the domain of $f'(x) = 1/(2\sqrt{x})$ is $(0, \infty)$, a subset of the original domain.]

Problem 18 Find $f'(x)$, the derivative of f at x, for $f(x) = x^{-1}$.

■ Tangent Lines

In the last section we defined the slope of the tangent line to the graph of $y = f(x)$ at $(x_1, f(x_1))$ to be

$$\lim_{\Delta x \to 0} \frac{f(x_1 + \Delta x) - f(x_1)}{\Delta x}$$

if the limit exists. This, of course, is $f'(x_1)$, the derivative of f at $x = x_1$. To find the equation of a tangent line to the graph of $y = f(x)$ at $(x_1, f(x_1))$, we use the point–slope form for the equation of a line, $y - y_1 = m(x - x_1)$, and the facts that $m = f'(x_1)$ and $y_1 = f(x_1)$ to obtain:

Tangent Line

The equation of the tangent line to the graph of $y = f(x)$ at $x = x_1$ is

$$y - y_1 = m(x - x_1)$$

where $y_1 = f(x_1)$ and $m = f'(x_1)$.
[*Note:* (x_1, y_1) is on the graph of f.]

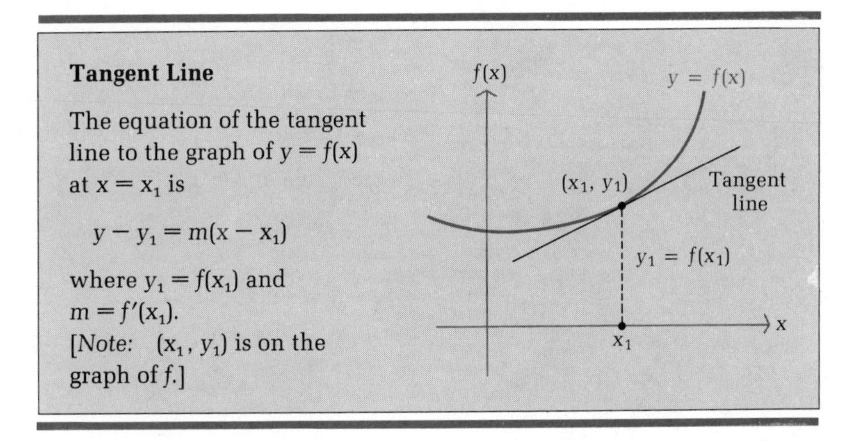

More generally,

$$f'(x) = \lim_{\Delta x \to 0} \frac{f(x + \Delta x) - f(x)}{\Delta x}$$

gives us the slope of the graph of f at *any* point $(x, f(x))$ on the graph of f for which the limit exists.

Example 19 In Example 17 we started with the function specified by $f(x) = 4x - x^2$ and found the derivative of f at x to be $f'(x) = 4 - 2x$. Thus, the slope of the graph of f at any point $(x, f(x))$ on the graph of f is

$$f'(x) = 4 - 2x$$

We will use this derivative in the following problems.

(A) Find the slopes of the graph of f at $x = 0$, 2, and 3.
(B) Find the equations of the tangent lines at $x = 0$, 2, and 3.
(C) Sketch the tangent lines to the graph of $y = 4x - x^2$ at $x = 0$, 2, and 3.

Solutions (A) Using $f'(x) = 4 - 2x$, we have:

$$f'(0) = 4 - 2(0) = 4$$
$$f'(2) = 4 - 2(2) = 0$$
$$f'(3) = 4 - 2(3) = -2$$

(B) Tangent line at $x = 0$:

$y - y_1 = m(x - x_1)$ $\qquad y_1 = f(x_1) = f(0) = 4(0) - (0)^2 = 0$
$\qquad m = f'(x_1) = f'(0) = 4$ (see part A)

$y - 0 = 4(x - 0)$

$y = 4x$ $\qquad$ Tangent line at $x = 0$

Tangent line at $x = 2$:

$y - y_1 = m(x - x_1)$ $\qquad y_1 = f(x_1) = f(2) = 4(2) - (2)^2 = 4$
$\qquad m = f'(x_1) = f'(2) = 0$ (see part A)

$y - 4 = 0(x - 2)$

$y = 4$ $\qquad$ Tangent line at $x = 2$

Tangent line at $x = 3$:

$y - y_1 = m(x - x_1)$ $\qquad y_1 = f(x_1) = f(3) = 4(3) - (3)^2 = 3$
$\qquad m = f'(x_1) = f'(3) = -2$ (see part A)

$y - 3 = -2(x - 3)$

$y = -2x + 9$ $\qquad$ Tangent line at $x = 3$

(C)

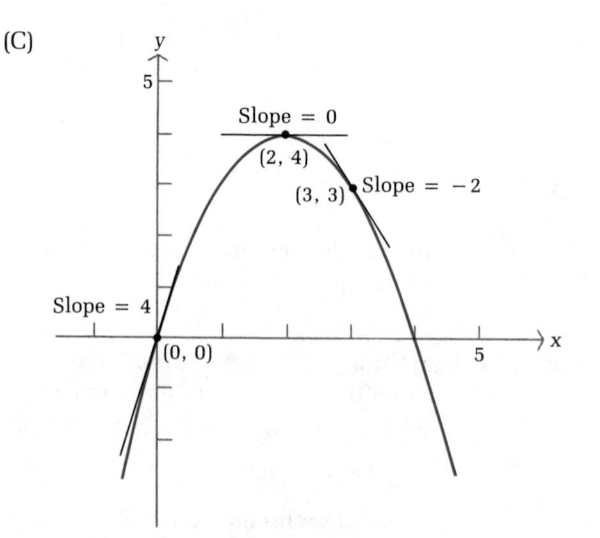

Problem 19 In Problem 17 we started with the function specified by $f(x) = 8x - 2x^2$ and found the derivative of f at x to be $f'(x) = 8 - 4x$.

(A) Find the slopes of the graph of f at $x = 1$, 2, and 4.
(B) Find the equations of the tangent lines at $x = 1$, 2, and 4.
(C) Sketch the tangent lines to the graph of $y = 8x - 2x^2$ at $x = 1$, 2, and 4.

■ Nonexistence of the Derivative

The existence of a derivative at $x = a$ depends on the existence of a limit at $x = a$; that is, on the existence of

$$f'(a) = \lim_{\Delta x \to 0} \frac{f(a + \Delta x) - f(a)}{\Delta x} \tag{2}$$

If the limit does not exist at $x = a$, we say that the function f is **nondifferentiable at $x = a$** or **$f'(a)$ does not exist.**

How can we recognize the points on the graph of f where $f'(a)$ does not exist? It is impossible to describe all the ways that the limit in (2) can fail to exist. However, we can illustrate some common situations where $f'(a)$ does fail to exist:

1. If f is not continuous at $x = a$, then $f'(a)$ does not exist (see Fig. 16A). It can be shown that **if f is differentiable at $x = a$, then f must be continuous at $x = a$.**
2. If the graph of f has a sharp corner at $x = a$, then $f'(a)$ does not exist and the graph has no tangent line at $x = a$ (see Fig. 16B).
3. If the graph of f has a vertical tangent line at $x = a$, then $f'(a)$ does not exist (see Figs. 16C and 16D).

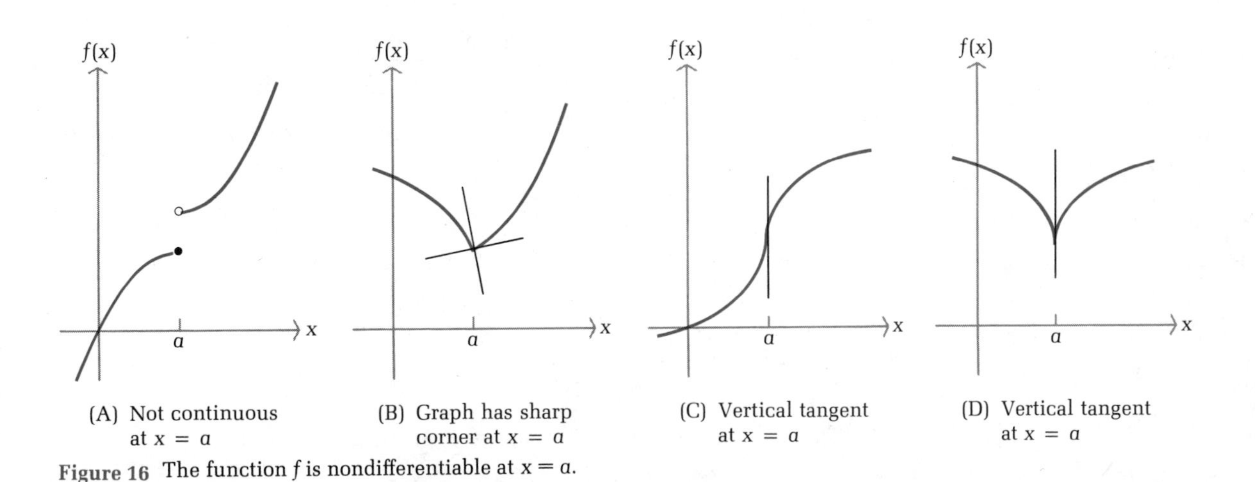

(A) Not continuous at $x = a$ (B) Graph has sharp corner at $x = a$ (C) Vertical tangent at $x = a$ (D) Vertical tangent at $x = a$

Figure 16 The function f is nondifferentiable at $x = a$.

If f is differentiable on the interval (a, b), then none of the situations in Figure 16 can occur. Thus, **the graph of a differentiable function is a continuous curve with no corners and no vertical tangent lines.**

■ Instantaneous Rates of Change

From the definition of instantaneous rate of change of $f(x)$ at x given in Section 1-3, we see that the instantaneous rate of change is simply the derivative of f at x—that is, $f'(x)$.

Example 20 Velocity

Refer to Example 15 in Section 1-3. Find a function that will give the instantaneous velocity, v, of the falling steel ball at any time x. Find the velocity at $x = 2$, 3, and 5 seconds.

Solution

Recall that the distance y (in feet) that the ball falls in x seconds is given by

$$y = f(x) = 16x^2$$

The instantaneous velocity function is $v = f'(x)$; thus,

$$v = f'(x) = \lim_{\Delta x \to 0} \frac{f(x + \Delta x) - f(x)}{\Delta x}$$

We will find $f'(x)$ using the two-step process described in Example 17.

Step 1. Find $[f(x + \Delta x) - f(x)]/\Delta x$ and simplify.

$$\begin{aligned}\frac{f(x + \Delta x) - f(x)}{\Delta x} &= \frac{[16(x + \Delta x)^2] - (16x^2)}{\Delta x} \\ &= \frac{16x^2 + 32x\Delta x + 16(\Delta x)^2 - 16x^2}{\Delta x} \\ &= \frac{32x\Delta x + 16(\Delta x)^2}{\Delta x} \\ &= \frac{\Delta x}{\Delta x}(32x + 16\Delta x) = 32x + 16\Delta x \qquad \Delta x \neq 0\end{aligned}$$

Step 2. Find the limit of the result of step 1.

$$\begin{aligned}\lim_{\Delta x \to 0} \frac{f(x + \Delta x) - f(x)}{\Delta x} &= \lim_{\Delta x \to 0} (32x + 16\Delta x) \\ &= 32x\end{aligned}$$

Thus,

$$v = f'(x) = 32x$$

The instantaneous velocities at $x = 2$, 3, and 5 seconds are

$$f'(2) = 32(2) = 64 \text{ feet per second}$$
$$f'(3) = 32(3) = 96 \text{ feet per second}$$
$$f'(5) = 32(5) = 160 \text{ feet per second}$$

An instantaneous rate of 64 feet per second at the end of 2 seconds means that *if* the rate were to remain constant for the next second, the object would fall an additional 64 feet. If the object is accelerating or decelerating (that is, if the rate does not remain constant), then the instantaneous rate is an approximation of what actually happens during the next second.

Problem 20 A steel ball falls so that its distance y (in feet) at time x (in seconds) is given by $y = f(x) = 16x^2 - 4x$.

(A) Find a function that will give the instantaneous velocity v at time x.
(B) Find the velocity at $x = 2$, 4, and 6 seconds.

■ Marginal Cost

In business and economics one is often interested in the rate at which something is taking place. A manufacturer, for example, is not only interested in the total cost $C(x)$ at certain production levels x, but is also interested in the rate of change of costs at various production levels.

In economics the word **marginal** refers to a rate of change; that is, to a derivative. Thus, if

$$C(x) = \text{Total cost of producing } x \text{ units during some unit of time}$$

then

$$\begin{aligned} C'(x) &= \text{Marginal cost} \\ &= \text{Rate of change in cost per unit change in production at an output level of } x \text{ units} \end{aligned}$$

Just as with instantaneous velocity, $C'(x)$ is an instantaneous rate. It indicates the change in cost for a 1 unit change in production at a production level of x units *if* the rate were to remain constant for the next unit change in production. If the rate does *not* remain constant, then the instantaneous rate is an approximation of what actually happens during the next unit change in production. Example 21 should help to clarify these ideas.

Example 21
Marginal Cost

Suppose the total cost $C(x)$ in thousands of dollars for manufacturing x sailboats per week is given by the function shown in the figure at the top of the next page.

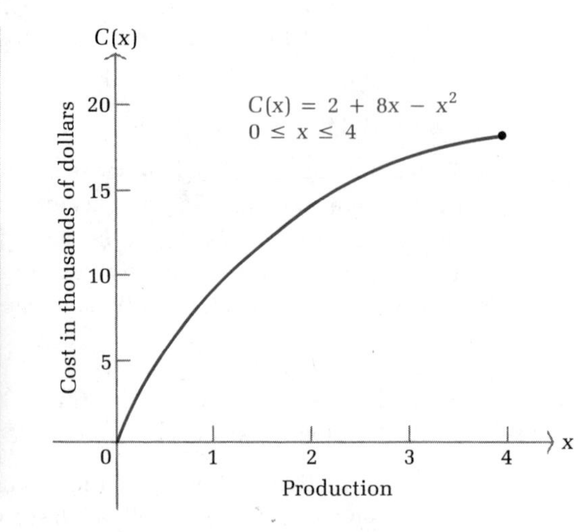

Find:

(A) The marginal cost at x
(B) The marginal cost at x = 1, 2, and 3 unit levels of production

Solutions (A) Marginal cost at x is

$$C'(x) = \lim_{\Delta x \to 0} \frac{C(x + \Delta x) - C(x)}{\Delta x}$$

which we find using the two-step process discussed in Example 17 (steps omitted here).

$$\text{Marginal cost} = C'(x) = 8 - 2x$$

(B) Marginal costs at x = 1, 2, and 3 unit levels of production are:

$C'(1) = 8 - 2(1) = 6$ $6,000 per unit increase in production
$C'(2) = 8 - 2(2) = 4$ $4,000 per unit increase in production
$C'(3) = 8 - 2(3) = 2$ $2,000 per unit increase in production

Notice that, as production goes up, the marginal cost goes down, as we might expect.

Let us now look at the marginal cost at the 1 unit level of production and interpret the result geometrically:

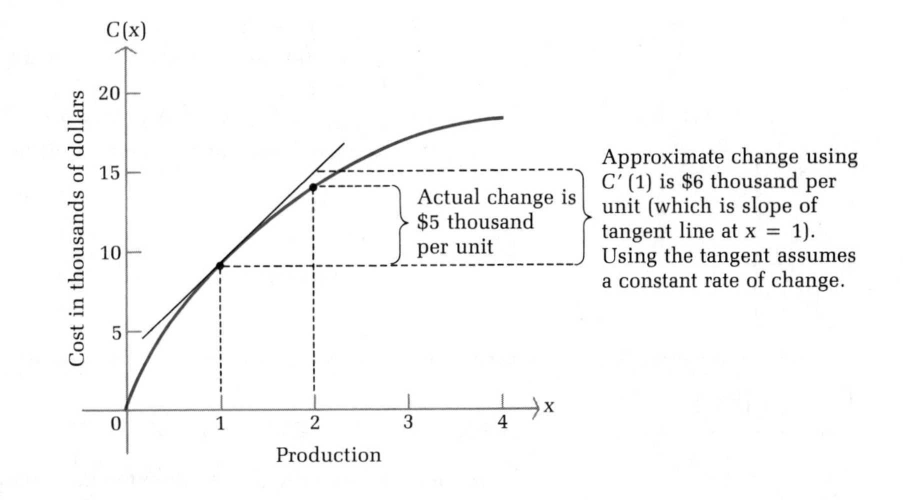

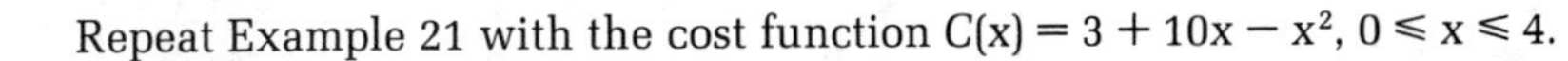

Problem 21 Repeat Example 21 with the cost function $C(x) = 3 + 10x - x^2$, $0 \leq x \leq 4$.

We will have more to say about marginal analysis in Section 1-8.

■ Summary

The concept of the derivative is a very powerful mathematical idea, and its applications are many and varied. In the next three sections we will develop formulas and general properties of derivatives that will enable us to find the derivatives of many functions without having to go through the (two-step) limiting process each time.

Answers to Matched Problems

17. $f'(x) = 8 - 4x$ 18. $f'(x) = -1/x^2$ or $-x^{-2}$
19. (A) $f'(1) = 4$, $f'(2) = 0$, $f'(4) = -8$
(B) At $x = 1$, $y = 4x + 2$; at $x = 2$, $y = 8$; at $x = 4$, $y = -8x + 32$
(C) f(x)
10
5
0 1 2 3 4 x

20. (A) $v = f'(x) = 32x - 4$
 (B) $f'(2) = 60$ feet per second, $f'(4) = 124$ feet per second, $f'(6) = 188$ feet per second

21. (A) Marginal cost $= C'(x) = 10 - 2x$
 (B) Marginal costs at 1, 2, and 3 unit levels of production are:

$C'(1) = 8$	\$8,000 per unit increase
$C'(2) = 6$	\$6,000 per unit increase
$C'(3) = 4$	\$4,000 per unit increase

Exercise 1-4

9–17 odd

A *Problems 1–8 refer to the function f in the following graph. Use the graph to determine if f′(x) exists at the indicated value of x.*

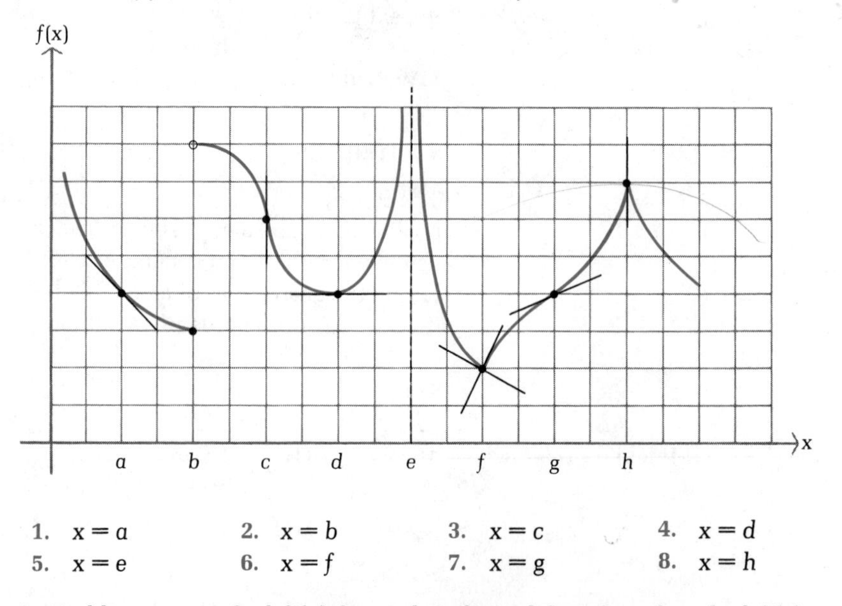

1. $x = a$
2. $x = b$
3. $x = c$
4. $x = d$
5. $x = e$
6. $x = f$
7. $x = g$
8. $x = h$

In Problems 9–18, find f′(x) for each indicated function; then find f′(1), f′(2), and f′(3).

9. $f(x) = 2x - 3$
10. $f(x) = 4x + 3$
11. $f(x) = 6x - x^2$
12. $f(x) = 8x - x^2$

B

13. $f(x) = \dfrac{1}{x + 1}$
14. $f(x) = \dfrac{1}{x - 5}$
15. $f(x) = \sqrt{x} - 3$
16. $f(x) = 2 - \sqrt{x}$
17. $f(x) = x^{-2}$
18. $f(x) = \dfrac{1}{x^2}$

19. If an object moves along a line so that it is at $y = f(x) = 4x^2 - 2x$ at time x (in seconds), find the instantaneous velocity function $v = f'(x)$ and find the velocity at times 1, 3, and 5 seconds (y is measured in feet).

20. Repeat Problem 19 with $f(x) = 8x^2 - 4x$.

21. Given $y = f(x) = x^2, \quad -3 \leqslant x \leqslant 3$:

(A) Find $f'(x)$.
(B) Find the slope of the tangent line to the graph of $y = x^2$ at $x = -2$, 0, and 2.
(C) Find the equations of the tangents at $x = -2$, 0, and 2.
(D) Sketch the tangent lines on the graph at $x = -2$, 0, and 2.

22. Repeat Problem 21 for $y = f(x) = x^2 + 1, \quad -3 \leqslant x \leqslant 3$.

C

23. For $f(x) = x^3 + 2x$, find:

(A) $f'(x)$ (B) $f'(1)$ and $f'(3)$

24. For $f(x) = x^2 - 3x^3$, find:

(A) $f'(x)$ (B) $f'(1)$ and $f'(2)$

In Problems 25 and 26, sketch the graph of f and determine where f is nondifferentiable.

25. $f(x) = \begin{cases} 2x & x < 1 \\ 2 & x \geqslant 1 \end{cases}$

26. $f(x) = \begin{cases} 2x & x < 2 \\ 6 - x & x \geqslant 2 \end{cases}$

In Problems 27–32, determine whether f is differentiable at x = 0 by considering

$$\lim_{\Delta x \to 0} \frac{f(\Delta x) - f(0)}{\Delta x}$$

27. $f(x) = |x|$

28. $f(x) = x + |x|$

29. $f(x) = x|x|$

30. $f(x) = x^{2/3}$

31. $f(x) = x^{1/3}$

32. $f(x) = x^3|x|$

Applications

Business & Economics

33. *Marginal cost.* The total cost per day, C(x) (in hundreds of dollars), for manufacturing x windsurfers is given by

$$C(x) = 3 + 10x - x^2 \qquad 0 \leqslant x \leqslant 5$$

(A) Find the marginal cost at x.
(B) Find the marginal cost at $x = 1$, 3, and 4 unit levels of production.

34. *Marginal cost.* Repeat Problem 33 for $C(x) = 5 + 12x - x^2$, $0 \leqslant x \leqslant 5$.

Life Sciences

35. *Negative growth.* A colony of bacteria was treated with a poison, and the number of survivors $N(t)$, in thousands, after t hours was found to be given approximately by

$$N(t) = t^2 - 8t + 16 \qquad 0 \leq t \leq 4$$

(A) Find $N'(t)$.
(B) Find the rate of change of the colony at $t = 1$, 2, and 3.

Social Sciences

36. *Learning.* A private foreign language school found that the average person learned $N(t)$ basic phrases in t continuous hours, as given approximately by

$$N(t) = 14t - t^2 \qquad 0 \leq t \leq 7$$

(A) Find $N'(t)$.
(B) Find the rate of learning at $t = 1$, 3, and 6 hours.

1-5 Derivatives of Constants, Power Forms, and Sums

- Derivative of a Constant
- Power Rule
- Derivative of a Constant Times a Function
- Derivatives of Sums and Differences
- Applications

In the last section we defined the derivative of f at x as

$$f'(x) = \lim_{\Delta x \to 0} \frac{f(x + \Delta x) - f(x)}{\Delta x}$$

(if the limit exists) and we used this definition and a two-step process to find the derivatives of a number of functions. In this and the next two sections we will develop some rules based on this definition that will enable us to determine the derivatives of a rather large class of functions without having to go through the two-step process each time.

Before starting on these rules, we list some symbols that are widely used to represent derivatives:

Derivative Notation

Given $y = f(x)$, then

$$f'(x) \qquad y' \qquad \frac{dy}{dx} \qquad D_x f(x)$$

all represent the derivative of f at x.

Each of these symbols for derivatives has its particular advantage in certain situations. All of them will become familiar to you after a little experience.

■ Derivative of a Constant

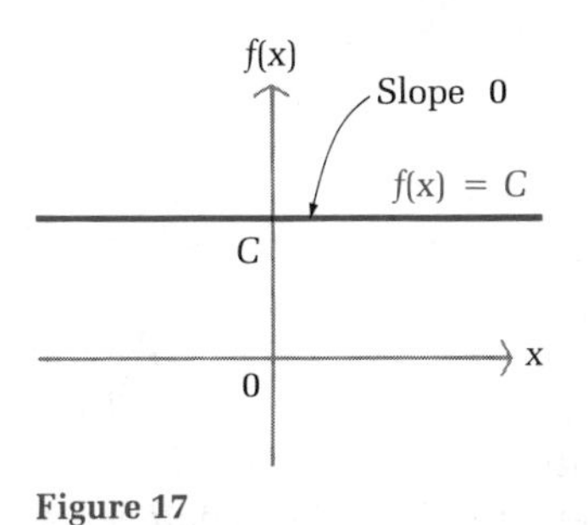

Figure 17

Suppose

$f(x) = C \qquad C$ a constant $\qquad$ A constant function

Geometrically, the graph of $f(x) = C$ is a horizontal straight line with slope 0 (see Figure 17); hence, we would expect $D_x C = 0$. We will show that this is actually the case using the definition of the derivative and the two-step process introduced in the last section. We want to find

$$f'(x) = \lim_{\Delta x \to 0} \frac{f(x + \Delta x) - f(x)}{\Delta x} \qquad \text{Definition of } f'(x)$$

Step 1. $\dfrac{f(x + \Delta x) - f(x)}{\Delta x} = \dfrac{C - C}{\Delta x} = \dfrac{0}{\Delta x} = 0 \qquad \Delta x \neq 0$

Step 2. $\lim\limits_{\Delta x \to 0} 0 = 0$

Thus,

$$D_x C = 0$$

And we conclude that **the derivative of any constant is 0.**

Derivative of a Constant

If $y = f(x) = C$, then

$$f'(x) = 0$$

Also, $y' = 0$, $dy/dx = 0$, and $D_x C = 0$.

Note: When we write $D_x C = 0$, we mean $D_x f(x) = 0$, where $f(x) = C$.

Example 22 (A) If $f(x) = 3$, then $f'(x) = 0$. (B) If $y = -1.4$, then $y' = 0$.
(C) If $y = \pi$, then $dy/dx = 0$. (D) $D_x(23) = 0$

Problem 22 Find:

(A) $f'(x)$ for $f(x) = -24$ (B) y' for $y = 12$
(C) dy/dx for $y = -\sqrt{7}$ (D) $D_x(-\pi)$

■ Power Rule

Using the definition of derivative and the two-step process introduced in the last section, we can show that:

$$\text{If } f(x) = x, \quad \text{then } f'(x) = 1$$
$$\text{If } f(x) = x^2, \quad \text{then } f'(x) = 2x$$
$$\text{If } f(x) = x^3, \quad \text{then } f'(x) = 3x^2$$
$$\text{If } f(x) = x^4, \quad \text{then } f'(x) = 4x^3$$

In general, for any positive integer n:

$$\text{If } f(x) = x^n, \quad \text{then } f'(x) = nx^{n-1} \tag{1}$$

In fact, more advanced techniques can be used to show that (1) holds for *any* real number n. We will assume this general result for the remainder of this book.

Power Rule

If $y = f(x) = x^n$, where n is a real number, then

$$f'(x) = nx^{n-1}$$

Example 23 (A) If $f(x) = x^5$, then $f'(x) = 5x^{5-1} = 5x^4$.
(B) If $y = x^{25}$, then $y' = 25x^{25-1} = 25x^{24}$.
(C) If $y = x^{-3}$, then $dy/dx = -3x^{-3-1} = -3x^{-4}$.
(D) $D_x x^{5/3} = \frac{5}{3}x^{(5/3)-1} = \frac{5}{3}x^{2/3}$

Problem 23 Find:

(A) $f'(x)$ for $f(x) = x^6$ (B) y' for $y = x^{30}$
(C) dy/dx for $y = x^{-2}$ (D) $D_x x^{3/2}$

In some cases, properties of exponents must be used to rewrite an expression before the power rule is applied.

Example 24 (A) If $f(x) = 1/x^4$, then we can write $f(x) = x^{-4}$ and

$$f'(x) = -4x^{-4-1} = -4x^{-5} \quad \text{or} \quad \frac{-4}{x^5}$$

(B) If $y = \sqrt{x}$, then we can write $y = x^{1/2}$ and

$$y' = \frac{1}{2}x^{(1/2)-1} = \frac{1}{2}x^{-1/2} \quad \text{or} \quad \frac{1}{2\sqrt{x}}$$

(C) $$D_x \frac{1}{\sqrt[3]{x}} = D_x x^{-1/3} = -\frac{1}{3}x^{(-1/3)-1} = -\frac{1}{3}x^{-4/3} \quad \text{or} \quad \frac{-1}{3\sqrt[3]{x^4}}$$

Problem 24 Find:

(A) $f'(x)$ for $f(x) = 1/x$ (B) y' for $y = \sqrt[3]{x^2}$ (C) $D_x(1/\sqrt{x})$

■ Derivative of a Constant Times a Function

Let $f(x) = 16x^2 = 16u(x)$, where $u(x) = x^2$. Using the power rule, we have

$$u'(x) = 2x$$

From Example 20 in the last section, $f'(x) = 32x$, which can be written in the form

$$f'(x) = 32x = 16(2x) = 16u'(x)$$

Using the definition of derivative and the two-step process introduced in the last section, we can show that, in general, if $f(x) = ku(x)$, where k is a constant and $u'(x)$ exists, then $f'(x)$ exists and $f'(x) = ku'(x)$. Thus, **the derivative of a constant times a differentiable function is the constant times the derivative of the function.**

Constant Times a Function Rule

If $y = f(x) = ku(x)$, then

$$f'(x) = ku'(x)$$

Also, $y' = ku'$, $dy/dx = k\,du/dx$, and $D_x ku(x) = kD_x u(x)$.

Example 25 (A) If $f(x) = 3x^2$, then $f'(x) = 3 \cdot 2x^{2-1} = 6x$.

(B) If $y = \frac{x^3}{6} = \frac{1}{6}x^3$, then $\frac{dy}{dx} = \frac{1}{6} \cdot 3x^{3-1} = \frac{1}{2}x^2$.

(C) If $y = \frac{1}{2x^4} = \frac{1}{2}x^{-4}$, then $y' = \frac{1}{2}(-4x^{-4-1}) = -2x^{-5}$ or $\frac{-2}{x^5}$.

(D) $D_x \frac{4}{\sqrt{x^3}} = D_x \frac{4}{x^{3/2}} = D_x\, 4x^{-3/2} = 4\left[-\frac{3}{2}x^{(-3/2)-1}\right]$

$$= -6x^{-5/2} \quad \text{or} \quad -\frac{6}{\sqrt{x^5}}$$

Problem 25 Find:

(A) $f'(x)$ for $f(x) = 4x^5$ (B) $\frac{dy}{dx}$ for $y = \frac{x^4}{12}$

(C) y' for $y = \frac{1}{3x^3}$ (D) $D_x \frac{9}{\sqrt[3]{x}}$

Derivatives of Sums and Differences

Let $f(x) = 4x - x^2 = u(x) - v(x)$, where $u(x) = 4x$ and $v(x) = x^2$. Using the power rule and the constant times a function rule, we have

$$u'(x) = 4 \quad \text{and} \quad v'(x) = 2x$$

From Example 17 in the last section, $f'(x) = 4 - 2x$, which can be written in the form

$$f'(x) = 4 - 2x = u'(x) - v'(x)$$

In general, we can show that if $f(x) = u(x) \pm v(x)$ and $u'(x)$ and $v'(x)$ exist, then $f'(x)$ exists and $f'(x) = u'(x) \pm v'(x)$. Thus, **the derivative of the sum of two differentiable functions is the sum of the derivatives and the derivative of the difference of two differentiable functions is the difference of the derivatives.** Together, we then have the sum and difference rule for differentiation.

Sum and Difference Rule

If $y = f(x) = u(x) \pm v(x)$, then

$$f'(x) = u'(x) \pm v'(x)$$

[*Note:* This rule generalizes to the sum and difference of any given number of functions.]

With this and the other rules stated previously, we will be able to compute the derivatives of all polynomials and a variety of other functions.

Example 26

(A) If $f(x) = 3x^2 + 2x$, then $f'(x) \boxed{= (3x^2)' + (2x)'} = 6x + 2$.

(B) If $y = 4 + 2x^3 - 3x^{-1}$, then $y' \boxed{= (4)' + (2x^3)' - (3x^{-1})'} = 6x^2 + 3x^{-2}$.

(C) If $y = \sqrt[3]{x} - 3x$, then $\dfrac{dy}{dx} = \dfrac{d}{dx}x^{1/3} - \dfrac{d}{dx}3x = \dfrac{1}{3}x^{-2/3} - 3$.

(D) $D_x\left(\dfrac{5}{3x^2} - \dfrac{2}{x^4} + \dfrac{x^3}{9}\right) \boxed{= D_x \dfrac{5}{3} x^{-2} - D_x 2x^{-4} + D_x \dfrac{1}{9} x^3}$

$$= \frac{5}{3}(-2)x^{-3} - 2(-4)x^{-5} + \frac{1}{9} \cdot 3x^2 = -\frac{10}{3x^3} + \frac{8}{x^5} + \frac{1}{3}x^2$$

Problem 26 Find:

(A) $f'(x)$ for $f(x) = 3x^4 - 2x^3 + x^2 - 5x + 7$
(B) y' for $y = 3 - 7x^{-2}$
(C) $\dfrac{dy}{dx}$ for $y = 5x^3 - \sqrt[4]{x}$
(D) $D_x\left(-\dfrac{3}{4x} + \dfrac{4}{x^3} - \dfrac{x^4}{8}\right)$

▪ Applications

Example 27
Instantaneous Velocity

An object moves along the y axis (marked in feet) so that its position at time x in seconds is

$$f(x) = x^3 - 6x^2 + 9x$$

(A) Find the instantaneous velocity function v.
(B) Find the velocity at $x = 2$ and $x = 5$ seconds.
(C) Find the time(s) when the velocity is 0.

Solutions

(A) $v = f'(x) = \boxed{(x^3)' - (6x^2)' + (9x)'} = 3x^2 - 12x + 9$

(B) $f'(2) = 3(2)^2 - 12(2) + 9 = -3$ feet per second

$f'(5) = 3(5)^2 - 12(5) + 9 = 24$ feet per second

(C) $v = f'(x) = 3x^2 - 12x + 9 = 0$

$$3(x^2 - 4x + 3) = 0$$
$$3(x - 1)(x - 3) = 0$$
$$x = 1, 3$$

Thus, $v = 0$ at $x = 1$ and $x = 3$ seconds.

Problem 27 Repeat Example 27 for $f(x) = x^3 - 15x^2 + 72x$.

Example 28 Tangents

Let $f(x) = x^4 - 8x^2 + 10$.

(A) Find $f'(x)$.
(B) Find the equation of the tangent line at $x = 1$.
(C) Find the values of x where the tangent line is horizontal.

Solutions

(A) $f'(x) = (x^4)' - (8x^2)' + (10)'$

$= 4x^3 - 16x$

(B) $y - y_1 = m(x - x_1)$ $\quad y_1 = f(x_1) = f(1) = (1)^4 - 8(1)^2 + 10 = 3$
$\quad m = f'(x_1) = f'(1) = 4(1)^3 - 16(1) = -12$

$y - 3 = -12(x - 1)$

$y = -12x + 15$ Tangent line at $x = 1$

(C) Since a horizontal line has 0 slope, we must solve $f'(x) = 0$ for x:

$$f'(x) = 4x^3 - 16x = 0$$
$$4x(x^2 - 4) = 0$$
$$4x(x - 2)(x + 2) = 0$$
$$x = 0, 2, -2$$

Thus, the tangent line to the graph of f will be horizontal at $x = -2$, $x = 0$, and $x = 2$. (In the next chapter, we will see how this information is used to help sketch the graph of f.)

Problem 28 Repeat Example 28 for $f(x) = x^4 - 4x^3 + 7$

Example 29 Marginal Cost

The total cost $C(x)$ in thousands of dollars for manufacturing x sailboats is given by

$$C(x) = 2 + 8x - x^2 \qquad 0 \leq x \leq 4$$

(A) The marginal cost at a production level of x is

$$C'(x) = (2)' + (8x)' - (x^2)' = 8 - 2x$$

(B) The marginal cost at $x = 1$ is

$C'(1) = 8 - 2(1) = 6$ $6,000 per unit increase in production

(C) The marginal cost at $x = 3$ is

$C'(3) = 8 - 2(3) - 2$ $2,000 per unit increase in production

Problem 29 Repeat Example 29 with the cost function $C(x) = 3 + 10x - x^2$, $0 \leq x \leq 4$.

Answers to Matched Problems

22. All are 0.
23. (A) $6x^5$ (B) $30x^{29}$ (C) $-2x^{-3}$ (D) $\frac{3}{2}x^{1/2}$
24. (A) $-x^{-2}$ or $-1/x^2$ (B) $\frac{2}{3}x^{-1/3}$ or $2/(3\sqrt[3]{x})$
(C) $-\frac{1}{2}x^{-3/2}$ or $-1/(2\sqrt{x^3})$

25. (A) $20x^4$ (B) $x^3/3$ (C) $-x^{-4}$ or $-1/x^4$
    (D) $-3x^{-4/3}$ or $-3/\sqrt[3]{x^4}$
26. (A) $12x^3 - 6x^2 + 2x - 5$ (B) $14x^{-3}$
    (C) $15x^2 - \frac{1}{4}x^{-3/4}$ (D) $3/(4x^2) - (12/x^4) - (x^3/2)$
27. (A) $v = 3x^2 - 30x + 72$
    (B) $f'(2) = 24$ feet per second; $f'(5) = -3$ feet per second
    (C) $x = 4$ seconds or $x = 6$ seconds
28. (A) $f'(x) = 4x^3 - 12x^2$ (B) $y = -8x + 12$ (C) $x = 0$ or $x = 3$
29. (A) Marginal cost $= C'(x) = 10 - 2x$
    (B) $C'(1) = 8$ \$8,000 per unit increase in production
    (C) $C'(3) = 4$ \$4,000 per unit increase in production

Exercise 1-5

Find each of the following:

A

1. $f'(x)$ for $f(x) = 12$
2. $\dfrac{dy}{dx}$ for $y = -\sqrt{3}$
3. $D_x 23$
4. y' for $y = \pi$
5. $\dfrac{dy}{dx}$ for $y = x^{12}$
6. $D_x x^5$
7. $f'(x)$ for $f(x) = x$
8. y' for $y = x^7$
9. y' for $y = x^{-7}$
10. $f'(x)$ for $f(x) = x^{-11}$
11. $\dfrac{dy}{dx}$ for $y = x^{5/2}$
12. $D_x x^{7/3}$
13. $D_x \dfrac{1}{x^5}$
14. $f'(x)$ for $f(x) = \dfrac{1}{x^9}$
15. $f'(x)$ for $f(x) = 2x^4$
16. $\dfrac{dy}{dx}$ for $y = -3x$
17. $D_x\left(\dfrac{1}{3}x^6\right)$
18. y' for $y = \dfrac{1}{2}x^4$
19. $\dfrac{dy}{dx}$ for $y = \dfrac{x^5}{15}$
20. $f'(x)$ for $f(x) = \dfrac{x^6}{24}$

B

21. $D_x(2x^{-5})$
22. y' for $y = -4x^{-1}$
23. $f'(x)$ for $f(x) = \dfrac{4}{x^4}$
24. $\dfrac{dy}{dx}$ for $y = \dfrac{-3}{x^6}$
25. $D_x \dfrac{-1}{2x^2}$
26. y' for $y = \dfrac{1}{6x^3}$
27. $f'(x)$ for $f(x) = -3x^{1/3}$
28. $\dfrac{dy}{dx}$ for $y = -8x^{1/4}$

29. $D_x(2x^2 - 3x + 4)$

30. y' for $y = 3x^2 + 4x - 7$

31. $\dfrac{dy}{dx}$ for $y = 3x^5 - 2x^3 + 5$

32. $f'(x)$ for $y = 2x^3 - 6x + 5$

33. $D_x(3x^{-4} + 2x^{-2})$

34. y' for $y = 2x^{-3} - 4x^{-1}$

35. $\dfrac{dy}{dx}$ for $y = \dfrac{1}{2x^4} - \dfrac{2}{3x^3}$

36. $f'(x)$ for $f(x) = \dfrac{3}{4x^3} + \dfrac{1}{2x^5}$

37. $D_x(3x^{2/3} - 5x^{1/3})$

38. $D_x(8x^{3/4} + 4x^{-1/4})$

39. $D_x\left(\dfrac{3}{x^{3/5}} - \dfrac{6}{x^{1/2}}\right)$

40. $D_y\left(\dfrac{5}{y^{1/5}} - \dfrac{8}{y^{3/2}}\right)$

41. $D_x \dfrac{1}{\sqrt[3]{x}}$

42. y' for $y = \dfrac{10}{\sqrt[5]{x}}$

43. $\dfrac{dy}{dx}$ for $y = \dfrac{12}{\sqrt{x}} - 3x^{-2} + x$

44. $f'(x)$ for $f(x) = 2x^{-3} - \dfrac{6}{\sqrt[3]{x^2}} + 7$

45. Given the equation $y = f(x) = 6x - x^2$, find:

(A) $f'(x)$
(B) The equation of the lines tangent to the graph at $x = 2$ and at $x = 4$

46. Repeat Problem 45 for $y = f(x) = 2x^2 + 8x$.
47. Repeat Problem 45 for $y = f(x) = x^3 - 3x^2 + 2$.
48. Repeat Problem 45 for $y = f(x) = 2x^3 - 3x^2 - 5$.
49. If an object moves along the y axis (marked in feet) so that its position at time x in seconds is given by $y = f(x) = 176x - 16x^2$, find:

(A) The instantaneous velocity function $v = f'(x)$
(B) The velocity at $x = 0$, 3, and 6 seconds
(C) The time(s) when $v = 0$

50. Repeat Problem 49 for $y = f(x) = 80x - 10x^2$.
51. Repeat Problem 49 for $y = f(x) = x^3 - 9x^2 + 15x$
52. Repeat Problem 49 for $y = f(x) = x^3 - 9x^2 + 24x$
53. Given the equation $y = f(x) = x^3 + 6x^2 - 15x$, find:

(A) $f'(x)$
(B) The values of x where the tangent line is horizontal

54. Repeat Problem 53 for $y = f(x) = x^3 - 9x^2 + 27x - 9$.
55. Repeat Problem 53 for $y = f(x) = 3x^4 - 4x^3 + 2$.
56. Repeat Problem 53 for $y = f(x) = x^4 - 32x^2 + 10$.

Find each of the following:

C

57. $f'(x)$ for $f(x) = \dfrac{10x + 20}{x}$

58. $\dfrac{dy}{dx}$ for $y = \dfrac{x^2 + 25}{x^2}$

59. $D_x \dfrac{x^4 - 3x^3 + 5}{x^2}$

60. y' for $y = \dfrac{2x^5 - 4x^3 + 2x}{x^3}$

In Problems 61–64, use the definition of derivative and the two-step process to verify each statement.

61. $D_x x^3 = 3x^2$

62. $D_x x^4 = 4x^3$

63. $D_x[kf(x)] = kD_x f(x)$

64. $D_x[u(x) + v(x)] = u'(x) + v'(x)$

Applications

Business & Economics

65. *Advertising.* Using past records it is estimated that a company will sell $N(x)$ units of a product after spending $\$x$ thousand on advertising, as given by

$$N(x) = 60x - x^2 \qquad 5 \leq x \leq 30$$

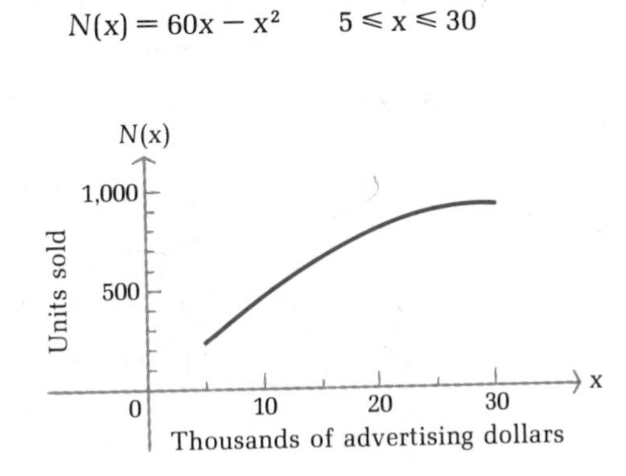

(A) Find $N'(x)$, the rate of change of sales per unit change in money spent on advertising at the $\$x$ thousand level.

(B) Find $N'(10)$ and $N'(20)$ and interpret.

66. *Marginal average cost.* (This topic is treated in detail in Section 1-8.) Economists often work with average costs—cost per unit output—rather than total costs. We would expect higher average costs, because of plant inefficiency, at low output levels and also at output levels near plant capacity. Therefore, we would expect the graph of an average cost function to be U-shaped. Suppose that for a given firm the total cost of producing x thousand units is given by

$$C(x) = x^3 - 6x^2 + 12x$$

Then the average cost $\overline{C}(x)$ is given by

$$\overline{C}(x) = \frac{C(x)}{x} = x^2 - 6x + 12$$

(A) Find the marginal average cost $\overline{C}'(x)$.

(B) Find the marginal average cost at $x = 2$, 3, and 4, and interpret.

Life Sciences

67. *Medicine.* A person x inches tall has a pulse rate of y beats per minute, as given approximately by

$$y = 590x^{-1/2} \qquad 30 \leq x \leq 75$$

What is the instantaneous rate of change of pulse rate at the:

(A) 36 inch level? (B) 64 inch level?

68. *Ecology.* A coal-burning electrical generating plant emits sulfur dioxide into the surrounding air. The concentration $C(x)$ in parts per million is given approximately by

$$C(x) = \frac{0.1}{x^2}$$

where x is the distance from the plant in miles. Find the (instantaneous) rate of change of concentration at:

(A) $x = 1$ mile (B) $x = 2$ miles

Social Sciences

69. *Learning.* Suppose a person learns y items in x hours, as given by

$$y = 50\sqrt{x} \qquad 0 \leq x \leq 9$$

Find the rate of learning at the end of:

(A) 1 hour (B) 9 hours

70. *Learning.* If a person learns y items in x hours, as given by

$$y = 21\sqrt[3]{x^2} \qquad 0 \leq x \leq 8$$

find the rate of learning at the end of:

(A) 1 hour (B) 8 hours

1-6 Derivatives of Products and Quotients

- Derivatives of Products
- Derivatives of Quotients

The derivative rules discussed in the last section added substantially to our ability to compute and apply derivatives to many practical problems. In this and the next section we will add a few more rules that will increase this ability even further.

Derivatives of Products

In the last section we found that the derivative of a sum is the sum of the derivatives. Is the derivative of a product the product of the derivatives? Let us take a look at a simple example. Consider

$$f(x) = u(x)v(x) = (x^2 - 3x)(2x^3 - 1) \quad (1)$$

where $u(x) = x^2 - 3x$ and $v(x) = 2x^3 - 1$. The product of the derivatives is

$$u'(x)v'(x) = (2x - 3)6x^2 = 12x^3 - 18x^2 \quad (2)$$

To see if this is equal to the derivative of the product, we multiply the right side of (1) and use derivative formulas from the last section:

$$f(x) = (x^2 - 3x)(2x^3 - 1) = 2x^5 - 6x^4 - x^2 + 3x$$

Thus,

$$f'(x) = 10x^4 - 24x^3 - 2x + 3 \quad (3)$$

Since (2) and (3) are not equal, we conclude that the derivative of a product is *not* the product of the derivatives. There is a product rule for derivatives, but it is slightly more complicated than you might expect.

Using the definition of derivative and the two-step process, we can show that **the derivative of a product is the first times the derivative of the second plus the second times the derivative of the first.**

Product Rule

If $y = f(x) = u(x)v(x)$, then

$$f'(x) = u(x)v'(x) + v(x)u'(x)$$

Also,

$$y' = uv' + vu'$$

$$\frac{dy}{dx} = u\frac{dv}{dx} + v\frac{du}{dx}$$

$$D_x[u(x)v(x)] = u(x)D_xv(x) + v(x)D_xu(x)$$

Example 30 Find $f'(x)$ for $f(x) = 2x^2(3x^4 - 2)$ two ways.

Solution Method I. Use the product rule:

$$\begin{aligned} f'(x) &= 2x^2(3x^4 - 2)' + (3x^4 - 2)(2x^2)' \\ &= 2x^2(12x^3) + (3x^4 - 2)(4x) \\ &= 24x^5 + 12x^5 - 8x \\ &= 36x^5 - 8x \end{aligned}$$

First times derivative of second plus second times derivative of first

Method II. Multiply first; then take derivatives:

$$f(x) = 2x^2(3x^4 - 2) = 6x^6 - 4x^2$$
$$f'(x) = 36x^5 - 8x$$

Problem 30 Find $f'(x)$ two ways for $f(x) = 3x^3(2x^2 - 3x + 1)$.

At this point, all the products we will encounter can be differentiated by either of the methods illustrated in Example 30. In the next and later sections, we will see that there are situations where the product rule must be used. Unless instructed otherwise, you should use the product rule to differentiate all products in this section to gain experience with the use of this important differentiation rule.

Example 31 Let $f(x) = (2x - 9)(x^2 + 6)$.

(A) Find the equation of the line tangent to the graph of $f(x)$ at $x = 3$.
(B) Find the values of x where the tangent line is horizontal.

Solutions (A) First find $f'(x)$:

$$f'(x) = (2x - 9)(x^2 + 6)' + (x^2 + 6)(2x - 9)'$$
$$= (2x - 9)(2x) + (x^2 + 6)(2)$$

Now find the equation of the tangent line at $x = 3$.

$$y - y_1 = m(x - x_1) \qquad y_1 = f(x_1) = f(3) = -45$$
$$m = f'(x_1) = f'(3) = 12$$
$$y - (-45) = 12(x - 3)$$
$$y = 12x - 81 \qquad \text{Tangent line at } x = 3$$

(B) The tangent line is horizontal at values of x such that $f'(x) = 0$, so

$$f'(x) = (2x - 9)2x + (x^2 + 6)2 = 0$$
$$6x^2 - 18x + 12 = 0$$
$$x^2 - 3x + 2 = 0$$
$$(x - 1)(x - 2) = 0$$
$$x = 1, 2$$

The tangent line is horizontal at $x = 1$ and at $x = 2$.

Problem 31 Repeat Example 31 for $f(x) = (2x + 9)(x^2 - 12)$.

As Example 31 illustrates, the way we write $f'(x)$ depends on what we want to do with it. If we are interested only in evaluating $f'(x)$ at specified values of x, the form in part A is sufficient. However, if we want to solve $f'(x) = 0$, we must multiply and collect like terms, as we did in part B.

▪ Derivatives of Quotients

As is the case with a product, the derivative of a quotient is *not* the quotient of the derivatives.

Let

$$f(x) = \frac{u(x)}{v(x)} \qquad \text{where} \qquad u'(x) \text{ and } v'(x) \text{ exist}$$

Starting with the definition of a derivative, you can show that

$$f'(x) = \frac{v(x)u'(x) - u(x)v'(x)}{[v(x)]^2}$$

Thus, **the derivative of a quotient is the denominator times the derivative of the numerator minus the numerator times the derivative of the denominator, all over the denominator squared.**

Quotient Rule

If

$$y = f(x) = \frac{u(x)}{v(x)}$$

then

$$f'(x) = \frac{v(x)u'(x) - u(x)v'(x)}{[v(x)]^2}$$

Also,

$$y' = \frac{vu' - uv'}{v^2}$$

$$\frac{dy}{dx} = \frac{v(du/dx) - u(dv/dx)}{v^2}$$

$$D_x \frac{u(x)}{v(x)} = \frac{v(x)D_x u(x) - u(x)D_x v(x)}{[v(x)]^2}$$

Example 32 (A) If

$$f(x) = \frac{x^2}{2x - 1}$$

find $f'(x)$.

(B) Find

$$D_x \frac{x^2 - x}{x^3 + 1}$$

(C) Find

$$D_x \frac{x^2 - 3}{x^2}$$

by using the quotient rule and also by splitting the fraction into two fractions.

Solutions (A)

$$f'(x) = \frac{(2x - 1)(x^2)' - x^2(2x - 1)'}{(2x - 1)^2}$$

The denominator times the derivative of the numerator minus the numerator times the derivative of the denominator, all over the square of the denominator

$$= \frac{(2x - 1)(2x) - x^2(2)}{(2x - 1)^2}$$

$$= \frac{4x^2 - 2x - 2x^2}{(2x - 1)^2}$$

$$= \frac{2x^2 - 2x}{(2x - 1)^2}$$

(B)

$$D_x \frac{x^2 - x}{x^3 + 1} = \frac{(x^3 + 1)D_x(x^2 - x) - (x^2 - x)D_x(x^3 + 1)}{(x^3 + 1)^2}$$

$$= \frac{(x^3 + 1)(2x - 1) - (x^2 - x)(3x^2)}{(x^3 + 1)^2}$$

$$= \frac{2x^4 - x^3 + 2x - 1 - 3x^4 + 3x^3}{(x^3 + 1)^2}$$

$$= \frac{-x^4 + 2x^3 + 2x - 1}{(x^3 + 1)^2}$$

(C) Method I. Use the quotient rule:

$$D_x \frac{x^2 - 3}{x^2} = \frac{x^2 D_x(x^2 - 3) - (x^2 - 3)D_x x^2}{(x^2)^2}$$

$$= \frac{x^2(2x) - (x^2 - 3)2x}{x^4}$$

$$= \frac{2x^3 - 2x^3 + 6x}{x^4} = \frac{6x}{x^4} = \frac{6}{x^3}$$

Method II. Split into two fractions:

$$\frac{x^2 - 3}{x^2} = \frac{x^2}{x^2} - \frac{3}{x^2} = 1 - 3x^{-2}$$

$$D_x(1 - 3x^{-2}) = 0 - 3(-2)x^{-3} = \frac{6}{x^3}$$

Comparing methods I and II, we see that it may sometimes pay to change an expression algebraically before blindly using a differentiation formula.

Problem 32 Find:

(A) $f'(x)$ for $f(x) = \dfrac{2x}{x^2 + 3}$

(B) y' for $y = \dfrac{x^3 - 3x}{x^2 - 4}$

(C) $D_x \dfrac{2 + x^3}{x^3}$ two ways

Example 33
Sales Analysis

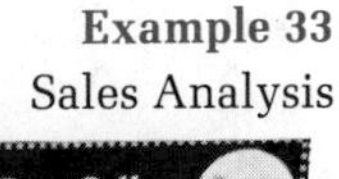

When a successful home video game is first introduced, the monthly sales generally increase rapidly for a period of time, and then begin to decrease. Suppose that the monthly sales $S(t)$ (in thousands of games) t months after the game is introduced are given by

$$S(t) = \frac{200t}{t^2 + 100}$$

(A) Find $S'(t)$.
(B) Find $S(5)$ and $S'(5)$ and interpret.
(C) Find $S(30)$ and $S'(30)$ and interpret.

Solutions

(A) $$S'(t) = \frac{(t^2 + 100)(200t)' - 200t(t^2 + 100)'}{(t^2 + 100)^2}$$

$$= \frac{(t^2 + 100)200 - 200t(2t)}{(t^2 + 100)^2}$$

$$= \frac{200t^2 + 20{,}000 - 400t^2}{(t^2 + 100)^2}$$

$$= \frac{20{,}000 - 200t^2}{(t^2 + 100)^2}$$

(B) $S(5) = \dfrac{200(5)}{5^2 + 100} = 8$ and $S'(5) = \dfrac{20{,}000 - 200(5)^2}{(5^2 + 100)^2} = 0.96$

The sales for the fifth month are 8,000 units. At this point in time, sales are increasing at the rate of $0.96(1000) = 960$ units per month.

(C) $S(30) = \dfrac{200(30)}{30^2 + 100} = 6$ and $S'(30) = \dfrac{20{,}000 - 200(30)^2}{(30^2 + 100)^2} = -0.16$

The sales for the thirtieth month are 6,000 units. At this point in time, sales are decreasing at the rate of $0.16(1{,}000) = 160$ units per month.

The function $S(t)$ in Example 33 is graphed in Figure 18. Notice that the maximum monthly sales seem to occur during the tenth month. In the next chapter, we will see how the derivative $S'(t)$ is used to help sketch the graph of $S(t)$ and to find the maximum monthly sales.

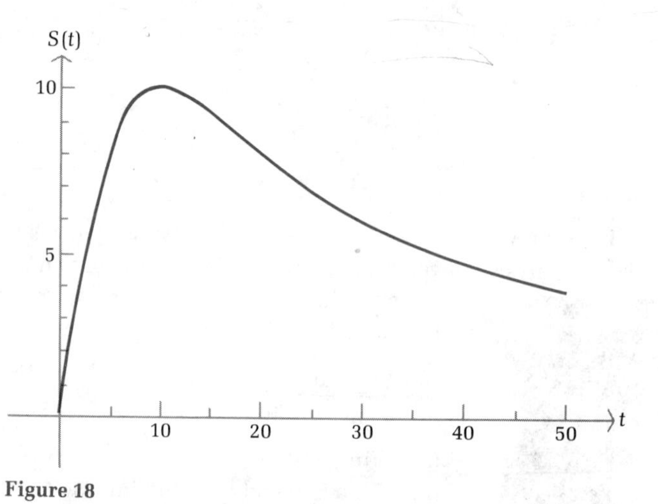

Figure 18

Problem 33 Refer to Example 33. Suppose that the monthly sales $S(t)$ (in thousands of games) t months after the game is introduced are given by

$$S(t) = \frac{200t}{t^2 + 64}$$

(A) Find $S'(t)$.
(B) Find $S(4)$ and $S'(4)$ and interpret.
(C) Find $S(24)$ and $S'(24)$ and interpret.

Answers to Matched Problems

30. $30x^4 - 36x^3 + 9x^2$ 31. (A) $y = 84x - 297$ (B) $x = -4$; $x = 1$

32. (A) $\dfrac{(x^2 + 3)2 - (2x)(2x)}{(x^2 + 3)^2} = \dfrac{6 - 2x^2}{(x^2 + 3)^2}$

(B) $\dfrac{(x^2 - 4)(3x^2 - 3) - (x^3 - 3x)(2x)}{(x^2 - 4)^2} = \dfrac{x^4 - 9x^2 + 12}{(x^2 - 4)^2}$

(C) $-\dfrac{6}{x^4}$

33. (A) $S'(t) = \dfrac{12{,}800 - 200t^2}{(t^2 + 64)^2}$

(B) $S(4) = 10$; $S'(4) = 1.5$; at $t = 4$ months, monthly sales are 10,000 and increasing at 1,500 games per month.

(C) $S(24) = 7.5$; $S'(24) = -0.25$; at $t = 24$ months, monthly sales are 7,500 and decreasing at 250 games per month.

Exercise 1-6

A *For f(x) as given, find f′(x) and simplify.*

1. $f(x) = 2x^3(x^2 - 2)$
2. $f(x) = 5x^2(x^3 + 2)$
3. $f(x) = (x - 3)(2x - 1)$
4. $f(x) = (3x + 2)(4x - 5)$
5. $f(x) = \dfrac{x}{x - 3}$
6. $f(x) = \dfrac{3x}{2x + 1}$
7. $f(x) = \dfrac{2x + 3}{x - 2}$
8. $f(x) = \dfrac{3x - 4}{2x + 3}$
9. $f(x) = (x^2 + 1)(2x - 3)$
10. $f(x) = (3x + 5)(x^2 - 3)$
11. $f(x) = \dfrac{x^2 + 1}{2x - 3}$
12. $f(x) = \dfrac{3x + 5}{x^2 - 3}$
13. $f(x) = (x^2 + 2)(x^2 - 3)$
14. $f(x) = (x^2 - 4)(x^2 + 5)$
15. $f(x) = \dfrac{x^2 + 2}{x^2 - 3}$
16. $f(x) = \dfrac{x^2 - 4}{x^2 + 5}$

B *Find each of the following and simplify:*

17. $f'(x)$ for $f(x) = (2x + 1)(x^2 - 3x)$
18. y' for $y = (x^3 + 2x^2)(3x - 1)$
19. $\dfrac{dy}{dx}$ for $y = (2x - x^2)(5x + 2)$
20. $D_x[(3 - x^3)(x^2 - x)]$
21. y' for $y = \dfrac{5x - 3}{x^2 + 2x}$
22. $f'(x)$ for $f(x) = \dfrac{3x^2}{2x - 1}$
23. $D_x \dfrac{x^2 - 3x + 1}{x^2 - 1}$
24. $\dfrac{dy}{dx}$ for $y = \dfrac{x^4 - x^3}{3x - 1}$

In Problems 25–28, find $f'(x)$ and find the equation of the line tangent to the graph of f at $x = 2$.

25. $f(x) = (1 + 3x)(5 - 2x)$
26. $f(x) = (7 - 3x)(1 + 2x)$
27. $f(x) = \dfrac{x - 8}{3x - 4}$
28. $f(x) = \dfrac{2x - 5}{2x - 3}$

In Problems 29–32, find $f'(x)$ and find the values of x where $f'(x) = 0$.

29. $f(x) = (2x - 15)(x^2 + 18)$
30. $f(x) = (2x - 3)(x^2 - 6)$
31. $f(x) = \dfrac{x}{x^2 + 1}$
32. $f(x) = \dfrac{x}{x^2 + 9}$

In Problems 33–36, find $f'(x)$ two ways: by using the product or quotient rule and by simplifying first.

33. $f(x) = x^3(x^4 - 1)$
34. $f(x) = x^4(x^3 - 1)$
35. $f(x) = \dfrac{x^3 + 9}{x^3}$
36. $f(x) = \dfrac{x^4 + 4}{x^4}$

C *Find each of the following. Do not simplify.*

37. $f'(x)$ for $f(x) = (2x^4 - 3x^3 + x)(x^2 - x + 5)$
38. $\dfrac{dy}{dx}$ for $y = (x^2 - 3x + 1)(x^3 + 2x^2 - x)$
39. $D_x \dfrac{3x^2 - 2x + 3}{4x^2 + 5x - 1}$
40. y' for $y = \dfrac{x^3 - 3x + 4}{2x^2 + 3x - 2}$
41. $\dfrac{dy}{dx}$ for $y = 9x^{1/3}(x^3 + 5)$
42. $D_x[(4x^{1/2} - 1)(3x^{1/3} + 2)]$
43. $f'(x)$ for $f(x) = \dfrac{6\sqrt[3]{x}}{x^2 - 3}$
44. y' for $y = \dfrac{2\sqrt{x}}{x^2 - 3x + 1}$
45. $D_x \dfrac{x^3 - 2x^2}{\sqrt[3]{x^2}}$
46. $\dfrac{dy}{dx}$ for $y = \dfrac{x^2 - 3x + 1}{\sqrt[4]{x}}$
47. $f'(x)$ for $f(x) = \dfrac{(2x^2 - 1)(x^2 + 3)}{x^2 + 1}$
48. y' for $y = \dfrac{2x - 1}{(x^3 + 2)(x^2 - 3)}$

Applications

Business & Economics

49. *Sales analysis.* The monthly sales S (in thousands) for a record album are given by

$$S(t) = \frac{200t}{t^2 + 36}$$

where t is the number of months since the release of the album.

(A) Find $S'(t)$, the rate of change of monthly sales with respect to time.
(B) Find $S(2)$ and $S'(2)$ and interpret.
(C) Find $S(8)$ and $S'(8)$ and interpret.

50. *Sales analysis.* A communications company has installed a cable television system in a city. The total number N (in thousands) of subscribers t months after the installation of the system is given by

$$N(t) = \frac{200t}{t + 5}$$

(A) Find $N'(t)$, the rate of change of total number of subscribers with respect to time.
(B) Find $N(5)$ and $N'(5)$ and interpret.
(C) Find $N(15)$ and $N'(15)$ and interpret.

51. *Price–demand function.* According to classical economic theory, the demand $d(x)$ for a commodity in a free market decreases as the price x increases. Suppose that the number $d(x)$ of transistor radios people are willing to buy per week in a given city at a price \$$x$ is given by

$$d(x) = \frac{50{,}000}{x^2 + 10x + 25} \qquad \$4 \leq x \leq \$15$$

(A) Find $d'(x)$, the rate of change of demand with respect to price change.
(B) Find $d'(5)$ and $d'(10)$ and interpret.

52. *Employee training.* A company producing computer components has established that on the average a new employee can assemble $N(t)$ components per day after t days of on-the-job training, as given by

$$N(t) = \frac{100t}{t + 9}$$

(A) Find $N'(t)$, the rate of change of units assembled with respect to time.
(B) Find $N'(1)$ and $N'(11)$ and interpret.

Life Sciences

53. *Medicine.* A drug is injected into the bloodstream of a patient through her right arm. The concentration of the drug in the bloodstream of the left arm t hours after the injection is given by

$$C(t) = \frac{0.14t}{t^2 + 1}$$

(A) Find $C'(t)$, the rate of change of drug concentration with respect to time.
(B) Find $C'(0.5)$ and $C'(3)$ and interpret.

54. *Drug sensitivity.* One hour after x milligrams of a particular drug are given to a person, the change in body temperature $T(x)$ in degrees Fahrenheit is given approximately by

$$T(x) = x^2\left(1 - \frac{x}{9}\right) \qquad 0 \leq x \leq 7$$

The rate at which T changes with respect to the size of the dosage x, $T'(x)$, is called the *sensitivity* of the body to the dosage.

(A) Find $T'(x)$, using the product rule.
(B) Find $T'(1)$, $T'(3)$, and $T'(6)$.

Social Sciences

55. *Learning.* In the early days of quantitative learning theory (around 1917), L. L. Thurstone found that a given person successfully accomplished $N(x)$ acts after x practice acts, as given by

$$N(x) = \frac{100x + 200}{x + 32}$$

(A) Find the rate of change of learning, $N'(x)$, with respect to the number of practice acts x.
(B) Find $N'(4)$ and $N'(68)$.

1-7 General Power Rule

- General Power Rule
- Combining Rules of Differentiation

General Power Rule

We have already made extensive use of the power rule:

$$D_x x^n = nx^{n-1} \qquad n \text{ any real number} \tag{1}$$

Now we want to generalize this rule so that we can differentiate functions of the form $[u(x)]^n$. Is (1) still valid if we replace x with a function $u(x)$? We begin by considering a simple example. Let $u(x) = 2x$ and $n = 4$. Then

$$[u(x)]^n = (2x)^4 = 2^4x^4 = 16x^4$$

and

$$D_x[u(x)]^n = D_x 16x^4 = 64x^3 \tag{2}$$

But

$$n[u(x)]^{n-1} = 4(2x)^3 = 32x^3 \tag{3}$$

Comparing (2) and (3), we see that

$$D_x[u(x)]^n \neq n[u(x)]^{n-1}$$

for this particular choice of u(x) and n. (In fact, it can be shown that the only time this last equation is valid is if u(x) = x.) Thus, we cannot generalize the power rule by simply substituting u(x) for x in (1).

How can we find a formula for $D_x[u(x)]^n$ where u(x) is an arbitrary differentiable function? Let us first find $D_x[u(x)]^2$ and $D_x[u(x)]^3$ to see if a general pattern emerges. Since $[u(x)]^2 = u(x)u(x)$, we use the product rule with v(x) = u(x) to write

$$\begin{aligned} D_x[u(x)]^2 &= D_x u(x)u(x) = u(x)u'(x) + u'(x)u(x) \\ &= 2u(x)u'(x) \qquad (4) \end{aligned}$$

Since $[u(x)]^3 = [u(x)]^2u(x)$, we now use the product rule with $v(x) = [u(x)]^2$ and (4) to write

$$\begin{aligned} D_x[u(x)]^3 &= D_x[u(x)]^2u(x) = [u(x)]^2D_xu(x) + u(x)D_x[u(x)]^2 \\ &= [u(x)]^2u'(x) + u(x)[2u(x)u'(x)] \\ &= 3[u(x)]^2u'(x) \end{aligned}$$

Continuing in this fashion, it can be shown that

$$D_x[u(x)]^n = n[u(x)]^{n-1}u'(x) \qquad n \text{ a positive integer} \qquad (5)$$

Using more advanced techniques, the formula in (5) can be established for all real numbers n. Thus, we have the *general power rule.*

General Power Rule

If n is any real number, then

$$D_x[u(x)]^n = n[u(x)]^{n-1}u'(x)$$

provided u′(x) exists. This rule is often written more compactly as

$$D_xu^n = u^{n-1}\frac{du}{dx} \qquad u = u(x)$$

Example 34 Find $f'(x)$:

(A) $f(x) = (3x + 1)^4$

(B) $f(x) = (x^3 + 4)^7$

(C) $f(x) = \dfrac{1}{(x^2 + x + 4)^3}$

(D) $f(x) = \sqrt{3 - x}$

Solutions

(A) $f(x) = (3x + 1)^4$ — Let $u = 3x + 1$, $n = 4$.

$$f'(x) = 4(3x+1)^3 D_x(3x+1) \qquad nu^{n-1}\frac{du}{dx}$$

$$= 4(3x+1)^3 3 \qquad \frac{du}{dx} = 3$$

$$= 12(3x+1)^3$$

(B) $f(x) = (x^3 + 4)^7$ — Let $u = (x^3 + 4)$, $n = 7$.

$$f'(x) = 7(x^3+4)^6 D_x(x^3+4) \qquad nu^{n-1}\frac{du}{dx}$$

$$= 7(x^3+4)^6 3x^2 \qquad \frac{du}{dx} = 3x^2$$

$$= 21x^2(x^3+4)^6$$

(C) $f(x) = \dfrac{1}{(x^2+x+4)^3} = (x^2+x+4)^{-3}$ — Let $u = x^2 + x + 4$, $n = -3$.

$$f'(x) = -3(x^2+x+4)^{-4} D_x(x^2+x+4) \qquad nu^{n-1}\frac{du}{dx}$$

$$= -3(x^2+x+4)^{-4}(2x+1) \qquad \frac{du}{dx} = 2x+1$$

$$= \frac{-3(2x+1)}{(x^2+x+4)^4}$$

(D) $f(x) = \sqrt{3-x} = (3-x)^{1/2}$ — Let $u = 3 - x$, $n = \frac{1}{2}$.

$$f'(x) = \frac{1}{2}(3-x)^{-1/2} D_x(3-x) \qquad nu^{n-1}\frac{du}{dx}$$

$$= \frac{1}{2}(3-x)^{-1/2}(-1) \qquad \frac{du}{dx} = -1$$

$$= -\frac{1}{2(3-x)^{1/2}} \quad \text{or} \quad -\frac{1}{2\sqrt{3-x}}$$

Problem 34 Find $f'(x)$:

(A) $f(x) = (5x + 2)^3$ (B) $f(x) = (x^4 - 5)^5$

(C) $f(x) = \dfrac{1}{(x^2+4)^2}$ (D) $f(x) = \sqrt{4 - x}$

Notice that we used two steps to differentiate each function in Example 34: First, we applied the general power rule; then we found du/dx. As you gain experience with the general power rule, you may want to combine these two steps. If you do this, be certain to multiply by du/dx. For example,

$D_x(x^5 + 1)^4 = 4(x^5 + 1)^3 5x^4$ — Correct

~~$D_x(x^5 + 1)^4 = 4(x^5 + 1)^3$~~ — Incorrect, $du/dx = 5x^4$ is missing

If we let $u(x) = x$, then $du/dx = 1$ and the general power rule reduces to the (ordinary) power rule discussed in Section 1-5. Compare the following:

$D_x x^n = nx^{n-1}$ Yes—power rule

$D_x u^n = nu^{n-1}\dfrac{du}{dx}$ Yes—general power rule

$D_x u^n = nu^{n-1}$ No, unless $u(x) = x$ and $du/dx = 1$

■ Combining Rules of Differentiation

The following examples illustrate the use of the general power rule in combination with other rules of differentiation.

Example 35 Find the line tangent to the graph of f at $x = 2$ for

$$f(x) = x^2\sqrt{2x + 12}$$

Solution

$$f(x) = x^2\sqrt{2x+12}$$

$$= x^2(2x+12)^{1/2}$$

Apply the product rule with $u = x^2$ and $v = (2x+12)^{1/2}$.

$$f'(x) = x^2 D_x(2x+12)^{1/2} + (2x+12)^{1/2}D_x x^2$$

Use the general power rule to differentiate $(2x+12)^{1/2}$ and the ordinary power rule to differentiate x^2.

$$= x^2\tfrac{1}{2}(2x+12)^{-1/2}(2) + (2x+12)^{1/2}(2x)$$

$$= \frac{x^2}{\sqrt{2x+12}} + 2x\sqrt{2x+12}$$

$$f'(2) = \frac{4}{\sqrt{16}} + 4\sqrt{16} = 1 + 16 = 17$$

$$f(2) = 4\sqrt{16} = 16$$

$(x_1, y_1) = (2, f(2)) = (2, 16)$ Point

$m = f'(2) = 17$ Slope

$y - 16 = 17(x - 2)$

$y = 17x - 18$ Tangent line

Problem 35 Find the line tangent to the graph of f at $x = 3$ for

$$f(x) = x\sqrt{15 - 2x}$$

Example 36 Find the values of x where the tangent line is horizontal for

$$f(x) = \frac{x^3}{(2 - 3x)^5}$$

Solution Use the quotient rule with $u = x^3$ and $v = (2 - 3x)^5$

$$f'(x) = \frac{(2-3x)^5 D_x x^3 - x^3 D_x (2-3x)^5}{[(2-3x)^5]^2}$$

Use the ordinary power rule to differentiate x^3 and the general power rule to differentiate $(2 - 3x)^5$.

$$= \frac{(2-3x)^5 3x^2 - x^3 5(2-3x)^4(-3)}{(2-3x)^{10}}$$

$$= \frac{(2-3x)^4 3x^2[(2-3x) + 5x]}{(2-3x)^{10}}$$

$$= \frac{3x^2(2+2x)}{(2-3x)^6} = \frac{6x^2(x+1)}{(2-3x)^6}$$

Since a fraction is 0 when the numerator is zero and the denominator is not, we see that $f'(x) = 0$ at $x = -1$ and $x = 0$. Thus, the graph of f will have horizontal tangent lines at $x = -1$ and $x = 0$.

Problem 36 Find the values of x where the tangent line is horizontal for

$$f(x) = \frac{x^3}{(3x-2)^2}$$

Example 37 Starting with the function f in Example 36, write f as a product and then differentiate.

Solution

$$f(x) = \frac{x^3}{(2-3x)^5} = x^3(2-3x)^{-5}$$

$$f'(x) = x^3 D_x(2-3x)^{-5} + (2-3x)^{-5} D_x x^3$$

$$= x^3(-5)(2-3x)^{-6}(-3) + (2-3x)^{-5} 3x^2$$

$$= 15x^3(2-3x)^{-6} + 3x^2(2-3x)^{-5}$$

At this point, we have an unsimplified form for $f'(x)$. This may be satisfactory for some purposes, but not for others. For example, if we need to solve the equation $f'(x) = 0$, we must perform the following algebraic simplifications.

$$f'(x) = \frac{15x^3}{(2-3x)^6} + \frac{3x^2}{(2-3x)^5}$$

$$= \frac{15x^3}{(2-3x)^6} + \frac{3x^2(2-3x)}{(2-3x)^6}$$

$$= \frac{15x^3 + 3x^2(2-3x)}{(2-3x)^6}$$

$$= \frac{3x^2(5x + 2 - 3x)}{(2-3x)^6}$$

$$= \frac{3x^2(2+2x)}{(2-3x)^6} = \frac{6x^2(1+x)}{(2-3x)^6}$$

Problem 37 Refer to the function f in Problem 36. Write f as a product and then differentiate. Do not simplify.

As Example 37 illustrates, any quotient can be converted to a product and differentiated by the product rule. However, if the derivative must be simplified, it is usually easier to use the quotient rule. (Compare the algebraic simplifications in Example 37 with those in Example 36.) There is one special case where using negative exponents is the preferred method —a fraction whose numerator is a constant.

Example 38 Find $f'(x)$ two ways for $f(x) = \frac{4}{(x^2+9)^3}$.

Solution Method I: Use the quotient rule:

$$f'(x) = \frac{(x^2+9)^3 D_x 4 - 4D_x(x^2+9)^3}{[(x^2+9)^3]^2}$$

$$= \frac{(x^2+9)^3(0) - 4[3(x^2+9)^2(2x)]}{(x^2+9)^6}$$

$$= \frac{-24x(x^2+9)^2}{(x^2+9)^6} = \frac{-24x}{(x^2+9)^4}$$

Method II: Rewrite, then use the general power rule:

$$f(x) = \frac{4}{(x^2+9)^3} = 4(x^2+9)^{-3}$$

$$f'(x) = 4(-3)(x^2+9)^{-4}(2x)$$

$$= \frac{-24x}{(x^2+9)^4}$$

Which method do you prefer?

Problem 38 Find $f'(x)$ two ways for $f(x) = \frac{5}{(x^3+1)^2}$.

Answers to Matched Problems

34. (A) $15(5x+2)^2$ (B) $20x^3(x^4-5)^4$
(C) $-4x/(x^2+4)^3$ (D) $-1/(2\sqrt{4-x})$
35. $y = 2x + 3$ 36. $x = 0, x = 2$
37. $-6x^3(3x-2)^{-3} + 3x^2(3x-2)^{-2}$
38. $-30x^2/(x^3+1)^3$

Exercise 1-7

A *Find $f'(x)$ using the general power rule.*

1. $f(x) = (2x+5)^3$ $3(2x+5)^2(2)$
2. $f(x) = (3x-7)^5$
3. $f(x) = (5-2x)^4$
4. $f(x) = (9-5x)^2$

5. $f(x) = (3x^2 + 5)^5$
6. $f(x) = (5x^2 - 3)^6$
7. $f(x) = (x^3 - 2x^2 + 2)^8$
8. $f(x) = (2x^2 + x + 1)^7$
9. $f(x) = (2x - 5)^{1/2}$ $\frac{1}{2}(2x-5)^{-1/2}(2)$
10. $f(x) = (4x + 3)^{1/2}$
11. $f(x) = (x^4 + 1)^{-2}$
12. $f(x) = (x^5 + 2)^{-3}$

B *Find dy/dx using the general power rule.*

13. $y = 3(x^2 - 2)^4$
14. $y = 2(x^3 + 6)^5$
15. $y = 2(x^2 + 3x)^{-3}$
16. $y = 3(x^3 + x^2)^{-2}$
17. $y = \sqrt{x^2 + 8}$
18. $y = \sqrt[3]{3x - 7}$
19. $y = \sqrt[3]{3x + 4}$
20. $y = \sqrt{2x - 5}$
21. $y = (x^2 - 4x + 2)^{1/2}$
22. $y = (2x^2 + 2x - 3)^{1/2}$
23. $y = \dfrac{1}{2x + 4}$
24. $y = \dfrac{1}{3x - 7}$
25. $y = \dfrac{1}{(x^3 + 4)^5}$
26. $y = \dfrac{1}{(x^2 - 3)^6}$
27. $y = \dfrac{1}{4x^2 - 4x + 1}$
28. $y = \dfrac{1}{2x^2 - 3x + 1}$
29. $y = \dfrac{4}{\sqrt{x^2 - 3x}}$
30. $y = \dfrac{3}{\sqrt[3]{x - x^2}}$

In Problems 31–36, find f′(x) and find the equation of the line tangent to the graph of f at the indicated value of x.

31. $f(x) = x(4 - x)^3, \quad x = 2$
32. $f(x) = x^2(1 - x)^4, \quad x = 2$
33. $f(x) = \dfrac{x}{(2x - 5)^3}, \quad x = 3$
34. $f(x) = \dfrac{x^4}{(3x - 8)^2}, \quad x = 4$
35. $f(x) = x\sqrt{2x + 2}, \quad x = 1$
36. $f(x) = x\sqrt{x - 6}, \quad x = 7$

In Problems 37–42, find f′(x) and find the values of x where the tangent line is horizontal.

37. $f(x) = x^2(x - 5)^3$
38. $f(x) = x^3(x - 7)^4$
39. $f(x) = \dfrac{x}{(2x + 5)^2}$
40. $f(x) = \dfrac{x - 1}{(x - 3)^3}$
41. $f(x) = \sqrt{x^2 - 8x + 20}$
42. $f(x) = \sqrt{x^2 + 4x + 5}$

C *Find each derivative and simplify.*

43. $D_x[3x(x^2 + 1)^3]$
44. $D_x[2x^2(x^3 - 3)^4]$
45. $D_x \dfrac{(x^3 - 7)^4}{2x^3}$
46. $D_x \dfrac{3x^2}{(x^2 + 5)^3}$
47. $D_x[(2x - 3)^2(2x^2 + 1)^3]$
48. $D_x[(x^2 - 1)^3(x^2 - 2)^2]$
49. $D_x[4x^2\sqrt{x^2 - 1}]$
50. $D_x[3x\sqrt{2x^2 + 3}]$

51. $D_x \dfrac{2x}{\sqrt{x-3}}$

52. $D_x \dfrac{x^2}{\sqrt{x^2+1}}$

53. $D_x \sqrt{(2x-1)^3(x^2+3)^4}$

54. $D_x \sqrt{\dfrac{4x+1}{2x^2+1}}$

Applications

Business & Economics

55. *Marginal average cost.* A manufacturer of skis finds that the average cost $\overline{C}(x)$ per pair of skis at an output level of x thousand skis is

$$\overline{C}(x) = (2x-8)^2 + 25$$

(A) Find the marginal average cost $\overline{C}'(x)$ using the general power rule.
(B) Find $\overline{C}'(2)$, $\overline{C}'(4)$, and $\overline{C}'(6)$.

56. *Compound interest.* If \$100 is invested at an interest rate of i compounded semiannually, the amount in the account at the end of 5 years is given by

$$A = 100\left(1 + \frac{1}{2}i\right)^{10}$$

Find dA/di.

Life Sciences

57. *Bacteria growth.* The number y of bacteria in a certain colony after x days is given approximately by

$$y = (3 \times 10^6)\left(1 - \frac{1}{\sqrt[3]{(x^2-1)^2}}\right)$$

Find dy/dx.

58. *Pollution.* A small lake in a resort area became contaminated with harmful bacteria because of excessive septic tank seepage. After treating the lake with a bactericide, the Department of Public Health estimated the bacteria concentration (number per cubic centimeter) after t days to be given by

$$C(t) = 500(8-t)^2 \qquad 0 \leq t \leq 7$$

(A) Find $C'(t)$ using the general power rule.
(B) Find $C'(1)$ and $C'(6)$, and interpret.

Social Sciences

59. *Learning.* In 1930, L. L. Thurstone developed the following formula to indicate how learning time T depends on the length of a list n:

$$T = f(n) = \frac{c}{k} n\sqrt{n-a}$$

where a, c, and k are empirical constants. Suppose for a particular person, time T in minutes for learning a list of length n is

$$T = f(n) = 2n\sqrt{n-2}$$

(A) Find dT/dn, the rate of change in time with respect to n.
(B) Find $f'(11)$ and $f'(27)$, and interpret.

1-8 Marginal Analysis in Business and Economics

- Marginal Cost, Revenue, and Profit
- Application
- Marginal Average Cost, Revenue, and Profit

Marginal Cost, Revenue, and Profit

One important use of calculus in business and economics is in marginal analysis. We introduced the concept of marginal cost earlier. There is no reason to stop there. Economists also talk about **marginal revenue** and **marginal profit.** Recall that the word *marginal* refers to a rate of change—that is, a derivative. Thus, we define the following:

Marginal Cost, Revenue, and Profit

If x is the number of units of product produced in some time interval, then

$$\begin{aligned} \text{Total cost} &= C(x) \\ \text{Marginal cost} &= C'(x) \\ \text{Total revenue} &= R(x) \\ \text{Marginal revenue} &= R'(x) \\ \text{Total profit} &= P(x) = R(x) - C(x) \\ \text{Marginal profit} &= P'(x) = R'(x) - C'(x) \\ &= (\text{Marginal revenue}) - (\text{Marginal cost}) \end{aligned}$$

Marginal functions have several important economic interpretations. We will discuss these interpretations in terms of the marginal cost function. Similar statements can be made for marginal revenue and marginal profit.

We have already seen that marginal cost is the rate of change of total cost per unit change in production at a given level of production. As Figure 19 illustrates, this implies that **the marginal cost approximates the change in total cost that results from a unit change in production.**

Since $C(x)$ is the total cost of producing x units and $C(x + 1)$ is the total cost of producing $x + 1$ units, the change in total cost

$$\begin{matrix} & \begin{pmatrix}\text{Cost for}\\ x+1 \text{ units}\end{pmatrix} & - & \begin{pmatrix}\text{Cost for}\\ x \text{ units}\end{pmatrix} \\ \Delta C = & C(x+1) & - & C(x)\end{matrix}$$

is also the cost of producing the $(x + 1)$st item. Thus, **the marginal cost $C'(x)$ also approximates the cost of producing the $(x + 1)$st item.**

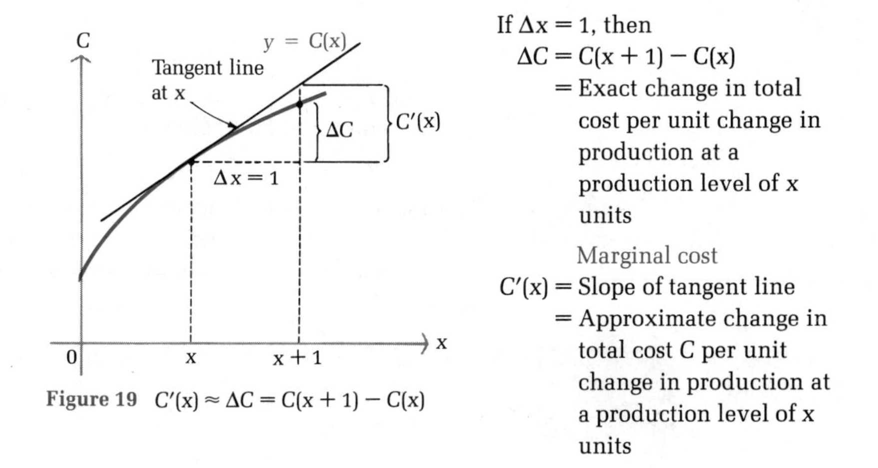

Figure 19 $C'(x) \approx \Delta C = C(x + 1) - C(x)$

Example 39 Marginal Cost

A small machine shop manufactures drill bits used in the petroleum industry. The shop manager estimates that the total daily cost in dollars of producing x bits is

$$C(x) = 1{,}000 + 25x - \frac{x^2}{10}$$

(A) Find $C'(10)$ and interpret.
(B) Find $C(11) - C(10)$ and interpret.

Solutions

(A)
$$C'(x) = 25 - \frac{x}{5} \qquad \text{Marginal cost function}$$

$$C'(10) = 25 - 2 \qquad \text{Marginal cost at a production level of 10 bits}$$

$$= 23$$

At a production level of 10 bits, a unit increase in production will increase total production costs by approximately \$23. Also, the cost of producing the 11th bit is approximately \$23.

(B) $C(10) = \$1{,}240$ Total cost of producing 10 bits

$C(11) = \$1{,}262.90$ Total cost of producing 11 bits

$\Delta C = C(11) - C(10) = \$22.90$

At a production level of 10 bits, a unit increase in production will increase total production costs by exactly \$22.90. Also, the cost of producing the 11th bit is exactly \$22.90.

Problem 39 Refer to the total cost function in Example 39.

(A) Find $C'(20)$ and interpret.
(B) Find $C(21) - C(20)$ and interpret.

▪ Application

We now present an example in market research to show how marginal cost, revenue, and profit are tied together.

Example 40
Production Strategy

The market research department of a company recommends that the company manufacture and market a new transistor radio. After suitable test marketing, the research department presents the following **demand equation:**

$$x = 10{,}000 - 1{,}000p \qquad x \text{ is demand at \$p per radio} \tag{1}$$

or

$$p = 10 - \frac{x}{1{,}000} \tag{2}$$

where x is the number of radios retailers are likely to buy per week at \$p per radio. Equation (2) is simply equation (1) solved for p in terms of x. Notice that as price goes up, demand goes down.

The financial department provides the following **cost equation:**

$$C(x) = 7{,}000 + 2x \tag{3}$$

where \$7,000 is the estimated fixed costs (tooling and overhead) and \$2 is the estimated variable costs (cost per unit for materials, labor, marketing, transportation, storage, etc.).

The **marginal cost** is

$$C'(x) = 2$$

Since this is a constant, it costs an additional \$2 to produce one more radio at all production levels.

The **revenue** (the amount of money R received by the company for manufacturing and selling x units at \$p per unit) is

$$R = (\text{Number of units sold})(\text{Price per unit}) = xp$$

In general, the revenue R can be expressed in terms of p by using equation (1) or in terms of x by using equation (2). In marginal analysis (problems involving marginal cost, marginal revenue, or marginal profit), cost, revenue, and profit must be expressed in terms of the number of units x. Thus, the **revenue equation** in terms of x is

$$R(x) = xp = x\left(10 - \frac{x}{1{,}000}\right) \qquad \text{Using equation (2)} \qquad (4)$$

$$= 10x - \frac{x^2}{1{,}000}$$

The **marginal revenue** is

$$R'(x) = 10 - \frac{x}{500}$$

For production levels of $x = 2{,}000$, 5,000, and 7,000, we have

$$R'(2{,}000) = 6 \qquad R'(5{,}000) = 0 \qquad R'(7{,}000) = -4$$

This means that at production levels of 2,000, 5,000, and 7,000, the respective approximate changes in revenue per unit change in production are \$6, \$0, and −\$4. That is, at the 2,000 output level revenue increases as production increases; at the 5,000 output level revenue does not change with a "small" change in production; and at the 7,000 output level revenue decreases with an increase in production.

When we graph $R(x)$ and $C(x)$ in the same coordinate system, we obtain Figure 20.

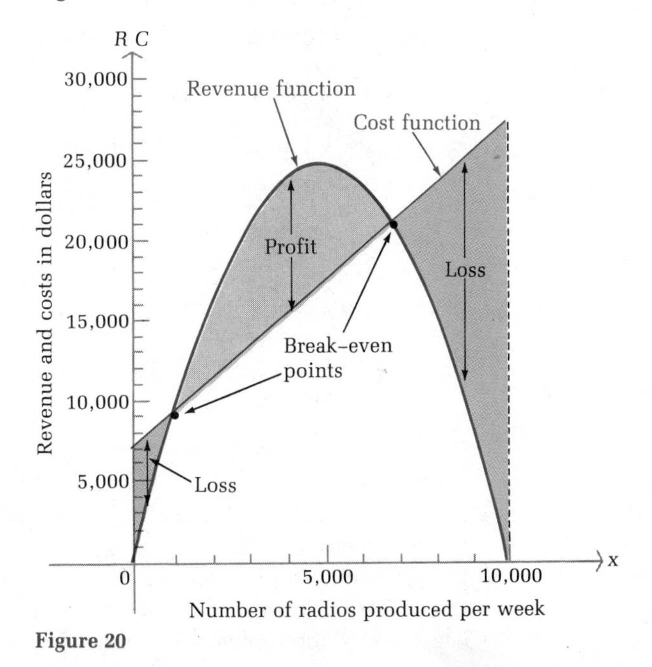

Figure 20

The break-even points (the points where revenue equals cost) are obtained as follows:

$$C(x) = R(x)$$

$$7{,}000 + 2x = 10x - \frac{x^2}{1{,}000}$$

$$\frac{x^2}{1{,}000} - 8x + 7{,}000 = 0$$

$$x^2 - 8{,}000x + 7{,}000{,}000 = 0$$ Solve using the quadratic formula. (See Section 0-3.)

$$\begin{aligned} x &= \frac{8{,}000 \pm \sqrt{8{,}000^2 - 4(7{,}000{,}000)}}{2} \\ &= \frac{8{,}000 \pm \sqrt{36{,}000{,}000}}{2} \\ &= \frac{8{,}000 \pm 6{,}000}{2} \\ &= 1{,}000, \quad 7{,}000 \end{aligned}$$

$$R(1{,}000) = 10(1{,}000) - \frac{1{,}000^2}{1{,}000} = 9{,}000$$

$$C(1{,}000) = 7{,}000 + 2(1{,}000) = 9{,}000$$

$$R(7{,}000) = 10(7{,}000) - \frac{7{,}000^2}{1{,}000} = 21{,}000$$

$$C(7{,}000) = 7{,}000 + 2(7{,}000) = 21{,}000$$

Thus, the break-even points are (1,000, 9,000) and (7,000, 21,000).

The **profit equation** is

$$\begin{aligned} P(x) &= R(x) - C(x) \\ &= \left(10x - \frac{x^2}{1{,}000}\right) - (7{,}000 + 2x) \\ &= -\frac{x^2}{1{,}000} + 8x - 7{,}000 \end{aligned}$$

The graph in Figure 20 also provides some useful information concerning the profit equation. At a production level of 1,000 or 7,000, revenue equals cost; hence, profit is 0 and the company will break even. For any production level between 1,000 and 7,000, revenue is greater than cost; hence, $P(x)$ is positive and the company will make a profit. For production levels less than 1,000 or greater than 7,000, revenue is less than cost; hence, $P(x)$ is negative and the company will have a loss.

The **marginal profit** is

$$P'(x) = -\frac{x}{500} + 8$$

For production levels of 1,000, 4,000, and 6,000, we have

$$P'(1{,}000) = 6 \qquad P'(4{,}000) = 0 \qquad P'(6{,}000) = -4$$

This means that at production levels of 1,000, 4,000, and 6,000, the respective approximate changes in profit per unit change in production are \$6, \$0, and −\$4. That is, at the 1,000 output level profit will be increased if production is increased; at the 4,000 output level profit does not change for "small" changes in production; and at the 6,000 output level profits will decrease if production is increased. It seems the best production level to produce a maximum profit is 4,000. [In the next chapter we will develop a systematic procedure for finding the production level (and, using the demand equation, the selling price) that will maximize profit.] This example warrants careful study, since a number of important ideas in economics and calculus are involved.

Problem 40 Refer to the revenue and profit equations in Example 40.

(A) Find $R'(3{,}000)$ and $R'(6{,}000)$ and interpret.
(B) Find $P'(2{,}000)$ and $P'(7{,}000)$ and interpret.

■ Marginal Average Cost, Revenue, and Profit

Sometimes it is desirable to carry out marginal analysis relative to **average cost (cost per unit), average revenue (revenue per unit), and average profit (profit per unit).** The relevant definitions are summarized in the following box:

Marginal Average Cost, Revenue, and Profit

If x is the number of units of a product produced in some time interval, then

$$\text{Average cost} = \overline{C}(x) = \frac{C(x)}{x} \qquad \text{Cost per unit}$$

$$\text{Marginal average cost} = \overline{C}'(x) = D_x\overline{C}(x)$$

$$\text{Average revenue} = \overline{R}(x) = \frac{R(x)}{x} \qquad \text{Revenue per unit}$$

$$\text{Marginal average revenue} = \overline{R}'(x) = D_x\overline{R}(x)$$

$$\text{Average profit} = \overline{P}(x) = \frac{P(x)}{x} \qquad \text{Profit per unit}$$

$$\text{Marginal average profit} = \overline{P}'(x) = D_x\overline{P}(x)$$

As was the case with marginal cost, **the marginal average cost approximates the change in average cost that results from a unit increase in production.** Similar statements can be made for marginal average revenue and marginal average profit.

Example 41 Referring to Example 39, we have

$$C(x) = 1{,}000 + 25x - \frac{x^2}{10}$$ Total cost function

$$\overline{C}(x) = \frac{C(x)}{x} = \frac{1{,}000}{x} + 25 - \frac{x}{10}$$ Average cost function

$$\overline{C}'(x) = D_x\overline{C}(x) = -\frac{1{,}000}{x^2} - \frac{1}{10}$$ Marginal average cost function

$$\overline{C}(10) = \frac{1{,}000}{10} + 25 - \frac{10}{10} = \$124$$ Average cost per unit if 10 units are produced

$$\overline{C}'(10) = -\frac{1{,}000}{100} - \frac{1}{10} = -\$10.10$$ A unit increase in production will decrease the average cost per unit by approximately \$10.10 at a production level of 10 units.

Problem 41 Let $C(x) = 7{,}000 + 2x$ be the cost function considered in Example 40.

(A) Find $\overline{C}(x)$ and $\overline{C}'(x)$.
(B) Find $\overline{C}(1{,}000)$ and $\overline{C}'(1{,}000)$ and interpret.

Answers to Matched Problems

39. (A) $C'(20) = 21$. At a production level of 20 bits, a unit increase in production will increase total production costs by approximately \$21. Also, the cost of producing the 21st bit is approximately \$21.
    (B) $C(21) - C(20) = 20.90$. At a production level of 20 bits, a unit increase in production will increase total production costs by exactly \$20.90. Also, the cost of producing the 21st bit is exactly \$20.90.
40. (A) $R'(3{,}000) = 4$. At a production level of 3,000, a unit increase in production will increase revenue by approximately \$4. $R'(6{,}000) = -2$. At a production level of 6,000, a unit increase in production will decrease revenue by approximately \$2.
    (B) $P'(2{,}000) = 4$. At a production level of 2,000, a unit increase in production will increase profit by approximately \$4. $P'(7{,}000) = -6$. At a production level of 7,000, a unit increase in production will decrease profit by approximately \$6.
41. (A) $\overline{C}(x) = \frac{7{,}000}{x} + 2,\quad \overline{C}'(x) = -\frac{7{,}000}{x^2}$

(B) $\overline{C}(1{,}000) = 9$. At a production level of 1,000, the average cost per unit is \$9. $\overline{C}'(1{,}000) = -0.007$. At a production level of 1,000, a unit increase in production will decrease the average cost per unit by approximately 0.7¢.

Exercise 1-8

Applications

1. *Cost analysis.* The total cost in dollars of producing x food processors is

$$C(x) = 2{,}000 + 50x - \frac{x^2}{2}$$

(A) Find the exact cost of producing the 21st food processor.
(B) Use the marginal cost to approximate the cost of producing the 21st food processor.

2. *Cost analysis.* The total cost in dollars of producing x electric guitars is

$$C(x) = 1{,}000 + 100x - \frac{x^2}{4}$$

(A) Find the exact cost of producing the 51st guitar.
(B) Use the marginal cost to approximate the cost of producing the 51st guitar.

3. *Cost analysis.* The total cost in dollars of manufacturing x auto body frames is

$$C(x) = 60{,}000 + 300x$$

(A) Find the average cost per unit if 500 frames are produced.
(B) Find the marginal average cost at a production level of 500 units and interpret.

4. *Cost analysis.* The total cost in dollars of printing x dictionaries is

$$C(x) = 20{,}000 + 10x$$

(A) Find the average cost per unit if 1,000 dictionaries are produced.
(B) Find the marginal average cost at a production level of 1,000 units and interpret.

5. *Revenue analysis.* The total revenue in dollars from the sale of x clock

radios is

$$R(x) = 100x - \frac{x^2}{40}$$

Evaluate the marginal revenue at the given values of x and interpret the results.

(A) $x = 1{,}600$ (B) $x = 2{,}500$

6. *Revenue analysis.* The total revenue in dollars from the sale of x steam irons is

$$R(x) = 50x - \frac{x^2}{20}$$

Evaluate the marginal revenue at the given values of x and interpret the results.

(A) $x = 400$ (B) $x = 650$

7. *Profit analysis.* The total profit in dollars from the sale of x skateboards is

$$P(x) = 30x - \frac{x^2}{2} - 250$$

(A) Find the exact profit from the sale of the 26th skateboard.
(B) Use the marginal profit to approximate the profit from the sale of the 26th skateboard.

8. *Profit analysis.* The total profit in dollars from the sale of x portable stereos is

$$P(x) = 22x - \frac{x^2}{10} - 400$$

(A) Find the exact profit from the sale of the 41st stereo.
(B) Use the marginal profit to approximate the profit from the sale of the 41st stereo.

9. *Profit analysis.* The total profit in dollars from the sale of x video cassettes is

$$P(x) = 5x - \frac{x^2}{200} - 450$$

Evaluate the marginal profit at the given values of x and interpret the results.

(A) $x = 450$ (B) $x = 750$

10. *Profit analysis.* The total profit in dollars from the sale of x cameras is

$$P(x) = 12x - \frac{x^2}{50} - 1{,}000$$

Evaluate the marginal profit at the given values of x and interpret the results.

(A) $x = 200$ (B) $x = 350$

11. *Profit analysis.* Refer to the profit equation in Problem 9.

(A) Find the average profit per unit if 150 cassettes are produced.
(B) Find the marginal average profit at a production level of 150 units and interpret.

12. *Profit analysis.* Refer to the profit equation in Problem 10.

(A) Find the average profit per unit if 200 cameras are produced.
(B) Find the marginal average profit at a production level of 200 units and interpret.

13. *Revenue, cost, and profit.* In Example 40, suppose we have the demand equation

$$x = 6{,}000 - 30p \qquad \text{or} \qquad p = 200 - \frac{x}{30}$$

and the cost equation

$$C(x) = 72{,}000 + 60x$$

(A) Find the marginal cost.
(B) Find the revenue equation in terms of x.
(C) Find the marginal revenue.
(D) Find $R'(1{,}500)$ and $R'(4{,}500)$, and interpret.
(E) Graph the cost function and the revenue function on the same coordinate system for $0 \leq x \leq 6{,}000$. Find the break-even points, and indicate regions of loss and profit.
(F) Find the profit equation in terms of x.
(G) Find the marginal profit.
(H) Find $P'(1{,}500)$ and $P'(3{,}000)$, and interpret.

14. *Revenue, cost, and profit.* In Example 40, suppose we have the demand equation

$$x = 9{,}000 - 30p \qquad \text{or} \qquad p = 300 - \frac{x}{30}$$

and the cost equation

$$C(x) = 150{,}000 + 30x$$

(A) Find the marginal cost.
(B) Find the revenue equation in terms of x.
(C) Find the marginal revenue.
(D) Find $R'(3{,}000)$ and $R'(6{,}000)$, and interpret.

(E) Graph the cost function and the revenue function on the same coordinate system for $0 \leq x \leq 9{,}000$. Find the break-even points, and indicate regions of loss and profit.
(F) Find the profit equation in terms of x.
(G) Find the marginal profit.
(H) Find $P'(1{,}500)$ and $P'(4{,}500)$, and interpret.

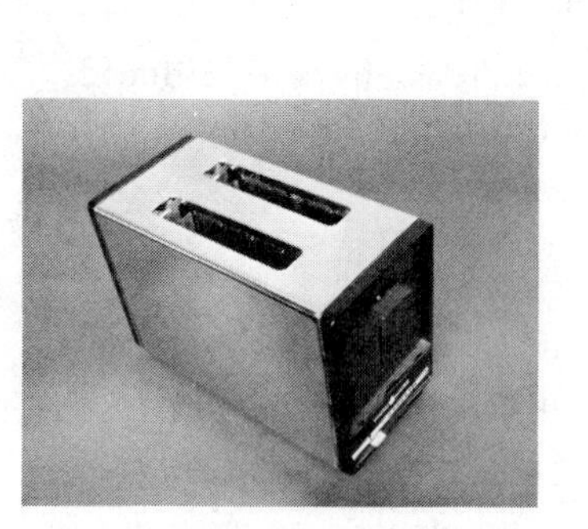

15. *Revenue, cost, and profit.* A company is planning to manufacture and market a new two-slice electric toaster. After conducting extensive market surveys, the research department provides the following estimates: a weekly demand of 200 toasters at a price of $16 per toaster and a weekly demand of 300 toasters at a price of $14 per toaster. The financial department estimates that weekly fixed costs will be $1,400 and the variable costs (cost per unit) will be $4.

(A) Assume that the demand equation is a linear equation of the form $p = mx + b$ and use the research department's estimates to find m and b.
(B) Find the revenue equation in terms of x.
(C) Assume that the cost equation is a linear equation of the form $C(x) = mx + b$ and use the financial department's estimates to find m and b.
(D) Graph the cost function and the revenue function on the same coordinate system for $0 \leq x \leq 1{,}000$. Find the break-even points and indicate regions of loss and profit.
(E) Find the profit equation in terms of x.
(F) Evaluate the marginal profit at $x = 250$ and $x = 475$, and interpret the results.

16. *Revenue, cost, and profit.* The company in Problem 15 is also planning to manufacture and market a four-slice toaster. For this toaster, the research department's estimates are a weekly demand of 300 toasters at a price of $25 per toaster and a weekly demand of 400 toasters at a price of $20. The financial department's estimates are fixed weekly costs of $5,000 and variable costs of $5 per toaster. Assume the demand and cost equations are linear. (See Problem 15, parts A and C.)

(A) Use the research department's estimates to find the demand equation.
(B) Find the revenue equation in terms of x.
(C) Use the financial department's estimates to find the cost equation in terms of x.
(D) Graph the cost function and the revenue function on the same coordinate system for $0 \leq x \leq 800$. Find the break-even points and indicate regions of loss and profit.
(E) Find the profit equation in terms of x.

(F) Evaluate the marginal profit at $x = 325$ and $x = 425$, and interpret the results.

1-9 Chapter Review

Important Terms and Symbols

1-1 *Limits.* Tangent line, secant line, limit of $f(x)$ as x approaches c, vertical asymptote, polynomial function, limit of $f(x)$ as x approaches infinity, horizontal asymptote, rational function, $\lim_{x \to c} f(x)$, $\lim_{x \to \pm\infty} f(x)$

1-2 *Continuity.* Continuous curve, discontinuous curve, continuity at a point, continuity on an open interval, discontinuity at a point, continuity properties, solving nonlinear inequalities, test number, sign chart

1-3 *Increments, tangent lines, and rates of change.* Increments, slope, tangent line, slope of graph at a point, average rate of change, instantaneous rate of change, difference quotient, average velocity, instantaneous velocity, Δx, Δy, average rate $= \Delta y/\Delta x$, instantaneous rate $= \lim_{\Delta x \to 0} \Delta y/\Delta x$

1-4 *The derivative.* The derivative of f at x, tangent line, differentiable function, nonexistence of the derivative, nondifferentiable at $x = a$, instantaneous rates of change, marginal cost, $f'(x)$

1-5 *Derivatives of constants, power forms, and sums.* Derivative notation, derivative of a constant, power rule, derivative of a constant times a function, derivatives of sums and differences, $f'(x)$, y', dy/dx, $D_x f(x)$

1-6 *Derivatives of products and quotients.* Derivatives of products, product rule, derivatives of quotients, quotient rule

1-7 *General power rule.* General power rule, combining rules of differentiation

1-8 *Marginal analysis in business and economics.* Demand equation, cost equation, marginal cost, revenue equation, marginal revenue, break-even points, profit equation, marginal profit, average cost, marginal average cost, average revenue, marginal average revenue, average profit, marginal average profit, $C'(x)$, $\overline{C}'(x)$, $R'(x)$, $\overline{R}'(x)$, $P'(x)$, $\overline{P}'(x)$

Summary of Rules of Differentiation

$$D_x C = 0$$

$$D_x x^n = nx^{n-1}$$

$$D_x kf(x) = kf'(x)$$

$$D_x[u(x) \pm v(x)] = u'(x) \pm v'(x)$$

$$D_x[u(x)v(x)] = u(x)v'(x) + v(x)u'(x)$$

$$D_x \frac{u(x)}{v(x)} = \frac{v(x)u'(x) - u(x)v'(x)}{[v(x)]^2}$$

$$D_x[u(x)]^n = n[u(x)]^{n-1}u'(x)$$

Exercise 1-9 Chapter Review

Work through all the problems in this chapter review and check your answers in the back of the book. (Answers to all review problems are there.) Where weaknesses show up, review appropriate sections in the text.

A *In Problems 1–10 find $f'(x)$ for $f(x)$ as given.*

1. $f(x) = 3x^4 - 2x^2 + 1$
2. $f(x) = 2x^{1/2} - 3x$
3. $f(x) = 5$
4. $f(x) = \frac{1}{2x^2} + \frac{x^2}{2}$
5. $f(x) = (2x - 1)(3x + 2)$
6. $f(x) = (x^2 - 1)(x^3 - 3)$
7. $f(x) = \frac{2x}{x^2 + 2}$
8. $f(x) = \frac{1}{3x + 2}$
9. $f(x) = (2x - 3)^3$
10. $f(x) = (x^2 + 2)^{-2}$

B *In Problems 11–18 find the indicated derivatives.*

11. $\frac{dy}{dx}$ for $y = 3x^4 - 2x^{-3} + 5$
12. y' for $y = (2x^2 - 3x + 2)(x^2 + 2x - 1)$
13. $f'(x)$ for $f(x) = \frac{2x - 3}{(x - 1)^2}$
14. y' for $y = 2\sqrt{x} + \frac{4}{\sqrt{x}}$
15. $D_x[(x^2 - 1)(2x + 1)^2]$
16. $D_x\sqrt[3]{x^3 - 5}$
17. $\frac{dy}{dx}$ for $y = \frac{3x^2 + 4}{x^2}$
18. $D_x\frac{(x^2 + 2)^4}{2x - 3}$
19. For $y = f(x) = x^2 + 4$, find:
    (A) The slope of the graph at $x = 1$
    (B) The equation of the tangent line at $x = 1$ in the form $y = mx + b$
20. Repeat Problem 19 for $f(x) = x^3(x + 1)^2$.

In Problems 21–24, find the values of x where the tangent line is horizontal.

21. $f(x) = 10x - x^2$
22. $f(x) = (x + 3)(x^2 - 45)$
23. $f(x) = \frac{x}{x^2 + 4}$
24. $f(x) = x^2(2x - 15)^3$

25. If an object moves along the y axis (scale in feet) so that it is at $y = f(x) = 16x^2 - 4x$ at time x (in seconds), find:

    (A) The instantaneous velocity function
    (B) The velocity at time $x = 3$ seconds

26. An object moves along the y axis (scale in feet) so that at time x (in seconds) it is at $y = f(x) = 96x - 16x^2$. Find:

    (A) The instantaneous velocity function
    (B) The time(s) when the velocity is 0

Solve each inequality. Express the answer in inequality notation.

27. $x^2 - x - 12 < 0$

28. $x^2 - 2x - 8 \geqslant 0$

29. $x^2 + 25 \geqslant 10x$

30. $\dfrac{x-5}{x^2+3x} \leqslant 0$

31. $\dfrac{x(x-2)^2}{(x-5)^3} < 0$

32. $(x+2)^3(x-1)^2(x-3) > 0$

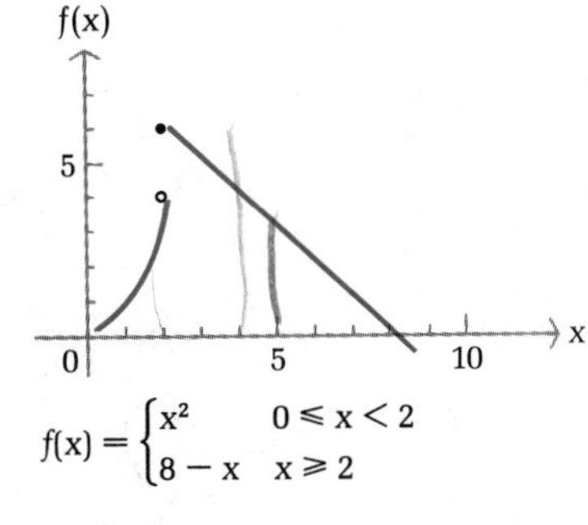

$$f(x) = \begin{cases} x^2 & 0 \leqslant x < 2 \\ 8 - x & x \geqslant 2 \end{cases}$$

Problems 33 and 34 refer to the function f described in the figure.

33. (A) $\lim_{x \to 2} f(x) = ?$ (B) $f(2) = ?$
    (C) Is f continuous at $x = 2$?

34. (A) $\lim_{x \to 5} f(x) = ?$ (B) $f(5) = ?$
    (C) Is f continuous at $x = 5$?

In Problems 35–40, determine where f is continuous. Express the answer in interval notation.

35. $f(x) = 2x^2 - 3x + 1$

36. $f(x) = \dfrac{1}{x+5}$

37. $f(x) = \dfrac{x-3}{x^2-x-6}$

38. $f(x) = \sqrt{x-3}$

39. $f(x) = \sqrt[3]{1-x^2}$

40. $f(x) = \sqrt{20 + x - x^2}$

In Problems 41–52, find each limit if it exists.

41. $\lim_{x \to 3} \dfrac{2x-3}{x+5}$

42. $\lim_{x \to 3} (2x^2 - x + 1)$

43. $\lim_{x \to 0} \dfrac{2x}{3x^2 - 2x}$

44. $\lim_{\Delta x \to 0} \dfrac{f(2 + \Delta x) - f(2)}{\Delta x}$ for $f(x) = x^2 + 4$

45. $\lim_{x\to 3} \frac{x-3}{x^2-9}$

46. $\lim_{x\to -3} \frac{x-3}{x^2-9}$

47. $\lim_{x\to 7} \frac{\sqrt{x}-\sqrt{7}}{x-7}$

48. $\lim_{x\to -2} \sqrt{\frac{x^2+4}{2-x}}$

49. $\lim_{x\to\infty} \left(3+\frac{1}{x^{1/3}}+\frac{2}{x^3}\right)$

50. $\lim_{x\to\infty} \frac{2x^2+3}{3x^2+2}$

51. $\lim_{x\to\infty} \frac{2x+3}{3x^2+2}$

52. $\lim_{x\to\infty} \frac{2x^2+3}{3x+2}$

In Problems 53 and 54, use the definition of the derivative to find $f'(x)$.

53. $f(x) = x^2 - x$

54. $f(x) = \sqrt{x} - 3$

C *Problems 55–58 refer to the function f in the figure. Determine if f is differentiable at the indicated value of x.*

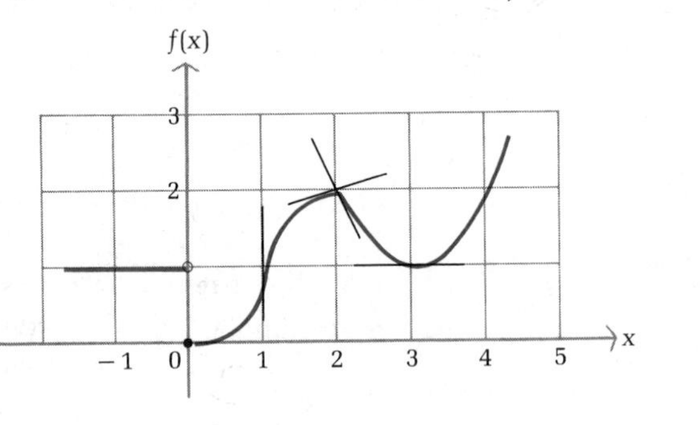

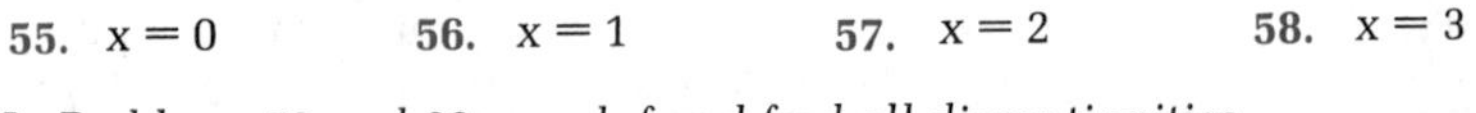

55. $x = 0$

56. $x = 1$

57. $x = 2$

58. $x = 3$

In Problems 59 and 60, graph f and find all discontinuities.

59. $f(x) = \begin{cases} 4 - x^2 & x < 0 \\ 2 + x^2 & x \geq 0 \end{cases}$

60. $f(x) = \begin{cases} 4 - x^2 & x < 1 \\ 2 + x^2 & x \geq 1 \end{cases}$

Problems 61–63 refer to

$$f(x) = \frac{2x^2 - 3x - 2}{3x^2 - 4x - 4}$$

61. (A) $\lim_{x\to 2} f(x) = ?$ (B) $f(2) = ?$ (C) Is f continuous at $x = 2$?

62. (A) $\lim_{x\to 0} f(x) = ?$ (B) $f(0) = ?$ (C) Is f continuous at $x = 0$?

63. Find all points of discontinuity for f.

In Problems 64–67, find $f'(x)$ and simplify.

64. $f(x) = (x-4)^4(x+3)^3$

65. $f(x) = \frac{x^5}{(2x+1)^4}$

66. $f(x) = \frac{\sqrt{x^2-1}}{x}$

67. $f(x) = \frac{x}{\sqrt{x^2+4}}$

Applications

Business & Economics

68. *Profit/loss analysis.* Let

$$p = 14 - x \quad \text{and} \quad C = 2x + 20 \qquad 0 \leq x \leq 14$$

be the demand equation and the cost equation, respectively, for a certain commodity.

(A) Express the cost C in terms of the price p.
(B) Express the revenue R in terms of the price p.
(C) Solve the inequality $R > C$ to find the range of prices that will result in a profit.

69. *Marginal analysis.* Let

$$p = 20 - x \quad \text{and} \quad C(x) = 2x + 56 \qquad 0 \leq x \leq 20$$

be the demand equation and the cost function, respectively, for a certain commodity.

(A) Find the marginal cost, average cost, and marginal average cost functions.
(B) Express the revenue in terms of x and find the marginal revenue, average revenue, and marginal average revenue functions.
(C) Find the profit, marginal profit, average profit, and marginal average profit functions.
(D) Find the break-even point(s).
(E) Evaluate the marginal profit at $x = 7$, 9, and 11, and interpret.
(F) Graph $R = R(x)$ and $C = C(x)$ on the same axes and locate regions of profit and loss.

70. *Employee training.* A company producing computer components has established that on the average, a new employee can assemble N(t) components per day after t days of on-the-job training, as given by

$$N(t) = \frac{40t}{t + 2}$$

(A) Find the average rate of change of N(t) from 3 days to 6 days.
(B) Find the instantaneous rate of change of N(t) at 3 days.
(C) Find $\lim_{t\to\infty} N(t)$.

Life Sciences

71. *Pollution.* A sewage treatment plant disposes of its effluent through a pipeline that extends 1 mile toward the center of a large lake. The concentration of effluent C(x), in parts per million, x meters from the end of the pipe is given approximately by

$$C(x) = 500(x + 1)^{-2}$$

What is the instantaneous rate of change of concentration at 9 meters? At 99 meters?

Social Sciences

72. *Learning.* If a person learns N items in t hours, as given by

$$N(t) = 20\sqrt{t}$$

find the rate of learning after:

(A) 1 hour (B) 4 hours

2 Additional Derivative Topics

CHAPTER 2 Contents

2-1 First Derivative and Graphs

- Increasing and Decreasing Functions
- Critical Values and Local Extrema
- First-Derivative Test

Since the derivative is associated with the slope of the graph of a function at a point, we might expect that it is also associated with other properties of a graph. As we will see in this and the next section, the derivative can tell us a great deal about the shape of the graph of a function. In addition, this investigation will lead to methods for finding absolute maximum and minimum values for functions that do not require graphing. Companies can use these methods to find production levels that will minimize cost or maximize profit. Pharmacologists can use them to find levels of drug dosages that will produce maximum sensitivity to a drug. And so on.

A brief review of Section 1-2, where we discussed the use of continuity and sign charts in solving inequalities, will prove useful in this section.

■ Increasing and Decreasing Functions

Graphs of functions generally have rising or falling sections as we move from left to right. It would be an aid to graphing if we could figure out where these sections occur. Suppose the graph of a function f is as indicated in Figure 1. As we move from left to right, we see that on the interval (a, b) the graph of f is rising, $f(x)$ is increasing,* and the slope of the graph is positive $[f'(x) > 0]$. On the other hand, on the interval (b, c) the graph of f is falling, $f(x)$ is decreasing, and the slope of the graph is negative $[f'(x) < 0]$. At $x = b$ the graph of f changes direction (from rising to falling), $f(x)$ changes from increasing to decreasing, the slope of the graph is 0 $[f'(b) = 0]$, and the tangent line is horizontal.

In general, we can prove that if $f'(x) > 0$ (is positive) on the interval (a, b),

* Formally, we say that $f(x)$ is *increasing* on an interval (a, b) if $f(x_2) > f(x_1)$ whenever $a < x_1 < x_2 < b$; f is decreasing on (a, b) if $f(x_2) < f(x_1)$ whenever $a < x_1 < x_2 < b$.

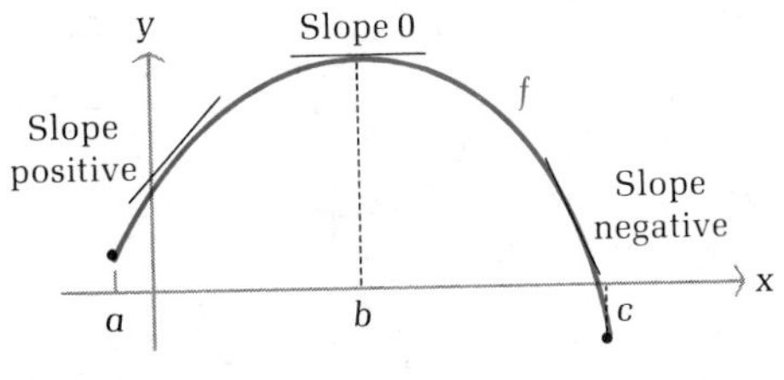

Figure 1

then $f(x)$ increases (↗) and the graph of f rises as we move from left to right over the interval; if $f'(x) < 0$ (is negative) on an interval (a, b), then $f(x)$ decreases (↘) and the graph of f falls as we move from left to right over the interval. We summarize these important results in the box.

Increasing and Decreasing Functions

For the interval (a, b):

$f'(x)$	$f(x)$	Graph of f	Examples
+	Increases ↗	Rises ↗	
–	Decreases ↘	Falls ↘	

Example 1 Given $f(x) = 8x - x^2$:

(A) Which values of x correspond to horizontal tangent lines?
(B) For which values of x is $f(x)$ increasing? Decreasing?
(C) Sketch a graph of f. Add horizontal tangent lines.

Solutions (A) $f'(x) = 8 - 2x = 0$

$$x = 4$$

Thus, a horizontal tangent line exists at $x = 4$ only.

(B) Construct a sign chart for $f'(x)$ to determine which values of x make $f'(x) > 0$ and which values make $f'(x) < 0$. (Solving inequalities by use of continuity and sign charts was discussed in Section 1-2.)

Sign chart for $f'(x) = 8 - 2x$:

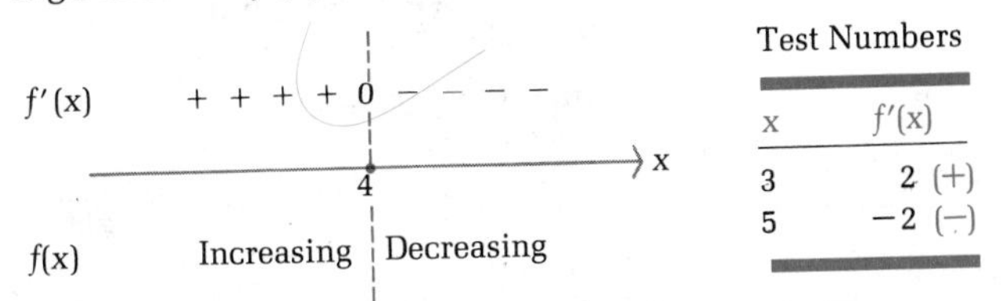

Test Numbers

x	$f'(x)$
3	2 (+)
5	−2 (−)

Thus, $f(x)$ is increasing on $(-\infty, 4)$ and decreasing on $(4, \infty)$.

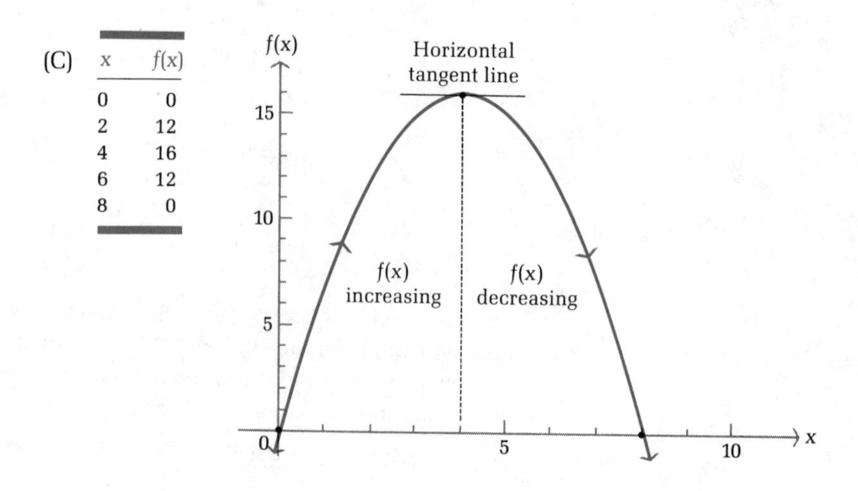

(C)

x	f(x)
0	0
2	12
4	16
6	12
8	0

Problem 1 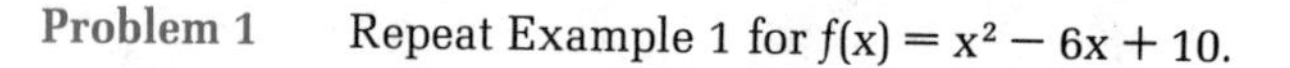
Repeat Example 1 for $f(x) = x^2 - 6x + 10$.

Example 2 Determine the intervals where f is increasing and those where f is decreasing for:

(A) $f(x) = 1 + x^3$ (B) $f(x) = (1 - x)^{1/3}$ (C) $f(x) = \dfrac{1}{x - 2}$

Solutions (A) $f(x) = 1 + x^3$

$$f'(x) = 3x^2 = 0$$
$$x = 0$$

Sign chart for $f'(x) = 3x^2$:

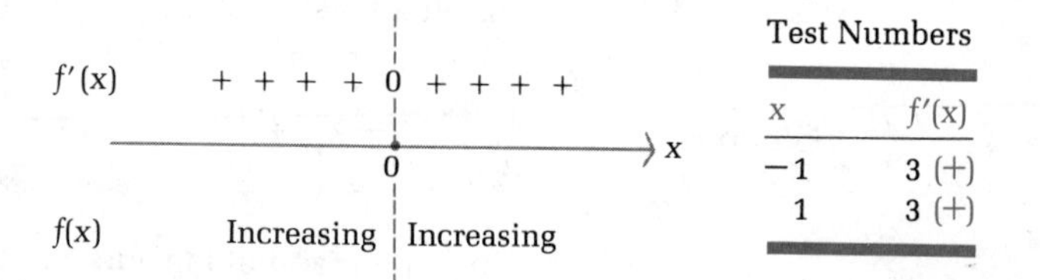

Test Numbers

x	$f'(x)$
−1	3 (+)
1	3 (+)

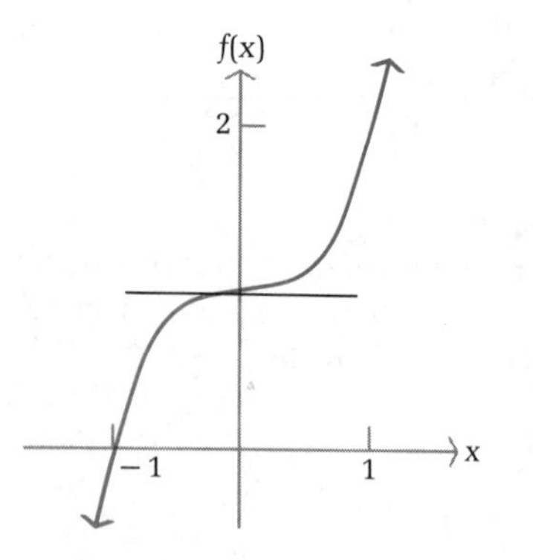

The sign chart indicates that $f(x)$ is increasing on $(-\infty, 0)$ and $(0, \infty)$. Since f is continuous at $x = 0$, it follows that $f(x)$ is increasing for all x. Thus, **a continuous function can be increasing (or decreasing) on an interval containing values of x where $f'(x) = 0$.** The graph of f is shown in the margin.

(B) $f(x) = (1 - x)^{1/3}$

$$f'(x) = -\frac{1}{3}(1 - x)^{-2/3} = \frac{-1}{3(1 - x)^{2/3}}$$

To construct a sign chart for a fraction, we plot the zeros of both the numerator and the denominator on a number line. In this case, the numerator is a constant, so the only point plotted is $x = 1$. We use the abbreviation ND to emphasize that $f'(x)$ is not defined at $x = 1$.

Sign chart for $f'(x) = -1/[3(1 - x)^{2/3}]$:

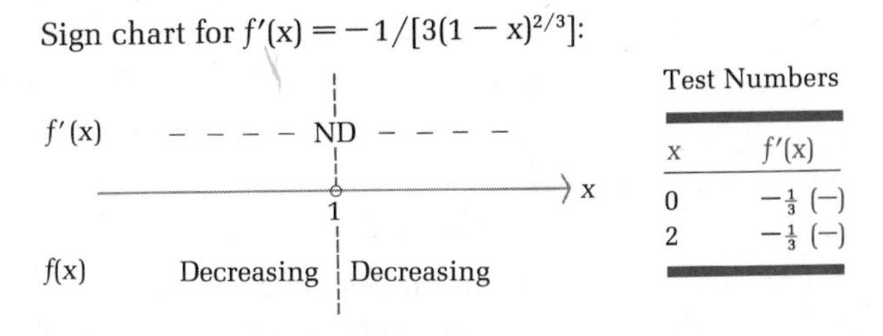

Test Numbers

x	$f'(x)$
0	$-\frac{1}{3}$ (−)
2	$-\frac{1}{3}$ (−)

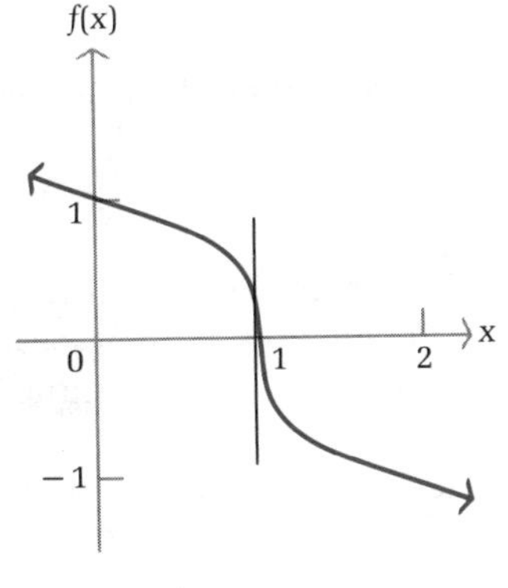

The sign chart indicates that f is decreasing on $(-\infty, 1)$ and $(1, \infty)$. Since f is continuous at $x = 1$, it follows that $f(x)$ is decreasing for all x. Thus, **a continuous function can be decreasing (or increasing) on an interval containing values of x where $f'(x)$ does not exist.** The graph of f is shown in the margin. Notice that the undefined derivative at $x = 1$ results in a vertical tangent line at $x = 1$. In general, **a vertical tangent will occur at $x = c$ if f is continuous at $x = c$ and $|f'(x)|$ becomes larger and larger as x approaches c.**

(C) $f(x) = \dfrac{1}{x - 2}$

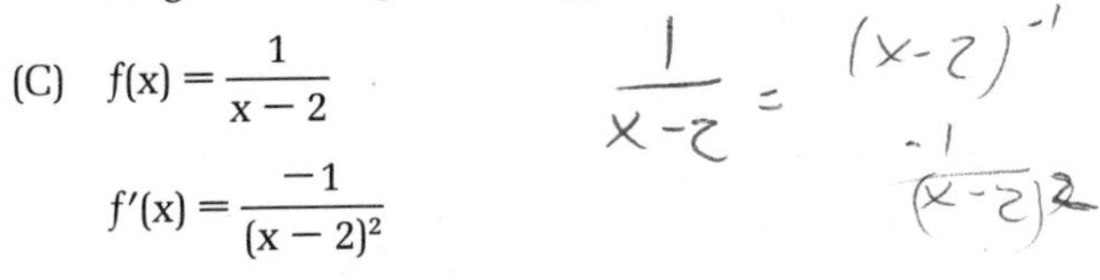

$f'(x) = \dfrac{-1}{(x - 2)^2}$

Sign chart for $f'(x) = -1/(x - 2)^2$:

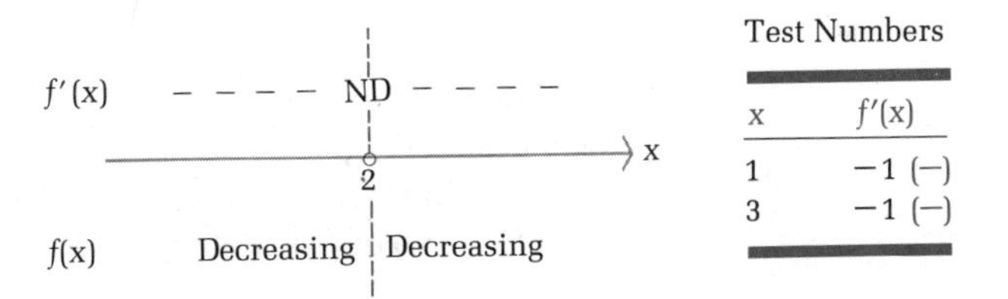

Test Numbers

x	$f'(x)$
1	−1 (−)
3	−1 (−)

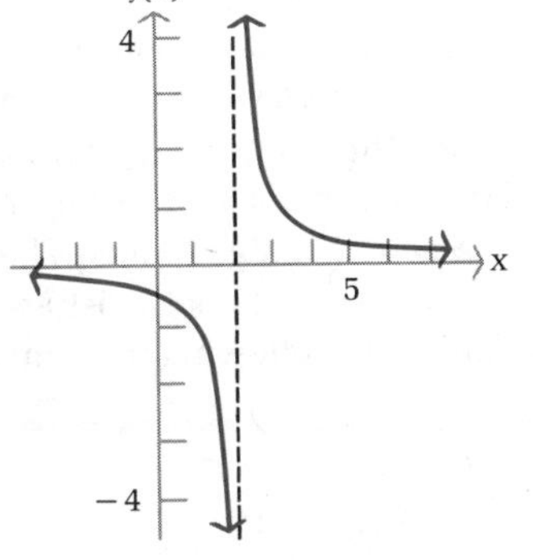

Thus, f is decreasing on $(-\infty, 2)$ and $(2, \infty)$. Since f is not continuous at $x = 2$, we cannot conclude that $f(x)$ is decreasing for all x, or even for all x except $x = 2$. (See the graph of f in the margin.) **The values where a function is increasing or decreasing must always be expressed in terms of open intervals that are subsets of the domain of the function.**

Problem 2 Determine the intervals where f is increasing and those where f is decreasing for:

(A) $f(x) = 1 - x^3$ (B) $f(x) = (1 + x)^{1/3}$ (C) $f(x) = \dfrac{1}{x}$

Critical Values and Local Extrema

When the graph of a continuous function changes from rising to falling, a high point or *local maximum* occurs, and when the graph changes from falling to rising, a low point or *local minimum* occurs. In Figure 2, high points occur at c_3 and c_6, and low points occur at c_2 and c_4. In general, we call $f(c)$ a **local maximum** if there exists an interval (m, n) containing c such that

$$f(x) \leq f(c)$$

for all x in (m, n). The quantity $f(c)$ is called a **local minimum** if there exists an interval (m, n) containing c such that

$$f(x) \geq f(c)$$

for all x in (m, n). The quantity $f(c)$ is called a **local extremum** if it is either a local maximum or a local minimum. Thus, in Figure 2 we see that local maxima occur at c_3 and c_6, local minima occur at c_2 and c_4, and all four of these points produce local extrema.

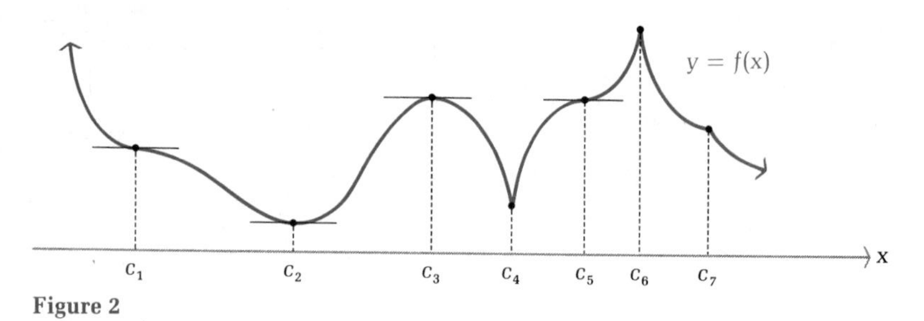

Figure 2

How can we locate local maxima and minima if we are given the equation for a function and not its graph? Figure 2 suggests an approach. It appears that local maxima and minima occur among those values of x such that $f'(x) = 0$ or $f'(x)$ does not exist; that is, among the values c_1, c_2, c_3, c_4, c_5, c_6, and c_7. [Recall from Section 1-4 that $f'(x)$ is not defined at points on the graph of f where there is a sharp corner or a vertical tangent line.] The values of x in the domain of f where $f'(x) = 0$ or $f'(x)$ does not exist are called the **critical values** of f. It is possible to prove the following theorem:

Theorem 1

Existence of Local Extrema

If f is continuous on the interval (a, b) and $f(c)$ is a local extremum, then either $f'(c) = 0$ or $f'(c)$ does not exist (is not defined).

Theorem 1 implies that a local extremum can occur only at a critical value, but it does not imply that every critical value produces a local extremum. In Figure 2, c_1 and c_5 are critical values (the slope is 0), but the function does not have a local maximum or local minimum at either of these values.

Our strategy for finding local extrema is now clear. We find all critical values for f and test each one to see if it produces a local maximum, a local minimum, or neither.

▪ First-Derivative Test

If $f'(x)$ exists on both sides of a critical value c, then the sign of $f'(x)$ can be used to determine if the point $(c, f(c))$ is a local maximum, a local minimum, or neither. The various possibilities are summarized in the box below and illustrated in Figure 3 on the next page.

First-Derivative Test for Local Extrema

Let c be a critical value of f [$f(c)$ is defined and either $f'(c) = 0$ or $f'(c)$ is not defined]. Construct a sign chart for $f'(x)$ close to and on either side of c.

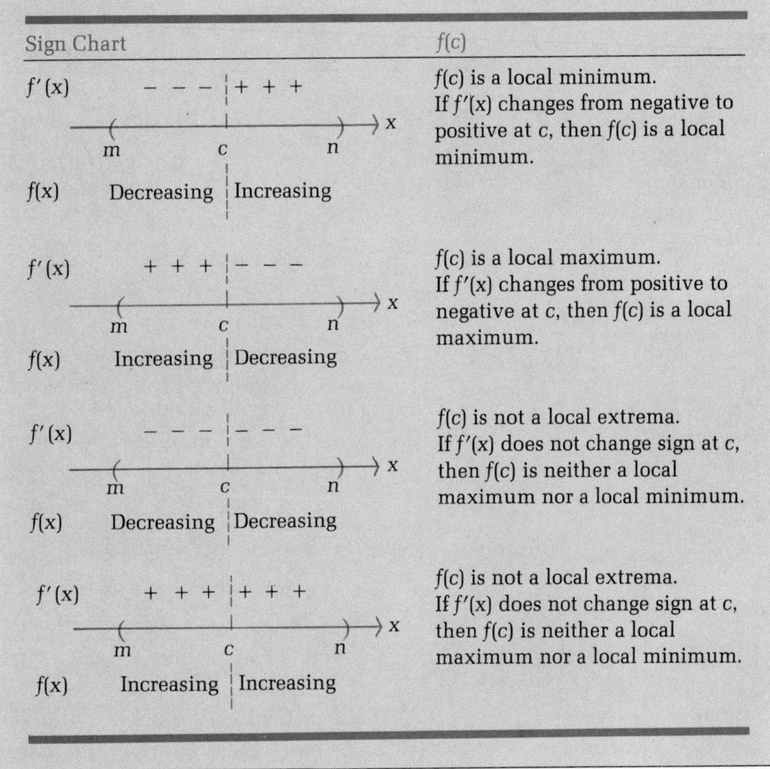

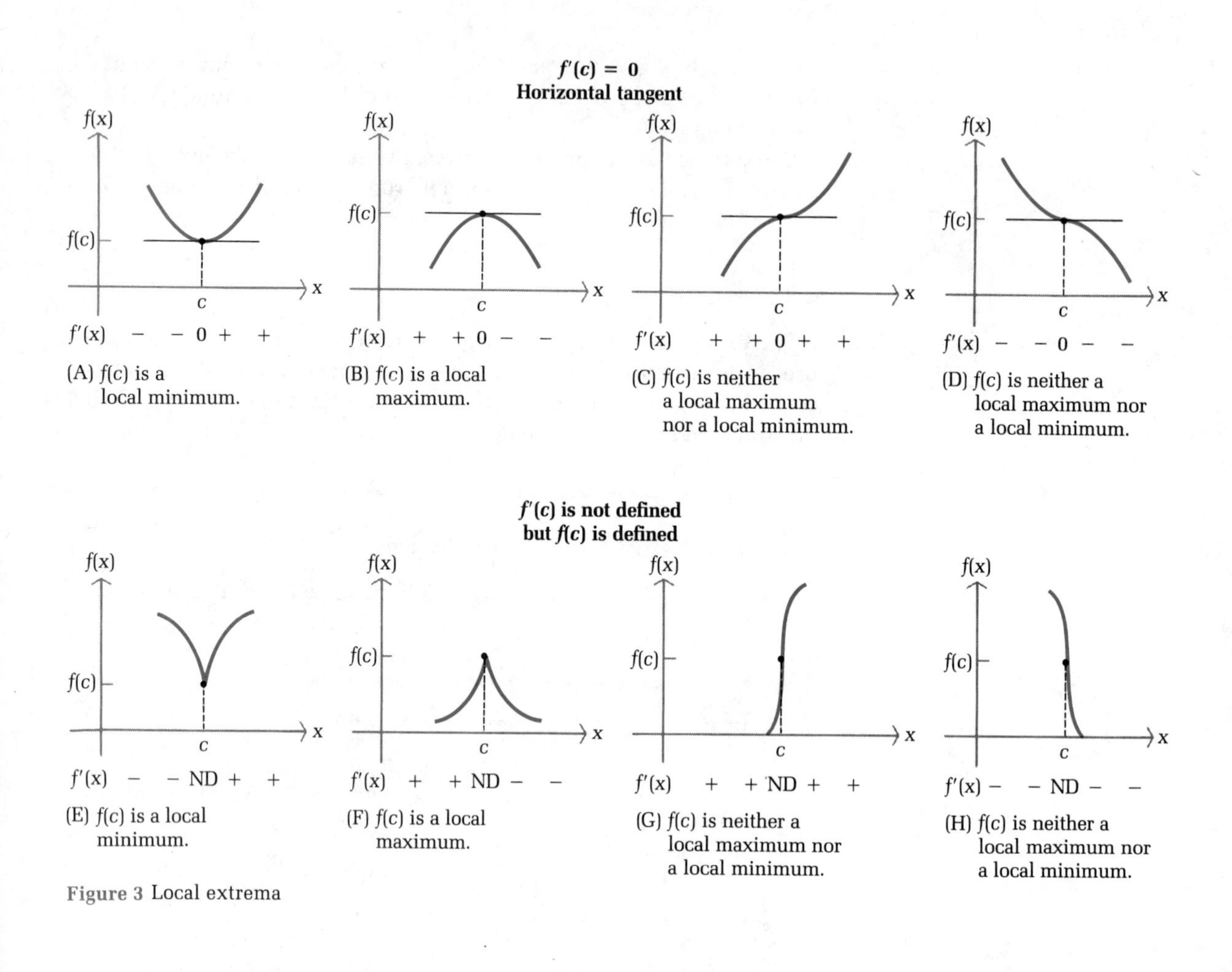

Figure 3 Local extrema

Example 3 Given $f(x) = x^3 - 6x^2 + 9x + 1$:

(A) Find the critical values of f.
(B) Find the local maxima and minima.
(C) Sketch the graph of f.

Solutions (A) $f'(x) = 3x^2 - 12x + 9 = 0$

$$3(x^2 - 4x + 3) = 0$$

$$3(x - 1)(x - 3) = 0$$

$$x = 1 \quad \text{or} \quad x = 3$$

The critical values are $x = 1$ and $x = 3$.

(B) The easiest way to apply the first-derivative test is to construct a sign chart for $f'(x)$ for all x.

Sign chart for $f'(x) = 3(x - 1)(x - 3)$:

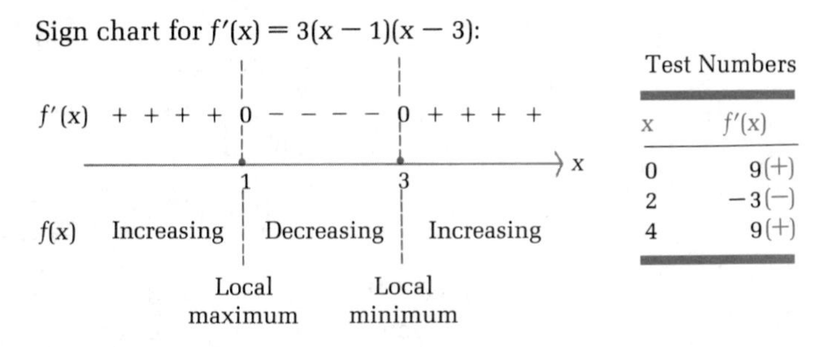

Test Numbers

x	$f'(x)$
0	9(+)
2	−3(−)
4	9(+)

The sign chart indicates that f increases on $(-\infty, 1)$, has a local maximum at $x = 1$, decreases on $(1, 3)$, has a local minimum at $x = 3$, and increases on $(3, \infty)$. These facts are summarized in the following table:

x	$f'(x)$	$f(x)$	Graph of f
$x < 1$	+	Increasing	Rising
$x = 1$	0	Local maximum	Horizontal tangent
$1 < x < 3$	−	Decreasing	Falling
$x = 3$	0	Local minimum	Horizontal tangent
$3 < x$	+	Increasing	Rising

(C) We sketch a graph of f using the information from part B and point-by-point plotting.

x	$f(x)$
0	1
1	5
2	3
3	1
4	5

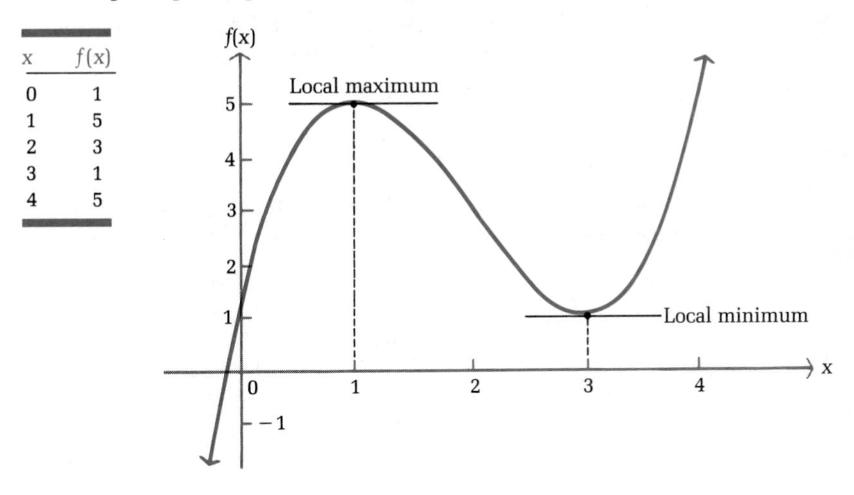

Problem 3 Given $f(x) = x^3 - 9x^2 + 24x - 10$:

(A) Find the critical values of f.
(B) Find the local maxima and minima.
(C) Sketch a graph of f.

In Example 3, the function f had local extrema at both of its critical values. However, as was noted earlier, not every critical value of a function will produce a local extremum. For example, consider the function discussed in Example 2B:

$$f(x) = (1 - x)^{1/3} \quad \text{and} \quad f'(x) = \frac{-1}{3(1 - x)^{2/3}}$$

Since $f(1)$ exists and $f'(1)$ does not exist, $x = 1$ is a critical value for this function. However, the sign chart for $f'(x)$ shows that $f'(x)$ does not change sign at $x = 1$:

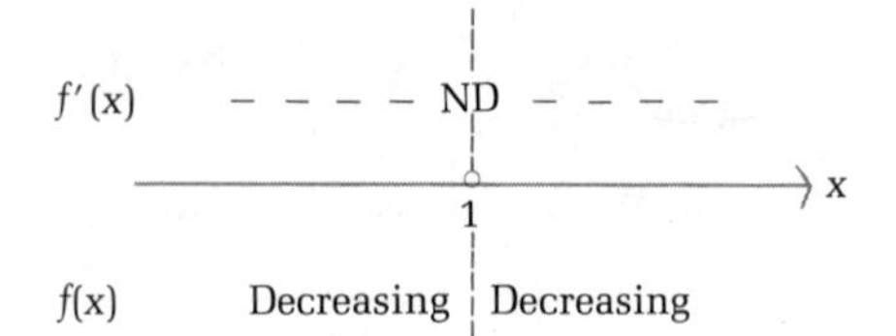

Thus, f does not have a local maximum or a local minimum at $x = 1$.

Finally, it is important to remember that a critical value must be in the domain of the function. Refer to the function discussed in Example 2C:

$$f(x) = \frac{1}{x - 2} \quad \text{and} \quad f'(x) = \frac{-1}{(x - 2)^2}$$

The derivative is not defined at $x = 2$, but neither is the function. Thus, $x = 2$ is not a critical value for f. In fact, this function does not have any critical values.

Answers to Matched Problems

1. (A) Horizontal tangent line at $x = 3$
   (B) Decreasing on $(-\infty, 3)$, increasing on $(3, \infty)$
   (C) $f(x)$

2. (A) Decreasing for all x (B) Increasing for all x
   (C) Decreasing on $(-\infty, 0)$ and $(0, \infty)$
3. (A) Critical values: $x = 2$, $x = 4$
   (B) Local maximum at $x = 2$, local minimum at $x = 4$

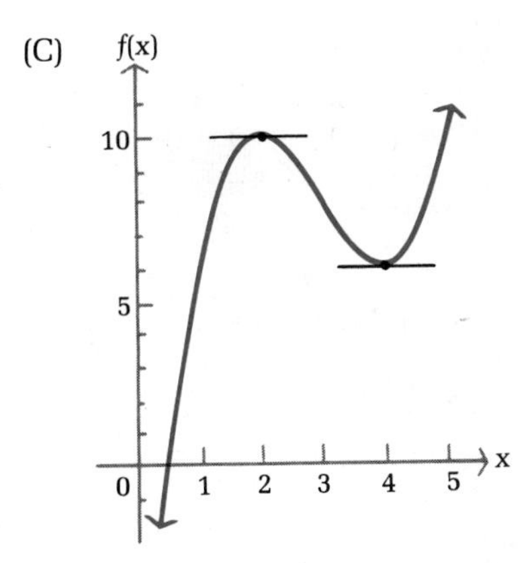

Exercise 2-1

A

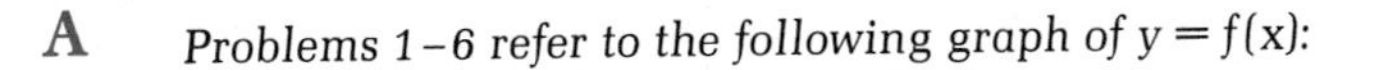

Problems 1–6 refer to the following graph of $y = f(x)$:

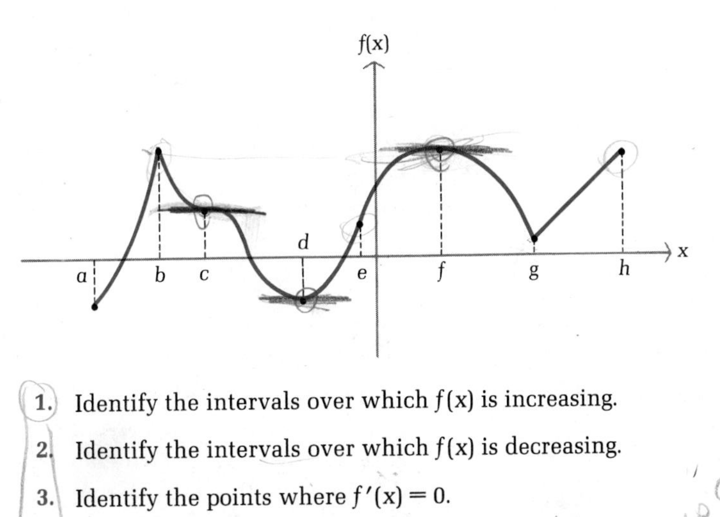

1. Identify the intervals over which $f(x)$ is increasing.
2. Identify the intervals over which $f(x)$ is decreasing.
3. Identify the points where $f'(x) = 0$.
4. Identify the points where $f'(x)$ does not exist.
5. Identify the points where f has a local maximum.
6. Identify the points where f has a local minimum.

B

Find the intervals where f(x) is increasing, the intervals where f(x) is decreasing, and the local extrema.

7. $f(x) = x^2 - 16x + 12$
8. $f(x) = x^2 + 6x + 7$
9. $f(x) = 4 + 10x - x^2$
10. $f(x) = 5 + 8x - 2x^2$

11. $f(x) = 2x^3 + 4$
12. $f(x) = 2 - 3x^3$
13. $f(x) = 2 - 6x - 2x^3$
14. $f(x) = x^3 + 9x + 7$
15. $f(x) = x^3 - 12x + 8$
16. $f(x) = 3x - x^3$
17. $f(x) = x^3 - 3x^2 - 24x + 7$
18. $f(x) = x^3 + 3x^2 - 9x + 5$
19. $f(x) = 2x^2 - x^4$
20. $f(x) = x^4 - 8x^2 + 3$

Find the intervals where f(x) is increasing, the intervals where f(x) is decreasing, and sketch the graph. Add horizontal tangent lines.

21. $f(x) = 4 + 8x - x^2$
22. $f(x) = 2x^2 - 8x + 9$
23. $f(x) = x^3 - 3x + 1$
24. $f(x) = x^3 - 12x + 2$
25. $f(x) = 10 - 12x + 6x^2 - x^3$
26. $f(x) = x^3 + 3x^2 + 3x$

C *Find the critical values, the intervals where f(x) is increasing, the intervals where f(x) is decreasing, and the local extrema. Do not graph.*

27. $f(x) = \dfrac{x-1}{x+2}$
28. $f(x) = \dfrac{x+2}{x-3}$
29. $f(x) = x + \dfrac{4}{x}$
30. $f(x) = \dfrac{9}{x} + x$
31. $f(x) = 1 + \dfrac{1}{x} + \dfrac{1}{x^2}$
32. $f(x) = 3 - \dfrac{4}{x} - \dfrac{2}{x^2}$
33. $f(x) = \dfrac{x^2}{x-2}$
34. $f(x) = \dfrac{x^2}{x+1}$
35. $f(x) = x^4(x-6)^2$
36. $f(x) = x^3(x-5)^2$
37. $f(x) = 3(x-2)^{2/3} + 4$
38. $f(x) = 6(4-x)^{2/3} + 4$
39. $f(x) = 2\sqrt{x} - x, \quad x > 0$
40. $f(x) = x - 4\sqrt{x}, \quad x > 0$

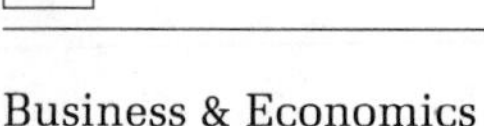

Applications

Business & Economics

41. *Average cost.* A manufacturer has the following costs in producing x toasters in one day, $0 < x < 150$: fixed costs, \$320; unit production cost, \$20 per toaster; equipment maintenance and repairs, $x^2/20$ dollars. Thus, the cost of manufacturing x toasters in one day is given by

$$C(x) = \frac{x^2}{20} + 20x + 320 \qquad 0 < x < 150$$

and the average cost per toaster is given by

$$\overline{C}(x) = \frac{C(x)}{x} = \frac{x}{20} + 20 + \frac{320}{x} \qquad 0 < x < 150$$

Find the critical values for $\overline{C}(x)$, the intervals where the average cost per toaster is decreasing, the intervals where the average cost per toaster is increasing, and the local extrema. Do not graph.

42. *Average cost.* A manufacturer has the following costs in producing x blenders in one day, $0 < x < 200$: fixed costs, \$450; unit production cost, \$60 per blender; equipment maintenance and repairs, $x^2/18$ dollars.

    (A) What is the average cost $\overline{C}(x)$ per blender if x blenders are produced in one day?
    (B) Find the critical values for $\overline{C}(x)$, the intervals where the average cost per blender is decreasing, the intervals where the average cost per blender is increasing, and the local extrema. Do not graph.

43. *Marginal analysis.* Show that profit will be increasing over production intervals (a, b) for which marginal revenue is greater than marginal cost. [*Hint:* $P(x) = R(x) - C(x)$.]
44. *Marginal analysis.* Show that profit will be decreasing over production intervals (a, b) for which marginal revenue is less than marginal cost.

Life Sciences

45. *Medicine.* A drug is injected into the bloodstream of a patient through the right arm. The concentration of the drug in the bloodstream of the left arm t hours after the injection is approximated by

$$C(t) = \frac{0.14t}{t^2 + 1} \qquad 0 < t < 24$$

Find the critical values for $C(t)$, the intervals where the concentration of the drug is increasing, the intervals where the concentration of the drug is decreasing, and the local extrema. Do not graph.

46. *Medicine.* The concentration $C(t)$ in milligrams per cubic centimeter of a particular drug in a patient's bloodstream is given by

$$C(t) = \frac{0.16t}{t^2 + 4t + 4} \qquad 0 < t < 12$$

where t is the number of hours after the drug is taken orally. Find the critical values for $C(t)$, the intervals where the concentration of the drug is increasing, the intervals where the concentration of the drug is decreasing, and the local extrema. Do not graph.

Social Sciences

47. *Politics.* Public awareness of a Congressional candidate before and after a successful campaign was approximated by

$$P(t) = \frac{8.4t}{t^2 + 49} + 0.1 \qquad 0 < t < 24$$

where t is time in months after the campaign started and $P(t)$ is the fraction of people in the Congressional district who could recall the candidate's (and later, Congressman's) name. Find the critical values for $P(t)$, the time intervals where the fraction is increasing, the time intervals where the fraction is decreasing, and the local extrema. Do not graph.

2-2 Second Derivative and Graphs

- Concavity
- Inflection Points
- Second-Derivative Test
- Application

In the preceding section we saw that the derivative can be used to determine when a graph is rising and falling. Now we want to see what the second derivative (the derivative of the derivative) can tell us about the shape of a graph.

Concavity

Consider the functions

$$f(x) = x^2 \qquad \text{and} \qquad g(x) = \sqrt{x}$$

for x in the interval $(0, \infty)$. Since

$$f'(x) = 2x > 0 \qquad \text{for } 0 < x < \infty$$

and

$$g'(x) = \frac{1}{2\sqrt{x}} > 0 \qquad \text{for } 0 < x < \infty$$

both functions are increasing on $(0, \infty)$.

Notice the different shapes of the graphs of f and g (see Figure 4). Even though the graph of each function is rising and each graph starts at (0, 0) and goes through (1, 1), the graphs are quite dissimilar. The graph of f opens upward while the graph of g opens downward. We say that the graph of f is *concave upward* and the graph of g is *concave downward*. It will help us draw graphs if we can determine the concavity of the graph before we draw it. How can we find a mathematical formulation of concavity?

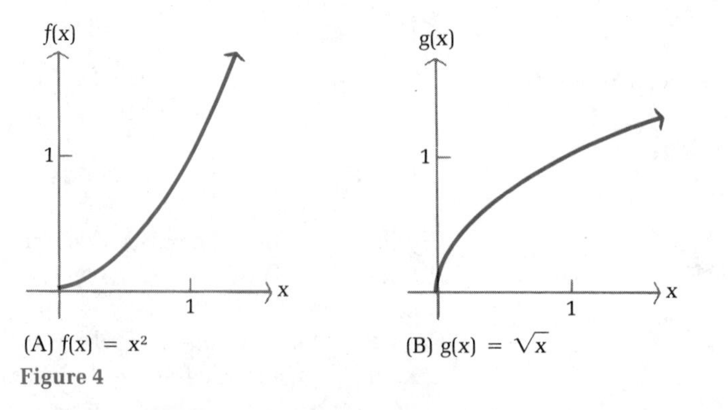

(A) $f(x) = x^2$ (B) $g(x) = \sqrt{x}$

Figure 4

It will be instructive to examine the slopes of f and g at various points on their graphs (see Figure 5). There are two observations we can make about each graph. Looking at the graph of f in Figure 5A, we see that $f'(x)$ (the slope of the tangent line) is *increasing* and that the graph lies *above* each tangent line. Looking at Figure 5B, we see that $g'(x)$ is *decreasing* and that the graph lies *below* each tangent line. With these ideas in mind, we state the general definition of concavity: The graph of a function f is **concave upward on the interval (*a*, *b*)** if $f'(x)$ is *increasing* on (a, b) and is **concave downward on the interval (*a*, *b*)** if $f'(x)$ is *decreasing* on (a, b). Geometrically, the graph is concave upward on (a, b) if it lies above its tangent lines in (a, b) and is concave downward on (a, b) if it lies below its tangent lines in (a, b).

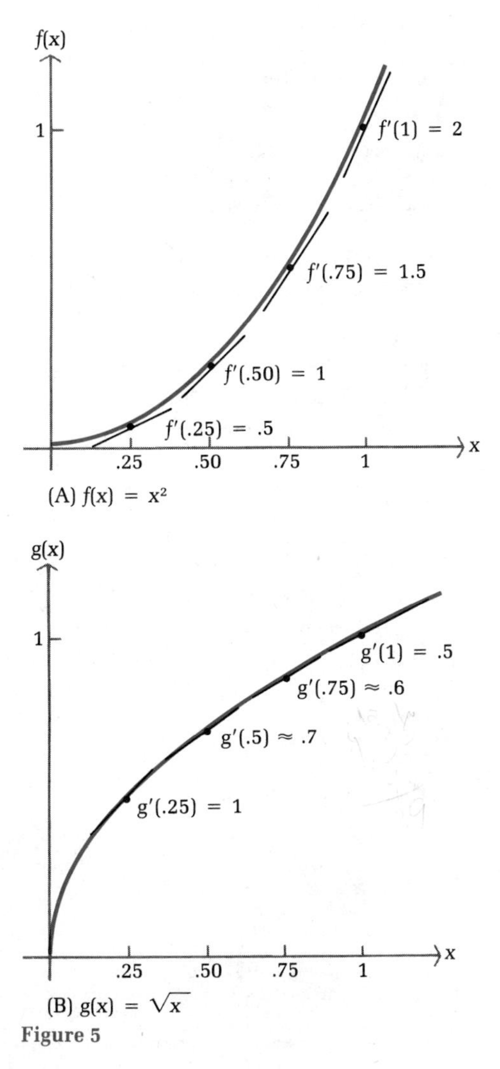

Figure 5

How can we determine when $f'(x)$ is increasing or decreasing? In the last section we used the derivative of a function to determine when that function is increasing or decreasing. Thus, to determine when the function $f'(x)$ is increasing or decreasing, we use the derivative of $f'(x)$. The derivative of the derivative of a function is called the *second derivative* of the function. Various notations for the second derivative are given in the following box:

Second Derivative

For $y = f(x)$, the **second derivative** of f is

$$f''(x) = D_x f'(x)$$

Other notations for $f''(x)$ are

$$\frac{d^2y}{dx^2} \qquad y'' \qquad D_x^2 f(x)$$

Returning to the functions f and g discussed at the beginning of this section, we have

$$\begin{aligned} f(x) &= x^2 & g(x) &= \sqrt{x} = x^{1/2} \\ f'(x) &= 2x & g'(x) &= \frac{1}{2}x^{-1/2} = \frac{1}{2\sqrt{x}} \\ f''(x) &= D_x 2x & g''(x) &= D_x \frac{1}{2}x^{-1/2} \\ &= 2 & &= \frac{1}{4}x^{-3/2} = -\frac{1}{4\sqrt{x^3}} \end{aligned}$$

For $x > 0$ we see that $f''(x) > 0$; thus, $f'(x)$ is increasing and the graph of f is concave upward (see Fig. 5A). For $x > 0$ we also see that $g''(x) < 0$; thus, $g'(x)$ is decreasing and the graph of g is concave downward (see Fig. 5B). These ideas are summarized in the following box:

Concavity

For the interval (a, b)

$f''(x)$	$f'(x)$	Graph of $y = f(x)$	Example
$+$	Increasing	Concave upward	
$-$	Decreasing	Concave downward	

Be careful not to confuse concavity with falling and rising. As Figure 6 illustrates, a graph that is concave upward on an interval may be falling, rising, or both falling and rising on that interval. A similar statement holds for a graph that is concave downward.

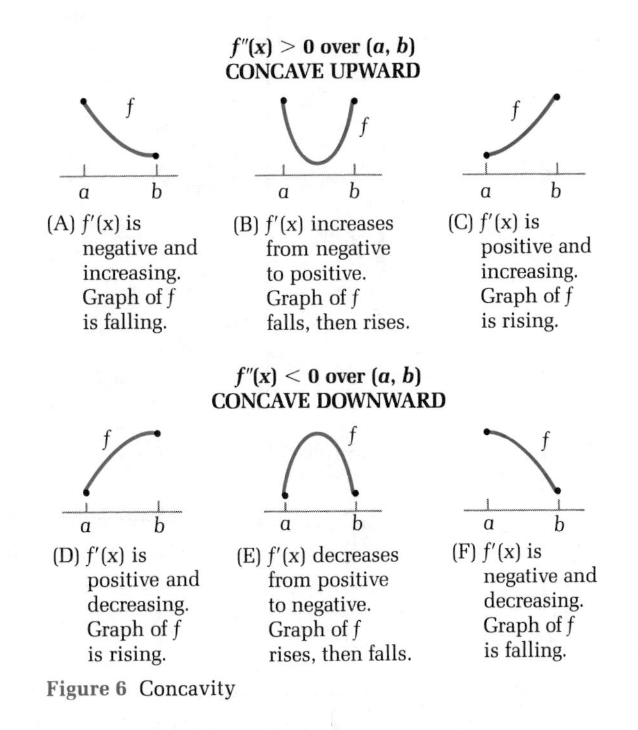

Figure 6 Concavity

Example 4 Let $f(x) = x^3$. Find the intervals where the graph of f is concave upward and the intervals where the graph of f is concave downward. Sketch a graph of f.

Solution To determine concavity, we must determine the sign of $f''(x)$.

$$f(x) = x^3$$
$$f'(x) = 3x^2$$
$$f''(x) = 6x$$

Sign chart for $f''(x) = 6x$:

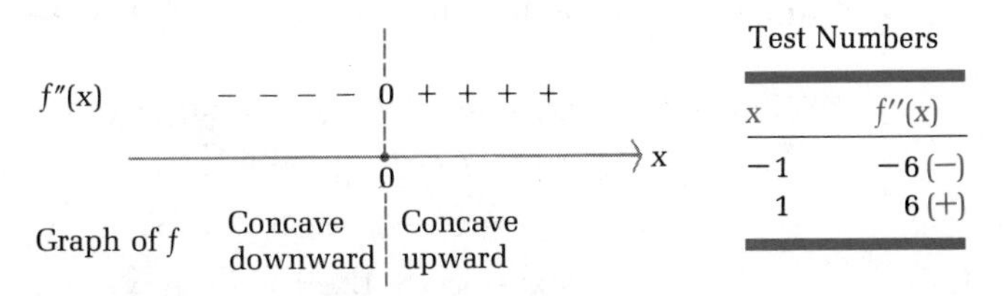

Test Numbers

x	$f''(x)$
-1	$-6\,(-)$
1	$6\,(+)$

Thus, the graph of f is concave downward on $(-\infty, 0)$ and concave upward on $(0, \infty)$. The graph of f (without going through other graphing details) is shown in the figure.

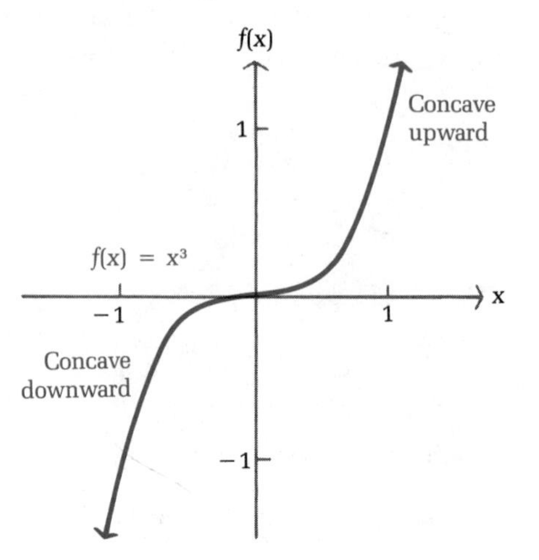

Problem 4 Repeat Example 4 for $f(x) = 1 - x^3$.

The graph in Example 4 changes from concave downward to concave upward at the point (0, 0). This point is called an *inflection point*.

Inflection Points

In general, an **inflection point** is a point on the graph of a function where the concavity changes (from upward to downward or from downward to upward). In order for the concavity to change at a point, $f''(x)$ must change sign at that point. Reasoning as we did in the previous section, we conclude that the inflection points must occur at points where $f''(x) = 0$ or $f''(x)$ does not exist [but $f(x)$ must exist]. Figure 7 illustrates several typical cases.

If $f'(c)$ exists and $f''(x)$ changes sign at $x = c$, then the tangent line at an inflection point $(c, f(c))$ will always lie below the graph on the side that is concave upward and above the graph on the side that is concave downward (see Figs. 7A, B, and C).

Example 5 Find the inflection points of $f(x) = x^3 - 6x^2 + 9x + 1$.

Solution Since inflection points occur at values of x where $f''(x)$ changes sign, we construct a sign chart for $f''(x)$.

$$f(x) = x^3 - 6x^2 + 9x + 1$$
$$f'(x) = 3x^2 - 12x + 9$$
$$f''(x) = 6x - 12 = 6(x - 2)$$

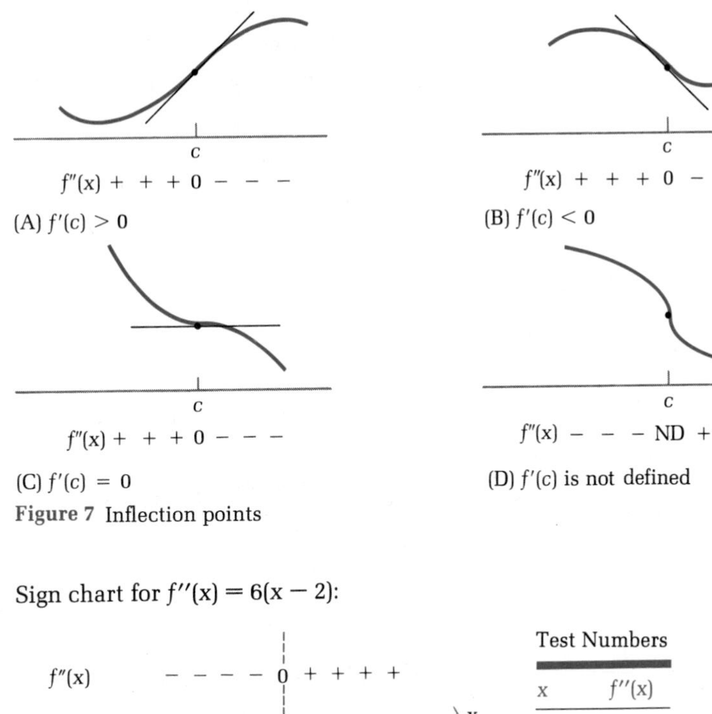

Figure 7 Inflection points

Sign chart for $f''(x) = 6(x - 2)$:

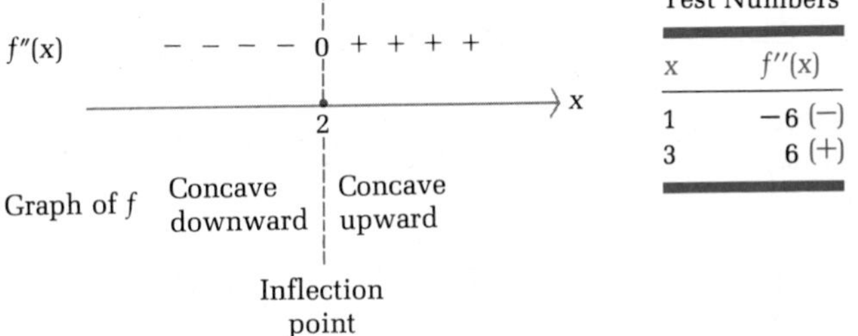

Test Numbers

x	$f''(x)$
1	-6 (−)
3	6 (+)

From the sign chart, we see that the graph of f has an inflection point at $x = 2$. The graph of f is shown in the figure. (See Example 3 in Section 2-1.)

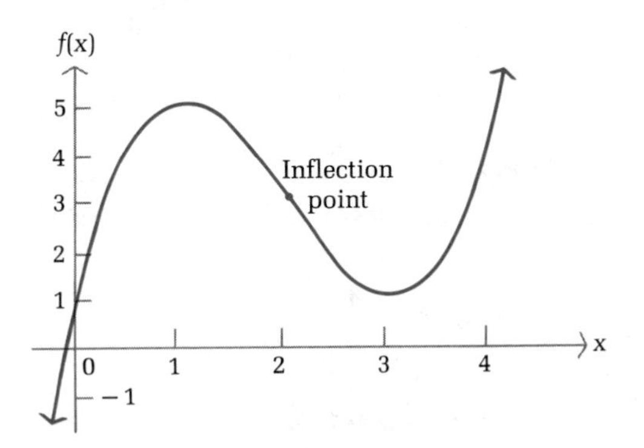

Problem 5 Find the inflection points of $f(x) = x^3 - 9x^2 + 24x - 10$. (See Problem 3 in Section 2-1 for the graph of f.)

It is important to remember that the values of x where $f''(x) = 0$ [or $f''(x)$ does not exist] are only candidates for inflection points. The second derivative must change sign at $x = c$ in order for the graph of f to have an inflection point at $x = c$. For example, consider

$$f(x) = x^4, \quad f'(x) = 4x^3, \quad \text{and} \quad f''(x) = 12x^2$$

The second derivative is 0 at $x = 0$, but $f''(x) > 0$ for all other values of x. Since $f''(x)$ does not change sign at $x = 0$, the graph of f does not have an inflection point at $x = 0$, as illustrated in Figure 8.

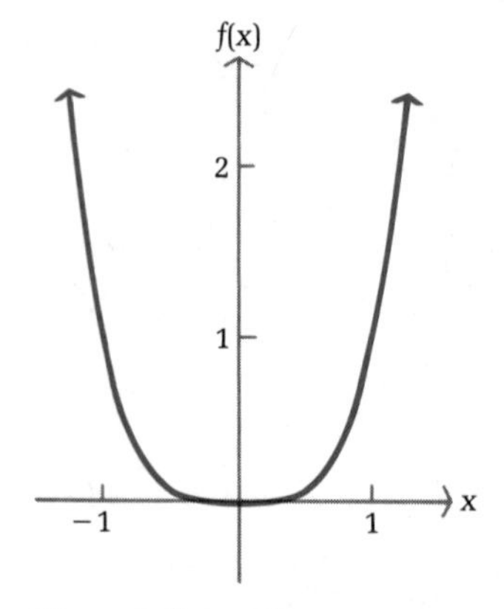

Figure 8 $f(x) = x^4$

▪ Second-Derivative Test

Now we want to see how the second derivative can be used to find local extrema. Suppose f is a function satisfying $f'(c) = 0$ and $f''(c) > 0$. First, note that if $f''(c) > 0$, then it follows from the properties of limits* that $f''(x) > 0$ in some interval (m, n) containing c. Thus, the graph of f must be concave upward in this interval. But this implies that $f'(x)$ is increasing in this interval. Since $f'(c) = 0$, $f'(x)$ must change from negative to positive at $x = c$ and $f(c)$ is a local minimum (see Figure 9). Reasoning in the same fashion, we conclude that if $f'(c) = 0$ and $f''(c) < 0$, then $f(c)$ is a local maximum. Of course, it is possible that both $f'(c) = 0$ and $f''(c) = 0$. In this case the second derivative cannot be used to determine the shape of the graph around $x = c$; $f(c)$ may be a local minimum, a local maximum, or neither.

* Actually, we are assuming that $f''(x)$ is continuous in an interval containing c. It is very unlikely that we will encounter a function for which $f''(c)$ exists, but $f''(x)$ is not continuous in an interval containing c.

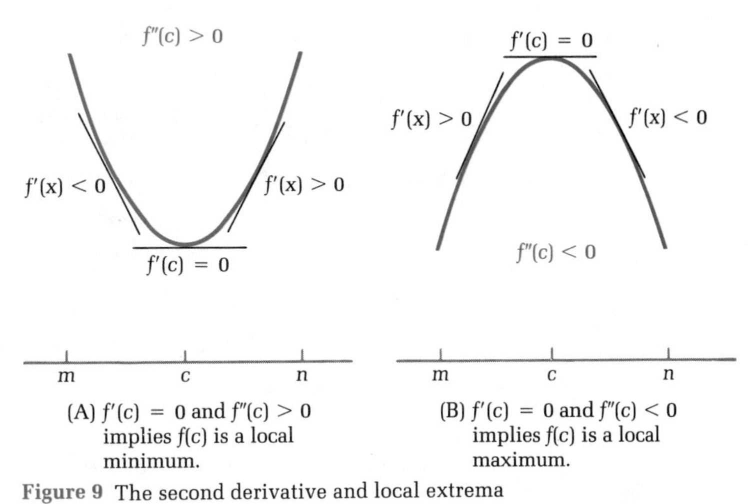

(A) $f'(c) = 0$ and $f''(c) > 0$ implies $f(c)$ is a local minimum.

(B) $f'(c) = 0$ and $f''(c) < 0$ implies $f(c)$ is a local maximum.

Figure 9 The second derivative and local extrema

The sign of the second derivative thus provides a simple test for identifying local maxima and minima. This test is most useful when we do not want to draw the graph of the function. If we are interested in drawing the graph and have already constructed the sign chart for $f'(x)$, then the first-derivative test can be used to identify the local extrema.

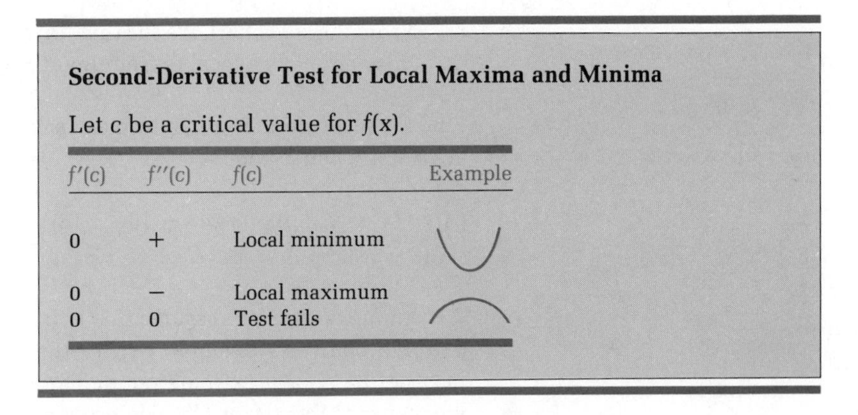

Second-Derivative Test for Local Maxima and Minima

Let c be a critical value for $f(x)$.

$f'(c)$	$f''(c)$	$f(c)$	Example
0	+	Local minimum	
0	−	Local maximum	
0	0	Test fails	

Example 6 Find the local maxima and minima of each function. Use the second-derivative test when it applies.

(A) $f(x) = x^3 - 6x^2 + 9x + 1$ (B) $f(x) = \frac{1}{6}x^6 - 4x^5 + 25x^4$

Solutions (A) Take first and second derivatives and find critical values:

$$f(x) = x^3 - 6x^2 + 9x + 1$$
$$f'(x) = 3x^2 - 12x + 9 = 3(x-1)(x-3)$$
$$f''(x) = 6x - 12 = 6(x-2)$$

Critical values are $x = 1, 3$.

$f''(1) = -6 < 0$ $\quad$ f has a local maximum at $x = 1$.

$f''(3) = 6 > 0$ $\quad$ f has a local minimum at $x = 3$.

(B) $f(x) = \frac{1}{6}x^6 - 4x^5 + 25x^4$

$f'(x) = x^5 - 20x^4 + 100x^3 = x^3(x - 10)^2$

$f''(x) = 5x^4 - 80x^3 + 300x^2$

Critical values are $x = 0$ and $x = 10$.

$f''(0) = 0$ $\quad$ The second-derivative test fails at both critical

$f''(10) = 0$ $\quad$ values, so the first-derivative test must be used.

Sign chart for $f'(x) = x^3(x - 10)^2$:

	$(-\infty, 0)$	0	$(0, 10)$	10	$(10, \infty)$
$f'(x)$	− − − −	0	+ + + +	0	+ + + +
$f(x)$	Decreasing		Increasing		Increasing

Test Numbers

x	$f'(x)$
−1	−121 (−)
1	81 (+)
11	1,331 (+)

From the sign chart, we see that $f(x)$ has a local minimum at $x = 0$ and does not have a local extremum at $x = 10$.

Problem 6 Find the local maxima and minima of each function. Use the second-derivative test when it applies.

(A) $f(x) = x^3 - 9x^2 + 24x - 10$ $\quad$ (B) $f(x) = 10x^6 - 24x^5 + 15x^4$

A common error is to assume that $f''(c) = 0$ implies that $f(c)$ is not a local extreme point. As Example 6B illustrates, if $f''(c) = 0$, then $f(c)$ may or may not be a local extreme point. The first-derivative test *must* be used whenever $f''(c) = 0$ [or $f''(c)$ does not exist].

Application

Example 7
Maximum Rate of Change

Using past records, a company estimates that it will sell $N(x)$ units of a product after spending $\$x$ thousand on advertising, as given by

$$N(x) = 2{,}000 - 2x^3 + 60x^2 - 450x \qquad 5 \le x \le 15$$

When is the rate of change of sales per unit (thousand dollars) change in advertising increasing? Decreasing? What is the maximum rate of change? Graph N and N' on the same axes and interpret.

Solution The rate of change of sales per unit (thousand dollars) change in advertising expenditure is

$$N'(x) = -6x^2 + 120x - 450 = -6(x - 5)(x - 15)$$

To determine when this rate is increasing and decreasing, we find $N''(x)$, the derivative of $N'(x)$:

$$N''(x) = -12x + 120 = 12(10 - x)$$

The information obtained by analyzing the signs of $N'(x)$ and $N''(x)$ is summarized in the table (sign charts are omitted).

x	$N''(x)$	$N'(x)$	$N'(x)$	$N(x)$
$5 < x < 10$	+	+	Increasing	Increasing, concave upward
$x = 10$	0	+	Local maximum	Inflection point
$10 < x < 15$	−	+	Decreasing	Increasing, concave downward

Thus, we see that $N'(x)$, the rate of change of sales, is increasing on (5, 10) and decreasing on (10, 15). Both N and N' are graphed in the figure. An examination of the graph of $N'(x)$ shows that the maximum rate of change is $N'(10) = 150$. Notice that $N'(x)$ has a local maximum and $N(x)$ has an inflection point at $x = 10$. This value of x is referred to as the **point of diminishing returns** since the rate of change of sales begins to decrease at this point.

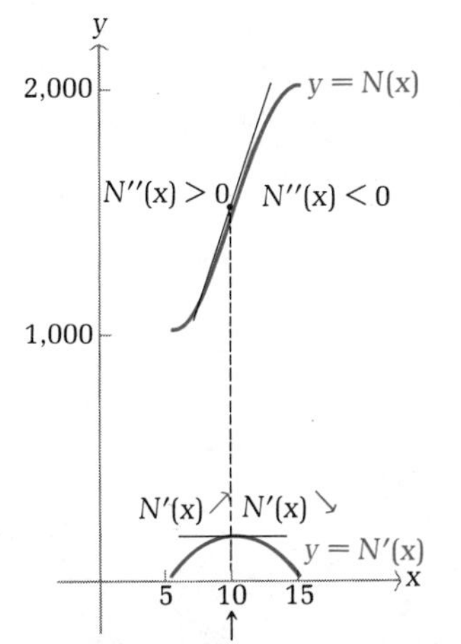

Problem 7 Repeat Example 7 for

$$N(x) = 5{,}000 - x^3 + 60x^2 - 900x \qquad 10 \leq x \leq 30$$

Answers to Matched Problems

4. Concave upward on $(-\infty, 0)$; concave downward on $(0, \infty)$

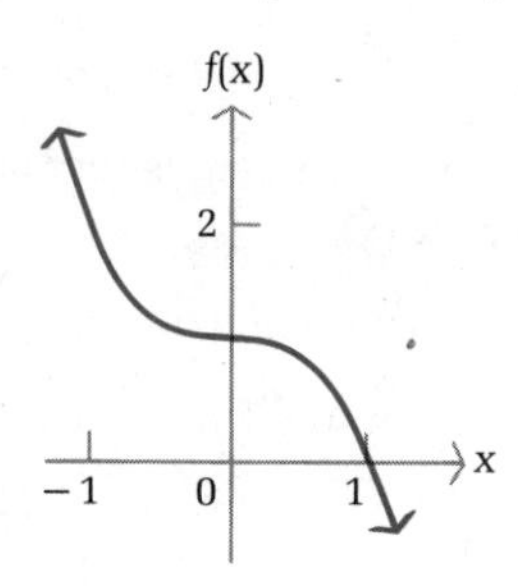

5. Inflection point at $x = 3$

6. (A) $f(2)$ is a local maximum; $f(4)$ is a local minimum
   (B) $f(0)$ is a local minimum; no local extremum at $x = 1$

7. $N'(x)$ is increasing on $(10, 20)$, decreasing on $(20, 30)$; maximum rate of change is $N'(20) = 300$; $x = 20$ is point of diminishing returns

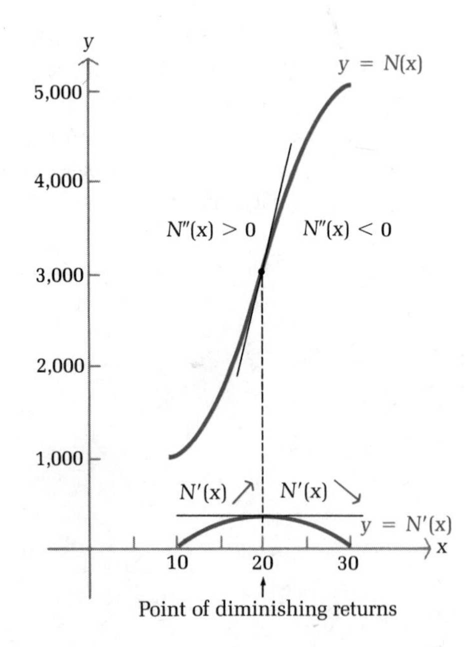

Exercise 2-2

A *Problems 1–4 refer to the following graph of $y = f(x)$:*

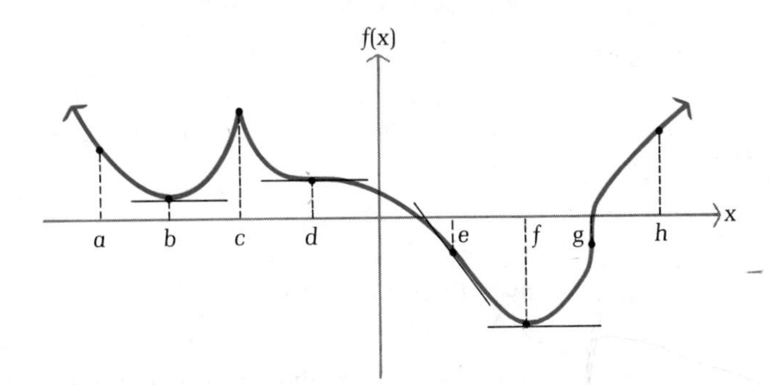

1. Identify intervals over which the graph of f is concave upward.
2. Identify intervals over which the graph of f is concave downward.
3. Identify inflection points.
4. Identify local extrema.

Find the indicated derivative for each function.

5. $f''(x)$ for $f(x) = x^3 - 2x^2 - 1$
6. $g''(x)$ for $g(x) = x^4 - 3x^2 + 5$
7. d^2y/dx^2 for $y = 2x^5 - 3$
8. d^2y/dx^2 for $y = 3x^4 - 7x$
9. $D_x^2(1 - 2x + x^3)$
10. $D_x^2(3x^2 - x^3)$
11. y'' for $y = (x^2 - 1)^3$
12. y'' for $y = (x^2 + 4)^4$
13. $f''(x)$ for $f(x) = 3x^{-1} + 2x^{-2} + 5$
14. $f''(x)$ for $f(x) = x^2 - x^{1/3}$

B *Find all local maxima and minima using the second-derivative test whenever it applies (do not graph). If the second-derivative test fails, use the first-derivative test.*

15. $f(x) = 2x^2 - 8x + 6$
16. $f(x) = 6x - x^2 + 4$
17. $f(x) = 2x^3 - 3x^2 - 12x - 5$
18. $f(x) = 2x^3 + 3x^2 - 12x - 1$
19. $f(x) = 3 - x^3 + 3x^2 - 3x$
20. $f(x) = x^3 + 6x^2 + 12x + 2$
21. $f(x) = x^4 - 8x^2 + 10$
22. $f(x) = x^4 - 18x^2 + 50$
23. $f(x) = x^6 + 3x^4 + 2$
24. $f(x) = 4 - x^6 - 6x^4$
25. $f(x) = x + \dfrac{16}{x}$
26. $f(x) = x + \dfrac{25}{x}$

Find the intervals where the graph of f is concave upward, the intervals where the graph is concave downward, and the inflection points.

27. $f(x) = x^2 - 4x + 5$
28. $f(x) = 9 + 3x - 4x^2$
29. $f(x) = x^3 - 18x^2 + 10x - 11$
30. $f(x) = x^3 + 24x^2 + 15x - 12$
31. $f(x) = x^4 - 24x^2 + 10x - 5$
32. $f(x) = x^4 + 6x^2 + 9x + 11$
33. $f(x) = -x^4 + 4x^3 + 3x + 7$
34. $f(x) = -x^4 - 2x^3 + 12x^2 + 15$

Find local maxima, local minima, and inflection points. Sketch the graph of each function. Include tangent lines at each local extreme point and inflection point.

35. $f(x) = x^3 - 6x^2 + 16$
36. $f(x) = x^3 - 9x^2 + 15x + 10$
37. $f(x) = x^3 + x + 2$
38. $f(x) = 1 - 3x - x^3$
39. $f(x) = (2 - x)^3 + 1$
40. $f(x) = (1 + x)^3 - 1$
41. $f(x) = x^3 - 12x$
42. $f(x) = 27x - x^3$

C *Find the inflection points. Do not graph.*

43. $f(x) = \dfrac{1}{x^2 + 12}$
44. $f(x) = \dfrac{x^2}{x^2 + 12}$
45. $f(x) = \dfrac{x}{x^2 + 12}$
46. $f(x) = \dfrac{x^3}{x^2 + 12}$

Applications

Business & Economics

47. *Revenue.* The marketing research department for a computer company used a large city to test market their new product. They found that a relationship between price p (dollars per unit) and the demand x (units per week) was given approximately by

$$p = 1{,}296 - 0.12x^2 \qquad 0 < x < 80$$

Thus, the weekly revenue can be approximated by

$$R(x) = xp = 1{,}296x - 0.12x^3 \qquad 0 < x < 80$$

(A) Find the local extrema for the revenue function.
(B) Over which intervals is the graph of the revenue function concave upward? Concave downward?

48. *Profit.* If the cost equation for the company in the preceding problem is

$$C(x) = 830 + 396x$$

(A) Find the local extrema for the profit function.
(B) Over which intervals is the graph of the profit function concave upward? Concave downward?

49. *Advertising.* A company estimates that it will sell N(x) units of a

product after spending \$x thousand on advertising, as given by

$$N(x) = -3x^3 + 225x^2 - 3{,}600x + 17{,}000 \qquad 10 \leq x \leq 40$$

(A) When is the rate of change of sales $N'(x)$ increasing? Decreasing?
(B) Find the inflection points for the graph of N.
(C) Graph N and N' on the same axes.
(D) What is the maximum rate of change of sales?

50. *Advertising.* A company estimates that it will sell $N(x)$ units of a product after spending \$x thousand on advertising, as given by

$$N(x) = -2x^3 + 90x^2 - 750x + 2{,}000 \qquad 5 \leq x \leq 25$$

(A) When is the rate of change of sales $N'(x)$ increasing? Decreasing?
(B) Find the inflection points for the graph of N.
(C) Graph N and N' on the same axes.
(D) What is the maximum rate of change of sales?

Life Sciences

51. *Population growth—bacteria.* A drug that stimulates reproduction is introduced into a colony of bacteria. After t minutes, the number of bacteria is given approximately by

$$N(t) = 1{,}000 + 30t^2 - t^3 \qquad 0 \leq t \leq 20$$

(A) When is the rate of growth $N'(t)$ increasing? Decreasing?
(B) Find the inflection points for the graph of N.
(C) Sketch the graph of N and N' on the same axes.
(D) What is the maximum rate of growth?

52. *Drug sensitivity.* One hour after x milligrams of a particular drug are given to a person, the change in body temperature $T(x)$ in degrees Fahrenheit is given by

$$T(x) = x^2\left(1 - \frac{x}{9}\right) \qquad 0 \leq x \leq 6$$

The rate at which $T(x)$ changes with respect to the size of the dosage x, $T'(x)$, is called the *sensitivity* of the body to the dosage.

(A) When is $T'(x)$ increasing? Decreasing?
(B) Where does the graph of T have inflection points?
(C) Sketch the graph of T and T' on the same axes.
(D) What is the maximum value of $T'(x)$?

Social Sciences

53. *Learning.* The time T in minutes it takes a person to learn a list of length n is

$$T(n) = \frac{2}{25}n^3 - \frac{6}{5}n^2 + 6n \qquad 0 \leq n$$

(A) When is the rate of change of T with respect to the length of the list increasing? Decreasing?

(B) Where does the graph of T have inflection points? Graph T and T' on the same axes.
(C) What is the minimum value of $T'(n)$?

2-3 Optimization; Absolute Maxima and Minima

- Absolute Maxima and Minima
- Applications

We are now ready to consider one of the most important applications of the derivative, namely, the use of derivatives to find the *absolute maximum* or *minimum* value of a function. As we mentioned earlier, an economist may be interested in the price or production level of a commodity that will bring a maximum profit; a doctor may be interested in the time it takes for a drug to reach its maximum concentration in the bloodstream after an injection; and a city planner might be interested in the location of heavy industry in a city to produce minimum pollution in residential and business areas. Before we launch an attack on problems of this type, we have to say a few words about the procedures needed to find absolute maximum and absolute minimum values of functions. We have most of the tools we need from the previous sections.

Absolute Maxima and Minima

First, what do we mean by *absolute maximum* and *absolute minimum*? We say that $f(c)$ is an **absolute maximum** of f if

$$f(c) \geq f(x)$$

for all x in the domain of f. Similarly, $f(c)$ is called an **absolute minimum** of f if

$$f(c) \leq f(x)$$

for all x in the domain of f. Figure 10 illustrates several typical examples.

In many practical problems, the domain of a function is restricted because of practical or physical considerations. If the domain is restricted to some closed interval, as is often the case, then Theorem 2 can be proved.

Theorem 2

A function f continuous on a closed interval $[a, b]$ assumes both an absolute maximum and an absolute minimum on that interval.

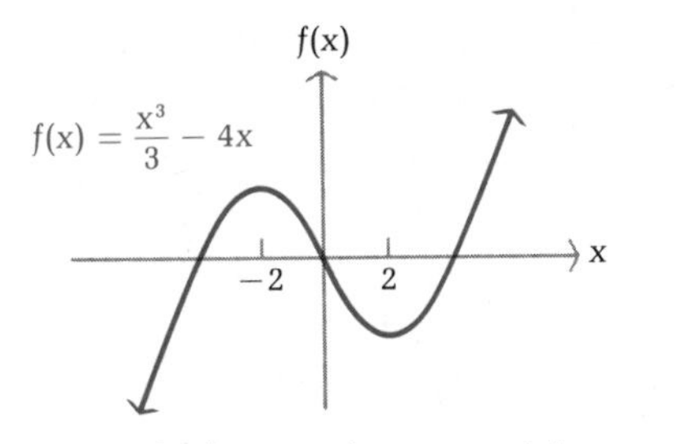

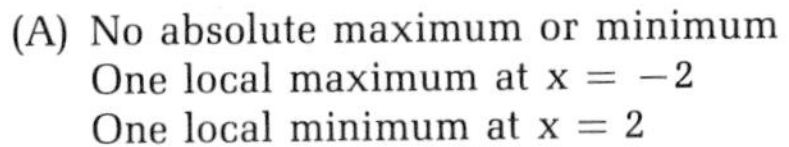

(A) No absolute maximum or minimum
One local maximum at $x = -2$
One local minimum at $x = 2$

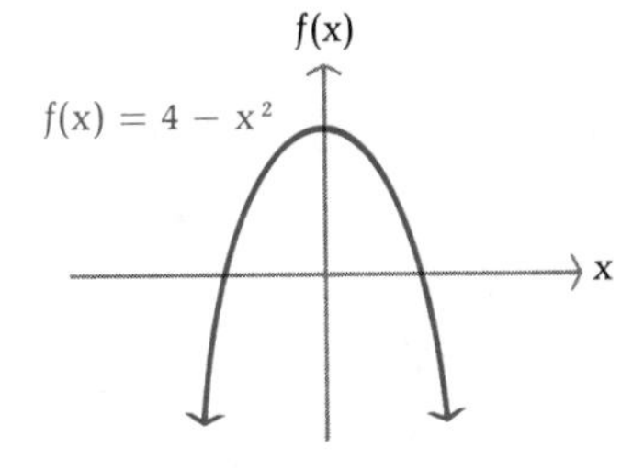

(B) Absolute maximum at $x = 0$
No absolute minimum

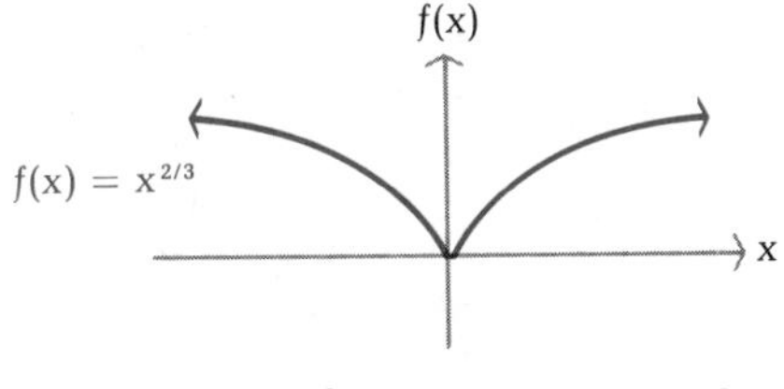

(C) Absolute minimum at $x = 0$
No absolute maximum

Figure 10

It is important to understand that the absolute maximum and minimum depend on both the function f and the interval $[a, b]$ (see Fig. 11, on the next page). However, in all four cases illustrated in Figure 11, the absolute maximum and the absolute minimum both occur at a critical value or an end point. It can be proved that absolute extrema (if they exist) must always occur at critical values or end points. Thus, to find the absolute maximum or minimum value of a continuous function on a closed interval, we simply identify the end points and the critical values in the interval, evaluate each, and then choose the largest and smallest values out of this group.

Steps in Finding Absolute Maximum and Minimum Values of Continuous Functions

1. Check to make certain that f is continuous over $[a, b]$.
2. Find the critical values in the interval (a, b).
3. Evaluate f at the end points and at the critical values found in step 2.
4. The absolute maximum $f(x)$ on $[a, b]$ is the largest of the values found in step 3.
5. The absolute minimum $f(x)$ on $[a, b]$ is the smallest of the values found in step 3.

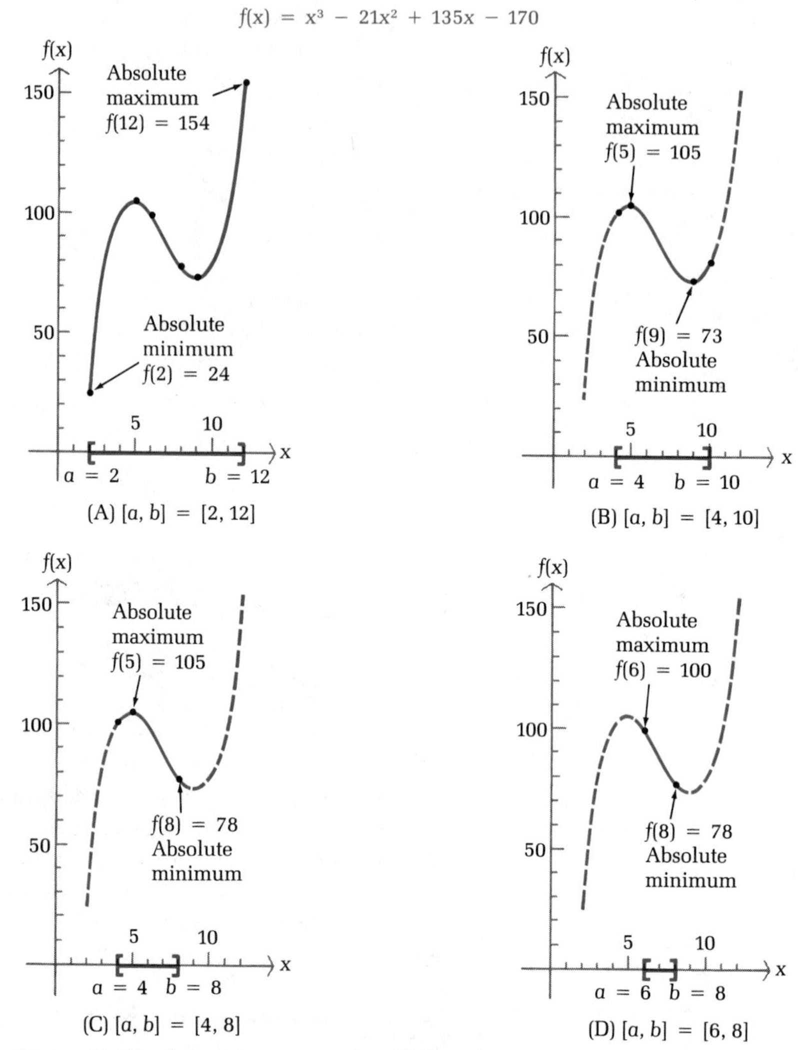

Figure 11 Absolute extrema on a closed interval

Example 8 Find the absolute maximum and absolute minimum values of

$$f(x) = x^3 + 3x^2 - 9x - 7$$

on each of the following intervals:

(A) $[-6, 4]$ (B) $[-4, 2]$ (C) $[-2, 2]$

Solutions (A) The function is continuous for all values of x.

$$f'(x) = 3x^2 + 6x - 9 = 3(x - 1)(x + 3)$$

Thus, $x = -3$ and $x = 1$ are critical values in the interval $(-6, 4)$. Evaluate f at the end points and critical values, $-6, -3, 1$, and 4, and choose the maximum and minimum from these:

$f(-6) = -61$ Absolute minimum

$f(-3) = 20$

$f(1) = -12$

$f(4) = 69$ Absolute maximum

(B) Interval: $[-4, 2]$

x	$f(x)$	
-4	13	
-3	20	Absolute maximum
1	-12	Absolute minimum
2	-5	

(C) Interval: $[-2, 2]$

x	$f(x)$	
-2	15	Absolute maximum
1	-12	Absolute minimum
2	-5	

The critical value $x = -3$ is not included in this table because it is not in the interval $[-2, 2]$.

Problem 8 Find the absolute maximum and absolute minimum values of

$$f(x) = x^3 - 12x$$

on each of the following intervals:

(A) $[-5, 5]$ (B) $[-3, 3]$ (C) $[-3, 1]$

Now suppose we want to find the absolute maximum or minimum value of a function that is continuous on an interval that is not closed. Since Theorem 2 no longer applies, we cannot be certain that the absolute maximum or minimum value exists. Figure 12 (on the next page) illustrates several ways that functions can fail to have absolute extrema.

In general, the best procedure to follow when the interval is not a closed interval (that is, is not of the form $[a, b]$) is to sketch the graph of the function. However, there is one special case that occurs frequently in applications and that can be analyzed without drawing a graph. It often happens that f is continuous on an interval I and has only one critical value c in the interval I (here I can be any type of interval—open, closed, or

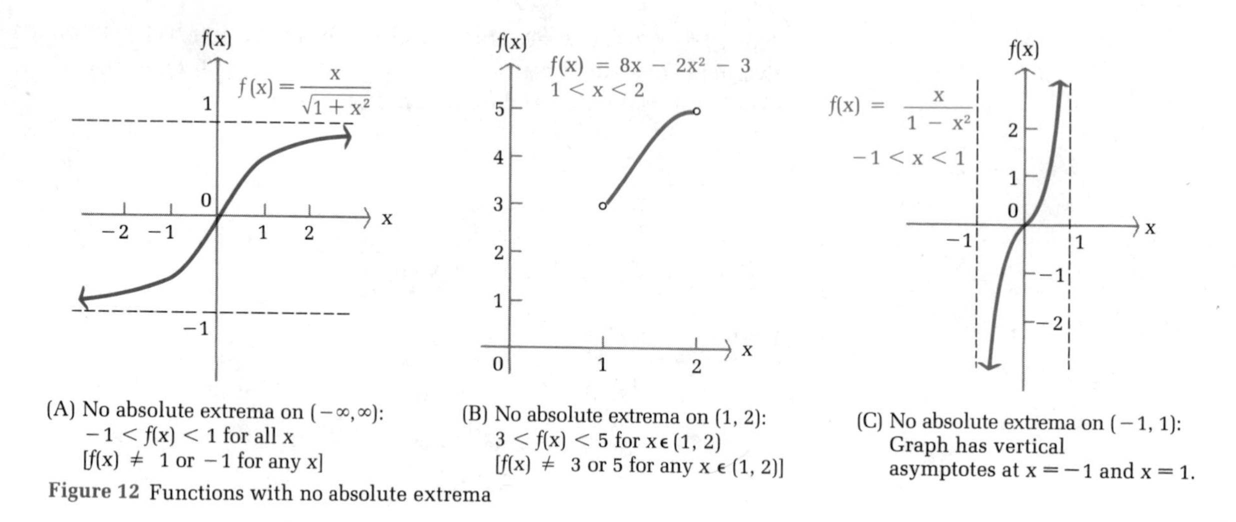

(A) No absolute extrema on $(-\infty, \infty)$:
$-1 < f(x) < 1$ for all x
$[f(x) \neq 1$ or -1 for any x$]$

(B) No absolute extrema on (1, 2):
$3 < f(x) < 5$ for $x \in (1, 2)$
$[f(x) \neq 3$ or 5 for any $x \in (1, 2)]$

(C) No absolute extrema on $(-1, 1)$:
Graph has vertical asymptotes at $x = -1$ and $x = 1$.

Figure 12 Functions with no absolute extrema

half-closed). If this is the case and if $f''(c)$ exists, then we have the second-derivative test for absolute extrema given in the box below.

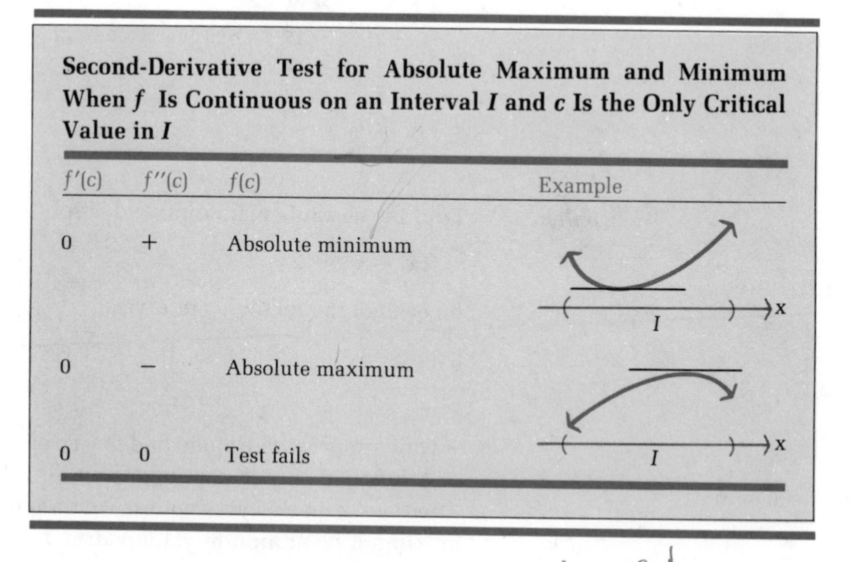

Second-Derivative Test for Absolute Maximum and Minimum When f Is Continuous on an Interval I and c Is the Only Critical Value in I

$f'(c)$	$f''(c)$	$f(c)$	Example
0	+	Absolute minimum	
0	−	Absolute maximum	
0	0	Test fails	

Example 9 Find the absolute minimum value of

$$f(x) = x + \frac{4}{x}$$

on the interval $(0, \infty)$.

Solution

$$f'(x) = 1 - \frac{4}{x^2} = \frac{x^2 - 4}{x^2} = \frac{(x-2)(x+2)}{x^2} \qquad f''(x) = \frac{8}{x^3}$$

The only critical value in the interval $(0, \infty)$ is $x = 2$. Since $f''(2) = 1 > 0$, $f(2) = 4$ is the absolute minimum value of f on $(0, \infty)$.

Problem 9 Find the absolute maximum value of

$$f(x) = 12 - x - \frac{9}{x}$$

on the interval $(0, \infty)$.

■ Applications

Now we want to solve some applied problems that involve absolute extrema. Before beginning, we outline in the next box the steps to follow in solving this type of problem. The first step is the most difficult one. The techniques used to construct the model are best illustrated through a series of examples.

A Strategy for Solving Applied Optimization Problems

Step 1. Introduce variables and construct a mathematical model of the form

Maximize (or minimize) $f(x)$ on the interval I

Step 2. Find the absolute maximum (or minimum) value of $f(x)$ on the interval I and the value(s) of x where this occurs.

Step 3. Use the solution to the mathematical model to answer the questions asked in the application.

Example 10 Cost–Demand

A company manufactures and sells x transistor radios per week. If the weekly cost and demand equations are

$$C(x) = 5{,}000 + 2x$$

$$p = 10 - \frac{x}{1{,}000} \qquad 0 \leq x \leq 8{,}000$$

find for each week:

(A) The maximum revenue

(B) The maximum profit, the production level that will realize the maximum profit, and the price that the company should charge for each radio

Solutions (A) The revenue received for selling x radios at $p per radio is

$$R(x) = xp$$
$$= x\left(10 - \frac{x}{1{,}000}\right)$$
$$= 10x - \frac{x^2}{1{,}000}$$

Thus, the mathematical model is

$$\text{Maximize} \quad R(x) = 10x - \frac{x^2}{1{,}000} \qquad 0 \leq x \leq 8{,}000$$

$$R'(x) = 10 - \frac{x}{500}$$

$$10 - \frac{x}{500} = 0$$

$$x = 5{,}000 \qquad \text{Only critical value}$$

Use the second-derivative test for absolute extrema:

$$R''(x) = -\frac{1}{500} < 0 \qquad \text{for all } x$$

Thus, the maximum revenue is

$$\text{Max } R(x) = R(5{,}000) = \$25{,}000$$

(B) Profit = Revenue − Cost

$$P(x) = R(x) - C(x)$$
$$= 10x - \frac{x^2}{1{,}000} - 5{,}000 - 2x$$
$$= 8x - \frac{x^2}{1{,}000} - 5{,}000$$

The mathematical model is

$$\text{Maximize} \quad P(x) = 8x - \frac{x^2}{1{,}000} - 5{,}000 \qquad 0 \leq x \leq 8{,}000$$

$$P'(x) = 8 - \frac{x}{500}$$

$$8 - \frac{x}{500} = 0$$

$$x = 4{,}000$$

$$P''(x) = -\frac{1}{500} < 0 \qquad \text{for all } x$$

Since $x = 4{,}000$ is the only critical value and $P''(x) < 0$,

$$\text{Max } P(x) = P(4{,}000) = \$11{,}000$$

Using the price–demand equation with $x = 4{,}000$, we find

$$p = 10 - \frac{4{,}000}{1{,}000} = \$6$$

Thus, a maximum profit of \$11,000 per week is realized when 4,000 radios are produced weekly and sold for \$6 each. Notice that this is not the same level of production that produces the maximum revenue.

All the results in this example are illustrated in Figure 13. We also note that profit is maximum when

$$P'(x) = R'(x) - C'(x) = 0$$

that is, when the marginal revenue is equal to the marginal cost (the rate of increase in revenue is the same as the rate of increase in cost at the 4,000 output level—notice that the slopes of the two curves are the same at this point).

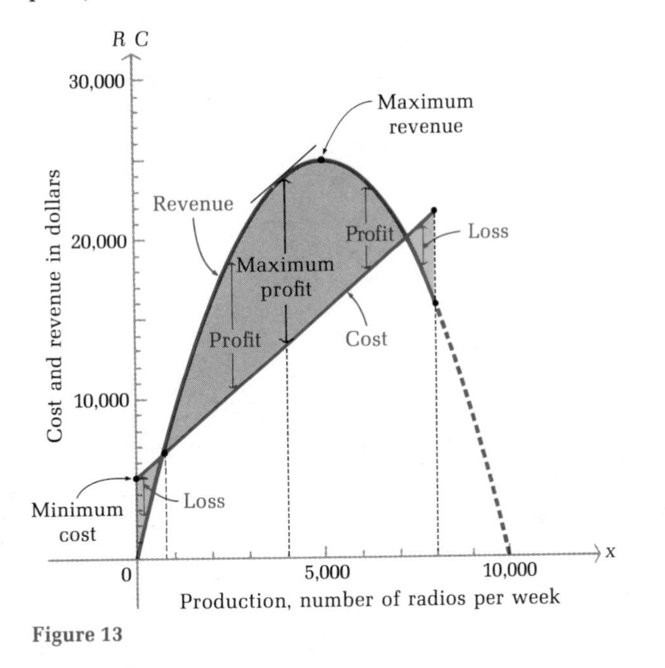

Figure 13

Problem 10 Repeat Example 10 for

$$C(x) = 90{,}000 + 30x$$

$$p = 300 - \frac{x}{30} \qquad 0 \le x \le 9{,}000$$

Example 11 Profit

In Example 10 the government has decided to tax the company \$2 for each radio produced. Taking into account this additional cost, how many radios should the company manufacture each week in order to maximize its weekly profit? What is the maximum weekly profit? How much should it charge for the radios?

Solution

The tax of \$2 per unit changes the company's cost equation:

$$\begin{aligned} C(x) &= \text{Original cost} + \text{Tax} \\ &= 5{,}000 + 2x + 2x \\ &= 5{,}000 + 4x \end{aligned}$$

The new profit function is

$$\begin{aligned} P(x) &= R(x) - C(x) \\ &= 10x - \frac{x^2}{1{,}000} - 5{,}000 - 4x \\ &= 6x - \frac{x^2}{1{,}000} - 5{,}000 \end{aligned}$$

Thus, we must solve the following:

$$\text{Maximize} \quad P(x) = 6x - \frac{x^2}{1{,}000} - 5{,}000 \qquad 0 \leq x \leq 8{,}000$$

$$P'(x) = 6 - \frac{x}{500}$$

$$6 - \frac{x}{500} = 0$$

$$x = 3{,}000$$

$$P''(x) = -\frac{1}{500} < 0$$

$$\text{Max } P(x) = P(3{,}000) = \$4{,}000$$

Using the price–demand equation with $x = 3{,}000$, we find

$$p = 10 - \frac{3{,}000}{1{,}000} = \$7$$

Thus, the company's maximum profit is \$4,000 when 3,000 radios are produced and sold weekly at a price of \$7.

Even though the tax caused the company's cost to increase by \$2 per radio, the price that the company should charge to maximize its profit increases by only \$1. The company must absorb the other \$1 with a resulting decrease of \$7,000 in maximum profit.

Problem 11 Repeat Example 11 if

$$C(x) = 90{,}000 + 30x$$

$$p = 300 - \frac{x}{30} \qquad 0 \leq x \leq 9{,}000$$

and the government decides to tax the company \$20 for each unit produced. Compare the results with the results in Problem 10B.

Example 12
Maximize Yield

A walnut grower estimates from past records that if twenty trees are planted per acre, each tree will average 60 pounds of nuts per year. If for each additional tree planted per acre (up to fifteen) the average yield per tree drops 2 pounds, how many trees should be planted to maximize the yield per acre? What is the maximum yield?

Solution Let x be the number of additional trees planted per acre. Then

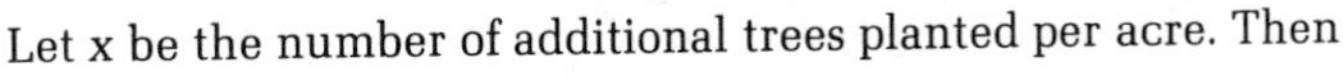

$$20 + x = \text{Total number of trees per acre}$$

$$60 - 2x = \text{Yield per tree}$$

$$\text{Yield per acre} = (\text{Total number of trees per acre})(\text{Yield per tree})$$

$$Y(x) = (20 + x)(60 - 2x)$$

$$= 1{,}200 + 20x - 2x^2 \qquad 0 \leq x \leq 15$$

Thus, we must solve the following:

$$\text{Maximize} \quad Y(x) = 1{,}200 + 20x - 2x^2 \qquad 0 \leq x \leq 15$$

$$Y'(x) = 20 - 4x$$

$$20 - 4x = 0$$

$$x = 5$$

$$Y''(x) = -4 < 0 \qquad \text{for all } x$$

Hence,

$$\text{Max } Y(x) = Y(5) = 1{,}250 \text{ pounds per acre}$$

Thus, a maximum yield of 1,250 pounds of nuts per acre is realized if twenty-five trees are planted per acre.

Problem 12 Repeat Example 12 starting with thirty trees per acre and a reduction of 1 pound per tree for each additional tree planted.

Example 13
Maximize Area

A farmer wants to construct a rectangular pen next to a barn 60 feet long, using all of the barn as part of one side of the pen. Find the dimensions of the pen with the largest area that the farmer can build if:

(A) 160 feet of fencing material is available
(B) 250 feet of fencing material is available

Solutions (A) We begin by constructing and labeling a figure:

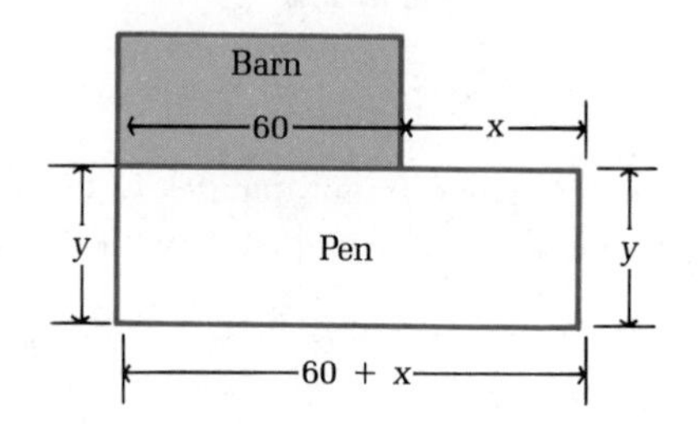

The area of the pen is

$$A = (x + 60)y$$

Before we can maximize the area, we must determine a relationship between x and y in order to express A as a function of one variable. In this case, x and y are related to the total amount of available fencing material:

$$x + y + 60 + x + y = 160$$
$$2x + 2y = 100$$
$$y = 50 - x$$

Thus,

$$A(x) = (x + 60)(50 - x)$$

Now we need to determine the permissible values of x. Since the farmer wants to use all of the barn as part of one side of the pen, x cannot be negative. Since y is the other dimension of the pen, y cannot be negative. Thus,

$$y = 50 - x \geqslant 0$$
$$50 \geqslant x$$

Thus, we must solve the following:

$$\text{Maximize} \quad A(x) = (x + 60)(50 - x) \qquad 0 \leqslant x \leqslant 50$$

$$A(x) = 3{,}000 - 10x - x^2$$
$$A'(x) = -10 - 2x$$
$$-10 - 2x = 0$$
$$x = -5$$

Since $x = -5$ is not in the interval [0, 50], there are no critical points in the interval. $A(x)$ is continuous on [0, 50], so the absolute maximum must occur at one of the end points.

$$A(0) = 3{,}000 \qquad \text{Maximum area}$$
$$A(50) = 0$$

If $x = 0$, then $y = 50$. Thus, the dimensions of the pen with largest area are 60 feet by 50 feet.

(B) If there is 250 feet of fencing material available, then

$$x + y + x + 60 + y = 250$$
$$2x + 2y = 190$$
$$y = 95 - x$$

The model becomes

$$\text{Maximize} \quad A(x) = (x + 60)(95 - x) \qquad 0 \leq x \leq 100$$

$$A(x) = 5{,}700 + 35x - x^2$$
$$A'(x) = 35 - 2x$$
$$35 - 2x = 0$$
$$x = \tfrac{35}{2} = 17.5 \qquad \text{The only critical value}$$
$$A''(x) = -2 < 0$$
$$\text{Max } A(x) = A(17.5) = 6{,}006.25$$
$$y = 95 - 17.5 = 77.5$$

This time the dimensions of the pen with the largest area are 77.5 feet by 77.5 feet.

Problem 13 Repeat Example 13 if the barn is 80 feet long.

Example 14
Inventory Control

A record company anticipates that there will be a demand for 20,000 copies of a certain album during the following year. It costs the company \$0.50 to store a record for one year. Each time it must press additional records, it costs \$200 to set up the equipment. How many records should the company press during each production run in order to minimize its total storage and set-up costs?

Solution

This type of problem is called an **inventory control problem.** One of the basic assumptions made in such problems is that the demand is uniform. For example, if there are 250 working days in a year, then the daily demand would be $20{,}000/250 = 800$ records. The company could decide to produce all 20,000 records at the beginning of the year. This would certainly minimize the set-up costs, but would result in very large storage costs. At the other extreme, it could produce 800 records each day. This would minimize the storage costs, but would result in very large set-up costs. Somewhere between these two extremes is the optimal solution that will minimize the total storage and set-up costs. Let

x = Number of records pressed during each production run

y = Number of production runs

It is easy to see that the total set-up cost for the year is $200y$, but what is the total storage cost? If the demand is uniform, then the number of records in storage between production runs will decrease from x to 0 and the average number in storage each day is $x/2$. This result is illustrated in the following figure:

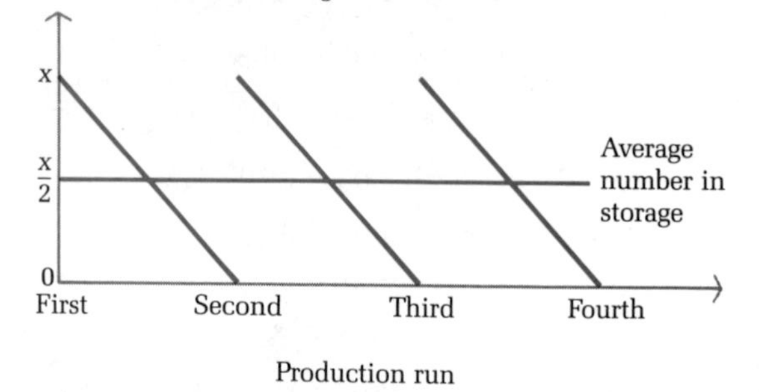

Since it costs \$0.50 to store one record for a year, the total storage cost is $0.5(x/2) = 0.25x$ and the total cost is

$$\text{Total cost} = \text{Set-up cost} + \text{Storage cost}$$

$$C = 200y + 0.25x$$

If the company produces x records in each of y production runs, then the total number of records produced is xy. Thus,

$$xy = 20{,}000$$

$$y = \frac{20{,}000}{x}$$

Certainly, x must be at least 1 and cannot exceed 20,000. Thus, we must solve the following:

$$\text{Minimize} \quad C(x) = 200\left(\frac{20{,}000}{x}\right) + 0.25x \qquad 1 \leq x \leq 20{,}000$$

$$C(x) = \frac{4{,}000{,}000}{x} + 0.25x$$

$$C'(x) = -\frac{4{,}000{,}000}{x^2} + 0.25$$

$$-\frac{4{,}000{,}000}{x^2} + 0.25 = 0$$

$$x^2 = \frac{4{,}000{,}000}{0.25}$$

$$x^2 = 16{,}000{,}000$$

$$x = 4{,}000 \qquad -4{,}000 \text{ is not a critical value since } 1 \leq x \leq 20{,}000$$

$$C''(x) = \frac{8{,}000{,}000}{x^3} > 0 \qquad \text{for } x \in (1, 20{,}000)$$

Thus,

$$\text{Min } C(x) = C(4{,}000) = 2{,}000$$

$$y = \frac{20{,}000}{4{,}000} = 5$$

The company will minimize its total cost by pressing 4,000 records five times during the year.

Problem 14 Repeat Example 14 if it costs $250 to set up a production run and $0.40 to store one record for a year.

Answers to Matched Problems

8. (A) Absolute maximum: $f(5) = 65$; absolute minimum: $f(-5) = -65$
   (B) Absolute maximum: $f(-2) = 16$; absolute minimum: $f(2) = -16$
   (C) Absolute maximum: $f(-2) = 16$; absolute minimum: $f(1) = -11$
9. $f(3) = 6$
10. (A) Max $R(x) = R(4{,}500) = \$675{,}000$
    (B) Max $P(x) = P(4{,}050) = \$456{,}750$; $p = \$165$
11. Max $P(x) = P(3{,}750) = \$378{,}750$; $p = \$175$; price increases $10, profit decreases $78,000
12. Max $Y(x) = Y(15) = 2{,}025$ pounds per acre
13. (A) 80 feet by 40 feet (B) 82.5 feet by 82.5 feet
14. Press 5,000 records four times during the year

Exercise 2-3

A *Find the absolute maximum and absolute minimum, if either exists, for each function.*

1. $f(x) = x^2 - 4x + 5$
2. $f(x) = x^2 + 6x + 7$
3. $f(x) = 10 + 8x - x^2$
4. $f(x) = 6 - 8x - x^2$
5. $f(x) = 1 - x^3$
6. $f(x) = 1 - x^4$

B *Find the indicated extrema of each function.*

7. Absolute maximum value of $f(x) = 24 - 2x - \frac{8}{x}, \quad x > 0$

8. Absolute minimum value of $f(x) = 3x + \frac{27}{x}$, $x > 0$

9. Absolute minimum value of $f(x) = 5 + 3x + \frac{12}{x^2}$, $x > 0$

10. Absolute maximum value of $f(x) = 10 - 2x - \frac{27}{x^2}$, $x > 0$

Find the absolute maximum and minimum, if either exists, of each function on the indicated intervals.

11. $f(x) = x^3 - 6x^2 + 9x - 6$

    (A) $[-1, 5]$ (B) $[-1, 3]$ (C) $[2, 5]$

12. $f(x) = 2x^3 - 3x^2 - 12x + 24$

    (A) $[-3, 4]$ (B) $[-2, 3]$ (C) $[-2, 1]$

13. $f(x) = (x - 1)(x - 5)^3 + 1$

    (A) $[0, 3]$ (B) $[1, 7]$ (C) $[3, 6]$

14. $f(x) = x^4 - 8x^2 + 16$

    (A) $[-1, 3]$ (B) $[0, 2]$ (C) $[-3, 4]$

Preliminary word problems:

C

15. How would you divide a 10 inch line so that the product of the two lengths is a maximum?
16. What quantity should be added to 5 and subtracted from 5 in order to produce the maximum product of the results?
17. Find two numbers whose difference is 30 and whose product is a minimum.
18. Find two positive numbers whose sum is 60 and whose product is a maximum.
19. Find the dimensions of a rectangle with perimeter 100 centimeters that has maximum area. Find the maximum area.
20. Find the dimensions of a rectangle of area 225 square centimeters that has the least perimeter. What is the perimeter?

Applications

Business & Economics

21. *Average costs.* If the average manufacturing cost (in dollars) per pair of sunglasses is given by

    $$\overline{C}(x) = x^2 - 6x + 12 \qquad 0 \leq x \leq 6$$

    where x is the number (in thousands) of pairs manufactured, how many pairs of glasses should be manufactured to minimize the average cost per pair? What is the minimum average cost per pair?

22. *Maximum revenue and profit.* A company manufactures and sells x television sets per month. The monthly cost and demand equations are

$$C(x) = 72{,}000 + 60x$$

$$p = 200 - \frac{x}{30} \qquad 0 \leq x \leq 6{,}000$$

(A) Find the maximum revenue.

(B) Find the maximum profit, the production level that will realize the maximum profit, and the price the company should charge for each television set.

(C) If the government decides to tax the company $5 for each set it produces, how many sets should the company manufacture each month in order to maximize its profit? What is the maximum profit? What should the company charge for each set?

23. *Car rental.* A car rental agency rents 200 cars per day at a rate of $30 per day. For each $1 increase in rate, five fewer cars are rented. At what rate should the cars be rented to produce the maximum income? What is the maximum income?

24. *Rental income.* A 300 room hotel in Las Vegas is filled to capacity every night at $80 a room. For each $1 increase in rent, three fewer rooms are rented. If each rented room costs $10 to service per day, how much should the management charge for each room to maximize gross profit? What is the maximum gross profit?

25. *Agriculture.* A commercial cherry grower estimates from past records that if thirty trees are planted per acre, each tree will yield an average of 50 pounds of cherries per season. If for each additional tree planted per acre (up to twenty) the average yield per tree is reduced by 1 pound, how many trees should be planted per acre to obtain the maximum yield per acre? What is the maximum yield?

26. *Agriculture.* A commercial pear grower must decide on the optimum time to have fruit picked and sold. If the pears are picked now, they will bring 30¢ per pound, with each tree yielding an average of 60 pounds of salable pears. If the average yield per tree increases 6 pounds per tree per week for the next 4 weeks, but the price drops 2¢ per pound per week, when should the pears be picked to realize the maximum return per tree? What is the maximum return?

27. *Manufacturing.* A candy box is to be made out of a piece of cardboard that measures 8 by 12 inches. Squares of equal size will be cut out of each corner, and then the ends and sides will be folded up to form a rectangular box. What size square should be cut from each corner to obtain a maximum volume?

28. *Packaging.* A parcel delivery service will deliver a package only if the length plus girth (distance around) does not exceed 108 inches.

(A) Find the dimensions of a rectangular box with square ends that satisfies the delivery service's restriction and has maximum volume. What is the maximum volume?

(B) Find the dimensions (radius and height) of a cylindrical container that meets the delivery service's requirement and has maximum volume. What is the maximum volume?

29. *Construction costs.* A fence is to be built to enclose a rectangular area of 800 square feet. The fence along three sides is to be made of material that costs \$2 per foot. The material for the fourth side costs \$6 per foot. Find the dimensions of the rectangle that will allow the most economical fence to be built.

30. *Construction costs.* The owner of a retail lumber store wants to construct a fence to enclose an outdoor storage area adjacent to the store as indicated in the accompanying figure. Find the dimensions that will enclose the largest area if:

(A) 240 feet of fencing material is used.

(B) 400 feet of fencing material is used.

31. *Inventory control.* A publishing company sells 50,000 copies of a certain book each year. It costs the company \$1.00 to store a book for one year. Each time it must print additional copies, it costs the company \$1,000 to set up the presses. How many books should the company produce during each printing in order to minimize its total storage and set-up costs?

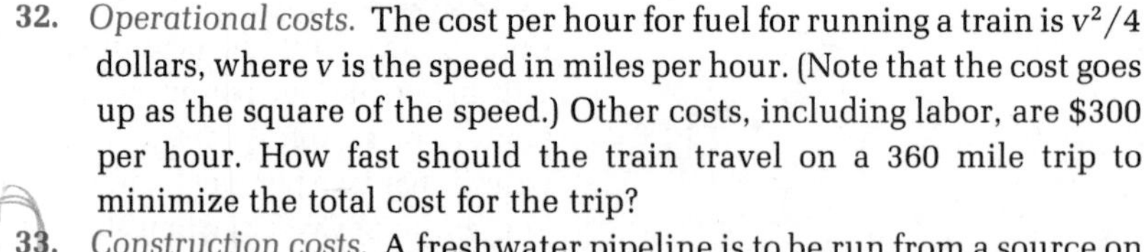

32. *Operational costs.* The cost per hour for fuel for running a train is $v^2/4$ dollars, where v is the speed in miles per hour. (Note that the cost goes up as the square of the speed.) Other costs, including labor, are \$300 per hour. How fast should the train travel on a 360 mile trip to minimize the total cost for the trip?

33. *Construction costs.* A freshwater pipeline is to be run from a source on the edge of a lake to a small resort community on an island 5 miles off-shore, as indicated in the figure at the top of the next page.

(A) If it costs 1.4 times as much to lay the pipe in the lake as it does on land, what should x be (in miles) to minimize the total cost of the project?

(B) If it costs only 1.1 times as much to lay the pipe in the lake as it does on land, what should x be to minimize the total cost of the project? [*Note:* Compare with Problem 38.]

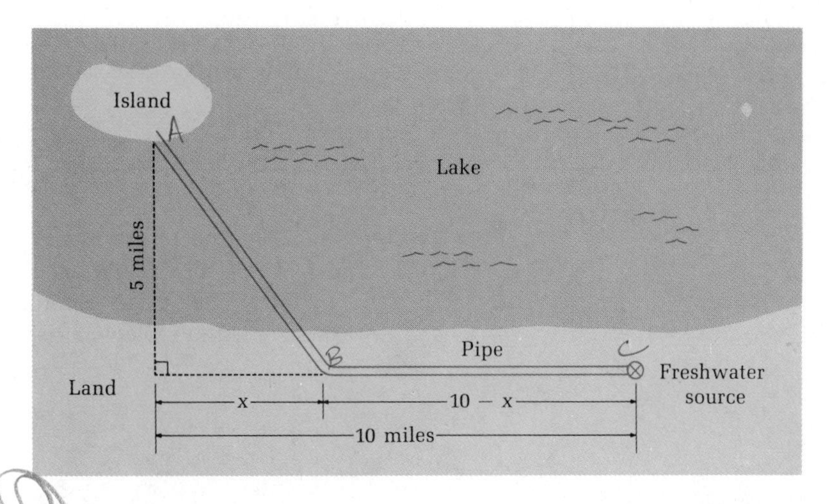

34. *Manufacturing costs.* A manufacturer wants to produce cans that will hold 12 ounces (approximately 22 cubic inches) in the form of a right circular cylinder. Find the dimensions (radius of an end and height) of the can that will use the smallest amount of material. Assume the circular ends are cut out of squares, with the corner portions wasted, and the sides are made from rectangles, with no waste.

Life Sciences

35. *Bacteria control.* A recreational swimming lake is treated periodically to control harmful bacteria growth. Suppose t days after a treatment, the concentration of bacteria per cubic centimeter is given by

$$C(t) = 30t^2 - 240t + 500 \qquad 0 \le t \le 8$$

How many days after a treatment will the concentration be minimal? What is the minimum concentration?

36. *Drug concentration.* The concentration $C(t)$ in milligrams per cubic centimeter of a particular drug in a patient's bloodstream is given by

$$C(t) = \frac{0.16t}{t^2 + 4t + 4}$$

where t is the number of hours after the drug is taken. How many hours after the drug is given will the concentration be maximum? What is the maximum concentration?

37. *Laboratory management.* A laboratory uses 500 white mice each year for experimental purposes. It costs \$4.00 to feed a mouse for one year. Each time mice are ordered from a supplier, there is a service charge of \$10 for processing the order. How many mice should be ordered each time in order to minimize the total cost of feeding the mice and of placing the orders for the mice?

38. *Bird flights.* Some birds tend to avoid flights over large bodies of water during daylight hours. It is speculated that more energy is required to fly over water than land because air generally rises over land and falls over water during the day. Suppose an adult bird with these tenden-

cies is taken from its nesting area on the edge of a large lake to an island 5 miles off-shore and is then released (see the accompanying figure).

(A) If it takes 1.4 times as much energy to fly over water as land, how far up-shore (x, in miles) should the bird head in order to minimize the total energy expended in returning to the nesting area?

(B) If it takes only 1.1 times as much energy to fly over water as land, how far up-shore should the bird head in order to minimize the total energy expended in returning to the nesting area? [*Note:* Compare with Problem 33.]

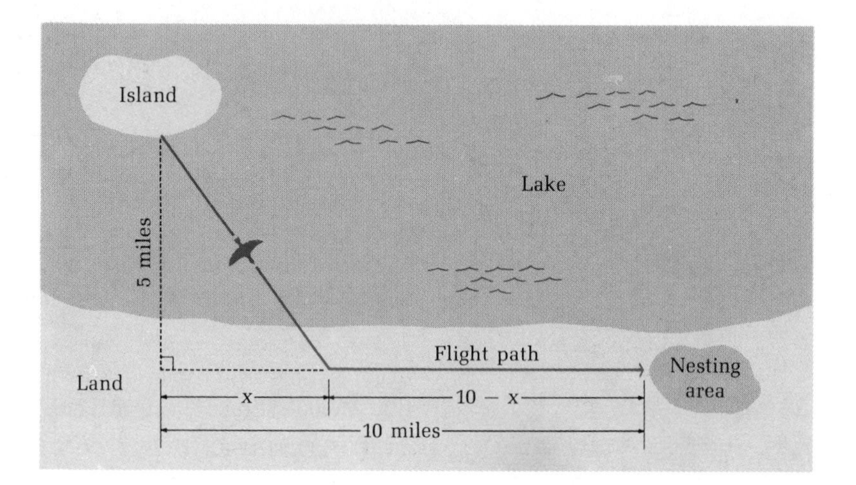

39. *Botany.* If it is known from past experiments that the height in feet of a given plant after t months is given approximately by

$$H(t) = 4t^{1/2} - 2t \qquad 0 \leq t \leq 2$$

how long, on the average, will it take a plant to reach its maximum height? What is the maximum height?

40. *Pollution.* Two heavy industrial areas are located 10 miles apart, as indicated in the figure at the top of the next page. If the concentration of particulate matter in parts per million decreases as the reciprocal of the square of the distance from the source, and area A_1 emits eight times the particulate matter as A_2, then the concentration of particulate matter at any point between the two areas is given by

$$C(x) = \frac{8k}{x^2} + \frac{k}{(10 - x)^2} \qquad 0.5 \leq x \leq 9.5, \quad k > 0$$

How far from A_1 will the concentration of particulate matter be at a minimum?

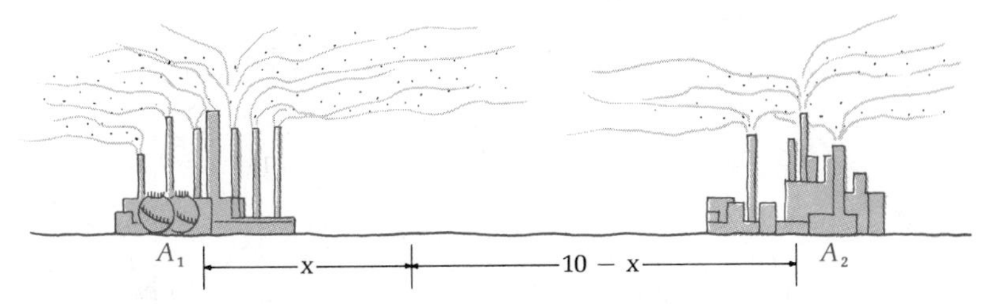

Social Sciences

41. *Politics.* In a newly incorporated city it is estimated that the voting population (in thousands) will increase according to

$$N(t) = 30 + 12t^2 - t^3 \qquad 0 \leq t \leq 8$$

where t is time in years. When will the rate of increase be most rapid?

42. *Learning.* A large grocery chain found that, on the average, a checker can memorize $P\%$ of a given price list in x continuous hours, as given approximately by

$$P(x) = 96x - 24x^2 \qquad 0 \leq x \leq 3$$

How long should a checker plan to take to memorize the maximum percentage? What is the maximum?

2-4 Curve Sketching Techniques: Unified and Extended

- Asymptotes
- Graphing Strategy
- Using the Strategy
- Application

In this section we will apply, in a systematic way, all the graphing concepts discussed in Sections 2-1 and 2-2. Before outlining a graphing strategy and considering the graphs of specific functions, we need to discuss one additional graphing concept, *asymptotes.*

Asymptotes

Horizontal and vertical asymptotes were introduced in Section 1-1. To review these concepts, consider the function f whose graph is shown in Figure 14 (on the next page). The lines $y = L$ and $y = M$ are horizontal asymptotes since $\lim_{x\to\infty} f(x) = L$ and $\lim_{x\to-\infty} f(x) = M$. The lines $x = a$ and $x = b$ are vertical asymptotes since the values of $f(x)$ become very large in absolute value for x near a and x near b.

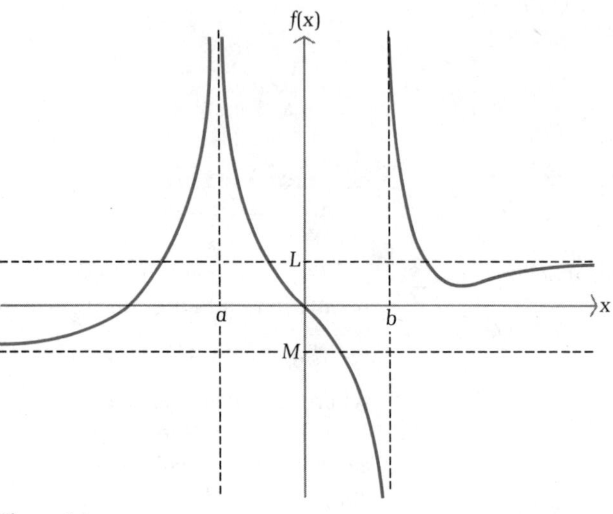

Figure 14

How can we find the asymptotes of a function before we draw its graph? Horizontal asymptotes are located by evaluating

$$\lim_{x\to\infty} f(x) \quad \text{and} \quad \lim_{x\to-\infty} f(x) \tag{1}$$

If f is a rational function (a ratio of polynomials), and either of the limits in (1) exists, then the other also exists and has the same value (see Theorem 3 in Section 1-1). Thus, **a rational function has at most one horizontal asymptote.** (Notice that this implies that the function graphed in Figure 14 is not a rational function, since it has two horizontal asymptotes.)

Theorem 3 provides a simple method for locating vertical asymptotes.

Theorem 3

Let $f(x) = p(x)/q(x)$, where both $p(x)$ and $q(x)$ are continuous at $x = c$. If $q(c) = 0$ and $p(c) \neq 0$, then the line $x = c$ is a vertical asymptote for the graph of f.

Since polynomial functions are continuous for all values of x, Theorem 3 can be applied to any rational function. Thus, **a rational function has vertical asymptotes at all values of x where the denominator equals 0, provided that the numerator is nonzero at that value of x.**

The next example shows how Theorem 3 and the limit techniques discussed in Section 1-1 can be used to locate asymptotes. Later in this section, we will discuss graphing techniques for functions with asymptotes.

Example 15 Find horizontal and vertical asymptotes for:

(A) $f(x) = \dfrac{6x + 5}{2x - 4}$

(B) $f(x) = \dfrac{x}{x^2 + 1}$

(C) $f(x) = \dfrac{x^2 - 4}{x}$

Solutions

(A) $\lim_{x \to \infty} \dfrac{6x + 5}{2x - 4} = \dfrac{6}{2} = 3$ Theorem 3 in Section 1-1

Thus, the line $y = 3$ is a horizontal asymptote. Now let $q(x) = 2x - 4$ and $p(x) = 6x + 5$. Since $q(2) = 0$ and $p(2) = 17 \neq 0$, Theorem 3 (on the facing page) implies that the line $x = 2$ is a vertical asymptote.

(B) $\lim_{x \to \infty} \dfrac{x}{x^2 + 1} = 0$ Theorem 3 in Section 1-1

Thus, the line $y = 0$ (the x axis) is a horizontal asymptote. Since the denominator is never 0, there are no vertical asymptotes.

(C) $\lim_{x \to \infty} \dfrac{x^2 - 4}{x}$ does not exist   Theorem 3 in Section 1-1

Thus, there is no horizontal asymptote. Let $q(x) = x$ and $p(x) = x^2 - 4$. Since $q(0) = 0$ and $p(0) = -4 \neq 0$, the line $x = 0$ (the y axis) is a vertical asymptote.

Problem 15 Find horizontal and vertical asymptotes for:

(A) $f(x) = \dfrac{3x + 5}{x + 2}$

(B) $f(x) = \dfrac{x + 1}{x^2}$

(C) $f(x) = \dfrac{x^3}{x^2 + 4}$

■ Graphing Strategy

We now have powerful tools to determine the shape of a graph of a function, even before we plot any points. We can accurately sketch the graphs of many functions using these tools and point-by-point plotting as necessary (often, very little point-by-point plotting is necessary). We organize these tools in the graphing strategy summarized in the box on the next page.

A Graphing Strategy

[Omit any of the following steps if procedures involved are too difficult or impossible (what may seem too difficult now, with a little practice, will become less so).]

Step 1. Use the first derivative. Construct a sign chart for $f'(x)$, determine the intervals where $f(x)$ is increasing and decreasing, and find local maxima and minima.

Step 2. Use the second derivative. Construct a sign chart for $f''(x)$, determine the intervals where the graph of f is concave upward and downward, and find any inflection points.

Step 3. Find horizontal and vertical asymptotes. Find any horizontal asymptotes by calculating $\lim_{x\to\pm\infty} f(x)$. Find any vertical asymptotes by using Theorem 3.

Step 4. Find intercepts. Find the y intercept by evaluating $f(0)$, if it exists. Find x intercepts by solving the equation $f(x) = 0$ for x, if possible. This equation may be too difficult to solve and the x intercepts are omitted.

Step 5. Sketch the graph of f. Draw asymptotes and locate intercepts, local maxima and minima, and inflection points. Sketch in what you know from steps 1–4. Use point-by-point plotting to complete the graph in regions of uncertainty.

Using the Strategy

Several examples will illustrate the use of the graphing strategy.

Example 16 Graph $f(x) = x^4 - 2x^3 + 2$ using the graphing strategy.

Solution Step 1. Use the first derivative:

$$f'(x) = 4x^3 - 6x^2 = 4x^2(x - \tfrac{3}{2})$$

Sign chart for $f'(x) = 4x^2(x - \frac{3}{2})$:

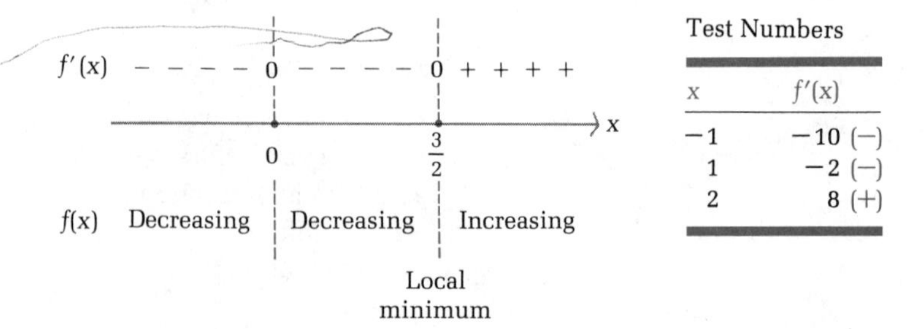

Test Numbers

x	$f'(x)$
−1	−10 (−)
1	−2 (−)
2	8 (+)

Thus, $f(x)$ is decreasing on $(-\infty, \frac{3}{2})$, increasing on $(\frac{3}{2}, \infty)$, and has a local minimum at $x = \frac{3}{2}$.

Step 2. Use the second derivative:

$$f''(x) = 12x^2 - 12x = 12x(x - 1)$$

Sign chart for $f''(x) = 12x(x - 1)$:

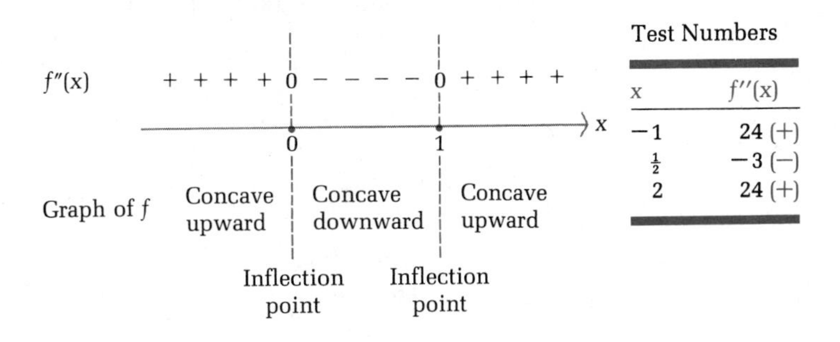

x	$f''(x)$
-1	$24\ (+)$
$\frac{1}{2}$	$-3\ (-)$
2	$24\ (+)$

Thus, the graph of f is concave upward on $(-\infty, 0)$ and $(1, \infty)$, concave downward on $(0, 1)$, and has inflection points at $x = 0$ and $x = 1$.

Step 3. Find horizontal and vertical asymptotes. Since $f(x)$ is a polynomial, there are no asymptotes.

Step 4. Find intercepts. The y intercept is $f(0) = 2$. Since the equation $f(x) = 0$ cannot be solved easily, we do not try to find x intercepts.

Step 5. Sketch the graph of f:

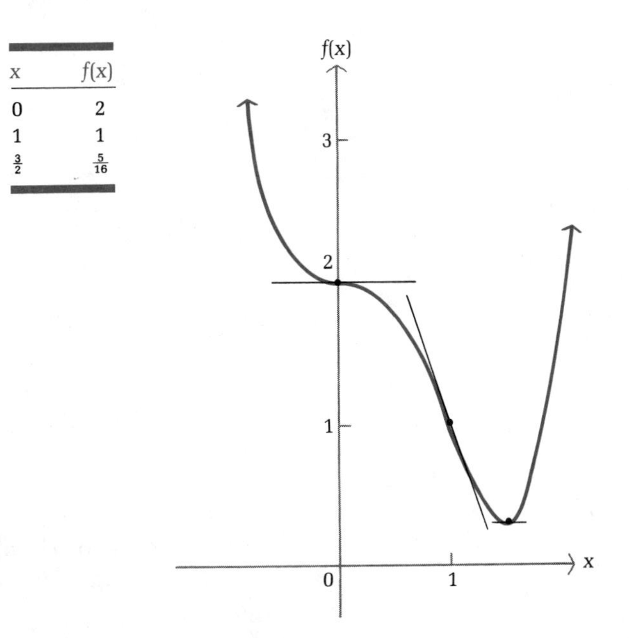

x	$f(x)$
0	2
1	1
$\frac{3}{2}$	$\frac{5}{16}$

Problem 16 Graph $f(x) = x^4 + 4x^3 + 10$ using the graphing strategy.

Example 17 Graph

$$f(x) = \frac{x-1}{x-2}$$

using the graphing strategy.

Solution Step 1. Use the first derivative:

$$f'(x) = \frac{(x-2)(1) - (x-1)(1)}{(x-2)^2} = \frac{-1}{(x-2)^2}$$

Sign chart for $f'(x) = -1/(x-2)^2$:

$f'(x)$ − − − − ND − − − −

x-axis, with 2 marked

$f(x)$ Decreasing | Decreasing

Test Numbers

x	$f'(x)$
1	-1 (−)
3	-1 (−)

Thus, $f(x)$ is decreasing on $(-\infty, 2)$ and $(2, \infty)$. There are no local extrema.

Step 2. Use the second derivative:

$$f''(x) = \frac{2}{(x-2)^3}$$

Sign chart for $f''(x) = 2/(x-2)^3$:

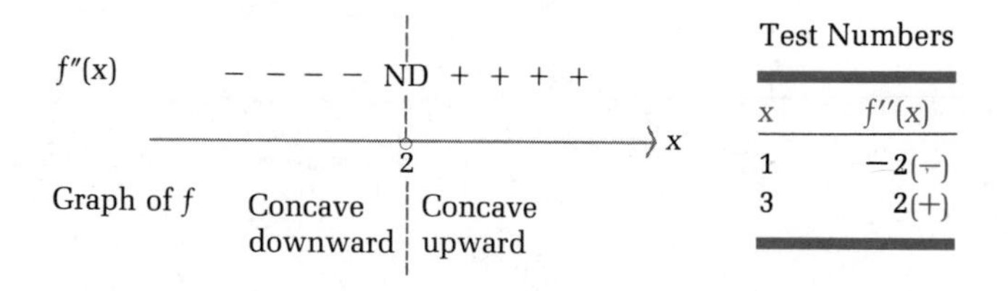

x	$f''(x)$
1	-2 (−)
3	2 (+)

Thus, the graph of f is concave downward on $(-\infty, 2)$ and concave upward on $(2, \infty)$. Since $f(2)$ is not defined, there is no inflection point at $x = 2$, even though $f''(x)$ changes sign at $x = 2$.

Step 3. Find horizontal and vertical asymptotes:

$$\lim_{x\to\infty} \frac{x-1}{x-2} = \frac{1}{1} = 1$$

Thus, $y = 1$ is a horizontal asymptote. Let $p(x) = x - 1$ and $q(x) = x - 2$. Since $q(2) = 0$ and $p(2) = 1 \neq 0$, $x = 2$ is a vertical asymptote.

Step 4. Find intercepts. The y intercept is

$$y = f(0) = \frac{0-1}{0-2} = \frac{1}{2}$$

Since a fraction is 0 when its numerator is 0, the x intercept is $x = 1$.

Step 5. Sketch the graph of f. Draw the asymptotes and plot points. (For functions with asymptotes, plotting additional points is often helpful.) Then sketch the graph.

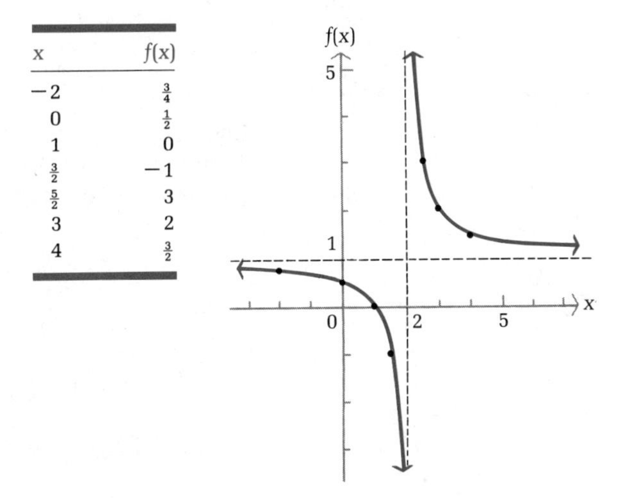

x	f(x)
−2	$\frac{3}{4}$
0	$\frac{1}{2}$
1	0
$\frac{3}{2}$	−1
$\frac{5}{2}$	3
3	2
4	$\frac{3}{2}$

Problem 17 Graph

$$f(x) = \frac{2x}{1-x}$$

using the graphing strategy.

■ Application

Example 18 Average Cost

Given the cost function $C(x) = 5{,}000 + \frac{1}{2}x^2$, where x is the number of units produced, graph the average cost function and the marginal cost function on the same set of coordinate axes.

Solution Let

$$\overline{C}(x) = \frac{C(x)}{x} = \frac{5{,}000}{x} + \frac{1}{2}x \qquad x > 0$$

and use the graphing strategy.

Step 1. Use the first derivative:

$$\overline{C}'(x) = -\frac{5{,}000}{x^2} + \frac{1}{2} = \frac{x^2 - 10{,}000}{2x^2}$$

Sign chart for $\overline{C}'(x) = (x^2 - 10{,}000)/(2x^2)$:

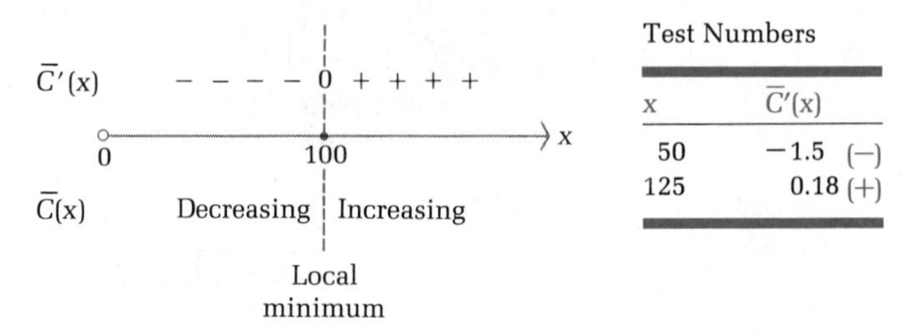

Test Numbers

x	$\overline{C}'(x)$
50	−1.5 (−)
125	0.18 (+)

Thus, $\overline{C}(x)$ is decreasing on (0, 100), increasing on (100, ∞), and has a local minimum at $x = 100$.

Step 2. Use the second derivative:

$$\overline{C}''(x) = \frac{10{,}000}{x^3} > 0 \qquad x > 0$$

Thus, the graph of $\overline{C}(x)$ is concave upward on (0, ∞).

Step 3. Find horizontal and vertical asymptotes. To determine asymptotes, we first write $\overline{C}(x)$ as a single fraction:

$$\overline{C}(x) = \frac{5{,}000}{x} + \frac{1}{2}x = \frac{10{,}000 + x^2}{2x}$$

Let $q(x) = 2x$ and $p(x) = 10{,}000 + x^2$. Since $q(0) = 0$ and $p(0) = 10{,}000 \neq 0$, $x = 0$ is a vertical asymptote.

$$\lim_{x \to \infty} \frac{10{,}000 + x^2}{2x} \quad \text{does not exist} \qquad \text{Theorem 3, Section 1-1}$$

Thus, there is no horizontal asymptote. However, consider the following limit:

$$\lim_{x \to \infty} \left[\overline{C}(x) - \frac{1}{2}x \right] = \lim_{x \to \infty} \frac{5{,}000}{x} = 0$$

This implies that the graph of $\overline{C}(x)$ approaches the line $y = \frac{1}{2}x$ as x approaches ∞. This line is called an **oblique asymptote** for the graph of $\overline{C}(x)$.

Step 4. Find intercepts. Since $x > 0$, there is no *y* intercept. Since $p(x)$ is never 0, there are no x intercepts.

Step 5. Sketch the graph of $\overline{C}$. The graph of $\overline{C}$ is shown in the figure. The marginal cost function is $C'(x) = x$. The graph of this linear function is also shown in the figure. This graph illustrates an important principle in economics: The minimal average cost occurs when the average cost is equal to the marginal cost.

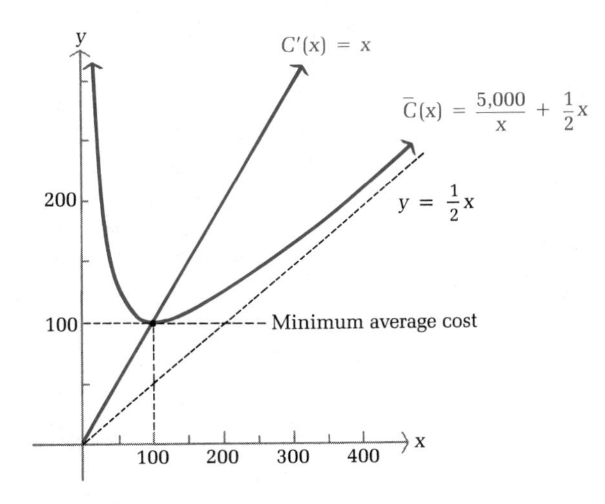

Problem 18 Given the cost function $C(x) = 1{,}600 + \frac{1}{4}x^2$:

(A) Find the minimum average cost.
(B) Find the marginal cost function.
(C) Graph the average cost function and the marginal cost function on the same axes. Include any oblique asymptotes.

Answers to Matched Problems

15. (A) Horizontal asymptote: $y = 3$; vertical asymptote: $x = -2$
    (B) Horizontal asymptote: $y = 0$ (x axis); vertical asymptote: $x = 0$ (y axis)
    (C) No horizontal asymptote; no vertical asymptotes
16. Increasing on $(-3, \infty)$
    Decreasing on $(-\infty, -3)$
    Local minimum at $x = -3$
    Concave upward on $(-\infty, -2)$ and $(0, \infty)$
    Concave downward on $(-2, 0)$
    Inflection points at $x = -2$ and $x = 0$
    y intercept: $f(0) = 10$

x	f(x)
−3	−17
−2	−6
0	10

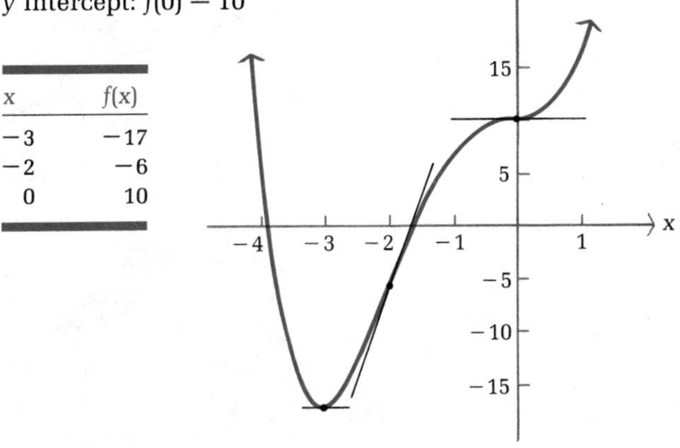

17. Increasing on $(-\infty, 1)$ and $(1, \infty)$
Concave upward on $(-\infty, 1)$
Concave downward on $(1, \infty)$
Horizontal asymptote: $y = -2$
Vertical asymptote: $x = 1$
x (and y) intercept: $f(0) = 0$

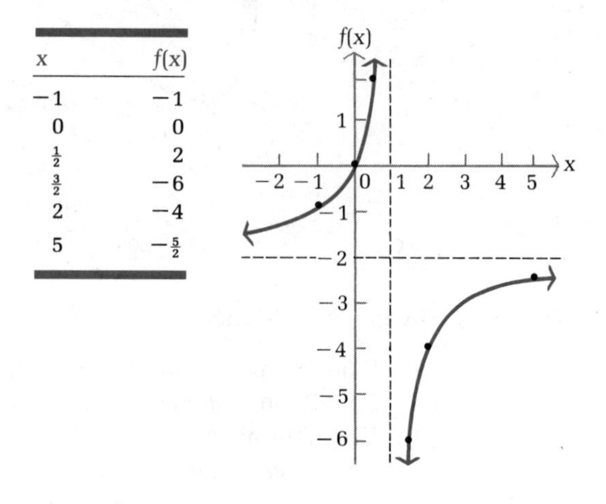

x	f(x)
−1	−1
0	0
$\frac{1}{2}$	2
$\frac{3}{2}$	−6
2	−4
5	$-\frac{5}{2}$

18. (A) Minimal average cost is 40 at $x = 80$
(B) $C'(x) = \frac{1}{2}x$
(C)

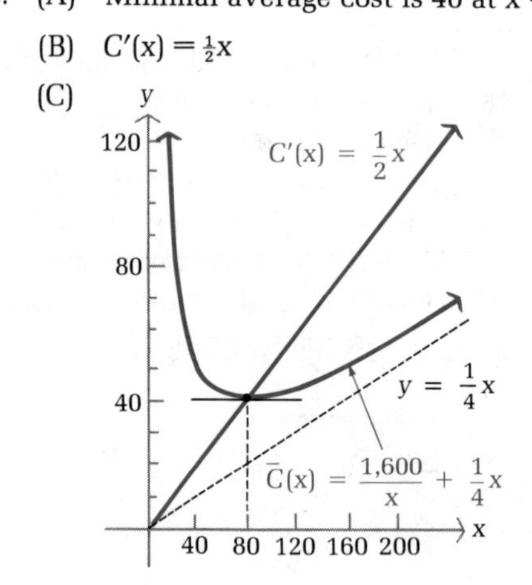

Exercise 2-4

A *Problems 1–10 refer to the following graph of $y = f(x)$:*

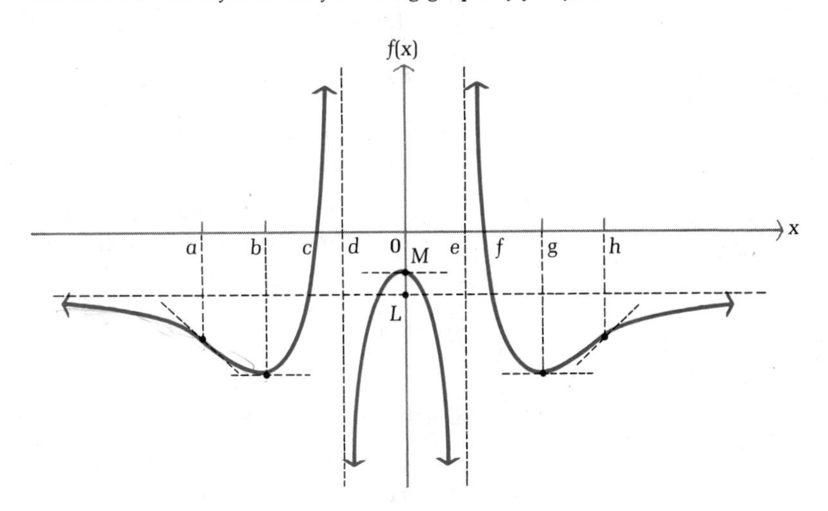

1. Identify the intervals over which $f(x)$ is increasing.
2. Identify the intervals over which $f(x)$ is decreasing.
3. Identify the points where $f(x)$ has a local maximum.
4. Identify the points where $f(x)$ has a local minimum.
5. Identify the intervals over which the graph of f is concave upward.
6. Identify the intervals over which the graph of f is concave downward.
7. Identify the inflection points.
8. Identify the horizontal asymptotes.
9. Identify the vertical asymptotes.
10. Identify the x and y intercepts.

B *Find the horizontal and vertical asymptotes.*

11. $f(x) = \dfrac{2x}{x + 2}$
12. $f(x) = \dfrac{3x + 2}{x - 4}$
13. $f(x) = \dfrac{x^2 + 1}{x^2 - 1}$
14. $f(x) = \dfrac{x^2 - 1}{x^2 + 2}$
15. $f(x) = \dfrac{x^3}{x^2 + 6}$
16. $f(x) = \dfrac{x}{x^2 - 4}$
17. $f(x) = \dfrac{x}{x^2 + 4}$
18. $f(x) = \dfrac{x^2 + 9}{x}$
19. $f(x) = \dfrac{x^2}{x - 3}$
20. $f(x) = \dfrac{x + 5}{x^2}$

Sketch a graph of $y = f(x)$ using the graphing strategy.

21. $f(x) = x^2 - 6x + 5$
22. $f(x) = 3 + 2x - x^2$
23. $f(x) = x^3 - 6x^2$
24. $f(x) = 3x^2 - x^3$
25. $f(x) = (x + 4)(x - 2)^2$
26. $f(x) = (2 - x)(x + 1)^2$
27. $f(x) = 8x^3 - 2x^4$
28. $f(x) = x^4 - 4x^3$
29. $f(x) = \dfrac{x + 3}{x - 3}$
30. $f(x) = \dfrac{2x - 4}{x + 2}$
31. $f(x) = \dfrac{x}{x - 2}$
32. $f(x) = \dfrac{2 + x}{3 - x}$

C

In Problems 33 and 34, show that the line $y = x$ is an oblique asymptote for the graph of $y = f(x)$ and then use the graphing strategy to sketch a graph of $y = f(x)$.

33. $f(x) = x + \dfrac{1}{x}$
34. $f(x) = x - \dfrac{1}{x}$

Sketch a graph of $y = f(x)$ using the graphing strategy.

35. $f(x) = x^3 - x$
36. $f(x) = x^3 + x$
37. $f(x) = (x^2 + 3)(9 - x^2)$
38. $f(x) = (x^2 + 3)(x^2 - 1)$
39. $f(x) = (x^2 - 4)^2$
40. $f(x) = (x^2 - 1)(x^2 - 5)$
41. $f(x) = 2x^6 - 3x^5$
42. $f(x) = 3x^5 - 5x^4$
43. $f(x) = \dfrac{x}{x^2 - 4}$
44. $f(x) = \dfrac{1}{x^2 - 4}$
45. $f(x) = \dfrac{1}{1 + x^2}$
46. $f(x) = \dfrac{x^2}{1 + x^2}$

Applications

Business & Economics

47. Revenue. The marketing research department for a computer company used a large city to test market their new product. They found that a relationship between price p (dollars per unit) and the demand x (units per week) was given approximately by

$$p = 1{,}296 - 0.12x^2 \qquad 0 < x < 80$$

Thus, the weekly revenue can be approximated by

$$R(x) = xp = 1{,}296x - 0.12x^3 \qquad 0 < x < 80$$

Graph the revenue function R.

48. Profit. If the cost equation for the company in the preceding problem is

$$C(x) = 830 + 396x$$

(A) Write an equation for the profit $P(x)$.
(B) Graph the profit function P.

49. *Pollution.* In Silicon Valley (California), a number of computer related manufacturing firms were found to be contaminating underground water supplies with toxic chemicals stored in leaking underground containers. A water quality control agency ordered the companies to take immediate corrective action and to contribute to a monetary pool for testing and cleanup of the underground contamination. Suppose the required monetary pool (in millions of dollars) for the testing and cleanup is estimated to be given by

$$P(x) = \frac{2x}{1-x} \qquad 0 \leq x < 1$$

where x is the percentage (expressed as a decimal fraction) of the total contaminant removed.

(A) Where is $P(x)$ increasing? Decreasing?
(B) Where is the graph of P concave upward? Downward?
(C) Find the horizontal and vertical asymptotes.
(D) Find the x and y intercepts.
(E) Sketch a graph of P.

50. *Employee training.* A company producing computer components has established that on the average a new employee can assemble $N(t)$ components per day after t days of on-the-job training, as given by

$$N(t) = \frac{100t}{t+9} \qquad t \geq 0$$

(A) Where is $N(t)$ increasing? Decreasing?
(B) Where is the graph of N concave upward? Downward?
(C) Find the horizontal and vertical asymptotes.
(D) Find the intercepts.
(E) Sketch a graph of N.

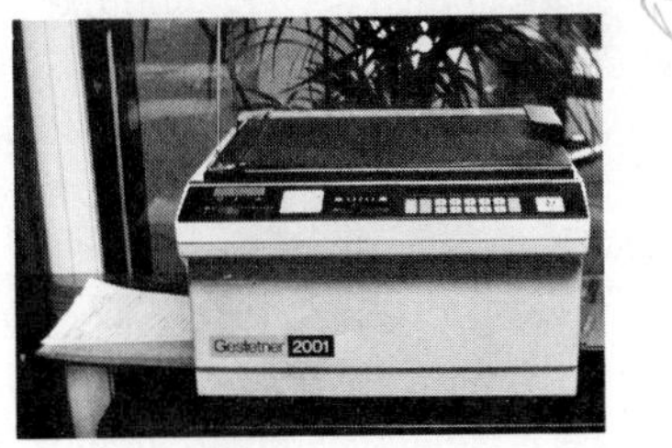

51. *Replacement time.* An office copier has an initial price of \$3,200. A maintenance/service contract costs \$300 for the first year and increases \$100 per year thereafter. It can be shown that the total cost of the copier after n years is given by

$$C(n) = 3{,}200 + 250n + 50n^2$$

(A) Write an expression for the average cost per year, $\overline{C}(n)$, for n years.
(B) When is the average cost per year minimum? (This is frequently referred to as the **replacement time** for this piece of equipment.)
(C) Graph the average cost function found in part A.

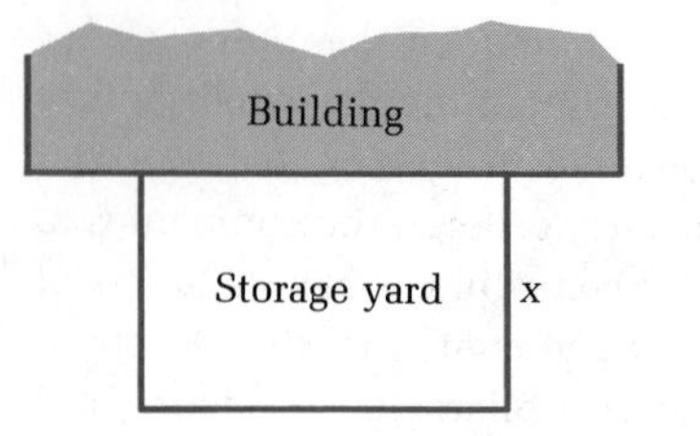

52. *Construction costs.* The management of a manufacturing plant wishes to add a fenced-in rectangular storage yard of 20,000 square feet, using the plant building as one side of the yard. If x is the distance from the building to the fence parallel to the building, then show that the length of the fence required for the yard is given by

$$L(x) = 2x + \frac{20{,}000}{x} \qquad x > 0$$

(A) What are the dimensions of the rectangle requiring the least amount of fencing?
(B) Graph L.

53. *Average and marginal costs.* The cost of producing x units of a certain product is given by

$$C(x) = 1{,}000 + 5x + \tfrac{1}{10}x^2$$

(A) Find the minimum average cost.
(B) Sketch the graph of the average cost function and the marginal cost function on the same set of coordinate axes. Include any oblique asymptotes.

54. Repeat Problem 53 for $C(x) = 500 + 2x + \frac{1}{5}x^2$.

Life Sciences

55. *Medicine.* A drug is injected into the bloodstream of a patient through her right arm. The concentration of the drug in the bloodstream of the left arm t hours after the injection is given by

$$C(t) = \frac{0.14t}{t^2 + 1}$$

Graph C.

56. *Physiology.* In a study on the speed of muscle contraction in frogs under various loads, researchers W. O. Fems and J. Marsh found that the speed of contraction decreases with increasing loads. More precisely, they found that the relationship between speed of contraction S (in centimeters per second) and load w (in grams) is given approximately by

$$S(w) = \frac{26 + 0.06w}{w} \qquad w \geq 5$$

Graph S.

Social Sciences

57. *Psychology—retention.* An experiment on retention is conducted in a psychology class. Each student in the class is given one day to memorize the same list of thirty special characters. The lists are turned in at the end of the day, and for each succeeding day for thirty days each student is asked to turn in a list of as many of the symbols as can be recalled. Averages are taken and it is found that

$$N(t) = \frac{5t + 20}{t} \qquad t \geq 1$$

provides a good approximation of the average number of symbols, $N(t)$, retained after t days. Graph N.

2-5 Differentials

- The Differential
- Approximations Using Differentials
- Applications

The Differential

In Chapter 1 we introduced the concept of increment. Recall that for a function defined by

$$y = f(x)$$

we said that Δx represents a change in the independent variable x; that is,

$$\Delta x = x_2 - x_1 \qquad \text{or} \qquad x_2 = x_1 + \Delta x$$

And Δy represents the corresponding change in the dependent variable y; that is,

$$\Delta y = f(x_1 + \Delta x) - f(x_1)$$

We then defined the derivative of f at x_1 to be

$$\frac{dy}{dx} = \lim_{\Delta x \to 0} \frac{\Delta y}{\Delta x}$$

If the limit exists, then it follows that

$$\frac{\Delta y}{\Delta x} \approx \frac{dy}{dx} \qquad \text{for small } \Delta x$$

or

$$\Delta y \approx \frac{dy}{dx} \Delta x \tag{1}$$

We used dy/dx as an alternate symbol for $f'(x)$. We will now give dy and dx special meaning, and we will show how dy can be used to approximate Δy. This turns out to be quite useful, since a number of practical problems require the computation of Δy, and we will be able to use the more readily computed dy. The symbols dy and dx are called **differentials** and are defined in the box on the next page.

Differentials

If $y = f(x)$ defines a differentiable function, then:

1. The **differential** dx of the independent variable x is an arbitrary real number.
2. The **differential** dy of the dependent variable y is defined as the product of $f'(x)$ and dx; that is,

$$dy = f'(x)\,dx \tag{2}$$

The differential dy is actually a function involving two independent variables, x and dx—a change in either one or both will affect dy.

Example 19 Find dy for $f(x) = x^2 + 3x$. Evaluate dy for $x = 2$ and $dx = 0.1$, for $x = 3$ and $dx = 0.1$, and for $x = 1$ and $dx = 0.02$.

Solution

$$\begin{aligned} dy &= f'(x)\,dx \\ &= (2x + 3)\,dx \end{aligned}$$

When $x = 2$ and $dx = 0.1$,

$$dy = [2(2) + 3]0.1 = 0.7$$

When $x = 3$ and $dx = 0.1$,

$$dy = [2(3) + 3]0.1 = 0.9$$

When $x = 1$ and $dx = 0.02$,

$$dy = [2(1) + 3]0.02 = 0.1$$

Problem 19 Find dy for $f(x) = \sqrt{x} + 3$. Evaluate dy for $x = 4$ and $dx = 0.1$, for $x = 9$ and $dx = 0.12$, and for $x = 1$ and $dx = 0.01$.

We now have two interpretations of the symbol dy/dx. Referring to the function $y = f(x) = x^2 + 3x$ in Example 19 with $x = 2$ and $dx = 0.1$, we have

$$\frac{dy}{dx} = f'(2) = 7 \qquad \text{Derivative}$$

and

$$\frac{dy}{dx} = \frac{0.7}{0.1} = 7 \qquad \text{Ratio of differentials}$$

Since differentials are defined in terms of derivatives, the derivative rules discussed in Chapter 1 lead to the differential rules in the box.

Differential Rules

If u and v are differentiable functions and c is a constant, then:

1. $dc = 0$
2. $du^n = nu^{n-1}\, du$
3. $d(u \pm v) = du \pm dv$
4. $d(uv) = u\, dv + v\, du$
5. $d\left(\frac{u}{v}\right) = \frac{v\, du - u\, dv}{v^2}$

Example 20 (A) Find dy for $y = \frac{x^2}{4 + x^2}$. (B) Find du for $u = (7 + x^3)^5$.

Solutions (A)

$$dy = \frac{(4 + x^2)\, d(x^2) - x^2\, d(4 + x^2)}{(4 + x^2)^2}$$

$$= \frac{(4 + x^2)2x\, dx - x^2 2x\, dx}{(4 + x^2)^2}$$

$$= \frac{8x\, dx}{(4 + x^2)^2}$$

(B)

$$du = 5(7 + x^3)^4\, d(7 + x^3)$$

$$= 5(7 + x^3)^4 3x^2\, dx$$

$$= 15x^2(7 + x^3)^4\, dx$$

Problem 20 (A) Find dy for $y = x^2(2x - 5)^3$. (B) Find du for $u = \frac{1}{4x + x^2}$.

■ Approximations Using Differentials

The differential of a function $y = f(x)$ can be used to approximate the change in y and the values of $f(x)$. If we let $dx = \Delta x$, then from (1) and (2) we have

$$\Delta y \approx \frac{dy}{dx}\Delta x = f'(x)\, dx = dy$$

Thus, dy can be used to approximate Δy. To interpret this result geometrically, we need to recall a basic property of slope. The vertical change in a line is equal to the product of the slope and the horizontal change, as shown in Figure 15 (on the next page).

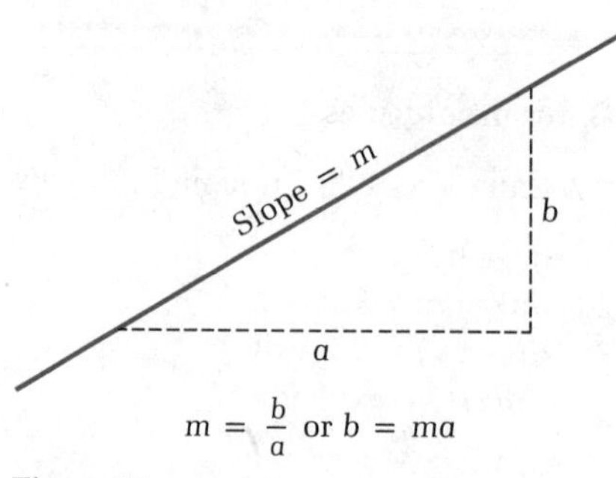

Figure 15

Now consider the line tangent to the graph of $y = f(x)$, as shown in Figure 16. Since $f'(x)$ is the slope of the tangent line and dx is the horizontal change in the tangent line, it follows that the vertical change in the tangent line is given by $dy = f'(x)\,dx$, as indicated in Figure 16.

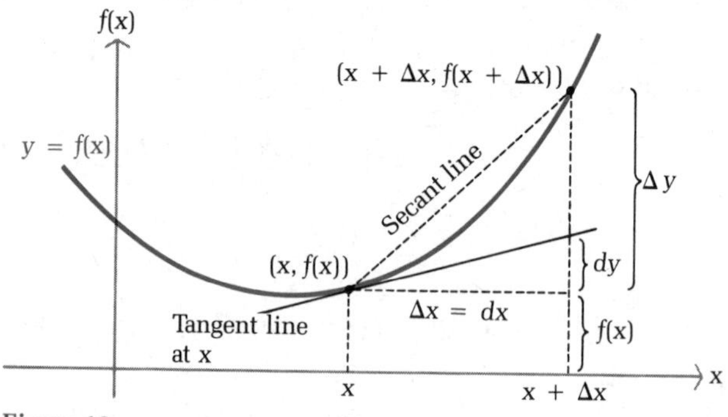

Figure 16

To see how differentials can be used to approximate the values of a function, we write

$$\Delta y \approx dy$$
$$f(x + \Delta x) - f(x) \approx f'(x)\,dx$$
$$f(x + \Delta x) \approx f(x) + f'(x)\,dx$$

These relationships are summarized in the following box.

Differential Approximation

If $f'(x)$ exists and $dx = \Delta x$, then for small Δx

$$\Delta y \approx dy$$

and

$$f(x + \Delta x) \approx f(x) + f'(x)\,dx$$

We will use these relationships in the examples that follow. (Before proceeding, however, it should be mentioned that even though differentials can be used to approximate certain quantities, the error can be substantial in certain cases.)

Example 21 Find Δy and dy for $f(x) = 6x - x^2$ when $x = 2$ and $\Delta x = dx = 0.1$.

Solution

$$\begin{aligned} \Delta y &= f(x + \Delta x) - f(x) \\ &= f(2.1) - f(2) \\ &= [6(2.1) - (2.1)^2] - [6(2) - 2^2] \\ &= 8.19 - 8 \\ &= 0.19 \end{aligned}$$

$$\begin{aligned} dy &= f'(x)\, dx \\ &= (6 - 2x)\, dx \\ &= [6 - 2(2)](0.1) \\ &= 0.2 \end{aligned}$$

Notice that dy and Δy differ by only 0.01 in this case.

Problem 21 Repeat Example 21 for $x = 4$ and $\Delta x = dx = 0.2$.

Example 22 Use differentials to approximate $\sqrt[3]{27.54}$.

Solution Even though the problem is trivial using a hand calculator, its solution using differentials will help increase the understanding of this concept. Form the function

$$y = f(x) = \sqrt[3]{x} = x^{1/3}$$

and note that we can compute $f(27)$ and $f'(27)$ exactly. Thus, if we let $x = 27$ and $dx = \Delta x = 0.54$ and use

$$\begin{aligned} f(x + \Delta x) &= f(x) + \Delta y \\ &\approx f(x) + dy \\ &= f(x) + f'(x)\, dx \end{aligned}$$

we will obtain an approximation for $f(27.54) = \sqrt[3]{27.54}$ that is easy to compute.

$$f(x + \Delta x) \approx f(x) + f'(x)\, dx$$

$$(x + \Delta x)^{1/3} \approx x^{1/3} + \frac{1}{3x^{2/3}}\, dx$$

$$(27 + 0.54)^{1/3} \approx 27^{1/3} + \frac{1}{3(27)^{2/3}}(0.54)$$

Thus,

$$\sqrt[3]{27.54} \approx 3 + \frac{0.54}{27} = 3.02 \qquad \text{(Calculator value} = 3.0199)$$

Problem 22 Use differentials to approximate $\sqrt{36.72}$.

■ Applications

Example 23
Weight–Height

A formula relating the approximate weight, W (in pounds), of an average person and their height, h (in inches), is

$$W = 0.0005h^3 \qquad 30 \leq h \leq 74$$

What is the approximate change in weight for a height increase from 40 to 42 inches?

Solution We are actually interested in finding ΔW, the change in weight brought about by the change in height from 40 to 42 inches ($\Delta h = 2$). We will use the differential dW to approximate ΔW, since Δh is small. The problem is now to find dW for $h = 40$ and $dh = \Delta h = 2$.

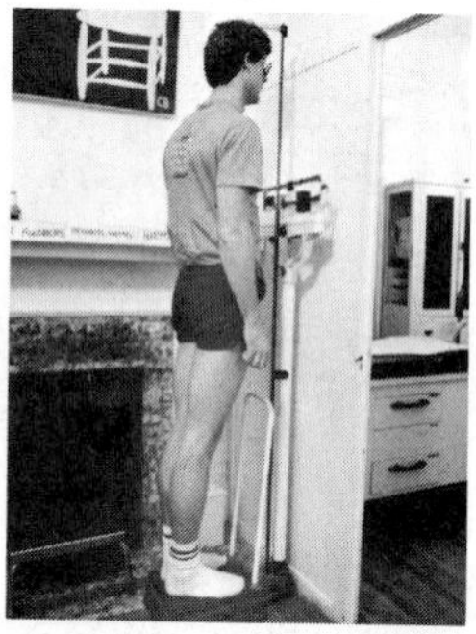

$$\begin{aligned} W(h) &= 0.0005h^3 \\ dW &= W'(h)\,dh \\ &= 0.0015h^2\,dh \\ &= 0.0015(40)^2(2) \\ &= 4.8 \text{ pounds} \end{aligned}$$

Thus, a child growing from 40 inches to 42 inches would expect to increase in weight by approximately 4.8 pounds. Notice that using the differential is somewhat easier than finding $\Delta W = W(42) - W(40)$.

Problem 23 Refer to Example 23. Approximate the change in weight resulting from a height increase from 70 to 72 inches.

Example 24
Cost–Revenue

A company manufactures and sells x transistor radios per week. If the weekly cost and revenue equations are

$$C(x) = 5{,}000 + 2x$$
$$R(x) = 10x - \frac{x^2}{1{,}000} \qquad 0 \leq x \leq 8{,}000$$

find the approximate changes in revenue and profit if production is increased from 2,000 to 2,010 units per week.

Solution We will approximate ΔR and ΔP with dR and dP, respectively, using $x = 2{,}000$ and $dx = \Delta x = 2{,}010 - 2{,}000 = 10$.

$$R(x) = 10x - \frac{x^2}{1{,}000}$$

$$\begin{aligned} dR &= R'(x)\,dx \\ &= \left(10 - \frac{x}{500}\right) dx \\ &= \left(10 - \frac{2{,}000}{500}\right) 10 \\ &= \$60 \text{ per week} \end{aligned}$$

$$\begin{aligned} P(x) &= R(x) - C(x) \\ &= 10x - \frac{x^2}{1{,}000} - 5{,}000 - 2x \\ &= 8x - \frac{x^2}{1{,}000} - 5{,}000 \end{aligned}$$

$$\begin{aligned} dP &= P'(x)\,dx \\ &= \left(8 - \frac{x}{500}\right) dx \\ &= \left(8 - \frac{2{,}000}{500}\right) 10 \\ &= \$40 \text{ per week} \end{aligned}$$

Problem 24 Repeat Example 24 with production increasing from 6,000 to 6,010.

Comparing the results in Example 24 and Problem 24, we see that an increase in production results in a revenue and profit increase at the 2,000 production level, but a revenue and profit loss at the 6,000 production level.

Answers to Matched Problems

19. $dy = \frac{1}{2\sqrt{x}}\,dx$; 0.025, 0.02, 0.005

20. (A) $dy = 2x(2x-5)^3\,dx + 6x^2(2x-5)^2\,dx$

    (B) $du = -\frac{(4+2x)\,dx}{(4x+x^2)^2}$

21. $\Delta y = -0.44$, $dy = -0.4$ 22. 6.06 23. 14.7 pounds

24. $dR = -\$20$ per week, $dP = -\$40$ per week

Exercise 2-5

A *Find dy for each function.*

1. $y = 30 + 12x^2 - x^3$

2. $y = 200x - \frac{x^2}{30}$

3. $y = x^2\left(1 - \frac{x}{9}\right)$

4. $y = x^3(60 - x)$

5. $y = f(x) = \frac{590}{\sqrt{x}}$

6. $y = 52\sqrt{x}$

7. $y = 75\left(1 - \frac{2}{x}\right)$

8. $y = 100\left(x - \frac{4}{x^2}\right)$

B

9. $y = (2x + 1)^3$

10. $y = (3x + 5)^5$

11. $y = \frac{x}{x^2 + 9}$

12. $y = \frac{x^2}{(x + 1)^2}$

Evaluate dy and Δy for each function at the indicated values.

13. $y = f(x) = x^2 - 3x + 2, \quad x = 5, \quad \Delta x = dx = 0.2$
14. $y = f(x) = 30 + 12x^2 - x^3, \quad x = 2, \quad \Delta x = dx = 0.1$
15. $y = f(x) = 75\left(1 - \frac{2}{x}\right), \quad x = 5, \quad dx = \Delta x = 0.5$
16. $y = f(x) = 100\left(x - \frac{4}{x^2}\right), \quad x = 2, \quad \Delta x = dx = 0.1$

Use differentials to approximate the indicated roots.

17. $\sqrt[4]{17}$
18. $\sqrt{83}$
19. $\sqrt[3]{28}$
20. $\sqrt[5]{34}$
21. A cube with sides 10 inches long is covered with a 0.2 inch thick coat of fiberglass. Use differentials to estimate the volume of the fiberglass shell.
22. A sphere with a radius of 5 centimeters is coated with ice 0.1 centimeter thick. Use differentials to estimate the volume of the ice [recall that $V = \frac{4}{3}\pi r^3$, $\pi \approx 3.14$].

C

23. Find dy if $y = \sqrt[3]{3x^2 - 2x + 1}$.
24. Find dy if $y = (2x^2 - 4)\sqrt{x + 2}$.
25. Find dy and Δy for $y = 52\sqrt{x}$, $x = 4$, and $\Delta x = dx = 0.3$.
26. Find dy and Δy for $y = 590/\sqrt{x}$, $x = 64$, and $\Delta x = dx = 1$.

Applications

Use differential approximations in the following problems.

Business & Economics

27. *Advertising.* Using past records, it is estimated that a company will sell N units of a product after spending x thousand dollars in advertising, as given by

$$N = 60x - x^2 \qquad 5 \leq x \leq 30$$

Approximately what increase in sales will result by increasing the advertising budget from \$10,000 to \$11,000? From \$20,000 to \$21,000?

28. *Price–demand.* Suppose in a grocery chain the daily demand in pounds for chocolate candy at $x per pound is given by

$$D = 1{,}000 - 40x^2 \qquad 1 \leq x \leq 5$$

If the price is increased from $3.00 per pound to $3.20 per pound, what is the approximate change in demand?

29. *Average cost.* For a company that manufactures tennis rackets, the average cost per racket, $\overline{C}$, is found to be

$$\overline{C} = x^2 - 20x + 110 \qquad 6 \leq x \leq 14$$

where x is the number of rackets produced per hour. What will the approximate change in cost per racket be if production is increased from seven per hour to eight per hour? From twelve per hour to thirteen per hour?

30. *Revenue and profit.* A company manufactures and sells x televisions per month. If the cost and revenue equations are

$$C(x) = 72{,}000 + 60x$$
$$R(x) = 200x - \frac{x^2}{30} \qquad 0 \leq x \leq 6{,}000$$

what will the approximate changes in revenue and profit be if production is increased from 1,500 to 1,501? From 4,500 to 4,501?

Life Sciences

31. *Pulse rate.* The average pulse rate y in beats per minute of a healthy person x inches tall is given approximately by

$$y = \frac{590}{\sqrt{x}} \qquad 30 \leq x \leq 75$$

Approximately how will the pulse rate change for a height change from 36 to 37 inches? From 64 to 65 inches?

32. *Measurement.* An egg of a particular bird is very nearly spherical. If the radius to the inside of the shell is 5 millimeters and the radius to the outside of the shell is 5.3 millimeters, approximately what is the volume of the shell? [Remember that $V = \frac{4}{3}\pi r^3$ and use $\pi \approx 3.14$.]

33. *Medicine.* A drug is given to a patient to dilate her arteries. If the radius of an artery is increased from 2 to 2.1 millimeters, approximately how much is a cross-sectional area increased? (Assume the cross-section of the artery is circular; $A = \pi r^2$ and $\pi \approx 3.14$.)

34. *Drug sensitivity.* One hour after x milligrams of a particular drug are given to a person, the change in body temperature T in degrees Fahrenheit is given by

$$T = x^2\left(1 - \frac{x}{9}\right) \qquad 0 \leq x \leq 6$$

Approximate the changes in body temperature produced by the following changes in drug dosages:

(A) From 2 to 2.1 milligrams
(B) From 3 to 3.1 milligrams
(C) From 4 to 4.1 milligrams

Social Sciences

35. *Learning.* A particular person learning to type has an achievement record given approximately by

$$N = 75\left(1 - \frac{2}{t}\right) \qquad 3 \leq t \leq 20$$

where N is the number of words per minute typed after t weeks of practice. What is the approximate improvement from 5 to 5.5 weeks of practice?

36. *Learning.* If a person learns y items in x hours, as given approximately by

$$y = 52\sqrt{x} \qquad 0 \leq x \leq 9$$

what is the approximate increase in the number of items learned when x changes from 1 to 1.1 hours? From 4 to 4.1 hours?

37. *Politics.* In a newly incorporated city it is estimated that the voting population (in thousands) will increase according to

$$N(t) = 30 + 12t^2 - t^3 \qquad 0 \leq t \leq 8$$

where t is time in years. Find the approximate change in votes for the following time changes:

(A) From 1 to 1.1 years
(B) From 4 to 4.1 years
(C) From 7 to 7.1 years

2-6 Chapter Review

Important Terms and Symbols

2-1 *First derivative and graphs.* Increasing function, decreasing function, rising, falling, critical value, local extrema, local maximum, local minimum, first-derivative test for local extrema

2-2 *Second derivative and graphs.* Concave upward, concave downward, second derivative, concavity and the second derivative, inflection point, second-derivative test for local maxima and minima, $f''(x)$, d^2y/dx^2, y'', $D_x^2 f(x)$

2-3 *Optimization; absolute maxima and minima.* Absolute maxima, absolute minima, absolute extrema of a function continuous on a closed interval, second-derivative test for absolute maximum and minimum

2-4 *Curve sketching techniques: unified and extended.* Locating asymptotes, graphing strategy, increasing, decreasing, local maxima and minima, concave upward, concave downward, inflection point, horizontal asymptote, vertical asymptote, y intercept, x intercept, oblique asymptote

2-5 *Differentials.* Differential dx, differential $dy = f'(x)\,dx$, differential approximation, $\Delta y \approx dy$, $f(x + \Delta x) \approx f(x) + f'(x)\,dx$, differential rules, $dc = 0$, $du^n = nu^{n-1}\,du$, $d(u \pm v) = du \pm dv$,

$$d(uv) = u\,dv + v\,du,\quad d\left(\frac{u}{v}\right) = \frac{v\,du - u\,dv}{v^2}$$

Exercise 2-6 Chapter Review

Work through all the problems in this chapter review and check your answers in the back of the book. (Answers to all review problems are there.) Where weaknesses show up, review appropriate sections in the text.

A *Problems 1–8 refer to the following graph of $y = f(x)$:*

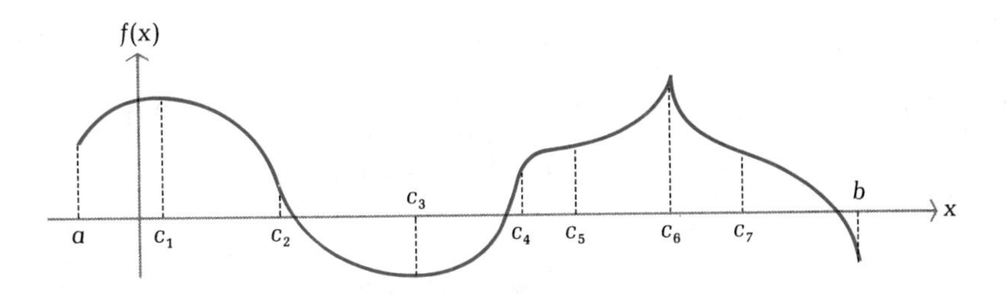

Identify the points or intervals on the x axis that produce the indicated behavior.

1. $f(x)$ is increasing
2. $f'(x) < 0$
3. Graph of f is concave downward
4. Local minima
5. Absolute maxima
6. $f'(x)$ appears to be 0
7. $f'(x)$ does not exist
8. Inflection points
9. Find $f''(x)$ for $f(x) = x^4 + 5x^3$.
10. Find y'' for $y = 3x + 4/x$.
11. Find dy for $y = f(x) = x^3 + 4x$.
12. Find dy for $y = f(x) = (3x^2 - 7)^3$.

B *Problems 13–18 refer to the function $y = f(x) = x^3 + 3x^2 - 24x - 3$.*

13. Identify critical values.
14. Find intervals over which $f(x)$ is increasing. Decreasing.

15. Find local maxima and minima.
16. Find intervals over which the graph of f is concave upward. Concave downward.
17. Identify inflection points.
18. Graph f.

Problems 19–23 refer to the function $y = f(x) = \dfrac{3x}{x+2}$.

19. Find horizontal asymptotes.
20. Find vertical asymptotes.
21. Find intervals over which $f(x)$ is increasing. Decreasing.
22. Find intervals over which the graph of f is concave upward. Concave downward.
23. Graph f.
24. Find the absolute maximum and minimum for

$$y = f(x) = x^3 - 12x + 12 \qquad -3 \leq x \leq 5$$

25. Find the absolute minimum for

$$y = f(x) = x^2 + \frac{16}{x^2} \qquad x > 0$$

Find horizontal and vertical asymptotes.

26. $f(x) = \dfrac{x}{x^2 + 9}$

27. $f(x) = \dfrac{x^3}{x^2 - 9}$

28. Find dy and Δy for $f(x) = x^3 - 2x + 1$, $x = 5$, and $\Delta x = dx = 0.1$.
29. Approximate $\sqrt{17}$ using differentials.

C

30. Find the absolute maximum for $f'(x)$ if

$$f(x) = 6x^2 - x^3 + 8$$

Graph f and f' on the same axes.

31. Find two positive numbers whose product is 400 and whose sum is a minimum. What is the minimum sum?
32. Sketch the graph of $f(x) = (x - 1)^3(x + 3)$ using the graphing strategy discussed in Section 2-4.
33. Find dy and Δy for $y = (2/\sqrt{x}) + 8$, $x = 16$, and $\Delta x = dx = 0.2$.

Applications

Business & Economics

34. *Profit.* The profit for a company manufacturing and selling x units per month is given by

$$P(x) = 150x - \frac{x^2}{40} - 50{,}000 \qquad 0 \leq x \leq 5{,}000$$

What production level will produce the maximum profit? What is the maximum profit?

35. *Average cost.* The total cost of producing x units per month is given by

$$C(x) = 4{,}000 + 10x + \tfrac{1}{10}x^2$$

Find the minimum average cost. Graph the average cost and the marginal cost functions on the same axes. Include any oblique asymptotes.

36. *Rental income.* A 200 room hotel in Fresno is filled to capacity every night at a rate of $40 per room. For each $1 increase in the nightly rate, four fewer rooms are rented. If each rented room costs $8 a day to service, how much should the management charge per room in order to maximize gross profit? What is the maximum gross profit?

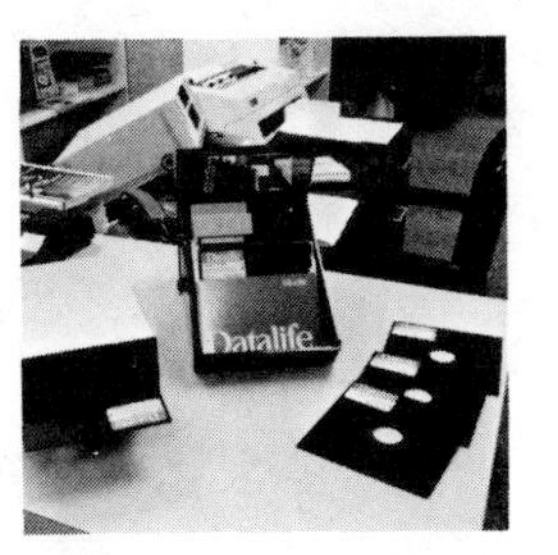

37. *Inventory control.* A computer store sells 7,200 boxes of floppy discs annually. It costs the store $0.20 to store a box of discs for one year. Each time it reorders discs, the store must pay a $5.00 service charge for processing the order. How many times during the year should the store order discs in order to minimize the total storage and reorder costs?

38. *Rate of change of revenue.* A company is manufacturing a new video game and can sell all it manufactures. The revenue (in dollars) is given by

$$R = 36x - \frac{x^2}{20}$$

where the production output in one day is x games. Use dR to approximate the change in revenue if production is increased from 250 to 260 games per day.

Life Sciences

39. *Bacteria control.* If t days after a treatment the bacteria count per cubic centimeter in a body of water is given by

$$C(t) = 20t^2 - 120t + 800 \qquad 0 \le t \le 9$$

in how many days will the count be a minimum?

Social Sciences

40. *Politics.* In a new suburb it is estimated that the number of registered voters will grow according to

$$N = 10 + 6t^2 - t^3 \qquad 0 \le t \le 5$$

where t is time in years and N is in thousands. When will the rate of increase be maximum?

Exponential and Logarithmic Functions

3

CHAPTER 3 Contents

We now know how to differentiate algebraic functions—that is, functions that can be defined using the algebraic operations of addition, subtraction, multiplication, division, powers, and roots. In this chapter we will discuss differentiation of forms that involve the exponential and logarithmic functions discussed in Sections 0-6 and 0-7. (You might want to review some of the properties of these functions before proceeding further.) We begin with a discussion of the irrational number e.

3-1 The Constant e and Continuous Compound Interest

- The Constant e
- Continuous Compound Interest

The Constant e

The special irrational number e is a particularly suitable base for both exponential and logarithmic functions. The reasons for choosing this number as a base will become clear as we develop differentiation formulas for the exponential function e^x and the natural logarithmic function $\ln x$.

In precalculus treatments, the number e is informally defined as an irrational number that can be approximated by the expression $[1 + (1/n)]^n$ by taking n sufficiently large. Now we will use the limit concept to formally define e as either of the following two limits:

The Number e

$$e = \lim_{n\to\infty}\left(1 + \frac{1}{n}\right)^n$$

or, alternately,

$$e = \lim_{s\to 0}(1 + s)^{1/s}$$

$$e = 2.718\ 281\ 8\ .\ .\ .$$

We will use both these forms. [*Note:* If $s = 1/n$, then as $n \to \infty$, $s \to 0$.]

The proof that the indicated limits exist and represent an irrational number between 2 and 3 is not easy and is omitted here. Many people reason (incorrectly) that the limits are 1, since "$(1 + s)$ approaches 1 as $s \to 0$, and 1 to any power is 1." A little experimentation with a pocket calculator can convince you otherwise. Consider the table of values for s and $f(s) = (1 + s)^{1/s}$ and the graph shown in Figure 1 for s close to 0.

s approaches 0 from the left $\to 0 \leftarrow$ s approaches 0 from the right

s	-0.5	-0.2	-0.1	$-0.01 \to 0 \leftarrow$	0.01	0.1	0.2	0.5
$(1 + s)^{1/s}$	4.000 0	3.051 8	2.868 0	2.732 0 $\to e \leftarrow$	2.704 8	2.593 7	2.488 3	2.250 0

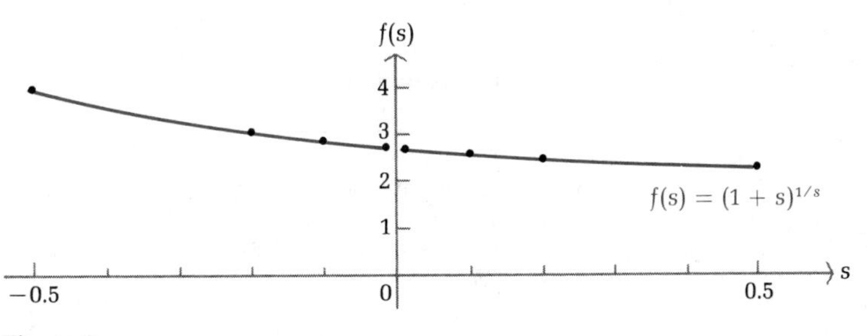

Figure 1

Compute some of the table values with a calculator yourself and also try several values of s even closer to 0. Note that the function is discontinuous at $s = 0$.

Exactly who discovered e is still being debated. It is named after the great mathematician Leonhard Euler (1707–1783), who computed e to twenty-three decimal places using $[1 + (1/n)]^n$.

▪ Continuous Compound Interest

Now we will see how e appears quite naturally in the important application of compound interest. Let us start with simple interest, move on to compound interest, and then to continuous compound interest.

If a principal P is borrowed at an annual rate of $100r\%$, then after t years at simple interest the borrower will owe the lender an amount A given by

$$A = P + Prt = P(1 + rt) \qquad \text{Simple interest} \tag{1}$$

On the other hand, if interest is compounded n times a year, then the borrower will owe the lender an amount A given by

$$A = P\left(1 + \frac{r}{n}\right)^{nt} \qquad \text{Compound interest} \tag{2}$$

where r/n is the interest rate per compounding period and nt is the number of compounding periods. Suppose P, r, and t in (2) are held fixed and n is increased. Will the amount A increase without bound or will it tend to some limiting value?

Let us perform a calculator experiment before we attack the general limit problem. If $P = \$100$, $r = 0.06$, and $t = 2$ years, then

$$A = 100\left(1 + \frac{0.06}{n}\right)^{2n}$$

We compute A for several values of n in Table 1. The biggest gain appears in the first step; then the gains slow down as n increases. In fact, it appears that A might be tending to something close to \$112.75 as n gets larger and larger.

Table 1

Compounding Frequency	n	$A = 100\left(1 + \frac{0.06}{n}\right)^{2n}$
Annually	1	\$112.3600
Semiannually	2	112.5509
Quarterly	4	112.6493
Weekly	52	112.7419
Daily	365	112.7486
Hourly	8,760	112.7491

Now we turn back to the general problem for a moment. Keeping P, r, and t fixed in equation (2), we compute the following limit and observe an interesting and useful result:

$$\lim_{n\to\infty} P\left(1 + \frac{r}{n}\right)^{nt} = P \lim_{n\to\infty}\left(1 + \frac{r}{n}\right)^{(n/r)rt} \qquad \text{Insert } r/r \text{ in the exponent and let } s = r/n.$$

$$= P[\lim_{s\to 0}(1 + s)^{1/s}]^{rt} \qquad \lim_{s\to 0}(1 + s)^{1/s} = e$$

$$= Pe^{rt}$$

The resulting formula is called the **continuous compound interest formula,** a very important and widely used formula in business and economics.

Continuous Compound Interest

$$A = Pe^{rt}$$

where

P = Principal

r = Annual nominal interest rate compounded continuously

t = Time in years

A = Amount at time t

Example 1 If \$100 is invested at an annual nominal rate of 6%, compounded continuously, what amount will be in the account after 2 years?

Solution

$$A = Pe^{rt}$$

$$= 100e^{(0.06)(2)} \qquad 6\% \text{ is equivalent to } r = 0.06.$$

$$\approx \$112.7497$$

(Compare this result with the values calculated in Table 1.)

Problem 1 What amount (to the nearest cent) will an account have after 5 years if \$100 is invested at an annual nominal rate of 8%, compounded annually? Semiannually? Continuously?

Example 2 If \$100 is invested at 12%, compounded continuously,* graph the amount in the account relative to time for a period of 10 years.

Solution We want to graph

$$A = 100e^{0.12t} \qquad 0 \le t \le 10$$

We construct a table of values using a calculator or Table I of Appendix B, graph the points from the table, and join the points with a smooth curve.

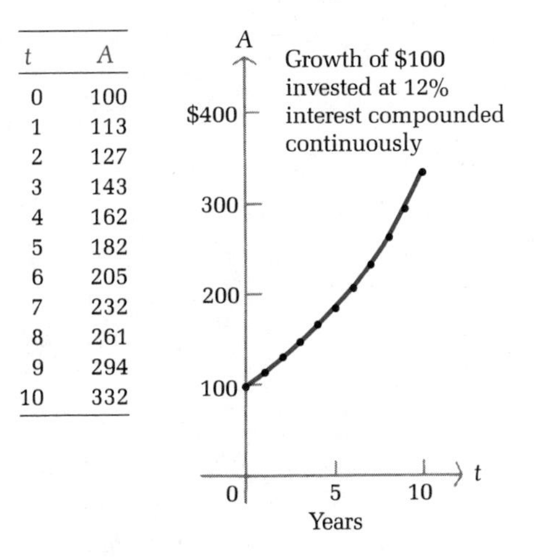

t	A
0	100
1	113
2	127
3	143
4	162
5	182
6	205
7	232
8	261
9	294
10	332

Problem 2 If \$5,000 is invested at 20%, compounded continuously, graph the amount in the account relative to time for a period of 10 years.

* Following common usage, we will often write the form "at 12%, compounded continuously," understanding that this means "at an annual nominal rate of 12%, compounded continuously."

Example 3 How long will it take money to double if it is invested at 18%, compounded continuously?

Solution Starting with the continuous compound interest formula $A = Pe^{rt}$, we must solve for t given $A = 2P$ and $r = 0.18$.

$$2P = Pe^{0.18t} \qquad \text{Divide both sides by } P.$$

$$e^{0.18t} = 2 \qquad \text{Take natural logs of both sides.}$$

$$\ln e^{0.18t} = \ln 2 \qquad \text{Recall that } \log_b b^x = x.$$

$$0.18t = \ln 2$$

$$t = \frac{\ln 2}{0.18}$$

$$t = 3.85 \text{ years}$$

Problem 3 How long will it take money to triple if it is invested at 12%, compounded continuously?

Answers to Matched Problems

1. \$146.93; \$148.02; \$149.18
2. $A = 5{,}000e^{0.2t}$

t	A
0	5,000
1	6,107
2	7,459
3	9,111
4	11,128
5	13,591
6	16,601
7	20,276
8	24,765
9	30,248
10	36,945

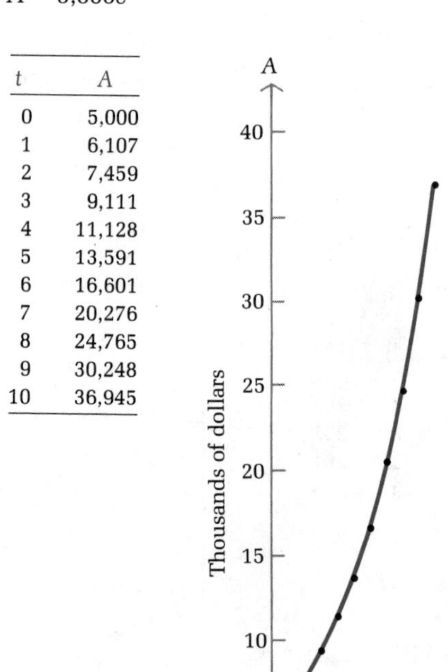

3. 9.16 years

Exercise 3-1

A *Use a calculator or table to evaluate A to the nearest cent in Problems 1–2.*

1. $A = \$1{,}000e^{0.1t}$ for $t = 2, 5,$ and 8
2. $A = \$5{,}000e^{0.08t}$ for $t = 1, 4,$ and 10

B *In Problems 3–8 solve for t or r to two decimal places.*

3. $2 = e^{0.06t}$
4. $2 = e^{0.03t}$
5. $3 = e^{0.1t}$
6. $3 = e^{0.25t}$
7. $2 = e^{5r}$
8. $3 = e^{10r}$

C *In Problems 9 and 10 complete each table to five decimal places using a hand calculator.*

9.

n	$(1 + 1/n)^n$
10	2.593 74
100	
1,000	
10,000	
100,000	
1,000,000	
10,000,000	
↓	↓
∞	$e = 2.718\ 281\ 8 \ldots$

10.

s	$(1 + s)^{1/s}$
0.01	2.704 81
−0.01	
0.001	
−0.001	
0.000 1	
−0.000 1	
0.000 01	
−0.000 01	
↓	↓
0	$e = 2.718\ 281\ 8 \ldots$

Applications

Business & Economics

11. *Continuous compound interest.* If \$20,000 is invested at an annual nominal rate of 12%, compounded continuously, how much will it be worth in 8.5 years?
12. *Continuous compound interest.* Assume \$1 had been invested at an annual nominal rate of 4%, compounded continuously, at the birth of Christ. What would be the value of the account in solid gold earths in the year 2000? (Assume that the earth weighs approximately 2.11×10^{26} ounces and that gold will be worth \$1,000 an ounce in the year 2000). What would be the value of the account in dollars at simple interest?
13. *Present value.* A note will pay \$20,000 at maturity 10 years from now. How much should you be willing to pay for the note now if money is worth 7%, compounded continuously?
14. *Present value.* A note will pay \$50,000 at maturity 5 years from now. How much should you be willing to pay for the note now if money is worth 8%, compounded continuously?

15. *Continuous compound interest.* An investor buys stock for $20,000. Four years later the stock is sold for $30,000. If interest is compounded continuously, what annual nominal rate of interest did the original $20,000 investment earn?

16. *Continuous compound interest.* A family paid $40,000 cash for a house. Ten years later, they sell the house for $100,000. If interest is compounded continuously, what annual nominal rate of interest did the original $40,000 investment earn?

17. *Present value.* Solving $A = Pe^{rt}$ for P, we obtain

$$P = Ae^{-rt}$$

which is the present value of the amount A due in t years if money is worth $100r\%$, compounded continuously.

(A) Graph $P = 10{,}000e^{-0.08t}$, $0 \leqslant t \leqslant 50$.
(B) $\lim_{x \to \infty} 10{,}000e^{-0.08t} = ?$ (Guess, using part A.)

(*Conclusion:* The further out that amount A is due, the smaller its present value, as we would expect.)

18. *Present value.* Referring to the preceding problem, in how many years will the $10,000 have to be due in order for its present value to be $5,000?

19. *Doubling time.* How long will it take money to double if invested at 25%, compounded continuously?

20. *Doubling time.* How long will it take money to double if invested at 5%, compounded continuously?

21. *Doubling rate.* At what nominal rate compounded continuously must money be invested to double in 5 years?

22. *Doubling rate.* At what nominal rate compounded continuously must money be invested to double in 3 years?

23. *Doubling time.* It is instructive to look at doubling times for money invested at various nominal rates of interest compounded continuously. Show that doubling time t at $100r\%$ interest compounded continuously is given by

$$t = \frac{\ln 2}{r}$$

24. *Doubling time.* Graph the doubling time equation from Problem 23 for $0 < r < 1.00$. Identify vertical and horizontal asymptotes.

Life Sciences

25. *World population.* A mathematical model for world population growth over short periods of time is given by

$$P = P_0e^{rt}$$

where

P_0 = Population at time $t = 0$
r = Continuous compound rate of growth

$t =$ Time

$P =$ Population at time t

How long will it take the earth's population to double if it continues to grow at its current continuous compound rate of 2% per year?

26. *World population.* Repeat Problem 25 under the assumption that the world population is growing at a continuous compound rate of 1% per year.

27. *Population growth.* Some underdeveloped nations have population doubling times of 20 years. At what continuous compound rate is the population growing? (Use the population growth model in Problem 25.)

28. *Population growth.* Some developed nations have population doubling times of 120 years. At what continuous compound rate is the population growing? (Use the population growth model in Problem 25.)

29. *Radioactive decay.* A mathematical model for the decay of radioactive substances is given by

$$Q = Q_0 e^{rt}$$

where

$Q_0 =$ Amount of the substance at time $t = 0$

$r =$ Continuous compound rate of decay

$t =$ Time

$Q =$ Amount of the substance at time t

If the continuous compound rate of decay of radium per year is $r = -0.000\,433\,2$, how long will it take an amount of radium to decay to half the original amount? (This period of time is called the half-life of the substance.)

30. *Radioactive decay.* The continuous compound rate of decay of carbon-14 per year is $r = -0.000\,123\,8$. How long will it take an amount of carbon-14 to decay to half the original amount? (Use the radioactive decay model in Problem 29.)

31. *Radioactive decay.* A cesium isotope has a half-life of 30 years. What is the continuous compound rate of decay? (Use the radioactive decay model in Problem 29.)

32. *Radioactive decay.* A strontium isotope has a half-life of 90 years. What is the continuous compound rate of decay? (Use the radioactive decay model in Problem 29.)

Social Sciences

33. *World population.* If the world population is now 4 billion (4×10^9) people and if it continues to grow at a continuous compound rate of 2% per year, how long will it be before there is only 1 square yard of land per person? (The earth has approximately 1.68×10^{14} square yards of land.)

3-2 Derivatives of Logarithmic and Exponential Functions

- Derivative Formulas
- Common Error
- Graphing Techniques
- Application

Derivative Formulas

We are now ready to derive a formula for the derivative of

$$f(x) = \ln x = \log_e x \qquad x > 0$$

using the definition of the derivative

$$f'(x) = \lim_{\Delta x \to 0} \frac{f(x + \Delta x) - f(x)}{\Delta x}$$

and the two-step process discussed in Section 1-4.

Step 1. Simplify the difference quotient first:

$$\begin{aligned}
\frac{f(x + \Delta x) - f(x)}{\Delta x} &= \frac{\ln(x + \Delta x) - \ln x}{\Delta x} \\
&= \frac{1}{\Delta x}[\ln(x + \Delta x) - \ln x] \\
&= \frac{1}{\Delta x} \ln \frac{x + \Delta x}{x} && \text{Property of logs} \\
&= \frac{1}{x}\left(\frac{x}{\Delta x}\right) \ln\left(1 + \frac{\Delta x}{x}\right) && \text{Multiply by } \frac{x}{x} = 1. \\
&= \frac{1}{x} \ln\left(1 + \frac{\Delta x}{x}\right)^{x/\Delta x} && \text{Property of logs}
\end{aligned}$$

Step 2. Find the limit. Let $s = \Delta x/x$. For x fixed, if $\Delta x \to 0$, then $s \to 0$. Thus,

$$\begin{aligned}
D_x \ln x &= \lim_{\Delta x \to 0} \frac{f(x + \Delta x) - f(x)}{\Delta x} \\
&= \lim_{\Delta x \to 0} \frac{1}{x} \ln\left(1 + \frac{\Delta x}{x}\right)^{x/\Delta x} && \text{Let } s = \frac{\Delta x}{x}. \\
&= \lim_{s \to 0} \frac{1}{x} \ln(1 + s)^{1/s} \\
&= \frac{1}{x} \ln[\lim_{s \to 0}(1 + s)^{1/s}] && \text{Properties of limits and continuity of log functions}
\end{aligned}$$

$$D_x \ln x = \frac{1}{x} \ln e \qquad \text{Definition of } e$$

$$= \frac{1}{x} \qquad \ln e = \log_e e = 1$$

Thus,

$$D_x \ln x = \frac{1}{x}$$

If we apply the two-step process to the exponential function $f(x) = e^x$, we can show that (see Problems 51 and 52 at the end of this section)

$$D_x e^x = e^x$$

Thus, **the derivative of the exponential function is the function itself.** (This important property is the reason that, out of all the possible exponential functions, the exponential function to the base e is often referred to as *the* exponential function.)

These two new and important derivative formulas are restated in the box:

Derivatives of the Natural Logarithmic and Exponential Functions

$$D_x \ln x = \frac{1}{x} \qquad D_x e^x = e^x$$

These new derivative formulas can be combined with the rules of differentiation discussed in Chapter 1 to differentiate a wide variety of functions.

Example 4 Find $f'(x)$ for:

(A) $f(x) = 2e^x + 3 \ln x$ (B) $f(x) = \dfrac{e^x}{x^3}$

(C) $f(x) = (\ln x)^4$ (D) $f(x) = \ln x^4$

Solutions

(A) $f'(x) = 2D_x e^x + 3D_x \ln x$

$$= 2e^x + 3\left(\frac{1}{x}\right) = 2e^x + \frac{3}{x}$$

(B) $f'(x) = \dfrac{x^3 D_x e^x - e^x D_x x^3}{(x^3)^2}$ Quotient rule

$$= \frac{x^3 e^x - e^x 3x^2}{x^6} = \frac{x^2 e^x (x-3)}{x^6} = \frac{e^x(x-3)}{x^4}$$

(C) $D_x(\ln x)^4 = 4(\ln x)^3 D_x \ln x$ Power rule for functions

$$= 4(\ln x)^3\left(\frac{1}{x}\right) = \frac{4(\ln x)^3}{x}$$

(D) $D_x \ln x^4 = D_x\, 4 \ln x$ Property of logarithms

$$= 4\left(\frac{1}{x}\right) = \frac{4}{x}$$

Problem 4 Find $f'(x)$ for:

(A) $f(x) = 4 \ln x - 5e^x$ (B) $f(x) = x^2 e^x$
(C) $f(x) = \ln x^3$ (D) $f(x) = (\ln x)^3$

■ Common Error

~~$D_x e^x = x e^{x-1}$~~

The power rule cannot be used to differentiate the exponential function. The power rule applies to exponential forms x^n where the exponent is a constant and the base is a variable. In the exponential form e^x, the base is a constant and the exponent is a variable.

■ Graphing Techniques

Using the techniques discussed in Chapter 2, we can use first and second derivatives to give us useful information about the graphs of $y = \ln x$ and $y = e^x$. Using the derivative formulas given previously, we have the following:

$\ln x$		e^x	
$y = \ln x$	$x > 0$	$y = e^x$	$-\infty < x < \infty$
$y' = 1/x > 0$	$x > 0$	$y' = e^x > 0$	$-\infty < x < \infty$
$y'' = -1/x^2 < 0$	$x > 0$	$y'' = e^x > 0$	$-\infty < x < \infty$

Thus, we see that both functions are increasing throughout their respective domains, the graph of $y = \ln x$ is always concave downward, and the graph of $y = e^x$ is always concave upward. It can be shown that the y axis is a vertical asymptote for the graph of $y = \ln x$ and the x axis is a horizontal asymptote for the graph of $y = e^x$ ($\lim_{x \to -\infty} e^x = 0$). Both equations are graphed in Figure 2.

Notice that if we fold the page along the line $y = x$, the two graphs match exactly (see Section 0-7). Also notice that both graphs are unbounded as $x \to \infty$. Comparing each graph with the graph of $y = x$ (the dashed line), we

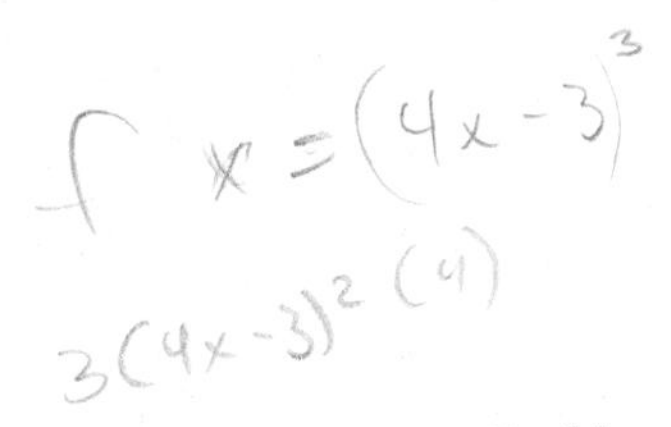

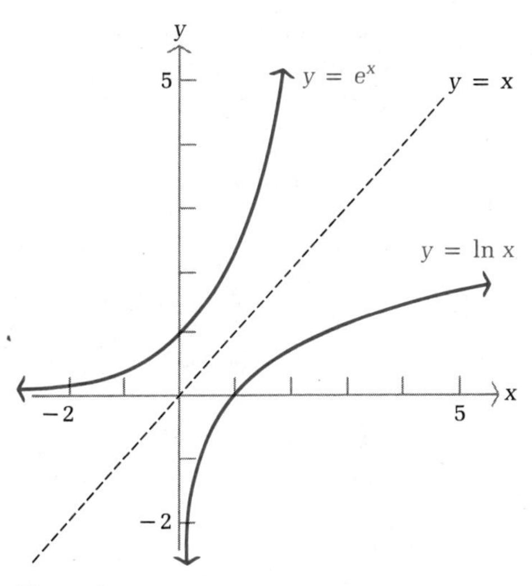

Figure 2

conclude that e^x grows more rapidly than x and ln x grows more slowly than x. In fact, the following limits can be established:

$$\lim_{x\to\infty} \frac{x^n}{e^x} = 0, \quad n > 0 \qquad \text{and} \qquad \lim_{x\to\infty} \frac{\ln x}{x^n} = 0, \quad n > 0$$

These limits indicate that e^x grows more rapidly than any positive power of x and ln x grows more slowly than any positive power of x.

Now let's apply graphing techniques to a slightly more complicated function.

Example 5 Sketch the graph of $f(x) = xe^x$.

Solution Step 1. Use the first derivative:

$$\begin{aligned} f'(x) &= xD_x e^x + e^x D_x x \\ &= xe^x + e^x \\ &= e^x(x+1) \end{aligned}$$

Since e^x is never 0 (see Figure 2), the only critical value is $x = -1$.

Sign chart for $f'(x) = e^x(x + 1)$:

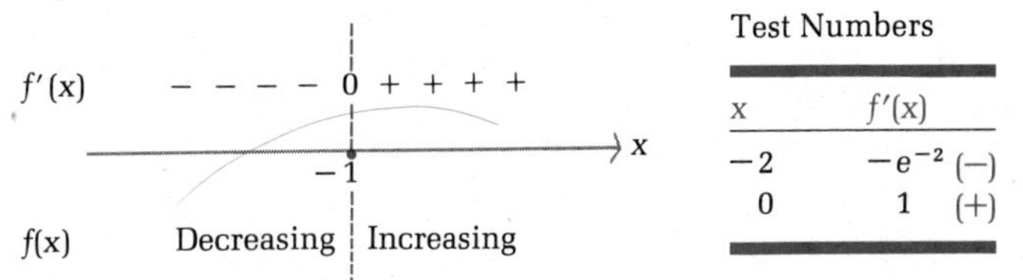

Test Numbers

x	f′(x)
−2	$-e^{-2}$ (−)
0	1 (+)

Thus, $f(x)$ decreases on $(-\infty, -1)$, has a local minimum at $x = -1$, and increases on $(-1, \infty)$. Since $e^x > 0$ for all x (see Figure 2), we do not have to evaluate e^{-2} to conclude that $-e^{-2} < 0$.

Step 2. Use the second derivative:

$$\begin{aligned} f''(x) &= e^x D_x(x+1) + (x+1)D_x e^x \\ &= e^x + (x+1)e^x \\ &= e^x(x+2) \end{aligned}$$

Sign chart for $f''(x) = e^x(x+2)$:

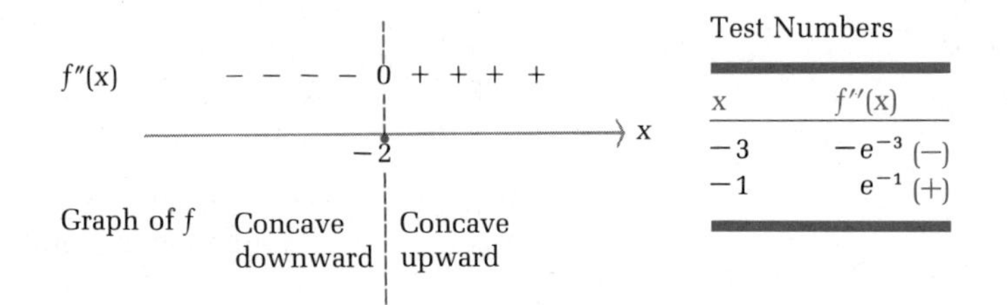

Thus, the graph of f is concave downward on $(-\infty, -2)$, has an inflection point at $x = -2$, and is concave upward on $(-2, \infty)$.

Step 3. Find intercepts. The y (and x) intercept is $f(0) = 0$. Since e^x is never 0, there are no other x intercepts.

Step 4. Investigate behavior as $x \to \infty$ and $x \to -\infty$. We have not developed limit techniques for functions of this type, but the following tables of values are sufficient to determine the nature of the graph as $x \to \infty$ and $x \to -\infty$:

x	1	5	10	$\to \infty$
$f(x)$	2.72	742.07	220,264.66	$\to \infty$

x	-1	-5	-10	$\to -\infty$
$f(x)$	-0.37	-0.03	$-0.000\ 45$	$\to 0$

Step 5. Sketch a graph of f, using the information from steps 1–4.

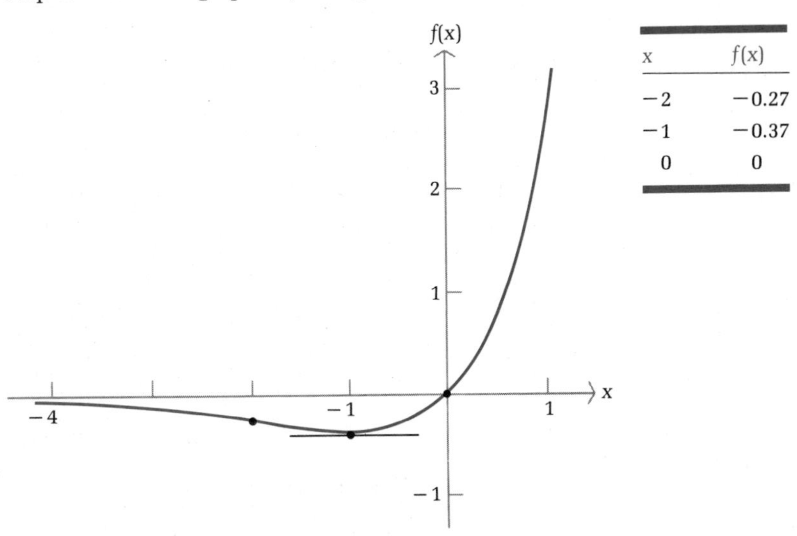

x	f(x)
−2	−0.27
−1	−0.37
0	0

Problem 5 Sketch the graph of $f(x) = x \ln x$.

▪ Application

Example 6
Maximum Profit

The market research department of a pet store chain test marketed their aquarium pumps (as well as other items) in several pet stores in a test city. They found that the weekly demand for aquarium pumps is given approximately by

$$p = 12 - 2 \ln x \qquad 0 < x < 90$$

where x is the number of pumps sold each week and \$$p$ is the price of one pump. If each pump costs the chain \$3, how should it be priced in order to maximize the weekly profit?

Solution Although we want to find the price that maximizes the weekly profit, it will be easier to first find the number of pumps that will maximize the weekly profit. The revenue equation is

$$R(x) = xp = 12x - 2x \ln x$$

The cost equation is

$$C(x) = 3x$$

and the profit equation is

$$P(x) = R(x) - C(x)$$
$$= 12x - 2x \ln x - 3x$$
$$= 9x - 2x \ln x$$

Thus, we must solve the following:

$$\text{Maximize } P(x) = 9x - 2x \ln x \qquad 0 < x < 90$$

$$P'(x) = 9 - 2x\left(\frac{1}{x}\right) - 2 \ln x$$
$$= 7 - 2 \ln x = 0$$
$$2 \ln x = 7$$
$$\ln x = 3.5$$
$$x = e^{3.5}$$

$$P''(x) = -2\left(\frac{1}{x}\right) = -\frac{2}{x}$$

Since $x = e^{3.5}$ is the only critical value and $P''(e^{3.5}) < 0$, the maximum weekly profit occurs when $x = e^{3.5} \approx 33$ and $p = 12 - 2 \ln e^{3.5} = \$5$.

Problem 6 Repeat Example 6 if each pump costs the chain $3.50.

Answers to Matched Problems

4. (A) $4/x - 5e^x$ (B) $xe^x(x + 2)$ (C) $3/x$ (D) $3(\ln x)^2/x$
5. Increasing on (e^{-1}, ∞)
   Decreasing on $(0, e^{-1})$
   Local minimum at $x = e^{-1}$
   Concave upward on $(0, \infty)$
   $f(1) = 0$

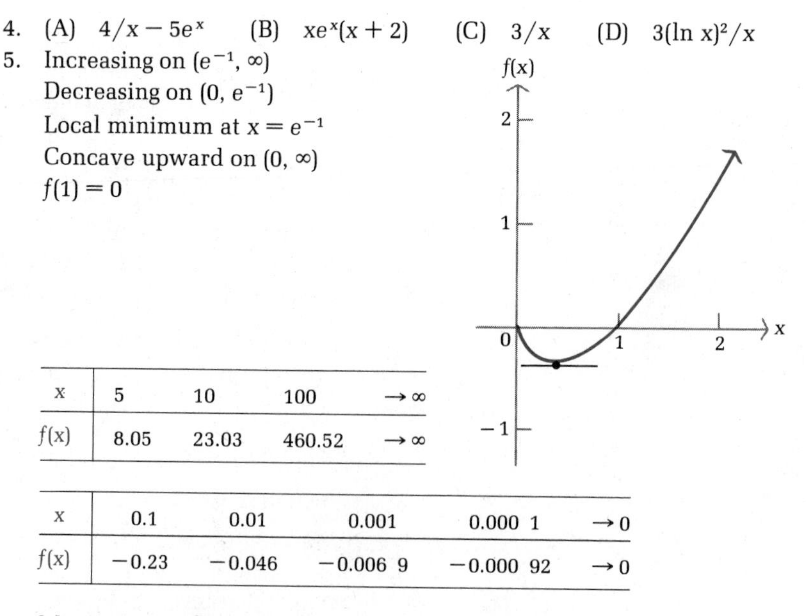

x	5	10	100	$\to \infty$
$f(x)$	8.05	23.03	460.52	$\to \infty$

x	0.1	0.01	0.001	0.000 1	$\to 0$
$f(x)$	−0.23	−0.046	−0.006 9	−0.000 92	$\to 0$

6. Maximum profit occurs for $x = e^{3.25} \approx 26$ and $p = \$5.50$

Exercise 3-2

A *Find $f'(x)$.*

1. $f(x) = 6e^x - 7 \ln x$
2. $f(x) = 4e^x + 5 \ln x$
3. $f(x) = 2x^e + 3e^x$
4. $f(x) = 4e^x - ex^e$
5. $f(x) = \ln x^5$
6. $f(x) = (\ln x)^5$
7. $f(x) = (\ln x)^2$
8. $f(x) = \ln x^2$

B

9. $f(x) = x^4 \ln x$
10. $f(x) = x^3 \ln x$
11. $f(x) = x^3e^x$
12. $f(x) = x^4e^x$
13. $f(x) = \dfrac{e^x}{x^2 + 9}$
14. $f(x) = \dfrac{e^x}{x^2 + 4}$
15. $f(x) = \dfrac{\ln x}{x^4}$
16. $f(x) = \dfrac{\ln x}{x^3}$
17. $f(x) = (x + 2)^3 \ln x$
18. $f(x) = (x - 1)^2 \ln x$
19. $f(x) = (x + 1)^3e^x$
20. $f(x) = (x - 2)^3e^x$
21. $f(x) = \dfrac{x^2 + 1}{e^x}$
22. $f(x) = \dfrac{x + 1}{e^x}$
23. $f(x) = x(\ln x)^3$
24. $f(x) = x(\ln x)^2$
25. $f(x) = (4 - 5e^x)^3$
26. $f(x) = (5 - \ln x)^4$
27. $f(x) = \sqrt{1 + \ln x}$
28. $f(x) = \sqrt{1 + e^x}$
29. $f(x) = xe^x - e^x$
30. $f(x) = x \ln x - x$
31. $f(x) = 2x^2 \ln x - x^2$
32. $f(x) = x^2e^x - 2xe^x + 2e^x$

Find the equation of the line tangent to the graph of $y = f(x)$ at the indicated value of x.

33. $f(x) = e^x; \quad x = 1$
34. $f(x) = e^x; \quad x = 2$
35. $f(x) = \ln x; \quad x = e$
36. $f(x) = \ln x; \quad x = 1$

C *Find the indicated extremum of each function for $x > 0$.*

37. Absolute maximum value of $f(x) = 4x - x \ln x$
38. Absolute minimum value of $f(x) = x \ln x - 3x$
39. Absolute minimum value of $f(x) = \dfrac{e^x}{x}$
40. Absolute maximum value of $f(x) = \dfrac{x^2}{e^x}$
41. Absolute maximum value of $f(x) = \dfrac{1 + 2 \ln x}{x}$
42. Absolute minimum value of $f(x) = \dfrac{1 - 5 \ln x}{x}$

Sketch the graph of $y = f(x)$.

43. $f(x) = 1 - e^x$
44. $f(x) = 1 - \ln x$
45. $f(x) = x - \ln x$
46. $f(x) = e^x - x$
47. $f(x) = (3 - x)e^x$
48. $f(x) = (x - 2)e^x$
49. $f(x) = x^2 \ln x$
50. $f(x) = \dfrac{\ln x}{x}$

Problems 51 and 52 refer to the function $f(x) = e^x$.

51. Show that

$$\lim_{\Delta x \to 0} \frac{f(x + \Delta x) - f(x)}{\Delta x} = e^x \lim_{\Delta x \to 0} \frac{e^{\Delta x} - 1}{\Delta x}$$

52. Use a calculator to estimate

$$\lim_{\Delta x \to 0} \frac{e^{\Delta x} - 1}{\Delta x}$$

and then use the result in Problem 51 to show that $f'(x) = e^x$.

Applications

Business & Economics

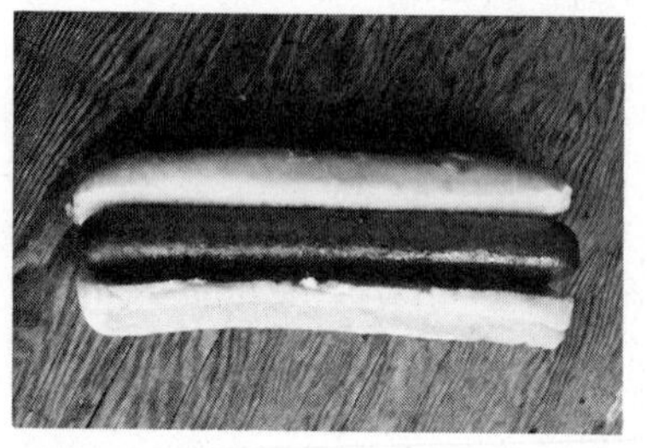

53. *Maximum profit.* A national food service runs food concessions for sporting events throughout the country. Their marketing research department chose a particular baseball/football stadium to test market a new jumbo hot dog. It was found that the demand for the new hot dog is given approximately by

$$p = 5 - \ln x \qquad 5 \le x \le 50$$

where x is the number of hot dogs (in thousands) that can be sold during one game at a price of $\$p$. If the concessionaire pays $1 for each hot dog, how should the hot dogs be priced to maximize the profit per game?

54. *Maximum profit.* On a national tour of a rock band, the demand for tee shirts is given by

$$p = 15 - 4 \ln x \qquad 5 \le x \le 40$$

where x is the number of shirts (in thousands) that can be sold during a single concert at a price of $\$p$. If the shirts cost the band $5 each, how should they be priced in order to maximize the profit per concert?

55. *Minimum average cost.* The cost of producing x units of a product is given by

$$C(x) = 600 + 100x - 100 \ln x \qquad x \ge 1$$

Find the minimal average cost.

56. *Minimum average cost.* The cost of producing x units of a product is

given by

$$C(x) = 1{,}000 + 200x - 200 \ln x \qquad x \geq 1$$

Find the minimal average cost.

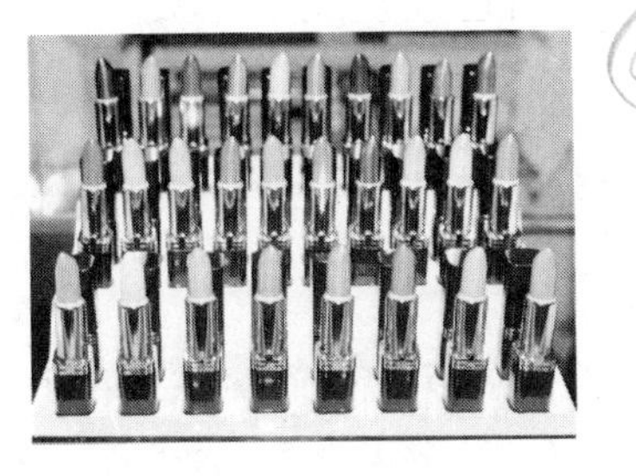

57. *Maximizing revenue.* A cosmetic company is planning the introduction and promotion of a new lipstick line. The marketing research department, after test marketing the new line in a large carefully selected city, found that the demand in that city is given approximately by

$$p = 10e^{-x} \qquad 0 \leq x \leq 2$$

where x thousand lipsticks were sold per week at a price of p dollars each.

(A) At what price will the weekly revenue $R(x) = xp$ be maximum? What is the maximum weekly revenue in the test city?

(B) Graph R for $0 \leq x \leq 2$.

58. *Maximizing revenue.* Repeat the preceding problem using the demand equation $p = 12e^{-x}$, $0 \leq x \leq 2$.

Life Sciences

59. *Blood pressure.* An experiment was set up to find a relationship between weight and systolic blood pressure in normal children. Using hospital records for 5,000 normal children, it was found that the systolic blood pressure was given approximately by

$$P(x) = 17.5(1 + \ln x) \qquad 10 \leq x \leq 100$$

where $P(x)$ is measured in millimeters of mercury and x is measured in pounds. What is the rate of change of blood pressure with respect to weight at the 40 pound weight level? At the 90 pound weight level?

60. *Blood pressure.* Graph the systolic blood pressure equation in the preceding problem.

61. *Drug concentration.* The concentration of a drug in the blood stream t hours after injection is given approximately by

$$C(t) = 4.35e^{-t} \qquad 0 \leq t \leq 5$$

where $C(t)$ is concentration in milligrams per milliliter.

(A) What is the rate of change of concentration after 1 hour? After 4 hours?

(B) Graph C.

62. *Water pollution.* The use of iodine crystals is a popular way of making small quantities of nondrinkable water drinkable. Crystals placed in a 1 ounce bottle of water will dissolve until the solution is saturated. After saturation, one-half of this solution is poured into a quart container of questionable water, then after about an hour the questionable water is usually drinkable. The half empty 1 ounce bottle is then refilled to be used again in the same way. Suppose the concentration of iodine in the 1 ounce bottle t minutes after the crystals are introduced can be approximated by

$$C(t) = 250(1 - e^{-t}) \qquad t \geq 0$$

where $C(t)$ is the concentration of iodine in micrograms per milliliter.

(A) What is the rate of change of the concentration after 1 minute? After 4 minutes?
(B) Graph C for $0 \leq t \leq 5$.

Social Sciences

63. *Psychology—stimulus/response.* In psychology the Weber-Fechner Law for stimulus response is

$$R = k \ln\left(\frac{S}{S_0}\right)$$

where R is the response, S is the stimulus, and S_0 is the lowest level of stimulus that can be detected. Find dR/dS.

64. *Psychology–learning.* A mathematical model for the average of a group of people learning to type is given by

$$N(t) = 10 + 6 \ln t \qquad t \geq 1$$

where $N(t)$ is the number of words per minute typed after t hours of instruction and practice (2 hours per day, 5 days per week). What is the rate of learning after 10 hours of instruction and practice? After 100 hours?

3-3 Chain Rule

- Composite Functions
- Chain Rule
- Generalized Derivative Rules
- Other Logarithmic and Exponential Functions

Suppose you were asked to find the derivative of

$$h(x) = \ln(2x + 1)$$

We have developed formulas for computing the derivatives of the natural logarithm function and polynomial functions separately, but not in the indicated combination. In this section we will discuss one of the most important derivative rules of all—the **chain rule.** This rule will enable us to determine the derivatives of some fairly complicated functions in terms of derivatives of more elementary functions. The chain rule is used to compute derivatives of functions that are compositions of more elementary functions whose derivatives are known.

▪ Composite Functions

Let us look at the given function h more closely:

$$h(x) = \ln(2x + 1)$$

The function h is a combination of the natural logarithm function and a linear function. To see this more clearly, let

$$y = f(u) = \ln u$$
$$u = g(x) = 2x + 1$$

Then we can express y as a function of x as follows:

$$y = f(u) = f[g(x)] = \ln(2x + 1) = h(x)$$

The function h is said to be the *composite* of the two simpler functions f and g. (Loosely speaking, we can think of h as a function of a function.) In general:

Composite Functions

A function h is a **composite** of functions f and g if

$$h(x) = f[g(x)]$$

The domain of h is the set of all numbers x such that x is in the domain of g and $g(x)$ is in the domain of f.

Example 7 Let $f(u) = e^u$ and $g(x) = 3x^2 + 1$. Find:

(A) $f[g(x)]$ (B) $g[f(u)]$

Solutions

(A) $f[g(x)] = e^{g(x)} = e^{3x^2+1}$

(B) $g[f(u)] = 3[f(u)]^2 + 1 = 3[e^u]^2 + 1 = 3e^{2u} + 1$

Problem 7 Let $f(u) = \ln u$ and $g(x) = 2x^3 + 4$. Find:

(A) $f[g(x)]$ (B) $g[f(u)]$

Example 8 Write each function as a composition of the natural logarithm or exponential function and a polynomial.

(A) $y = \ln(x^3 - 2x^2 + 1)$ (B) $y = e^{x^2+4}$

Solutions (A) Let

$$y = f(u) = \ln u$$
$$u = g(x) = x^3 - 2x^2 + 1$$

Check $y = f[g(x)] = \ln[g(x)] = \ln(x^3 - 2x^2 + 1)$

(B) Let

$$y = f(u) = e^u$$
$$u = g(x) = x^2 + 4$$

Check $y = f[g(x)] = e^{g(x)} = e^{x^2+4}$

Problem 8 Repeat Example 8 for:

(A) $y = e^{2x^3+7}$ (B) $y = \ln(x^4 + 10)$

▪ Chain Rule

The word "chain" comes from the fact that a function formed by composition (such as those in Example 7) involves a "chain" of functions—that is, "a function of a function." We now introduce the *chain rule*, which will enable us to compute the derivative of a composite function in terms of the derivatives of the functions making up the composition.

Suppose

$$y = h(x) = f[g(x)]$$

is a composite of f and g where

$$y = f(u) \qquad \text{and} \qquad u = g(x)$$

We would like to express the derivative dy/dx in terms of the derivatives of f and g. From the definition of a derivative we have

$$\frac{dy}{dx} = \lim_{\Delta x \to 0} \frac{h(x + \Delta x) - h(x)}{\Delta x}$$

$$= \lim_{\Delta x \to 0} \frac{\Delta y}{\Delta x} \qquad (1)$$

Noting that

$$\frac{\Delta y}{\Delta x} = \frac{\Delta y}{\Delta u} \frac{\Delta u}{\Delta x} \qquad (2)$$

we might be tempted to substitute (2) into (1) to obtain

$$\frac{dy}{dx} = \lim_{\Delta x \to 0} \frac{\Delta y}{\Delta u} \frac{\Delta u}{\Delta x}$$

and reason that $\Delta u \to 0$ as $\Delta x \to 0$ so that

$$\frac{dy}{dx} = \left(\lim_{\Delta u \to 0} \frac{\Delta y}{\Delta u}\right)\left(\lim_{\Delta x \to 0} \frac{\Delta u}{\Delta x}\right)$$

$$= \frac{dy}{du}\frac{du}{dx}$$

The result is correct under rather general conditions, and is called the *chain rule*, but our "derivation" is superficial, because it ignores a number of hidden problems. Since a formal proof of the **chain rule** is beyond the scope of this book, we simply state it as follows:

Chain Rule

If $y = f(u)$ and $u = g(x)$, define the composite function

$$y = h(x) = f[g(x)]$$

Then

$$\frac{dy}{dx} = \frac{dy}{du}\frac{du}{dx} \qquad \text{provided } \frac{dy}{du} \text{ and } \frac{du}{dx} \text{ exist}$$

Example 9 Find dy/dx, given:

(A) $y = \ln(x^2 - 4x + 2)$ (B) $y = e^{2x^3+5}$

Solutions (A) Let $y = \ln u$ and $u = x^2 - 4x + 2$. Then

$$\frac{dy}{dx} = \frac{dy}{du}\frac{du}{dx} \; ^*$$

$$= \frac{1}{u}(2x - 4)$$

$$= \frac{1}{x^2 - 4x + 2}(2x - 4) \qquad \text{Since } u = x^2 - 4x + 2$$

$$= \frac{2x - 4}{x^2 - 4x + 2}$$

(B) Let $y = e^u$ and $u = 2x^3 + 5$. Then

$$\frac{dy}{dx} = \frac{dy}{du}\frac{du}{dx}$$

$$= e^u(6x^2)$$

$$= 6x^2e^{2x^3+5} \qquad \text{Since } u = 2x^3 + 5$$

* After some experience with the chain rule, the steps in the dashed boxes are usually done mentally.

Problem 9 Find dy/dx, given:

(A) $y = e^{3x^4+6}$ (B) $y = \ln(x^2 + 9x + 4)$

The chain rule can be extended to compositions of three or more functions. For example, if $y = f(w)$, $w = g(u)$, and $u = h(x)$ then

$$\frac{dy}{dx} = \frac{dy}{dw}\frac{dw}{du}\frac{du}{dx}$$

Example 10 For $y = h(x) = e^{1+(\ln x)^2}$, find dy/dx.

Solution Note that h is of the form $y = e^w$ where $w = 1 + u^2$ and $u = \ln x$. Thus,

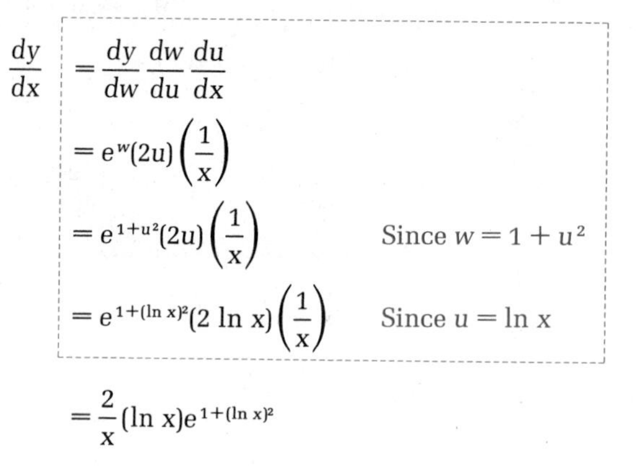

$$\frac{dy}{dx} = \frac{dy}{dw}\frac{dw}{du}\frac{du}{dx}$$

$$= e^w(2u)\left(\frac{1}{x}\right)$$

$$= e^{1+u^2}(2u)\left(\frac{1}{x}\right) \qquad \text{Since } w = 1 + u^2$$

$$= e^{1+(\ln x)^2}(2 \ln x)\left(\frac{1}{x}\right) \qquad \text{Since } u = \ln x$$

$$= \frac{2}{x}(\ln x)e^{1+(\ln x)^2}$$

Problem 10 For $y = h(x) = [\ln(1 + e^x)]^3$, find dy/dx.

■ Generalized Derivative Rules

In practice, it is not necessary to introduce additional variables when using the chain rule, as we did in Examples 9 and 10. Instead, the chain rule can be used to extend the derivative rules for specific functions to general derivative rules for compositions. This is what we did in Section 1-7 when we discussed the general power rule. In fact, the general power rule is a consequence of the chain rule. To see this, let $y = [f(x)]^n$ and $u = f(x)$. Applying the chain rule to $y = u^n$, we have

$$\frac{dy}{dx} = \frac{dy}{du}\frac{du}{dx} = nu^{n-1}f'(x) = n[f(x)]^{n-1}f'(x)$$

The same technique can be applied to functions of the form $y = e^{f(x)}$ and $y = \ln[f(x)]$ (see Problems 59 and 60 at the end of this section). The results are summarized in the following box:

General Derivative Rules

$$D_x[f(x)]^n = n[f(x)]^{n-1}f'(x) \quad (3)$$

$$D_x \ln[f(x)] = \frac{1}{f(x)} f'(x) \quad (4)$$

$$D_x e^{f(x)} = e^{f(x)}f'(x) \quad (5)$$

For power, natural logarithm, or exponential forms, we can either use the chain rule discussed earlier or these special differentiation formulas based on the chain rule. Use whichever is easier for you. In Example 11, we will use the general derivative rules.

Example 11

(A) $D_x e^{2x} = e^{2x} D_x 2x$ Using (5)

$= e^{2x}(2) = 2e^{2x}$

(B) $D_x \ln(x^2 + 9) = \frac{1}{x^2 + 9} D_x(x^2 + 9)$ Using (4)

$= \frac{1}{x^2 + 9} 2x = \frac{2x}{x^2 + 9}$

(C) $D_x(1 + e^{x^2})^3 = 3(1 + e^{x^2})^2 D_x(1 + e^{x^2})$ Using (3)

$= 3(1 + e^{x^2})^2 e^{x^2} D_x x^2$ Using (5)

$= 3(1 + e^{x^2})^2 e^{x^2}(2x)$

$= 6xe^{x^2}(1 + e^{x^2})^2$

Problem 11 Find:

(A) $D_x \ln(x^3 + 2x)$ (B) $D_x e^{3x^2+2}$ (C) $D_x(2 + e^{-x^2})^4$

▪ Other Logarithmic and Exponential Functions

In most applications involving logarithmic or exponential functions, the number e is the preferred base. However, there are situations where it is convenient to use a base other than e. Derivatives of $y = \log_b x$ and $y = b^x$ can be obtained by expressing these functions in terms of the natural logarithmic and exponential functions. We begin by finding a relationship between $\log_b x$ and $\ln x$ for any base b, $b > 0$ and $b \neq 1$.

$y = \log_b x$ Change to exponential form.

$b^y = x$ Take the natural log of both sides.

$\ln b^y = \ln x$ Recall that $\ln b^y = y \ln b$.

$y \ln b = \ln x$ Solve for y.

$$y = \frac{1}{\ln b} \ln x$$

Thus,

$$\log_b x = \frac{1}{\ln b} \ln x \qquad \text{Change of base formula*} \tag{6}$$

Differentiating both sides of (6), we have

$$D_x \log_b x = \frac{1}{\ln b} D_x \ln x = \frac{1}{\ln b} \frac{1}{x}$$

Example 12 Find $f'(x)$ for:

(A) $f(x) = \log_2 x$ (B) $f(x) = \log(1 + x^3)$

Solutions (A) $f(x) = \log_2 x = \frac{1}{\ln 2} \ln x$ Using (6)

$$f'(x) = \frac{1}{\ln 2} \frac{1}{x}$$

(B) $f(x) = \log(1 + x^3)$ Recall that $\log r = \log_{10} r$.

$$= \frac{1}{\ln 10} \ln(1 + x^3) \qquad \text{Using (6)}$$

$$f'(x) = \frac{1}{\ln 10} \frac{1}{1 + x^3} 3x^2 = \frac{1}{\ln 10} \frac{3x^2}{1 + x^3}$$

Problem 12 Find $f'(x)$ for:

(A) $f(x) = \log x$ (B) $f(x) = \log_3(x + x^2)$

Now we want to find a relationship between b^x and e^x for any base b, $b > 0$ and $b \neq 1$.

$$y = b^x \qquad \text{Take the natural log of both sides.}$$

$$\ln y = \ln b^x$$

$$= x \ln b \qquad \text{Change to exponential form.}$$

$$y = e^{x \ln b}$$

Thus,

$$b^x = e^{x \ln b} \tag{7}$$

Differentiating both sides of (7), we have

$$D_x b^x = e^{x \ln b} \ln b = b^x \ln b$$

Example 13 Find $f'(x)$ for:

(A) $f(x) = 2^x$ (B) $f(x) = 10^{x^5 + x}$

* Equation (6) is a special case of the **change of base** formula for logarithms (which can be derived in the same way): $\log_b x = \log_a x / \log_a b$.

Solutions (A) $f(x) = 2^x = e^{x \ln 2}$ Using (7)

$f'(x) = e^{x \ln 2} \ln 2 = 2^x \ln 2$

(B) $f(x) = 10^{x^5+x} = e^{(x^5+x)\ln 10}$ Using (7)

$f'(x) = e^{(x^5+x)\ln 10}(5x^4 + 1) \ln 10$

$= 10^{x^5+x}(5x^4 + 1) \ln 10$

Problem 13 Find $f'(x)$ for:

(A) $f(x) = 5^x$ (B) $f(x) = 4^{x^2+3x}$

Answers to Matched Problems

7. (A) $f[g(x)] = \ln(2x^3 + 4)$ (B) $g[f(u)] = 2(\ln u)^3 + 4$
8. (A) $y = f(u) = e^u$ and $u = g(x) = 2x^3 + 7$
   (B) $y = f(u) = \ln u$ and $u = g(x) = x^4 + 10$
9. (A) $12x^3e^{3x^4+6}$ (B) $\dfrac{2x + 9}{x^2 + 9x + 4}$
10. $\dfrac{dy}{dx} = \dfrac{3e^x[\ln(1 + e^x)]^2}{1 + e^x}$
11. (A) $\dfrac{3x^2 + 2}{x^3 + 2x}$ (B) $6xe^{3x^2+2}$ (C) $-8xe^{-x^2}(2 + e^{-x^2})^3$
12. (A) $\dfrac{1}{\ln 10}\dfrac{1}{x}$ (B) $\dfrac{1}{\ln 3}\dfrac{1 + 2x}{x + x^2}$
13. (A) $5^x \ln 5$ (B) $4^{x^2+3x}(2x + 3) \ln 4$

Exercise 3-3

A

Write each composite function in the form $y = f(u)$ and $u = g(x)$.

1. $y = (2x + 5)^3$
2. $y = (3x - 7)^5$
3. $y = \ln(2x^2 + 7)$
4. $y = \ln(x^2 - 2x + 5)$
5. $y = e^{x^2-2}$
6. $y = e^{3x^3+5x}$

Express y in terms of x. Use the chain rule to find dy/dx and then express dy/dx in terms of x.

7. $y = u^2, \quad u = 2 + e^x$
8. $y = u^3, \quad u = 3 - \ln x$
9. $y = e^u, \quad u = 2 - x^4$
10. $y = e^u, \quad u = x^6 + 5x^2$
11. $y = \ln u, \quad u = 4x^5 - 7$
12. $y = \ln u, \quad u = 2 + 3x^4$

Find each derivative.

13. $D_x \ln(x - 3)$
14. $D_w \ln(w + 100)$
15. $D_t \ln(3 - 2t)$
16. $D_y \ln(4 - 5y)$

17. $D_x 3e^{2x}$
18. $D_y 2e^{3y}$
19. $D_t 2e^{-4t}$
20. $D_r 6e^{-3r}$

B

21. $D_x 100e^{-0.03x}$
22. $D_t 1{,}000e^{0.06t}$
23. $D_x \ln(x+1)^4$
24. $D_x \ln(x+1)^{-3}$
25. $D_x(2e^{2x} - 3e^x + 5)$
26. $D_t(1 + e^{-t} - e^{-2t})$
27. $D_x e^{3x^2-2x}$
28. $D_x e^{x^3-3x^2+1}$
29. $D_t \ln(t^2 + 3t)$
30. $D_x \ln(x^3 - 3x^2)$
31. $D_x \ln(x^2+1)^{1/2}$
32. $D_x \ln(x^4+5)^{3/2}$
33. $D_t[\ln(t^2+1)]^4$
34. $D_w[\ln(w^3-1)]^2$
35. $D_x(e^{2x}-1)^4$
36. $D_x(e^{x^2}+3)^5$
37. $D_x \dfrac{e^{2x}}{x^2+1}$
38. $D_x \dfrac{e^{x+1}}{x+1}$
39. $D_x(x^2+1)e^{-x}$
40. $D_x(1-x)e^{2x}$
41. $D_x e^{-x} \ln x$
42. $D_x \dfrac{\ln x}{e^x + 1}$
43. $D_x \dfrac{1}{\ln(1+x^2)}$
44. $D_x \dfrac{1}{\ln(1-x^3)}$
45. $D_x \sqrt[3]{\ln(1-x^2)}$
46. $D_t \sqrt[5]{\ln(1-t^5)}$

C

Sketch the graph of $y = f(x)$.

47. $f(x) = 1 - e^{-x}$
48. $f(x) = 2 - 3e^{-2x}$
49. $f(x) = \ln(1-x)$
50. $f(x) = \ln(2x+4)$
51. $f(x) = e^{-(1/2)x^2}$
52. $f(x) = \ln(x^2+4)$

Express y in terms of x. Use the chain rule to find dy/dx. Express dy/dx in terms of x.

53. $y = 1 + w^2, \quad w = \ln u, \quad u = 2 + e^x$
54. $y = \ln w, \quad w = 1 + e^u, \quad u = x^2$

Find each derivative.

55. $D_x \log_2(3x^2 - 1)$
56. $D_x \log(x^3 - 1)$
57. $D_x 10^{x^2+x}$
58. $D_x 8^{1-2x^2}$
59. Use the chain rule to derive the formula

$$D_x \ln[f(x)] = \frac{1}{f(x)} f'(x)$$

60. Use the chain rule to derive the formula

$$D_x e^{f(x)} = e^{f(x)} f'(x)$$

Applications

Business & Economics

61. *Maximum revenue.* Suppose the price–demand equation for x units of a commodity is determined from empirical data to be

$$p = 100e^{-0.05x}$$

where x units are sold per day at a price of $p each. Find the production level and price that maximizes revenue. What is the maximum revenue?

62. *Maximum revenue.* Repeat the preceding problem using the price–demand equation

$$p = 10e^{-0.04x}$$

63. *Salvage value.* The salvage value S, in dollars, of a company airplane after t years is estimated to be given by

$$S(t) = 300{,}000e^{-0.1t}$$

What is the rate of depreciation in dollars per year after 1 year? 5 years? 10 years?

64. *Resale value.* The resale value R, in dollars, of a company car after t years is estimated to be given by

$$R(t) = 20{,}000e^{-0.15t}$$

What is the rate of depreciation in dollars per year after 1 year? 2 years? 3 years?

65. *Promotion and maximum profit.* A recording company has produced a new compact disc featuring a very popular recording group. Before launching a national sales campaign, the marketing research department chose to test market the disc in a bellwether city. Their interest is in determining the length of a sales campaign that will maximize total profits. From empirical data, the research department estimates that the proportion of a target group of 50,000 persons buying the disc after t days of television promotion is given by $1 - e^{-0.03t}$. If a $4 profit is realized on each disc sold, then the total revenue after t days of promotion will be approximated by

$$R(t) = (4)(50{,}000)(1 - e^{-0.03t}) \qquad t \geq 0$$

Television promotion costs are

$$C(t) = 4{,}000 + 3{,}000t \qquad t \geq 0$$

(A) How many days of television promotion should be used to maximize total profit? What is the maximum total profit? What percentage of the target market will have purchased the disc when the maximum profit is reached?

(B) Graph the profit function.

66. *Promotion and maximum profit.* Repeat the preceding problem using the revenue equation

$$R(t) = (3)(60{,}000)(1 - e^{-0.04t})$$

Life Sciences

67. *Blood pressure and age.* A research group using hospital records developed the following approximate mathematical model relating systolic blood pressure and age:

$$P(x) = 40 + 25\ln(x + 1) \qquad 0 \leqslant x \leqslant 65$$

where $P(x)$ is pressure measured in millimeters of mercury and x is age in years. What is the rate of change of pressure at the end of 10 years? At the end of 30 years? At the end of 60 years?

68. *Biology.* A yeast culture at room temperature (68°F) is placed in a refrigerator maintaining a constant temperature of 38°F. After t hours the temperature T of the culture is given approximately by

$$T = 30e^{-0.58t} + 38 \qquad t \geqslant 0$$

What is the rate of change of temperature of the culture at the end of 1 hour? At the end of 4 hours?

69. *Bacterial growth.* A single cholera bacterium divides every 0.5 hour to produce two complete cholera bacteria. If we start with a colony of 5,000 bacteria, then after t hours there will be

$$A(t) = 5{,}000 \cdot 2^{2t}$$

bacteria. Find $A'(t)$, $A'(1)$, and $A'(5)$ and interpret.

70. *Bacterial growth.* Repeat the preceding problem for a starting colony of 1,000 bacteria where a single bacterium divides every 0.25 hour.

Social Sciences

71. *Sociology.* Daniel Lowenthal, a sociologist at Columbia University, made a 5 year study on the sale of popular records relative to their position in the top 20. He found that the average number of sales $N(n)$ of the nth ranking record was given approximately by

$$N(n) = N_1e^{-0.09(n-1)} \qquad 1 \leqslant n \leqslant 20$$

where N_1 was the number of sales of the number one record on the list at a given time. Graph N for $N_1 = 1{,}000{,}000$ records.

72. *Political science.* Thomas W. Casstevens, a political scientist at Oakland University, has studied legislative turnover. He (with others) found that the number $N(t)$ of continuously serving members of an elected legislative body remaining t years after an election is given approximately by a function of the form

$$N(t) = N_0e^{-ct}$$

In particular, for the 1965 election for the U.S. House of Representatives, it was found that

$$N(t) = 434e^{-0.0866t}$$

What is the rate of change after 2 years? After 10 years?

3-4 Chapter Review

Important Terms and Symbols

3-1 *The constant e and continuous compound interest.* Definition of e, continuous compound interest

3-2 *Derivatives of logarithmic and exponential functions.* Derivative formulas for the natural logarithmic and exponential functions, graph properties of $y = \ln x$ and $y = e^x$

3-3 *Chain rule.* Composite functions, chain rule, general derivative formulas, derivative formulas for $y = \log_b x$ and $y = b^x$

Additional Rules of Differentiation

$$D_x \ln x = \frac{1}{x}$$

$$D_x e^x = e^x$$

$$D_x \ln[f(x)] = \frac{1}{f(x)} f'(x)$$

$$D_x e^{f(x)} = e^{f(x)} f'(x)$$

$$D_x \log_b x = D_x \frac{1}{\ln b} \ln x = \frac{1}{\ln b} \frac{1}{x}$$

$$D_x b^x = D_x e^{x \ln b} = e^{x \ln b} \ln b = b^x \ln b$$

$$\frac{dy}{dx} = \frac{dy}{du}\frac{du}{dx}, \quad \frac{dy}{dx} = \frac{dy}{dw}\frac{dw}{du}\frac{du}{dx}, \quad \text{and so on}$$

Exercise 3-4 Chapter Review

Work through all the problems in this chapter review and check your answers in the back of the book. (Answers to all review problems are there.) Where weaknesses show up, review appropriate sections in the text.

A

1. Use a calculator to evaluate $A = 2{,}000e^{0.09t}$ to the nearest cent for $t = 5$, 10, and 20.

Find the indicated derivatives in Problems 2–4.

2. $D_x(2 \ln x + 3e^x)$
3. $D_x e^{2x-3}$
4. y' for $y = \ln(2x + 7)$
5. Let $y = \ln u$ and $u = 3 + e^x$.

   (A) Express y in terms of x.

   (B) Use the chain rule to find dy/dx and then express dy/dx in terms of x.

B

6. Graph $y = 100e^{-0.1x}$.

Find the indicated derivatives in Problems 7–12.

7. $D_z[(\ln z)^7 + \ln z^7]$
8. $D_x x^6 \ln x$
9. $D_x \dfrac{e^x}{x^6}$
10. y' for $y = \ln(2x^3 - 3x)$
11. $f'(x)$ for $f(x) = e^{x^3 - x^2}$
12. dy/dx for $y = e^{-2x} \ln 5x$
13. Find the equation of the line tangent to the graph of $y = f(x) = 1 + e^{-x}$ at $x = 0$. At $x = -1$.

C

In Problems 14 and 15, find the absolute maximum value of $f(x)$ for $x > 0$.

14. $f(x) = 11x - 2x \ln x$
15. $f(x) = 10xe^{-2x}$

Sketch the graph of $y = f(x)$ in Problems 16 and 17.

16. $f(x) = 5 - 5e^{-x}$
17. $f(x) = x^3 \ln x$
18. Let $y = w^3$, $w = \ln u$, and $u = 4 - e^x$.

    (A) Express y in terms of x.
    (B) Use the chain rule to find dy/dx and then express dy/dx in terms of x.

Find the indicated derivatives in Problems 19–21.

19. y' for $y = 5^{x^2-1}$
20. $D_x \log_5(x^2 - x)$
21. $D_x \sqrt{\ln(x^2 + x)}$

Applications

Business & Economics

22. *Doubling time.* How long will it take money to double if it is invested at 5% interest compounded

    (A) Annually? (B) Continuously?

23. *Continuous compound interest.* If \$100 is invested at 10% interest compounded continuously, the amount (in dollars) at the end of t years is given by

    $$A = 100e^{0.1t}$$

    Find $A'(t)$, $A'(1)$, and $A'(10)$.

24. *Marginal analysis.* If the price–demand equation for x units of a commodity is

    $$p(x) = 1{,}000e^{-0.02x}$$

    find the marginal revenue equation.

25. *Maximum revenue.* For the price–demand equation in the preceding problem, find the production level and price per unit that produces the maximum revenue. What is the maximum revenue?

26. *Maximum revenue.* Graph the revenue function from the preceding two problems for $0 \leq x \leq 100$.

27. *Minimal average cost.* The cost of producing x units of a product is given by

$$C(x) = 200 + 50x - 50 \ln x \qquad x \geq 1$$

Find the minimal average cost.

Life Sciences

28. *Drug concentration.* The concentration of a drug in the bloodstream t hours after injection is given approximately by

$$C(t) = 5e^{-0.3t}$$

where $C(t)$ is concentration in milligrams per milliliter. What is the rate of change of concentration after 1 hour? After 5 hours?

Social Sciences

29. *Psychology—learning.* In a computer assembly plant, a new employee on the average is able to assemble

$$N(t) = 10(1 - e^{-0.4t})$$

units after t days of on-the-job training.

(A) What is the rate of learning after 1 day? After 5 days?

(B) Graph N for $0 \leq t \leq 10$.

Integration

4

CHAPTER 4 Contents

The last three chapters dealt with differential calculus. We now begin the development of the second main part of calculus, called *integral calculus.* Two types of integrals will be introduced, the *indefinite integral* and the *definite integral;* each is quite different from the other. But through the remarkable *fundamental theorem of calculus,* we will show that not only are the two integral forms intimately related, but both are intimately related to differentiation.

4-1 Antiderivatives and Indefinite Integrals

- Antiderivatives
- Indefinite Integrals
- Indefinite Integrals Involving Algebraic Functions
- Indefinite Integrals Involving Exponential and Logarithmic Functions
- Applications

■ Antiderivatives

Many operations in mathematics have reverses—compare addition and subtraction, multiplication and division, and powers and roots. The function $f(x) = \frac{1}{3}x^3$ has the derivative $f'(x) = x^2$. Reversing this process is referred to as *antidifferentiation.* Thus,

$$\frac{x^3}{3} \quad \text{is an antiderivative of} \quad x^2$$

since

$$D_x\left(\frac{x^3}{3}\right) = x^2$$

In general, we say that $F(x)$ is an **antiderivative** of $f(x)$ if

$$F'(x) = f(x)$$

Note that

$$D_x\left(\frac{x^3}{3} + 2\right) = x^2 \qquad D_x\left(\frac{x^3}{3} - \pi\right) = x^2 \qquad D_x\left(\frac{x^3}{3} + \sqrt{5}\right) = x^2$$

Hence,

$$\frac{x^3}{3} + 2 \qquad \frac{x^3}{3} - \pi \qquad \frac{x^3}{3} + \sqrt{5}$$

are also antiderivatives of x^2, since each has x^2 as a derivative. In fact, it appears that

$$\frac{x^3}{3} + C$$

for any real number C, is an antiderivative of x^2, since

$$D_x\left(\frac{x^3}{3} + C\right) = x^2$$

Thus, antidifferentiation of a given function does not, in general, lead to a unique function, but to a whole set of functions.

Does the expression

$$\frac{x^3}{3} + C$$

with C any real number, include all antiderivatives of x^2? Theorem 1 (which we state without proof) indicates that the answer is yes.

Theorem 1

If F and G are differentiable functions on the interval (a, b) and $F'(x) = G'(x)$, then $F(x) = G(x) + k$ for some constant k.

■ Indefinite Integrals

In words, Theorem 1 states that **if the derivatives of two functions are equal, then the functions differ by at most a constant.** We use the symbol

$$\int f(x)\, dx$$

called the **indefinite integral,** to represent all antiderivatives of $f(x)$, and we write

$$\int f(x)\, dx = F(x) + C \qquad \text{where } F'(x) = f(x)$$

that is, if $F(x)$ is any antiderivative of $f(x)$. The symbol $\int$ is called an **integral sign** and $f(x)$ is called the **integrand.** (We will have more to say about the symbol dx later.) The arbitrary constant C is called the **constant of integration.**

▪ Indefinite Integrals Involving Algebraic Functions

Just as with differentiation, we can develop formulas and special properties that will enable us to find indefinite integrals of many frequently encountered functions. To start, we list some formulas that can be established using the definitions of antiderivative and indefinite integral, and the properties of derivatives considered in Chapter 1.

Indefinite Integral Formulas and Properties

For k and C constants:

1. $\int k\,dx = kx + C$
2. $\int x^n\,dx = \frac{x^{n+1}}{n+1} + C \qquad n \neq -1$
3. $\int kf(x)\,dx = k\int f(x)\,dx$
4. $\int [f(x) \pm g(x)]\,dx = \int f(x)\,dx \pm \int g(x)\,dx$

We will establish formula 2 and property 3 here (the others may be shown to be true in a similar manner). To establish formula 2, we simply differentiate the right side to obtain the integrand on the left side. Thus,

$$D_x\left(\frac{x^{n+1}}{n+1} + C\right) = \frac{(n+1)x^n}{n+1} + 0 = x^n \qquad n \neq -1$$

(The case when $n = -1$ will be considered later in this section.) To establish property 3, let F be a function such that $F'(x) = f(x)$. Then

$$k\int f(x)\,dx = k\int F'(x)\,dx = k[F(x) + C_1] = kF(x) + kC_1$$

and since $(kF(x))' = kF'(x) = kf(x)$, we have

$$\int kf(x)\,dx = \int kF'(x)\,dx = kF(x) + C_2$$

But $kF(x) + kC_1$ and $kF(x) + C_2$ describe the same set of functions, since C_1 and C_2 are arbitrary real numbers. It is important to remember that property 3 states that a constant factor can be moved across an integral sign; a variable factor cannot be moved across an integral sign.

Correct *Incorrect*

$$\int 5x^{1/2}\,dx = 5\int x^{1/2}\,dx \qquad \cancel{\int xx^{1/2}\,dx = x\int x^{1/2}\,dx}$$

Now let us put the formulas and properties to use.

Example 1

(A) $\int 5\,dx = 5x + C$

(B) $\int x^4\,dx = \dfrac{x^{4+1}}{4+1} + C = \dfrac{x^5}{5} + C$

(C) $\int 5x^7\,dx = 5\int x^7\,dx = 5\dfrac{x^8}{8} + C = \dfrac{5}{8}x^8 + C$

(D)
$$\begin{aligned}\int (4x^3 + 2x - 1)\,dx &= \int 4x^3\,dx + \int 2x\,dx - \int dx \\ &= 4\int x^3\,dx + 2\int x\,dx - \int dx \\ &= \frac{4x^4}{4} + \frac{2x^2}{2} - x + C \\ &= x^4 + x^2 - x + C\end{aligned}$$

Property 4 can be extended to the sum and difference of an arbitrary number of functions.

(E) $\int \dfrac{3\,dx}{x^2} = \int 3x^{-2}\,dx = \dfrac{3x^{-2+1}}{-2+1} + C = -3x^{-1} + C$

(F)
$$\begin{aligned}\int 5\sqrt[3]{x^2}\,dx &= 5\int x^{2/3}\,dx = 5\frac{x^{(2/3)+1}}{\frac{2}{3}+1} + C \\ &= 5\frac{x^{5/3}}{\frac{5}{3}} + C = 3x^{5/3} + C\end{aligned}$$

To check any of these, we differentiate the final result to obtain the integrand in the original indefinite integral. When you evaluate an indefinite integral, do not forget to include the arbitrary constant C.

Problem 1 Find each of the following:

(A) $\int dx$ (B) $\int 3x^4\,dx$ (C) $\int (2x^5 - 3x^2 + 1)\,dx$

(D) $\int 4\sqrt[5]{x^3}\,dx$ (E) $\int \left(2x^{2/3} - \dfrac{3}{x^4}\right)dx$

Example 2

(A)
$$\int \frac{x^3 - 3}{x^2}\,dx = \int \left(\frac{x^3}{x^2} - \frac{3}{x^2}\right) dx$$
$$= \int (x - 3x^{-2})\,dx$$
$$= \int x\,dx - 3\int x^{-2}\,dx$$
$$= \frac{x^{1+1}}{1+1} - 3\frac{x^{-2+1}}{-2+1} + C$$
$$= \tfrac{1}{2}x^2 + 3x^{-1} + C$$

(B)
$$\int \left(\frac{2}{\sqrt[3]{x}} - 6\sqrt{x}\right) dx = \int (2x^{-1/3} - 6x^{1/2})\,dx$$
$$= 2\int x^{-1/3}\,dx - 6\int x^{1/2}\,dx$$
$$= 2\frac{x^{(-1/3)+1}}{-\frac{1}{3}+1} - 6\frac{x^{(1/2)+1}}{\frac{1}{2}+1} + C$$
$$= 2\frac{x^{2/3}}{\frac{2}{3}} - 6\frac{x^{3/2}}{\frac{3}{2}} + C$$
$$= 3x^{2/3} - 4x^{3/2} + C$$

Problem 2 Find each indefinite integral.

(A) $\displaystyle\int \frac{x^4 - 8x^3}{x^2}\,dx$ (B) $\displaystyle\int \left(8\sqrt[3]{x} - \frac{6}{\sqrt{x}}\right) dx$

▪ Indefinite Integrals Involving Exponential and Logarithmic Functions

Formula 5 (in the next box) follows immediately from the derivative formula for the exponential function discussed in the last chapter. Because of the absolute value, formula 6 does not follow directly from the derivative formula for the natural logarithm function. Let us show that

$$D_x \ln|x| = \frac{1}{x} \qquad x \neq 0$$

We consider two cases:

Case 1. $x > 0$

$$D_x \ln|x| = D_x \ln x \qquad \text{Since } |x| = x \text{ for } x > 0$$
$$= \frac{1}{x}$$

Indefinite Integral Formulas

5. $\int e^x \, dx = e^x + C$

6. $\int \frac{dx}{x} = \ln|x| + C \qquad x \neq 0$

Case 2. $x < 0$

$$\begin{aligned} D_x \ln|x| &= D_x \ln(-x) \qquad \text{Since } |x| = -x \text{ for } x < 0 \\ &= \frac{1}{-x} D_x(-x) \\ &= \frac{-1}{-x} = \frac{1}{x} \end{aligned}$$

Thus,

$$D_x \ln|x| = \frac{1}{x} \qquad x \neq 0$$

Hence,

$$\int \frac{1}{x} \, dx = \ln|x| + C \qquad x \neq 0$$

What about the indefinite integral of ln x? We postpone a discussion of $\int \ln x \, dx$ until Section 5-3, where we will be able to find it using a technique called *integration by parts.*

Example 3

$$\begin{aligned} \int \left(2e^x + \frac{3}{x}\right) dx &= 2\int e^x \, dx + 3\int \frac{1}{x} \, dx \\ &= 2e^x + 3 \ln|x| + C \end{aligned}$$

Problem 3 Find $\int \left(\frac{5}{x} - 4e^x\right) dx$.

Let us now consider some applications of the indefinite integral to see why we are interested in finding antiderivatives of functions.

■ Applications

Example 4 Curves

Find the equation of the curve that passes through (2, 5) if its slope is given by $dy/dx = 2x$ at any point x.

Solution We are interested in finding a function $y = f(x)$ such that

$$\frac{dy}{dx} = 2x \tag{1}$$

and

$$y = 5 \qquad \text{when } x = 2 \tag{2}$$

If

$$\frac{dy}{dx} = 2x$$

then

$$\begin{aligned} y &= \int 2x\,dx \\ &= x^2 + C \end{aligned} \tag{3}$$

Since $y = 5$ when $x = 2$, we determine the *particular* value of C so that

$$5 = 2^2 + C$$

Thus,

$$C = 1$$

and

$$y = x^2 + 1$$

is the particular antiderivative out of all those possible from (3) that satisfies both (1) and (2). See Figure 1.

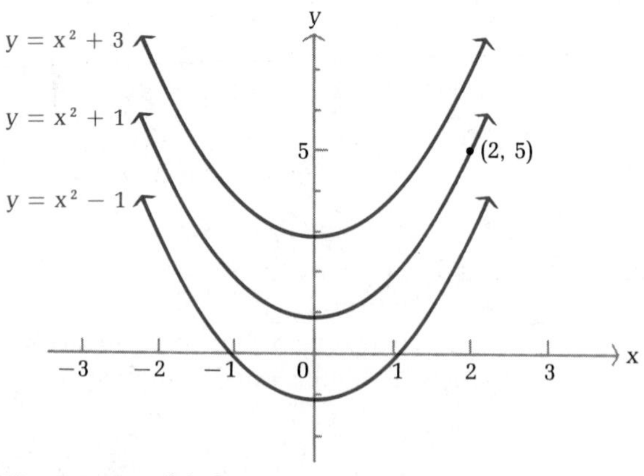

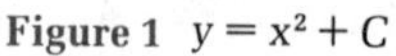
Figure 1 $y = x^2 + C$

Problem 4 Find the equation of the curve that passes through (2, 6) if the slope of the curve at any point x is given by $dy/dx = 3x^2$.

In certain situations it is easier to determine the rate at which something happens than how much of it has happened in a given length of time (e.g., population growth rates, business growth rates, rate of healing of a wound, rates of learning or forgetting). If a rate function (derivative) is given and we know the value of the dependent variable for a given value of the independent variable, then—if the rate function is not too complicated—we can often find the original function by integration.

Example 5 Cost Function

If the marginal cost of producing x units is given by

$$C'(x) = 0.3x^2 + 2x$$

and the fixed cost is \$2,000, find the cost function $C(x)$ and the cost of producing twenty units.

Solution

Recall that marginal cost is the derivative of the cost function and that fixed cost is cost at a zero production level. Thus, the mathematical problem is to find $C(x)$ given

$$C'(x) = 0.3x^2 + 2x \qquad C(0) = 2{,}000$$

We now find the indefinite integral of $0.3x^2 + 2x$ and determine the arbitrary integration constant using $C(0) = 2{,}000$.

$$C'(x) = 0.3x^2 + 2x$$

$$C(x) = \int (0.3x^2 + 2x)\,dx$$

$$= 0.1x^3 + x^2 + K$$

Since C represents the cost, we use K for the constant of integration.

But

$$C(0) = (0.1)0^3 + 0^2 + K = 2{,}000$$

Thus,

$$K = 2{,}000$$

and the particular cost function is

$$C(x) = 0.1x^3 + x^2 + 2{,}000$$

We now find $C(20)$, the cost of producing twenty units:

$$C(20) = (0.1)20^3 + 20^2 + 2{,}000$$

$$= \$3{,}200$$

Problem 5

Find the revenue function $R(x)$ when the marginal revenue is

$$R'(x) = 400 - 0.4x$$

and no revenue results at a zero production level. What is the revenue at a production level of 1,000 units?

Answers to Matched Problems

1. (A) $x + C$ (B) $\frac{3}{5}x^5 + C$ (C) $(x^6/3) - x^3 + x + C$
   (D) $\frac{5}{2}x^{8/5} + C$ (E) $\frac{6}{5}x^{5/3} + x^{-3} + C$
2. (A) $\frac{1}{3}x^3 - 4x^2 + C$ (B) $6x^{4/3} - 12x^{1/2} + C$
3. $5 \ln|x| - 4e^x + C$
4. $y = x^3 - 2$
5. $R(x) = 400x - 0.2x^2$; $R(1{,}000) = \$200{,}000$

Exercise 4-1

A *Find each indefinite integral. (Check by differentiating.)*

1. $\int 7\,dx$
2. $\int \pi\,dx$
3. $\int x^6\,dx$
4. $\int x^3\,dx$
5. $\int 8t^3\,dt$
6. $\int 10t^4\,dt$
7. $\int (2u + 1)\,du$
8. $\int (1 - 2u)\,du$
9. $\int (3x^2 + 2x - 5)\,dx$
10. $\int (2 + 4x - 6x^2)\,dx$
11. $\int (s^4 - 8s^5)\,ds$
12. $\int (t^5 + 6t^3)\,dt$
13. $\int 3e^t\,dt$
14. $\int 2e^t\,dt$
15. $\int 2z^{-1}\,dz$
16. $\int \frac{3}{s}\,ds$

Find all the antiderivatives for each derivative.

17. $\frac{dy}{dx} = 200x^4$
18. $\frac{dx}{dt} = 42t^5$
19. $\frac{dP}{dx} = 24 - 6x$
20. $\frac{dy}{dx} = 3x^2 - 4x^3$
21. $\frac{dy}{du} = 2u^5 - 3u^2 - 1$
22. $\frac{dA}{dt} = 3 - 12t^3 - 9t^5$
23. $\frac{dy}{dx} = e^x + 3$
24. $\frac{dy}{dx} = x - e^x$
25. $\frac{dx}{dt} = 5t^{-1} + 1$
26. $\frac{du}{dv} = \frac{4}{v} + \frac{v}{4}$

B *Find each indefinite integral. (Check by differentiation.)*

27. $\int 6x^{1/2}\,dx$

28. $\int 8t^{1/3}\,dt$

29. $\int 8x^{-3}\,dx$

30. $\int 12u^{-4}\,du$

31. $\int \frac{du}{\sqrt{u}}$

32. $\int \frac{dt}{\sqrt[3]{t}}$

33. $\int \frac{dx}{4x^3}$

34. $\int \frac{6\,dm}{m^2}$

35. $\int \frac{du}{2u^5}$

36. $\int \frac{dy}{3y^4}$

37. $\int \left(3x^2 - \frac{2}{x^2}\right) dx$

38. $\int \left(4x^3 + \frac{2}{x^3}\right) dx$

39. $\int \left(10x^4 - \frac{8}{x^5} - 2\right) dx$

40. $\int \left(\frac{6}{x^4} - \frac{2}{x^3} + 1\right) dx$

41. $\int \left(3\sqrt{x} + \frac{2}{\sqrt{x}}\right) dx$

42. $\int \left(\frac{2}{\sqrt[3]{x}} - \sqrt[3]{x^2}\right) dx$

43. $\int \left(\sqrt[3]{x^2} - \frac{4}{x^3}\right) dx$

44. $\int \left(\frac{12}{x^5} - \frac{1}{\sqrt[3]{x^2}}\right) dx$

45. $\int \frac{e^x - 3x}{4}\,dx$

46. $\int \frac{e^x - 3x^2}{2}\,dx$

47. $\int (2z^{-3} + z^{-2} + z^{-1})\,dz$

48. $\int (3x^{-2} - x^{-1})\,dx$

In Problems 49–58, find the particular antiderivative of each derivative that satisfies the given condition.

49. $\frac{dy}{dx} = 2x - 3, \quad y(0) = 5$

50. $\frac{dy}{dx} = 5 - 4x, \quad y(0) = 20$

51. $C'(x) = 6x^2 - 4x, \quad C(0) = 3{,}000$

52. $R'(x) = 600 - 0.6x, \quad R(0) = 0$

53. $\frac{dx}{dt} = \frac{20}{\sqrt{t}}, \quad x(1) = 40$

54. $\frac{dR}{dt} = \frac{100}{t^2}, \quad R(1) = 400$

55. $\frac{dy}{dx} = 2x^{-2} + 3x^{-1} - 1, \quad y(1) = 0$

56. $\frac{dy}{dx} = 3x^{-1} + x^{-2}, \quad y(1) = 1$

57. $\frac{dx}{dt} = 4e^t - 2, \quad x(0) = 1$

58. $\frac{dy}{dt} = 5e^t - 4, \quad y(0) = -1$

59. Find the equation of the curve that passes through (2, 3) if its slope is given by

$$\frac{dy}{dx} = 4x - 3$$

for each x.

60. Find the equation of the curve that passes through (1, 3) if its slope is given by

$$\frac{dy}{dx} = 12x^2 - 12x$$

for each x.

C *Find each indefinite integral.*

61. $\int \frac{2x^4 - x}{x^3}\, dx$

62. $\int \frac{x^{-1} - x^4}{x^2}\, dx$

63. $\int \frac{x^5 - 2x}{x^4}\, dx$

64. $\int \frac{1 - 3x^4}{x^2}\, dx$

65. $\int \frac{x^2e^x - 2x}{x^2}\, dx$

66. $\int \frac{1 - xe^x}{x}\, dx$

Find the antiderivative of each of the derivatives that satisfies the given condition.

67. $\frac{dM}{dt} = \frac{t^2 - 1}{t^2}, \quad M(4) = 5$

68. $\frac{dR}{dx} = \frac{1 - x^4}{x^3}, \quad R(1) = 4$

69. $\frac{dy}{dx} = \frac{5x + 2}{\sqrt[3]{x}}, \quad y(1) = 0$

70. $\frac{dx}{dt} = \frac{\sqrt{t^3} - t}{\sqrt{t^3}}, \quad x(9) = 4$

71. $p'(x) = -\frac{10}{x^2}, \quad p(1) = 20$

72. $p'(x) = \frac{10}{x^3}, \quad p(1) = 15$

Applications

Business & Economics

73. *Profit function.* If the marginal profit for producing x units is given by

$$P'(x) = 50 - 0.04x \qquad P(0) = 0$$

where P(x) is the profit in dollars, find the profit function P and the profit on 100 units of production.

74. *Natural resources.* The world demand for wood is increasing. In 1975 the demand was 12.6 billion cubic feet, and the rate of increase in demand is given approximately by

$$d'(t) = 0.009t$$

where t is time in years after 1975 (data from the U.S. Department of Agriculture and Forest Service). Noting that $d(0) = 12.6$, find $d(t)$. Also find $d(25)$, the demand in the year 2000.

75. *Revenue function.* The marginal revenue for producing x digital sports watches is given by

$$R'(x) = 100 - \tfrac{1}{5}x \qquad R(0) = 0$$

where $R(x)$ is the revenue in dollars. Find the revenue function and the price–demand equation. What is the price when the demand is 700 units?

76. *Cost function.* The marginal average cost for producing x digital sports watches is given by

$$\overline{C}'(x) = -\frac{1{,}000}{x^2} \qquad \overline{C}(100) = 25$$

where $\overline{C}(x)$ is the average cost in dollars. Find the average cost function and the cost function. What are the fixed costs?

77. *Labor costs and learning.* A defense contractor is starting production on a new missile control system. On the basis of data collected while assembling the first 16 control systems, the production manager obtained the following function describing the rate of labor use:

$$g(x) = 2{,}400x^{-1/2}$$

where $g(x)$ is the number of labor-hours required to assemble the xth unit of the control system. For example, after assembling 16 units, the rate of assembly is 600 labor-hours per unit, and after assembling 25 units, the rate of assembly is 480 labor-hours per unit. The more units assembled, the more efficient the process because of learning. If 19,200 labor-hours are required to assemble the first 16 units, how many labor-hours, $L(x)$, will be required to assemble the first x units? The first 25 units?

78. *Labor costs and learning.* If the rate of labor use in Problem 77 is

$$g(x) = 2{,}000x^{-1/3}$$

and if the first 8 control units require 12,000 labor-hours, how many labor-hours, $L(x)$, will be required for the first x control units? The first 27 control units?

Life Sciences

79. *Weight–height.* The rate of change of an average person's weight with respect to their height h (in inches) is given approximately by

$$\frac{dW}{dh} = 0.0015h^2$$

Find $W(h)$ if $W(60) = 108$ pounds. Also find the average weight for a person who is 5 feet 10 inches tall.

80. *Wound healing.* If the area of a healing wound changes at a rate given approximately by

$$\frac{dA}{dt} = -4t^{-3} \qquad 1 \leq t \leq 10$$

where t is in days and $A(1) = 2$ square centimeters, what will the area of the wound be in 10 days?

Social Sciences

81. *Urban growth.* A suburban area of Chicago incorporated into a city. The growth rate t in years after incorporation is estimated to be

$$\frac{dN}{dt} = 400 + 600\sqrt{t} \qquad 0 \leq t \leq 9$$

If the current population is 5,000, what will the population be 9 years from now?

82. *Learning.* A beginning high school language class was chosen for an experiment in learning. Using a list of 50 words, the experiment involved measuring the rate of vocabulary memorization at different times during a continuous 5 hour study session. It was found that the average rate of learning for the whole class was inversely proportional to the time spent studying and was given approximately by

$$V'(t) = \frac{15}{t} \qquad 1 \leq t \leq 5$$

If the average number of words memorized after 1 hour of study was 15 words, what was the average number of words learned after t hours of study, $1 \leq t \leq 5$? After 4 hours of study? (Round answer to the nearest whole number.)

4-2 Integration by Substitution

- General Integral Formulas
- Integration by Substitution
- Application
- Common Errors

General Integral Formulas

In Section 3-3, we saw that the chain rule extends the derivative formulas for x^n, e^x, and $\ln x$ to derivative formulas for $[f(x)]^n$, $e^{f(x)}$, and $\ln[f(x)]$. The chain rule can also be used to extend the integral formulas discussed in Section 4-1. The general formulas are summarized in the following box:

General Integral Formulas

1. $\displaystyle\int [f(x)]^n f'(x)\,dx = \frac{[f(x)]^{n+1}}{n+1} + C \qquad n \neq -1$

2. $\displaystyle\int e^{f(x)} f'(x)\,dx = e^{f(x)} + C$

3. $\displaystyle\int \frac{1}{f(x)} f'(x)\,dx = \ln|f(x)| + C$

Each of these formulas can be verified by using the chain rule to show that the derivative of the function on the right is the integrand on the left. For example,

$$D_x[e^{f(x)} + C] = e^{f(x)} f'(x)$$

verifies formula 2.

Example 6

(A) $\displaystyle\int (3x+4)^{10} 3\,dx = \frac{(3x+4)^{11}}{11} + C$ Formula 1 with $f(x) = 3x + 4$ and $f'(x) = 3$

Check $\displaystyle D_x \frac{(3x+4)^{11}}{11} = 11\frac{(3x+4)^{10}}{11} D_x(3x+4) = (3x+4)^{10} 3$

(B) $\displaystyle\int e^{x^2} 2x\,dx = e^{x^2} + C$ Formula 2 with $f(x) = x^2$ and $f'(x) = 2x$

Check $D_x e^{x^2} = e^{x^2} D_x x^2 = e^{x^2} 2x$

(C) $\displaystyle\int \frac{1}{1+x^3} 3x^2\,dx = \ln|1+x^3| + C$ Formula 3 with $f(x) = 1 + x^3$ and $f'(x) = 3x^2$

Check $\displaystyle D_x \ln|1+x^3| = \frac{1}{1+x^3} D_x(1+x^3) = \frac{1}{1+x^3} 3x^2$

Problem 6 Find each indefinite integral.

(A) $\displaystyle\int (2x^3 - 3)^{20} 6x^2\,dx$ (B) $\displaystyle\int e^{5x} 5\,dx$ (C) $\displaystyle\int \frac{1}{4+x^2} 2x\,dx$

▪ Integration by Substitution

The key step in using formulas 1, 2, and 3 is recognizing the form of the integrand. Some people find it difficult to identify $f(x)$ and $f'(x)$ in these formulas and prefer to use a *substitution* to simplify the integrand. The

method of substitution, which we now discuss, becomes increasingly useful as one progresses in studies of integration.

Example 7 Find $\int (x^2 + 2x + 5)^5(2x + 2)\, dx$.

Solution If

$$u = x^2 + 2x + 5$$

then the differential of u is (see Section 2-5)

$$du = (2x + 2)\, dx$$

Notice that du is one of the factors in the integrand. Substitute u for $x^2 + 2x + 5$ and du for $(2x + 2)\, dx$ to obtain

$$\begin{aligned}\int (x^2 + 2x + 5)^5(2x + 2)\, dx &= \int u^5\, du\\ &= \frac{u^6}{6} + C\\ &= \tfrac{1}{6}(x^2 + 2x + 5)^6 + C \qquad \text{Since } u = x^2 + 2x + 5\end{aligned}$$

Check

$$\begin{aligned}D_x\tfrac{1}{6}(x^2 + 2x + 5)^6 &= \tfrac{1}{6}(6)(x^2 + 2x + 5)^5 D_x(x^2 + 2x + 5)\\ &= (x^2 + 2x + 5)^5(2x + 2)\end{aligned}$$

Problem 7 Find $\int (x^2 - 3x + 7)^4(2x - 3)\, dx$ by substitution.

Substituting $u = f(x)$ and $du = f'(x)\, dx$ in formulas 1, 2, and 3 produces the formulas in the next box. These formulas are valid if u is an independent variable or if u is a function and du is the differential of u.

Integration Formulas

4. $\displaystyle\int u^n\, du = \frac{u^{n+1}}{n+1} + C \qquad n \neq -1$

5. $\displaystyle\int e^u\, du = e^u + C$

6. $\displaystyle\int \frac{1}{u}\, du = \ln|u| + C$

In some cases, the integrand will have to be modified before making a substitution and using one of the integration formulas. Example 8 illustrates this process.

Example 8 Integrate:

(A) $\displaystyle\int \frac{1}{4x + 7}\, dx$ (B) $\displaystyle\int xe^{-x^2}\, dx$ (C) $\displaystyle\int 4x^2\sqrt{x^3 + 5}\, dx$

Solutions (A) If $u = 4x + 7$, then $du = 4\,dx$. We are missing a factor of 4 in the integrand to match formula 6 exactly. Recalling that a constant factor can be moved across an integral sign, we proceed as follows:

$$\int \frac{1}{4x+7}\,dx = \int \frac{1}{4x+7}\frac{4}{4}\,dx$$

$$= \frac{1}{4}\int \frac{1}{4x+7}4\,dx \qquad \text{Substitute } u = 4x+7 \text{ and } du = 4\,dx.$$

$$= \frac{1}{4}\int \frac{1}{u}\,du \qquad \text{Use formula 6.}$$

$$= \tfrac{1}{4}\ln|u| + C$$

$$= \tfrac{1}{4}\ln|4x+7| + C \qquad \text{Since } u = 4x+7$$

Check $$D_x\frac{1}{4}\ln|4x+7| = \frac{1}{4}\frac{1}{4x+7}D_x(4x+7) = \frac{1}{4}\frac{1}{4x+7}(4) = \frac{1}{4x+7}$$

(B) If $u = -x^2$, then $du = -2x\,dx$. Proceed as in part A:

$$\int xe^{-x^2}\,dx = \int e^{-x^2}\frac{-2}{-2}x\,dx$$

$$= -\frac{1}{2}\int e^{-x^2}(-2x)\,dx \qquad \text{Substitute } u = -x^2 \text{ and } du = -2x\,dx.$$

$$= -\frac{1}{2}\int e^u\,du \qquad \text{Use formula 5.}$$

$$= -\tfrac{1}{2}e^u + C$$

$$= -\tfrac{1}{2}e^{-x^2} + C \qquad \text{Since } u = -x^2$$

Check $$D_x(-\tfrac{1}{2}e^{-x^2}) = -\tfrac{1}{2}e^{-x^2}D_x(-x^2) = -\tfrac{1}{2}e^{-x^2}(-2x) = xe^{-x^2}$$

(C) $$\int 4x^2\sqrt{x^3+5}\,dx = 4\int \sqrt{x^3+5}\,x^2\,dx \qquad \text{Move the 4 across the integral sign and proceed as before.}$$

$$= 4\int \sqrt{x^3+5}\,\frac{3}{3}x^2\,dx$$

$$= \frac{4}{3}\int \sqrt{x^3+5}\,3x^2\,dx \qquad \text{Substitute } u = x^3+5 \text{ and } du = 3x^2.$$

$$= \frac{4}{3}\int \sqrt{u}\,du$$

$$= \frac{4}{3}\int u^{1/2}\,du \qquad \text{Use formula 4.}$$

$$= \frac{4}{3}\frac{u^{3/2}}{\frac{3}{2}} + C$$

$$= \tfrac{8}{9}u^{3/2} + C$$

$$= \tfrac{8}{9}(x^3+5)^{3/2} + C \qquad \text{Since } u = x^3+5$$

Check $D_x[\frac{8}{9}(x^3+5)^{3/2}] = \frac{4}{3}(x^3+5)^{1/2}D_x(x^3+5)$
$= \frac{4}{3}(x^3+5)^{1/2}3x^2 = 4x^2\sqrt{x^3+5}$

Problem 8 Integrate:

(A) $\int e^{-3x}\,dx$ (B) $\int \frac{x}{x^2-9}\,dx$ (C) $\int 5x^2(x^3+4)^{-2}\,dx$

■ Application

Example 9 Price–Demand

The market research department for a supermarket chain has determined that for one store the marginal price $p'(x)$ at x tubes per week for a certain brand of toothpaste is given by

$$p'(x) = -0.015e^{-0.01x}$$

Find the price–demand equation if the weekly demand is 50 when the price of a tube is \$2.35. Find the weekly demand when the price of a tube is \$1.89.

Solution

$$p(x) = \int -0.015e^{-0.01x}\,dx$$
$$= -0.015\int e^{-0.01x}\,dx$$
$$= -0.015\int e^{-0.01x}\frac{-0.01}{-0.01}\,dx$$
$$= \frac{-0.015}{-0.01}\int e^{-0.01x}(-0.01)\,dx \qquad \text{Substitute } u = -0.01x \text{ and } du = -0.01\,dx.$$
$$= 1.5\int e^u\,du$$
$$= 1.5e^u + C$$
$$= 1.5e^{-0.01x} + C \qquad \text{Since } u = -0.01x$$

We find C by noting that

$$p(50) = 1.5e^{-0.01(50)} + C = \$2.35$$
$$C = \$2.35 - 1.5e^{-0.5} \qquad \text{Use a calculator or a table.}$$
$$C = \$2.35 - 0.91$$
$$C = \$1.44$$

Thus,

$$p(x) = 1.5e^{-0.01x} + 1.44$$

To find the demand when the price is \$1.89, we solve $p(x) = \$1.89$ for x:

$$1.5e^{-0.01x} + 1.44 = 1.89$$
$$1.5e^{-0.01x} = 0.45$$
$$e^{-0.01x} = 0.3$$
$$-0.01x = \ln 0.3$$
$$x = -100 \ln 0.3 \approx 120 \text{ tubes}$$

Problem 9 The marginal price $p'(x)$ at a supply level of x tubes per week for a certain brand of toothpaste is given by

$$p'(x) = 0.001e^{0.01x}$$

Find the price–supply equation if the supplier is willing to supply 100 tubes per week at a price of $1.65 each. How many tubes would the supplier be willing to supply at a price of $1.98 each?

■ Common Errors

1. $$\int 2(x^2-3)^{3/2}\,dx = \int (x^2-3)^{3/2} 2\frac{x}{x}\,dx$$
$$\cancel{= \frac{1}{x}\int (x^2-3)^{3/2}(2x)\,dx}$$

A variable cannot be moved across an integral sign! This integral requires techniques that are beyond the scope of this book.

2. $$\int \frac{2x^2}{(x^2-3)^2}\,dx = \int (x^2-3)^{-2}2x^2\,dx$$
$$\cancel{= x\int (x^2-3)^{-2}(2x)\,dx}$$

No, for the same reason as in illustration 1.

3. $$\int \frac{1}{x^2+9}\,dx = \int \frac{1}{u}\,dx \qquad u = x^2+9$$
$$\cancel{= \ln|u|}$$

An integrand must be expressed entirely in terms of u and du before formulas 4, 5, and 6 on page 318 can be used.

A constant factor can be moved back and forth across an integral sign, but a variable factor cannot.

Yes	No
$\int kf(x)\,dx = k\int f(x)\,dx$	$\cancel{\int f(x)g(x)\,dx = f(x)\int g(x)\,dx}$
(k a constant factor)	[f(x) a variable factor]

Answers to Matched Problems

6. (A) $\frac{1}{21}(2x^3 - 3)^{21} + C$ (B) $e^{5x} + C$
   (C) $\ln|4 + x^2| + C$ or $\ln(4 + x^2) + C$, since $4 + x^2 > 0$
7. $\frac{1}{5}(x^2 - 3x + 7)^5 + C$
8. (A) $-\frac{1}{3}e^{-3x} + C$ (B) $\frac{1}{2}\ln|x^2 - 9| + C$ (C) $-\frac{5}{3}(x^3 + 4)^{-1} + C$
9. $p(x) = 0.1e^{0.01x} + 1.38$; 179 tubes

Exercise 4-2

A *Find each indefinite integral and check the result by differentiating.*

1. $\int (x^2 - 4)^5 2x\, dx$
2. $\int (x^3 + 1)^4 3x^2\, dx$
3. $\int e^{4x} 4\, dx$
4. $\int e^{-3x}(-3)\, dx$
5. $\int \frac{1}{2t + 3} 2\, dt$
6. $\int \frac{1}{5t - 7} 5\, dt$

B

7. $\int (3x - 2)^7\, dx$
8. $\int (5x + 3)^9\, dx$
9. $\int (x^2 + 3)^7 x\, dx$
10. $\int (x^3 - 5)^4 x^2\, dx$
11. $\int 10e^{-0.5t}\, dt$
12. $\int 4e^{0.01t}\, dt$
13. $\int \frac{1}{10x + 7}\, dx$
14. $\int \frac{1}{100 - 3x}\, dx$
15. $\int xe^{2x^2}\, dx$
16. $\int x^2 e^{4x^3}\, dx$
17. $\int \frac{x^2}{x^3 + 4}\, dx$
18. $\int \frac{x}{x^2 - 2}\, dx$
19. $\int \frac{t}{(3t^2 + 1)^4}\, dt$
20. $\int \frac{t^2}{(t^3 - 2)^5}\, dt$
21. $\int \frac{x^2}{(4 - x^3)^2}\, dx$
22. $\int \frac{x}{(5 - 2x^2)^5}\, dx$
23. $\int e^{2x}(1 + e^{2x})^3\, dx$
24. $\int e^{-x}(1 - e^{-x})^4\, dx$
25. $\int \frac{1 + x}{4 + 2x + x^2}\, dx$
26. $\int \frac{x^2 - 1}{x^3 - 3x + 7}\, dx$
27. $\int (2x + 1)e^{x^2+x+1}\, dx$
28. $\int (x^2 + 2x)e^{x^3+3x^2}\, dx$

29. $\int (e^x - 2x)^3(e^x - 2)\, dx$

30. $\int (x^2 - e^x)^4(2x - e^x)\, dx$

31. $\int \frac{x^3 + x}{(x^4 + 2x^2 + 1)^4}\, dx$

32. $\int \frac{x^2 - 1}{(x^3 - 3x + 7)^2}\, dx$

C

33. $\int x\sqrt{3x^2 + 7}\, dx$

34. $\int x^2\sqrt{2x^3 + 1}\, dx$

35. $\int \frac{x^3}{\sqrt{2x^4 + 3}}\, dx$

36. $\int \frac{x^2}{\sqrt{4x^3 - 1}}\, dx$

37. $\int \frac{(\ln x)^3}{x}\, dx$

38. $\int \frac{e^x}{1 + e^x}\, dx$

39. $\int \frac{1}{x^2} e^{-1/x}\, dx$

40. $\int \frac{1}{x \ln x}\, dx$

Find the antiderivative of each derivative.

41. $\frac{dx}{dt} = 7t^2(t^3 + 5)^6$

42. $\frac{dm}{dn} = 10n(n^2 - 8)^7$

43. $\frac{dy}{dt} = \frac{3t}{\sqrt{t^2 - 4}}$

44. $\frac{dy}{dx} = \frac{5x^2}{(x^3 - 7)^4}$

45. $\frac{dp}{dx} = \frac{e^x + e^{-x}}{(e^x - e^{-x})^2}$

46. $\frac{dm}{dt} = \frac{\ln(t - 5)}{t - 5}$

Applications

Business & Economics

47. *Price–demand equation.* The marginal price for a weekly demand of x bottles of baby shampoo in a drug store is given by

$$p'(x) = \frac{-6{,}000}{(3x + 50)^2}$$

Find the price–demand equation if the weekly demand is 150 when the price of a bottle of shampoo is \$4. What is the weekly demand when the price is \$2.50?

48. *Price–supply equation.* The marginal price at a supply level of x bottles of baby shampoo per week is given by

$$p'(x) = \frac{300}{(3x + 25)^2}$$

Find the price–supply equation if the distributor of the shampoo is willing to supply 75 bottles a week at a price of \$1.60 per bottle. How many bottles would the supplier be willing to supply at a price of \$1.75 per bottle?

49. *Cost function.* The weekly marginal cost of producing x pairs of tennis

shoes is given by

$$C'(x) = 12 + \frac{500}{x + 1}$$

where $C(x)$ is cost in dollars. If the fixed costs are $2,000 per week, find the cost function. What is the average cost per pair of shoes if 1,000 pairs of shoes are produced each week?

50. *Revenue function.* The weekly marginal revenue from the sale of x pairs of tennis shoes is given by

$$R'(x) = 40 - 0.02x + \frac{200}{x + 1} \qquad R(0) = 0$$

where $R(x)$ is revenue in dollars. Find the revenue function. Find the revenue from the sale of 1,000 pairs of shoes.

51. *Marketing.* An automobile company is ready to introduce a new line of cars with a national sales campaign. After test marketing the line in a carefully selected city, the marketing research department estimates that sales (in millions of dollars) will increase at the monthly rate of

$$S'(t) = 10 - 10e^{-0.1t} \qquad 0 \leqslant t \leqslant 24$$

t months after the national campaign has started. What will be the total sales, $S(t)$, t months after the beginning of the national campaign if we assume zero sales at the beginning of the campaign? What is the estimated total sales for the first 12 months of the campaign?

52. *Marketing.* Repeat Problem 51 if the monthly rate of increase in sales is found to be approximated by

$$S'(t) = 20 - 20e^{-0.05t} \qquad 0 \leqslant t \leqslant 24$$

53. *Oil production.* Using data from the first 3 years' production as well as geological studies, the management of an oil company estimates that oil will be pumped from a producing field at a rate given by

$$R(t) = \frac{100}{t + 1} + 5 \qquad 0 \leqslant t \leqslant 20$$

where $R(t)$ is the rate of production in thousands of barrels per year t years after pumping begins. How many barrels of oil, $Q(t)$, will the field produce the first t years if $Q(0) = 0$? How many barrels will be produced the first 9 years?

54. *Oil production.* In Problem 53, if the rate is found to be

$$R(t) = \frac{120t}{t^2 + 1} + 3 \qquad 0 \leqslant t \leqslant 20$$

how many barrels of oil, $Q(t)$, will the field produce the first t years if $Q(0) = 0$? How many barrels will be produced the first 5 years?

Life Sciences

55. *Biology.* A yeast culture is growing at the rate of $W'(t) = 0.2e^{0.1t}$ grams per hour. If the starting culture weighs 2 grams, what will be the weight of the culture, $W(t)$, after t hours? After 8 hours?

56. *Medicine.* The rate of healing for a skin wound (in square centimeters per day) is approximated by $A'(t) = -0.9e^{0.1t}$. If the initial wound has an area of 9 square centimeters, what will its area, $A(t)$, be after t days? After 5 days?

57. *Pollution.* A contaminated lake is treated with a bactericide. The rate of decrease in harmful bacteria t days after the treatment is given by

$$\frac{dN}{dt} = -\frac{2{,}000t}{1 + t^2} \qquad 0 \leq t \leq 10$$

where $N(t)$ is the number of bacteria per milliliter of water. If the initial count was 5,000 bacteria per milliliter, find $N(t)$ and then find the bacteria count after 10 days.

58. *Pollution.* An oil tanker aground on a reef is losing oil and producing an oil slick that is radiating outward at a rate given approximately by

$$\frac{dR}{dt} = \frac{60}{\sqrt{t + 9}} \qquad t \geq 0$$

where R is the radius in feet of the circular slick after t minutes. Find the radius of the slick after 16 minutes if the radius is 0 when $t = 0$.

Social Sciences

59. *Learning.* In a particular business college, it was found that an average student enrolled in an advanced typing class progressed at a rate of $N'(t) = 6e^{-0.1t}$ words per minute per week, t weeks after enrolling in a 15 week course. If at the beginning of the course a student could type 40 words per minute, how many words per minute, $N(t)$, would the student be expected to type t weeks into the course? After completing the course?

60. *Learning.* In the same business college, it was also found that an average student enrolled in a beginning shorthand class progressed at a rate of $N'(t) = 12e^{-0.06t}$ words per minute per week, t weeks after enrolling in a 15 week course. If at the beginning of the course a student could take dictation in shorthand at 0 words per minute, how many words per minute, $N(t)$, would the student be expected to handle t weeks into the course? After completing the course?

61. *College enrollment.* The projected rate of increase in enrollment in a new college is estimated by

$$\frac{dE}{dt} = 5{,}000(t + 1)^{-3/2} \qquad t \geq 0$$

where $E(t)$ is the projected enrollment in t years. If enrollment is 2,000 when $t = 0$, find the projected enrollment 15 years from now.

4-3 Differential Equations—Growth and Decay

- Differential Equations
- Continuous Compound Interest Revisited
- Exponential Growth Law
- Population Growth, Radioactive Decay, Learning
- A Comparison of Exponential Growth Phenomena

Differential Equations

In the last section we considered equations of the form

$$\frac{dy}{dx} = 6x^2 - 4x \qquad p'(x) = -400e^{-0.04x}$$

These are examples of differential equations. In general an equation is a **differential equation** if it involves an unknown function (often denoted by y) and one or more of its derivatives. Other examples of differential equations are

$$\frac{dy}{dx} = ky \qquad y'' - xy' + x^2 = 5$$

Finding solutions to different types of differential equations (functions that satisfy the equation) is the subject matter for whole books and courses on the subject. Here we will consider only a few very special but very important types of equations that have immediate and significant application. We start by considering the problem of continuous compound interest from another point of view, which will enable us to generalize the concept and apply the results to problems from a number of different fields.

Continuous Compound Interest Revisited

Let P be the initial amount of money deposited in an account and let A be the amount in the account at any time t. Instead of assuming that the money in the account earns a particular rate of interest, suppose we say that *the rate of growth of the amount of money in the account at any time t is proportional to the amount present at that time.* Since dA/dt is the rate of growth of A with respect to t, we have

$$\frac{dA}{dt} = rA \qquad A(0) = P \qquad A, P > 0 \tag{1}$$

where r is an appropriate constant. We would like to find a function $A = A(t)$ that satisfies these conditions. Multiplying both sides of equation

(1) by $1/A$, we obtain

$$\frac{1}{A}\frac{dA}{dt} = r$$

Now we integrate each side with respect to t,

$$\int \frac{1}{A}\frac{dA}{dt}\,dt = \int r\,dt$$ Use formula 3 in Section 4-2 to evaluate the left side.

$$\ln|A| = rt + C$$ $|A| = A$ since $A > 0$

$$\ln A = rt + C$$

and convert this last equation into the equivalent exponential form

$$A = e^{rt+C}$$ From Section 0-7, $y = \ln x$ if and only if $x = e^y$.

$$= e^C e^{rt}$$ Property of exponents: $b^m b^n = b^{m+n}$

Since $A(0) = P$, we evaluate $A(t) = e^C e^{rt}$ at $t = 0$ and set it equal to P:

$$A(0) = e^C e^0 = e^C = P$$

Hence, $e^C = P$, and we can rewrite $A = e^C e^{rt}$ in the form

$$A = Pe^{rt}$$

This is the same continuous compound interest formula obtained in Section 3-1, where the principal P is invested at an annual nominal rate of $100r\%$ compounded continuously for t years.

■ Exponential Growth Law

In general, if the rate of change with respect to time of a quantity Q is proportional to the amount present and $Q(0) = Q_0$, then proceeding in exactly the same way as above, we obtain the following:

Exponential Growth Law

If $\dfrac{dQ}{dt} = rQ$ and $Q(0) = Q_0$, then $Q = Q_0 e^{rt}$.

$Q_0 =$ Amount at $t = 0$

$r =$ Continuous compound growth rate

$t =$ Time

$Q =$ Quantity at time t

The constant r in the exponential growth law is sometimes called the **growth constant** or even the **growth rate.** This last term can be misleading, since the rate of growth of Q with respect to time is dQ/dt, not r. Notice that if $r < 0$, then $dQ/dt < 0$ and Q is decreasing. This type of growth is called **exponential decay.**

Once we know that the rate of growth of something is proportional to the amount present, then we know it has exponential growth and we can use the results summarized in the box without having to solve the involved differential equation each time. The exponential growth law applies not only to money invested at interest compounded continuously, but also to many other types of problems—population growth, radioactive decay, natural resource depletion, and so on.

▪ Population Growth, Radioactive Decay, Learning

The world population is growing at an ever-increasing rate, as illustrated in Figure 2. **Population growth** over certain periods of time can often be approximated by the exponential growth law described above.

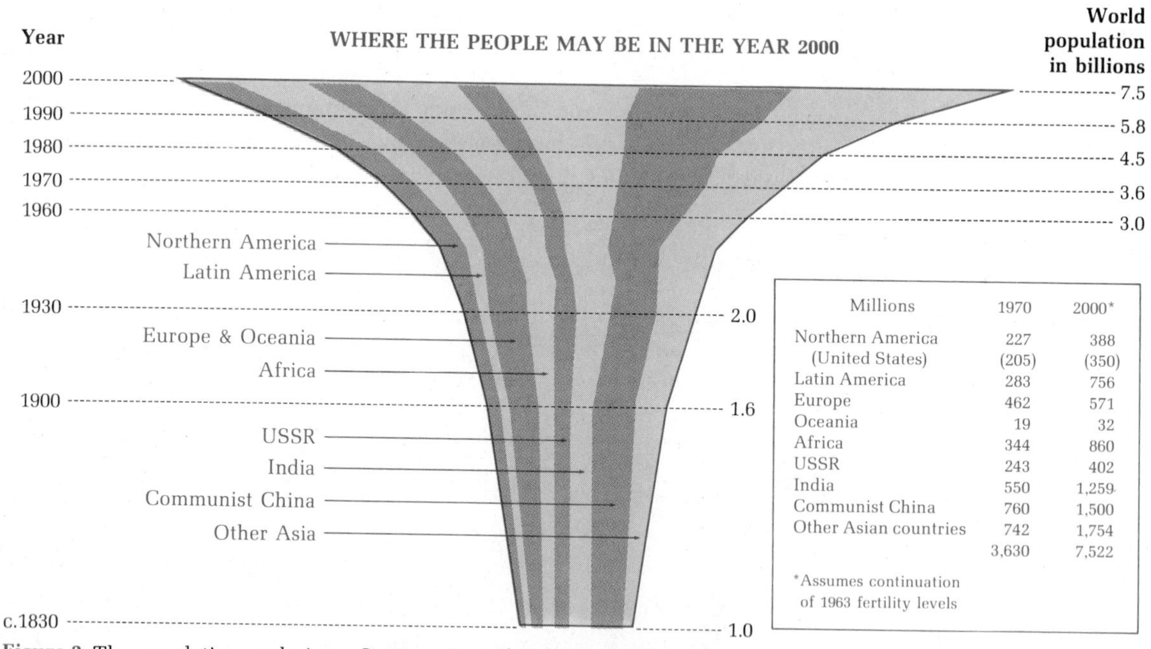

Millions	1970	2000*
Northern America	227	388
(United States)	(205)	(350)
Latin America	283	756
Europe	462	571
Oceania	19	32
Africa	344	860
USSR	243	402
India	550	1,259
Communist China	760	1,500
Other Asian countries	742	1,754
	3,630	7,522

*Assumes continuation of 1963 fertility levels

Figure 2 The population explosion. *Source:* United States State Department

Example 10
Population Growth

India had a population of 500 million people in 1966 ($t = 0$) and a growth rate of 3% per year (which we will assume is compounded continuously). If

P is the population in millions t years after 1966, and the same growth rate continues, then

$$\frac{dP}{dt} = 0.03P \qquad P(0) = 500$$

Thus, using the exponential growth law, we obtain

$$P = 500e^{0.03t}$$

With this result, we can estimate the population of India in 1986 ($t = 20$) to be

$$\begin{aligned} P(20) &= 500e^{0.03(20)} \\ &\approx 911 \text{ million people} \end{aligned}$$

Problem 10 Assuming the same continuous compound growth rate, what will India's population be in the year 2001?

Example 11
Population Growth

If the exponential growth law applies to Russia's population growth, at what continuous compound growth rate will the population double over the next 100 years?

Solution The problem is to find r, given $P = 2P_0$ and $t = 100$:

$$\begin{aligned} P &= P_0e^{rt} \\ 2P_0 &= P_0e^{100r} \\ 2 &= e^{100r} \qquad \text{Take ln of both sides and reverse equation.} \\ 100r &= \ln 2 \\ r &= \frac{\ln 2}{100} \\ &\approx 0.0069 \quad \text{or} \quad 0.69\% \end{aligned}$$

Problem 11 If the exponential growth law applies to population growth in Mexico, find the doubling time of the population if it continues to grow at 3.2% per year compounded continuously.

We now turn to another type of exponential growth—**radioactive decay.** In 1946, Willard Libby (who later received a Nobel Prize in chemistry) found that as long as a plant or animal is alive, radioactive carbon-14 is maintained at a constant level in its tissues. Once the plant or animal is dead, however, the radioactive carbon-14 diminishes by radioactive decay at a rate proportional to the amount present. Thus,

$$\frac{dQ}{dt} = rQ \qquad Q(0) = Q_0$$

and we have another example of the exponential growth law. The continuous compound rate of decay for radioactive carbon-14 is found to be 0.000 123 8; thus, $r = -0.000\ 123\ 8$, since decay implies a negative continuous compound growth rate.

Example 12 Archaeology

A piece of human bone was found at an archaeological site in Africa. If 10% of the original amount of radioactive carbon-14 was present, estimate the age of the bone.

Solution Using the exponential growth law for

$$\frac{dQ}{dt} = -0.000\ 123\ 8Q \qquad Q(0) = Q_0$$

we find that

$$Q = Q_0 e^{-0.0001238t}$$

and our problem is to find t so that $Q = 0.1Q_0$ (the amount of carbon-14 present now is 10% of the amount present, Q_0, at the death of the person). Thus,

$$0.1Q_0 = Q_0 e^{-0.0001238t}$$

$$0.1 = e^{-0.0001238t}$$

$$\ln 0.1 = \ln e^{-0.0001238t}$$

$$t = \frac{\ln 0.1}{-0.0001238} \approx 18{,}600 \text{ years}$$

Problem 12 Estimate the age of the bone in Example 12 if 50% of the original amount of carbon-14 is present.

In learning certain skills such as typing and swimming, a mathematical model often used is one that assumes there is a maximum skill attainable, say M, and the rate of improving is proportional to the difference between that achieved, y, and that attainable, M. Mathematically,

$$\frac{dy}{dt} = k(M - y) \qquad y(0) = 0$$

We solve this using the same technique that was used to obtain the exponential growth law. First, multiply both sides of the first equation by $1/(M - y)$ to obtain

$$\frac{1}{M - y}\frac{dy}{dt} = k$$

and then integrate both sides:

$$\int \frac{1}{M-y}\frac{dy}{dt}\,dt = \int k\,dt$$

$$-\int \frac{1}{M-y}\left(-\frac{dy}{dt}\right)dt = \int k\,dt$$ Use formula 3 in Section 4-2 to evaluate the left side.

$$-\ln(M-y) = kt + C$$ No absolute value signs required. (Why?)

Change this last equation to equivalent exponential form:

$$M - y = e^{-kt-C}$$
$$M - y = e^{-C}e^{-kt}$$
$$y = M - e^{-C}e^{-kt}$$

Now $y(0) = 0$; hence,

$$y(0) = M - e^{-C}e^{0} = 0$$

Solving for e^{-C}, we obtain

$$e^{-C} = M$$

and our final solution is

$$y = M - Me^{-kt} = M(1 - e^{-kt})$$

Example 13 Learning

For a particular person who is learning to swim, it is found that the distance y (in feet) the person is able to swim in 1 minute after t hours of practice is given approximately by

$$y = 50(1 - e^{-0.04t})$$

What is the rate of improvement after 10 hours of practice?

Solution

$$y = 50 - 50e^{-0.04t}$$
$$y'(t) = 2e^{-0.04t}$$
$$y'(10) = 2e^{-0.04(10)} \approx 1.34 \text{ feet per hour of practice}$$

Problem 13 In Example 13, what is the rate of improvement after 50 hours of practice?

▪ A Comparison of Exponential Growth Phenomena

The graphs and equations given in Table 1 (on the next page) compare several widely used growth models. These are divided basically into two groups: unlimited growth and limited growth. Following each graph and equation is a short, incomplete list of areas in which the models are used. This only touches on a subject that has been extensively developed and which you are likely to encounter in greater depth in the future.

Table 1 Exponential Growth

Description	Model	Solution	Graph	Uses
Unlimited growth Rate of growth is proportional to the amount present.	$\frac{dy}{dt} = ky$ $k, t > 0$ $y(0) = c$	$y = ce^{kt}$	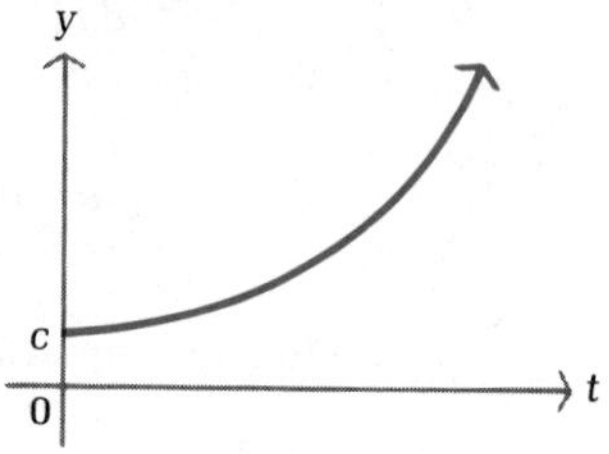	• Short-term population growth (people, bacteria, etc.) • Growth of money at continuous compound interest • Price–supply curves • Depletion of natural resources
Exponential decay Rate of growth is proportional to the amount present.	$\frac{dy}{dt} = -ky$ $k, t > 0$ $y(0) = c$	$y = ce^{-kt}$	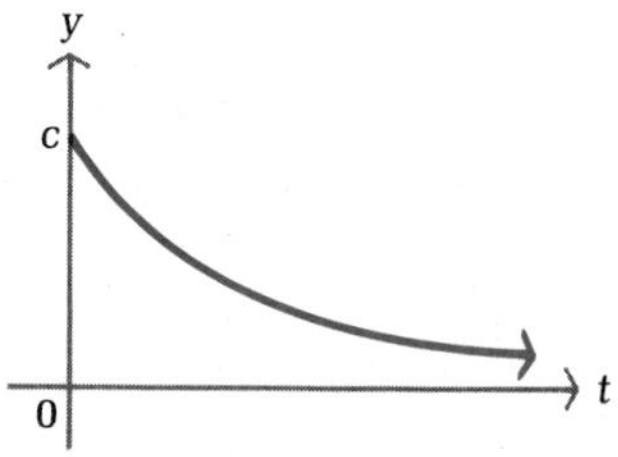	• Radioactive decay • Light absorption in water • Price–demand curves • Atmospheric pressure (t is altitude)
Limited growth Rate of growth is proportional to the difference between the amount present and a fixed limit.	$\frac{dy}{dt} = k(M - y)$ $k, t > 0$ $y(0) = 0$	$y = c(1 - e^{-kt})$	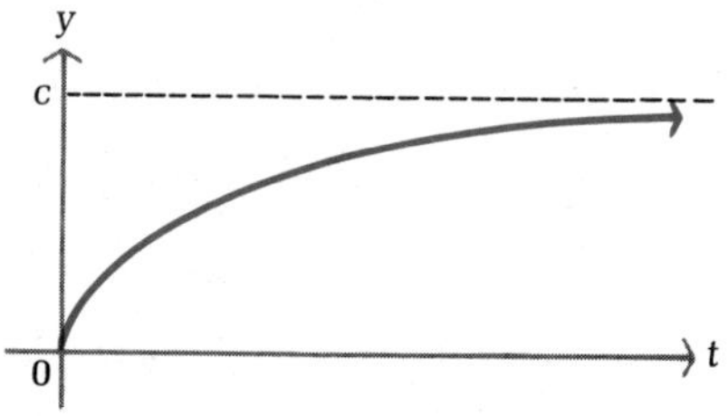	• Learning • Sales fads (e.g., skateboards) • Depreciation of equipment • Company growth
Logistic growth Rate of growth is proportional to the amount present and to the difference between the amount present and a fixed amount.	$\frac{dy}{dt} = ky(M - y)$ $k, t > 0$ $y(0) = \frac{M}{1 + c}$	$y = \frac{M}{1 + ce^{-kMt}}$	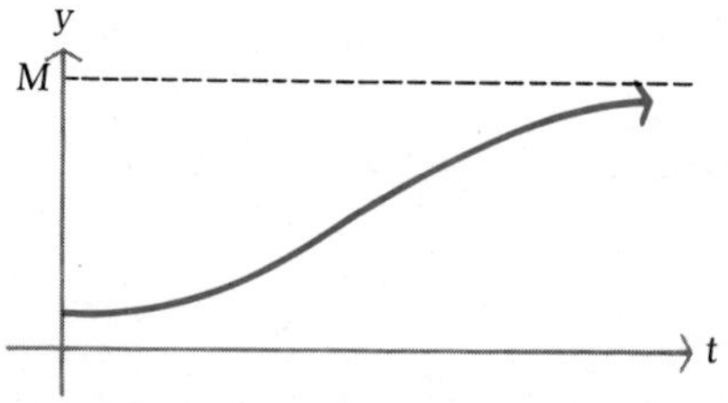	• Learning • Long-term population growth • Epidemics • Sales of new products • Company growth

Answers to Matched Problems

10. 1,429 million people
11. Approx. 22 years
12. Approx. 5,600 years
13. Approx. 0.27 foot per hour

Exercise 4-3

Applications

Business & Economics

1. *Continuous compound interest.* Find the amount A in an account after t years if

$$\frac{dA}{dt} = 0.08A \qquad \text{and} \qquad A(0) = 1{,}000$$

2. *Continuous compound interest.* Find the amount A in an account after t years if

$$\frac{dA}{dt} = 0.12A \qquad \text{and} \qquad A(0) = 5{,}250$$

3. *Continuous compound interest.* Find the amount A in an account after t years if

$$\frac{dA}{dt} = rA \qquad A(0) = 8{,}000 \quad A(2) = 9{,}020$$

4. *Continuous compound interest.* Find the amount A in an account after t years if

$$\frac{dA}{dt} = rA \qquad A(0) = 5{,}000 \quad A(5) = 7{,}460$$

5. *Price-demand.* If the marginal price dp/dx at x units of demand per week is proportional to the price p, and if at $100 there is no weekly demand [$p(0) = 100$], and if at $77.88 there is a weekly demand of 5 units [$p(5) = 77.88$], find the price–demand equation.
6. *Price–supply.* If the marginal price dp/dx at x units of supply per day is proportional to the price p, and if at a price of $10 there is no daily supply [$p(0) = 10$], and if at a price of $12.84 there is a daily supply of 50 units [$p(50) = 12.84$], find the price–supply equation.
7. *Advertising.* A company is trying to expose a new product to as many people as possible through television advertising. Suppose the rate of exposure to new people is proportional to the number of those who have not seen the product out of L possible viewers. If no one is aware of the product at the start of the campaign and after 10 days 40% of L are aware of the product, solve

$$\frac{dN}{dt} = k(L - N) \qquad N(0) = 0 \quad N(10) = 0.4L$$

for $N = N(t)$, the number of people who are aware of the product after t days of advertising.

8. *Advertising.* Repeat Problem 7 for

$$\frac{dN}{dt} = k(L - N) \qquad N(0) = 0 \quad N(10) = 0.1L$$

Life Sciences

9. *Ecology.* For relatively clear bodies of water, light intensity is reduced according to

$$\frac{dI}{dx} = -kI \qquad I(0) = I_0$$

where I is the intensity of light at x feet below the surface. For the Sargasso Sea off the West Indies, $k = 0.009\,42$. Find I in terms of x and find the depth at which the light is reduced to half of that at the surface.

10. *Blood pressure.* It can be shown under certain assumptions that blood pressure P in the largest artery in the human body (the aorta) changes between beats with respect to time t according to

$$\frac{dP}{dt} = -aP \qquad P(0) = P_0$$

where a is a constant. Find $P = P(t)$ that satisfies both conditions.

11. *Drug concentrations.* A single injection of a drug is administered to a patient. The amount Q in the body then decreases at a rate proportional to the amount present, and for this particular drug the rate is 4% per hour. Thus,

$$\frac{dQ}{dt} = -0.04Q \qquad Q(0) = Q_0$$

where t is time in hours. If the initial injection is 3 milliliters [$Q(0) = 3$], find $Q = Q(t)$ that satisfies both conditions. How many milliliters of the drug are still in the body after 10 hours?

12. *Simple epidemic.* A community of 1,000 individuals is assumed to be homogeneously mixed. One individual who has just returned from another community has influenza. Assume the home community has not had influenza shots and all are susceptible. One mathematical model for an influenza epidemic assumes that influenza tends to spread at a rate in direct proportion to the number who have it, N, and to the number who have not contracted it, in this case, $1{,}000 - N$. Mathematically,

$$\frac{dN}{dt} = kN(1{,}000 - N) \qquad N(0) = 1$$

where N is the number of people who have contracted influenza after t days. For $k = 0.0004$, it can be shown that $N(t)$ is given by

$$N(t) = \frac{1{,}000}{1 + 999e^{-0.4t}}$$

See Table 1 (logistic growth) for the characteristic graph.

(A) How many people have contracted influenza after 10 days? After 20 days?

(B) How many days will it take until half the community has contracted influenza?

(C) Find $\lim_{t \to \infty} N(t)$.

Social Sciences

13. *Archaeology.* A skull from an ancient tomb was discovered and was found to have 5% of the original amount of radioactive carbon-14 present. Estimate the age of the skull. (See Example 12.)

14. *Learning.* For a particular person learning to type, it was found that the number of words per minute, N, the person was able to type after t hours of practice was given approximately by

$$N = 100(1 - e^{-0.02t})$$

See Table 1 (limited growth) for a characteristic graph. What is the rate of improvement after 10 hours of practice? After 40 hours of practice?

15. *Small group analysis.* In a study on small group dynamics, sociologists Stephan and Mischler found that, when the members of a discussion group of ten were ranked according to the number of times each participated, the number of times $N(k)$ the kth-ranked person participated was given approximately by

$$N(k) = N_1e^{-0.11(k-1)} \qquad 1 \leq k \leq 10$$

where N_1 is the number of times the first-ranked person participated in the discussion. If, in a particular discussion group of ten people, $N_1 = 180$, estimate how many times the sixth-ranked person participated. The tenth-ranked person.

16. *Perception.* One of the oldest laws in mathematical psychology is the Weber–Fechner law (discovered in the middle of the nineteenth century). It concerns a person's sensed perception of various strengths of stimulation involving weights, sound, light, shock, taste, and so on. One form of the law states that the rate of change of sensed sensation S with respect to stimulus R is inversely proportional to the strength of the stimulus R. Thus,

$$\frac{dS}{dR} = \frac{k}{R}$$

where k is a constant. If we let R_0 be the threshold level at which the stimulus R can be detected (the least amount of sound, light, weight, and so on that can be detected), then it is appropriate to write

$$S(R_0) = 0$$

Find a function S in terms of R that satisfies the above conditions.

17. *Rumor spread.* A group of 400 parents, relatives, and friends are waiting anxiously at Kennedy Airport for a student charter to return after a year in Europe. It is stormy and the plane is late. A particular parent thought he had heard that the plane's radio had gone out and related this news to some friends, who in turn passed it on to others, and so on. Sociologists have studied rumor propagation and have found that a rumor tends to spread at a rate in direct proportion to the number who have heard it, x, and to the number who have not, $P - x$, where P is the total population. Mathematically, for our case, $P = 400$ and

$$\frac{dx}{dt} = 0.001x(400 - x) \qquad x(0) = 1$$

where t is time in minutes. From this, it can be shown that

$$x(t) = \frac{400}{1 + 399e^{-0.4t}}$$

See Table 1 (logistic growth) for a characteristic graph.

(A) How many people have heard the rumor after 5 minutes? 20 minutes?
(B) Find $\lim_{t\to\infty} x(t)$.

4-4 Definite Integral

- Definite Integral
- Properties
- Definite Integrals and Substitution
- Applications
- Common Errors

Definite Integral

We start this discussion with a simple example, out of which will evolve a new integral form, called the *definite integral*. Our approach in this section will be intuitive and informal; these concepts will be made more precise in Section 5-1.

Suppose a manufacturing company's marginal cost equation for a given product is given by

$$C'(x) = 2 - 0.2x \qquad 0 \leq x \leq 8$$

where the marginal cost is in thousands of dollars and production is x units per day. What is the total change in cost per day going from a production level of 2 units per day to 6 units per day? If $C = C(x)$ is the cost function, then

$$\begin{pmatrix}\text{Total net change in cost}\\ \text{between } x = 2 \text{ and } x = 6\end{pmatrix} = C(6) - C(2) = C(x)\Big|_2^6 \qquad (1)$$

The special symbol $C(x)|_2^6$ is a convenient way of representing the center expression that will prove useful to us later.

To evaluate (1), we need to find the antiderivative of $C'(x)$; that is,

$$C(x) = \int (2 - 0.2x)\, dx = 2x - 0.1x^2 + K \tag{2}$$

Thus, we are within a constant of knowing the original cost function. However, we do not need to know the constant K to solve the original problem (1). We compute $C(6) - C(2)$ for $C(x)$ found in (2):

$$\begin{aligned} C(6) - C(2) &= [2(6) - 0.1(6)^2 + K] - [2(2) - 0.1(2)^2 + K] \\ &= 12 - 3.6 + K - 4 + 0.4 - K \\ &= \$4.8 \text{ thousand per day increase in costs for a production} \\ &\qquad \text{increase from 2 to 6 units per day} \end{aligned}$$

The unknown constant K canceled out! Thus, we conclude that any antiderivative of $C'(x) = 2 - 0.2x$ will do, since antiderivatives of a given function can differ by at most a constant (see Section 4-1). Thus, we really do not have to find the constant in the original cost function to solve the problem.

Since $C(x)$ is an antiderivative of $C'(x)$, the above discussion suggests the following notation:

$$C(6) - C(2) = C(x)|_2^6 = \int_2^6 C'(x)\, dx \tag{3}$$

The integral form on the right in (3) is called a *definite integral*—it represents the number found by evaluating an antiderivative of the integrand at 6 and 2 and taking the difference as indicated.

Definite Integral

The **definite integral** of a continuous function f over an interval from $x = a$ to $x = b$ is the net change of an antiderivative of f over the interval. Symbolically, if $F(x)$ is an antiderivative of $f(x)$, then

$$\int_a^b f(x)\, dx = F(x)|_a^b = F(b) - F(a) \qquad \text{where} \quad F'(x) = f(x)$$

Integrand: $f(x)$ **Upper limit:** b **Lower limit:** a

In Section 5-1 we will formally define a definite integral as a limit of a special sum. Then the relationship in the box turns out to be the most important theorem in calculus—the fundamental theorem of calculus. Our intent in this and the next section is to give you some intuitive experience with the definite integral concept and its use. You will then be better able to understand a formal definition and to appreciate the significance of the fundamental theorem.

Example 14 Evaluate $\int_{-1}^{2} (3x^2 - 2x)\,dx$.

Solution We choose the simplest antiderivative of $(3x^2 - 2x)$, namely $(x^3 - x^2)$, since any antiderivative will do (see discussion at beginning of section).

$$\int_{-1}^{2} (3x^2 - 2x)\,dx = (x^3 - x^2)|_{-1}^{2}$$

$$= (2^3 - 2^2) - [(-1)^3 - (-1)^2]$$

$$= 4 - (-2) = 6$$

Be careful of sign errors here.

Problem 14 Evaluate $\int_{-2}^{2} (2x - 1)\,dx$.

Remark

Do not confuse a definite integral with an indefinite integral. The definite integral $\int_a^b f(x)\,dx$ is a real number; the indefinite integral $\int f(x)\,dx$ is a whole set of functions—all the antiderivatives of $f(x)$.

■ Properties

In the next box we state several useful properties of the definite integral. You will note that some of these parallel the properties for the indefinite integral listed in Section 4-1.

Definite Integral Properties

1. $\int_a^a f(x)\,dx = 0$
2. $\int_a^b f(x)\,dx = -\int_b^a f(x)\,dx$
3. $\int_a^b Kf(x)\,dx = K\int_a^b f(x)\,dx$ K a constant
4. $\int_a^b [f(x) \pm g(x)]\,dx = \int_a^b f(x)\,dx \pm \int_a^b g(x)\,dx$
5. $\int_a^b f(x)\,dx = \int_a^c f(x)\,dx + \int_c^b f(x)\,dx$

These properties are justified as follows: If $F'(x) = f(x)$, then

1. $$\int_a^a f(x)\,dx = F(x)\Big|_a^a = F(a) - F(a) = 0$$

2. $$\int_a^b f(x)\,dx = F(x)\Big|_a^b = F(b) - F(a) = -[F(a) - F(b)] = -\int_b^a f(x)\,dx$$

3. $$\int_a^b Kf(x)\,dx = KF(x)\Big|_a^b = KF(b) - KF(a) = K[F(b) - F(a)]$$
$$= K\int_a^b f(x)\,dx$$

and so on.

Example 15 Evaluate $\displaystyle\int_1^2 \left(2x + 3e^x - \frac{4}{x}\right) dx.$

Solution

$$\int_1^2 \left(2x + 3e^x - \frac{4}{x}\right) dx = 2\int_1^2 x\,dx + 3\int_1^2 e^x\,dx - 4\int_1^2 \frac{1}{x}\,dx$$
$$= 2\,\frac{x^2}{2}\Big|_1^2 + 3e^x\Big|_1^2 - 4\ln|x|\Big|_1^2$$
$$= (2^2 - 1^2) + (3e^2 - 3e^1) - (4\ln 2 - 4\ln 1)$$
$$= 3 + 3e^2 - 3e - 4\ln 2 \approx 14.24$$

Problem 15 Evaluate $\displaystyle\int_1^3 \left(4x - 2e^x + \frac{5}{x}\right) dx.$

▪ Definite Integrals and Substitution

The evaluation of a definite integral is a two-step process: First find an antiderivative and then find the net change in that antiderivative. The next example illustrates several different ways these steps can be performed when substitution is involved.

Example 16 Evaluate:

(A) $\displaystyle\int_2^4 e^{-0.5x}\,dx$ (B) $\displaystyle\int_1^2 (2x - 1)^3\,dx$ (C) $\displaystyle\int_0^5 \frac{x}{x^2 + 10}\,dx$

Solutions (A) First find an antiderivative:

$$\int e^{-0.5x}\,dx = \int e^{-0.5x}\,\frac{-0.5}{-0.5}\,dx$$
$$= \frac{1}{-0.5}\int e^{-0.5x}(-0.5)\,dx$$ Substitute $u = -0.5x$ and $du = -0.5\,dx$.
$$= -2\int e^u\,du$$

$$\int e^{-0.5x}\,dx = -2e^{u} + C$$
$$= -2e^{-0.5x} + C \qquad \text{Since } u = -0.5x$$

Now find the net change in an antiderivative:

$$\int_2^4 e^{-0.5x}\,dx = -2e^{-0.5x}\Big|_2^4 \qquad \text{Choose } C = 0.$$
$$= [-2e^{-0.5(4)}] - [-2e^{-0.5(2)}]$$
$$= -2e^{-2} + 2e^{-1} \approx 0.465$$

(B) Perform the substitution mentally:

$$\int_1^2 (2x-1)^3\,dx = \frac{1}{2}\int_1^2 (2x-1)^3 2\,dx$$
$$= \frac{1}{2}\,\frac{(2x-1)^4}{4}\Big|_1^2$$
$$= \frac{(2\cdot 2-1)^4}{8} - \frac{(2\cdot 1-1)^4}{8}$$
$$= \frac{3^4}{8} - \frac{1^4}{8} = \frac{81}{8} - \frac{1}{8} = \frac{80}{8} = 10$$

The integrand has the form $u^3\,du$. The antiderivative is $u^4/4 = (2x-1)^4/4$.

(C) Substitute directly in the definite integral, changing the limits of integration:

$$\int_0^5 \frac{x}{x^2+10}\,dx = \frac{1}{2}\int_0^5 \frac{1}{x^2+10}\,2x\,dx$$
$$= \frac{1}{2}\int_{10}^{35} \frac{1}{u}\,du$$
$$= \frac{1}{2}\ln|u|\Big|_{10}^{35}$$
$$= \tfrac{1}{2}\ln 35 - \tfrac{1}{2}\ln 10 \approx 0.626$$

If $u = x^2 + 10$, then $x = 0$ implies $u = 10$ and $x = 5$ implies $u = 35$.

Problem 16 Evaluate:

(A) $\displaystyle\int_{10}^{30} e^{-0.1x}\,dx$ (B) $\displaystyle\int_0^1 \frac{1}{2x+4}\,dx$ (C) $\displaystyle\int_{-1}^{2} x^2(1+x^3)^2\,dx$

■ Applications

Example 17 Pollution

A large factory on the Mississippi River discharges pollutants into the river at a rate that is estimated by a water quality control agency to be

$$P'(t) = R(t) = t\sqrt{t^2+1} \qquad 0 \le t \le 5$$

where $P(t)$ is the total number of tons of pollutants discharged into the river after t years of operation. What quantity of pollutants will be discharged into the river during the first 3 years of operation?

Solution

$$
\begin{aligned}
P(3) - P(0) &= \int_0^3 t\sqrt{t^2+1}\,dt \\
&= \int_0^3 (t^2+1)^{1/2} t\,dt \\
&= \frac{1}{2}\int_0^3 (t^2+1)^{1/2} 2t\,dt \\
&= \frac{1}{2} \cdot \frac{(t^2+1)^{3/2}}{\frac{3}{2}} \Bigg|_0^3 \\
&= \frac{1}{3}(t^2+1)^{3/2} \Bigg|_0^3 \\
&= \frac{1}{3}(3^2+1)^{3/2} - \frac{1}{3}(0^2+1)^{3/2} \\
&= \frac{1}{3}(10^{3/2} - 1) \approx 10.2 \text{ tons}
\end{aligned}
$$

Problem 17 Repeat Example 17 for the time interval from 3 to 5 years.

Example 18
Useful Life

An amusement company maintains records for each video game it installs in an arcade. Suppose that $C(t)$ and $R(t)$ represent the total accumulated costs and revenues (in thousands of dollars), respectively, t years after a particular game has been installed and that

$$C'(t) = 2$$
$$R'(t) = 9e^{-0.5t}$$

The value of t for which $C'(t) = R'(t)$ is called the **useful life** of the game.

(A) Find the useful life of the game to the nearest year.
(B) Find the total profit accumulated during the useful life of the game.

Solutions

(A)

$$
\begin{aligned}
R'(t) &= C'(t) \\
9e^{-0.5t} &= 2 \\
e^{-0.5t} &= \tfrac{2}{9} \\
-0.5t &= \ln \tfrac{2}{9} \\
t &= -2 \ln \tfrac{2}{9} \approx 3 \text{ years}
\end{aligned}
$$

Thus, the game has a useful life of 3 years.

(B) The total profit accumulated during the useful life of the game is

$$P(3) - P(0) = \int_0^3 P'(t)\,dt$$
$$= \int_0^3 [R'(t) - C'(t)]\,dt$$
$$= \int_0^3 (9e^{-0.5t} - 2)\,dt$$
$$= (-18e^{-0.5t} - 2t)\Big|_0^3$$
$$= (-18e^{-1.5} - 6) - (-18e^0 - 0)$$
$$= 12 - 18e^{-1.5} \approx 7.984 \text{ or } \$7{,}984$$

Problem 18 Repeat Example 18 if $C'(t) = 1$ and $R'(t) = 7.5e^{-0.5t}$.

■ Common Errors

1. $$\int_0^2 e^x\,dx = e^x\Big|_0^2 = \cancel{e^2}$$

Do not forget to evaluate the antiderivative at both the upper and lower limits of integration and do not assume that the antiderivative is 0 just because the lower limit is 0. The correct procedure for this problem is

$$\int_0^2 e^x\,dx = e^x\Big|_0^2 = e^2 - e^0 = e^2 - 1$$

2. $$\int_0^5 \frac{x}{x^2 + 10}\,dx = \frac{1}{2}\int_{\cancel{0}}^{\cancel{5}} \frac{1}{u}\,du \qquad \begin{aligned} u &= x^2 + 10 \\ du &= 2x\,dx \end{aligned}$$

If a substitution is made in a definite integral, the limits of integration also must be changed. The new limits are determined by the particular substitution used in the integral. (See Example 16C for the correct procedure for this integral.)

Answers to Matched Problems

14. -4 15. $16 + 2e - 2e^3 + 5 \ln 3 \approx -13.241$
16. (A) $10(e^{-1} - e^{-3}) \approx 3.181$ (B) $(\ln 6 - \ln 4)/2 \approx 0.203$
(C) 81
17. $\frac{1}{3}(26^{3/2} - 10^{3/2}) \approx 33.7$ tons
18. (A) $-2 \ln(\frac{2}{15}) \approx 4$ years (B) $11 - 15e^{-2} \approx 8.970$ or \$8,970

Exercise 4-4

Evaluate.

A 1. $\int_2^3 2x\,dx$ 2. $\int_1^2 3x^2\,dx$

3. $\int_{3}^{4} 5\,dx$

4. $\int_{12}^{20} dx$

5. $\int_{1}^{3} (2x - 3)\,dx$

6. $\int_{1}^{3} (6x + 5)\,dx$

7. $\int_{0}^{4} (3x^2 - 4)\,dx$

8. $\int_{0}^{2} (6x^2 - 2x)\,dx$

9. $\int_{-3}^{4} (4 - x^2)\,dx$

10. $\int_{-1}^{2} (x^2 - 4x)\,dx$

11. $\int_{0}^{1} 24x^{11}\,dx$

12. $\int_{0}^{2} 30x^5\,dx$

13. $\int_{0}^{1} e^{2x}\,dx$

14. $\int_{-1}^{1} e^{5x}\,dx$

15. $\int_{1}^{3.5} 2x^{-1}\,dx$

16. $\int_{1}^{2} \frac{dx}{x}$

B

17. $\int_{1}^{2} (2x^{-2} - 3)\,dx$

18. $\int_{1}^{2} (5 - 16x^{-3})\,dx$

19. $\int_{1}^{4} 3\sqrt{x}\,dx$

20. $\int_{4}^{25} \frac{2}{\sqrt{x}}\,dx$

21. $\int_{2}^{3} 12(x^2 - 4)^5 x\,dx$

22. $\int_{0}^{1} 32(x^2 + 1)^7 x\,dx$

23. $\int_{3}^{9} \frac{1}{x - 1}\,dx$

24. $\int_{2}^{8} \frac{1}{x + 1}\,dx$

25. $\int_{0}^{1} (e^{2x} - 2x)^2(e^{2x} - 1)\,dx$

26. $\int_{0}^{1} \frac{2e^{4x} - 3}{e^{2x}}\,dx$

27. $\int_{-2}^{-1} (x^{-1} + 2x)\,dx$

28. $\int_{-3}^{-1} (-3x^{-2} + x^{-1})\,dx$

C

29. $\int_{2}^{3} x\sqrt{2x^2 - 3}\,dx$

30. $\int_{0}^{1} x\sqrt{3x^2 + 2}\,dx$

31. $\int_{0}^{1} \frac{x - 1}{x^2 - 2x + 3}\,dx$

32. $\int_{1}^{2} \frac{x + 1}{2x^2 + 4x + 4}\,dx$

33. $\int_{-1}^{1} \frac{e^{-x} - e^{x}}{(e^{-x} + e^{x})^2}\,dx$

34. $\int_{6}^{7} \frac{\ln(t - 5)}{t - 5}\,dt$

Applications

Business & Economics

35. *Salvage value.* A new piece of industrial equipment will depreciate in value rapidly at first, then less rapidly as time goes on. Suppose the rate (in dollars per year) at which the book value of a new milling machine changes is given approximately by

$$V'(t) = f(t) = 500(t - 12) \qquad 0 \leqslant t \leqslant 10$$

where $V(t)$ is the value of the machine after t years. What is the total loss in value of the machine in the first 5 years? In the second 5 years? Set up appropriate integrals and solve.

36. *Maintenance costs.* Maintenance costs for an apartment house generally increase as the building gets older. From past records, a managerial service determines that the rate of increase in maintenance costs (in dollars per year) for a particular apartment complex is given approximately by

$$M'(x) = f(x) = 90x^2 + 5{,}000$$

where x is the age of the apartment in years and $M(x)$ is the total (accumulated) cost of maintenance for x years. Write a definite integral that will give the total maintenance costs from 2 to 7 years after the apartment house was built, and evaluate it.

37. *Useful life.* The total accumulated costs $C(t)$ and revenues $R(t)$ (in thousands of dollars), respectively, for a coin-operated photocopying machine satisfy

$$C'(t) = \tfrac{1}{11}t \quad \text{and} \quad R'(t) = 5te^{-t^2}$$

where t is time in years. Find the useful life of the machine to the nearest year. What is the total profit accumulated during the useful life of the machine?

38. *Useful life.* The total accumulated costs $C(t)$ and revenues $R(t)$ (in thousands of dollars), respectively, for a coal mine satisfy

$$C'(t) = 3 \quad \text{and} \quad R'(t) = 15e^{-0.1t}$$

where t is the number of years the mine has been in operation. Find the useful life of the mine to the nearest year. What is the total profit accumulated during the useful life of the mine?

39. *Labor costs and learning.* A defense contractor is starting production on a new missile control system. On the basis of data collected while assembling the first 16 control systems, the production manager obtained the following function for rate of labor use:

$$g(x) = 2{,}400x^{-1/2}$$

where $g(x)$ is the number of labor-hours required to assemble the xth unit of a control system. Approximately how many labor-hours will be required to assemble the 17th through the 25th control units? [*Hint:* Let $a = 16$ and $b = 25$.]

40. *Labor costs and learning.* If the rate of labor use in Problem 39 is

$$g(x) = 2{,}000x^{-1/3}$$

approximately how many labor-hours will be required to assemble the 9th through the 27th control units? [*Hint:* Let $a = 8$ and $b = 27$.]

41. *Oil production.* Using data from the first 3 years' production as well as geological studies, the management of an oil company estimates that oil will be pumped from a producing field at a rate given by

$$R(t) = \frac{100}{t+1} + 5 \qquad 0 \leq t \leq 20$$

where $R(t)$ is the rate of production in thousands of barrels per year t years after pumping begins. Approximately how many barrels of oil will the field produce during the first 10 years of production? From the end of the 10th year to the end of the 20th year of production?

42. *Oil production.* In Problem 41, if the rate is found to be

$$R(t) = \frac{120t}{t^2+1} + 3 \qquad 0 \leq t \leq 20$$

approximately how many barrels of oil will the field produce during the first 5 years of production? The second 5 years of production?

43. *Marketing.* An automobile company is ready to introduce a new line of cars with a national sales campaign. After test marketing the line in a carefully selected city, the marketing research department estimates that sales (in millions of dollars) will increase at the monthly rate of

$$S'(t) = 10 - 10e^{-0.1t} \qquad 0 \leq t \leq 24$$

t months after the national campaign is started. What will be the approximate total sales during the first 12 months of the campaign? The second 12 months of the campaign?

44. *Marketing.* Repeat Problem 43 if the monthly rate of increase in sales is found to be approximated by

$$S'(t) = 20 - 20e^{-0.05t} \qquad 0 \leq t \leq 24$$

Life Sciences

45. *Natural resource depletion.* The instantaneous rate of change of demand for wood in the United States since 1970 ($t = 0$) in billions of cubic feet per year is estimated to be given by

$$Q'(t) = 12 + 0.006t^2 \qquad 0 \leq t \leq 50$$

where $Q(t)$ is the total amount of wood consumed in billions of cubic feet t years after 1970. How many billions of cubic feet of wood will be consumed from 1980 to 1990?

46. *Natural resource depletion.* Repeat Problem 45 for the time interval from 1990 to 2000.

47. *Biology.* A yeast culture weighing 2 grams is removed from a refrigerator unit and is expected to grow at the rate of $W'(t) = 0.2e^{0.1t}$ grams per hour at a higher controlled temperature. How much will the weight of the culture increase during the first 8 hours of growth? How much will the weight of the culture increase from the end of the 8th hour to the end of the 16th hour of growth?

48. *Medicine.* The rate of healing for a skin wound (in square centimeters per day) is approximated by $A'(t) = -0.9e^{-0.1t}$. The initial wound has an area of 9 square centimeters. How much will the area change during the first 5 days? The second 5 days?

Social Sciences

49. *Learning.* In a particular business college, it was found that an average student enrolled in an advanced typing class progressed at a rate of $N'(t) = 6e^{-0.1t}$ words per minute per week, t weeks after enrolling in a 15 week course. At the beginning of the course an average student could type 40 words per minute. How much improvement would be expected during the first 5 weeks of the course? The second 5 weeks of the course? The last 5 weeks of the course?

50. *Learning.* In the same business college, it was also found that an average student enrolled in a beginning shorthand class progressed at a rate of $N'(t) = 12e^{-0.06t}$ words per minute per week, t weeks after enrolling in a 15 week course. At the beginning of the course none of the students could take any dictation by shorthand. How much improvement would be expected during the first 5 weeks of the course? The second 5 weeks of the course? The last 5 weeks of the course?

4-5 Area and the Definite Integral

- Area under a Curve
- Area between a Curve and the x Axis
- Area between Two Curves
- Application: Distribution of Income

Area under a Curve

Consider the graph of $f(x) = x$ from $x = 0$ to $x = 4$ (Fig. 3). We can easily compute the area of the triangle bounded by $f(x) = x$, the x axis ($y = 0$), and the line $x = 4$, using the formula for the area of a triangle:

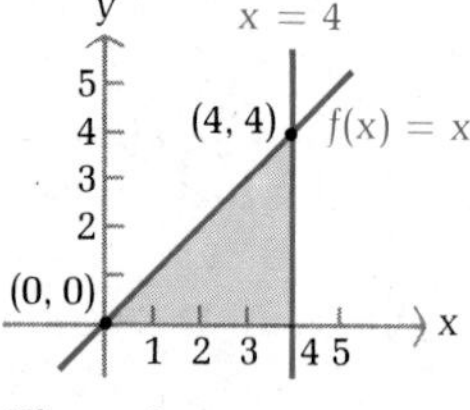

Figure 3

$$A = \frac{bh}{2} = \frac{4 \cdot 4}{2} = 8$$

Let us integrate $f(x) = x$ from $x = 0$ to $x = 4$:

$$\int_0^4 x\,dx = \frac{x^2}{2}\Big|_0^4 = \frac{4^2}{2} - \frac{0^2}{2} = 8$$

We get the same result! It turns out that this is not a coincidence. In general, we can prove the following:

Area under a Curve

If f is continuous and $f(x) \geq 0$ over the interval $[a, b]$, then the area bounded by $y = f(x)$, the x axis $(y = 0)$, and the vertical lines $x = a$ and $x = b$ is given exactly by

$$A = \int_a^b f(x)\,dx$$

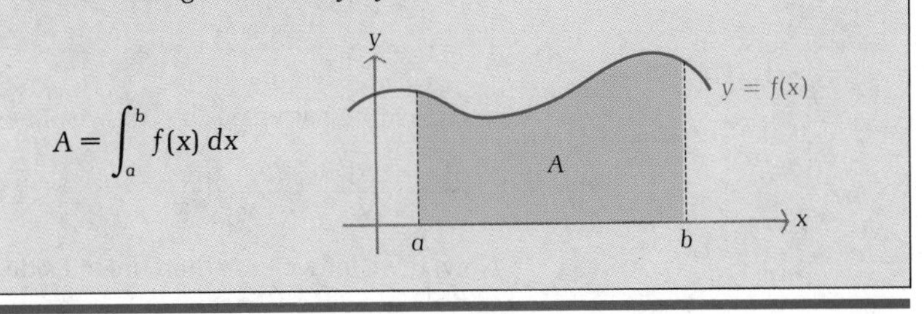

Let us see why the definite integral gives us the area exactly. Let $A(x)$ be the area under the graph of $y = f(x)$ from a to x, as indicated in Figure 4.

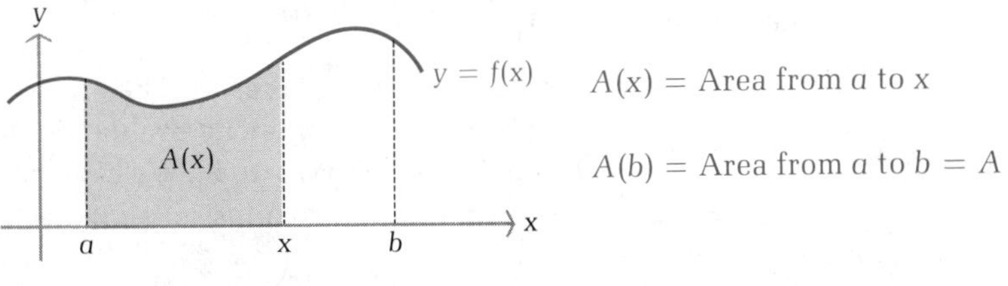

Figure 4

If we can show that $A(x)$ is an antiderivative of $f(x)$, then we can write

$$\begin{aligned}\int_a^b f(x)\,dx &= A(x)\big|_a^b = A(b) - A(a)\\ &= \begin{pmatrix}\text{Area from}\\ x = a \text{ to } x = b\end{pmatrix} - \begin{pmatrix}\text{Area from}\\ x = a \text{ to } x = a\end{pmatrix}\\ &= A - 0 = A\end{aligned}$$

To show that $A(x)$ is an antiderivative of $f(x)$—that is, $A'(x) = f(x)$—we use the definition of a derivative (Section 1-4) and write

$$A'(x) = \lim_{\Delta x \to 0} \frac{A(x + \Delta x) - A(x)}{\Delta x}$$

Geometrically, $A(x + \Delta x) - A(x)$ is the area from x to $x + \Delta x$ (see Fig. 5 on the next page). This area is given approximately by the area of the rectangle $\Delta x \cdot f(x)$, and the smaller Δx is, the better the approximation. Using

$$A(x + \Delta x) - A(x) \approx \Delta x \cdot f(x)$$

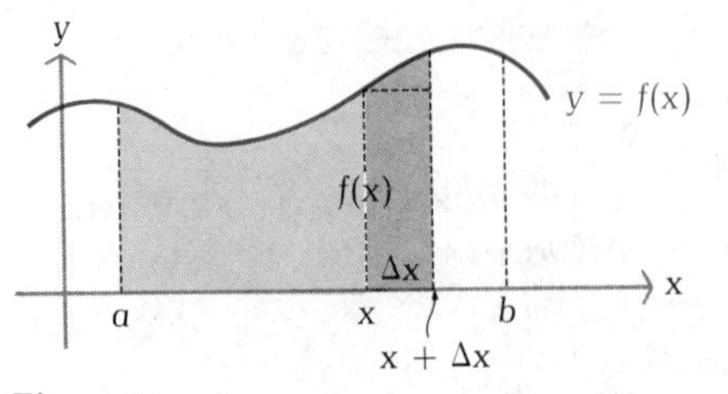

Figure 5

and dividing both sides by Δx, we obtain

$$\frac{A(x+\Delta x) - A(x)}{\Delta x} \approx f(x)$$

Now, if we let $\Delta x \to 0$, then the left side has $A'(x)$ as a limit, which is equal to the right side. Hence,

$$A'(x) = f(x)$$

that is, $A(x)$ is an antiderivative of $f(x)$. Thus,

$$\int_a^b f(x)\, dx = A(x)|_a^b = A(b) - A(a) = A - 0 = A$$

This is a remarkable result: The area under the graph of $y = f(x)$, $f(x) \geqslant 0$, can be obtained simply by evaluating the antiderivative of $f(x)$ at the end points of the interval $[a, b]$. We have now solved, at least in part, the third basic problem of calculus stated in Section 1-1.

Example 19 Find the area bounded by $f(x) = 6x - x^2$ and $y = 0$ for $1 \leqslant x \leqslant 4$.

Solution We sketch a graph of the region first:

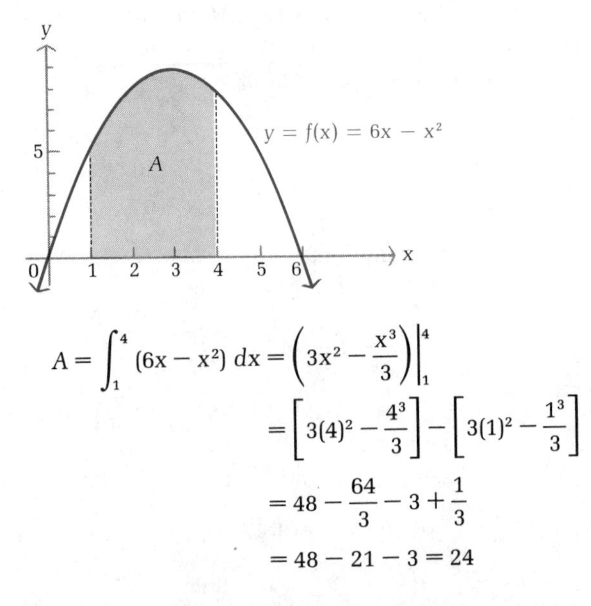

$$A = \int_1^4 (6x - x^2)\, dx = \left(3x^2 - \frac{x^3}{3}\right)\Bigg|_1^4$$

$$= \left[3(4)^2 - \frac{4^3}{3}\right] - \left[3(1)^2 - \frac{1^3}{3}\right]$$

$$= 48 - \frac{64}{3} - 3 + \frac{1}{3}$$

$$= 48 - 21 - 3 = 24$$

Problem 19 Find the area bounded by $f(x) = x^2 + 1$ and $y = 0$ for $-1 \leqslant x \leqslant 3$.

■ Area between a Curve and the x Axis

The condition $f(x) \geqslant 0$ is essential to the relationship between an area under a graph and the definite integral. How can we find the area between the graph of f and the x axis if $f(x) \leqslant 0$ on $[a, b]$ or if $f(x)$ is both positive and negative on $[a, b]$? To begin, suppose $f(x) \leqslant 0$ and A is the area between the graph of f and the x axis for $a \leqslant x \leqslant b$, as illustrated in Figure 6A. If we let $g(x) = -f(x)$, then A is also the area between the graph of g and the x axis for $a \leqslant x \leqslant b$ (see Fig. 6B). Since $g(x) \geqslant 0$ for $a \leqslant x \leqslant b$, we can use the definite integral of g to find A:

$$A = \int_a^b g(x)\,dx = \int_a^b [-f(x)]\,dx$$

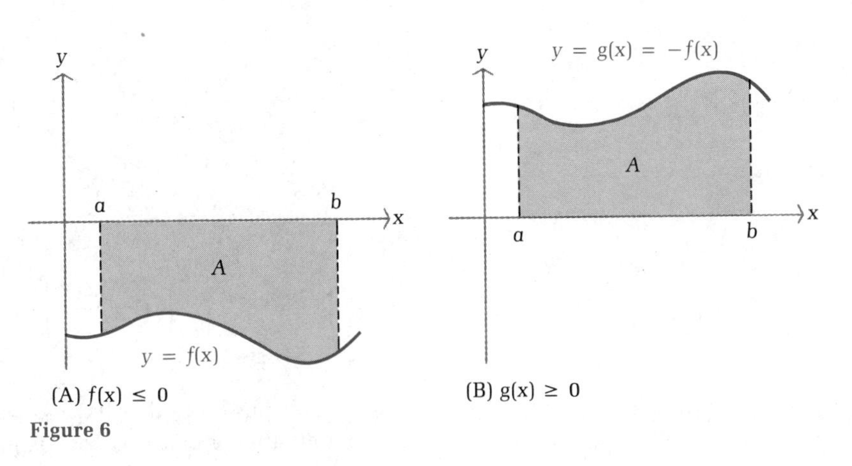

Figure 6

Thus, **the area between the graph of a negative function and the x axis is equal to the definite integral of the negative of the function.** Finally, if $f(x)$ is positive for some values of x and negative for others, the area between the graph of f and the x axis can be obtained by dividing $[a, b]$ into intervals over which f is always positive or always negative.

Example 20 Find the area between the graph of $f(x) = x^2 - 2x$ and the x axis over the indicated intervals:

(A) $[1, 2]$ (B) $[-1, 1]$

Solutions We begin by sketching the graph of f. (The solution of every area problem should begin with a sketch.)

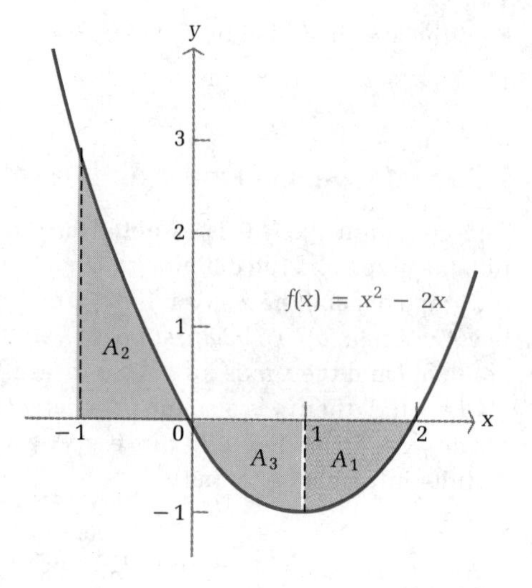

(A) From the graph we see that $f(x) \leqslant 0$ for $1 \leqslant x \leqslant 2$, so we integrate $-f(x)$:

$$\begin{aligned} A_1 &= \int_1^2 [-f(x)]\, dx \\ &= \int_1^2 (2x - x^2)\, dx \\ &= \left(x^2 - \frac{x^3}{3}\right)\Big|_1^2 \\ &= \left[(2)^2 - \frac{(2)^3}{3}\right] - \left[(1)^2 - \frac{(1)^3}{3}\right] \\ &= 4 - \tfrac{8}{3} - 1 + \tfrac{1}{3} = \tfrac{2}{3} \end{aligned}$$

(B) Since the graph shows that $f(x) \geqslant 0$ on $[-1, 0]$ and $f(x) \leqslant 0$ on $[0, 1]$, the computation of this area will require two integrals:

$$\begin{aligned} A &= A_2 + A_3 \\ &= \int_{-1}^0 f(x)\, dx + \int_0^1 [-f(x)]\, dx \\ &= \int_{-1}^0 (x^2 - 2x)\, dx + \int_0^1 (2x - x^2)\, dx \\ &= \left(\frac{x^3}{3} - x^2\right)\Big|_{-1}^0 + \left(x^2 - \frac{x^3}{3}\right)\Big|_0^1 \\ &= \tfrac{4}{3} + \tfrac{2}{3} = 2 \end{aligned}$$

Problem 20 Find the area between the graph of $f(x) = x^2 - 9$ and the x axis over the indicated intervals:

(A) $[0, 2]$ (B) $[2, 4]$

■ Area between Two Curves

Consider the area bounded by $y = f(x)$ and $y = g(x)$, $f(x) \geq g(x) \geq 0$, for $a \leq x \leq b$, as indicated in Figure 7.

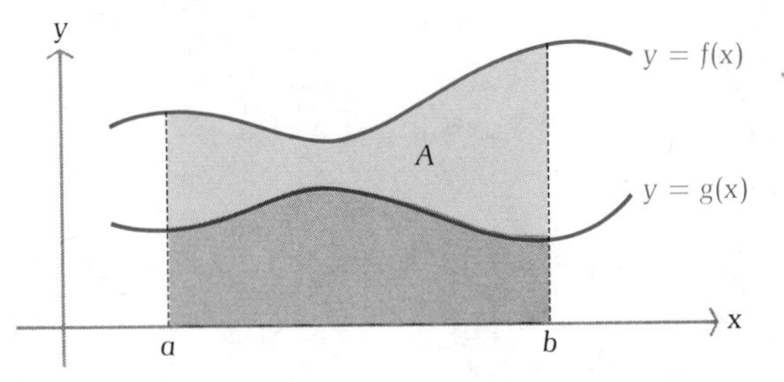

Figure 7

$$\begin{pmatrix}\text{Area } A \text{ between}\\ f(x) \text{ and } g(x)\end{pmatrix}$$

$$= \begin{pmatrix}\text{Area under}\\ f(x)\end{pmatrix} - \begin{pmatrix}\text{Area under}\\ g(x)\end{pmatrix}$$ Areas are from $x = a$ to $x = b$ above the x axis.

$$= \int_a^b f(x)\,dx - \int_a^b g(x)\,dx$$ Use definite integral property 4 (Section 4-4).

$$= \int_a^b [f(x) - g(x)]\,dx$$

It can be shown that the above result does not require $f(x)$ or $g(x)$ to remain positive over the interval $[a, b]$. A more general result is stated in the box:

Area between Two Curves

If f and g are continuous and $f(x) \geq g(x)$ over the interval $[a, b]$, then the area bounded by $y = f(x)$ and $y = g(x)$ for $a \leq x \leq b$ is given exactly by

$$A = \int_a^b [f(x) - g(x)]\,dx$$

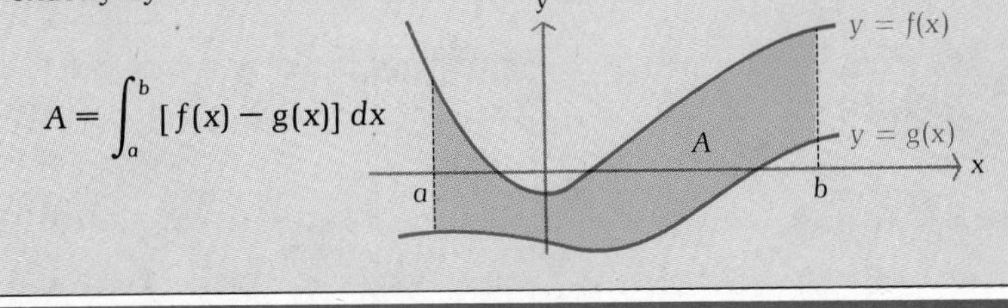

Example 21 Find the area bounded by $f(x) = \frac{1}{2}x + 3$, $g(x) = -x^2 + 1$, $x = -2$, and $x = 1$.

Solution We first sketch the area, then set up and evaluate an appropriate definite integral:

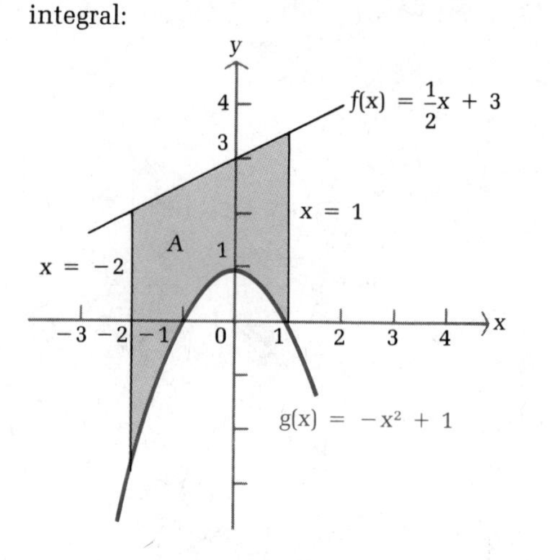

We observe from the graph that $f(x) \geqslant g(x)$ for $-2 \leqslant x \leqslant 1$, so

$$A = \int_{-2}^{1} [f(x) - g(x)]\, dx = \int_{-2}^{1} \left[\left(\frac{x}{2} + 3\right) - (-x^2 + 1)\right] dx$$

$$= \int_{-2}^{1} \left(x^2 + \frac{x}{2} + 2\right) dx$$

$$= \left(\frac{x^3}{3} + \frac{x^2}{4} + 2x\right)\bigg|_{-2}^{1} = \left(\frac{1}{3} + \frac{1}{4} + 2\right) - \left(\frac{-8}{3} + \frac{4}{4} - 4\right) = \frac{33}{4}$$

Problem 21 Find the area bounded by $f(x) = x^2 - 1$, $g(x) = -\frac{1}{2}x - 3$, $x = -1$, and $x = 2$.

Example 22 Find the area bounded by $f(x) = 5 - x^2$ and $g(x) = 2 - 2x$.

Solution

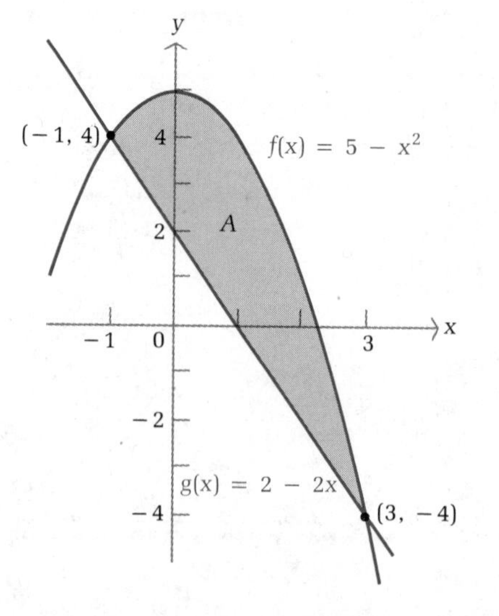

The graph of f is a parabola and the graph of g is a line, as shown in the figure. To find the points of intersection (hence the upper and lower limits of integration), we solve $y = 5 - x^2$ and $y = 2 - 2x$ simultaneously by setting $5 - x^2$ equal to $2 - 2x$ (substitution method):

$$5 - x^2 = 2 - 2x$$
$$x^2 - 2x - 3 = 0$$
$$x = -1, 3$$

The figure shows that $f(x) \geqslant g(x)$ over the interval $[-1, 3]$, so

$$\begin{aligned} A &= \int_{-1}^{3} [f(x) - g(x)]\,dx = \int_{-1}^{3} [5 - x^2 - (2 - 2x)]\,dx \\ &= \int_{-1}^{3} (3 + 2x - x^2)\,dx \\ &= \left(3x + x^2 - \frac{x^3}{3}\right)\Bigg|_{-1}^{3} \\ &= \left[3(3) + (3)^2 - \frac{(3)^3}{3}\right] - \left[3(-1) + (-1)^2 - \frac{(-1)^3}{3}\right] \\ &= 9 + 9 - 9 + 3 - 1 - \tfrac{1}{3} = \tfrac{32}{3} \end{aligned}$$

Problem 22 Find the area bounded by $f(x) = 6 - x^2$ and $g(x) = x$.

▪ Application: Distribution of Income

Economists use a cumulative distribution called a **Lorenz curve** to describe the distribution of income between different households in a given country. A typical Lorenz curve is shown in Figure 8. The points on this curve are determined by ranking all households by income and then computing

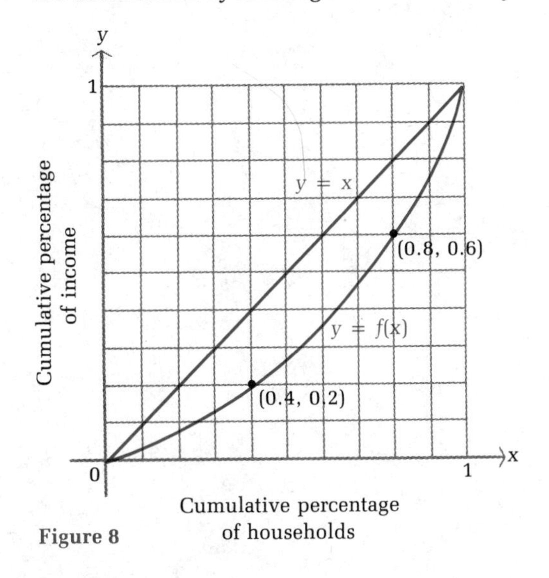

Figure 8

the percentage of households whose income is less than or equal to a given percentage of the total income for the country. For example, the point (0.4, 0.2) on the Lorenz curve in Figure 8 indicates that the bottom 40% of the households receive 20% of the income, the point (0.8, 0.6) indicates that the bottom 80% of the households receive 60% of the income, and so on. **Absolute equality** of income distribution would occur if every household received the same income. This is represented by the line $y = x$ in Figure 8. The area between the Lorenz curve and the line $y = x$ is used to indicate how much the income distribution for a given country differs from absolute equality. More precisely, the **coefficient of inequality** of income distribution is defined to be the ratio of the area between the Lorenz curve and the line $y = x$ to the area under the line $y = x$. Since the area under the line $y = x$ from $x = 0$ to $x = 1$ is $\frac{1}{2}$, it follows that *the coefficient of inequality is simply twice the area between the two curves.*

If we are given a function f whose graph is a Lorenz curve, then we can use a definite integral to find the coefficient of inequality.

Coefficient of Inequality for a Lorenz Curve

If $y = f(x)$ is the equation of a Lorenz curve, then

$$\text{Coefficient of inequality} = 2\int_0^1 [x - f(x)]\,dx$$

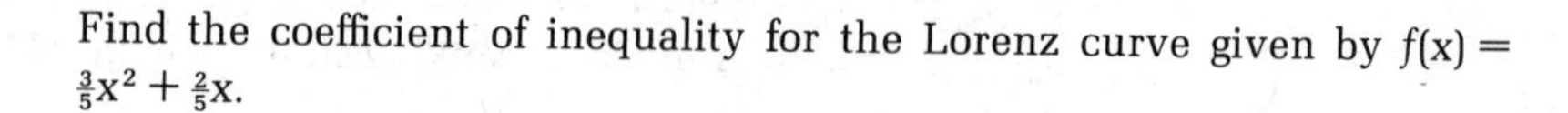

Example 23 Find the coefficient of inequality for the Lorenz curve given by $f(x) = \frac{3}{5}x^2 + \frac{2}{5}x$.

Solution

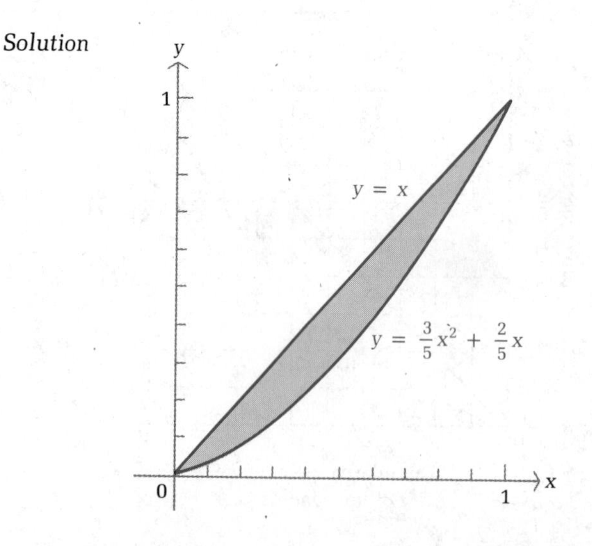

$$2\int_0^1 [x - f(x)]\,dx = 2\int_0^1 \left[x - \left(\frac{3}{5}x^2 + \frac{2}{5}x\right)\right] dx$$
$$= 2\int_0^1 \left(\frac{3}{5}x - \frac{3}{5}x^2\right) dx$$
$$= 2\left(\frac{3}{10}x^2 - \frac{1}{5}x^3\right)\Big|_0^1$$
$$= 2(\tfrac{3}{10} - \tfrac{1}{5}) = \tfrac{1}{5}$$

Thus, the coefficient of inequality is $\frac{1}{5}$ or 0.2. This number provides a relative measure of the income distribution in the country. For example, if the coefficient of inequality for a second country is 0.3, then we would conclude that income is less equally distributed in this second country.

Problem 23 Find the coefficient of inequality for the Lorenz curve given by $f(x) = \frac{9}{10}x^2 + \frac{1}{10}x$.

Answers to Matched Problems

19. $A = \int_{-1}^{3} (x^2 + 1)\,dx = \frac{40}{3}$
20. (A) $A = \int_0^2 (9 - x^2)\,dx = \frac{46}{3}$
    (B) $A = \int_2^3 (9 - x^2)\,dx + \int_3^4 (x^2 - 9)\,dx = 6$
21. $A = \int_{-1}^{2} [(x^2 - 1) - (-x/2 - 3)]\,dx = \frac{39}{4}$
22. $A = \int_{-3}^{2} [(6 - x^2) - x]\,dx = \frac{125}{6}$ 23. 0.3

Exercise 4-5

Find the area bounded by the graphs of the indicated equations.

A

1. $y = 2x + 4, \quad y = 0, \quad 1 \le x \le 3$
2. $y = -2x + 6, \quad y = 0, \quad 0 \le x \le 2$
3. $y = 3x^2, \quad y = 0, \quad 1 \le x \le 2$
4. $y = 4x^3, \quad y = 0, \quad 1 \le x \le 2$
5. $y = x^2 + 2, \quad y = 0, \quad -1 \le x \le 0$
6. $y = 3x^2 + 1, \quad y = 0, \quad -2 \le x \le 0$
7. $y = 4 - x^2, \quad y = 0, \quad -1 \le x \le 2$
8. $y = 12 - 3x^2, \quad y = 0, \quad -2 \le x \le 1$
9. $y = e^x, \quad y = 0, \quad -1 \le x \le 2$
10. $y = e^{-x}, \quad y = 0, \quad -2 \le x \le 1$
11. $y = \dfrac{1}{t}, \quad y = 0, \quad 0.5 \le t \le 1$
12. $y = \dfrac{1}{t}, \quad y = 0, \quad 0.1 \le t \le 1$

B

13. $y = 12, \quad y = -2x + 8, \quad -1 \le x \le 2$
14. $y = 3, \quad y = 2x + 6, \quad -1 \le x \le 2$

15. $y = 3x^2, \quad y = 12$
16. $y = x^2, \quad y = 9$
17. $y = 4 - x^2, \quad y = -5$
18. $y = x^2 - 1, \quad y = 3$
19. $y = x^2 + 1, \quad y = 2x - 2, \quad -1 \leq x \leq 2$
20. $y = x^2 - 1, \quad y = x - 2, \quad -2 \leq x \leq 1$
21. $y = -x, \quad y = 0, \quad -2 \leq x \leq 1$
22. $y = -x + 1, \quad y = 0, \quad -1 \leq x \leq 2$
23. $y = e^{0.5x}, \quad y = \frac{-1}{x}, \quad 1 \leq x \leq 2$
24. $y = \frac{1}{x}, \quad y = -e^x, \quad 0.5 \leq x \leq 1$

C

25. $y = x^2 - 4, \quad y = 0, \quad 0 \leq x \leq 3$
26. $y = 4\sqrt[3]{x}, \quad y = 0, \quad -1 \leq x \leq 8$
27. $y = x^2 + 2x + 3, \quad y = 2x + 4$
28. $y = 8 + 4x - x^2, \quad y = x^2 - 2x$
29. $y = x^2 - 4x - 10, \quad y = 14 - 2x - x^2$
30. $y = 6 + 6x - x^2, \quad y = 13 - 2x$
31. $y = x^3, \quad y = 4x$
32. $y = x^3 + 1, \quad y = x + 1$

Applications

Business & Economics

33. *Oil production.* Using data from the first 3 years' production as well as geological studies, the management of an oil company estimates that oil will be pumped from a producing field at a rate given by

$$R(t) = \frac{100}{t + 10} + 10 \qquad 0 \leq t \leq 15$$

where $R(t)$ is the rate of production in thousands of barrels per year t years after pumping begins. Find the area between the graph of R and the t axis over the interval [5, 10] and interpret.

34. *Oil production.* In Problem 33, if the rate is found to be

$$R(t) = \frac{100t}{t^2 + 25} + 4 \qquad 0 \leq t \leq 25$$

find the area between the graph of R and the t axis over the interval [5, 15] and interpret.

35. *Useful life.* An amusement company maintains records for each video game it installs in an arcade. Suppose that $C(t)$ and $R(t)$ represent the total accumulated costs and revenues (in thousands of dollars), respectively, t years after a particular game has been installed. If

$$C'(t) = 2 \quad \text{and} \quad R'(t) = 9e^{-0.3t}$$

find the area between the graphs of C' and R' over the interval on the t axis from 0 to the useful life of the game and interpret.

36. *Useful life.* Repeat Problem 35 if

$$C'(t) = 2t \quad \text{and} \quad R'(t) = 5te^{-0.1t^2}$$

37. *Income distribution.* The income distribution for a certain country is represented by the Lorenz curve with the equation

$$f(x) = \tfrac{9}{25}x^2 + \tfrac{16}{25}x$$

(A) What is the percentage of total income received by the bottom 25% of the households? By the bottom 50% of the households?
(B) Find the coefficient of inequality.

38. *Income distribution.* Repeat Problem 37 for

$$f(x) = \tfrac{3}{25}x^2 + \tfrac{22}{25}x$$

39. *Distribution of wealth.* Lorenz curves can be used to provide a relative measure of the distribution of the total assets of a country. A report by the U.S. Congressional Joint Economic Committee stated that in 1963, the bottom 90% of the households in the United States controlled 35% of the country's assets. Show that the Lorenz curve given by

$$y = f(x) = x^{9.964}$$

passes through the point (0.9, 0.35). Find the coefficient of inequality.

40. *Distribution of wealth.* The congressional report discussed in Problem 39 also stated that by 1983, the percentage of total assets controlled by the bottom 90% of the households in the United States had declined to 28%. Show that the Lorenz curve given by

$$y = f(x) = x^{12.082}$$

passes through the point (0.9, 0.28). Find the coefficient of inequality.

Life Sciences

41. *Biology.* A yeast culture is growing at a rate of $W'(t) = 0.3e^{0.1t}$ grams per hour. Find the area between the graph of W' and the t axis over the interval [0, 10] and interpret.

42. *Natural resource depletion.* The instantaneous rate of change of the demand for lumber in the United States since 1970 ($t = 0$) in billions of cubic feet per year is estimated to be given by

$$Q'(t) = 12 + 0.006t^2 \qquad 0 \leq t \leq 50$$

Find the area between the graph of Q' and the t axis over the interval [15, 20] and interpret.

Social Sciences

43. *Learning.* A beginning high school language class was chosen for an experiment on learning. Using a list of 50 words, the experiment involved measuring the rate of vocabulary memorization at different

times during a continuous 5 hour study session. It was found that the average rate of learning for the whole class was inversely proportional to the time spent studying and was given approximately by

$$V'(t) = \frac{15}{t} \qquad 1 \leq t \leq 5$$

Find the area between the graph of V' and the t axis over the interval [2, 4] and interpret.

4-6 Chapter Review

Important Terms and Symbols

4-1 *Antiderivatives and indefinite integrals.* Antiderivative, indefinite integral, integral sign, integrand, constant of integration, $\int f(x)\,dx$

4-2 *Integration by substitution.* General integral formulas, method of substitution

4-3 *Differential equations—growth and decay.* Differential equation, continuous compound interest, exponential growth law, $dQ/dt = rQ$, $Q = Q_0e^{rt}$

4-4 *Definite integral.* Definite integral, integrand, upper limit, lower limit, $\int_a^b f(x)\,dx$

4-5 *Area and the definite integral.* Area under a curve, area between a curve and the x axis, area between two curves, distribution of income

Integration Formulas

$$\int k\,dx = kx + C$$

$$\int kf(x)\,dx = k\int f(x)\,dx$$

$$\int [f(x) \pm g(x)]\,dx = \int f(x)\,dx \pm \int g(x)\,dx$$

$$\int u^n\,du = \frac{u^{n+1}}{n+1} + C \qquad n \neq -1$$

$$\int e^u\,du = e^u + C$$

$$\int \frac{1}{u}\,du = \ln|u| + C$$

Exercise 4-6 Chapter Review

Work through all the problems in this chapter review and check your answers in the back of the book. (Answers to all review problems are there.) Where weaknesses show up, review appropriate sections in the text.

A *Find each integral in Problems 1–6.*

1. $\int (3t^2 - 2t)\,dt$
2. $\int_2^5 (2x - 3)\,dx$
3. $\int (3t^{-2} - 3)\,dt$
4. $\int_1^4 x\,dx$
5. $\int e^{-0.5x}\,dx$
6. $\int_1^5 \frac{2}{u}\,du$
7. Find a function $y = f(x)$ that satisfies both conditions:

$$\frac{dy}{dx} = 3x^2 - 2 \qquad f(0) = 4$$

8. Find the area bounded by the graphs of $y = 3x^2 + 1$, $y = 0$, $x = -1$, and $x = 2$.

B *Find each integral in Problems 9–14.*

9. $\int \sqrt[3]{6x - 5}\,dx$
10. $\int_0^1 10(2x - 1)^4\,dx$
11. $\int \left(\frac{2}{x^2} - 2xe^{x^2}\right) dx$
12. $\int_0^4 \sqrt{x^2 + 4}\,x\,dx$
13. $\int (e^{-2x} + x^{-1})\,dx$
14. $\int_0^{10} 10e^{-0.02x}\,dx$
15. Find a function $y = f(x)$ that satisfies both conditions:

$$\frac{dy}{dx} = 3x^{-1} - x^{-2} \qquad f(1) = 5$$

16. Find the equation of the curve that passes through (2, 10) if its slope is given by

$$\frac{dy}{dx} = 6x + 1$$

for each x.

C 17. Find the area bounded by the graphs of $y = x^2 - 4$, $y = 0$, $x = -2$, and $x = 4$.

Find each integral in Problems 18–23.

18. $\int_0^3 \frac{x}{1 + x^2}\,dx$
19. $\int_0^3 \frac{x}{(1 + x^2)^2}\,dx$
20. $\int x^3(2x^4 + 5)^5\,dx$
21. $\int \frac{e^{-x}}{e^{-x} + 3}\,dx$
22. $\int \frac{e^x}{(e^x + 2)^2}\,dx$
23. $\int \frac{(\ln x)^2}{x}\,dx$

24. Find a function $y = f(x)$ that satisfies both conditions:

$$\frac{dy}{dx} = 9x^2e^{x^3} \qquad f(0) = 2$$

25. Solve the differential equation:

$$\frac{dN}{dt} = 0.06N \qquad N(0) = 800 \qquad N > 0$$

26. Find the area bounded by the graphs of $y = 6 - x^2$, $y = x^2 - 2$, $x = 0$, and $x = 3$. Be careful!

Applications

Business & Economics

27. *Profit function.* If the marginal profit for producing x units per day is given by

$$P'(x) = 100 - 0.02x \qquad P(0) = 0$$

where $P(x)$ is the profit in dollars, find the profit function P and the profit on ten units of production per day.

28. *Resource depletion.* An oil well starts out producing oil at the rate of 60,000 barrels of oil per year, but the production rate is expected to decrease by 4,000 barrels per year. Thus, if $P(t)$ is the total production (in thousands of barrels) in t years, then

$$P'(t) = f(t) = 60 - 4t \qquad 0 \leqslant t \leqslant 15$$

Write a definite integral that will give the total production after 15 years of operation. Evaluate it.

29. *Profit and production.* The weekly marginal profit for an output of x units is given approximately by

$$P'(x) = 150 - \frac{x}{10} \qquad 0 \leqslant x \leqslant 40$$

What is the total change in profit for a production change from ten units per week to forty units? Set up a definite integral and evaluate it.

30. *Useful life.* The total accumulated costs $C(t)$ and revenues $R(t)$ (in thousands of dollars), respectively, for a coal mine satisfy

$$C'(t) = 3 \quad \text{and} \quad R'(t) = 20e^{-0.1t}$$

where t is the number of years the mine has been in operation. Find the useful life of the mine to the nearest year. What is the total profit accumulated during the useful life of the mine?

31. *Marketing.* The marketing research department for an automobile company estimates that sales (in millions of dollars) will increase at

the monthly rate of

$$S'(t) = 15 - 15e^{-0.03t} \qquad 0 \leq t \leq 24$$

t months after the beginning of a national sales campaign. What will be the approximate total sales the first 12 months of the campaign?

32. *Income distribution.* Find the coefficient of inequality for the income distribution represented by the Lorenz curve with the equation

$$f(x) = \tfrac{1}{2}x + \tfrac{1}{2}x^2$$

Life Sciences

33. *Wound healing.* The area of a small, healing surface wound changes at a rate given approximately by

$$\frac{dA}{dt} = -5t^{-2} \qquad 1 \leq t \leq 5$$

where t is in days and $A(1) = 5$ square centimeters. What will the area of the wound be in 5 days?

34. *Pollution.* An environmental protection agency estimates that the rate of seepage of toxic chemicals from a waste dump in gallons per year is given by

$$R(t) = \frac{1{,}000}{(1 + t)^2}$$

where t is time in years since the discovery of the seepage. Find the total amount of toxic chemicals that seep from the dump during the first 4 years after the seepage is discovered.

35. *Population.* The population of the United States in 1980 was approximately 226 million. The continuous compound growth rate for the decade from 1970 to 1980 was 1.1%. (Data from the 1980 census.)

    (A) If the population continues to grow at the same continuous compound growth rate, what will be the population in 1990?
    (B) How long will it take the population to double at this continuous compound growth rate?

Social Sciences

36. *Archaeology.* The continuous compound rate of decay for carbon-14 is $r = -0.000\ 123\ 8$. A piece of animal bone found at an archaeological site contains 4% of the original amount of carbon-14. Estimate the age of the bone.

37. *Learning.* In a particular business college, it was found that an average student enrolled in a typing class progressed at a rate of $N'(t) = 7e^{-0.1t}$ words per minute t weeks after enrolling in a 15 week course. If at the beginning of the course a student could type 25 words per minute, how many words per minute, $N(t)$, would the student be expected to type t weeks into the course? After completing the course?

5 Additional Integration Topics

CHAPTER 5 Contents

5-1 Definite Integral as a Limit of a Sum

- Rectangle Rule for Approximating Definite Integrals
- Definite Integral as a Limit of a Sum
- Recognizing a Definite Integral
- Average Value of a Continuous Function

Up to this point, in order to evaluate a definite integral

$$\int_a^b f(x)\, dx$$

we need to find an antiderivative of the function f so that we can write

$$\int_a^b f(x)\, dx = F(x)\Big|_a^b = F(b) - F(a) \qquad F'(x) = f(x)$$

But suppose we cannot find an antiderivative of f (it may not even exist in a convenient or closed form). For example, how would you evaluate the following?

$$\int_2^8 \sqrt{x^3 + 1}\, dx \qquad \int_1^5 \left(\frac{x}{x+1}\right)^3 dx \qquad \int_0^5 e^{-x^2} dx$$

We now introduce the *rectangle rule* for approximating definite integrals, and out of this discussion will evolve a new way of looking at definite integrals.

■ Rectangle Rule for Approximating Definite Integrals

In Section 4-5 we saw that any definite integral of a positive continuous function f over an interval $[a, b]$ can always be interpreted as the area bounded by $y = f(x)$, $y = 0$, $x = a$, and $x = b$ (see Fig. 1). What we need is a way of approximating such areas, given $y = f(x)$ and an interval $[a, b]$.

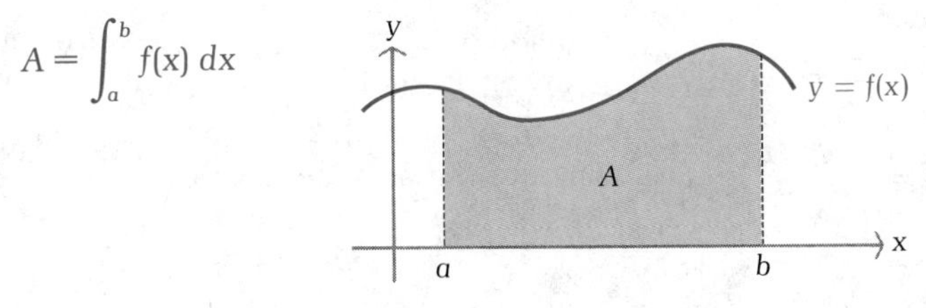

Figure 1

Let us start with a concrete example and generalize from the experience. We will start with a simple definite integral we can evaluate exactly:

$$\int_1^5 (x^2 + 3)\,dx = \left(\frac{x^3}{3} + 3x\right)\Bigg|_1^5$$
$$= \left[\frac{5^3}{3} + 3(5)\right] - \left[\frac{1^3}{3} + 3(1)\right]$$
$$= \left(\frac{125}{3} + 15\right) - \left(\frac{1}{3} + 3\right)$$
$$= \frac{160}{3} = 53\tfrac{1}{3}$$

This integral represents the area bounded by $y = x^2 + 3$, $y = 0$, $x = 1$, and $x = 5$, as indicated in Figure 2.

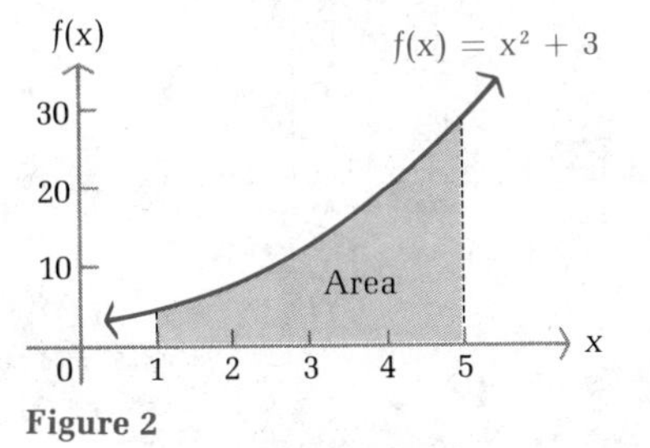

Figure 2

Since areas of rectangles are easy to compute, we cover the area in Figure 2 with rectangles (see Fig. 3) so that the top of each rectangle has a point in

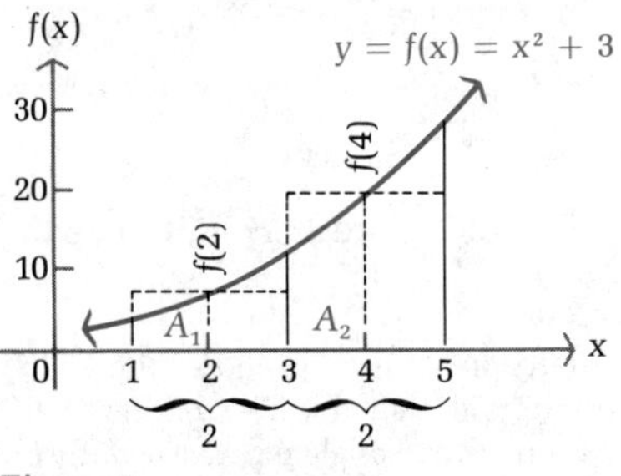

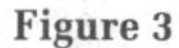

Figure 3

common with the graph of $y = f(x)$. As our first approximation, we divide the interval [1, 5] into two equal subintervals, each with length $(b - a)/2 = (5 - 1)/2 = 2$, and use the midpoint of each subinterval to compute the altitude of the rectangle sitting on top of that subinterval (see Fig. 3).

$$\begin{aligned}\int_1^5 (x^2 + 3)\, dx &\approx A_1 + A_2 \\ &= f(2) \cdot 2 + f(4) \cdot 2 \\ &= 2[f(2) + f(4)] \\ &= 2(7 + 19) = 52\end{aligned}$$

This approximation is less than 3% off of the exact area we found above ($53\frac{1}{3}$).

Now let us divide the interval [1, 5] into four equal subintervals, each of length $(b - a)/4 = (5 - 1)/4 = 1$, and use the midpoint* of each subinterval to compute the altitude of the rectangle corresponding to that subinterval (see Fig. 4).

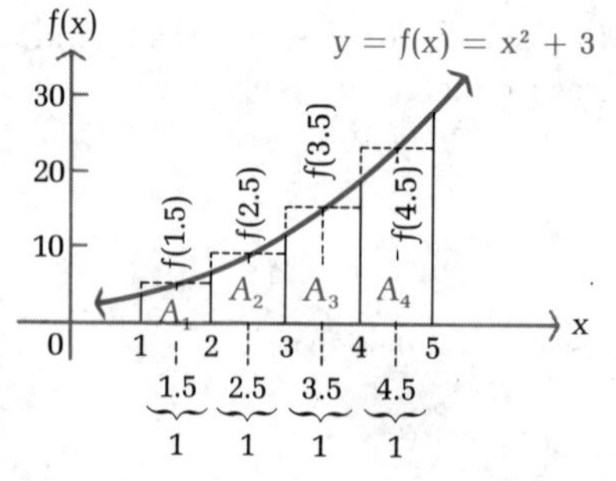

Figure 4

$$\begin{aligned}\int_1^5 (x^2 + 3)\, dx &\approx A_1 + A_2 + A_3 + A_4 \\ &= f(1.5) \cdot 1 + f(2.5) \cdot 1 + f(3.5) \cdot 1 + f(4.5) \cdot 1 \\ &= f(1.5) + f(2.5) + f(3.5) + f(4.5) \\ &= 5.25 + 9.25 + 15.25 + 23.25 \\ &= 53\end{aligned}$$

Now we are less than 1% off of the exact area ($53\frac{1}{3}$).

* We actually do not need to choose the midpoint of each subinterval; any point from each subinterval will do. The midpoint is often a convenient point to choose, because then the rectangle tops are usually above part of the graph and below part of the graph. This tends to cancel some of the error that occurs.

We would expect the approximations to continue to improve as we use more and more rectangles with smaller and smaller bases. We now state the *rectangle rule* for approximating definite integrals of a continuous function f over the interval from $x = a$ to $x = b$.

Rectangle Rule

Divide the interval from $x = a$ to $x = b$ into n equal subintervals of length $\Delta x = (b - a)/n$. Let c_k be any point in the kth subinterval. Then

$$\begin{aligned}\int_a^b f(x)\,dx &\approx f(c_1)\Delta x + f(c_2)\Delta x + \cdots + f(c_n)\Delta x \\ &= \Delta x[f(c_1) + f(c_2) + \cdots + f(c_n)]\end{aligned}$$

The rectangle rule is valid for any continuous function f. However, if f is not positive on $[a, b]$, then neither the definite integral nor the approximating sum represent areas.

Example 1 Use the rectangle rule to approximate

$$\int_2^{10} \frac{x}{x+1}\,dx$$

using $n = 4$ and c_k the midpoint of each subinterval. Compute the approximation to three decimal places.

Solution Step 1. Find Δx, the length of each subinterval:

$$\Delta x = \frac{b-a}{n} = \frac{10-2}{4} = \frac{8}{4} = 2$$

Step 2. Use the midpoint of each subinterval for c_k:

$$\begin{array}{rcccc} \text{Subintervals:} & [2, 4], & [4, 6], & [6, 8], & [8, 10] \\ \text{Midpoints:} & c_1 = 3, & c_2 = 5, & c_3 = 7, & c_4 = 9 \end{array}$$

Step 3. Use the rectangle rule with $n = 4$:

$$\begin{aligned}\int_a^b f(x)\,dx &\approx f(c_1)\Delta x + f(c_2)\Delta x + f(c_3)\Delta x + f(c_4)\Delta x \\ &= \Delta x[f(c_1) + f(c_2) + f(c_3) + f(c_4)] \\ &= 2[f(3) + f(5) + f(7) + f(9)] \\ &= 2(0.750 + 0.833 + 0.875 + 0.900) \\ &= 2(3.358) = 6.716 \qquad \text{To 3 decimal places}\end{aligned}$$

Problem 1 Use the rectangle rule to approximate

$$\int_2^{14} \frac{x}{x-1}\, dx$$

using $n = 4$ and c_k the midpoint of each subinterval. Compute the approximation to two decimal places.

One important application of the rectangle rule is the approximation of definite integrals involving **tabular functions**—that is, functions defined by tables rather than by formulas. The following example illustrates this approach.

Example 2 Real Estate

A developer is interested in estimating the area of the irregularly shaped property shown in Figure 5A. A surveyor used the straight horizontal road at the bottom of the property as the x axis and measured the vertical distance across the property at 400 foot intervals, starting at 200 (see Fig. 5B). These distances can be viewed as the values of the continuous function f whose graph forms the top of the property. Use the rectangle rule to approximate the area of the property.

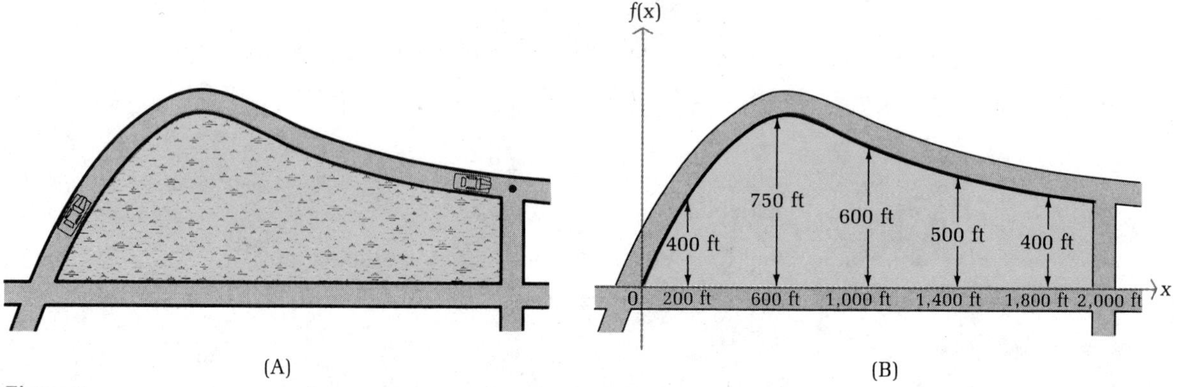

Figure 5

Solution List the values of the function f in a table:

x	200	600	1,000	1,400	1,800
$f(x)$	400	750	600	500	400

The area of the property is given by

$$A = \int_0^{2,000} f(x)\, dx$$

Using the rectangle rule with $n = 5$ and $\Delta x = 400$, and the values of f in the table, we have

$$\begin{aligned} A &\approx \Delta x[f(200) + f(600) + f(1{,}000) + f(1{,}400) + f(1{,}800)] \\ &= 400[400 + 750 + 600 + 500 + 400] \\ &= 1{,}060{,}000 \text{ square feet} \end{aligned}$$

Problem 2 To obtain a more accurate approximation of the area of the property shown in Figure 5A, the surveyor measured the vertical distances at 200 foot intervals, starting at 100. The results are listed in the table. Use these values and the rectangle rule with $n = 10$ and $\Delta x = 200$ to approximate the area of the property.

x	100	300	500	700	900	1,100	1,300	1,500	1,700	1,900
$f(x)$	225	500	700	725	650	575	525	475	425	375

■ Definite Integral as a Limit of a Sum

In using the rectangle rule to approximate a definite integral, one might expect

$$\lim_{\Delta x \to 0} [f(c_1)\Delta x + f(c_2)\Delta x + \cdots + f(c_n)\Delta x] = \int_a^b f(x)\,dx$$

This idea motivates the formal definition of a definite integral that we referred to in Section 4-4.

Definition of a Definite Integral

Let f be a continuous function defined on the closed interval $[a, b]$, and let

1. $a = x_0 \leqslant x_1 \leqslant \cdots \leqslant x_{n-1} \leqslant x_n = b$
2. $\Delta x_k = x_k - x_{k-1}$ for $k = 1, 2, \ldots, n$
3. $\Delta x_k \to 0$ as $n \to \infty$
4. $x_{k-1} \leqslant c_k \leqslant x_k$ for $k = 1, 2, \ldots, n$

Then

$$\int_a^b f(x)\,dx = \lim_{n\to\infty} [f(c_1)\Delta x_1 + f(c_2)\Delta x_2 + \cdots + f(c_n)\Delta x_n]$$

is called a **definite integral.**

In the definition of a definite integral, we divide the closed interval $[a, b]$ into n subintervals of arbitrary lengths in such a way that the length of each subinterval $\Delta x_k = x_k - x_{k-1}$ tends to 0 as n increases without bound. From each of the n subintervals we then select a point c_k and form the sum

$$f(c_1)\Delta x_1 + f(c_2)\Delta x_2 + \cdots + f(c_n)\Delta x_n$$

which is called a **Riemann sum** [named after the celebrated German mathematician Georg Riemann (1826–1866)].

Under the conditions stated in the definition, it can be shown that the limit of the Riemann sum always exists and it is a real number. The limit is independent of the nature of the subdivisions of $[a, b]$ as long as condition 3 holds, and it is independent of the choice of the c_k as long as condition 4 holds.

In a more formal treatment of the subject, we would then prove the remarkable **fundamental theorem of calculus,** which shows that the limit in the definition of a definite integral can be determined exactly by evaluating an antiderivative of $f(x)$, if it exists, at the end points of the interval $[a, b]$ and taking the difference.

Theorem 1

Fundamental Theorem of Calculus

Under the conditions stated in the definition of a definite integral

(Definition)

$$\int_a^b f(x)\,dx = \lim_{n\to\infty} [f(c_1)\Delta x_1 + f(c_2)\Delta x_2 + \cdots + f(c_n)\Delta x_n]$$

(Theorem)

$$= F(b) - F(a) \qquad \text{where } F'(x) = f(x)$$

Now we are free to evaluate a definite integral by using the fundamental theorem if an antiderivative of $f(x)$ can be found; otherwise, we can approximate it using the formal definition in the form of the rectangle rule.

Recognizing a Definite Integral

Recall that the derivative of a function f was defined by

$$f'(x) = \lim_{\Delta x\to 0} \frac{f(x + \Delta x) - f(x)}{\Delta x}$$

a form that is generally not easy to compute directly, but is easy to recognize in certain practical problems (slope, instantaneous velocity, rates of change, etc.). Once it is recognized that we are dealing with a derivative, we then proceed to try to compute it using derivative formulas and rules.

Similarly, evaluating a definite integral using the definition

$$\int_a^b f(x)\,dx = \lim_{n\to\infty} [f(c_1)\Delta x_1 + f(c_2)\Delta x_2 + \cdots + f(c_n)\Delta x_n] \tag{1}$$

is generally not easy; but the form on the right occurs naturally in many practical problems. We can use the fundamental theorem to evaluate the integral (once it is recognized) if an antiderivative can be found; otherwise, we will approximate it using the rectangle rule. We will now illustrate these points by finding the average value of a continuous function.

▪ Average Value of a Continuous Function

Suppose the temperature T (in degrees Fahrenheit) in the middle of a small shallow lake from 8 AM ($t = 0$) to 6 PM ($t = 10$) during the month of May is given approximately as shown in Figure 6.

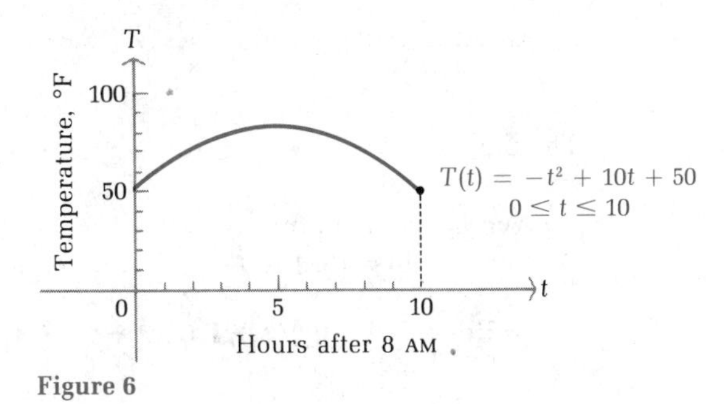

Figure 6

How can we compute the average temperature from 8 AM to 6 PM? We know that the average of a finite number of values

$$a_1, a_2, \ldots, a_n$$

is given by

$$\text{Average} = \frac{a_1 + a_2 + \cdots + a_n}{n}$$

But how can we handle a continuous function with infinitely many values? It would seem reasonable to divide the time interval [0, 10] into n equal subintervals, compute the temperature at a point in each subinterval, and then use the average of these values as an approximation of the average value of the continuous function $T = T(t)$ over [0, 10]. We would expect the approximations to improve as n increases. In fact, we would be inclined to define the limit of the average for n values as $n \to \infty$ as *the average value of T* over [0, 10], if the limit exists. This is exactly what we will do:

$$\begin{pmatrix}\text{Average temperature} \\ \text{for } n \text{ values}\end{pmatrix} = \frac{1}{n}[T(t_1) + T(t_2) + \cdots + T(t_n)] \tag{2}$$

where t_k is a point in the kth subinterval. We will call the limit of (2) as $n \to \infty$ the *average temperature over the time interval* [0, 10].

Form (2) looks sort of like form (1), but we are missing the Δt_k. We take care of this by multiplying (2) by $(b - a)/(b - a)$, which will change the form of (2) without changing its value:

$$\begin{aligned}
&\frac{b-a}{b-a} \cdot \frac{1}{n}[T(t_1) + T(t_2) + \cdots + T(t_n)] \\
&= \frac{1}{b-a} \cdot \frac{b-a}{n}[T(t_1) + T(t_2) + \cdots + T(t_n)] \\
&= \frac{1}{b-a} \cdot \left[T(t_1)\frac{b-a}{n} + T(t_2)\frac{b-a}{n} + \cdots + T(t_n)\frac{b-a}{n}\right] \\
&= \frac{1}{b-a}[T(t_1)\Delta t + T(t_2)\Delta t + \cdots + T(t_n)\Delta t]
\end{aligned}$$

Thus,

$$\begin{aligned}
&\begin{pmatrix}\text{Average temperature} \\ \text{over } [a, b] = [0, 10]\end{pmatrix} \\
&= \lim_{n\to\infty} \frac{1}{b-a}[T(t_1)\Delta t + T(t_2)\Delta t + \cdots + T(t_n)\Delta t] \\
&= \frac{1}{b-a}\left\{\lim_{n\to\infty}[T(t_1)\Delta t + T(t_2)\Delta t + \cdots + T(t_n)\Delta t]\right\}
\end{aligned}$$

Now the part in the braces is of form (1)—that is, a definite integral. Thus,

$$\begin{aligned}
&\begin{pmatrix}\text{Average temperature} \\ \text{over } [a, b] = [0, 10]\end{pmatrix} \\
&= \frac{1}{b-a}\int_a^b T(t)\, dt \\
&= \frac{1}{10-0}\int_0^{10} (-t^2 + 10t + 50)\, dt \\
&= \frac{1}{10}\left(-\frac{t^3}{3} + 5t^2 + 50t\right)\Bigg|_0^{10} \\
&= \frac{200}{3} \approx 67^\circ\text{F}
\end{aligned}$$

We now evaluate the definite integral using the fundamental theorem.

In general, proceeding as above for an arbitrary continuous function f over an interval $[a, b]$, we obtain the general formula:

Average Value of a Continuous Function *f* over [*a*, *b*]

$$\frac{1}{b-a}\int_a^b f(x)\,dx$$

Example 3 Find the average value of $f(x) = x - 3x^2$ over the interval $[-1, 2]$.

Solution

$$\frac{1}{b-a}\int_a^b f(x)\,dx = \frac{1}{2-(-1)}\int_{-1}^{2} (x - 3x^2)\,dx$$

$$= \frac{1}{3}\left(\frac{x^2}{2} - x^3\right)\Bigg|_{-1}^{2} = -\frac{5}{2}$$

Problem 3 Find the average value of $g(t) = 6t^2 - 2t$ over the interval $[-2, 3]$.

Example 4 Average Price

Given the demand function

$$p = D(x) = 100e^{-0.05x}$$

find the average price (in dollars) over the demand interval $[40, 60]$.

Solution

$$\text{Average price} = \frac{1}{b-a}\int_a^b D(x)\,dx$$

$$= \frac{1}{60-40}\int_{40}^{60} 100e^{-0.05x}\,dx$$

$$= \frac{100}{20}\int_{40}^{60} e^{-0.05x}\,dx$$

$$= -\frac{5}{0.05}e^{-0.05x}\Bigg|_{40}^{60}$$

$$= 100(e^{-2} - e^{-3}) \approx \$8.55$$

Problem 4 Given the supply equation

$$p = S(x) = 10e^{0.05x}$$

find the average price (in dollars) over the supply interval $[20, 30]$.

Example 5 Advertising

A metropolitan newspaper currently has a daily circulation of 50,000 papers (weekdays and Sunday). The management of the paper decides to initiate an aggressive advertising campaign to increase circulation. Suppose that the daily circulation (in thousands of papers) t days after the beginning of the campaign is given by

$$S(t) = 100 - 50e^{-0.01t}$$

What is the average daily circulation during the first 30 days of the campaign? During the second 30 days of the campaign?

Solution

$$\begin{aligned}\begin{pmatrix}\text{Average daily circulation}\\ \text{over } [a, b] = [0, 30]\end{pmatrix} &= \frac{1}{b-a}\int_a^b S(t)\,dt\\ &= \frac{1}{30}\int_0^{30} (100 - 50e^{-0.01t})\,dt\\ &= \frac{1}{30}(100t + 5{,}000e^{-0.01t})\Big|_0^{30}\\ &= \tfrac{1}{30}(3{,}000 + 5{,}000e^{-0.3} - 5{,}000)\\ &\approx 56.8 \text{ or } 56{,}800 \text{ papers}\end{aligned}$$

$$\begin{aligned}\begin{pmatrix}\text{Average daily circulation}\\ \text{over } [a, b] = [30, 60]\end{pmatrix} &= \frac{1}{60-30}\int_{30}^{60} (100 - 50e^{-0.01t})\,dt\\ &= \frac{1}{30}(100t + 5{,}000e^{-0.01t})\Big|_{30}^{60}\\ &= \tfrac{1}{30}(6{,}000 + 5{,}000e^{-0.6} - 3{,}000 - 5{,}000e^{-0.3})\\ &\approx 68.0 \text{ or } 68{,}000 \text{ papers}\end{aligned}$$

Problem 5 Refer to Example 5. Satisfied with the increase in circulation, management decides to terminate the advertising campaign. Suppose that the daily circulation (in thousands of papers) t days after the end of the advertising campaign is given by

$$S(t) = 65 + 8e^{-0.02t}$$

What is the average daily circulation during the first 30 days after the end of the campaign? During the second 30 days after the end of the campaign?

Answers to Matched Problems

1. 14.36
2. 1,035,000 square feet
3. 13
4. \$35.27
5. $\dfrac{2{,}350 - 400e^{-0.6}}{30} \approx 71.0$ or 71,000 papers;

   $\dfrac{1{,}950 - 400e^{-1.2} + 400e^{-0.6}}{30} \approx 68.3$ or 68,300 papers

Exercise 5-1

For Problems 1–8:

(A) Use the rectangle rule to approximate each definite integral for the indicated number of subintervals n. Choose c_k as the midpoint of each subinterval.

(B) Evaluate each integral exactly using an antiderivative.

A

1. $\int_1^5 3x^2\,dx,\quad n = 2$
2. $\int_2^6 x^2\,dx,\quad n = 2$
3. $\int_1^5 3x^2\,dx,\quad n = 4$
4. $\int_2^6 x^2\,dx,\quad n = 4$

B

5. $\int_0^4 (4 - x^2)\,dx,\quad n = 2$
6. $\int_0^4 (3x^2 - 12)\,dx,\quad n = 2$
7. $\int_0^4 (4 - x^2)\,dx,\quad n = 4$
8. $\int_0^4 (3x^2 - 12)\,dx,\quad n = 4$

In Problems 9–12, use the rectangle rule with n = 4 and the values of f in the given table to approximate the indicated definite integral. Choose c_k as the midpoint of each interval.

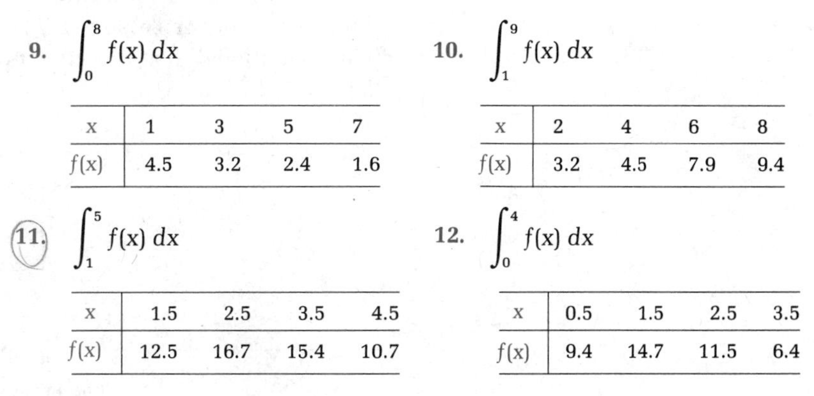

9. $\int_0^8 f(x)\,dx$

x	1	3	5	7
f(x)	4.5	3.2	2.4	1.6

10. $\int_1^9 f(x)\,dx$

x	2	4	6	8
f(x)	3.2	4.5	7.9	9.4

11. $\int_1^5 f(x)\,dx$

x	1.5	2.5	3.5	4.5
f(x)	12.5	16.7	15.4	10.7

12. $\int_0^4 f(x)\,dx$

x	0.5	1.5	2.5	3.5
f(x)	9.4	14.7	11.5	6.4

Find the average value of each function over the indicated interval.

13. $f(x) = 500 - 50x,\quad [0, 10]$
14. $g(x) = 2x + 7,\quad [0, 5]$
15. $f(t) = 3t^2 - 2t,\quad [-1, 2]$
16. $g(t) = 4t - 3t^2,\quad [-2, 2]$
17. $f(x) = \sqrt[3]{x},\quad [1, 8]$
18. $g(x) = \sqrt{x + 1},\quad [3, 8]$
19. $f(x) = 4e^{-0.2x},\quad [0, 10]$
20. $f(x) = 64e^{0.08x},\quad [0, 10]$

Use the rectangle rule to approximate (to three decimal places) each quantity in Problems 21–24. Use n = 4 and choose c_k as the midpoint of each subinterval.

21. The average value of $f(x) = (x + 1)/(x^2 + 1)$ for $[-1, 1]$

22. The average value of $f(x) = x/(x+1)$ for $[0, 4]$
23. The area under the graph of $f(x) = \ln(1 + x^3)$ from $x = 0$ to $x = 2$
24. The area under the graph of $f(x) = 1/(2 + x^3)$ from $x = -1$ to $x = 1$

C

In Problems 25–28, use the rectangle rule to approximate (to three decimal places) each definite integral for the indicated number of subintervals n. Choose c_k as the midpoint of each subinterval.

25. $\int_0^1 e^{-x^2}\,dx, \quad n = 5$

26. $\int_0^1 e^{x^2}\,dx, \quad n = 5$

27. $\int_0^1 e^{-x^2}\,dx, \quad n = 10$

28. $\int_0^1 e^{x^2}\,dx, \quad n = 10$

29. Find the average value of $f'(x)$ over the interval $[a, b]$ for any differentiable function f.
30. Show that the average value of $f(x) = Ax + B$ over the interval $[a, b]$ is

$$f\left(\frac{a+b}{2}\right)$$

Applications

Business & Economics

31. *Inventory.* A store orders 600 units of a product every 3 months. If the product is steadily depleted to zero by the end of each 3 months, the inventory on hand, I, at any time t during the year is illustrated as follows:

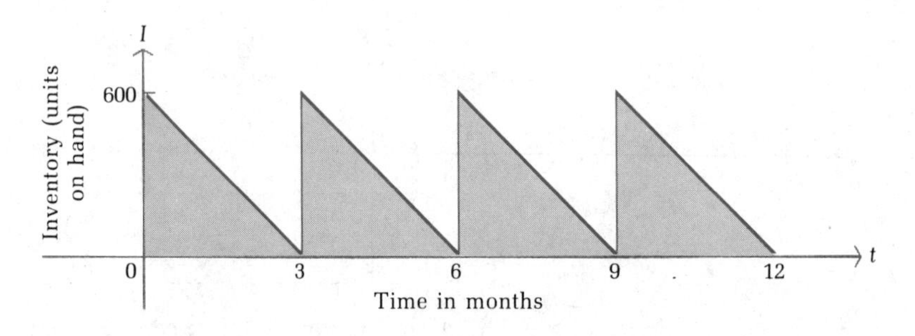

(A) Write an inventory function (assume it is continuous) for the first 3 months. [The graph is a straight line joining (0, 600) and (3, 0).]

(B) What is the average number of units on hand for a 3 month period?

32. Repeat Problem 31 with an order of 1,200 units every 4 months.
33. *Cash reserves.* Suppose cash reserves (in thousands of dollars) are approximated by

$$C(x) = 1 + 12x - x^2 \qquad 0 \le x \le 12$$

where x is the number of months after the first of the year. What is the average cash reserve for the first quarter?

34. Repeat Problem 33 for the second quarter.

35. *Average cost.* The total cost in dollars of manufacturing x auto body frames is

$$C(x) = 60{,}000 + 300x$$

(A) Find the average cost per unit if 500 frames are produced. [*Hint:* Recall that $\overline{C}(x)$ is the average cost per unit.]

(B) Find the average value of the cost function over the interval [0, 500].

36. *Average cost.* The total cost in dollars of printing x dictionaries is

$$C(x) = 20{,}000 + 10x$$

(A) Find the average cost per unit if 1,000 dictionaries are produced.

(B) Find the average value of the cost function over the interval [0, 1,000].

37. *Continuous compound interest.* If $100 is deposited in an account that earns interest at an annual nominal rate of 8% compounded continuously, find the amount in the account after 5 years and the average amount in the account during this 5 year period. (Continuous compound interest is discussed in Section 3-1.)

38. *Continuous compound interest.* If $500 is deposited in an account that earns interest at an annual nominal rate of 12% compounded continuously, find the amount in the account after 10 years and the average amount in the account during this 10 year period.

39. *Supply function.* Given the supply function

$$p = S(x) = 10(e^{0.02x} - 1)$$

find the average price (in dollars) over the supply interval [20, 30].

40. *Demand function.* Given the demand function

$$p = D(x) = \frac{1{,}000}{x}$$

find the average price (in dollars) over the demand range [400, 600].

41. *Advertising.* The number of hamburgers (in thousands) sold each day by a chain of restaurants t days after the beginning of an advertising campaign is given by

$$S(t) = 20 - 10e^{-0.1t}$$

What is the average number of hamburgers sold each day during the first week of the advertising campaign? During the second week of the campaign?

42. *Advertising.* The number of hamburgers (in thousands) sold each day by a chain of restaurants t days after the end of an advertising campaign is given by

$$S(t) = 10 + 8e^{-0.2t}$$

What is the average number of hamburgers sold each day during the first week after the end of the advertising campaign? During the second week after the end of the campaign?

43. *Real estate.* A surveyor produced the table below by measuring the vertical distance (in feet) across a piece of real estate at 600 foot intervals, starting at 300 (see the figure). Use these values and the rectangle rule to estimate the area of the property.

x	300	900	1,500	2,100
f(x)	900	1,700	1,700	900

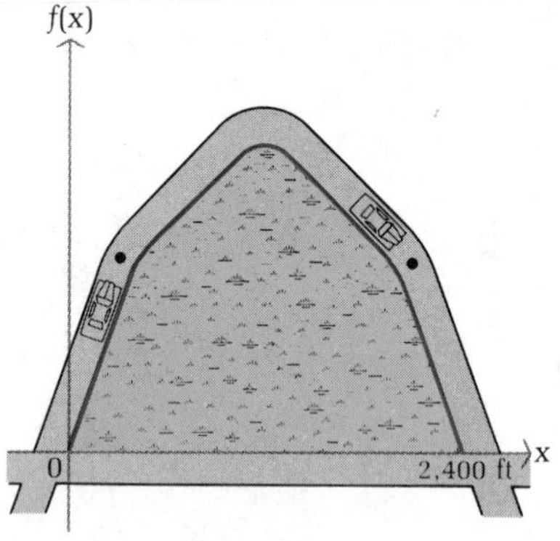

44. *Real estate.* Repeat Problem 43 for the following table of measurements:

x	200	600	1,000	1,400	1,800	2,200
f(x)	600	1,400	1,800	1,800	1,400	600

Life Sciences

45. *Temperature.* If the temperature $C(t)$ in an artificial habitat was made to change according to

$$C(t) = t^3 - 2t + 10 \qquad 0 \leq t \leq 2$$

(in degrees Celsius) over a 2 hour period, what is the average temperature over this period?

46. *Medicine.* A drug is injected into the bloodstream of a patient through her right arm. The concentration of the drug in the bloodstream of the left arm t hours after the injection is given by

$$C(t) = \frac{0.14t}{t^2 + 1}$$

What is the average concentration of the drug in the bloodstream of the left arm during the first hour after the injection? During the first 2 hours after the injection?

Social Sciences

47. *Politics.* Public awareness of a Congressional candidate before and after a successful campaign was approximated by

$$P(t) = \frac{8.4t}{t^2 + 49} + 0.1 \qquad 0 \leq t \leq 24$$

where t is time in months after the campaign started and $P(t)$ is the fraction of people in the Congressional district who could recall the candidate's name. What is the average fraction of people who could recall the candidate's name during the first 7 months after the campaign began? During the first 2 years after the campaign began?

48. *Population composition.* Because of various factors (such as birth rate expansion, then contraction; family flights from urban areas; etc.), the number of children in a large city was found to increase and then decrease rather drastically. If the number of children over a 6 year period was found to be given approximately by

$$N(t) = -\tfrac{1}{4}t^2 + t + 4 \qquad 0 \leq t \leq 6$$

what was the average number of children in the city over the 6 year time period? [Assume $N = N(t)$ is continuous.]

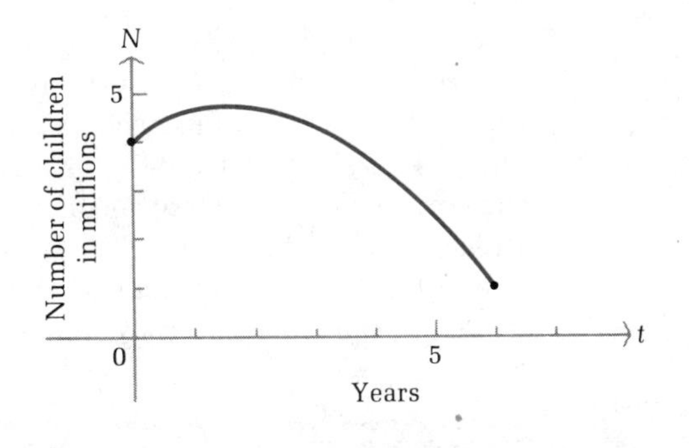

5-2 Applications in Business and Economics

- Continuous Income Stream
- Present Value of a Continuous Income Stream
- Consumers' and Producers' Surplus

▪ Continuous Income Stream

We start this discussion with an example.

Example 6 The rate of change of the income produced by a vending machine (in dollars per year) is given by

$$f(t) = 1{,}000e^{0.04t}$$

where t is time in years since the installation of the machine. Find the total income produced by the machine during the first 5 years of operation.

Solution

Since we have been given the rate of change of income, we can find the total income by using a definite integral:

$$\begin{aligned}
\text{Total income} &= \int_0^5 1{,}000e^{0.04t}\,dt \\
&= 25{,}000e^{0.04t}\Big|_0^5 \\
&= 25{,}000e^{0.04(5)} - 25{,}000e^{0.04(0)} \\
&= 30{,}535 - 25{,}000 \\
&= \$5{,}535
\end{aligned}$$

Rounded to the nearest dollar

Thus, the vending machine produces a total income of \$5,535 during the first 5 years of operation.

Problem 6 Refer to Example 6. Find the total income produced during the second 5 years of operation.

In reality, income from a vending machine is not received as a single payment at the end of the 5 year period. Instead, the income is collected on a regular basis, perhaps daily or weekly. In problems of this type, it is convenient to assume that income is actually received in a **continuous stream.** That is, we assume that income is a continuous function of time. The rate of change of income is called the **rate of flow** of the continuous income stream. In general, we have:

Total Income for a Continuous Income Stream

If $f(t)$ is the rate of flow of a continuous income stream, then the **total income** produced during the time period from $t = a$ to $t = b$ is

$$\text{Total income} = \int_a^b f(t)\,dt$$

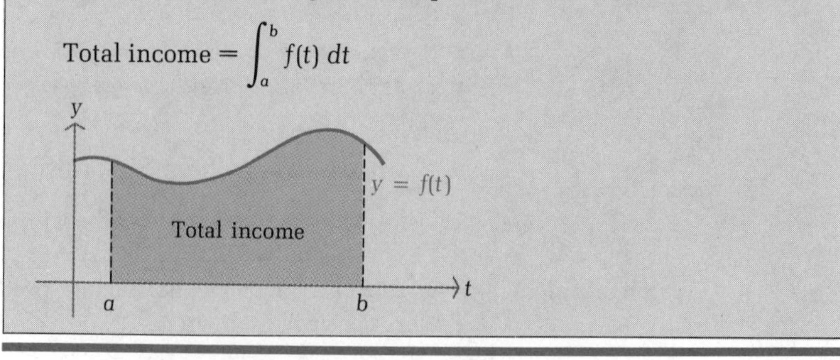

▪ Present Value of a Continuous Income Stream

In Section 3-1, we discussed continuous compound interest. Recall that if P dollars are invested at an annual nominal rate of $100r\%$, compounded continuously, then the amount A after t years is

$$A = Pe^{rt}$$

Solving this equation for P, we have

$$P = Ae^{-rt}$$

which is referred to as the present value of A. That is, P is the amount that would have to be invested at time $t = 0$ (the present) in order to receive A dollars t years from now (the future) at $100r\%$ compounded continuously. For example, if a bond pays \$10,000 in 5 years and money is worth 12% compounded continuously, then the present value is (to the nearest dollar)

$$P = 10{,}000e^{-0.12(5)} = \$5{,}488$$

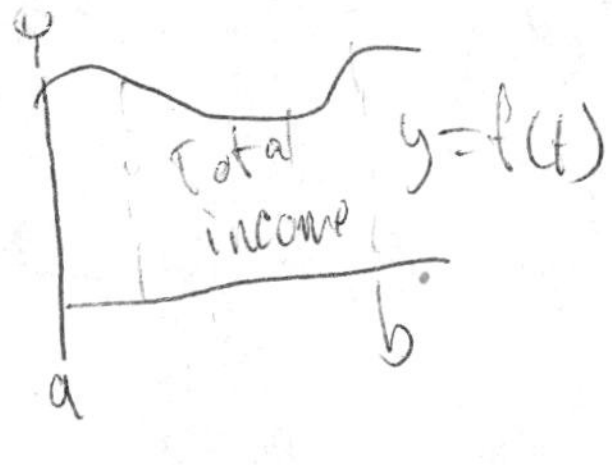

Thus, it is easy to compute the present value of an investment that returns a single payment. Now we want to consider investments that return payments to the investor on a regular basis.

In order to generalize the concept of present value to continuous income streams, we assume that the income produced is instantaneously invested at an annual rate of $100r\%$, compounded continuously. (In general, the value of money is determined by the currently available interest rate.) Suppose $f(t)$ is the rate of flow of a continuous income stream. To find the present value of this continuous income stream at $100r\%$, compounded continuously for T years, we begin by dividing the interval $[0, T]$ into n equal subintervals of length Δt and choose an arbitrary point c_k in each subinterval, as illustrated in Figure 7. The total income produced during the time period from $t = t_{k-1}$ to $t = t_k$ is equal to the area under the graph of $f(t)$ over this subinterval and is approximately equal to $f(c_k)\Delta t$, the area of the shaded rectangle in Figure 7.

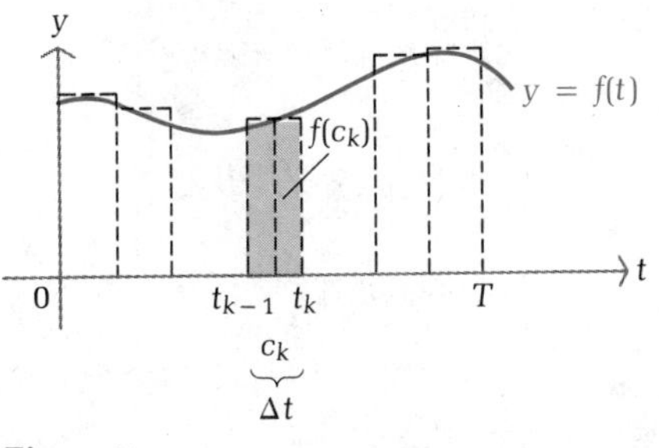

Figure 7

Using the present value formula $P = Ae^{-rt}$ with $A = f(c_k)\Delta t$ and $t = c_k$, the present value of the income received during the time period from

$t = t_{k-1}$ to $t = t_k$ is approximately equal to

$$f(c_k)\Delta t e^{-rc_k}$$

The total present value for the time period from $t = 0$ to $t = T$ is approximately equal to

$$f(c_1)\Delta t e^{-rc_1} + f(c_2)\Delta t e^{-rc_2} + \cdots + f(c_n)\Delta t e^{-rc_n}$$

which has the form of a Riemann sum. And in the limit we have a definite integral. (See the definition of definite integral in Section 5-1.) Thus, we have:

Present Value of a Continuous Income Stream

If $f(t)$ is the rate of flow of a continuous income stream, then the **present value,** *PV*, at 100r% compounded continuously for *T* years is

$$\text{PV} = \int_0^T f(t)e^{-rt}\,dt$$

Example 7 Let $f(t) = 1{,}000e^{0.04t}$ be the rate of flow of the income produced by the vending machine in Example 6.

(A) Find the present value of this income stream at 12% compounded continuously for 5 years.

(B) Find the amount (income plus interest) produced by the vending machine over this 5 year period.

(C) Find the interest earned during this 5 year period.

Solutions (A)

$$\begin{aligned} \text{PV} &= \int_0^T f(t)e^{-rt}\,dt \\ &= \int_0^5 1{,}000e^{0.04t}e^{-0.12t}\,dt \\ &= 1{,}000\int_0^5 e^{-0.08t}\,dt \\ &= -12{,}500e^{-0.08t}\Big|_0^5 \\ &= -12{,}500e^{-0.4} + 12{,}500 \\ &= \$4{,}121 \end{aligned}$$

(B) The total amount produced by a continuous income stream is the same as the amount that would result if the present value were deposited in an account earning 12% interest compounded continuously for 5 years. Using the present value in the formula for continuous compound interest, we have

$$A = PVe^{rt} = 4{,}121e^{0.12(5)} = \$7{,}509$$

(C) In Example 6, we saw that the income produced by this vending machine was \$5,535. Since the amount is the total of all money produced (income plus interest), the difference between the amount and the income is interest. Thus,

$$7{,}509 - 5{,}535 = \$1{,}974$$

is the interest earned by the income produced during the 5 year period.

Problem 7 Repeat Example 7 if the interest rate is 9% compounded continuously.

■ Consumers' and Producers' Surplus

Let $p = D(x)$ be the price–demand equation for a product, let $\bar{p}$ be the current price, and let $\bar{x}$ be the corresponding demand.* The demand curve in Figure 8 shows that some of these $\bar{x}$ units could have been sold at a higher price. The consumers who would have been willing to pay higher prices have saved money. We want to determine the amount of money saved by these consumers.

To do this, consider the interval $[c_k, c_k + \Delta x]$ where $c_k < \bar{x}$. If the price remained constant over this interval, then the savings on each unit would

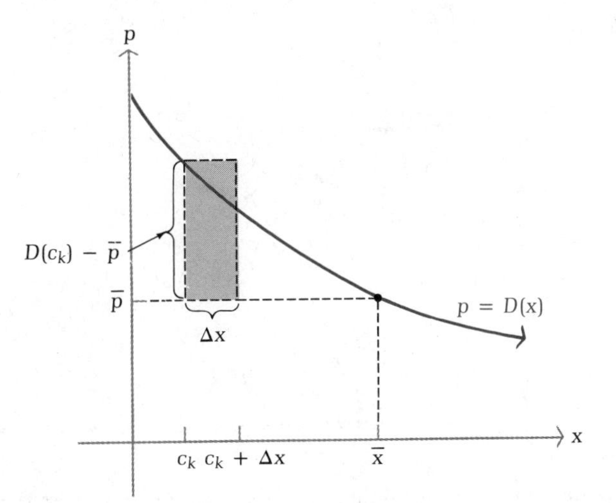

Figure 8

* Unless otherwise specified, we will assume that price is in dollars and demand is in number of units.

be the difference between $D(c_k)$, the price consumers are willing to pay, and $\bar{p}$, the price they actually pay. The total savings to consumers over this interval is approximately equal to

$$[D(c_k) - \bar{p}]\Delta x$$

which is the area of the shaded rectangle shown in Figure 8. If we divide the interval $[0, \bar{x}]$ into n equal subintervals, then the total savings to consumers is approximately equal to

$$[D(c_1) - \bar{p}]\Delta x + [D(c_2) - \bar{p}]\Delta x + \cdots + [D(c_n) - \bar{p}]\Delta x$$

which we recognize as a Riemann sum for the integral

$$\int_0^{\bar{x}} [D(x) - \bar{p}]\, dx$$

Thus, we define the *consumers' surplus* to be this integral.

Consumers' Surplus

If $(\bar{x}, \bar{p})$ is a point on the graph of the price–demand equation $p = D(x)$, then the **consumers' surplus,** CS, at a price level of $\bar{p}$ is

$$CS = \int_0^{\bar{x}} [D(x) - \bar{p}]\, dx$$

which is the area between $p = \bar{p}$ and $p = D(x)$ from $x = 0$ to $x = \bar{x}$.

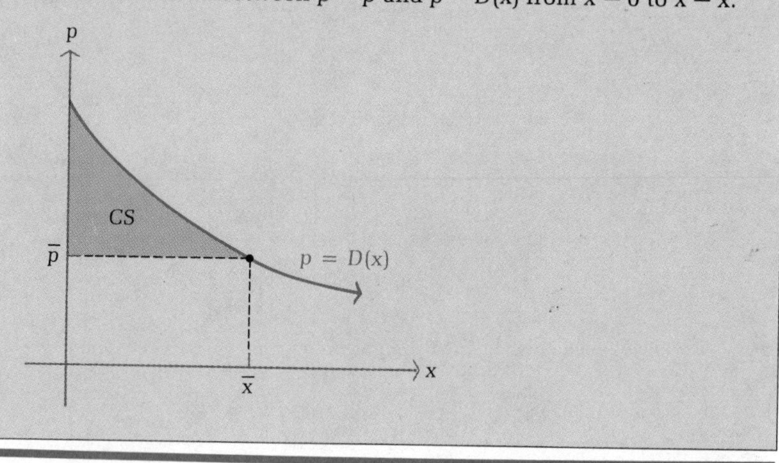

Example 8 Find the consumers' surplus at a price level of \$8 for the price–demand equation

$$p = D(x) = 20 - \frac{1}{20}x$$

Solution Step 1. Find $\bar{x}$, the demand when the price is $\bar{p} = 8$:

$$\bar{p} = 20 - \frac{1}{20}\bar{x}$$

$$8 = 20 - \frac{1}{20}\bar{x}$$

$$\frac{1}{20}\bar{x} = 12$$

$$\bar{x} = 240$$

Step 2. Sketch a graph:

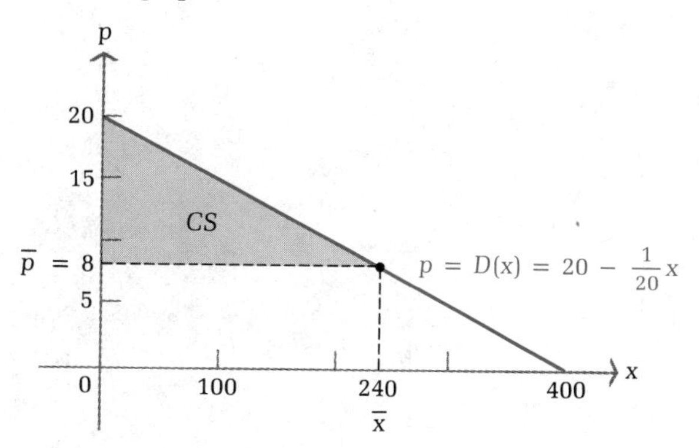

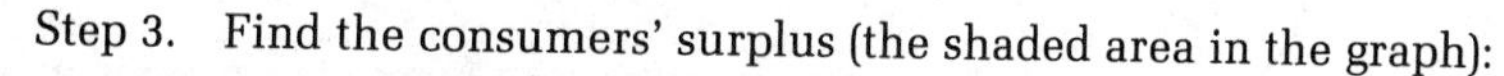

Step 3. Find the consumers' surplus (the shaded area in the graph):

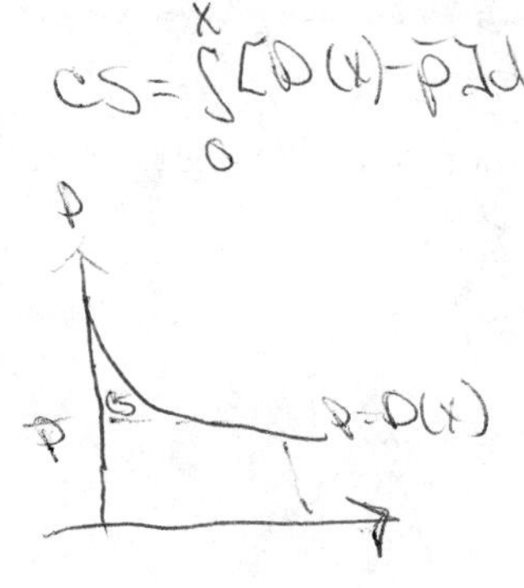

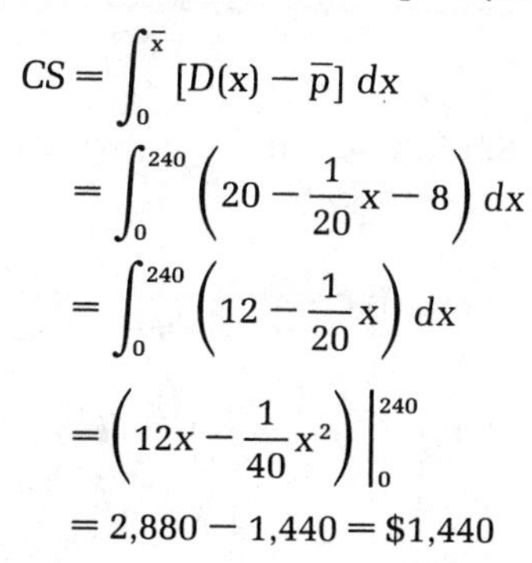

$$CS = \int_0^{\bar{x}} [D(x) - \bar{p}]\, dx$$

$$= \int_0^{240} \left(20 - \frac{1}{20}x - 8\right) dx$$

$$= \int_0^{240} \left(12 - \frac{1}{20}x\right) dx$$

$$= \left(12x - \frac{1}{40}x^2\right)\Big|_0^{240}$$

$$= 2{,}880 - 1{,}440 = \$1{,}440$$

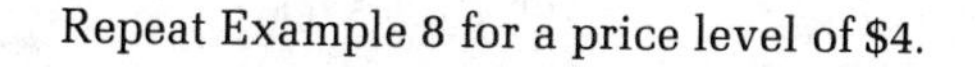

Problem 8 Repeat Example 8 for a price level of \$4.

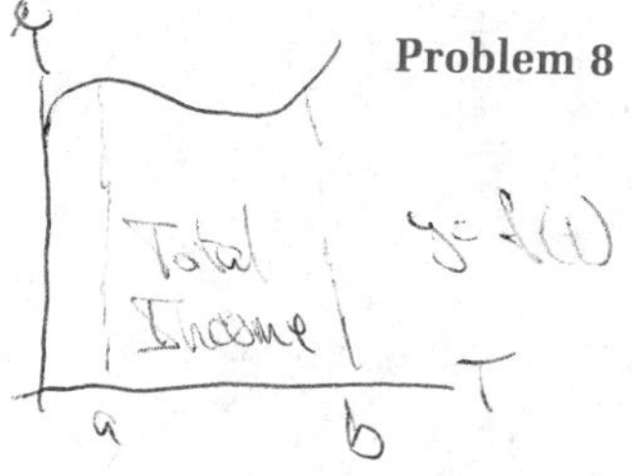

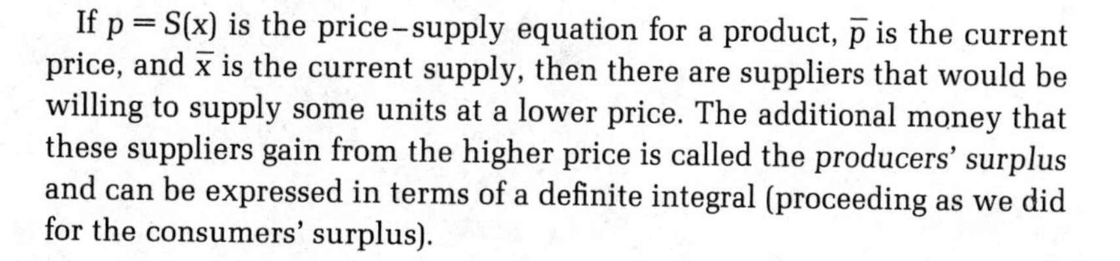

If $p = S(x)$ is the price–supply equation for a product, $\bar{p}$ is the current price, and $\bar{x}$ is the current supply, then there are suppliers that would be willing to supply some units at a lower price. The additional money that these suppliers gain from the higher price is called the *producers' surplus* and can be expressed in terms of a definite integral (proceeding as we did for the consumers' surplus).

Producers' Surplus

If $(\bar{x}, \bar{p})$ is a point on the graph of the price–supply equation $p = S(x)$, then the **producers' surplus,** PS, at a price level of $\bar{p}$ is

$$PS = \int_0^{\bar{x}} [\bar{p} - S(x)]\, dx$$

which is the area between $p = \bar{p}$ and $p = S(x)$ from $x = 0$ to $x = \bar{x}$.

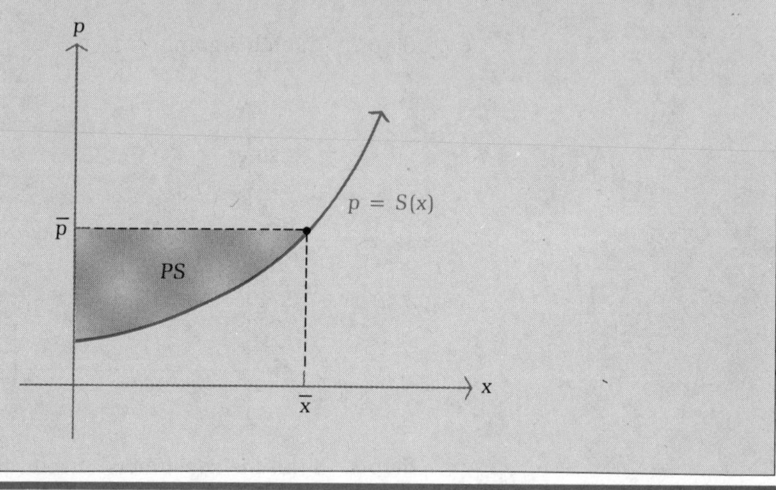

Example 9 Find the producers' surplus at a price level of \$20 for the price–supply equation

$$p = S(x) = 2 + \frac{1}{5{,}000}x^2$$

Solution Step 1. Find $\bar{x}$, the supply when the price is $\bar{p} = 20$:

$$\bar{p} = 2 + \frac{1}{5{,}000}\bar{x}^2$$

$$20 = 2 + \frac{1}{5{,}000}\bar{x}^2$$

$$\frac{1}{5{,}000}\bar{x}^2 = 18$$

$$\bar{x}^2 = 90{,}000$$

$$\bar{x} = 300 \qquad \text{There is only one solution since } \bar{x} \geq 0.$$

Step 2. Sketch a graph:

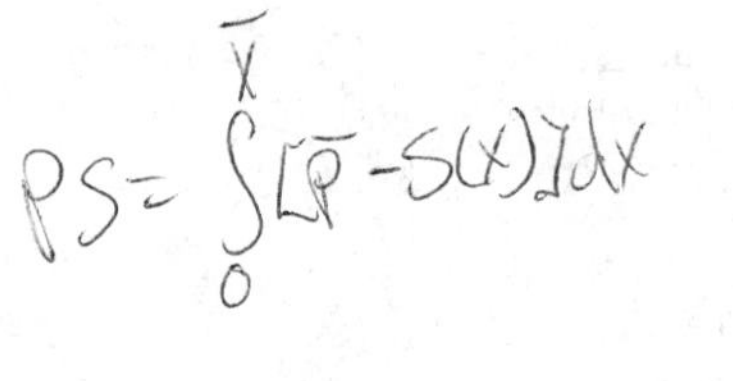

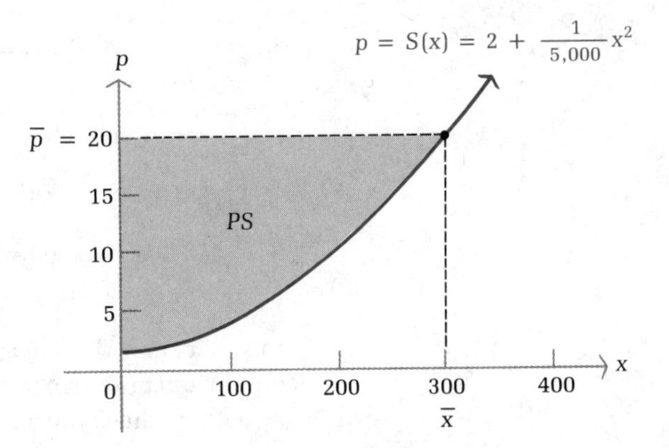

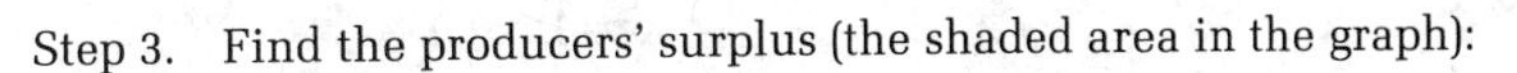

Step 3. Find the producers' surplus (the shaded area in the graph):

$$PS = \int_0^{\bar{x}} [\bar{p} - S(x)]\, dx$$

$$= \int_0^{300} \left[20 - \left(2 + \frac{1}{5,000}x^2\right)\right] dx$$

$$= \int_0^{300} \left(18 - \frac{1}{5,000}x^2\right) dx$$

$$= \left(18x - \frac{1}{15,000}x^3\right)\Big|_0^{300}$$

$$= 5,400 - 1,800 = \$3,600$$

Problem 9 Repeat Example 9 for a price level of \$4.

In a competitive market, the price of a product is determined by the relationship between supply and demand. If $p = D(x)$ and $p = S(x)$ are the price–demand and price–supply equations, respectively, for a product, then the price at which supply equals demand $[S(x) = D(x)]$ is called the **equilibrium price.** If the price stabilizes at the equilibrium price, then both consumers and producers benefit.

Example 10 Find the equilibrium price and then find the consumers' surplus and the producers' surplus at the equilibrium price level if

$$p = D(x) = 20 - \frac{1}{20}x \quad \text{and} \quad p = S(x) = 2 + \frac{1}{5,000}x^2$$

Solution Step 1. Find the equilibrium point. Set $D(x)$ equal to $S(x)$ and solve:

$$D(x) = S(x)$$

$$20 - \frac{1}{20}x = 2 + \frac{1}{5{,}000}x^2$$

$$\frac{1}{5{,}000}x^2 + \frac{1}{20}x - 18 = 0$$

$$x^2 + 250x - 90{,}000 = 0$$

$$x = 200, -450$$

Since x cannot be negative, the only solution is $x = 200$. The equilibrium price can be determined by using $D(x)$ or $S(x)$. We will use both to check our work:

$$\bar{p} = D(200) \qquad\qquad \bar{p} = S(200)$$

$$= 20 - \frac{1}{20}(200) = 10 \qquad\qquad = 2 + \frac{1}{5{,}000}(200)^2 = 10$$

Thus, the equilibrium price is $\bar{p} = 10$, and the corresponding demand and supply is $\bar{x} = 200$.

Step 2. Sketch a graph:

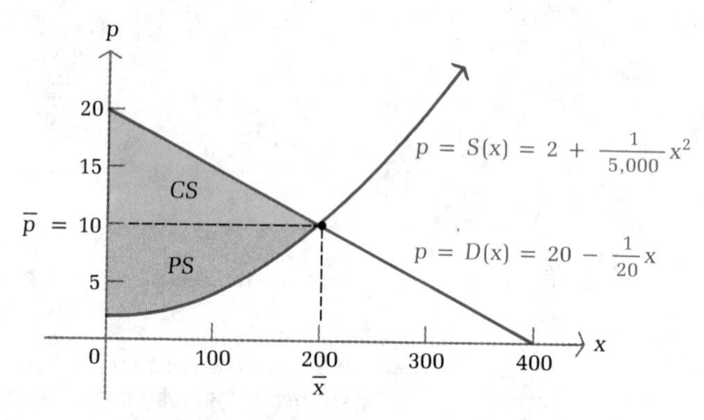

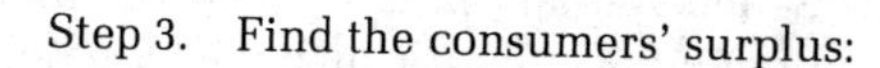
Step 3. Find the consumers' surplus:

$$CS = \int_0^{\bar{x}} [D(x) - \bar{p}]\, dx$$

$$= \int_0^{200} \left(20 - \frac{1}{20}x - 10\right) dx$$

$$= \int_0^{200} \left(10 - \frac{1}{20}x\right) dx$$

$$= \left(10x - \frac{1}{40}x^2\right)\Big|_0^{200}$$

$$= 2{,}000 - 1{,}000 = \$1{,}000$$

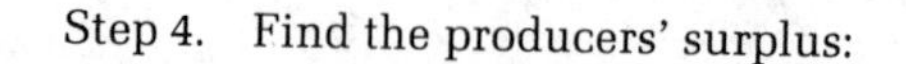
Step 4. Find the producers' surplus:

$$
\begin{aligned}
PS &= \int_0^{\bar{x}} [\bar{p} - S(x)]\, dx \\
&= \int_0^{200} \left[10 - \left(2 + \frac{1}{5,000}x^2\right)\right] dx \\
&= \int_0^{200} \left(8 - \frac{1}{5,000}x^2\right) dx \\
&= \left(8x - \frac{1}{15,000}x^3\right)\Big|_0^{200} \\
&= 1,600 - \frac{1,600}{3} \approx \$1,067 \qquad \text{Rounded to the nearest dollar}
\end{aligned}
$$

Problem 10 Repeat Example 10 for

$$p = D(x) = 25 - \frac{1}{1,000}x^2 \quad \text{and} \quad p = S(x) = 5 + \frac{1}{10}x$$

Answers to Matched Problems 6. \$6,761 7. (A) \$4,424 (B) \$6,938 (C) \$1,403 8. \$2,560 9. \$133 10. $\bar{p} = 15$; $CS = \$667$; $PS = \$500$

Exercise 5-2

Applications

Business & Economics

1. Find the total income produced by a continuous income stream in the first 5 years if the rate of flow is $f(t) = 2,500$.
2. Repeat Problem 1 if the rate of flow is $f(t) = 3,000$.
3. Find the total income produced by a continuous income stream in the first 3 years if the rate of flow is $f(t) = 400e^{0.05t}$.
4. Repeat Problem 3 if the rate of flow is $f(t) = 600e^{-0.06t}$.
5. Find the present value at 8% interest compounded continuously for 10 years for the continuous income stream with rate of flow $f(t) = 800$.
6. Repeat Problem 5 if the rate of flow is $f(t) = 1,200$.
7. Find the present value at 10% interest compounded continuously for 4 years for the continuous income stream with rate of flow $f(t) = 1,500e^{-0.02t}$.
8. Repeat Problem 7 if the rate of flow is $f(t) = 2,000e^{0.06t}$.
9. Find the consumers' surplus at a price level of $\bar{p} = 150$ for the price–demand equation $p = D(x) = 400 - \frac{1}{20}x$.
10. Find the consumers' surplus at a price level of $\bar{p} = 120$ for the price–demand equation $p = D(x) = 200 - \frac{1}{50}x$.

11. Find the producers' surplus at a price level of $\overline{p} = 65$ for the price–supply equation

$$p = S(x) = 10 + \frac{1}{10}x + \frac{1}{3,600}x^2$$

12. Find the producers' surplus at a price level of $\overline{p} = 55$ for the price–supply equation

$$p = S(x) = 15 + \frac{1}{10}x + \frac{3}{1,000}x^2$$

13. Find the equilibrium price, and then find the consumers' surplus and the producers' surplus for

$$p = D(x) = 50 - \frac{1}{10}x \quad \text{and} \quad p = S(x) = 11 + \frac{1}{20}x$$

14. Repeat Problem 13 for

$$p = D(x) = 20 - \frac{1}{600}x^2 \quad \text{and} \quad p = S(x) = 2 + \frac{1}{300}x^2$$

15. Find the equilibrium price (rounded to the nearest dollar), and then find the consumers' surplus and the producers' surplus for

$$p = D(x) = 80e^{-0.001x} \quad \text{and} \quad p = S(x) = 30e^{0.001x}$$

16. Repeat Problem 15 for

$$p = D(x) = 185e^{-0.005x} \quad \text{and} \quad p = S(x) = 25e^{0.005x}$$

17. Find the amount (income plus interest) produced by a continuous income stream at 14% compounded continuously for 2 years if the rate of flow is $f(t) = 1,000$.
18. Repeat Problem 17 if the rate of flow is $f(t) = 1,000e^{0.04t}$.
19. Find the interest earned at 12% compounded continuously for 3 years by a continuous income stream with rate of flow $f(t) = 2,000e^{0.07t}$.
20. Repeat Problem 19 if the rate of flow is $f(t) = 2,000$.
21. An investor is presented with a choice of two investments, an established clothing store and a new computer store. Each choice requires the same initial investment and each produces a continuous income stream at 10% compounded continuously. The rate of flow of income from the clothing store is $f(t) = 12,000$ and the rate of flow of income from the computer store is $g(t) = 10,000e^{0.05t}$. Compare the present values of these investments to determine which is the better choice over the next 5 years.
22. Refer to Problem 21. Which investment is the better choice over the next 10 years?

Problems 23 and 24 refer to a continuous income stream with rate of flow $f(t) = k$, where k is a constant.

23. Find the present value at 100r% compounded continuously for T years.

24. Find the amount after T years at 100r% compounded continuously.

5-3 Integration by Parts

In Section 4-1 we said that we would return to the indefinite integral

$$\int \ln x \, dx$$

later, since none of the integration techniques considered up to that time could be used to find an antiderivative for ln x. We will now develop a very useful technique, called *integration by parts,* that will not only enable us to find the above integral, but also many others, including integrals such as

$$\int x \ln x \, dx \qquad \text{and} \qquad \int xe^x \, dx$$

The integration by parts technique is also used to derive many integration formulas that are tabulated in mathematical handbooks.

The method of integration by parts is based on the product formula for derivatives. If f and g are differentiable functions, then

$$D_x[f(x)g(x)] = f(x)g'(x) + g(x)f'(x)$$

which can be written in the equivalent form

$$f(x)g'(x) = D_x[f(x)g(x)] - g(x)f'(x)$$

Integrating both sides, we obtain

$$\int f(x)g'(x) \, dx = \int D_x[f(x)g(x)] \, dx - \int g(x)f'(x) \, dx$$

The first integral to the right of the equal sign is $f(x)g(x) + C$. (Why?) We will leave out the constant of integration for now, since we can add it after integrating the second integral to the right of the equal sign. So we have

$$\int f(x)g'(x) \, dx = f(x)g(x) - \int g(x)f'(x) \, dx$$

This last form can be transformed into a more convenient form by letting $u = f(x)$ and $v = g(x)$; then $du = f'(x) \, dx$ and $dv = g'(x) \, dx$. Making these substitutions, we obtain the **integration by parts formula:**

Integration by Parts Formula

$$\int u\,dv = uv - \int v\,du$$

This formula can be very useful when the integral on the left is difficult to integrate using standard formulas. If u and dv are chosen with care, then the integral on the right side may be easier to integrate than the one on the left. Several examples will demonstrate the use of the formula.

Example 11 Find $\int x \ln x\, dx$, $x > 0$, using integration by parts.

Solution First, write the integration by parts formula

$$\int u\,dv = uv - \int v\,du$$

Then try to identify u and dv in $\int x \ln x\, dx$ (this is the key step) so that when $\int u\,dv$ is written in the form $uv - \int v\,du$, the new integral will be easier to integrate.

Suppose we choose

$$u = x \qquad \text{and} \qquad dv = \ln x\, dx$$

Then

$$du = dx \qquad v = ?$$

We do not know an antiderivative of ln x yet, so we change our choice for u and dv to

$$u = \ln x \qquad dv = x\, dx$$

Then

$$du = \frac{1}{x}\, dx \qquad v = \frac{x^2}{2}$$

Any constant may be added to v (we choose 0 for simplicity). There are cases where it is convenient to add a constant other than 0, but in most cases 0 will do. The general arbitrary constant of integration will be added at the end of the process.

Using the chosen u, du, dv, and v in the integration by parts formula, we obtain

$$\int u \quad dv = u \quad v - \int v \quad du$$

$$\int (\ln x)x\,dx = (\ln x)\left(\frac{x^2}{2}\right) - \int \left(\frac{x^2}{2}\right)\frac{1}{x}\,dx$$

$$= \frac{x^2}{2}\ln x - \int \frac{x}{2}\,dx$$

This new integral is easy to integrate.

$$= \frac{x^2}{2}\ln x - \frac{x^2}{4} + C$$

To check this result, show that

$$D_x\left(\frac{x^2}{2}\ln x - \frac{x^2}{4} + C\right) = x \ln x$$

which is the integrand in the original integral.

Problem 11 Find $\int x \ln 2x\,dx$.

Example 12 Find $\int xe^x\,dx$.

Solution We write the integration by parts formula

$$\int u\,dv = uv - \int v\,du$$

and choose

$$u = e^x \qquad dv = x\,dx$$

Then

$$du = e^x\,dx \qquad v = \frac{x^2}{2}$$

and

$$\int u \quad dv = u \quad v - \int v \quad du$$

$$\int e^x x\,dx = e^x\left(\frac{x^2}{2}\right) - \int \left(\frac{x^2}{2}\right)e^x\,dx$$

$$= \frac{x^2}{2}e^x - \frac{1}{2}\int x^2e^x\,dx$$

This new integral is more complicated than the original one.

This time the integration by parts formula leads to a new integral that is more complicated than the one we started with. This does not mean that there is an error in our calculations or in the formula. It simply means that our first choice for u and dv did not change the original problem into one that we can solve. Thus, we must make a different selection. Suppose we choose

$$u = x \qquad dv = e^x\, dx$$

Then

$$du = dx \qquad v = e^x$$

and

$$\int u\, dv = u\, v - \int v\, du$$

$$\int xe^x\, dx = xe^x - \int e^x\, dx \qquad \text{This integral is one we can evaluate.}$$

$$= xe^x - e^x + C$$

Check this result by differentiation.

Problem 12 Find $\int xe^{2x}\, dx$.

Integration by Parts: Selection of *u* and *dv*

1. It must be possible to integrate dv (preferably by using standard formulas or simple substitutions).
2. The new integral, $\int v\, du$, should be simpler than the original integral, $\int u\, dv$.
3. For integrals involving $x^p(\ln x)^q$, try

   $$u = (\ln x)^q \qquad dv = x^p\, dx$$

4. For integrals involving $x^p e^{ax}$, try

   $$u = x^p \qquad dv = e^{ax}\, dx$$

5. For integrals involving $x^p(ax + b)^q$, try

   $$u = x^p \qquad dv = (ax + b)^q\, dx$$

Example 13 Find $\displaystyle\int x(x + 5)^9\, dx$.

Solution Following suggestion 5 in the box, we choose

$$u = x \qquad dv = (x + 5)^9\, dx$$

Then

$$du = dx \qquad v = \frac{(x + 5)^{10}}{10}$$

and

$$\int x(x+5)^9\,dx = x\frac{(x+5)^{10}}{10} - \int \frac{(x+5)^{10}}{10}\,dx$$

$$= \frac{1}{10}x(x+5)^{10} - \frac{1}{110}(x+5)^{11} + C$$

Problem 13 Find $\int x(x-4)^8\,dx$.

Example 14 Find $\int x^2e^{-x}\,dx$.

Solution Following suggestion 4 in the box, we choose

$$u = x^2 \qquad dv = e^{-x}\,dx$$

Then

$$du = 2x\,dx \qquad v = -e^{-x}$$

and

$$\int x^2e^{-x}\,dx = x^2(-e^{-x}) - \int (-e^{-x})2x\,dx$$

$$= -x^2e^{-x} + 2\int xe^{-x}\,dx \tag{1}$$

The new integral is not one we can evaluate by standard formulas, but it is simpler than the original integral. Applying the integration by parts formula to it will produce an even simpler integral. For the integral $\int xe^{-x}\,dx$, we choose

$$u = x \qquad dv = e^{-x}\,dx$$

Then

$$du = dx \qquad v = -e^{-x}$$

and

$$\int xe^{-x}\,dx = x(-e^{-x}) - \int (-e^{-x})\,dx$$

$$= -xe^{-x} + \int e^{-x}\,dx$$

$$= -xe^{-x} - e^{-x} \tag{2}$$

Substituting (2) into (1) and adding a constant of integration, we have

$$\int x^2e^{-x}\,dx = -x^2e^{-x} + 2(-xe^{-x} - e^{-x}) + C$$

$$= -x^2e^{-x} - 2xe^{-x} - 2e^{-x} + C$$

Problem 14 Find $\int x^2 e^{2x}\,dx$.

Example 15 Find $\int_1^e \ln x\,dx$.

Solution First, find $\int \ln x\,dx$; then return to the definite integral. Following suggestion 3 in the box (with $p = 0$), we choose

$$u = \ln x \qquad dv = dx$$

Then

$$du = \frac{1}{x}\,dx \qquad v = x$$

Hence,

$$\int \ln x\,dx = (\ln x)(x) - \int (x)\frac{1}{x}\,dx$$
$$= x \ln x - x + C$$

Thus,

$$\int_1^e \ln x\,dx = (x \ln x - x)\Big|_1^e$$
$$= (e \ln e - e) - (1 \ln 1 - 1)$$
$$= (e - e) - (0 - 1)$$
$$= 1$$

Problem 15 Find $\int_1^2 \ln 3x\,dx$.

Answers to Matched Problems

11. $\frac{x^2}{2} \ln 2x - \frac{x^2}{4} + C$ 12. $\frac{x}{2} e^{2x} - \frac{1}{4} e^{2x} + C$

13. $\frac{1}{9} x(x-4)^9 - \frac{1}{90}(x-4)^{10} + C$ 14. $\frac{x^2}{2} e^{2x} - \frac{x}{2} e^{2x} + \frac{1}{4} e^{2x} + C$

15. $2 \ln 6 - \ln 3 - 1 \approx 1.4849$

Exercise 5-3

A *Integrate using integration by parts. Assume $x > 0$ whenever the natural log function is involved.*

1. $\int xe^{3x}\,dx$
2. $\int xe^{4x}\,dx$
3. $\int x^2 \ln x\,dx$
4. $\int x^3 \ln x\,dx$
5. $\int x(x-6)^7\,dx$
6. $\int x(x+9)^6\,dx$

B *Problems 7–24 are mixed—some require integration by parts and others can be solved using techniques we have considered earlier. Integrate as indicated, assuming $x > 0$ whenever the natural log function is involved.*

7. $\int xe^{-x}\,dx$
8. $\int (x-1)e^{-x}\,dx$
9. $\int xe^{x^2}\,dx$
10. $\int xe^{-x^2}\,dx$
11. $\int x(2x-1)^5\,dx$
12. $\int (2x-1)^5\,dx$
13. $\int (3x+2)^6\,dx$
14. $\int x(3x+2)^6\,dx$
15. $\int_0^1 (x-3)e^x\,dx$
16. $\int_0^2 (x+5)e^x\,dx$
17. $\int_1^3 \ln 2x\,dx$
18. $\int_2^3 \ln 7x\,dx$
19. $\int \frac{2x}{x^2+1}\,dx$
20. $\int \frac{x^2}{x^3+5}\,dx$
21. $\int \frac{\ln x}{x}\,dx$
22. $\int \frac{e^x}{e^x+1}\,dx$
23. $\int \sqrt{x}\ln x\,dx$
24. $\int \frac{\ln x}{\sqrt{x}}\,dx$

C *Some of these problems may require using the integration by parts formula more than once. Assume $x > 0$ whenever the natural log function is involved.*

25. $\int x^2e^x\,dx$
26. $\int x^3e^x\,dx$
27. $\int xe^{ax}\,dx, \quad a \neq 0$
28. $\int \ln(ax)\,dx, \quad a > 0$
29. $\int_1^e \frac{\ln x}{x^2}\,dx$
30. $\int_1^2 x^3e^{x^2}\,dx$
31. $\int x^2(x+1)^7\,dx$
32. $\int x(x-7)^{-4}\,dx$
33. $\int x(x+6)^{-3}\,dx$
34. $\int x^2(x+2)^6\,dx$
35. $\int (\ln x)^2\,dx$
36. $\int x(\ln x)^2\,dx$
37. $\int (\ln x)^3\,dx$
38. $\int x(\ln x)^3\,dx$
39. $\int x\sqrt{x+4}\,dx$
40. $\int \frac{x}{\sqrt{x-2}}\,dx$

Find the area bounded by the graphs of the indicated equations.

41. $y = x - \ln x, \quad y = 0, \quad x = 1, \quad x = e$
42. $y = 1 - \ln x, \quad y = 0, \quad x = 1, \quad x = 4$
43. $y = (x - 2)e^x, \quad y = 0, \quad x = 0, \quad x = 3$
44. $y = (3 - x)e^x, \quad y = 0, \quad x = 0, \quad x = 3$

Applications

Business & Economics

45. *Profit.* If the rate of change of profit in millions of dollars per year is given by

$$P'(t) = 2t - te^{-t}$$

where t is time in years and the profit at time 0 is 0, find $P = P(t)$.

46. *Production.* An oil field is estimated to produce oil at a rate of $R(t)$ thousand barrels per month t months from now, as given by

$$R(t) = 10te^{-0.1t}$$

Estimate the total production in the first year of operation by use of an appropriate definite integral.

47. *Continuous income stream.* Find the present value at 8% compounded continuously for 5 years for the continuous income stream with rate of flow

$$f(t) = 1{,}000 - 200t$$

48. *Continuous income stream.* Find the amount (income plus interest) at 10% compounded continuously for 4 years for the continuous income stream with rate of flow

$$f(t) = 4{,}000 - 250t^2$$

49. *Income distribution.* Find the coefficient of inequality for the Lorenz curve with equation

$$y = \frac{1}{16}x(x + 1)^4$$

50. *Producers' surplus.* Find the producers' surplus at a price level of $\overline{p} = 26$ for the price–supply equation

$$p = S(x) = 5 \ln(x + 1)$$

[*Hint:* To find an antiderivative of $\ln(x + 1)$, first substitute $t = x + 1$, then use integration by parts.]

Life Sciences

51. *Pollution.* The concentration of particulate matter in parts per million t hours after a factory ceases operation for the day is given by

$$C(t) = \frac{20 \ln(t + 1)}{(t + 1)^2}$$

Find the average concentration for the time period from $t = 0$ to $t = 5$.

52. *Medicine.* After a person takes a pill, the drug contained in the pill is assimilated into the bloodstream. The rate of assimilation t minutes after taking the pill is

$$R(t) = te^{-0.2t}$$

Find the total amount of the drug that is assimilated into the bloodstream during the first 10 minutes after the pill is taken.

Social Sciences

53. *Politics.* The number of voters (in thousands) in a certain city is given by

$$N(t) = 20 + 4t - 5te^{-0.1t}$$

where t is the time in years. Find the average number of voters during the time period from $t = 0$ to $t = 5$.

5-4 Improper Integrals

- Improper Integrals
- Application: Capital Value
- Probability Density Functions

Improper Integrals

We are now going to consider an integral form that has wide application in probability studies as well as other areas. Earlier, when we introduced the idea of a definite integral,

$$\int_a^b f(x)\, dx \qquad (1)$$

we required f to be continuous over a closed interval $[a, b]$. Now we are going to extend the meaning of (1) so that the interval $[a, b]$ may become infinite in length.

Let us investigate a particular example that will motivate several general definitions. What would be a reasonable interpretation for the following expression?

$$\int_1^\infty \frac{dx}{x^2}$$

Sketching a graph of $f(x) = 1/x^2$, $x \geq 1$ (see Fig. 9 on the next page), we note that for any fixed $b > 1$, $\int_1^b f(x)\, dx$ is the area between the curve $y = 1/x^2$, the x axis, $x = 1$, and $x = b$.

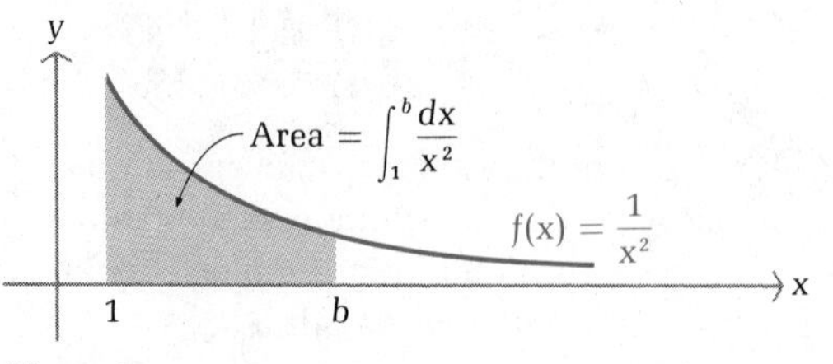

Figure 9

Let us see what happens when we let $b \to \infty$; that is, when we compute the following limit:

$$\lim_{b\to\infty} \int_1^b \frac{dx}{x^2} = \lim_{b\to\infty} \left[(-x^{-1}) \Big|_1^b \right]$$

$$= \lim_{b\to\infty} \left(-\frac{1}{b} + 1 \right) = 1$$

Did you expect this result? No matter how large b is taken, the area under the curve from $x = 1$ to $x = b$ never exceeds 1, and in the limit it is 1. This suggests that we write

$$\int_1^\infty \frac{dx}{x^2} = \lim_{b\to\infty} \int_1^b \frac{dx}{x^2} = 1$$

This integral is an example of an *improper integral*. In general, the forms

$$\int_{-\infty}^b f(x)\,dx \qquad \int_a^\infty f(x)\,dx \qquad \int_{-\infty}^\infty f(x)\,dx$$

where f is continuous over the indicated interval, are called **improper integrals.** (There are also other types of improper integrals that will not be considered here. These involve certain types of points of discontinuity within the interval of integration.) Each type of improper integral above is formally defined in the box:

Improper Integrals

If f is continuous over the indicated interval and the limit exists, then:

1. $\displaystyle\int_a^\infty f(x)\,dx = \lim_{b\to\infty} \int_a^b f(x)\,dx$
2. $\displaystyle\int_{-\infty}^b f(x)\,dx = \lim_{a\to-\infty} \int_a^b f(x)\,dx$
3. $\displaystyle\int_{-\infty}^\infty f(x)\,dx = \int_{-\infty}^c f(x)\,dx + \int_c^\infty f(x)\,dx$

   where c is any point on $(-\infty, \infty)$, provided *both* improper integrals on the right exist.

If the indicated limit exists, then the improper integral is said to exist or **converge;** if the limit does not exist, then the improper integral is said not to exist or to **diverge** (and no value is assigned to it).

Example 16 Evaluate $\int_2^\infty dx/x$ if it converges.

Solution

$$\int_2^\infty \frac{dx}{x} = \lim_{b\to\infty} \int_2^b \frac{dx}{x}$$

$$= \lim_{b\to\infty} \left[(\ln x)\Big|_2^b \right]$$

$$= \lim_{b\to\infty} (\ln b - \ln 2)$$

Since $\ln b \to \infty$ as $b \to \infty$, the limit does not exist. Hence, the improper integral diverges.

Problem 16 Evaluate $\int_3^\infty dx/(x-1)^2$ if it converges.

Example 17 Evaluate $\int_{-\infty}^2 e^x\, dx$ if it converges.

Solution

$$\int_{-\infty}^2 e^x\, dx = \lim_{a\to-\infty} \int_a^2 e^x\, dx$$

$$= \lim_{a\to-\infty} (e^x|_a^2)$$

$$= \lim_{a\to-\infty} (e^2 - e^a) = e^2 - 0 = e^2 \qquad \text{The integral converges.}$$

Problem 17 Evaluate $\int_{-\infty}^{-1} x^{-2}\, dx$ if it converges.

Example 18 Evaluate

$$\int_{-\infty}^\infty \frac{2x}{(1+x^2)^2}\, dx$$

if it converges.

Solution

$$\int_{-\infty}^\infty \frac{2x}{(1+x^2)^2}\, dx = \int_{-\infty}^0 (1+x^2)^{-2}2x\, dx + \int_0^\infty (1+x^2)^{-2}2x\, dx$$

$$= \lim_{a\to-\infty} \int_a^0 (1+x^2)^{-2}2x\, dx + \lim_{b\to\infty} \int_0^b (1+x^2)^{-2}2x\, dx$$

$$= \lim_{a\to-\infty} \left[\frac{(1+x^2)^{-1}}{-1}\Big|_a^0 \right] + \lim_{b\to\infty} \left[\frac{(1+x^2)^{-1}}{-1}\Big|_0^b \right]$$

$$= \lim_{a\to-\infty} \left[-1 + \frac{1}{1+a^2} \right] + \lim_{b\to\infty} \left[-\frac{1}{1+b^2} + 1 \right]$$

$$= -1 + 1 = 0 \qquad \text{The integral converges.}$$

Problem 18 Evaluate $\int_{-\infty}^\infty dx/e^x$ if it converges.

Example 19 Oil Production

It is estimated that an oil well will produce oil at a rate of $R(t)$ thousand barrels per month t months from now, as given by

$$R(t) = 50e^{-0.05t} - 50e^{-0.1t}$$

Estimate the total amount of oil produced by this well.

Solution The total amount of oil produced in T months of operation is

$$\int_0^T R(t)\,dt$$

At some point in time, the monthly production rate will become so low that it will no longer be economically feasible to operate the well. However, for the purpose of estimating the total production, it is convenient to assume that the well is operated indefinitely. Thus, the total amount of oil produced is

$$\begin{aligned}\int_0^\infty R(t)\,dt &= \lim_{T\to\infty}\int_0^T R(t)\,dt\\ &= \lim_{T\to\infty}\int_0^T (50e^{-0.05t} - 50e^{-0.1t})\,dt\\ &= \lim_{T\to\infty}\left[(-1{,}000e^{-0.05t} + 500e^{-0.1t})\Big|_0^T\right]\\ &= \lim_{T\to\infty}(-1{,}000e^{-0.05T} + 500e^{-0.1T} + 500)\\ &= 500 \text{ thousand barrels}\end{aligned}$$

Problem 19 Find the total amount of oil produced by a well whose monthly production rate (in thousands of barrels) is given by

$$R(t) = 100e^{-0.1t} - 25e^{-0.2t}$$

▪ Application: Capital Value

If $f(t)$ is the rate of flow of a continuous income stream at an annual nominal rate of $100r\%$, compounded continuously, then the **capital value** of the income stream is the present value over the time interval $[0, \infty)$. That is,

$$\text{Capital value} = \int_0^\infty f(t)e^{-rt}\,dt$$

Example 20 A family has leased the oil rights of a property to a petroleum company in return for a perpetual annual payment of \$1,200. Find the capital value of this lease at 10% compounded continuously.

Solution The annual payments from the oil company produce a continuous income stream with rate of flow $f(t) = 1{,}200$ that continues indefinitely. (It is common practice to treat a sequence of equal periodic payments as a continuous income stream with a constant rate of flow, even if the income is received only at the end of each period.) Thus, the capital value is

$$\begin{aligned}\int_0^\infty f(t)e^{-rt}\,dt &= \int_0^\infty 1{,}200e^{-0.1t}\,dt\\ &= \lim_{T\to\infty}\int_0^T 1{,}200e^{-0.1t}\,dt\\ &= \lim_{T\to\infty}\left[-12{,}000e^{-0.1t}\Big|_0^T\right]\\ &= \lim_{T\to\infty}(-12{,}000e^{-0.1T}+12{,}000) = \$12{,}000\end{aligned}$$

Problem 20 Repeat Example 20 if the interest rate is 8%, compounded continuously.

■ Probability Density Functions

We will now take a brief look at the use of improper integrals relative to probability density functions. The approach will be intuitive and informal.

Suppose an experiment is designed in such a way that any real number x on the interval $[a, b]$ is a possible outcome. For example, x may represent an IQ score, the height of a person in inches, or the life of a light bulb in hours.

In certain situations it is possible to find a function f with x as an independent variable that can be used to determine the probability that x will assume a value on a given subinterval of $(-\infty, \infty)$. Such a function, called a **probability density function,** must satisfy the following three conditions (see Fig. 10):

1. $f(x) \geq 0$ for all $x \in (-\infty, \infty)$
2. $\int_{-\infty}^{\infty} f(x)\,dx = 1$
3. If $[c, d]$ is a subinterval of $(-\infty, \infty)$, then

$$\text{Probability}(c \leq x \leq d) = \int_c^d f(x)\,dx$$

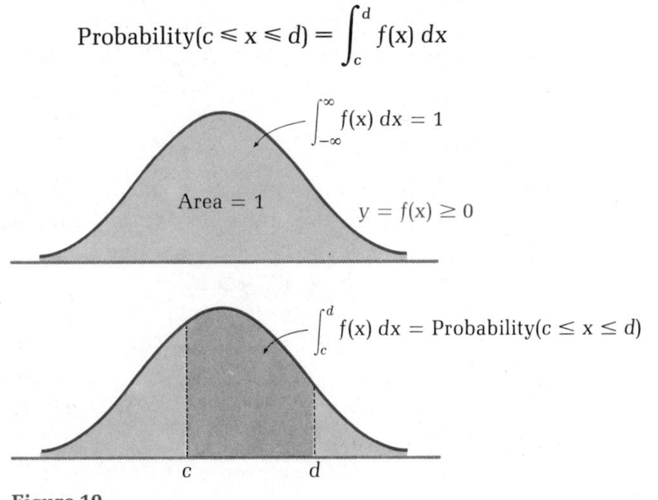

Figure 10

Example 21
Finish Time

A sailing club has a race over the same course twice a month. The races always start at 12 noon on Sunday, and the boats finish according to the probability density function (where x is hours after noon):

$$f(x) = \begin{cases} -\dfrac{x}{2} + 2 & 2 \leqslant x \leqslant 4 \\ 0 & \text{otherwise} \end{cases}$$

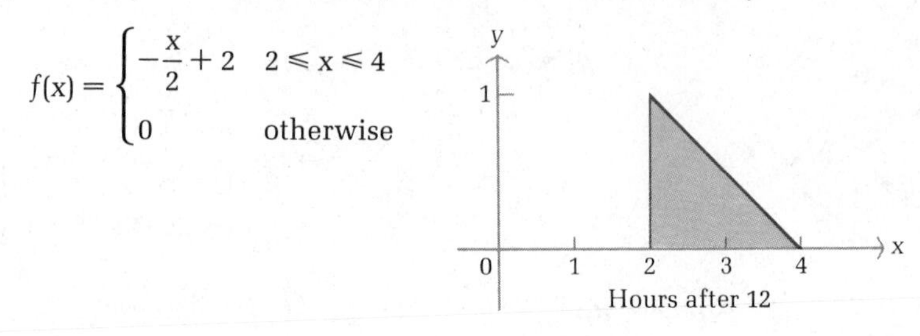

Note that

$$f(x) \geqslant 0$$

and

$$\int_{-\infty}^{\infty} f(x)\,dx = \int_2^4 \left(-\frac{x}{2} + 2\right) dx = \left(-\frac{x^2}{4} + 2x\right)\Big|_2^4 = 1$$

The probability that a boat selected at random from the sailing fleet will finish between 2 and 3 hours after the start is given by

$$\begin{aligned} \text{Probability}(2 \leqslant x \leqslant 3) &= \int_2^3 \left(-\frac{x}{2} + 2\right) dx \\ &= \left(-\frac{x^2}{4} + 2x\right)\Big|_2^3 = .75 \end{aligned}$$

which is the area under the curve from $x = 2$ to $x = 3$.

Problem 21 In Example 21, find the probability that a boat selected at random from the fleet will finish between 2:30 and 3:30 PM.

Example 22
Duration of
Telephone Calls

Suppose the length of telephone calls (in minutes) in a public telephone booth has the probability density function

$$f(t) = \begin{cases} \dfrac{1}{4} e^{-t/4} & t \geqslant 0 \\ 0 & \text{otherwise} \end{cases}$$

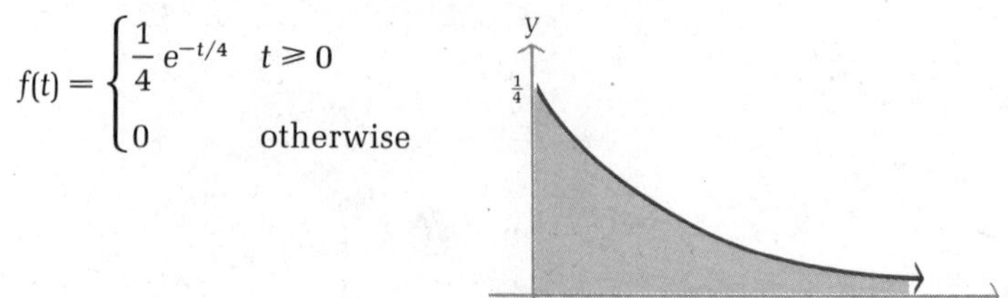

(A) Compute $\int_{-\infty}^{\infty} f(t)\,dt$.
(B) Determine the probability that a call selected at random will last between 2 and 3 minutes.

Solutions (A)
$$\int_{-\infty}^{\infty} f(t)\,dt = \int_{-\infty}^{0} f(t)\,dt + \int_{0}^{\infty} f(t)\,dt$$
$$= 0 + \int_{0}^{\infty} \frac{1}{4} e^{-t/4}\,dt$$
$$= \lim_{b\to\infty} \int_{0}^{b} \frac{1}{4} e^{-t/4}\,dt$$
$$= \lim_{b\to\infty} \left(-e^{-t/4}\Big|_{0}^{b}\right)$$
$$= \lim_{b\to\infty} (-e^{-b/4} + e^{0})$$
$$= \lim_{b\to\infty} \left(-\frac{1}{e^{b/4}} + 1\right)$$
$$= 0 + 1 = 1$$

(B)
$$\text{Probability}(2 \leq t \leq 3) = \int_{2}^{3} \frac{1}{4} e^{-t/4}\,dt$$
$$= (-e^{-t/4})\Big|_{2}^{3}$$
$$= -e^{-3/4} + e^{-1/2} \approx .13$$

Problem 22 In Example 22, find the probability that a call selected at random will last longer than 4 minutes.

The most important probability density function is the **normal probability density function** defined below and graphed in Figure 11.

$$f(x) = \frac{1}{\sigma\sqrt{2\pi}} e^{-(x-\mu)^2/2\sigma^2}$$

μ is the mean
σ is the standard deviation

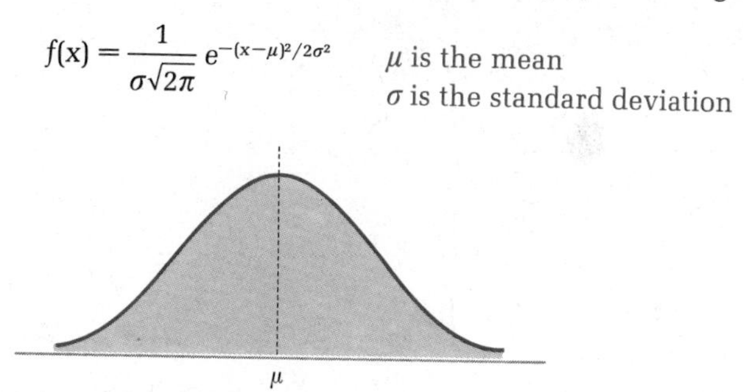

Figure 11 Normal curve

It can be shown, but not easily, that

$$\frac{1}{\sigma\sqrt{2\pi}} \int_{-\infty}^{\infty} e^{-(x-\mu)^2/2\sigma^2}\,dx = 1$$

Since $\int e^{-x^2}$ is nonintegrable in terms of elementary functions (that is, the antiderivative cannot be expressed as a finite combination of simple functions), probabilities such as

$$\text{Probability}(c \leq x \leq d) = \frac{1}{\sigma\sqrt{2\pi}} \int_c^d e^{-(x-\mu)^2/2\sigma^2}\, dx$$

are generally determined by making an appropriate substitution in the integrand and then using a table of areas under the standard normal curve (that is, the normal curve with $\mu = 0$ and $\sigma = 1$). Such tables are readily available in most mathematical handbooks. A table can be constructed by using the rectangle rule discussed in Section 5-1; however, digital computers that use refined techniques are generally used for this purpose. Some hand calculators have the capability of computing normal curve areas directly.

Answers to Matched Problems

16. $\frac{1}{2}$ 17. 1 18. Diverges 19. 875 thousand barrels
20. \$15,000 21. .5 22. $e^{-1} \approx .37$

Exercise 5-4

Find the value of each improper integral that converges.

A

1. $\int_1^\infty \frac{dx}{x^4}$
2. $\int_1^\infty \frac{dx}{x^3}$
3. $\int_0^\infty e^{-x/2}\, dx$
4. $\int_0^\infty e^{-x}\, dx$

B

5. $\int_1^\infty \frac{dx}{\sqrt{x}}$
6. $\int_1^\infty \frac{dx}{\sqrt[3]{x}}$
7. $\int_0^\infty \frac{dx}{(x+1)^2}$
8. $\int_0^\infty \frac{dx}{(x+1)^3}$
9. $\int_0^\infty \frac{dx}{(x+1)^{2/3}}$
10. $\int_0^\infty \frac{dx}{\sqrt{x+1}}$
11. $\int_1^\infty \frac{dx}{x^{0.99}}$
12. $\int_1^\infty \frac{dx}{x^{1.01}}$
13. $0.3\int_0^\infty e^{-0.3x}\, dx$
14. $0.01\int_0^\infty e^{-0.1x}\, dx$
15. In Example 21, find the probability that a randomly selected boat will finish before 3:30 PM.
16. In Example 21, find the probability that a randomly selected boat will finish after 2:30 PM.

17. In Example 22, find the probability that a telephone call selected at random will last longer than 1 minute.
18. In Example 22, find the probability that a telephone call selected at random will last less than 3 minutes.

C

Find the value of each improper integral that converges. Note that $\lim_{x\to\infty} x^n e^{-x} = 0$ *and* $\lim_{x\to\infty} x^{-n} \ln x = 0$ *for all positive integers n.*

19. $\int_0^\infty \frac{1}{k} e^{-x/k}\, dx,\ k > 0$
20. $\int_0^\infty xe^{-x}\, dx$
21. $\int_{-\infty}^0 \frac{dx}{\sqrt{1-x}}$
22. $\int_{-\infty}^\infty xe^{-x^2}\, dx$
23. $\int_0^\infty (e^{-x} - e^{-2x})\, dx$
24. $\int_0^\infty x^2 e^{-x}\, dx$
25. $\int_1^\infty \frac{\ln x}{x}\, dx$
26. $\int_1^\infty \frac{\ln x}{x^2}\, dx$

Applications

Business & Economics

Note that $\lim_{t\to\infty} t^n e^{-kt} = 0$ *for* $n > 0$ *and* $k > 0$*. Limits of this form will occur in several of the following applications.*

27. *Capital value.* The perpetual annual rent for a property is $6,000. Find the capital value at 12% compounded continuously.
28. *Capital value.* The perpetual annual rent for a property is $10,000. Find the capital value at 5% compounded continuously.
29. *Capital value.* A trust fund produces a perpetual stream of income with rate of flow

    $$f(t) = 1{,}500e^{0.04t}$$

    Find the capital value at 9% compounded continuously.
30. *Capital value.* A trust fund produces a perpetual stream of income with rate of flow

    $$f(t) = 1{,}000t$$

    Find the capital value at 10% compounded continuously.
31. *Production.* The rate of production of a natural gas well in millions of cubic feet per month is given by

    $$R(t) = te^{-0.4t}$$

    Assuming that the well is operated indefinitely, find the total production.
32. *Production.* Repeat Problem 31 for $R(t) = t^2 e^{-0.4t}$.

33. *Consumption.* The daily per capita use of water (in hundreds of gallons) for domestic purposes has a probability density function of the form

$$g(x) = \begin{cases} .05e^{-.05x} & x \geqslant 0 \\ 0 & \text{otherwise} \end{cases}$$

Find the probability that a person chosen at random will use at least 300 gallons of water per day.

34. *Warranty.* A manufacturer guarantees a product for 1 year. The time for failure of a new product after it is sold is given by the probability density function

$$f(t) = \begin{cases} .01e^{-.01t} & t \geqslant 0 \\ 0 & \text{otherwise} \end{cases}$$

where t is time in months. What is the probability that a buyer chosen at random will have a product failure during the warranty period?

Life Sciences

35. *Pollution.* It has been estimated that the rate of seepage of toxic chemicals from a waste dump is $R(t)$ gallons per year t years from now, where

$$R(t) = \frac{500}{(1+t)^2}$$

Assuming that this seepage continues indefinitely, find the total amount of toxic chemicals that seep from the dump.

36. *Drug assimilation.* When a person takes a drug, the body does not assimilate all of the drug. One way to determine the amount of the drug that is assimilated is to measure the rate at which the drug is eliminated from the body. If the rate of elimination of the drug (in milliliters per minute) is given by

$$R(t) = te^{-0.2t}$$

where t is the time in minutes since the drug was administered, how much of the drug is eliminated from the body?

37. *Medicine.* If the length of stay for people in a hospital has a probability density function

$$g(t) = \begin{cases} .2e^{-.2t} & t \geqslant 0 \\ 0 & \text{otherwise} \end{cases}$$

where t is time in days, find the probability that a patient chosen at random will stay in the hospital less than 5 days.

38. *Medicine.* For a particular disease, the length of time in days for recovery has a probability density function of the form

$$R(t) = \begin{cases} .03e^{-.03t} & t \geqslant 0 \\ 0 & \text{otherwise} \end{cases}$$

For a randomly selected person who contracts this disease, what is the probability that he or she will take at least 7 days to recover?

Social Sciences

39. *Politics.* In a particular election, the length of time each voter spent on campaigning for a candidate or issue was found to have a probability density function

$$F(x) = \begin{cases} \dfrac{1}{(x+1)^2} & x \geq 0 \\ 0 & \text{otherwise} \end{cases}$$

where x is time in minutes. For a voter chosen at random, what is the probability of his or her spending at least 9 minutes on the campaign?

40. *Psychology.* In an experiment on conditioning, pigeons were required to recognize on a light display one pattern of dots out of five possible patterns to receive a food pellet. After the ninth successful trial, it was found that the probability density function for the length of time in seconds until success on the tenth trial is given by

$$f(t) = \begin{cases} e^{-t} & t \geq 0 \\ 0 & \text{otherwise} \end{cases}$$

What is the probability that a pigeon selected at random from those having successfully completed nine trials will take 2 or more seconds to complete the tenth trial successfully?

5-5 Chapter Review

Important Terms and Symbols

5-1 *Definite integral as a limit of a sum.* Rectangle rule, tabular function, definite integral (as a limit of a Riemann sum), fundamental theorem of calculus, average value of a continuous function

5-2 *Applications in business and economics.* Continuous income stream, rate of flow, total income, consumers' surplus, producers' surplus, equilibrium price

5-3 *Integration by parts.* $\int u\,dv = uv - \int v\,du$

5-4 *Improper integrals.* Improper integral, converge, diverge, capital value, probability density function, normal probability density function,
$\int_a^\infty f(x)\,dx = \lim_{b\to\infty} \int_a^b f(x)\,dx$, $\int_{-\infty}^b f(x)\,dx = \lim_{a\to-\infty} \int_a^b f(x)\,dx$,
$\int_{-\infty}^\infty f(x)\,dx = \int_{-\infty}^c f(x)\,dx + \int_c^\infty f(x)\,dx$

Exercise 5-5 Chapter Review

Work through all the problems in this chapter review and check your answers in the back of the book. (Answers to all review problems are there.) Where weaknesses show up, review appropriate sections in the text.

A *Evaluate the indicated integrals.*

1. $\int xe^{4x}\,dx$
2. $\int x \ln x\,dx$
3. $\int x(x+7)^5\,dx$
4. $\int_0^{\infty} e^{-2x}\,dx$
5. $\int_0^{\infty} \frac{1}{x+1}\,dx$
6. $\int_1^{\infty} \frac{16}{x^3}\,dx$
7. Approximate $\int_1^5 (x^2+1)\,dx$ using the rectangle rule with $n = 2$ and c_k as the midpoint of each subinterval.
8. Use the table of values below and the rectangle rule with $n = 4$ and c_k as the midpoint of each subinterval to approximate $\int_1^{17} f(x)\,dx$.

x	3	7	11	15
$f(x)$	1.2	3.4	2.6	0.5

9. Find the average value of $f(x) = 6x^2 + 2x$ over the interval $[-1, 2]$.

B *Evaluate the indicated integrals.*

10. $\int_0^1 xe^x\,dx$
11. $\int_{-\infty}^0 e^x\,dx$
12. $\int_{-1}^1 x(x+1)^3\,dx$
13. $\int te^{-0.5t}\,dt$
14. $\int x^2 \ln x\,dx$
15. $\int_0^{\infty} \frac{1}{(x+3)^2}\,dx$
16. Find the area bounded by the graphs of $y = \ln x$, $y = 0$, and $x = e$.
17. Approximate $\int_0^1 e^{2x^2}\,dx$ to three decimal places using the rectangle rule with $n = 5$ and c_k as the midpoint of each subinterval.
18. Find the average value of $f(x) = 3x^{1/2}$ over the interval $[1, 9]$.

C *Evaluate the indicated integrals.*

19. $\int \frac{(\ln x)^2}{x}\,dx$
20. $\int x(\ln x)^2\,dx$

21. $\int xe^{-2x^2}\, dx$

22. $\int x^2e^{-2x}\, dx$

23. $\int x^2(x-1)^4\, dx$

24. $\int_{-\infty}^{\infty} \frac{x}{(1+x^2)^3}\, dx$

25. $\int_0^{\infty} (x+1)e^{-x}\, dx$ [*Hint:* Recall that $\lim_{x\to\infty} x^n e^{-x} = 0,\ n > 0$.]

26. $\int_1^{\infty} \frac{\ln x}{x^3}\, dx$ $\left[\textit{Hint:}\ \text{Recall that } \lim_{x\to\infty} \frac{\ln x}{x^n} = 0,\ n > 0.\right]$

Applications

Business & Economics

27. *Inventory.* Suppose the inventory of a certain item t months after the first of the year is given approximately by

$$I(t) = 10 + 36t - 3t^2 \qquad 0 \le t \le 12$$

What is the average inventory for the second quarter of the year?

28. *Supply function.* Given the supply function

$$p = S(x) = 8(e^{0.05x} - 1)$$

find the average price (in dollars) over the supply interval [40, 50].

29. *Continuous income stream.* Find the present value at 15% compounded continuously for 4 years for the continuous income stream with rate of flow

$$f(t) = 2{,}500e^{0.05t}$$

Find the amount, the total income, and the interest earned during this 4 year period.

30. *Capital value.* The perpetual annual rent for a property is $2,400. Find the capital value at 12% compounded continuously.

31. *Consumers' and producers' surplus.* Given the price–demand and price–supply equations

$$p = D(x) = 70 - \frac{1}{5}x \qquad p = S(x) = 13 + \frac{3}{2{,}500}x^2$$

(A) Find the consumers' surplus at a price level of $\bar{p} = 50$.
(B) Find the producers' surplus at a price level of $\bar{p} = 25$.
(C) Find the equilibrium price, and then find the consumers' surplus and the producers' surplus at the equilibrium price level.

32. *Production.* An oil field is estimated to produce oil at the rate of $R(t)$ thousand barrels per month t months from now, as given by

$$R(t) = 25te^{-0.05t}$$

How much oil is produced during the first 2 years of operation? If the well is operated indefinitely, what is the total amount of oil produced? [Recall that $\lim_{t\to\infty} t^n e^{-kt} = 0$ for $n > 0$ and $k > 0$.]

33. *Parts testing.* If in testing printed circuits for hand calculators, failures occur relative to time in hours according to the probability density function

$$F(t) = \begin{cases} .02e^{-.02t} & t \geq 0 \\ 0 & \text{otherwise} \end{cases}$$

what is the probability that a circuit chosen at random will fail in the first hour of testing?

Life Sciences

34. *Drug assimilation.* The rate at which the body eliminates a drug (in milliliters per hour) is given by

$$R(t) = 20t(t + 1)^{-3}$$

where t is the number of hours since the drug was administered. How much of the drug is eliminated in the first hour after it was administered? What is the total amount of the drug that is eliminated by the body?

35. *Medicine.* For a particular doctor, the length of time in hours spent with a patient per office visit has the probability density function

$$f(t) = \begin{cases} \dfrac{\frac{4}{3}}{(t + 1)^2} & 0 \leq t \leq 3 \\ 0 & \text{otherwise} \end{cases}$$

What is the probability that the doctor will spend more than 1 hour with a randomly selected patient?

Social Sciences

36. *Politics.* The rate of change of the voting population of a city, $N'(t)$, with respect to time t in years is estimated to be

$$N'(t) = \frac{100t}{(1 + t^2)^2}$$

where $N(t)$ is in thousands. If $N(0)$ is the current voting population, how much will this population increase during the next 3 years? If the population continues to grow at this rate indefinitely, what is the total increase in the voting population?

37. *Psychology.* Rats were trained to go through a maze by rewarding them with a food pellet upon successful completion. After the seventh successful run, it was found that the probability density function for length of time in minutes until success on the eighth trial is given by

$$f(t) = \begin{cases} .5e^{-.5t} & t \geq 0 \\ 0 & \text{otherwise} \end{cases}$$

What is the probability that a rat selected at random after seven successful runs will take 2 or more minutes to complete the eighth run successfully?

Multivariable Calculus

6

CHAPTER 6 Contents

6-1 Functions of Several Variables

- Functions of Two or More Independent Variables
- Examples of Functions of Several Variables
- Three-Dimensional Coordinate Systems

■ Functions of Two or More Independent Variables

In Section 0-5 we introduced the concept of a function with one independent variable. Now we will broaden the concept to include functions with more than one independent variable. We start with an example.

A small manufacturing company produces a standard type of surfboard and no other products. If fixed costs are \$500 per week and variable costs are \$70 per board produced, then the weekly cost function is given by

$$C(x) = 500 + 70x \tag{1}$$

where x is the number of boards produced per week. The cost function is a function of a single independent variable x. For each value of x from the domain of C there exists exactly one value of $C(x)$ in the range of C.

Now, suppose the company decides to add a high-performance competition board to its line. If the fixed costs for the competition board are \$200 per week and the variable costs are \$100 per board, then the cost function (1) must be modified to

$$C(x, y) = 700 + 70x + 100y \tag{2}$$

where $C(x, y)$ is the cost for weekly output of x standard boards and y competition boards. Equation (2) is an example of a function with two independent variables, x and y. Of course, as the company expands its product line even further, its weekly cost function must be modified to include more and more independent variables, one for each new product produced.

In general, an equation of the form

$$z = f(x, y)$$

will describe a **function of two independent variables** if for each ordered pair (x, y) from the domain of f there is one and only one value of z determined by $f(x, y)$ in the range of f. Unless otherwise stated, we will assume that the domain of a function specified by an equation of the form $z = f(x, y)$ is the set of all ordered pairs of real numbers (x, y) such that $f(x, y)$ is also a real number. It should be noted, however, that certain conditions in practical problems often lead to further restrictions of the domain of a function.

We can similarly define functions of three independent variables, $w = f(x, y, z)$; of four independent variables, $u = f(w, x, y, z)$; and so on. In this chapter, we will primarily concern ourselves with functions with two independent variables.

Example 1 For $C(x, y) = 700 + 70x + 100y$, find $C(10, 5)$.

Solution

$$C(10, 5) = 700 + 70(10) + 100(5)$$
$$= \$1{,}900$$

Problem 1 Find $C(20, 10)$ for the cost function in Example 1.

Example 2 For $f(x, y, z) = 2x^2 - 3xy + 3z + 1$, find $f(3, 0, -1)$.

Solution

$$f(3, 0, -1) = 2(3)^2 - 3(3)(0) + 3(-1) + 1$$
$$= 18 - 0 - 3 + 1$$
$$= 16$$

Problem 2 Find $f(-2, 2, 3)$ for f in Example 2.

■ **Example 3** Revenue, Cost, and Profit Functions

The surfboard company discussed previously has determined that the demand equations for the two types of boards they produce are given by

$$p = 210 - 4x + y$$
$$q = 300 + x - 12y$$

where p is the price of the standard board, q is the price of the competition board, x is the weekly demand for standard boards, and y is the weekly demand for competition boards.

(A) Find the weekly revenue function $R(x, y)$ and evaluate $R(20, 10)$.
(B) If the weekly cost function is

$$C(x, y) = 700 + 70x + 100y$$

find the weekly profit function $P(x, y)$ and evaluate $P(20, 10)$.

Solutions (A)

$$\text{Revenue} = \begin{pmatrix}\text{Demand for}\\ \text{standard}\\ \text{boards}\end{pmatrix} \times \begin{pmatrix}\text{Price of a}\\ \text{standard}\\ \text{board}\end{pmatrix} + \begin{pmatrix}\text{Demand for}\\ \text{competition}\\ \text{boards}\end{pmatrix} \times \begin{pmatrix}\text{Price of a}\\ \text{competition}\\ \text{board}\end{pmatrix}$$

$$\begin{aligned} R(x, y) &= xp + yq \\ &= x(210 - 4x + y) + y(300 + x - 12y) \\ &= 210x + 300y - 4x^2 + 2xy - 12y^2 \\ R(20, 10) &= 210(20) + 300(10) - 4(20)^2 + 2(20)(10) - 12(10)^2 \\ &= \$4{,}800 \end{aligned}$$

(B) Profit = Revenue − Cost

$$\begin{aligned} P(x, y) &= R(x, y) - C(x, y) \\ &= 210x + 300y - 4x^2 + 2xy - 12y^2 - 700 - 70x - 100y \\ &= 140x + 200y - 4x^2 + 2xy - 12y^2 - 700 \\ P(20, 10) &= 140(20) + 200(10) - 4(20)^2 + 2(20)(10) - 12(10)^2 - 700 \\ &= \$1{,}700 \end{aligned}$$

Problem 3 Repeat Example 3 if the demand and cost equations are given by

$$\begin{aligned} p &= 220 - 6x + y \\ q &= 300 + 3x - 10y \\ C(x, y) &= 40x + 80y + 1{,}000 \end{aligned}$$

▪ Examples of Functions of Several Variables

A number of concepts we have already considered can be thought of in terms of functions of two or more variables. We list a few of these below.

Area of a rectangle $\qquad A(x, y) = xy$

Volume of a box $\qquad V(x, y, z) = xyz$

Volume of a right circular cylinder $\qquad V(r, h) = \pi r^2 h$

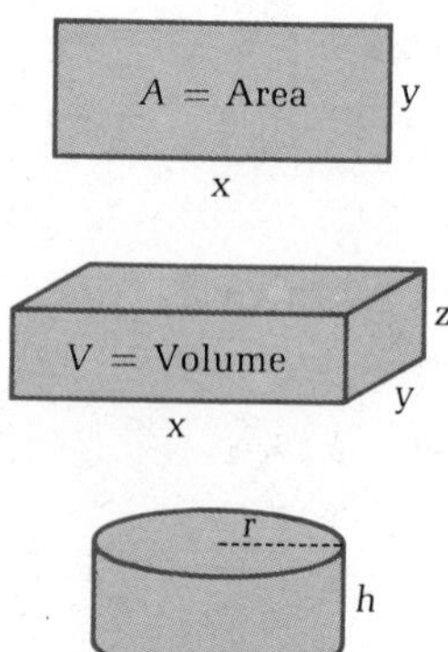

Simple interest	$A(P, r, t) = P(1 + rt)$	A = Amount P = Principal r = Annual rate t = Time in years
Compound interest	$A(P, r, t, n) = P\left(1 + \frac{r}{n}\right)^{nt}$	A = Amount P = Principal r = Annual rate t = Time in years n = Compound periods per year
IQ	$Q(M, C) = \frac{M}{C}(100)$	Q = IQ = Intelligence quotient M = MA = Mental age C = CA = Chronological age
Resistance for blood flow in a vessel	$R(L, r) = k\frac{L}{r^4}$	R = Resistance L = Length of vessel r = Radius of vessel k = Constant

Example 4 Package Design

A company uses a box with a square base and an open top for one of its products (see the figure). If x is the length in inches of each side of the base and y is the height in inches, find the total amount of material $M(x, y)$ required to construct one of these boxes and evaluate $M(5, 10)$.

Solution

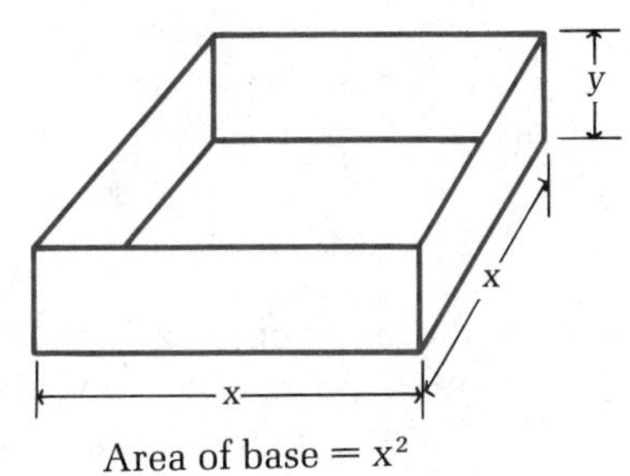

$$\text{Area of base} = x^2$$

$$\text{Area of one side} = xy$$

$$\text{Total material} = (\text{Area of base}) + 4(\text{Area of one side})$$

$$M(x, y) = x^2 + 4xy$$

$$M(5, 10) = 5^2 + 4(5)(10)$$

$$= 225 \text{ square inches}$$

Problem 4 For the box in Example 4, find the volume $V(x, y)$ and evaluate $V(5, 10)$.

▪ Three-Dimensional Coordinate Systems

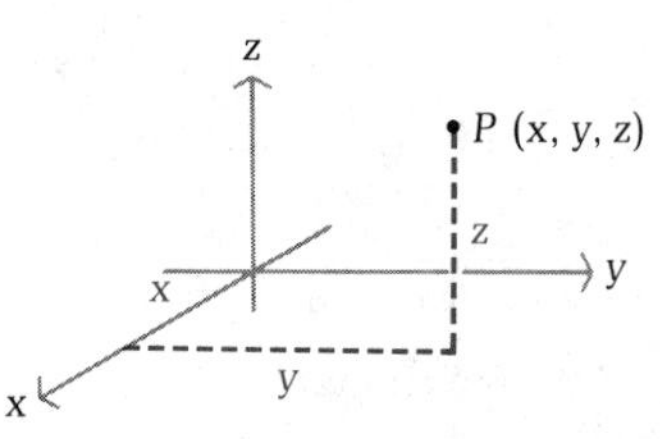

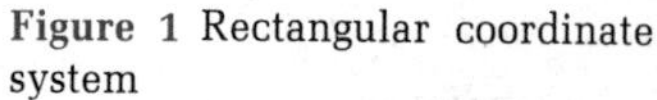
Figure 1 Rectangular coordinate system

We now take a brief look at some graphs of functions of two independent variables. Since functions of the form $z = f(x, y)$ involve two independent variables, x and y, and one dependent variable, z, we need a three-dimensional coordinate system for their graphs. We take three mutually perpendicular number lines intersecting at their origins to form a rectangular coordinate system in three-dimensional space (see Fig. 1). In such a system, every ordered triplet of numbers (x, y, z) can be associated with a unique point, and conversely.

Example 5 Locate $(-3, 5, 2)$ in a rectangular coordinate system.

Solution

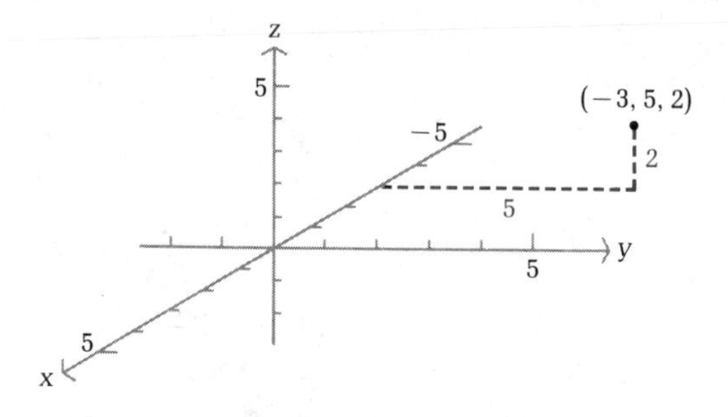

Problem 5 Find the coordinates of the corners *A*, *C*, *G*, and *D* of the rectangular box shown in the figure.

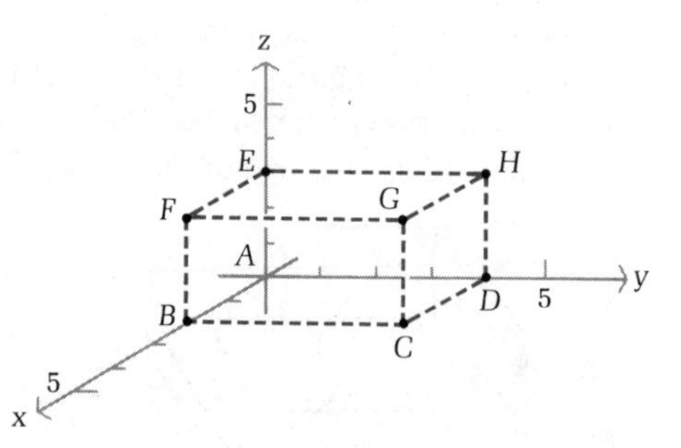

What does the graph of $z = x^2 + y^2$ look like? If we let $x = 0$ and graph $z = 0^2 + y^2 = y^2$ in the yz plane, we obtain a parabola; if we let $y = 0$ and graph $z = x^2 + 0^2 = x^2$ in the xz plane, we obtain another parabola. It can be shown that the graph of $z = x^2 + y^2$ is just one of these parabolas rotated around the z axis (see Fig. 2). This cup-shaped figure is a *surface* and is called a **paraboloid.**

In general, the graph of any function of the form $z = f(x, y)$ is called a **surface.** The graph of such a function is the graph of all ordered triplets of numbers (x, y, z) that satisfy the equation. Graphing functions of two independent variables is often a very difficult task, and the general process will not be dealt with in this book. We present only a few simple graphs to suggest extensions of earlier geometric interpretations of the derivative and local maxima and minima to functions of two variables. Note that $z = f(x, y) = x^2 + y^2$ appears (see Fig. 2) to have a local minimum at $(x, y) = (0, 0)$. Figure 3 shows a local maximum at $(x, y) = (0, 0)$.

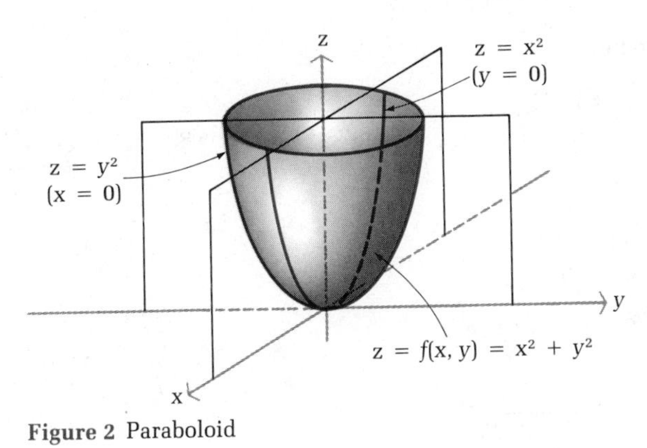

Figure 2 Paraboloid

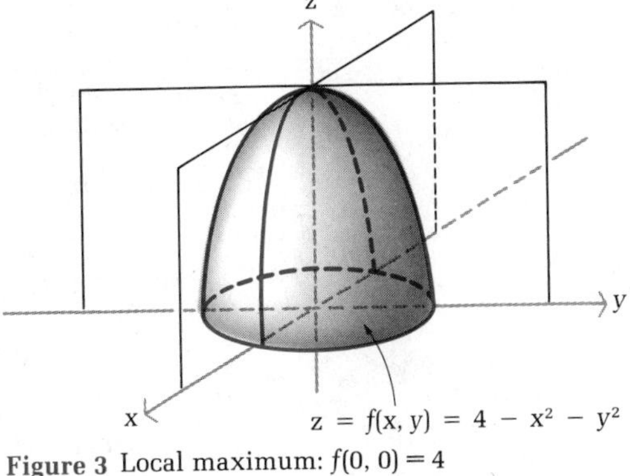

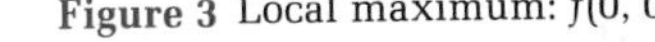

Figure 3 Local maximum: $f(0, 0) = 4$

Figure 4 shows a point at $(x, y) = (0, 0)$, called a **saddle point,** which is neither a local minimum nor a local maximum. More will be said about local maxima and minima in Section 6-3.

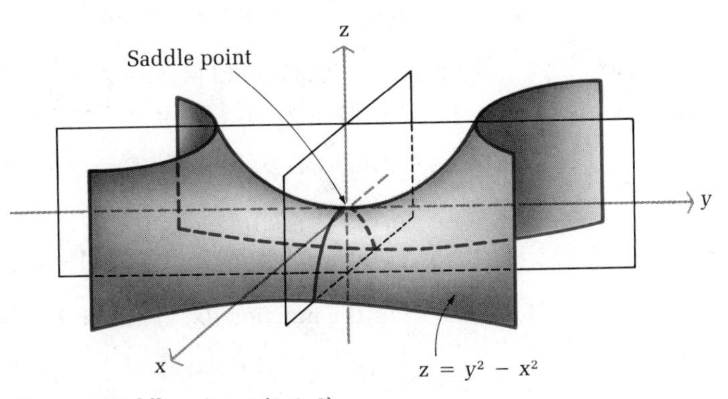

Figure 4 Saddle point at (0, 0, 0)

Answers to Matched Problems

1. \$3,100 2. 30
3. (A) $R(x, y) = 220x + 300y - 6x^2 + 4xy - 10y^2$; $R(20, 10) = \$4{,}800$
   (B) $P(x, y) = 180x + 220y - 6x^2 + 4xy - 10y^2 - 1{,}000$;
   $P(20, 10) = \$2{,}200$
4. $V(x, y) = x^2y$; $V(5, 10) = 250$ cubic inches
5. $A(0, 0, 0)$; $C(2, 4, 0)$; $G(2, 4, 3)$; $D(0, 4, 0)$

Exercise 6-1

A *For the functions*

$$f(x, y) = 10 + 2x - 3y \qquad g(x, y) = x^2 - 3y^2$$

find each of the following:

1. $f(0, 0)$ **2.** $f(2, 1)$ **3.** $f(-3, 1)$ **4.** $f(2, -7)$
5. $g(0, 0)$ **6.** $g(0, -1)$ **7.** $g(2, -1)$ **8.** $g(-1, 2)$

B *Find each of the following:*

9. $A(2, 3)$ for $A(x, y) = xy$
10. $V(2, 4, 3)$ for $V(x, y, z) = xyz$
11. $Q(12, 8)$ for $Q(M, C) = \frac{M}{C}(100)$
12. $T(50, 17)$ for $T(V, x) = \frac{33V}{x + 33}$
13. $V(2, 4)$ for $V(r, h) = \pi r^2 h$
14. $S(4, 2)$ for $S(x, y) = 5x^2y^3$
15. $R(1, 2)$ for $R(x, y) = -5x^2 + 6xy - 4y^2 + 200x + 300y$
16. $P(2, 2)$ for $P(x, y) = -x^2 + 2xy - 2y^2 - 4x + 12y + 5$
17. $R(6, 0.5)$ for $R(L, r) = 0.002\frac{L}{r^4}$
18. $L(2{,}000, 50)$ for $L(w, v) = (1.25 \times 10^{-5})wv^2$
19. $A(100, 0.06, 3)$ for $A(P, r, t) = P + Prt$
20. $A(10, 0.04, 3, 2)$ for $A(P, r, t, n) = P\left(1 + \frac{r}{n}\right)^{tn}$
21. $A(100, 0.08, 10)$ for $A(P, r, t) = Pe^{rt}$
22. $A(1{,}000, 0.06, 8)$ for $A(P, r, t) = Pe^{rt}$

C

23. For the function $f(x, y) = x^2 + 2y^2$, find:

$$\frac{f(x + \Delta x, y) - f(x, y)}{\Delta x}$$

24. For the function $f(x, y) = x^2 + 2y^2$, find:

$$\frac{f(x, y + \Delta y) - f(x, y)}{\Delta y}$$

25. For the function $f(x, y) = 2xy^2$, find:

$$\frac{f(x + \Delta x, y) - f(x, y)}{\Delta x}$$

26. For the function $f(x, y) = 2xy^2$, find:

$$\frac{f(x, y + \Delta y) - f(x, y)}{\Delta y}$$

27. Find the coordinates of E and F in the figure for Problem 5 in the text.

28. Find the coordinates of B and H in the figure for Problem 5 in the text.

Applications

Business & Economics

29. *Cost function.* A small manufacturing company produces two models of a surfboard: a standard model and a competition model. If the standard model is produced at a variable cost of \$70 each, the competition model at a variable cost of \$100 each, and the total fixed costs per month are \$2,000, then the monthly cost function is given by

$$C(x, y) = 2{,}000 + 70x + 100y$$

where x and y are the numbers of standard and competition models produced per month, respectively. Find $C(20, 10)$, $C(50, 5)$, and $C(30, 30)$.

30. *Advertising and sales.* A company spends x thousand dollars per week on newspaper advertising and y thousand dollars per week on television advertising. Its weekly sales were found to be given by

$$S(x, y) = 5x^2y^3$$

Find $S(3, 2)$ and $S(2, 3)$.

31. *Revenue function.* A supermarket sells two brands of coffee: brand A at \$p per pound and brand B at \$q per pound. The daily demand equations for brands A and B are, respectively,

$$x = 200 - 5p + 4q$$
$$y = 300 + 2p - 4q$$

(both in pounds). Find the daily revenue function $R(p, q)$. Evaluate $R(2, 3)$ and $R(3, 2)$.

32. *Revenue, cost, and profit functions.* A company manufactures ten-speed and three-speed bicycles. The weekly demand and cost equations are

$$p = 230 - 9x + y$$
$$q = 130 + x - 4y$$
$$C(x, y) = 200 + 80x + 30y$$

where \$p is the price of a ten-speed bicycle, \$q is the price of a three-speed bicycle, x is the weekly demand for ten-speed bicycles, y is the weekly demand for three-speed bicycles, and $C(x, y)$ is the cost function. Find the weekly revenue function $R(x, y)$ and the weekly profit function $P(x, y)$. Evaluate $R(10, 15)$ and $P(10, 15)$.

33. *Future value.* At the end of each year $2,000 is invested into an IRA account earning 9% compounded annually. How much will be in the account at the end of 30 years? Use the annuity formula

$$F(P, i, n) = P\frac{(1 + i)^n - 1}{i}$$

where

$P = PMT =$ Periodic payment

$i =$ Rate per period

$n =$ Number of payments (periods)

$F = FV =$ Amount or future value

34. *Package design.* The packaging department in a company has been asked to design a rectangular box with no top and a partition down the middle (see the accompanying figure). If x, y, and z are the dimensions in inches, find the total amount of material $M(x, y, z)$ used in constructing one of these boxes and evaluate $M(10, 12, 6)$.

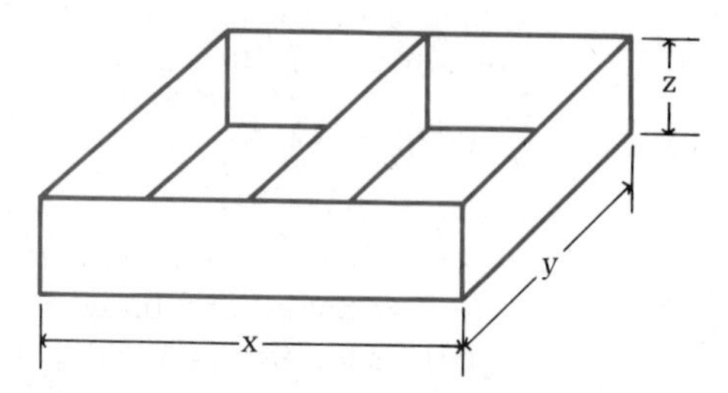

Life Sciences

35. *Marine biology.* In using scuba diving gear, a marine biologist estimates the time of a dive according to the equation

$$T(V, x) = \frac{33V}{x + 33}$$

where

$T =$ Time of dive in minutes

$V =$ Volume of air, at sea level pressure, compressed into tanks

$x =$ Depth of dive in feet

Find $T(70, 47)$ and $T(60, 27)$.

36. *Blood flow.* Poiseuille's law states that the resistance, R, for blood flowing in a blood vessel varies directly as the length of the vessel, L, and inversely as the fourth power of its radius, r. Stated as

an equation,

$$R(L, r) = k\frac{L}{r^4} \qquad k \text{ a constant}$$

Find $R(8, 1)$ and $R(4, 0.2)$.

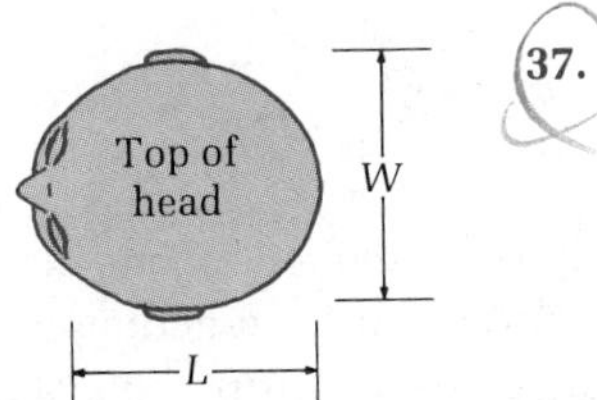

37. *Physical anthropology.* Anthropologists, in their study of race and human genetic groupings, often use an index called the *cephalic index*. The cephalic index, C, varies directly as the width, W, of the head, and inversely as the length, L, of the head (both viewed from the top). In terms of an equation,

$$C(W, L) = 100\frac{W}{L}$$

where

W = Width in inches

L = Length in inches

Find $C(6, 8)$ and $C(8.1, 9)$.

Social Sciences

38. *Safety research.* Under ideal conditions, if a person driving a car slams on the brakes and skids to a stop, the length of the skid marks is given by the formula

$$L(w, v) = kwv^2$$

where

k = Constant

w = Weight of car in pounds

v = Speed of car in miles per hour

For $k = 0.000\ 013\ 3$, find $L(2{,}000, 40)$ and $L(3{,}000, 60)$.

39. *Psychology.* Intelligence quotient (IQ) is defined to be the ratio of the mental age (MA), as determined by certain tests, and the chronological age (CA), multiplied by 100. Stated as an equation,

$$Q(M, C) = \frac{M}{C} \cdot 100$$

where

Q = IQ

M = MA

C = CA

Find $Q(12, 10)$ and $Q(10, 12)$.

6-2 Partial Derivatives

- Partial Derivatives
- Second-Order Partial Derivatives

Partial Derivatives

We know how to differentiate many kinds of functions of one independent variable and how to interpret the results. What about functions with two or more independent variables? Let us return to the surfboard example considered at the beginning of the chapter.

For the company producing only the standard board, the cost function was

$$C(x) = 500 + 70x$$

Differentiating with respect to x, we obtain the marginal cost function

$$C'(x) = 70$$

Since the marginal cost is constant, \$70 is the change in cost for one unit increase in production at any output level.

For the company producing two boards, a standard model and a competition model, the cost function was

$$C(x, y) = 700 + 70x + 100y$$

Now suppose we differentiate with respect to x, holding y fixed, and denote this by $C_x(x, y)$; or we differentiate with respect to y, holding x fixed, and denote this by $C_y(x, y)$. Differentiating in this way, we obtain

$$C_x(x, y) = 70 \qquad C_y(x, y) = 100$$

Both these are called **partial derivatives** and, in this example, both represent marginal costs. The first is the change in cost due to one unit increase in production of the standard board with the production of the competition model held fixed. The second is the change in cost due to one unit increase in production of the competition board with the production of the standard board held fixed.

In general, if $z = f(x, y)$, then the **partial derivative of f with respect to x,** denoted by $\partial z/\partial x$, f_x, $f_x(x, y)$, is defined by

$$\frac{\partial z}{\partial x} = \lim_{\Delta x \to 0} \frac{f(x + \Delta x, y) - f(x, y)}{\Delta x}$$

provided the limit exists. This is the ordinary derivative of f with respect to x, holding y constant. Thus, we are able to continue to use all the derivative rules and properties discussed in Chapters 1–3 for partials.

Similarly, the **partial derivative of f with respect to y,** denoted by $\partial z/\partial y$, f_y, or $f_y(x, y)$, is defined by

$$\frac{\partial z}{\partial y} = \lim_{\Delta y \to 0} \frac{f(x, y + \Delta y) - f(x, y)}{\Delta y}$$

which is the ordinary derivative with respect to y, holding x constant.

Parallel definitions and interpretations hold for functions with three or more independent variables.

Example 6 For $z = f(x, y) = 2x^2 - 3x^2y + 5y + 1$, find:

(A) $\dfrac{\partial z}{\partial x}$ (B) $f_x(2, 3)$

Solutions (A) $z = 2x^2 - 3x^2y + 5y + 1$

Differentiating with respect to x, holding y constant (that is, treating y as a constant), we obtain

$$\frac{\partial z}{\partial x} = 4x - 6xy$$

(B) $f(x, y) = 2x^2 - 3x^2y + 5y + 1$

First differentiate with respect to x (part A) to obtain

$$f_x(x, y) = 4x - 6xy$$

Then evaluate at (2, 3). Thus,

$$f_x(2, 3) = 4(2) - 6(2)(3) = -28$$

Problem 6 For f in Example 6, find:

(A) $\dfrac{\partial z}{\partial y}$ (B) $f_y(2, 3)$

Example 7 For $z = f(x, y) = e^{x^2+y^2}$, find:

(A) $\dfrac{\partial z}{\partial x}$ (B) $f_y(2, 1)$

Solutions (A) Using the chain rule [thinking of $z = e^u$, $u = u(x)$; y is held constant], we obtain

$$\frac{\partial z}{\partial x} = e^{x^2+y^2} \frac{\partial(x^2 + y^2)}{\partial x}$$
$$= 2xe^{x^2+y^2}$$

(B) $f_y(x, y) = 2ye^{x^2+y^2}$

$$f_y(2, 1) = 2(1)e^{2^2+1^2}$$
$$= 2e^5$$

Problem 7 For $z = f(x, y) = (x^2 + 2xy)^5$, find:

(A) $\dfrac{\partial z}{\partial y}$ (B) $f_x(1, 0)$

Example 8 Profit

The profit function for the surfboard company in Example 3 in Section 6-1 was

$$P(x, y) = 140x + 200y - 4x^2 + 2xy - 12y^2 - 700$$

Find $P_x(15, 10)$ and $P_x(30, 10)$, and interpret.

Solution

$$P_x(x, y) = 140 - 8x + 2y$$
$$P_x(15, 10) = 140 - 8(15) + 2(10) = 40$$
$$P_x(30, 10) = 140 - 8(30) + 2(10) = -80$$

At a production level of 15 standard and 10 competition boards per week, increasing the production of standard boards by 1 and holding the production of competition boards fixed at 10 will increase profit by approximately \$40. At a production level of 30 standard and 10 competition boards per week, increasing the production of standard boards by 1 unit and holding the production of competition boards fixed at 10 will decrease profit by approximately \$80.

Problem 8 For the profit function in Example 8, find $P_y(25, 10)$ and $P_y(25, 15)$, and interpret.

Partials have simple geometric interpretations, as indicated in Figure 5. If we hold x fixed, say $x = a$, then $f_y(a, y)$ is the slope of the curve obtained by intersecting the plane $x = a$ with the surface $z = f(x, y)$. A similar interpretation is given to $f_x(x, b)$.

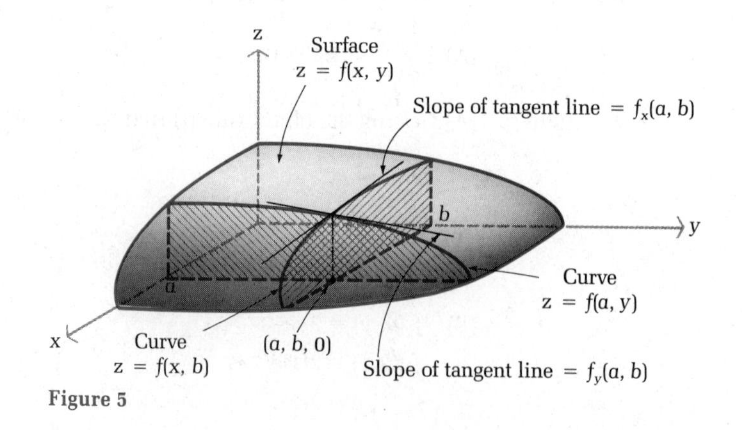

Figure 5

Second-Order Partial Derivatives

Just as there are second-order ordinary derivatives, there are second-order partials, and we will be using some of these in the next section when we discuss local maxima and minima. The following second-order partials will be useful:

Second-Order Partials

If $z = f(x, y)$, then

$$\frac{\partial^2 z}{\partial x^2} = \frac{\partial}{\partial x}\left(\frac{\partial z}{\partial x}\right) = f_{xx}(x, y) = f_{xx}$$

$$\frac{\partial^2 z}{\partial x\, \partial y} = \frac{\partial}{\partial x}\left(\frac{\partial z}{\partial y}\right) = f_{yx}(x, y) = f_{yx}$$

$$\frac{\partial^2 z}{\partial y\, \partial x} = \frac{\partial}{\partial y}\left(\frac{\partial z}{\partial x}\right) = f_{xy}(x, y) = f_{xy}$$

$$\frac{\partial^2 z}{\partial y^2} = \frac{\partial}{\partial y}\left(\frac{\partial z}{\partial y}\right) = f_{yy}(x, y) = f_{yy}$$

In the mixed partial $\partial^2 z/(\partial x\, \partial y) = f_{yx}$, we start with $z = f(x, y)$ and first differentiate with respect to y (holding x constant). Then we differentiate with respect to x (holding y constant). What is the order of differentiation for $\partial^2 z/(\partial y\, \partial x) = f_{xy}$? It can be shown that for the functions we will consider, $f_{xy}(x, y) = f_{yx}(x, y)$.

Example 9 For $z = f(x, y) = 3x^2 - 2xy^3 + 1$, find:

(A) $\dfrac{\partial^2 z}{\partial x\, \partial y}, \dfrac{\partial^2 z}{\partial y\, \partial x}$ (B) $\dfrac{\partial^2 z}{\partial x^2}$ (C) $f_{yx}(2, 1)$

Solutions (A) First differentiate with respect to y and then with respect to x:

$$\frac{\partial z}{\partial y} = -6xy^2 \qquad \frac{\partial^2 z}{\partial x\, \partial y} = \frac{\partial}{\partial x}\left(\frac{\partial z}{\partial y}\right) = \frac{\partial}{\partial x}(-6xy^2) = -6y^2$$

First differentiate with respect to x and then with respect to y:

$$\frac{\partial z}{\partial x} = 6x - 2y^3 \qquad \frac{\partial^2 z}{\partial y\, \partial x} = \frac{\partial}{\partial y}\left(\frac{\partial z}{\partial x}\right) = \frac{\partial}{\partial y}(6x - 2y^3) = -6y^2$$

(B) Differentiate with respect to x twice:

$$\frac{\partial z}{\partial x} = 6x - 2y^3 \qquad \frac{\partial^2 z}{\partial x^2} = \frac{\partial}{\partial x}\left(\frac{\partial z}{\partial x}\right) = 6$$

(C) First find $f_{yx}(x, y)$. Then evaluate at (2, 1). Again, remember that f_{yx} means to differentiate with respect to y first and then with respect to x. Thus,

$$f_y(x, y) = -6xy^2$$
$$f_{yx}(x, y) = -6y^2$$

and

$$f_{yx}(2, 1) = -6(1)^2 = -6$$

Problem 9 For the function in Example 9, find:

(A) $\dfrac{\partial^2 z}{\partial y\, \partial x}$ (B) $\dfrac{\partial^2 z}{\partial y^2}$ (C) $f_{xy}(2, 3)$ (D) $f_{yx}(2, 3)$

Answers to Matched Problems

6. (A) $\dfrac{\partial z}{\partial y} = -3x^2 + 5$ (B) $f_y(2, 3) = -7$
7. (A) $10x(x^2 + 2xy)^4$ (B) 10
8. $P_y(25, 10) = 10$: At a production level of $x = 25$ and $y = 10$, increasing y by 1 unit and holding x fixed at 25 will increase profit by approximately \$10; $P_y(25, 15) = -110$: At a production level of $x = 25$ and $y = 15$, increasing y by 1 unit and holding x fixed at 25 will decrease profit by approximately \$110.
9. (A) $-6y^2$ (B) $-12xy$ (C) -54 (D) -54

Exercise 6-2

A

For $z = f(x, y) = 10 + 3x + 2y$, find each of the following:

1. $\dfrac{\partial z}{\partial x}$
2. $\dfrac{\partial z}{\partial y}$
3. $f_y(1, 2)$
4. $f_x(1, 2)$

For $z = f(x, y) = 3x^2 - 2xy^2 + 1$, find each of the following:

5. $\dfrac{\partial z}{\partial y}$
6. $\dfrac{\partial z}{\partial x}$
7. $f_x(2, 3)$
8. $f_y(2, 3)$

For $S(x, y) = 5x^2y^3$, find each of the following:

9. $S_x(x, y)$
10. $S_y(x, y)$
11. $S_y(2, 1)$
12. $S_x(2, 1)$

B For $C(x, y) = x^2 - 2xy + 2y^2 + 6x - 9y + 5$, *find each of the following:*

13. $C_x(x, y)$
14. $C_y(x, y)$
15. $C_x(2, 2)$
16. $C_y(2, 2)$
17. $C_{xy}(x, y)$
18. $C_{yx}(x, y)$
19. $C_{xx}(x, y)$
20. $C_{yy}(x, y)$

For $z = f(x, y) = e^{2x+3y}$, *find each of the following:*

21. $\dfrac{\partial z}{\partial x}$
22. $\dfrac{\partial z}{\partial y}$
23. $\dfrac{\partial^2 z}{\partial x\, \partial y}$
24. $\dfrac{\partial^2 z}{\partial y\, \partial x}$
25. $f_{xy}(1, 0)$
26. $f_{yx}(0, 1)$
27. $f_{xx}(0, 1)$
28. $f_{yy}(1, 0)$

Find $f_x(x, y)$ *and* $f_y(x, y)$ *for each function f given by:*

29. $f(x, y) = (x^2 - y^3)^3$
30. $f(x, y) = \sqrt{2x - y^2}$
31. $f(x, y) = (3x^2y - 1)^4$
32. $f(x, y) = (3 + 2xy^2)^3$
33. $f(x, y) = \ln(x^2 + y^2)$
34. $f(x, y) = \ln(2x - 3y)$
35. $f(x, y) = y^2e^{xy^2}$
36. $f(x, y) = x^3e^{x^2y}$
37. $f(x, y) = \dfrac{x^2 - y^2}{x^2 + y^2}$
38. $f(x, y) = \dfrac{2x^2y}{x^2 + y^2}$

Find $f_{xx}(x, y)$, $f_{xy}(x, y)$, $f_{yx}(x, y)$, *and* $f_{yy}(x, y)$ *for each function f given by:*

39. $f(x, y) = x^2y^2 + x^3 + y$
40. $f(x, y) = x^3y^3 + x + y^2$
41. $f(x, y) = \dfrac{x}{y} - \dfrac{y}{x}$
42. $f(x, y) = \dfrac{x^2}{y} - \dfrac{y^2}{x}$
43. $f(x, y) = xe^{xy}$
44. $f(x, y) = x\ln(xy)$

C

45. For

$$P(x, y) = -x^2 + 2xy - 2y^2 - 4x + 12y - 5$$

find values of x and y such that

$$P_x(x, y) = 0 \quad \text{and} \quad P_y(x, y) = 0$$

simultaneously.

46. For

$$C(x, y) = 2x^2 + 2xy + 3y^2 - 16x - 18y + 54$$

find values of x and y such that

$$C_x(x, y) = 0 \quad \text{and} \quad C_y(x, y) = 0$$

simultaneously.

In Problems 47–48, show that the function f satisfies $f_{xx}(x, y) + f_{yy}(x, y) = 0$.

47. $f(x, y) = \ln(x^2 + y^2)$
48. $f(x, y) = x^3 - 3xy^2$
49. For $f(x, y) = x^2 + 2y^2$, find:

(A) $\lim_{\Delta x \to 0} \frac{f(x + \Delta x, y) - f(x, y)}{\Delta x}$ (B) $\lim_{\Delta y \to 0} \frac{f(x, y + \Delta y) - f(x, y)}{\Delta y}$

50. For $f(x, y) = 2xy^2$, find:

(A) $\lim_{\Delta x \to 0} \frac{f(x + \Delta x, y) - f(x, y)}{\Delta x}$ (B) $\lim_{\Delta y \to 0} \frac{f(x, y + \Delta y) - f(x, y)}{\Delta y}$

Applications

Business & Economics

51. *Profit function.* A firm produces two types of calculators, x thousand of type A and y thousand of type B per year. The revenue and cost functions for the year are (in thousands of dollars)

$$R(x, y) = 14x + 20y$$
$$C(x, y) = x^2 - 2xy + 2y^2 + 12x + 16y + 5$$

Find $P_x(1, 2)$ and $P_y(1, 2)$, and interpret.

52. *Advertising and sales.* A company spends x thousand dollars per week on newspaper advertising and y thousand dollars per week on television advertising. Its weekly sales were found to be given by

$$S(x, y) = 5x^2y^3$$

Find $S_x(3, 2)$ and $S_y(3, 2)$, and interpret.

53. *Demand equations.* A supermarket sells two brands of coffee, brand A at \$p per pound and brand B at \$q per pound. The daily demand equations for brands A and B are, respectively,

$$x = 200 - 5p + 4q$$
$$y = 300 + 2p - 4q$$

Find $\partial x/\partial p$ and $\partial y/\partial p$, and interpret.

54. *Revenue and profit functions.* A company manufactures ten-speed and three-speed bicycles. The weekly demand and cost functions are

$$p = 230 - 9x + y$$
$$q = 130 + x - 4y$$
$$C(x, y) = 200 + 80x + 30y$$

where \$p is the price of a ten-speed bicycle, \$q is the price of a three-speed bicycle, x is the weekly demand for ten-speed bicycles, y is the weekly demand for three-speed bicycles, and C(x, y) is the cost function. Find $R_x(10, 5)$ and $P_x(10, 5)$, and interpret.

*Problems 55 and 56 refer to the **Cobb–Douglas production function***

$$f(x, y) = kx^m y^n$$

*where k, m, and n are positive constants with $m + n = 1$. The function f measures the number of units of a finished product produced from the use of x units of labor and y units of capital (for equipment such as tools, machinery, buildings, and so on). The partial derivative $f_x(x, y)$ approximates the change in productivity per unit change in labor units and is called **marginal productivity of labor.** The partial derivative $f_y(x, y)$ approximates the change in productivity per unit change in capital units and is called **marginal productivity of capital.***

55. *Productivity.* The productivity of a third-world country is given approximately by the function

$$f(x, y) = 10x^{0.75}y^{0.25}$$

with the utilization of x units of labor and y units of capital.

(A) Find $f_x(x, y)$ and $f_y(x, y)$.
(B) If the country is now using 625 units of labor and 81 units of capital, find the marginal productivity of labor and the marginal productivity of capital.
(C) For the greatest increase in the country's productivity, should the government encourage increased use of labor or increased use of capital?

56. *Productivity.* The productivity of an automobile manufacturing company is given approximately by the function

$$f(x, y) = 50\sqrt{xy} = 50x^{0.5}y^{0.5}$$

with the utilization of x units of labor and y units of capital.

(A) Find $f_x(x, y)$ and $f_y(x, y)$.
(B) If the company is now using 256 units of labor and 144 units of capital, find the marginal productivity of labor and the marginal productivity of capital.
(C) For the greatest increase in the company's productivity, should the management encourage increased use of labor or increased use of capital?

*Problems 57–60 refer to the following: If a decrease in demand for one product results in an increase in demand for another product, then the two products are said to be **competitive** or **substitute products.** (Real whipping cream and imitation whipping cream are examples of competitive or substitute products.) If a decrease in demand for one product results in a decrease in demand for another product, then the two products are said to be **complementary products.** (Fishing boats and outboard motors are examples of* *complementary products.) Partial derivatives can be used to test whether two products are competitive, complementary, or neither. We start with demand functions for two products where the demand for either depends on the prices for both:*

$x = f(p, q)$ *Demand function for product A*

$y = g(p, q)$ *Demand function for product B*

The variables x and y represent the number of units demanded of products A and B, respectively, at a price p for 1 unit of product A and a price q for 1 unit of product B. Normally, if the price of A increases while the price of B is held constant, then the demand for A will decrease; that is, $f_p(p, q) < 0$. *Then, if A and B are competitive products, the demand for B will increase; that is,* $g_p(p, q) > 0$. *Similarly, if the price of B increases while the price of A is held constant, then the demand for B will decrease; that is,* $g_q(p, q) < 0$. *And if A and B are competitive products, then the demand for A will increase; that is,* $f_q(p, q) > 0$. *Reasoning similarly for complementary products, we arrive at the following test:*

Test for Competitive and Complementary Products

Partials	**Products *A* and *B***
$f_q(p, q) > 0$ and $g_p(p, q) > 0$	Competitive (Substitute)
$f_q(p, q) < 0$ and $g_p(p, q) < 0$	Complementary
$f_q(p, q) \geqslant 0$ and $g_p(p, q) \leqslant 0$	Neither
$f_q(p, q) \leqslant 0$ and $g_p(p, q) \geqslant 0$	Neither

57. *Product demand.* The weekly demand equations for the sale of butter and margarine in a supermarket are

$$x = f(p, q) = 8{,}000 - 0.09p^2 + 0.08q^2 \quad \text{Butter}$$
$$y = g(p, q) = 15{,}000 + 0.04p^2 - 0.3q^2 \quad \text{Margarine}$$

Determine whether the products are competitive, complementary, or neither.

58. *Product demand.* The daily demand equations for the sale of brand *A* coffee and brand *B* coffee in a supermarket are

$$x = f(p, q) = 200 - 5p + 4q \quad \text{Brand } A \text{ coffee}$$
$$y = g(p, q) = 300 + 2p - 4q \quad \text{Brand } B \text{ coffee}$$

Determine whether the two products are competitive, complementary, or neither.

59. *Product demand.* The monthly demand equations for the sale of skis and ski boots in a sporting goods store are

$$x = f(p, q) = 800 - 0.004p^2 - 0.003q^2 \quad \text{Skis}$$
$$y = g(p, q) = 600 - 0.003p^2 - 0.002q^2 \quad \text{Ski boots}$$

Determine whether the products are competitive, complementary, or neither.

60. *Product demand.* The monthly demand equations for the sale of tennis rackets and tennis balls in a sporting goods store are

$$x = f(p, q) = 500 - 0.5p - q^2 \qquad \text{Tennis rackets}$$
$$y = g(p, q) = 10{,}000 - 8p - 100q^2 \qquad \text{Tennis balls (cans)}$$

Determine whether the products are competitive, complementary, or neither.

Life Sciences

61. *Medicine.* The following empirical formula relates the surface area A (in square inches) of an average human body to its weight w (in pounds) and its height h (in inches):

$$A = f(w, h) = 15.64w^{0.425}h^{0.725}$$

Knowing the surface area of a human body is useful, for example, in studies pertaining to hypothermia (heat loss due to exposure).

(A) Find $f_w(w, h)$ and $f_h(w, h)$.
(B) For a 65 pound child who is 57 inches tall, find $f_w(65, 57)$ and $f_h(65, 57)$ and interpret.

62. *Blood flow.* Poiseuille's law states that the resistance, R, for blood flowing in a blood vessel varies directly as the length of the vessel, L, and inversely as the fourth power of its radius, r. Stated as an equation,

$$R(L, r) = k\frac{L}{r^4} \qquad k \text{ a constant}$$

Find $R_L(4, 0.2)$ and $R_r(4, 0.2)$, and interpret.

Social Sciences

63. *Physical anthropology.* Anthropologists, in their study of race and human genetic groupings, often use an index called the *cephalic index*. The cephalic index, C, varies directly as the width, W, of the head, and inversely as the length, L, of the head (both viewed from the top). In terms of an equation,

$$C(W, L) = 100\frac{W}{L}$$

where

W = Width in inches
L = Length in inches

Find $C_W(6, 8)$ and $C_L(6, 8)$, and interpret.

64. *Safety research.* Under ideal conditions, if a person driving a car slams on the brakes and skids to a stop, the length of the skid marks is given by the formula

$$L(w, v) = kwv^2$$

where

k = Constant

w = Weight of car in pounds

v = Speed of car in miles per hour

For $k = 0.000\ 013\ 3$, find $L_w(2{,}500, 60)$ and $L_v(2{,}500, 60)$, and interpret.

6-3 Maxima and Minima

We are now ready to undertake a brief but useful analysis of local maxima and minima for functions of the type $z = f(x, y)$. Basically, we are going to extend the second-derivative test developed for functions of a single independent variable. To start, we assume that all second-order partials exist for the function f in some circular region in the xy plane. This guarantees that the surface $z = f(x, y)$ has no sharp points, breaks, or ruptures. In other words, we are dealing only with smooth surfaces with no edges (like the edge of a box); or breaks (like an earthquake fault); or sharp points (like the bottom point of a golf tee). See Figure 6.

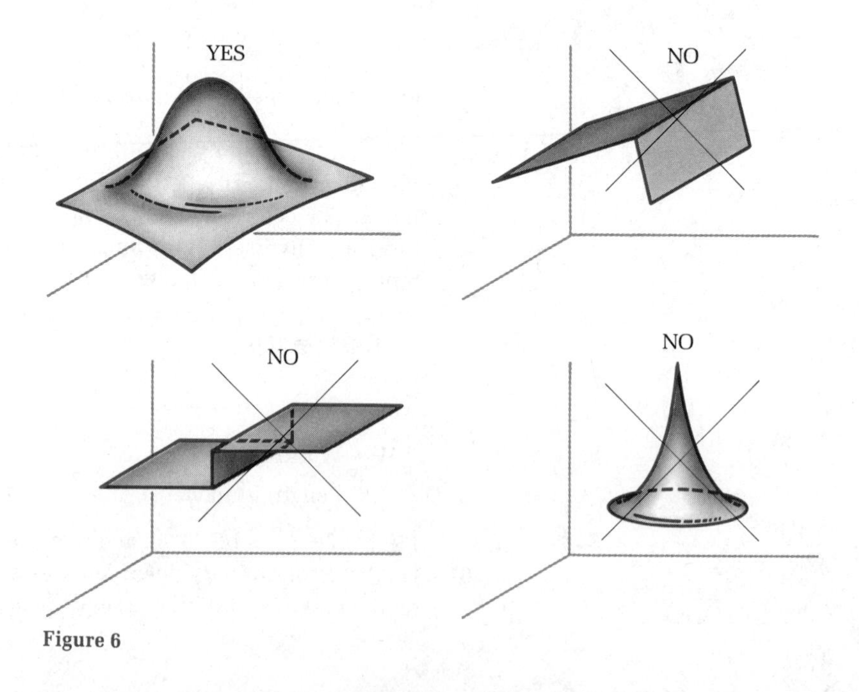

Figure 6

In addition, we will not concern ourselves with boundary points or absolute maxima–minima theory. In spite of these restrictions, the procedure we are now going to describe will help us solve a large number of useful problems.

What does it mean for $f(a, b)$ to be a local maximum or a local minimum? We say that ***f*(*a*, *b*) is a local maximum** if there exists a circular region in the domain of f with (a, b) as the center, such that

$$f(a, b) \geq f(x, y)$$

for all (x, y) in the region. Similarly, we say that ***f*(*a*, *b*) is a local minimum** if there exists a circular region in the domain of f with (a, b) as the center, such that

$$f(a, b) \leq f(x, y)$$

for all (x, y) in the region. Figure 7A illustrates a local maximum, Figure 7B a local minimum, and Figure 7C a saddle point, which is neither a local maximum nor a local minimum.

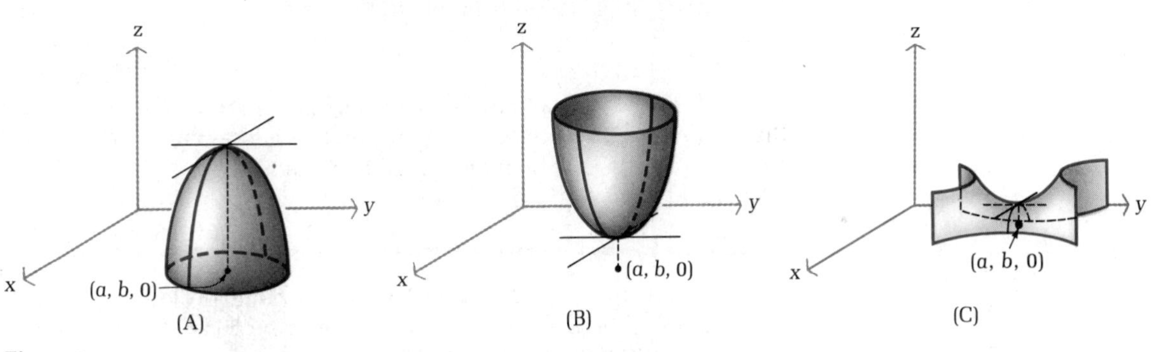

Figure 7

What happens to $f_x(a, b)$ and $f_y(a, b)$ if $f(a, b)$ is a local minimum or a local maximum and the partials of f exist in a circular region containing (a, b)? Figure 7 suggests that $f_x(a, b) = 0$ and $f_y(a, b) = 0$, since the tangents to the indicated curves are horizontal. Theorem 1 indicates that our intuitive reasoning is correct.

Theorem 1

Let $f(a, b)$ be an extreme (a local maximum or a local minimum) for the function f. If both f_x and f_y exist at (a, b), then

$$f_x(a, b) = 0 \quad \text{and} \quad f_y(a, b) = 0 \tag{1}$$

The converse of this theorem is false; that is, if $f_x(a, b) = 0$ and $f_y(a, b) = 0$, then $f(a, b)$ may or may not be a local extreme—the point $(a, b, f(a, b))$ may be a saddle point, for example (see Fig. 7C).

Theorem 1 gives us what are called *necessary* (but not *sufficient*) conditions for $f(a, b)$ to be a local extreme. We thus find all points (a, b) such that $f_x(a, b) = 0$ and $f_y(a, b) = 0$ and test these further to determine whether $f(a, b)$ is a local extreme or a saddle point. Points (a, b) such that (1) holds are called **critical points.** The next theorem, using second-derivative tests, gives us *sufficient* conditions for a local point to produce a local extreme or a saddle point. As was the case with Theorem 1, we state this theorem without proof.

Theorem 2

Second-Derivative Test for Local Extrema

If:

1. $z = f(x, y)$
2. $f_x(a, b) = 0$ and $f_y(a, b) = 0$ [(a, b) is a critical point]
3. All second-order partials of f exist in some circular region containing (a, b) as a center
4. $A = f_{xx}(a, b)$, $B = f_{xy}(a, b)$, $C = f_{yy}(a, b)$

Then:

1. If $AC - B^2 > 0$ and $A < 0$, then $f(a, b)$ is a local maximum.
2. If $AC - B^2 > 0$ and $A > 0$, then $f(a, b)$ is a local minimum.
3. If $AC - B^2 < 0$, then f has a saddle point at (a, b).
4. If $AC - B^2 = 0$, the test fails.

Know

To illustrate the use of Theorem 2, we will first find the local extrema for a very simple function whose solution is almost obvious: $z = f(x, y) = x^2 + y^2 + 2$. From the function f itself and its graph (Fig. 8), it is clear that a local minimum is found at (0, 0). Let us see how Theorem 2 confirms this observation.

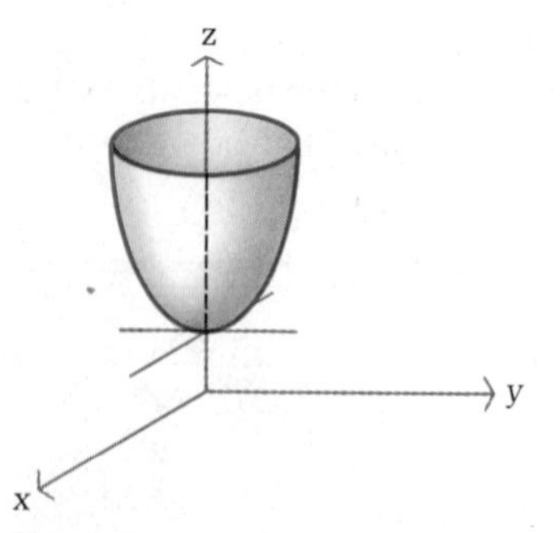

Figure 8

Step 1. Find critical points. Find (x, y) such that $f_x(x, y) = 0$ and $f_y(x, y) = 0$ simultaneously:

$$f_x(x, y) = 2x = 0$$
$$x = 0$$
$$f_y(x, y) = 2y = 0$$
$$y = 0$$

The only critical point is $(a, b) = (0, 0)$.

Step 2. Compute $A = f_{xx}(0, 0)$, $B = f_{xy}(0, 0)$, and $C = f_{yy}(0, 0)$:

$$f_{xx}(x, y) = 2, \quad \text{thus} \quad A = f_{xx}(0, 0) = 2$$
$$f_{xy}(x, y) = 0, \quad \text{thus} \quad B = f_{xy}(0, 0) = 0$$
$$f_{yy}(x, y) = 2, \quad \text{thus} \quad C = f_{yy}(0, 0) = 2$$

Step 3. Evaluate $AC - B^2$ and try to classify the critical point (0, 0) using Theorem 2:

$$AC - B^2 = 2 \cdot 2 - 0^2 = 4 > 0 \quad \text{and} \quad A = 2 > 0$$

Therefore, case 2 in Theorem 2 holds. That is, $f(0, 0) = 2$ is a local minimum.

We will now use Theorem 2 in the following examples to analyze extrema without the aid of graphs.

Example 10 Use Theorem 2 to find local extrema for

$$f(x, y) = -x^2 - y^2 + 6x + 8y - 21$$

Solution Step 1. Find critical points. Find (x, y) such that $f_x(x, y) = 0$ and $f_y(x, y) = 0$ simultaneously:

$$f_x(x, y) = -2x + 6 = 0$$
$$x = 3$$
$$f_y(x, y) = -2y + 8 = 0$$
$$y = 4$$

The only critical point is $(a, b) = (3, 4)$.

Step 2. Compute $A = f_{xx}(3, 4)$, $B = f_{xy}(3, 4)$, and $C = f_{yy}(3, 4)$:

$$f_{xx}(x, y) = -2, \quad \text{thus} \quad A = f_{xx}(3, 4) = -2$$
$$f_{xy}(x, y) = 0, \quad \text{thus} \quad B = f_{xy}(3, 4) = 0$$
$$f_{yy}(x, y) = -2, \quad \text{thus} \quad C = f_{yy}(3, 4) = -2$$

Step 3. Evaluate $AC - B^2$ and try to classify the critical point (3, 4) using Theorem 2:

$$AC - B^2 = (-2)(-2) - (0)^2 = 4 > 0 \quad \text{and} \quad A = -2 < 0$$

Therefore, case 1 in Theorem 2 holds. That is, $f(3, 4) = 4$ is a local maximum.

Problem 10 Use Theorem 2 to find local extrema for

$$f(x, y) = x^2 + y^2 - 10x - 2y + 36$$

Example 11 Use Theorem 2 to find local extrema for

$$f(x, y) = x^3 + y^3 - 6xy$$

Solution Step 1. Find critical points for $f(x, y) = x^3 + y^3 - 6xy$:

$$f_x(x, y) = 3x^2 - 6y = 0 \qquad \text{Solve for } y.$$

$$6y = 3x^2$$

$$y = \frac{1}{2}x^2 \tag{2}$$

$$f_y(x, y) = 3y^2 - 6x = 0$$

$$3y^2 = 6x \qquad \text{Use (2) to eliminate } y.$$

$$3\left(\frac{1}{2}x^2\right)^2 = 6x$$

$$\frac{3}{4}x^4 = 6x \qquad \text{Solve for } x.$$

$$3x^4 - 24x = 0$$

$$3x(x^3 - 8) = 0$$

$$x = 0 \quad \text{or} \quad x = 2$$

$$y = 0 \qquad y = \frac{1}{2}(2)^2 = 2$$

The critical points are (0, 0) and (2, 2). Since there are two critical points, steps 2 and 3 must be performed twice.

Test (0, 0) Step 2. Compute $A = f_{xx}(0, 0)$, $B = f_{xy}(0, 0)$, and $C = f_{yy}(0, 0)$:

$$f_{xx}(x, y) = 6x, \quad \text{thus} \quad A = f_{xx}(0, 0) = 0$$

$$f_{xy}(x, y) = -6, \quad \text{thus} \quad B = f_{xy}(0, 0) = -6$$

$$f_{yy}(x, y) = 6y, \quad \text{thus} \quad C = f_{yy}(0, 0) = 0$$

Step 3. Evaluate $AC - B^2$ and try to classify the critical point (0, 0) using Theorem 2:

$$AC - B^2 = (0)(0) - (-6)^2 = -36 < 0$$

Therefore, case 3 in Theorem 2 applies. That is, f has a saddle point at (0, 0).

Now we will consider the second critical point, (2, 2).

Test (2, 2) Step 2. Compute $A = f_{xx}(2, 2)$, $B = f_{xy}(2, 2)$, and $C = f_{yy}(2, 2)$:

$$f_{xx}(x, y) = 6x, \quad \text{thus} \quad A = f_{xx}(2, 2) = 12$$

$$f_{xy}(x, y) = -6, \quad \text{thus} \quad B = f_{xy}(2, 2) = -6$$

$$f_{yy}(x, y) = 6y, \quad \text{thus} \quad C = f_{yy}(2, 2) = 12$$

Step 3. Evaluate $AC - B^2$ and try to classify the critical point (2, 2) using Theorem 2:

$$AC - B^2 = (12)(12) - (-6)^2 = 108 > 0 \quad \text{and} \quad A = 12 > 0$$

Thus, case 2 in Theorem 2 applies and $f(2, 2) = -8$ is a local minimum.

Problem 11 Use Theorem 2 to find local extrema for

$$f(x, y) = x^3 + y^2 - 6xy$$

Example 12 Profit

Suppose the surfboard company discussed earlier has developed the yearly profit equation

$$P(x, y) = -2x^2 + 2xy - y^2 + 10x - 4y + 107$$

where x is the number (in thousands) of standard surfboards produced per year, y is the number (in thousands) of competition surfboards produced per year, and P is profit (in thousands of dollars). How many of each type of board should be produced per year to realize a maximum profit? What is the maximum profit?

Solution

Step 1. Find critical points:

$$P_x(x, y) = -4x + 2y + 10 = 0$$
$$P_y(x, y) = 2x - 2y - 4 = 0$$

Solving this system, we obtain (3, 1) as the only critical point.

Step 2. Compute $A = P_{xx}(3, 1)$, $B = P_{xy}(3, 1)$, and $C = P_{yy}(3, 1)$:

$$P_{xx}(x, y) = -4, \quad \text{thus} \quad A = P_{xx}(3, 1) = -4$$
$$P_{xy}(x, y) = 2, \quad \text{thus} \quad B = P_{xy}(3, 1) = 2$$
$$P_{yy}(x, y) = -2, \quad \text{thus} \quad C = P_{yy}(3, 1) = -2$$

Step 3. Evaluate $AC - B^2$ and try to classify the critical point (3, 1) using Theorem 2:

$$AC - B^2 = (-4)(-2) - (2)^2 = 8 - 4 = 4 > 0$$
$$A = -4 < 0$$

Therefore, case 1 in Theorem 2 applies. That is, $P(3, 1) = \$120{,}000$ is a local maximum. This is obtained by producing 3,000 standard boards and 1,000 competition boards per year.

Problem 12 Repeat Example 12 with

$$P(x, y) = -2x^2 + 4xy - 3y^2 + 4x - 2y + 77$$

Example 13 Package Design

The packaging department in a company has been asked to design a rectangular box with no top and a partition down the middle. The box must have a volume of 48 cubic inches. Find the dimensions that will minimize the amount of material used to construct the box.

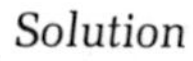

Solution

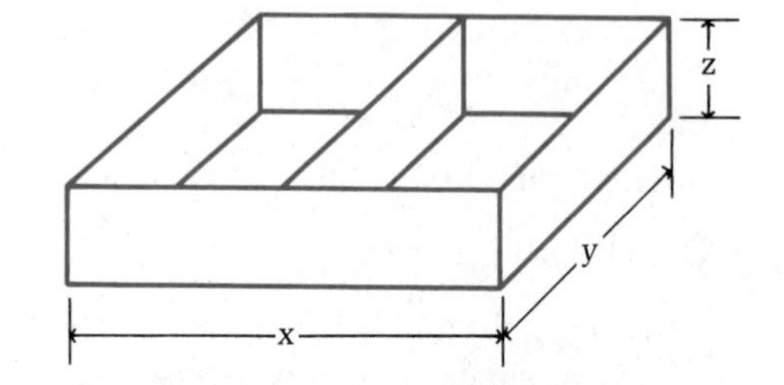

The amount of material used in constructing this box is

$$M = \underset{\text{Base}}{xy} + \underset{\substack{\text{Front}\\ \text{and}\\ \text{back}}}{2xz} + \underset{\substack{\text{Sides}\\ \text{and}\\ \text{partition}}}{3yz} \tag{3}$$

The volume of the box is

$$V = xyz = 48 \tag{4}$$

Since Theorem 2 applies only to functions with two independent variables, we must use (4) to eliminate one of the variables in (3).

$$M = xy + 2xz + 3yz \qquad \text{Substitute } z = \frac{48}{xy}$$

$$= xy + 2x\left(\frac{48}{xy}\right) + 3y\left(\frac{48}{xy}\right)$$

$$= xy + \frac{96}{y} + \frac{144}{x}$$

Thus, we must find the minimum value of

$$M(x, y) = xy + \frac{96}{y} + \frac{144}{x}$$

Step 1. Find critical points:

$$M_x(x, y) = y - \frac{144}{x^2} = 0$$

$$y = \frac{144}{x^2} \tag{5}$$

$$M_y(x, y) = x - \frac{96}{y^2} = 0$$

$$x = \frac{96}{y^2} \qquad \text{Solve for } y^2.$$

$$y^2 = \frac{96}{x} \qquad \text{Use (5) to eliminate } y \text{ and solve for } x.$$

$$\left(\frac{144}{x^2}\right)^2 = \frac{96}{x}$$

$$\frac{20{,}736}{x^4} = \frac{96}{x}$$

$$x^3 = \frac{20{,}736}{96} = 216$$

$$x = 6 \qquad \text{Use (5) to find } y.$$

$$y = \frac{144}{36} = 4$$

Step 2. Compute $A = M_{xx}(6, 4)$, $B = M_{xy}(6, 4)$, and $C = M_{yy}(6, 4)$:

$$M_{xx}(x, y) = \frac{288}{x^3}, \quad \text{thus} \quad A = M_{xx}(6, 4) = \frac{288}{216} = \frac{4}{3}$$

$$M_{xy} = 1, \quad \text{thus} \quad B = M_{xy}(6, 4) = 1$$

$$M_{yy}(x, y) = \frac{192}{y^3}, \quad \text{thus} \quad C = M_{yy}(6, 4) = \frac{192}{64} = 3$$

Step 3. Evaluate $AC - B^2$ and try to classify the critical point (6, 4) using Theorem 2:

$$AC - B^2 = \left(\frac{4}{3}\right)(3) - (1)^2 = 3 > 0 \quad \text{and} \quad A = \frac{4}{3} > 0$$

Therefore, case 2 in Theorem 2 applies; $M(x, y)$ has a local minimum at (6, 4). If $x = 6$ and $y = 4$, then

$$z = \frac{48}{xy} = \frac{48}{6(4)} = 2$$

Thus, the dimensions that will require the minimum amount of material are 6 inches by 4 inches by 2 inches.

Problem 13 If the box in Example 13 must have a volume of 384 cubic inches, find the dimensions that will require the least amount of material.

Answers to Matched Problems

10. $f(5, 1) = 10$ is a local minimum
11. f has a saddle point at (0, 0); $f(6, 18) = -108$ is a local minimum
12. Local maximum for $x = 2$ and $y = 1$; $P(2, 1) = \$80{,}000$
13. 12 inches by 8 inches by 4 inches

Exercise 6-3

Find local extrema using Theorem 2.

A

1. $f(x, y) = 6 - x^2 - 4x - y^2$
2. $f(x, y) = 3 - x^2 - y^2 + 6y$

3. $f(x, y) = x^2 + y^2 + 2x - 6y + 14$
4. $f(x, y) = x^2 + y^2 - 4x + 6y + 23$

B

5. $f(x, y) = xy + 2x - 3y - 2$
6. $f(x, y) = x^2 - y^2 + 2x + 6y - 4$
7. $f(x, y) = -3x^2 + 2xy - 2y^2 + 14x + 2y + 10$
8. $f(x, y) = -x^2 + xy - 2y^2 + x + 10y - 5$
9. $f(x, y) = 2x^2 - 2xy + 3y^2 - 4x - 8y + 20$
10. $f(x, y) = 2x^2 - xy + y^2 - x - 5y + 8$

C

11. $f(x, y) = e^{xy}$
12. $f(x, y) = x^2y - xy^2$
13. $f(x, y) = x^3 + y^3 - 3xy$
14. $f(x, y) = 2y^3 - 6xy - x^2$
15. $f(x, y) = 2x^4 + y^2 - 12xy$
16. $f(x, y) = 16xy - x^4 - 2y^2$
17. $f(x, y) = x^3 - 3xy^2 + 6y^2$
18. $f(x, y) = 2x^2 - 2x^2y + 6y^3$

Applications

Business & Economics

19. *Product mix for maximum profit.* A firm produces two types of calculators, x thousand of type A and y thousand of type B per year. If the revenue and cost equations for the year are (in millions of dollars)

$$R(x, y) = 2x + 3y$$
$$C(x, y) = x^2 - 2xy + 2y^2 + 6x - 9y + 5$$

find how many of each type of calculator should be produced per year to maximize profit. What is the maximum profit?

20. *Automation–labor mix for minimum cost.* The annual labor and automated equipment cost (in millions of dollars) for a company's production of television sets is given by

$$C(x, y) = 2x^2 + 2xy + 3y^2 - 16x - 18y + 54$$

where x is the amount spent per year on labor and y is the amount spent per year on automated equipment (both in millions of dollars). Determine how much should be spent on each per year to minimize this cost. What is the minimum cost?

21. *Maximizing profit.* A department store sells two brands of inexpensive calculators. The store pays \$6 for each brand A calculator and \$8 for each brand B calculator. The research department has estimated the following weekly demand equations for these two competitive products:

$$x = 116 - 30p + 20q \qquad \text{Demand equation for brand A}$$
$$y = 144 + 16p - 24q \qquad \text{Demand equation for brand B}$$

where p is the selling price for brand A and q is the selling price for brand B.

(A) Determine the demands x and y when $p = \$10$ and $q = \$12$; when $p = \$11$ and $q = \$11$.

(B) How should the store price each calculator to maximize weekly profits? What is the maximum weekly profit? [*Hint:* $C = 6x + 8y$, $R = px + qy$, and $P = R - C$.]

22. *Maximizing profit.* A store sells two brands of color print film. The store pays \$2 for each roll of brand A film and \$3 for each roll of brand B film. A consulting firm has estimated the following daily demand equations for these two competitive products:

$$x = 75 - 40p + 25q \qquad \text{Demand equation for brand A}$$

$$y = 80 + 20p - 30q \qquad \text{Demand equation for brand B}$$

where p is the selling price for brand A and q is the selling price for brand B.

(A) Determine the demands x and y when $p = \$4$ and $q = \$5$; when $p = \$4$ and $q = \$4$.

(B) How should the store price each brand of film to maximize daily profits? What is the maximum daily profit? [*Hint:* $C = 2x + 3y$, $R = px + qy$, and $P = R - C$.]

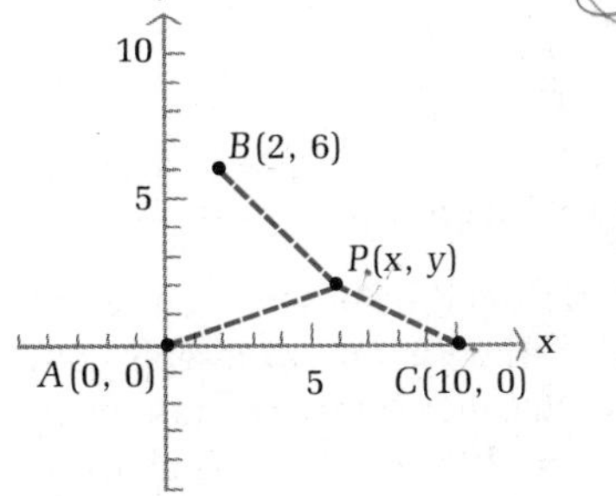

23. *Minimizing cost.* A satellite television reception station is to be located at $P(x, y)$ so that the sum of the squares of the distances from P to the three towns A, B, and C is minimum (see the figure). Find the coordinates of P. This location will minimize the cost of providing satellite cable television for all three towns.

24. *Minimizing cost.* Repeat Problem 23 replacing the coordinates of B with B(6, 9) and the coordinates of C with C(9, 0).

25. *Minimum material.* A rectangular box with no top and two parallel partitions (see accompanying figure) is to be made to hold 64 cubic inches. Find the dimensions that will require the least amount of material.

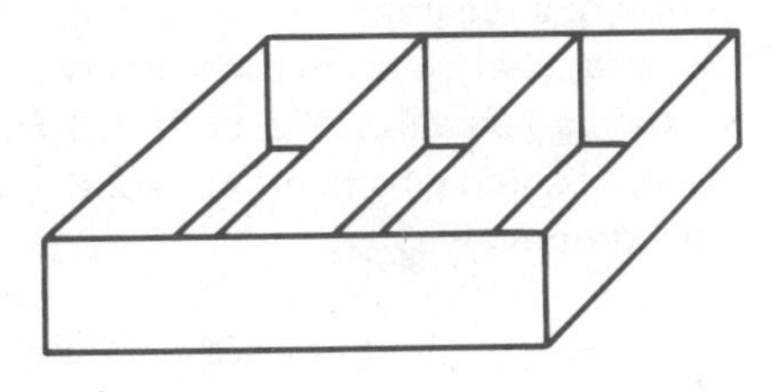

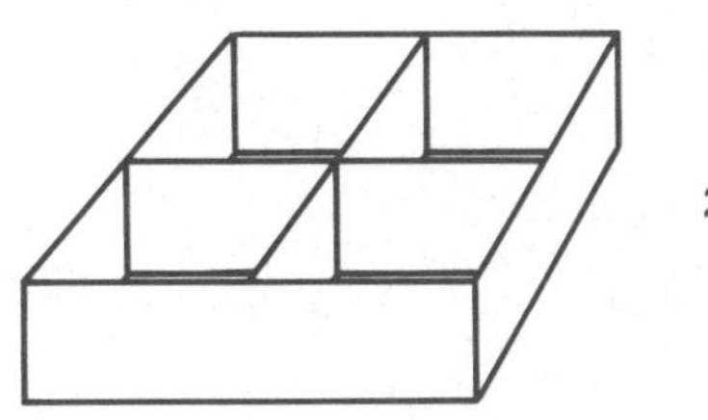

26. *Minimum material.* A rectangular box with no top and two intersecting partitions (see accompanying figure) is to be made to hold 72 cubic inches. What should its dimensions be in order to use the least amount of material in its construction?

27. *Maximum volume.* A mailing service states that a rectangular package shall have the sum of the length and girth not to exceed 120 inches (see the figure). What are the dimensions of the largest (in volume) mailing carton that can be constructed meeting these restrictions?

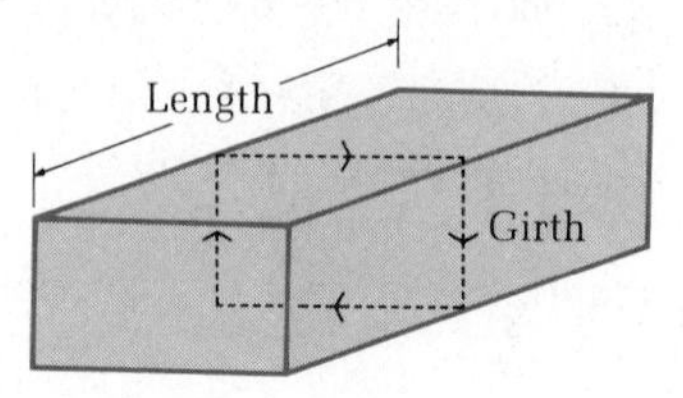

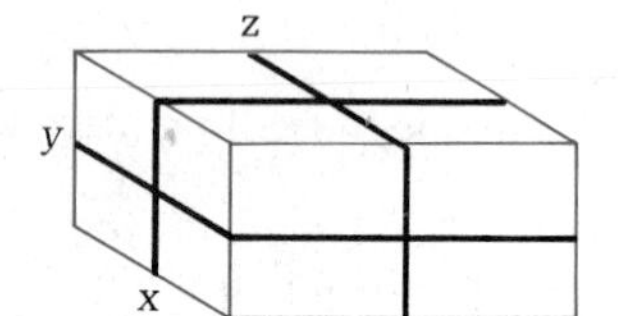

28. *Maximum shipping volume.* A shipping box is reinforced with steel bands in all three directions, as indicated in the figure. A total of 150 inches of steel tape are to be used, with 6 inches of waste because of a 2 inch overlap in each direction. Find the dimensions of the box with maximum volume that can be taped as indicated.

6-4 Maxima and Minima Using Lagrange Multipliers (Optional)

- Functions of Two Independent Variables
- Functions of Three Independent Variables

Functions of Two Independent Variables

We will now consider a particularly powerful method of solving a certain class of maxima–minima problems. The method is due to Joseph Louis Lagrange (1736–1813), an eminent eighteenth century French mathematician, and it is called the **method of Lagrange multipliers.** We introduce the method through an example; then we will formalize the discussion in the form of a theorem.

A rancher wants to construct two feeding pens of the same size along an existing fence (see Fig. 9). If 720 feet of fencing are available, how long should x and y be in order to obtain the maximum total area? What is the maximum area?

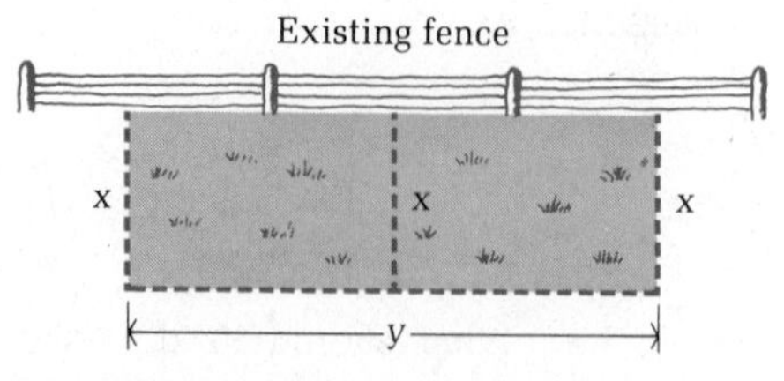

Figure 9

The total area is given by

$$f(x, y) = xy$$

which can be made as large as we like providing there are no restrictions on x and y. But there are restrictions on x and y, since we have only 720 feet of fencing. That is, x and y must be chosen so that

$$3x + y = 720$$

This restriction on x and y, also called a **constraint,** leads to the following maxima–minima problem:

$$\text{Maximize} \quad f(x, y) = xy \tag{1}$$

$$\text{Subject to} \quad 3x + y = 720 \quad \text{or} \quad 3x + y - 720 = 0 \tag{2}$$

This problem is a special case of a general class of problems of the form

$$\text{Maximize (or minimize)} \quad z = f(x, y) \tag{3}$$

$$\text{Subject to} \quad g(x, y) = 0 \tag{4}$$

Of course, we could try to solve (4) for y in terms of x, or for x in terms of y, then substitute the result into (3), and use methods developed in Section 2-3 for functions of a single variable. But what if (4) were more complicated than (2), and solving for one variable in terms of the other was either very difficult or impossible? In the method of Lagrange multipliers we work with $g(x, y)$ directly and avoid having to solve (4) for one variable in terms of the other. In addition, the method generalizes to functions of arbitrarily many variables subject to one or more constraints.

Now, to the method. We form a new function F, using functions f and g in (3) and (4), as follows:

$$F(x, y, \lambda) = f(x, y) + \lambda g(x, y) \tag{5}$$

where λ (lambda) is called a **Lagrange multiplier.** Theorem 3 forms the basis for the method.

Theorem 3

The relative maxima and minima of the function $z = f(x, y)$ subject to the constraint $g(x, y) = 0$ will be among those points (x_0, y_0) for which (x_0, y_0, λ_0) is a solution to the system

$$F_x(x, y, \lambda) = 0$$
$$F_y(x, y, \lambda) = 0$$
$$F_\lambda(x, y, \lambda) = 0$$

where $F(x, y, \lambda) = f(x, y) + \lambda g(x, y)$, provided all the partial derivatives exist.

We now solve the fence problem using the method of Lagrange multipliers.

Step 1. Formulate the problem in the form of equations (3) and (4):

$$\text{Maximize} \quad f(x, y) = xy$$
$$\text{Subject to} \quad g(x, y) = 3x + y - 720 = 0$$

Step 2. Form the function F, introducing the Lagrange multiplier λ:

$$\begin{aligned} F(x, y, \lambda) &= f(x, y) + \lambda g(x, y) \\ &= xy + \lambda(3x + y - 720) \end{aligned}$$

Step 3. Solve the system $F_x = 0$, $F_y = 0$, $F_\lambda = 0$. (Solutions are called **critical points** for F.)

$$F_x = y + 3\lambda = 0$$
$$F_y = x + \lambda = 0$$
$$F_\lambda = 3x + y - 720 = 0$$

From the first two equations, we see that

$$y = -3\lambda$$
$$x = -\lambda$$

Substitute these values for x and y into the third equation and solve for λ.

$$\begin{aligned} -3\lambda - 3\lambda &= 720 \\ -6\lambda &= 720 \\ \lambda &= -120 \end{aligned}$$

Thus,

$$y = -3(-120) = 360 \text{ feet}$$
$$x = -(-120) = 120 \text{ feet}$$

Step 4. Test the critical points for maxima and minima. The function F has only one critical point at (120, 360, −120), and since $f(x, y) = xy$ has a minimum at (0, 0), we conclude that (120, 360) produces a maximum for f. Hence,

$$\begin{aligned} \text{Max } f(x, y) &= f(120, 360) \\ &= (120)(360) = 43{,}200 \text{ square feet} \end{aligned}$$

The key steps in applying the method of Lagrange multipliers are listed in the following box:

Method of Lagrange Multipliers—Key Steps

1. Formulate the problem in the form

   Maximize (or minimize) $z = f(x, y)$

   Subject to $g(x, y) = 0$

2. Form the function F:

   $$F(x, y, \lambda) = f(x, y) + \lambda g(x, y)$$

3. Find the critical points for F; that is, solve the system

   $$F_x(x, y, \lambda) = 0$$
   $$F_y(x, y, \lambda) = 0$$
   $$F_\lambda(x, y, \lambda) = 0$$

4. Evaluate $z = f(x, y)$ at each point (x_0, y_0) such that (x_0, y_0, λ_0) satisfies the system in step 3. The maximum or minimum value of $f(x, y)$ will be among these values in the problems we consider.

Example 14 Minimize $f(x, y) = x^2 + y^2$ subject to $x + y = 10$.

Solution

Step 1. Minimize $f(x, y) = x^2 + y^2$

Subject to $g(x, y) = x + y - 10 = 0$

Step 2. $F(x, y, \lambda) = x^2 + y^2 + \lambda(x + y - 10)$

Step 3.

$$F_x = 2x + \lambda = 0$$
$$F_y = 2y + \lambda = 0$$
$$F_\lambda = x + y - 10 = 0$$

From the first two equations,

$$x = -\frac{\lambda}{2} \qquad y = -\frac{\lambda}{2}$$

Substituting these into the third equation, we obtain

$$-\frac{\lambda}{2} - \frac{\lambda}{2} = 10$$
$$-\lambda = 10$$
$$\lambda = -10$$

The critical point is $(5, 5, -10)$.

Step 4. $f(5, 5) = 5^2 + 5^2 = 50$

Checking other points on the line $x + y = 10$ near $(5, 5)$, we see that this is a minimum. (See Fig. 10 on the next page.)

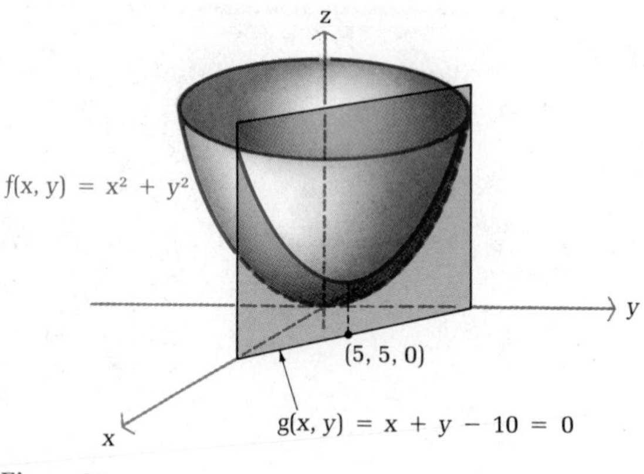

Figure 10

Problem 14 Maximize $f(x, y) = 25 - x^2 - y^2$ subject to $x + y = 4$. (See Fig. 11.)

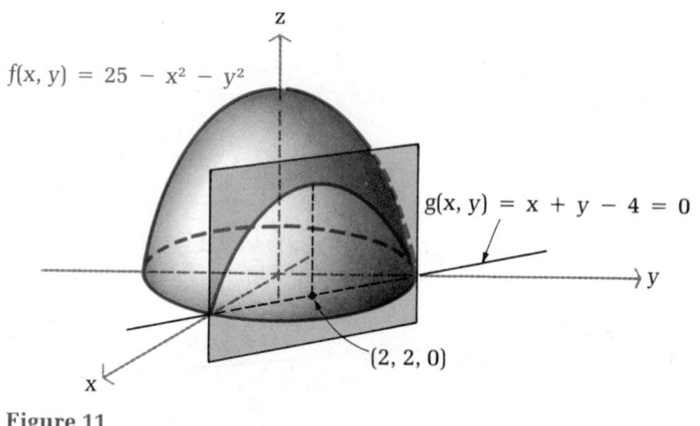

Figure 11

▪ Functions of Three Independent Variables

We have indicated that the method of Lagrange multipliers can be extended to functions with arbitrarily many independent variables with one or more constraints. We state a theorem for functions with three independent variables and one constraint and consider an example that will demonstrate the advantage of the method of Lagrange multipliers over the method used in Section 6-3.

Theorem 4

The relative maxima and minima of the function $w = f(x, y, z)$ subject to the constraint $g(x, y, z) = 0$ will be among the set of points (x_0, y_0, z_0) for which $(x_0, y_0, z_0, \lambda_0)$ is a solution to the system

$$F_x(x, y, z, \lambda) = 0$$
$$F_y(x, y, z, \lambda) = 0$$
$$F_z(x, y, z, \lambda) = 0$$
$$F_\lambda(x, y, z, \lambda) = 0$$

where $F(x, y, z, \lambda) = f(x, y, z) + \lambda g(x, y, z)$, provided all the partial derivatives exist.

Example 15
Package Design

A rectangular box with an open top and one partition is to be constructed from 162 square inches of cardboard. Find the dimensions that will result in a box with the largest possible volume.

Solution

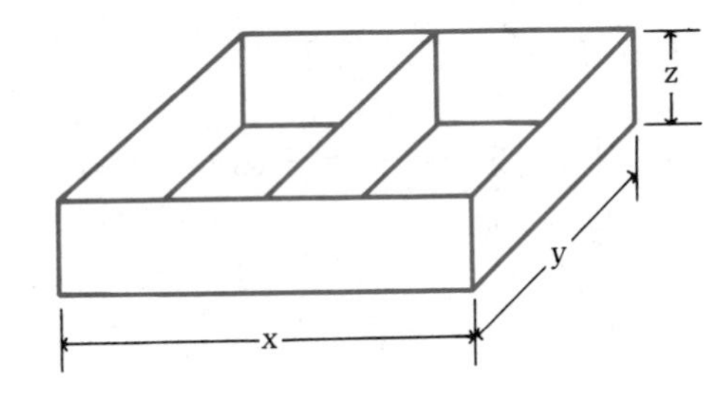

We must maximize

$$V(x, y, z) = xyz$$

subject to the constraint that the amount of material used is 162 square inches. Thus, x, y, and z must satisfy

$$xy + 2xz + 3yz = 162$$

Step 1. Maximize $V(x, y, z) = xyz$
Subject to $g(x, y, z) = xy + 2xz + 3yz - 162 = 0$

Step 2. $F(x, y, z, \lambda) = xyz + \lambda(xy + 2xz + 3yz - 162)$

Step 3. $F_x = yz + \lambda(y + 2z) = 0$
$F_y = xz + \lambda(x + 3z) = 0$
$F_z = xy + \lambda(2x + 3y) = 0$
$F_\lambda = xy + 2xz + 3yz - 162 = 0$

From the first two equations, we can write

$$\lambda = \frac{-yz}{y + 2z} \qquad \lambda = \frac{-xz}{x + 3z}$$

Eliminating λ, we have

$$\frac{-yz}{y + 2z} = \frac{-xz}{x + 3z}$$

$$-xyz - 3yz^2 = -xyz - 2xz^2$$

$$3yz^2 = 2xz^2 \qquad \text{We can assume } z \neq 0.$$

$$3y = 2x$$

$$x = \frac{3}{2}y$$

From the second and third equations,

$$\lambda = \frac{-xz}{x + 3z} \qquad \lambda = \frac{-xy}{2x + 3y}$$

Eliminating λ, we have

$$\frac{-xz}{x + 3z} = \frac{-xy}{2x + 3y}$$

$$-2x^2z - 3xyz = -x^2y - 3xyz$$

$$2x^2z = x^2y \qquad \text{We can assume } x \neq 0.$$

$$2z = y$$

$$z = \frac{1}{2}y$$

Substituting $x = \frac{3}{2}y$ and $z = \frac{1}{2}y$ in the fourth equation, we have

$$\left(\frac{3}{2}y\right)y + 2\left(\frac{3}{2}y\right)\left(\frac{1}{2}y\right) + 3y\left(\frac{1}{2}y\right) - 162 = 0$$

$$\frac{3}{2}y^2 + \frac{3}{2}y^2 + \frac{3}{2}y^2 = 162$$

$$y^2 = 36 \qquad \text{We can assume}$$

$$y = 6 \qquad y > 0.$$

$$x = \frac{3}{2}(6) = 9 \qquad \text{Using } x = \frac{3}{2}y$$

$$z = \frac{1}{2}(6) = 3 \qquad \text{Using } z = \frac{1}{2}y$$

Since (9, 6, 3) is the only critical point with x, y, and z all positive, the dimensions of the box with maximum volume are 9 inches by 6 inches by 3 inches.

Problem 15 Find the dimensions of the box of the type described in Example 15 with the largest volume that can be constructed from 288 square inches of cardboard.

Suppose we had decided to solve Example 15 by the method used in Section 6-3. First we would have to solve the material constraint for one of the variables, say z:

$$z = \frac{162 - xy}{2x + 3y}$$

Then we would eliminate z in the volume function and maximize

$$V(x, y) = xy\frac{162 - xy}{2x + 3y}$$

Using the method of Lagrange multipliers allows us to avoid the formidable task of finding the partial derivatives of V.

Answers to Matched Problems

14. Max $f(x, y) = f(2, 2) = 17$ (see Fig. 11)
15. 12 inches by 8 inches by 4 inches

Exercise 6-4

Use the method of Lagrange multipliers in the following problems:

A

1. Maximize $f(x, y) = 2xy$
   Subject to $x + y = 6$
2. Minimize $f(x, y) = 6xy$
   Subject to $y - x = 6$
3. Minimize $f(x, y) = x^2 + y^2$
   Subject to $3x + 4y = 25$
4. Maximize $f(x, y) = 25 - x^2 - y^2$
   Subject to $2x + y = 10$

B

5. Find the maximum and minimum of $f(x, y) = 2xy$ subject to $x^2 + y^2 = 18$.
6. Find the maximum and minimum of $f(x, y) = x^2 - y^2$ subject to $x^2 + y^2 = 25$.
7. Maximize the product of two numbers if their sum must be 10.
8. Minimize the product of two numbers if their difference must be 10.

C

9. Minimize $f(x, y, z) = x^2 + y^2 + z^2$
   Subject to $2x - y + 3z = -28$
10. Maximize $f(x, y, z) = xyz$
    Subject to $2x + y + 2z = 120$
11. Maximize and minimize $f(x, y, z) = x + y + z$
    Subject to $x^2 + y^2 + z^2 = 12$

12. Maximize and minimize $f(x, y, z) = 2x + 4y + 4z$
    Subject to $x^2 + y^2 + z^2 = 9$

Applications

Business & Economics

13. *Budgeting for least cost.* A manufacturing company produces two models of a television set, x units of model A and y units of model B per week, at a cost in dollars of

$$C(x, y) = 6x^2 + 12y^2$$

If it is necessary (because of shipping considerations) that

$$x + y = 90$$

how many of each type of set should be manufactured per week to minimize cost? What is the minimum cost?

14. *Budgeting for maximum production.* A manufacturing firm has budgeted $60,000 per month for labor and materials. If x thousand dollars is spent on labor and y thousand dollars is spent on materials, and if the monthly output in units is given by

$$N(x, y) = 4xy - 8x$$

how should the $60,000 be allocated to labor and materials in order to maximize N? What is the maximum N?

15. *Productivity.* A consulting firm for a manufacturing company arrived at the following Cobb–Douglas production function for a particular product:

$$N(x, y) = 50x^{0.8}y^{0.2}$$

where x is the number of units of labor and y is the number of units of capital required to produce $N(x, y)$ units of the product. If $400,000 is budgeted for production of the product, each unit of labor costs $40, and each unit of capital costs $80, determine the number of labor units and capital units required to maximize production.

16. *Productivity.* The research department for a manufacturing company arrived at the following Cobb–Douglas production function for a particular product:

$$N(x, y) = 10x^{0.6}y^{0.4}$$

where x is the number of units of labor and y is the number of units of capital required to produce $N(x, y)$ units of the product. If $300,000 is budgeted for production of the product, each unit of labor costs $30, and each unit of capital costs $60, determine the number of labor units and capital units required to maximize production.

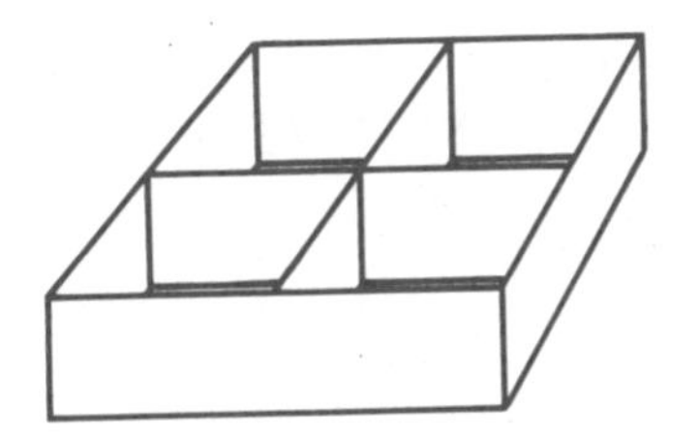

17. *Maximum volume.* A rectangular box with no top and two intersecting partitions is to be constructed from 192 square inches of cardboard (see accompanying figure). Find the dimensions that will maximize the volume.
18. *Maximum volume.* A mailing service states that a rectangular package shall have the sum of the length and girth not to exceed 120 inches (see the figure). What are the dimensions of the largest (in volume) mailing carton that can be constructed meeting these restrictions?

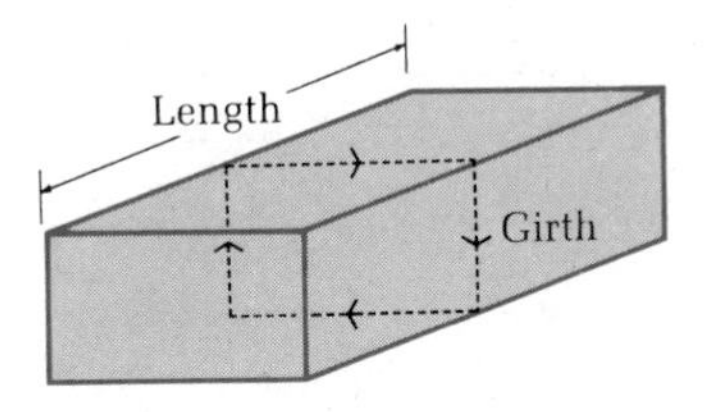

Life Sciences

19. *Agriculture.* Three pens of the same size are to be built along an existing fence (see the figure). If 400 feet of fencing are available, what length should x and y be to produce the maximum total area? What is the maximum area?

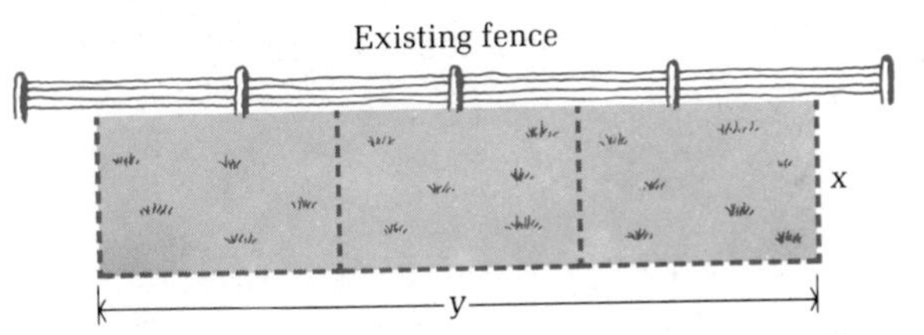

20. *Diet and minimum cost.* A group of guinea pigs is to receive 25,600 calories per week. Two available foods produce 200xy calories for a mixture of x kilograms of type M food and y kilograms of type N food. If type M costs \$1 per kilogram and type N costs \$2 per kilogram, how much of each type of food should be used to minimize weekly food costs? What is the minimum cost? [*Note:* $x \geq 0, y \geq 0$]

6-5 Method of Least Squares

- Least Squares Approximation
- Applications

Least Squares Approximation

In this section we will use the optimization techniques discussed in Section 6-3 to find the equation of a line which is a "best" approximation to a set of points in a rectangular coordinate system. This very popular method is known as **least squares approximation** or **linear regression.** Let us begin by considering a specific case.

A manufacturer wants to approximate the cost function for a product. The value of the cost function has been determined for certain levels of production, as listed in the table:

Number of Units x, in hundreds	Cost y, in thousands of dollars
2	4
5	6
6	7
9	8

Although these points do not all lie on a line (see Fig. 12), they are very close to being linear. The manufacturer would like to approximate the cost function by a linear function; that is, determine values m and d so that the line

$$y = mx + d$$

is, in some sense, the "best" approximation to the cost function.

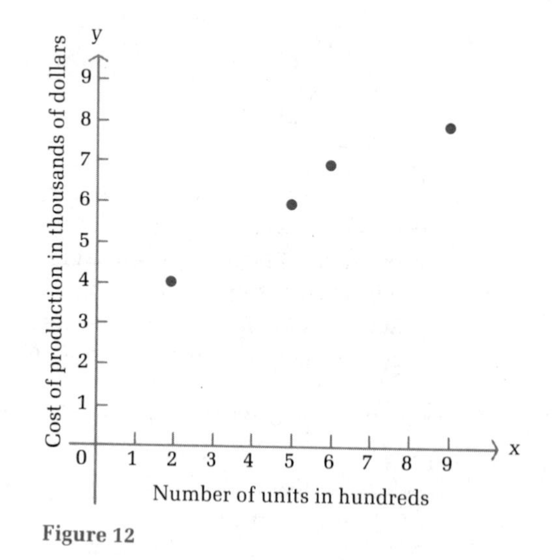

Figure 12

What do we mean by "best"? Since the line $y = mx + d$ will not go through all four points, it is reasonable to examine the differences between the y coordinates of the points listed in the table and the y coordinates of the corresponding points on the line. Each of these differences is called the **residual** at that point (see Fig. 13). For example, at $x = 2$ the point from the table is (2, 4) and the point on the line is $(2, 2m + d)$, so the residual is

$$4 - (2m + d) = 4 - 2m - d$$

All the residuals are listed in the table below:

x	y	mx + d	Residual
2	4	2m + d	4 − 2m − d
5	6	5m + d	6 − 5m − d
6	7	6m + d	7 − 6m − d
9	8	9m + d	8 − 9m − d

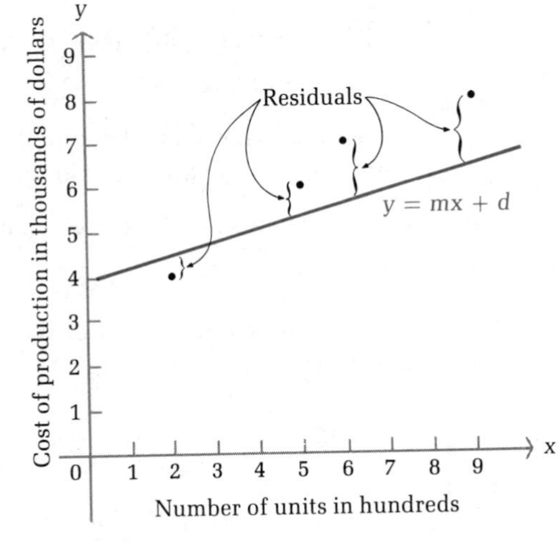

Figure 13

Our criterion for the "best" approximation is the following: Determine the values of m and d that *minimize* the sum of the squares of the residuals. The resulting line is called the **least squares line** or the **regression line.** To this end, we minimize

$$F(m, d) = (4 - 2m - d)^2 + (6 - 5m - d)^2 + (7 - 6m - d)^2 + (8 - 9m - d)^2$$

Step 1. Find critical points:

$$F_m(m, d) = 2(4 - 2m - d)(-2) + 2(6 - 5m - d)(-5) + 2(7 - 6m - d)(-6) + 2(8 - 9m - d)(-9)$$

$$= -304 + 292m + 44d = 0$$

$$F_d(m, d) = 2(4 - 2m - d)(-1) + 2(6 - 5m - d)(-1) + 2(7 - 6m - d)(-1) + 2(8 - 9m - d)(-1)$$

$$= -50 + 44m + 8d = 0$$

Solving the system

$$-304 + 292m + 44d = 0$$
$$-50 + 44m + 8d = 0$$

we obtain $(m, d) = (0.58, 3.06)$ as the only critical point.

Step 2. Compute $A = F_{mm}(m, d)$, $B = F_{md}(m, d)$, and $C = F_{dd}(m, d)$:

$$F_{mm}(m, d) = 292, \quad \text{thus} \quad A = F_{mm}(0.58, 3.06) = 292$$
$$F_{md}(m, d) = 44, \quad \text{thus} \quad B = F_{md}(0.58, 3.06) = 44$$
$$F_{dd}(m, d) = 8, \quad \text{thus} \quad C = F_{dd}(0.58, 3.06) = 8$$

Step 3. Evaluate $AC - B^2$ and try to classify the critical point (m, d) using Theorem 2 in Section 6-3:

$$AC - B^2 = (292)(8) - (44)^2 = 400 > 0$$
$$A = 292 > 0$$

Therefore, case 2 in Theorem 2 applies, and $F(m, d)$ has a local minimum at the critical point $(0.58, 3.06)$.

Thus, the least squares line for the given data is

$$y = 0.58x + 3.06 \qquad \text{Least squares line}$$

Note that the sum of the squares of the residuals is minimized for this choice of m and d (see Fig. 14).

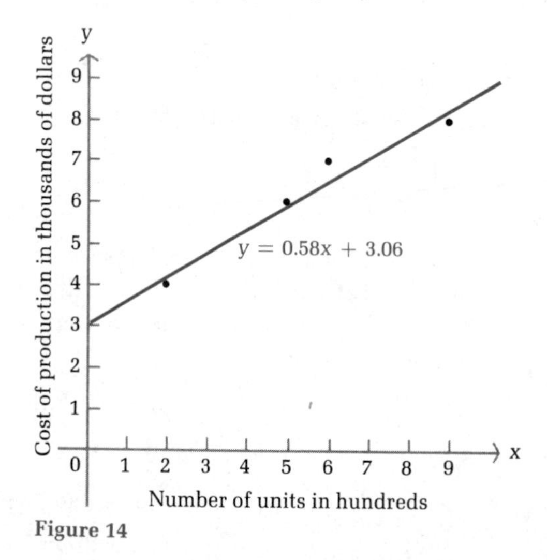

Figure 14

This linear function can now be used by the manufacturer to estimate any of the quantities normally associated with the cost function—such as costs, marginal costs, average costs, and so on. For example, the cost of producing 2,000 units is approximately

$$y = (0.58)(20) + 3.06 = 14.66 \quad \text{or} \quad \$14,660$$

The marginal cost function is

$$\frac{dy}{dx} = 0.58$$

The average cost function is

$$\bar{y} = \frac{0.58x + 3.06}{x}$$

In general, if we are given a set of n points $(x_1, y_1), (x_2, y_2), \ldots, (x_n, y_n)$, then it can be shown that the coefficients m and d of the least squares line $y = mx + d$ must satisfy the system of equations

$$\left(\sum_{k=1}^{n} x_k\right)m + nd = \sum_{k=1}^{n} y_k \tag{1}$$

$$\left(\sum_{k=1}^{n} x_k^2\right)m + \left(\sum_{k=1}^{n} x_k\right)d = \sum_{k=1}^{n} x_k y_k \tag{2}$$

Using the notation

$$\bar{x} = \frac{1}{n}\sum_{k=1}^{n} x_k \qquad \text{Average of the x coordinates}$$

$$\bar{y} = \frac{1}{n}\sum_{k=1}^{n} y_k \qquad \text{Average of the y coordinates}$$

to simplify the form of equations (1) and (2) and solving for m and d produces the formulas given in the box.

Least Squares Approximation

For a set of n points $(x_1, y_1), (x_2, y_2), \ldots, (x_n, y_n)$, the coefficients m and d of the least squares line

$$y = mx + d$$

are given by the formulas

$$m = \frac{\sum_{k=1}^{n} x_k y_k - n\bar{x}\bar{y}}{\sum_{k=1}^{n} x_k^2 - n\bar{x}^2} \tag{3}$$

$$d = \bar{y} - \bar{x}m \tag{4}$$

where

$$\bar{x} = \frac{1}{n}\sum_{k=1}^{n} x_k \qquad \text{Average of the x coordinates}$$

$$\bar{y} = \frac{1}{n}\sum_{k=1}^{n} y_k \qquad \text{Average of the y coordinates}$$

Since the value of m is used in equation (4) to compute the value of d, the value of m must always be computed first. Notice that equation (4) implies that the point $(\bar{x}, \bar{y})$ is always on the least squares line.

Applications

Example 16 Educational Testing

The table lists the midterm and final examination scores for ten students in a calculus course.

Midterm	Final	Midterm	Final
49	61	78	77
53	47	83	81
67	72	85	79
71	76	91	93
74	68	99	99

(A) Find the least squares line for the data given in the table.
(B) Use the least squares line to predict the final examination score for a student who scored 95 on the midterm examination.
(C) Graph the data and the least squares line on the same set of axes.

Solutions (A) A table is a convenient way to compute all the sums in the formulas for m and d:

	x_k	y_k	$x_k y_k$	x_k^2
	49	61	2,989	2,401
	53	47	2,491	2,809
	67	72	4,824	4,489
	71	76	5,396	5,041
	74	68	5,032	5,476
	78	77	6,006	6,084
	83	81	6,723	6,889
	85	79	6,715	7,225
	91	93	8,463	8,281
	99	99	9,801	9,801
Totals	750	753	58,440	58,496

Thus,

$$\bar{x} = \frac{1}{10}\sum_{k=1}^{10} x_k = \frac{1}{10}(750) = 75.0$$

$$\bar{y} = \frac{1}{10}\sum_{k=1}^{10} y_k = \frac{1}{10}(753) = 75.3$$

$$\sum_{k=1}^{10} x_k y_k = 58{,}440$$

$$\sum_{k=1}^{10} x_k^2 = 58{,}496$$

Substituting the appropriate values in equation (3),

$$m = \frac{\sum_{k=1}^{n} x_k y_k - n\bar{x}\bar{y}}{\sum_{k=1}^{n} x_k^2 - n\bar{x}^2}$$

$$= \frac{58{,}440 - 10(75.0)(75.3)}{58{,}496 - 10(75.0)^2} = \frac{1{,}965}{2{,}246} \approx 0.875$$

Then, using equation (4),

$$d = \bar{y} - \bar{x}m$$
$$\approx 75.3 - (75.0)(0.875) \approx 9.68$$

The least squares line is given (approximately) by

$$y = 0.875x + 9.68$$

(B) If $x = 95$, then the predicted score on the final examination is

$$y = 0.875(95) + 9.68$$
$$\approx 93 \qquad \text{Assuming that the score must be an integer}$$

(C)

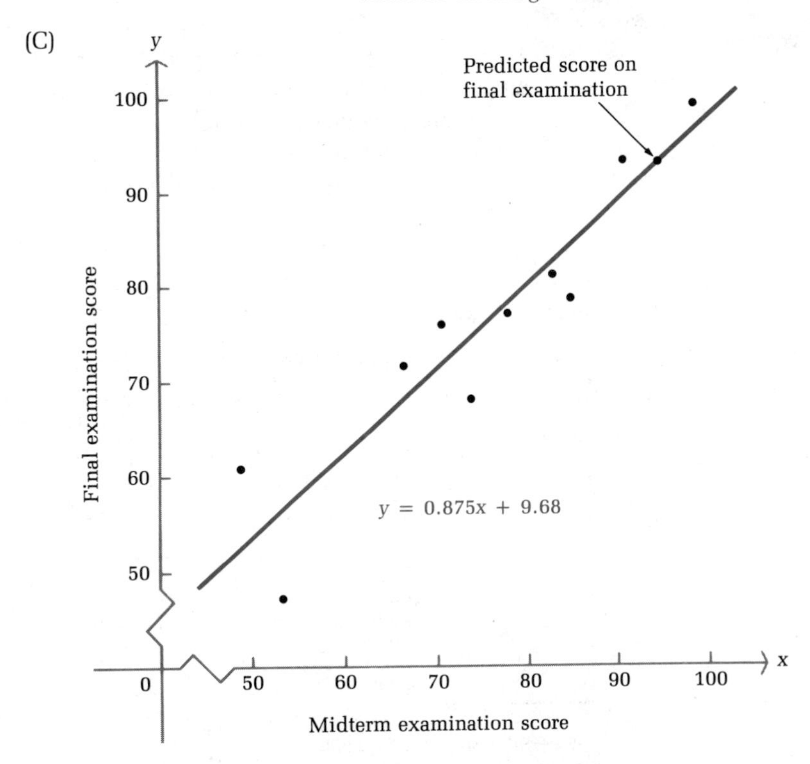

Problem 16 Repeat Example 16 for the following scores:

Midterm	Final	Midterm	Final
54	50	84	80
60	66	88	95
75	80	89	85
76	68	97	94
78	71	99	86

Example 17
Wool Production

Table 1 lists the annual production of wool throughout the world for the years 1970–1980. Use the data in the table to predict the worldwide wool production for 1981.

Table 1
World Wool Production

Year	Millions of Pounds	Year	Millions of Pounds
1970	6,107	1976	5,827
1971	5,972	1977	5,838
1972	5,560	1978	5,983
1973	5,474	1979	6,168
1974	5,769	1980	6,285
1975	5,911		

Solution Solving this problem by hand is certainly possible, but would require considerable effort. Instead, we used a computer to perform the necessary computations. (The program we used can be found in the computer supplement for this text. See the Preface.) The computer output is listed in Table 2.

Table 2

Input to Program

```
11 DATA POINTS HAVE BEEN ENTERED.

DO YOU WANT TO SEE THE POINTS (Y/N)?Y

 DATA POINTS
----------------------
0              6107
1              5972
2              5560
3              5474
4              5769
5              5911
6              5827
7              5838
8              5983
9              6168
10             6285
----------------------

PRESS RETURN TO CONTINUE?
```

Output from Program

```
<------- LEAST SQUARES LINE ------->

SLOPE: M = 33.9
Y INTERCEPT: D = 5729.96
EQUATION: Y = 33.9 X + 5729.96
-----------------------------------

TO COMPUTE AN ESTIMATED VALUE OF Y,
ENTER AN X VALUE. ENTER 999 TO STOP.
?11

X = 11   Y = 6102.85

ENTER AN X VALUE. ENTER 999 TO STOP.
?999
```

Notice that we used x = 0 for 1970, x = 1 for 1971, and so on. Examining the computer output in Table 2, we see that the least squares line is

$$y = 33.9x + 5{,}729.96$$

and the estimated worldwide wool production in 1981 is 6,102.85 million pounds.

Problem 17 Use the least squares line in Example 17 to estimate the worldwide wool production in 1982.

Answers to Matched Problems

16. (A) $y = 0.85x + 9.47$ (B) 90.2
(C)

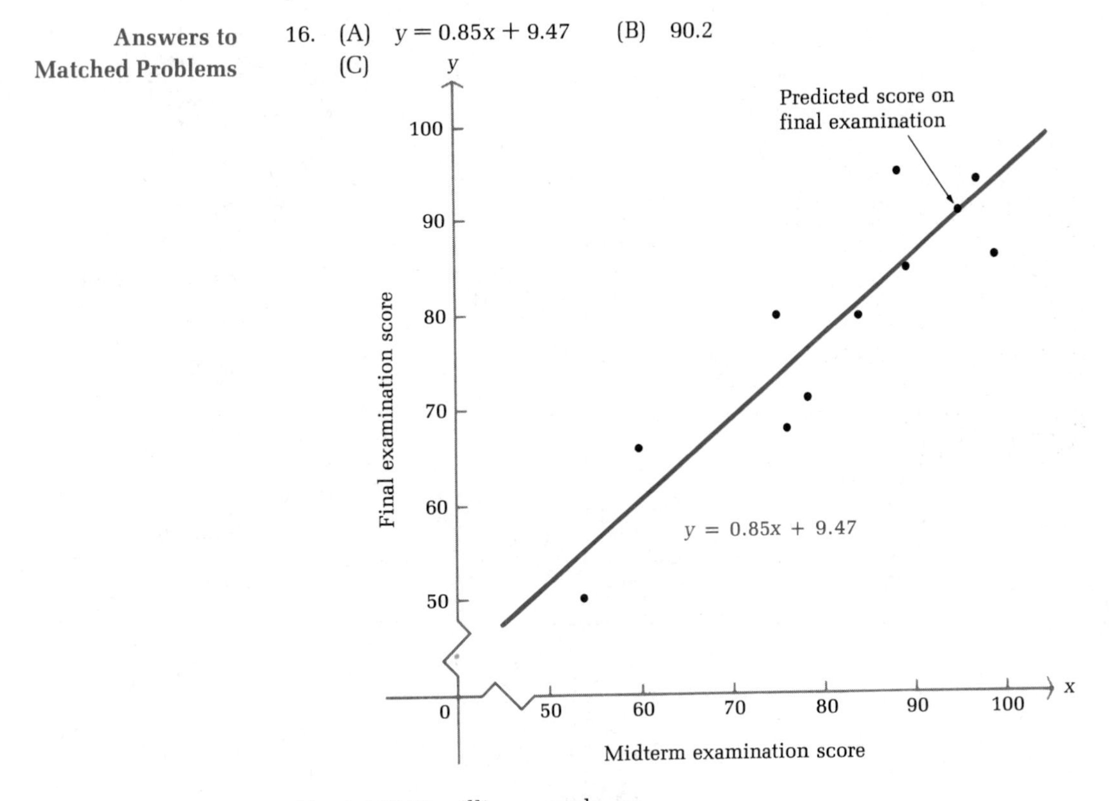

17. 6,136.76 million pounds

Exercise 6-5

A *Find the least squares line. Graph the data and the least squares line.*

1.

x	y
1	1
2	3
3	4
4	3

2.

x	y
1	−2
2	−1
3	3
4	5

3.

x	y
1	8
2	5
3	4
4	0

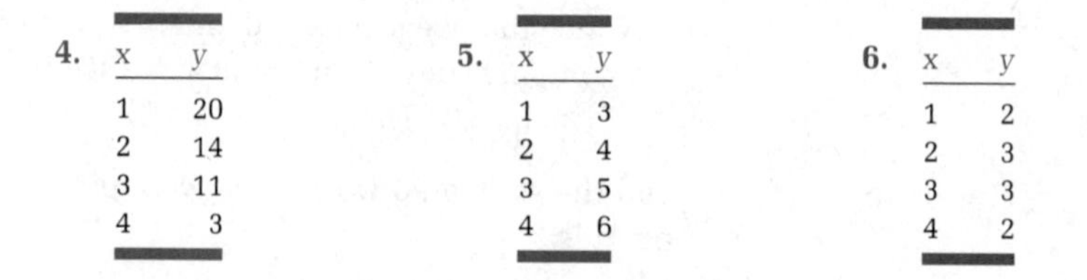

4.

x	y
1	20
2	14
3	11
4	3

5.

x	y
1	3
2	4
3	5
4	6

6.

x	y
1	2
2	3
3	3
4	2

B *Find the least squares line and use it to estimate y for the indicated value of* x.

7.

x	y
0	10
5	22
10	31
15	46
20	51

Estimate y when $x = 25$.

8.

x	y
−5	60
0	50
5	30
10	20
15	15

Estimate y when $x = 20$.

9.

x	y
−1	14
1	12
3	8
5	6
7	5

Estimate y when $x = 2$.

10.

x	y
2	−4
6	0
10	8
14	12
18	14

Estimate y for $x = 15$.

11.

x	y
0.5	25
2	22
3.5	21
5	21
6.5	18
9.5	12
11	11
12.5	8
14	5
15.5	1

Estimate y for $x = 8$.

12.

x	y
0	−15
2	−9
4	−7
6	−7
8	−1
12	11
14	13
16	19
18	25
20	33

Estimate y for $x = 10$.

C **13.** The method of least squares can be generalized to curves other than straight lines. To find the coefficients of the parabola

$$y = ax^2 + bx + c$$

that is the "best" fit for the points (1, 2), (2, 1), (3, 1), and (4, 3), minimize the sum of the squares of the residuals

$$F(a, b, c) = (a + b + c - 2)^2 + (4a + 2b + c - 1)^2 + (9a + 3b + c - 1)^2 + (16a + 4b + c - 3)^2$$

by solving the system

$$F_a(a, b, c) = 0 \qquad F_b(a, b, c) = 0 \qquad F_c(a, b, c) = 0$$

for a, b, and c. Graph the points and the parabola.

14. Repeat Problem 13 for the points $(-1, -2)$, $(0, 1)$, $(1, 2)$, and $(2, 0)$.

Applications

Business & Economics

15. *Cost.* The cost y in thousands of dollars for producing x units of a product at various times in the past is given in the table.

x	y
10	5
12	6
15	7
18	8
20	9

(A) Find the least squares line for the data.
(B) Use the least squares line to estimate the cost of producing 25 units.

x	y
4	100
5	120
6	150
7	190
8	240

16. *Advertising and sales.* A company spends x thousand dollars on advertising each month and has y thousand dollars in monthly sales. The data in the table were obtained by examining the past history of the company.

(A) Find the least squares line for the data.
(B) Use the least squares line to estimate the sales if $10,000 is spent on advertising.

x	y
5.0	2.0
5.5	1.8
6.0	1.4
6.5	1.2
7.0	1.1

17. *Maximizing profit.* The marketing research department for a drug store chain chose two summer resort areas to test market a new sun screen lotion packaged in 4 ounce plastic bottles. After a summer of varying the selling price and recording the monthly demand, the research department arrived at the given demand table, where y is the number of bottles purchased per month (in thousands) at x dollars per bottle.

(A) Find a demand equation using the method of least squares.
(B) If each bottle of sun screen costs the drug store chain $4, how should it be priced to achieve a maximum monthly profit? [*Hint:* Use the result of part A, with $C = 4y$, $R = xy$, and $P = R - C$.]

18. *Maximizing profit.* A marketing research consultant for a supermarket chain chose a large city to test market a new brand of mixed nuts packaged in 8 ounce cans. After a year of varying the selling price and recording the monthly demand, the consultant arrived at the given demand table, where y is the number of cans purchased per month (in thousands) at x dollars per can.

x	y
4.0	4.2
4.5	3.5
5.0	2.7
5.5	1.5
6.0	0.7

(A) Find a demand equation using the method of least squares.
(B) If each can of nuts costs the supermarket chain \$3, how should it be priced to achieve a maximum monthly profit?

Life Sciences

19. *Medicine.* If a person dives into cold water, a neural reflex response automatically shuts off blood circulation to the skin and muscles and reduces the pulse rate. A medical research team conducted an experiment using a group of ten 2-year-olds. A child's face was placed momentarily in cold water and the corresponding reduction in pulse rate recorded. The average reduction in heart rate for each temperature was summarized in the following table:

Water Temperature	Pulse Rate Reduction
50	15
55	13
60	10
65	6
70	2

(A) If T is water temperature in degrees Fahrenheit and P is pulse rate reduction in beats per minute, use the method of least squares to find a linear equation relating T and P.
(B) Use the equation found in part A to find P when $T = 57$.

20. *Biology.* In biology there is an approximate rule, called the *bioclimatic rule for temperate climates*, that has been known for a couple of hundred years. This rule states that in spring and early summer, periodic phenomena such as blossoming of flowers, appearance of insects, and ripening of fruit usually come about 4 days later for each 500 feet of altitude. Stated as a formula,

$$d = 8h \qquad 0 \leq h \leq 4$$

h	d
0	0
1	7
2	18
3	28
4	33

where d is the change in days and h is the altitude in thousands of feet. To test this rule, an experiment was set up to record the difference in blossoming time of the same type of apple tree at different altitudes. A summary of the results is given in the table in the margin.

(A) Use the method of least squares to find a linear equation relating h and d. Does the bioclimatic rule, $d = 8h$, appear to be approximately correct?
(B) How much longer will it take this type of apple tree to blossom at 3.5 thousand feet than at sea level? [Use the linear equation found in part A.]

Social Sciences

21. *Political science.* Association of economic class and party affiliation did not start with Roosevelt's New Deal; it goes back to the time of Andrew Jackson (1767–1845). Paul Lazarsfeld of Columbia University published an article in the November 1950 issue of *Scientific American* in which he discusses statistical investigations of the relationships between economic class and party affiliation. The data in the table are taken from this article.

Ward	**Average Assessed Value per Person (in $100)** 1836	**Percent Democratic Votes** 1836
12	1.7	51
3	2.1	49
1	2.3	53
5	2.4	36
2	3.6	65
11	3.7	35
10	4.7	29
4	6.2	40
6	7.1	34
9	7.4	29
8	8.7	20
7	11.9	23

(A) If A represents the average assessed value per person in a given ward in 1836 and D represents the percentage of people in that ward voting Democratic in 1836, use the method of least squares to find a linear equation relating A and D.
(B) If the average assessed value per person in a ward had been $300, what is the predicted percentage of people in that ward that would have voted Democratic?

22. *Education.* The table lists the high school grade-point averages of ten students and their college grade-point averages after one semester of college.

High School GPA	College GPA
2.0	1.5
2.2	1.5
2.4	1.6
2.7	1.8
2.9	2.1
3.0	2.3
3.1	2.5
3.3	2.9
3.4	3.2
3.7	3.5

(A) Find the least squares line for the data.
(B) Estimate the college GPA for a student with a high school GPA of 3.5.
(C) Estimate the high school GPA necessary for a college GPA of 2.7.

6-6 Double Integrals over Rectangular Regions

- Introduction
- Definition of the Double Integral
- Average Value over Rectangular Regions
- Volume and Double Integrals

Introduction

We have generalized the concept of differentiation to functions with two or more independent variables. How can we do the same with integration and how can we interpret the results? Let us first look at the operation of antidifferentiation. We can antidifferentiate a function of two or more variables with respect to one of the variables by treating all the other variables as though they were constants. Thus, this operation is the reverse operation of partial differentiation, just as ordinary antidifferentiation is the reverse operation of ordinary differentiation. We write $\int f(x, y)\,dx$ to indicate that we are to antidifferentiate $f(x, y)$ with respect to x, holding y fixed; we write $\int f(x, y)\,dy$ to indicate that we are to antidifferentiate $f(x, y)$ with respect to y, holding x fixed.

Example 18 Evaluate:

(A) $\displaystyle\int (6xy^2 + 3x^2)\,dy$ (B) $\displaystyle\int (6xy^2 + 3x^2)\,dx$

Solutions (A) Treating x as a constant and using the properties of antidifferentiation from Section 4-1, we have

$$\begin{aligned}\int (6xy^2 + 3x^2)\,dy &= \int 6xy^2\,dy + \int 3x^2\,dy\\ &= 6x\int y^2\,dy + 3x^2\int dy\\ &= 6x\left(\frac{y^3}{3}\right) + 3x^2(y) + C(x)\\ &= 2xy^3 + 3x^2y + C(x)\end{aligned}$$

The dy tells us we are looking for the antiderivative of $(6xy^2 + 3x^2)$ with respect to y only, holding x constant.

Notice that the constant of integration can actually be *any function of x alone*, since, for any such function, $\partial/\partial y\,[C(x)] = 0$. We can verify that our answer is correct by using partial differentiation:

$$\begin{aligned}\frac{\partial}{\partial y}[2xy^3 + 3x^2y + C(x)] &= 6xy^2 + 3x^2 + 0\\ &= 6xy^2 + 3x^2\end{aligned}$$

(B) Now we treat y as a constant:

$$\begin{aligned}\int (6xy^2 + 3x^2)\,dx &= \int 6xy^2\,dx + \int 3x^2\,dx\\ &= 6y^2\int x\,dx + 3\int x^2\,dx\\ &= 6y^2\left(\frac{x^2}{2}\right) + 3\left(\frac{x^3}{3}\right) + E(y)\\ &= 3x^2y^2 + x^3 + E(y)\end{aligned}$$

This time the antiderivative contains an arbitrary function E(y) of y alone.

Check

$$\begin{aligned}\frac{\partial}{\partial x}[3x^2y^2 + x^3 + E(y)] &= 6xy^2 + 3x^2 + 0\\ &= 6xy^2 + 3x^2\end{aligned}$$

Problem 18 Evaluate:

(A) $\displaystyle\int (4xy + 12x^2y^3)\,dy$ (B) $\displaystyle\int (4xy + 12x^2y^3)\,dx$

Now that we have extended the concept of antidifferentiation to functions with two variables, we can also evaluate definite integrals of the form

$$\int_a^b f(x, y)\, dx \qquad \text{or} \qquad \int_c^d f(x, y)\, dy$$

Example 19 Evaluate, substituting the limits of integration in y if dy is used and in x if dx is used:

(A) $\int_0^2 (6xy^2 + 3x^2)\, dy$

(B) $\int_0^1 (6xy^2 + 3x^2)\, dx$

Solutions (A) From Example 18A, we know that $\int (6xy^2 + 3x^2)\, dy = 2xy^3 + 3x^2y + C(x)$. According to the definition of the definite integral for a function of one variable, we can use any antiderivative to evaluate the definite integral. Thus, choosing $C(x) = 0$, we have

$$\begin{aligned}\int_0^2 (6xy^2 + 3x^2)\, dy &= (2xy^3 + 3x^2y)\Big|_{y=0}^{y=2} \\ &= [2x(2)^3 + 3x^2(2)] - [2x(0)^3 + 3x^2(0)] \\ &= 16x + 6x^2\end{aligned}$$

(B) From Example 18B, we know that $\int (6xy^2 + 3x^2)\, dx = 3x^2y^2 + x^3 + E(y)$. Thus, choosing $E(y) = 0$, we have

$$\begin{aligned}\int_0^1 (6xy^2 + 3x^2)\, dx &= (3x^2y^2 + x^3)\Big|_{x=0}^{x=1} \\ &= [3y^2(1)^2 + (1)^3] - [3y^2(0)^2 + (0)^3] \\ &= 3y^2 + 1\end{aligned}$$

Problem 19 Evaluate:

(A) $\int_0^1 (4xy + 12x^2y^3)\, dy$

(B) $\int_0^3 (4xy + 12x^2y^3)\, dx$

Notice that integrating and evaluating a definite integral, with integrand $f(x, y)$, with respect to y produces a function of x alone (or a constant). Likewise, integrating and evaluating a definite integral, with integrand $f(x, y)$, with respect to x produces a function of y alone (or a constant). Each of these results, involving at most one variable, can now be used as an integrand in a second definite integral.

Example 20 Evaluate:

(A) $\int_0^1 \left[\int_0^2 (6xy^2 + 3x^2)\, dy \right] dx$

(B) $\int_0^2 \left[\int_0^1 (6xy^2 + 3x^2)\, dx \right] dy$

Solutions (A) Example 19A showed that

$$\int_0^2 (6xy^2 + 3x^2)\, dy = 16x + 6x^2$$

Thus,

$$\begin{aligned}\int_0^1 \left[\int_0^2 (6xy^2 + 3x^2)\, dy \right] dx &= \int_0^1 (16x + 6x^2)\, dx \\ &= (8x^2 + 2x^3)\Big|_{x=0}^{x=1} \\ &= [8(1)^2 + 2(1)^3] - [8(0)^2 + 2(0)^3] \\ &= 10\end{aligned}$$

(B) Example 19B showed that

$$\int_0^1 (6xy^2 + 3x^2)\, dx = 3y^2 + 1$$

Thus,

$$\begin{aligned}\int_0^2 \left[\int_0^1 (6xy^2 + 3x^2)\, dx \right] dy &= \int_0^2 (3y^2 + 1)\, dy \\ &= (y^3 + y)\Big|_{y=0}^{y=2} \\ &= [(2)^3 + 2] - [(0)^3 + 0] \\ &= 10\end{aligned}$$

Problem 20 Evaluate:

(A) $\int_0^3 \left[\int_0^1 (4xy + 12x^2y^3)\, dy \right] dx$

(B) $\int_0^1 \left[\int_0^3 (4xy + 12x^2y^3)\, dx \right] dy$

▪ Definition of the Double Integral

Notice that the answers in Examples 20A and 20B are identical. This is not an accident. In fact, it is this property that enables us to define the **double integral.** (See the box at the top of the next page.)

Double Integral

The double integral of a function $f(x, y)$ over a rectangle $R = \{(x, y) | a \leq x \leq b, \quad c \leq y \leq d\}$ is

$$\iint_R f(x, y)\, dA = \int_a^b \left[\int_c^d f(x, y)\, dy \right] dx = \int_c^d \left[\int_a^b f(x, y)\, dx \right] dy$$

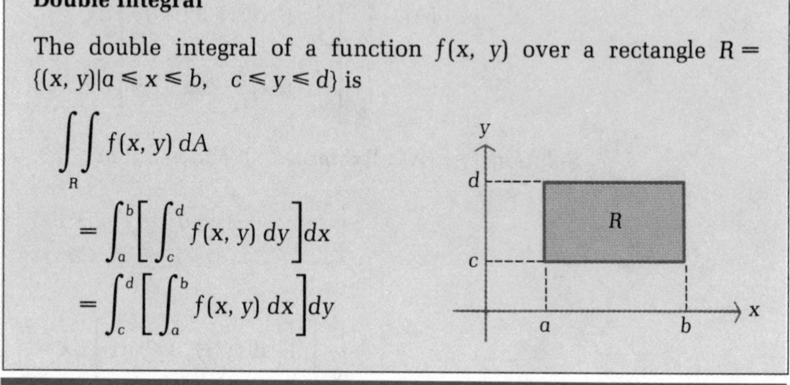

In the double integral $\iint_R f(x, y)\, dA$, $f(x, y)$ is called the **integrand** and R is called the **region of integration.** The expression dA indicates that this is an integral over a two-dimensional region. The integrals

$$\int_a^b \left[\int_c^d f(x, y)\, dy \right] dx \qquad \text{and} \qquad \int_c^d \left[\int_a^b f(x, y)\, dx \right] dy$$

are referred to as **iterated integrals** (the brackets are often omitted), and the order in which dx and dy are written indicates the order of integration. This is not the most general definition of the double integral over a rectangular region; however, it is equivalent to the general definition for all the functions we will consider.

Example 21 Evaluate $\iint_R (x + y)\, dA$ over $R = \{(x, y) | 1 \leq x \leq 3, \quad -1 \leq y \leq 2\}$.

Solution We can choose either order of iteration. As a check, we will evaluate the integral both ways:

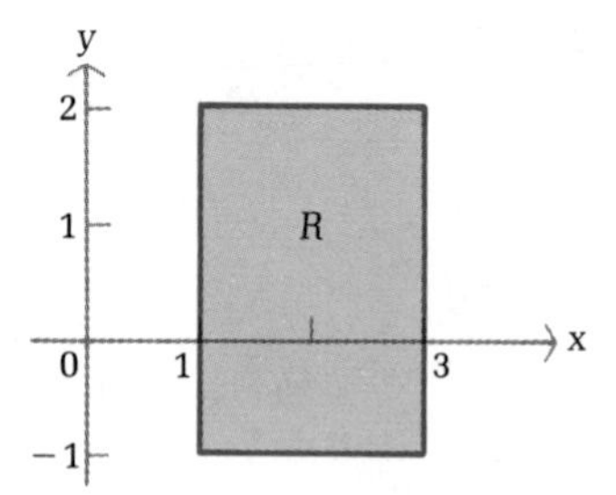

$$\begin{aligned}
\iint_R (x + y)\, dA &= \int_1^3 \int_{-1}^2 (x + y)\, dy\, dx \\
&= \int_1^3 \left[\left(xy + \frac{y^2}{2} \right) \Bigg|_{y=-1}^{y=2} \right] dx \\
&= \int_1^3 \left[(2x + 2) - \left(-x + \frac{1}{2} \right) \right] dx \\
&= \int_1^3 \left(3x + \frac{3}{2} \right) dx \\
&= \left(\frac{3}{2}x^2 + \frac{3}{2}x \right) \Bigg|_{x=1}^{x=3} \\
&= \left(\frac{27}{2} + \frac{9}{2} \right) - \left(\frac{3}{2} + \frac{3}{2} \right) \\
&= (18) - (3) = 15
\end{aligned}$$

$$\iint_R (x+y)\,dA = \int_{-1}^{2}\int_{1}^{3} (x+y)\,dx\,dy$$
$$= \int_{-1}^{2}\left[\left(\frac{x^2}{2}+xy\right)\Big|_{x=1}^{x=3}\right] dy$$
$$= \int_{-1}^{2}\left[\left(\frac{9}{2}+3y\right)-\left(\frac{1}{2}+y\right)\right] dy$$
$$= \int_{-1}^{2} (4+2y)\,dy$$
$$= (4y+y^2)\Big|_{y=-1}^{y=2}$$
$$= (8+4)-(-4+1)$$
$$= (12)-(-3) = 15$$

Problem 21 Evaluate both ways:

$$\iint_R (2x-y)\,dA \quad \text{over } R = \{(x,y)\,|\,-1 \leqslant x \leqslant 5,\quad 2 \leqslant y \leqslant 4\}$$

Example 22 Evaluate:

$$\iint_R 2xe^{x^2+y}\,dA \quad \text{over } R = \{(x,y)\,|\,0 \leqslant x \leqslant 1,\quad -1 \leqslant y \leqslant 1\}$$

Solution

$$\iint_R 2xe^{x^2+y}\,dA = \int_{-1}^{1}\int_{0}^{1} 2xe^{x^2+y}\,dx\,dy$$
$$= \int_{-1}^{1}\left[(e^{x^2+y})\Big|_{x=0}^{x=1}\right] dy$$
$$= \int_{-1}^{1} (e^{1+y}-e^{y})\,dy$$
$$= (e^{1+y}-e^{y})\Big|_{y=-1}^{y=1}$$
$$= (e^2-e)-(e^0-e^{-1})$$
$$= e^2-e-1+e^{-1}$$

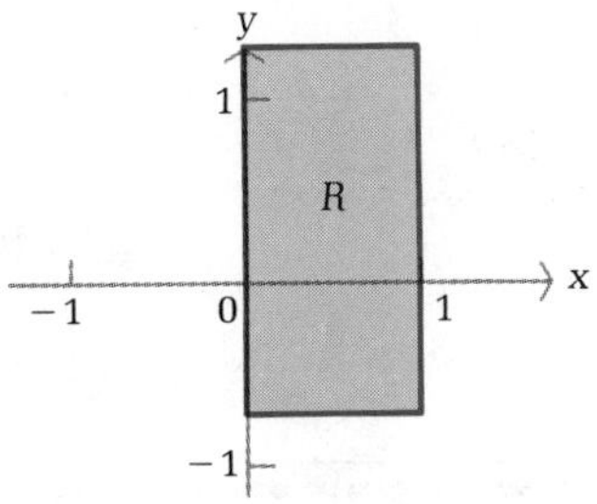

Problem 22 Evaluate:

$$\iint_R \frac{x}{y^2}e^{x/y}\,dA \quad \text{over } R = \{(x,y)\,|\,0 \leqslant x \leqslant 1,\quad 1 \leqslant y \leqslant 2\}$$

Average Value over Rectangular Regions

In Section 5-1 the average value of a function $f(x)$ over an interval $[a, b]$ was defined as

$$\frac{1}{b-a}\int_a^b f(x)\,dx$$

This definition is easily extended to functions of two variables over rectangular regions, as shown in the box. Notice that the denominator in the expression given in the box, $(b-a)(d-c)$, is simply the area of the rectangle R.

Average Value over Rectangular Regions

The **average value** of the function $f(x, y)$ over the rectangle $R = \{(x, y) | a \leq x \leq b, \quad c \leq y \leq d\}$ is

$$\frac{1}{(b-a)(d-c)}\iint_R f(x, y)\,dA$$

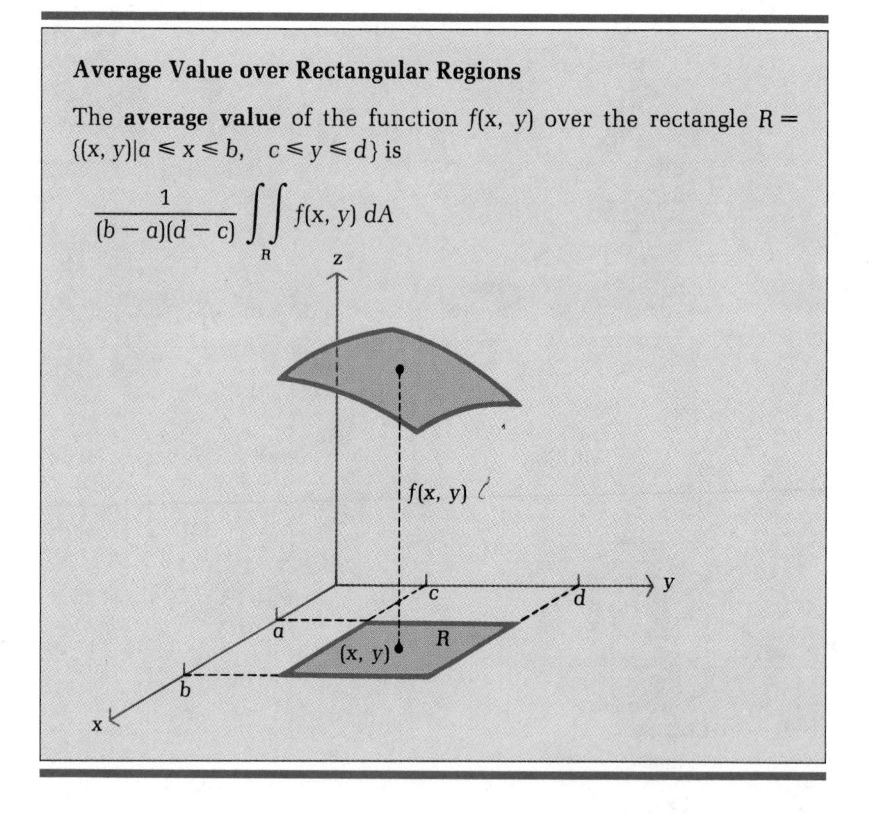

Example 23 Find the average value of $f(x, y) = 4 - \frac{1}{2}x - \frac{1}{2}y$ over the rectangle $R = \{(x, y) | 0 \leq x \leq 2, \quad 0 \leq y \leq 2\}$.

Solution

$$\frac{1}{(b-a)(d-c)}\iint_R f(x, y)\,dA = \frac{1}{(2-0)(2-0)}\iint_R \left(4 - \frac{1}{2}x - \frac{1}{2}y\right) dA$$

$$= \frac{1}{4}\int_0^2 \int_0^2 \left(4 - \frac{1}{2}x - \frac{1}{2}y\right) dy\,dx$$

$$= \frac{1}{4}\int_0^2 \left[\left(4y - \frac{1}{2}xy - \frac{1}{4}y^2\right)\Bigg|_{y=0}^{y=2}\right] dx$$

$$= \frac{1}{4}\int_0^2 (7 - x)\,dx$$

$$= \frac{1}{4}\left(7x - \frac{1}{2}x^2\right)\Bigg|_{x=0}^{x=2}$$

$$= \frac{1}{4}(12) = 3$$

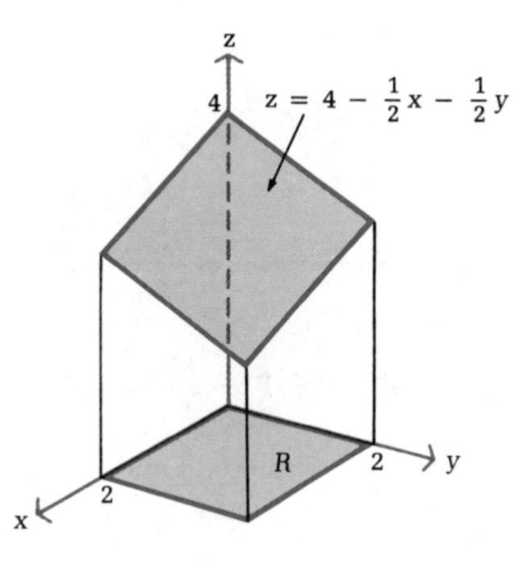

Problem 23 Find the average value of $f(x, y) = x + 2y$ over the rectangle $R = \{(x, y) \mid 0 \leq x \leq 2, \quad 0 \leq y \leq 1\}$.

▪ Volume and Double Integrals

One application of the definite integral of a function with one variable is the calculation of areas, so it is not surprising that the definite integral of a function of two variables can be used to calculate volumes of solids.

Volume under a Surface

If $f(x, y) \geqslant 0$ over a rectangle R, $R = \{(x, y) | a \leqslant x \leqslant b, \quad c \leqslant y \leqslant d\}$, then the volume of the solid formed by graphing f over the rectangle R is given by

$$V = \iint_R f(x, y)\, dA$$

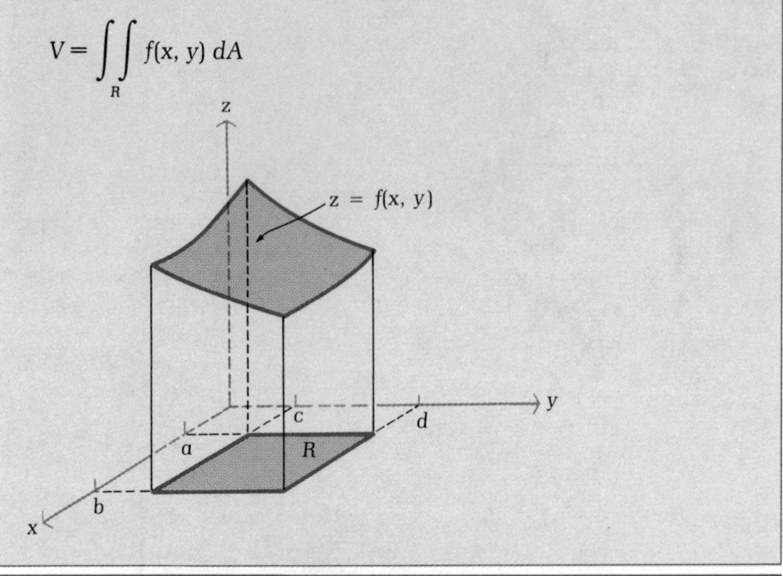

A proof of the statement in the box is left to a more advanced text.

Example 24 Find the volume of the solid under the graph of $f(x, y) = 1 + x^2 + y^2$ over the rectangle $R = \{(x, y) | 0 \leqslant x \leqslant 1, \quad 0 \leqslant y \leqslant 1\}$.

Solution

$$\begin{aligned} V &= \iint_R (1 + x^2 + y^2)\, dA \\ &= \int_0^1 \int_0^1 (1 + x^2 + y^2)\, dx\, dy \\ &= \int_0^1 \left[\left(x + \frac{1}{3}x^3 + xy^2 \right) \Bigg|_{x=0}^{x=1} \right] dy \\ &= \int_0^1 \left(\frac{4}{3} + y^2 \right) dy \\ &= \left(\frac{4}{3}y + \frac{1}{3}y^3 \right) \Bigg|_{y=0}^{y=1} = \frac{5}{3} \text{ cubic units} \end{aligned}$$

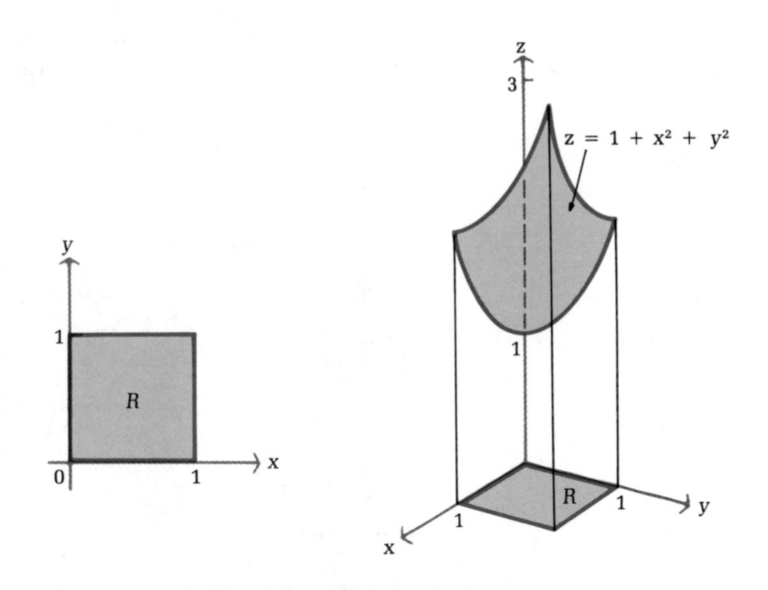

Problem 24 Find the volume of the solid under the graph of $f(x, y) = 1 + x + y$ over the rectangle $R = \{(x, y)|0 \leq x \leq 1, \quad 0 \leq y \leq 2\}$.

Answers to Matched Problems

18. (A) $2xy^2 + 3x^2y^4 + C(x)$ (B) $2x^2y + 4x^3y^3 + E(y)$
19. (A) $2x + 3x^2$ (B) $18y + 108y^3$
20. (A) 36 (B) 36
21. 12
22. $e - 2e^{1/2} + 1$
23. 2
24. 5 cubic units

Exercise 6-6

A *Find each antiderivative. Then use the antiderivative to evaluate the definite integral.*

1. (A) $\int 12x^2y^3\,dy$ (B) $\int_0^1 12x^2y^3\,dy$
2. (A) $\int 12x^2y^3\,dx$ (B) $\int_{-1}^{2} 12x^2y^3\,dx$
3. (A) $\int (4x + 6y + 5)\,dx$ (B) $\int_{-2}^{3} (4x + 6y + 5)\,dx$
4. (A) $\int (4x + 6y + 5)\,dy$ (B) $\int_1^4 (4x + 6y + 5)\,dy$
5. (A) $\int \frac{x}{\sqrt{y + x^2}}\,dx$ (B) $\int_0^2 \frac{x}{\sqrt{y + x^2}}\,dx$

6. (A) $\displaystyle\int \frac{x}{\sqrt{y+x^2}}\,dy$ (B) $\displaystyle\int_1^5 \frac{x}{\sqrt{y+x^2}}\,dy$

B *Evaluate each iterated integral. (See the indicated problem for the evaluation of the inner integral.)*

7. $\displaystyle\int_{-1}^{2}\int_0^1 12x^2y^3\,dy\,dx$

(see Problem 1)

8. $\displaystyle\int_0^1\int_{-1}^{2} 12x^2y^3\,dx\,dy$

(see Problem 2)

9. $\displaystyle\int_1^4\int_{-2}^{3} (4x+6y+5)\,dx\,dy$

(see Problem 3)

10. $\displaystyle\int_{-2}^{3}\int_1^4 (4x+6y+5)\,dy\,dx$

(see Problem 4)

11. $\displaystyle\int_1^5\int_0^2 \frac{x}{\sqrt{y+x^2}}\,dx\,dy$

(see Problem 5)

12. $\displaystyle\int_0^2\int_1^5 \frac{x}{\sqrt{y+x^2}}\,dy\,dx$

(see Problem 6)

Use both orders of iteration to evaluate each double integral.

13. $\displaystyle\iint_R xy\,dA;\quad R=\{(x,y)|0\le x\le 2,\quad 0\le y\le 4\}$

14. $\displaystyle\iint_R \sqrt{xy}\,dA;\quad R=\{(x,y)|1\le x\le 4,\quad 1\le y\le 9\}$

15. $\displaystyle\iint_R (x+y)^5\,dA;\quad R=\{(x,y)|-1\le x\le 1,\quad 1\le y\le 2\}$

16. $\displaystyle\iint_R xe^y\,dA;\quad R=\{(x,y)|-2\le x\le 3,\quad 0\le y\le 2\}$

Find the average value of each function over the given rectangle.

17. $f(x,y)=(x+y)^2;\quad R=\{(x,y)|1\le x\le 5,\quad -1\le y\le 1\}$
18. $f(x,y)=x^2+y^2;\quad R=\{(x,y)|-1\le x\le 2,\quad 1\le y\le 4\}$
19. $f(x,y)=\dfrac{x}{y};\quad R=\{(x,y)|1\le x\le 4,\quad 2\le y\le 7\}$
20. $f(x,y)=x^2y^3;\quad R=\{(x,y)|-1\le x\le 1,\quad 0\le y\le 2\}$

Find the volume of the solid under the graph of each function over the given rectangle.

21. $f(x,y)=2-x^2-y^2;\quad R=\{(x,y)|0\le x\le 1,\quad 0\le y\le 1\}$
22. $f(x,y)=5-x;\quad R=\{(x,y)|0\le x\le 5,\quad 0\le y\le 5\}$
23. $f(x,y)=4-y^2;\quad R=\{(x,y)|0\le x\le 2,\quad 0\le y\le 2\}$
24. $f(x,y)=e^{-x-y};\quad R=\{(x,y)|0\le x\le 1,\quad 0\le y\le 1\}$

C Evaluate each double integral. Select the order of integration carefully—each problem is easy to do one way and difficult the other.

25. $\iint_R xe^{xy}\, dA$; $R = \{(x, y) | 0 \leq x \leq 1,\ \ 1 \leq y \leq 2\}$

26. $\iint_R xye^{x^2y}\, dA$; $R = \{(x, y) | 0 \leq x \leq 1,\ \ 1 \leq y \leq 2\}$

27. $\iint_R \frac{2y + 3xy^2}{1 + x^2}\, dA$; $R = \{(x, y) | 0 \leq x \leq 1,\ \ -1 \leq y \leq 1\}$

28. $\iint_R \frac{2x + 2y}{1 + 4y + y^2}\, dA$; $R = \{(x, y) | 1 \leq x \leq 3,\ \ 0 \leq y \leq 1\}$

Applications

Business & Economics

29. *Economics–multiplier principle.* Suppose Congress enacts a one-time-only 10% tax rebate that is expected to infuse y billion dollars, $5 \leq y \leq 7$, into the economy. If every individual and corporation is expected to spend a proportion x, $0.6 \leq x \leq 0.8$, of each dollar received, then by the **multiplier principle** in economics (using the sum of an infinite geometric progression—see Appendix A), the total amount of spending S (in billions of dollars) generated by this tax rebate is given by

$$S(x, y) = \frac{y}{1 - x}$$

What is the average total amount of spending for the indicated ranges of the values of x and y? Set up a double integral and evaluate.

30. *Economics—multiplier principle.* Repeat Problem 29 if $6 \leq y \leq 10$ and $0.7 \leq x \leq 0.9$.

31. *Economics—Cobb–Douglas production function.* If an industry invests x thousand labor-hours, $10 \leq x \leq 20$, and y million dollars, $1 \leq y \leq 2$, in the production of N thousand units of a certain item, then N is given by

$$N(x, y) = x^{0.75}y^{0.25}$$

What is the average number of units produced for the indicated ranges of x and y? Set up a double integral and evaluate.

32. *Economics—Cobb–Douglas production function.* Repeat Problem 31 for

$$N(x, y) = x^{0.5}y^{0.5}$$

where $10 \leq x \leq 30$ and $1 \leq y \leq 3$.

Life Sciences

33. *Population distribution.* In order to study the population distribution of a certain species of insects, a biologist has constructed an artificial habitat in the shape of a rectangle 16 feet long and 12 feet wide. The only food available to the insects in this habitat is located at its center. The biologist has determined that the concentration C of insects per square foot at a point d units from the food supply (see the figure) is given approximately by

$$C = 10 - \tfrac{1}{10}d^2$$

What is the average concentration of insects throughout the habitat? Express C as a function of x and y, set up a double integral, and evaluate.

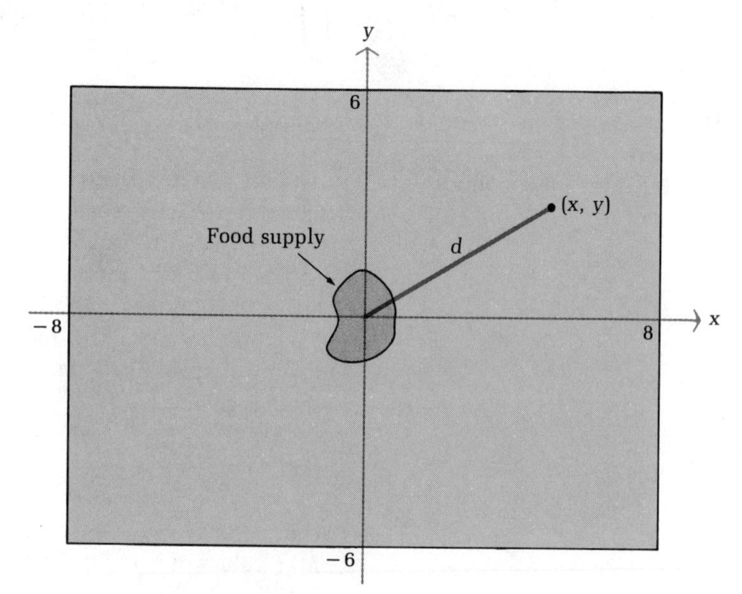

34. *Population distribution.* Repeat Problem 33 for a square habitat that measures 12 feet on each side, where the insect concentration is given by

$$C = 8 - \tfrac{1}{10}d^2$$

35. *Pollution.* A heavy industrial plant located in the center of a small town emits particulate matter into the atmosphere. Suppose the concentration of particulate matter in parts per million at a point d miles from the plant is given by

$$C = 100 - 15d^2$$

If the boundaries of the town form a rectangle 4 miles long and 2 miles wide, what is the average concentration of particulate matter throughout the city? Express C as a function of x and y, set up a double integral, and evaluate.

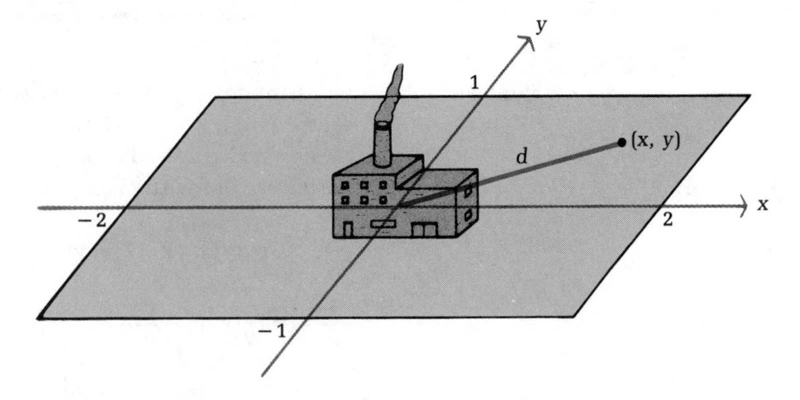

36. *Pollution.* Repeat Problem 35 if the boundaries of the town form a rectangle 8 miles long and 4 miles wide and the concentration of particulate matter is given by

$$C = 100 - 3d^2$$

Social Sciences

37. *Safety research.* Under ideal conditions, if a person driving a car slams on the brakes and skids to a stop, the length of the skid marks (in feet) is given by the formula

$$L = 0.000\ 013\ 3xy^2$$

where x is the weight of the car in pounds and y is the speed of the car in miles per hour. What is the average length of the skid marks for cars weighing between 2,000 and 3,000 pounds and traveling at speeds between 50 and 60 miles per hour? Set up a double integral and evaluate.

38. *Safety research.* Repeat Problem 37 for cars weighing between 2,000 and 2,500 pounds and traveling at speeds between 40 and 50 miles per hour.

39. *Psychology.* The intelligence quotient Q for an individual with mental age x and chronological age y is given by

$$Q(x, y) = 100\frac{x}{y}$$

In a group of sixth graders, the mental age varies between 8 and 16 years and the chronological age varies between 10 and 12 years. What is the average intelligence quotient for this group? Set up a double integral and evaluate.

40. *Psychology.* Repeat Problem 39 for a group with mental ages between 6 and 14 years and chronological ages between 8 and 10 years.

6-7 Chapter Review

Important Terms and Symbols

6-1 *Functions of several variables.* Functions of two independent variables, functions of several independent variables, surface, paraboloid, saddle point, $z = f(x, y)$, $w = f(x, y, z)$

6-2 *Partial derivatives.* Partial derivative of f with respect to x, partial derivative of f with respect to y, second-order partials,

$$\frac{\partial z}{\partial x}, \quad \frac{\partial z}{\partial y}, \quad f_x(x, y), \quad f_y(x, y), \quad \frac{\partial^2 z}{\partial x^2} = f_{xx}(x, y), \quad \frac{\partial^2 z}{\partial x\, \partial y} = f_{yx}(x, y),$$

$$\frac{\partial^2 z}{\partial y\, \partial x} = f_{xy}(x, y), \quad \frac{\partial^2 z}{\partial y^2} = f_{yy}(x, y)$$

6-3 *Maxima and minima.* Local maximum, local minimum, critical point, second-derivative test

6-4 *Maxima and minima using Lagrange multipliers (optional).* Constraint, Lagrange multiplier, method of Lagrange multipliers for functions of two variables, method of Lagrange multipliers for functions of three variables

6-5 *Method of least squares.* Least squares approximation, linear regression, residual, least squares line, regression line, estimation, approximation

6-6 *Double integrals over rectangular regions.* Double integral, iterated integral, average value over rectangular regions, volume under a surface, $\iint_R f(x, y)\, dA = \int_a^b \left[\int_c^d f(x, y)\, dy\right] dx = \int_c^d \left[\int_a^b f(x, y)\, dx\right] dy$

Exercise 6-7 Chapter Review

Work through all the problems in this chapter review and check your answers in the back of the book. (Answers to all review problems are there.) Where weaknesses show up, review appropriate sections in the text.

A

1. For $f(x, y) = 2{,}000 + 40x + 70y$, find $f(5, 10)$, $f_x(x, y)$, and $f_y(x, y)$.
2. For $z = x^3y^2$, find $\partial^2 z/\partial x^2$ and $\partial^2 z/\partial x\, \partial y$.
3. Evaluate: $\int (6xy^2 + 4y)\, dy$
4. Evaluate: $\int (6xy^2 + 4y)\, dx$
5. Evaluate: $\int_0^1 \int_0^1 4xy\, dy\, dx$

B

6. For $f(x, y) = 3x^2 - 2xy + y^2 - 2x + 3y - 7$, find $f(2, 3)$, $f_y(x, y)$, and $f_y(2, 3)$.

7. For $f(x, y) = -4x^2 + 4xy - 3y^2 + 4x + 10y + 81$, find

$$[f_{xx}(2, 3)][f_{yy}(2, 3)] - [f_{xy}(2, 3)]^2$$

x	y
2	12
4	10
6	7
8	3

8. Use the least squares line for the data in the table to estimate y when $x = 10$.
9. For $R = \{(x, y) | -1 \leq x \leq 1, \ 1 \leq y \leq 2\}$, evaluate the following two ways:

$$\iint_R (4x + 6y)\, dA$$

C

10. For $f(x, y) = e^{x^2+2y}$, find f_x, f_y, and f_{xy}.
11. For $f(x, y) = (x^2 + y^2)^5$, find f_x and f_{xy}.
12. Find all critical points and test for extrema for

$$f(x, y) = x^3 - 12x + y^2 - 6y$$

13. Find the least squares line for the data in the table:

x	y	x	y
10	50	60	80
20	45	70	85
30	50	80	90
40	55	90	90
50	65	100	110

14. Find the average value of $f(x, y) = x^{2/3}y^{1/3}$ over the rectangle

$$R = \{(x, y) | -8 \leq x \leq 8, \quad 0 \leq y \leq 27\}$$

15. Find the volume of the solid under the graph of

$$z = 3x^2 + 3y^2$$

over the rectangle

$$R = \{(x, y) | 0 \leq x \leq 1, \quad -1 \leq y \leq 1\}$$

Applications

Business & Economics

16. *Maximizing profit.* A company produces x units of product A and y units of product B (both in hundreds per month). The monthly profit equation (in thousands of dollars) is found to be

$$P(x, y) = -4x^2 + 4xy - 3y^2 + 4x + 10y + 81$$

(A) Find $P_x(1, 3)$ and interpret.
(B) How many of each product should be produced each month to maximize profit? What is the maximum profit?

17. *Minimizing material.* A rectangular box with no top and six compartments (see the figure) is to have a volume of 96 cubic inches. Find the dimensions that will require the least amount of material.

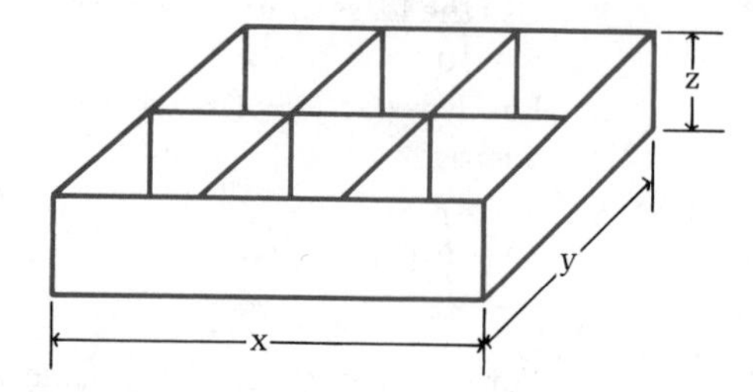

Year	Profit
1	2
2	2.5
3	3.1
4	4.2
5	4.3

18. *Profit.* A company's annual profit (in millions of dollars) over a 5 year period is given in the table. Use the least squares line to estimate the profit for the sixth year.

19. *Economics—Cobb–Douglas production function.* The Cobb–Douglas production function for an industry is

$$N(x, y) = x^{0.8}y^{0.2}$$

where x is the number of labor-hours (in thousands) and y is the amount of money (in millions) invested in the production of N thousand units of a certain item. If $10 \leq x \leq 12$ and $1 \leq y \leq 3$, find the average number of units produced. Set up a definite integral and evaluate.

Life Sciences

20. *Marine biology.* The function used for timing dives with scuba gear is

$$T(V, x) = \frac{33V}{x + 33}$$

where T is the time of the dive in minutes, V is the volume of air (at sea level pressure) compressed into tanks, and x is the depth of the dive in feet. Find $T_x(70, 17)$ and interpret.

21. *Pollution.* A heavy industrial plant located in the center of a small town emits particulate matter into the atmosphere. Suppose the concentration of particulate matter in parts per million at a point d miles from the plant is given by

$$C = 100 - 24d^2$$

If the boundaries of the town form a square 4 miles long and 4 miles wide, what is the average concentration of particulate matter throughout the city? Express C as a function of x and y, set up a double integral, and evaluate.

Social Sciences

22. *Sociology.* Joseph Cavanaugh, a sociologist, found that the number of long-distance telephone calls, n, between two cities in a given period of time varied (approximately) jointly as the populations P_1 and P_2 of the two cities, and varied inversely as the distance, d, between the two cities. In terms of an equation for a time period of 1 week,

$$n(P_1, P_2, d) = 0.001\frac{P_1P_2}{d}$$

Find $n(100{,}000, 50{,}000, 100)$.

23. *Education.* At the beginning of the semester, students in a foreign language course are given a proficiency exam. The same exam is given at the end of the semester. The results for five students are given in the table. Use the least squares line to estimate the score on the second exam for a student who scored 40 on the first exam.

First Exam	Second Exam
30	60
50	75
60	80
70	85
90	90

Trigonometric Functions

7

CHAPTER 7 Contents

Up until now we have restricted our attention to algebraic, logarithmic, and exponential functions. These functions were used to model many real-life situations from business, economics, and the life and social sciences. Now we turn our attention to another important class of functions, called the **trigonometric functions.** These functions are particularly useful in describing periodic phenomena; that is, phenomena that repeat in cycles. Consider the sunrise times for a 2 year period starting January 1, as pictured in Figure 1. We see that the cycle repeats after 1 year. Business cycles, blood pressure in the aorta, seasonal growth, water waves, and amounts of pollution in the atmosphere are often periodic and can be illustrated with similar types of graphs.

We assume the reader has had a course in trigonometry. The next section provides a brief review of those topics that are most important for our purposes.

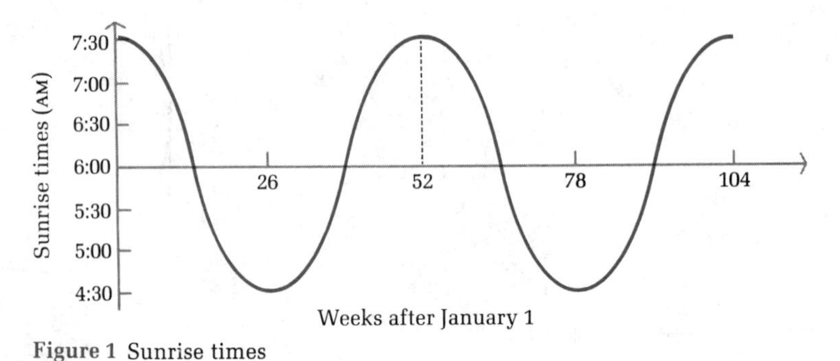

Figure 1 Sunrise times

7-1 Trigonometric Functions Review

- Angles; Degree–Radian Measure
- Trigonometric Functions
- Graphs of the Sine and Cosine Functions
- Four Other Trigonometric Functions

▪ Angles; Degree–Radian Measure

We start our discussion of trigonometry with the concept of **angle.** A point P on a line divides the line into two parts. Either part together with the end point is called a **half-line.** A geometric figure consisting of two half-lines with a common end point is called an **angle.** One of the half-lines is called

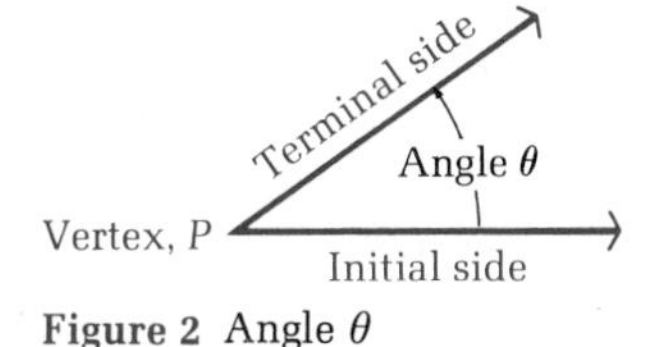

Figure 2 Angle θ

the **initial side** and the other half-line is called the **terminal side.** The common end point is called the **vertex.** Figure 2 illustrates these concepts.

There are two widely used measures of angles—the **degree** and the **radian.** A central angle of a circle subtended by an arc $\frac{1}{360}$ of the circumference of the circle is said to have **degree measure 1,** written **1°** (see Fig. 3). It follows that a central angle subtended by an arc $\frac{1}{4}$ the circumference has degree measure 90; $\frac{1}{2}$ the circumference has degree measure 180; and the whole circumference of a circle has degree measure 360.

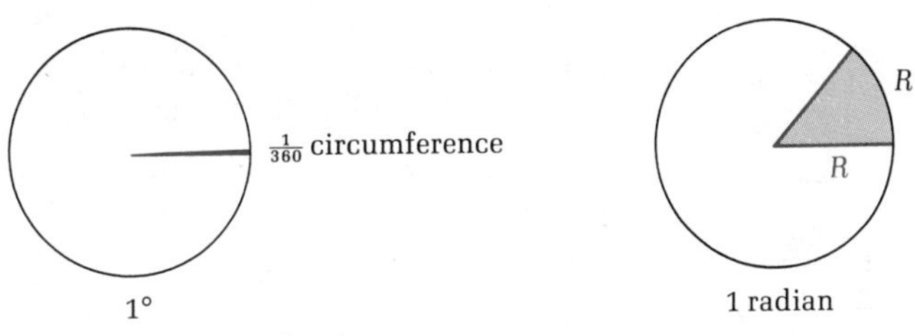

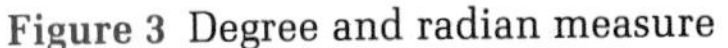

Figure 3 Degree and radian measure

The other measure of angles, which we will use extensively when we discuss the calculus of the trigonometric functions in the next two sections, is radian measure. A central angle subtended by an arc of length equal to the radius (R) of the circle is said to have **radian measure 1,** written **1 radian** (see Fig. 3). In general, a central angle subtended by an arc of length s has radian measure determined as follows:

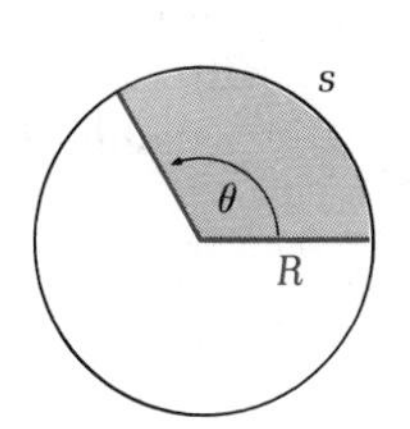

$$\theta^{r} = \text{Radian measure of } \theta = \frac{\text{Arc length}}{\text{Radius}} = \frac{s}{R}$$

[*Note:* If $R = 1$, then $\theta^{r} = s$.]

What is the radian measure of an angle of 180°? A central angle of 180° is subtended by an arc of $\frac{1}{2}$ the circumference of a circle. Thus,

$$s = \frac{C}{2} = \frac{2\pi R}{2} = \pi R$$

and

$$\theta^{r} = \frac{s}{R} = \frac{\pi R}{R} = \pi \text{ radians}$$

The following proportion can be used to convert degree measure to radian measure and vice versa:

Degree–Radian Conversion Formula

$$\frac{\theta^{\circ}}{180^{\circ}} = \frac{\theta^{r}}{\pi \text{ radians}}$$

Example 1 Find the radian measure of 1°.

Solution

$$\frac{1°}{180°} = \frac{\theta^r}{\pi \text{ radians}}$$

$$\theta^r = \frac{\pi}{180} \text{ radians} \approx 0.0175 \text{ radian}$$

Problem 1 Find the degree measure of 1 radian.

A comparison of degree and radian measures for a few important angles is given in the following table:

Radian	0	$\pi/6$	$\pi/4$	$\pi/3$	$\pi/2$	π	2π
Degree	0	30°	45°	60°	90°	180°	360°

Before defining trigonometric functions, we will generalize the notion of angle defined above. Starting with a rectangular coordinate system and two half-lines coinciding with the nonnegative x axis, the initial side of the angle remains fixed and the terminal side rotates until it reaches its terminal position. When the terminal side is rotated counterclockwise, the angle formed is considered positive (see Figs. 4A and 4C). When it is rotated clockwise, the angle formed is considered negative (see Fig. 4B). Angles located in a coordinate system in this manner are said to be in a **standard position.**

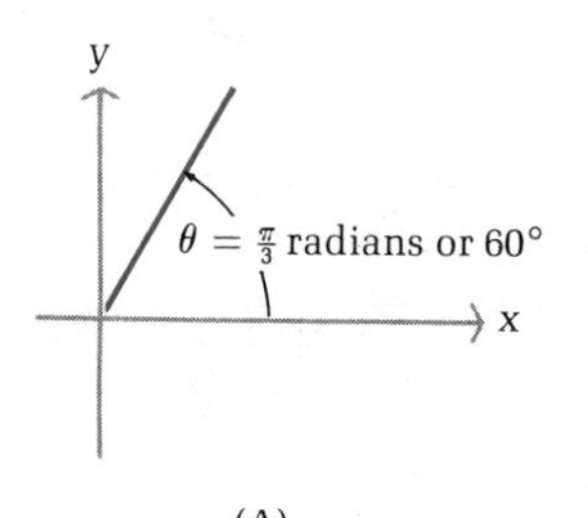

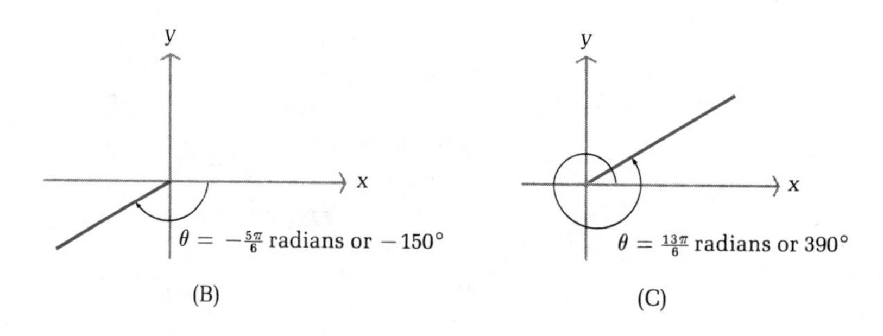

Figure 4 Generalized angles

▪ Trigonometric Functions

Let us locate a unit circle (radius 1) in a coordinate system with center at the origin (Fig. 5). The terminal side of any angle in standard position will pass through this circle at some point P. The abscissa of this point P is called the **cosine of θ** (abbreviated **cos θ**), and the ordinate of the point is the **sine of θ** (abbreviated **sin θ**). Thus, the set of all ordered pairs of the form $(\theta, \cos\theta)$, and the set of all ordered pairs of the form $(\theta, \sin\theta)$ constitute,

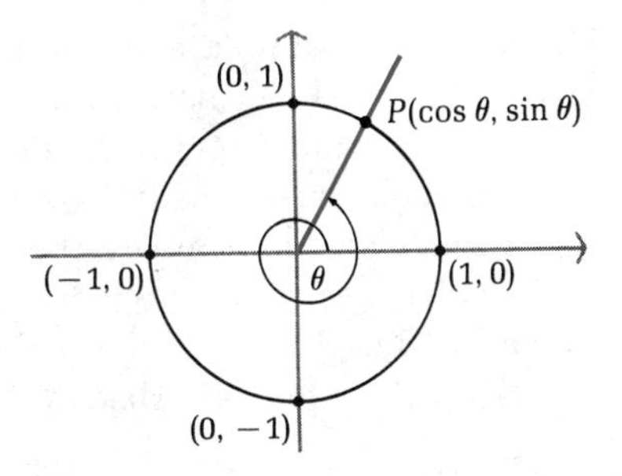

Figure 5

respectively, the **cosine** and **sine functions.** The **domain** of these two functions is the set of all angles, positive or negative, with measure either in degrees or radians. The **range** is a subset of the set of real numbers.

It is desirable, and necessary for our work in calculus, to define these two trigonometric functions in terms of real number domains. This is easily done as follows:

Sine and Cosine Functions with Real Number Domains

For any real number x,

$$\sin x = \sin(x \text{ radians})$$
$$\cos x = \cos(x \text{ radians})$$

Example 2 Referring to Figure 5, find:

(A) $\cos 90°$ (B) $\sin(-\pi/2 \text{ radians})$ (C) $\cos \pi$

Solutions

(A) The terminal side of an angle of degree measure 90 passes through (0, 1) on the unit circle. This point has abscissa 0. Thus,

$$\cos 90° = 0$$

(B) The terminal side of an angle of radian measure $-\pi/2$ $(-90°)$ passes through $(0, -1)$ on the unit circle. This point has ordinate -1. Thus,

$$\sin\left(-\frac{\pi}{2} \text{ radians}\right) = -1$$

(C) $\cos \pi = \cos(\pi \text{ radians}) = -1$, since the terminal side of an angle of radian measure π (180°) passes through $(-1, 0)$ on the unit circle and this point has abscissa -1.

Problem 2 Referring to Figure 5, find:

(A) $\sin 180°$ (B) $\cos(2\pi \text{ radians})$ (C) $\sin(-\pi)$

To find the value of either the sine or the cosine function for any angle or any real number by direct use of the definition is not easy. Tables are available, but hand calculators with SIN and COS buttons are even more convenient. Calculators generally have degree and radian options, so we can use the calculator to evaluate these functions for most of the real numbers in which we might have an interest. The following table includes a few values produced by a hand calculator. The x value is entered and then the SIN or COS button is pushed to obtain the desired value in display.

x	1	-7	35.26	-105.9
$\sin x$	0.8415	-0.6570	-0.6461	0.7920
$\cos x$	0.5403	0.7539	-0.7632	0.6105

Exact values of the sine and cosine functions can be obtained for multiples of the special angles in the triangles in Figure 6, since the triangles can be used to find the coordinate of the intersection of the terminal side of each angle with the unit circle. [*Note:* We now drop the word "radian" after $\pi/4$ and interpret $\pi/4$ as the radian measure of an angle or simply as a real number, depending on the context.]

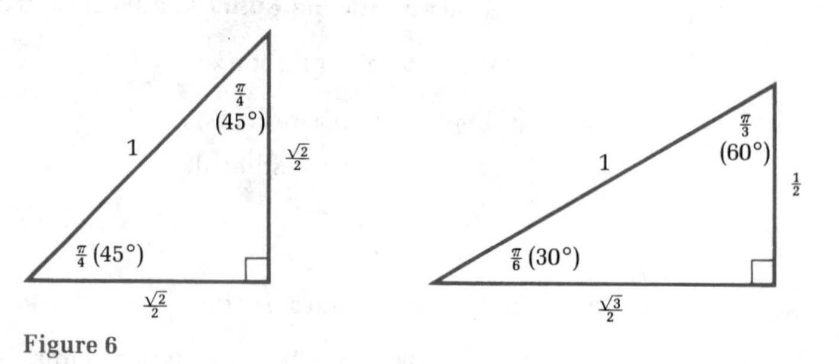

Figure 6

Example 3 Find the exact value of each of the following using Figure 6:

(A) $\cos \dfrac{\pi}{4}$ (B) $\sin \dfrac{\pi}{6}$ (C) $\sin\left(-\dfrac{\pi}{6}\right)$

Solutions (A) $\cos \dfrac{\pi}{4} = \dfrac{\sqrt{2}}{2}$ (B) $\sin \dfrac{\pi}{6} = \dfrac{1}{2}$

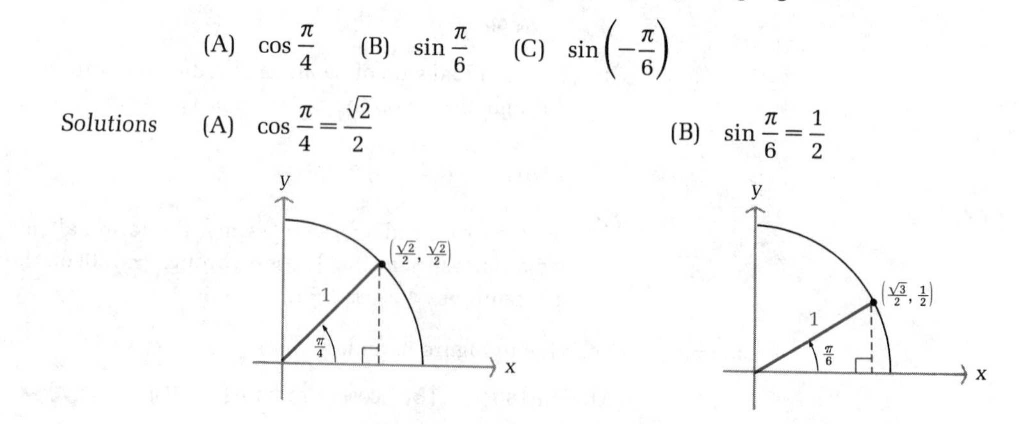

(C) $\sin\left(-\frac{\pi}{6}\right) = -\frac{1}{2}$

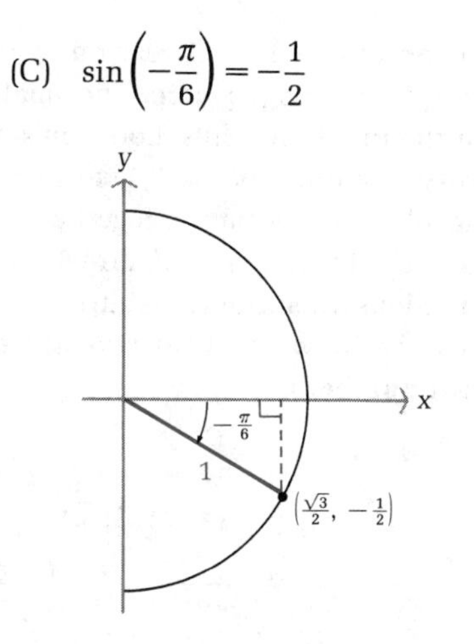

Problem 3 Find the exact value of each of the following using Figure 6:

(A) $\sin\frac{\pi}{4}$ (B) $\cos\frac{\pi}{3}$ (C) $\cos\left(-\frac{\pi}{3}\right)$

▪ Graphs of the Sine and Cosine Functions

To graph $y = \sin x$ or $y = \cos x$ for x a real number, we could use a calculator to produce a table, and then plot the ordered pairs from the table in a coordinate system. However, we can speed up the process by returning to basic definitions. Referring to Figure 7, we see that as x increases and P moves around the unit circle in a counterclockwise (positive) direction, both sin x and cos x behave in uniform ways.

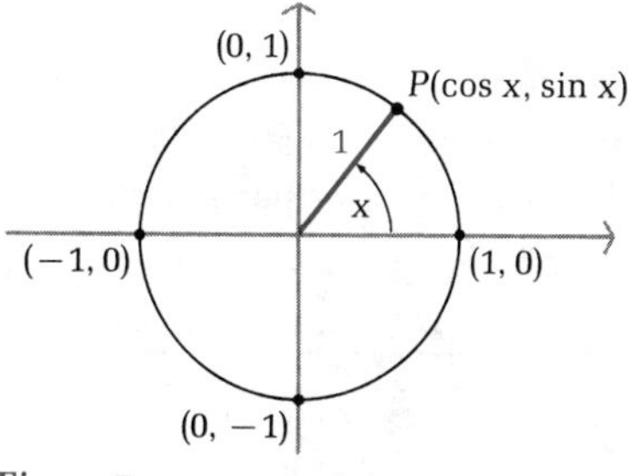

Figure 7

As x increases from	$y = \sin x$	$y = \cos x$
0 to $\pi/2$	Increases from 0 to 1	Decreases from 1 to 0
$\pi/2$ to π	Decreases from 1 to 0	Decreases from 0 to -1
π to $3\pi/2$	Decreases from 0 to -1	Increases from -1 to 0
$3\pi/2$ to 2π	Increases from -1 to 0	Increases from 0 to 1

Note that P has completed one revolution and is back at its starting place. If we let x continue to increase, then the second and third columns in the table will be repeated every 2π units. In general, it can be shown that

$$\sin(x + 2\pi) = \sin x$$
$$\cos(x + 2\pi) = \cos x$$

for all real numbers x. Functions such that

$$f(x + p) = f(x)$$

for some constant p and all real numbers x for which the functions are defined are said to be **periodic.** The smallest such value of p is called the **period** of the function. Thus, both the sine and cosine functions are periodic (a very important property) with period 2π.

Putting all this information together, and, perhaps adding a few values obtained from a calculator or Figure 6, we obtain the graphs of the sine and cosine functions illustrated in Figure 8. Notice that these curves are continuous. It can be shown that **the sine and cosine functions are continuous for all real numbers.**

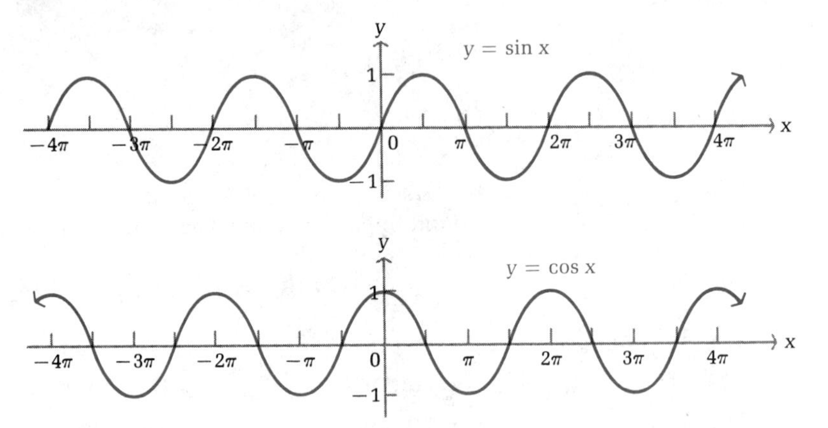

Figure 8

■ Four Other Trigonometric Functions

The sine and cosine functions are only two of six trigonometric functions. They are, however, the most important of the six for many applications. We define the other four trigonometric functions below. Exercises pertaining to these functions may be found in Exercise 7-1, part C.

Four Other Trigonometric Functions

$$\tan x = \frac{\sin x}{\cos x} \qquad \cos x \neq 0$$

$$\cot x = \frac{\cos x}{\sin x} \qquad \sin x \neq 0$$

$$\sec x = \frac{1}{\cos x} \qquad \cos x \neq 0$$

$$\csc x = \frac{1}{\sin x} \qquad \sin x \neq 0$$

Answers to Matched Problems

1. $180/\pi \approx 57.3°$ 2. (A) 0 (B) 1 (C) 0
3. (A) $\sqrt{2}/2$ (B) $\frac{1}{2}$ (C) $\frac{1}{2}$

Exercise 7-1

A

Recall that 180° corresponds to π radians. Mentally convert each degree measure to radian measure in terms of π.

1. 60° 2. 90° 3. 45°
4. 360° 5. 30° 6. 120°

In Problems 7–12, indicate the quadrant in which the terminal side of each angle lies.

II I
III IV

7. 150° 8. −190° 9. $-\frac{\pi}{3}$ radians
10. $\frac{7\pi}{6}$ radians 11. 400° 12. −250°

Use Figure 5 to find the exact value of each of the following:

13. $\cos 0°$ 14. $\sin 90°$ 15. $\sin \pi$
16. $\cos \frac{\pi}{2}$ 17. $\cos(-\pi)$ 18. $\sin \frac{3\pi}{2}$

B

Recall that π radians corresponds to 180°. Mentally convert each radian measure to degree measure.

19. $\frac{\pi}{3}$ radians 20. 2π radians 21. $\frac{\pi}{4}$ radian
22. $\frac{\pi}{2}$ radians 23. $\frac{\pi}{6}$ radian 24. $\frac{5\pi}{6}$ radians

Use Figure 6 to find the exact value of each of the following:

25. $\cos 30°$ 26. $\sin(-45°)$ 27. $\sin \frac{\pi}{6}$
28. $\sin\left(-\frac{\pi}{3}\right)$ 29. $\cos \frac{5\pi}{6}$ 30. $\cos(-120°)$

Use a hand calculator to find the value (to four decimal places) of each of the following:

31. $\sin 3$ 32. $\cos 13$ 33. $\cos 33.74$
34. $\sin 325.9$ 35. $\sin(-43.06)$ 36. $\cos(-502.3)$

C Convert to radian measure.

37. 27° **38.** 18°

Convert to degree measure.

39. $\frac{\pi}{12}$ radian **40.** $\frac{\pi}{60}$ radian

Use Figure 6 to find the exact value of each of the following:

41. $\tan 45°$ **42.** $\cot 45°$ **43.** $\sec \frac{\pi}{3}$

44. $\csc \frac{\pi}{6}$ **45.** $\cot \frac{\pi}{3}$ **46.** $\tan \frac{\pi}{6}$

47. Refer to Figure 5 and use the Pythagorean theorem to show that

$$(\sin x)^2 + (\cos x)^2 = 1$$

for all x.

48. Use the results of Problem 47 and basic definitions to show that

(A) $(\tan x)^2 + 1 = (\sec x)^2$ (B) $1 + (\cot x)^2 = (\csc x)^2$

Applications

Business & Economics

49. *Seasonal business cycle.* Suppose profit in the sale of swimming suits in a department store over a 2 year period is given approximately by

$$P(t) = 5 - 5 \cos \frac{\pi t}{26} \qquad 0 \leq t \leq 104$$

where P is profit in hundreds of dollars for a week of sales t weeks after January 1. The graph of the profit function is shown in the figure.

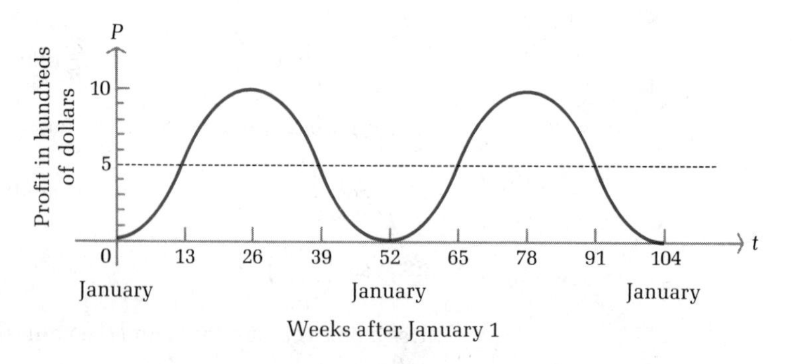

(A) Find the exact value of $P(13)$, $P(26)$, $P(39)$, and $P(52)$ by evaluating $P(t) = 5 - 5 \cos(\pi t/26)$ without tables or a calculator.

(B) Use a calculator to find $P(30)$ and $P(100)$.

50. *Seasonal business cycle.* A soft drink company has revenues from sales over a 2 year period as given approximately by

$$R(t) = 4 - 3\cos\frac{\pi t}{6} \qquad 0 \leqslant t \leqslant 24$$

where $R(t)$ is revenue in millions of dollars for a month of sales t months after February 1. The graph of the revenue function is shown in the figure.

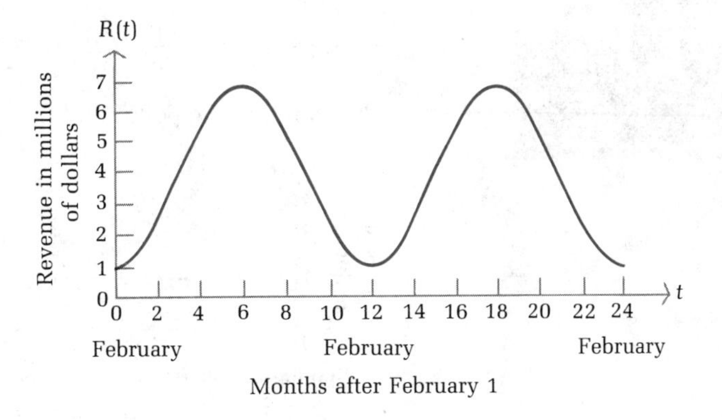

(A) Find the exact values of $R(0)$, $R(2)$, $R(3)$, and $R(18)$ without tables or a calculator.
(B) Use a calculator to find $R(5)$ and $R(23)$.

Life Sciences

51. *Physiology.* A normal seated adult breathes in and exhales about 0.8 liter of air every 4 seconds. The volume of air $V(t)$ in the lungs t seconds after exhaling is given approximately by

$$V(t) = 0.45 - 0.35\cos\frac{\pi t}{2} \qquad 0 \leqslant t \leqslant 8$$

The graph for two complete respirations is shown in the figure.

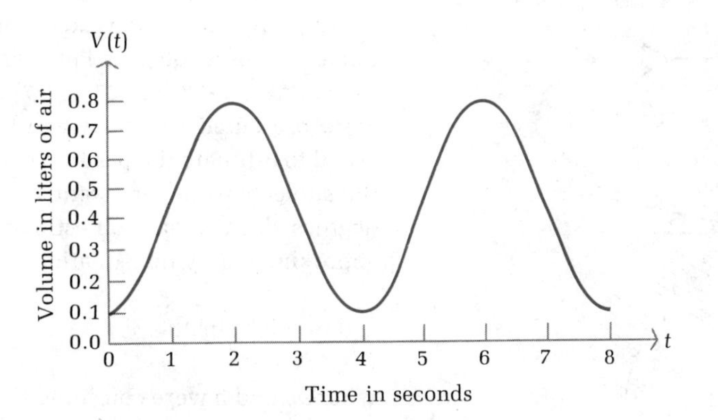

(A) Find the exact value of $V(0)$, $V(1)$, $V(2)$, $V(3)$, and $V(7)$ without using a table or a calculator.
(B) Use a calculator to find $V(3.5)$ and $V(5.7)$.

52. *Pollution.* In a large city the amount of sulfur dioxide pollutant released into the atmosphere due to the burning of coal and oil for heating purposes varies seasonally. Suppose the number of tons of pollutant released into the atmosphere during the nth week after January 1 is given approximately by

$$P(n) = 1 + \cos \frac{\pi n}{26} \qquad 0 \leqslant n \leqslant 104$$

The graph of the pollution function is shown in the figure.

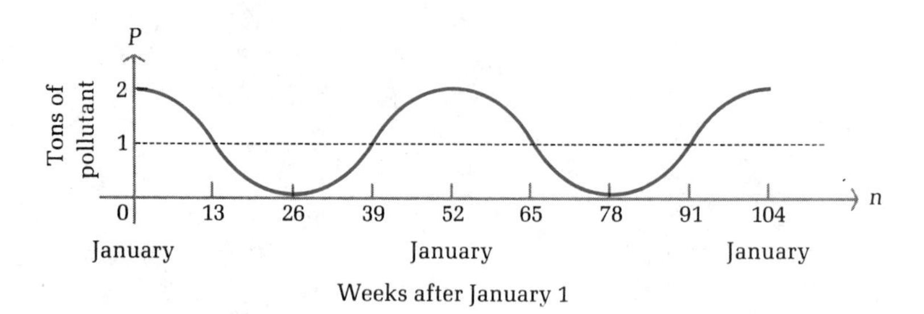

(A) Find the exact value of $P(0)$, $P(39)$, $P(52)$, and $P(65)$ by evaluating $P(n) = 1 + \cos(\pi n/26)$ without tables or a calculator.
(B) Use a calculator to find $P(10)$ and $P(95)$.

Social Sciences

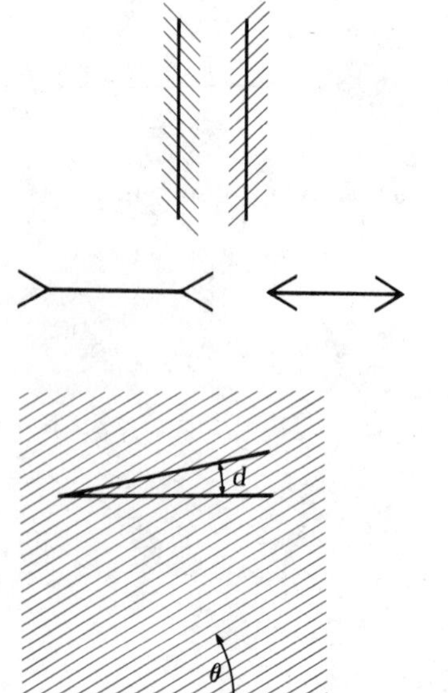

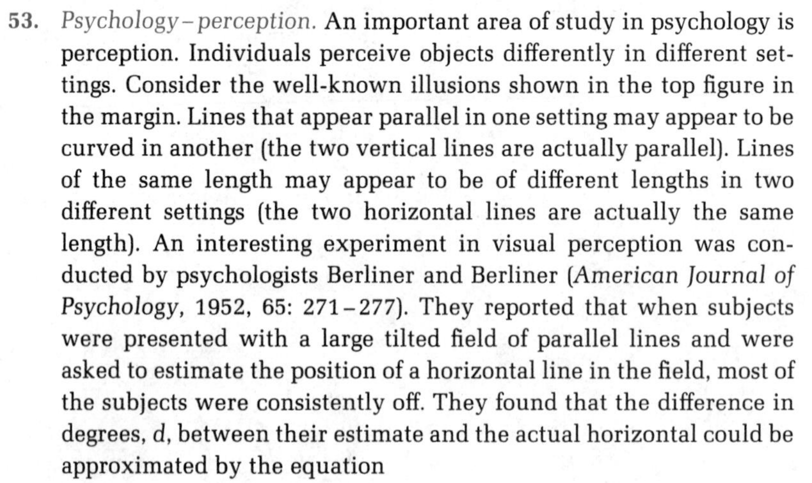

53. *Psychology–perception.* An important area of study in psychology is perception. Individuals perceive objects differently in different settings. Consider the well-known illusions shown in the top figure in the margin. Lines that appear parallel in one setting may appear to be curved in another (the two vertical lines are actually parallel). Lines of the same length may appear to be of different lengths in two different settings (the two horizontal lines are actually the same length). An interesting experiment in visual perception was conducted by psychologists Berliner and Berliner (*American Journal of Psychology*, 1952, 65: 271–277). They reported that when subjects were presented with a large tilted field of parallel lines and were asked to estimate the position of a horizontal line in the field, most of the subjects were consistently off. They found that the difference in degrees, d, between their estimate and the actual horizontal could be approximated by the equation

$$d = a + b \sin 4\theta$$

where a and b were constants associated with a particular individual and θ was the angle of tilt of the visual field in degrees. Suppose for a given individual $a = -2.1$ and $b = -4$. Find d if:

(A) $\theta = 30°$ (B) $\theta = 10°$

7-2 Derivatives of Trigonometric Functions

- Derivative Formulas
- Application

Derivative Formulas

In this section we will discuss derivative formulas for the sine and cosine functions. Once we have these formulas, we will automatically have integral formulas for the same functions, which we will discuss in the next section.

From the definition of derivative (Section 1-4),

$$D_x \sin x = \lim_{\Delta x \to 0} \frac{\sin(x + \Delta x) - \sin x}{\Delta x}$$

Using trigonometric identities and some special trigonometric limits, it can be shown that the limit on the right is $\cos x$. Similarly, it can be shown that $D_x \cos x = -\sin x$.

We now add the following two important derivative formulas to our list of derivative formulas from preceding chapters:

Derivative Formulas

$$\frac{d \sin u}{dx} = \cos u \frac{du}{dx}$$

$$\frac{d \cos u}{dx} = -\sin u \frac{du}{dx}$$

Example 4

(A) $D_x \sin x^2 = \cos x^2 \, D_x x^2 = 2x \cos x^2$

(B) $D_x \cos(2x - 5) = -\sin(2x - 5) \, D_x(2x - 5) = -2 \sin(2x - 5)$

(C) $D_x(3x^2 - x) \cos x = (3x^2 - x) \, D_x \cos x + (\cos x) \, D_x(3x^2 - x)$

$$= -(3x^2 - x) \sin x + (6x - 1) \cos x$$

$$= (x - 3x^2) \sin x + (6x - 1) \cos x$$

Problem 4 Find each of the following derivatives:

(A) $D_x \cos x^3$ (B) $D_x \sin(5 - 3x)$ (C) $D_x \dfrac{\sin x}{x}$

Example 5 Find the slope of the graph of $f(x) = \sin x$ at $(\pi/2, 1)$, and sketch in the tangent line to the graph at this point.

Solution Slope at $(\pi/2, 1) = f'(\pi/2) = \cos(\pi/2) = 0$.

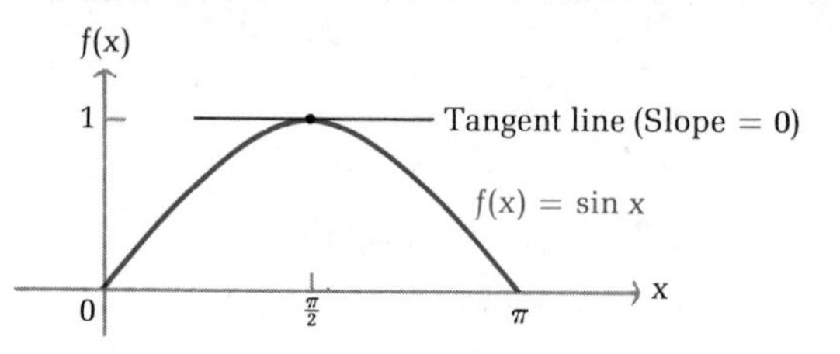

Problem 5 Find the slope of the graph of $f(x) = \cos x$ at $(\pi/6, \sqrt{3}/2)$.

Example 6 Find $D_x \sec x$.

Solution

$$\begin{aligned} D_x \sec x &= D_x \frac{1}{\cos x} && \text{Since } \sec x = \frac{1}{\cos x} \\ &= D_x(\cos x)^{-1} \\ &= -(\cos x)^{-2} D_x \cos x \\ &= -(\cos x)^{-2}(-\sin x) \\ &= \frac{\sin x}{(\cos x)^2} = \left(\frac{\sin x}{\cos x}\right)\left(\frac{1}{\cos x}\right) \\ &= \tan x \sec x && \text{Since } \tan x = \frac{\sin x}{\cos x} \end{aligned}$$

Problem 6 Find $D_x \csc x$.

▪ Application

Example 7 **Revenue** A sporting goods store has revenue from the sale of ski jackets that is given approximately by

$$R(t) = 1.55 + 1.45 \cos \frac{\pi t}{26} \qquad 0 \leq t \leq 104$$

where $R(t)$ is revenue in thousands of dollars for a week of sales t weeks after January 1.

(A) What is the rate of change of revenue t weeks after the first of the year?
(B) What is the rate of change of revenue 10 weeks after the first of the year? 26 weeks after the first of the year? 40 weeks after the first of the year?
(C) Find all local maxima and minima for $0 < t < 104$.
(D) Find the absolute maximum and minimum for $0 \leq t \leq 104$.

Solutions

(A) $R'(t) = -\frac{1.45\pi}{26} \sin \frac{\pi t}{26} \qquad 0 \leq t \leq 104$

(B) $R'(10) \approx -\$0.164$ thousand ($-\$164$ per week)
$R'(26) = \$0$ per week
$R'(40) \approx \$0.174$ thousand ($\$174$ per week)

(C) Find the critical values:

$$R'(t) = -\frac{1.45\pi}{26} \sin \frac{\pi t}{26} = 0 \qquad 0 < t < 104$$

$$\sin \frac{\pi t}{26} = 0$$

$$\frac{\pi t}{26} = \pi, 2\pi, 3\pi \qquad \textit{Note: } 0 < t < 104 \text{ implies } 0 < \pi t/26 < 4\pi.$$

$$t = 26, 52, 78$$

Use the second-derivative test:

$$R''(t) = -\frac{1.45\pi^2}{26^2} \cos \frac{\pi t}{26}$$

t	$R''(t)$	Graph of R
26	+	Local minimum
52	−	Local maximum
78	+	Local minimum

(D) Evaluate $R(t)$ at end points $t = 0$ and $t = 104$ and at the critical values found in part C:

t	$R(t)$	
0	\$3,000	Absolute maximum
26	\$100	Absolute minimum
52	\$3,000	Absolute maximum
78	\$100	Absolute minimum
104	\$3,000	Absolute maximum

The results above can be visualized in the graph of R, as shown on the next page.

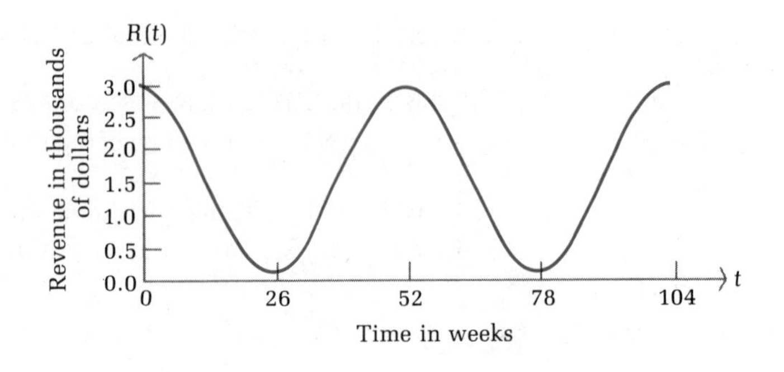

Problem 7 Suppose that in Example 7 revenue from the sale of ski jackets is given approximately by

$$R(t) = 6.2 + 5.8 \cos \frac{\pi t}{6} \qquad 0 \leq t \leq 24$$

where $R(t)$ is revenue in thousands of dollars for a month of sales t months after January 1.

(A) What is the rate of change of revenue t months after the first of the year?
(B) What is the rate of change of revenue 2 months after the first of the year? 12 months after the first of the year? 23 months after the first of the year?
(C) Find all local maxima and minima for $0 < t < 24$.
(D) Find the absolute maximum and minimum for $0 \leq t \leq 24$.

Answers to Matched Problems

4. (A) $-3x^2 \sin x^3$ (B) $-3 \cos(5 - 3x)$ (C) $\dfrac{x \cos x - \sin x}{x^2}$

5. $-\frac{1}{2}$ 6. $-\cot x \csc x$

7. (A) $R'(t) = -\dfrac{5.8\pi}{6} \sin \dfrac{\pi t}{6}$, $0 < t < 24$

(B) $R'(2) \approx -\$2.630$ thousand ($-\$2{,}630$ per month); $R'(12) = \$0$ per month; $R'(23) \approx \$1.518$ thousand ($1,518 per month)

(C) Local minima at $t = 6$ and $t = 18$; local maximum at $t = 12$

(D)

	t	$R(t)$	
End point	0	$12,000	Absolute maximum
	6	$400	Absolute minimum
	12	$12,000	Absolute maximum
	18	$400	Absolute minimum
End point	24	$12,000	Absolute maximum

Exercise 7-2

Find the following derivatives:

A

1. $D_t \cos t$
2. $D_w \sin w$
3. $D_x \sin x^3$
4. $D_x \cos(x^2 - 1)$

B

5. $D_t\, t \sin t$
6. $D_u\, u \cos u$
7. $D_x \sin x \cos x$
8. $D_x \dfrac{\sin x}{\cos x}$
9. $D_x(\sin x)^5$
10. $D_x(\cos x)^8$
11. $D_x \sqrt{\sin x}$
12. $D_x \sqrt{\cos x}$
13. $D_x \cos \sqrt{x}$
14. $D_x \sin \sqrt{x}$
15. Find the slope of the graph of $f(x) = \sin x$ at $x = \pi/6$.
16. Find the slope of the graph of $f(x) = \cos x$ at $x = \pi/4$.

C *Find the following derivatives:*

17. $D_x \tan x$
18. $D_x \cot x$
19. $D_x \sin \sqrt{x^2 - 1}$
20. $D_x \cos \sqrt{x^4 - 1}$

Applications

Business & Economics

21. *Profit.* Suppose profit in the sale of swimming suits in a department store is given approximately by

$$P(t) = 5 - 5 \cos \frac{\pi t}{26} \qquad 0 \le t \le 104$$

where $P(t)$ is profit in hundreds of dollars for a week of sales t weeks after January 1.

(A) What is the rate of change of profit t weeks after the first of the year?
(B) What is the rate of change of profit 8 weeks after the first of the year? 26 weeks after the first of the year? 50 weeks after the first of the year?
(C) Find all local maxima and minima for $0 < t < 104$.
(D) Find the absolute maximum and minimum for $0 \le t \le 104$.

22. *Revenue.* A soft drink company has revenues from sales over a 2 year period as given approximately by

$$R(t) = 4 - 3 \cos \frac{\pi t}{6} \qquad 0 \le t \le 24$$

where $R(t)$ is revenue in millions of dollars for a month of sales t months after February 1.

(A) What is the rate of change of revenue t months after February 1?
(B) What is the rate of change of revenue 1 month after February 1? 6 months after February 1? 11 months after February 1?
(C) Find all local maxima and minima for $0 < t < 24$.
(D) Find the absolute maximum and minimum for $0 \leq t \leq 24$.

Life Sciences

23. *Physiology.* A normal seated adult breathes in and exhales about 0.8 liter of air every 4 seconds. The volume of air $V(t)$ in the lungs t seconds after exhaling is given approximately by

$$V(t) = 0.45 - 0.35 \cos \frac{\pi t}{2} \qquad 0 \leq t \leq 8$$

(A) What is the rate of flow of air t seconds after exhaling?
(B) What is the rate of flow of air 3 seconds after exhaling? 4 seconds after exhaling? 5 seconds after exhaling?
(C) Find all local maxima and minima for $0 < t < 8$.
(D) Find the absolute maximum and minimum for $0 \leq t \leq 8$.

24. *Pollution.* In a large city the amount of sulfur dioxide pollutant released into the atmosphere due to the burning of coal and oil for heating purposes varies seasonally. Suppose the number of tons of pollutant released into the atmosphere during the nth week after January 1 is given approximately by

$$P(n) = 1 + \cos \frac{\pi n}{26} \qquad 0 \leq n \leq 104$$

(A) What is the rate of change of pollutant n weeks after the first of the year?
(B) What is the rate of change of pollutant 13 weeks after the first of the year? 26 weeks after the first of the year? 30 weeks after the first of the year?
(C) Find all local maxima and minima for $0 < t < 104$.
(D) Find the absolute maximum and minimum for $0 \leq t \leq 104$.

7-3 Integration of Trigonometric Functions

- Integral Formulas
- Application

Integral Formulas

Now that we know that

$$D_x \sin x = \cos x \qquad \text{and} \qquad D_x \cos x = -\sin x$$

then from the definition of the indefinite integral of a function (Section 4-1), we automatically have the two integral formulas

$$\int \cos x\, dx = \sin x + C \qquad \text{and} \qquad \int \sin x\, dx = -\cos x + C$$

Example 8 Find the area under the sine curve $y = \sin x$ from 0 to π.

Solution

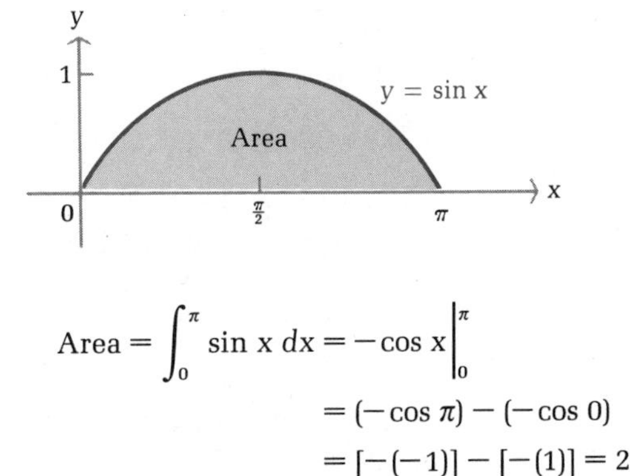

$$\text{Area} = \int_0^\pi \sin x\, dx = -\cos x \Big|_0^\pi$$
$$= (-\cos \pi) - (-\cos 0)$$
$$= [-(-1)] - [-(1)] = 2$$

Problem 8 Find the area under the cosine curve $y = \cos x$ from 0 to $\pi/2$.

From the facts that

$$D_x \sin u = \cos u \frac{du}{dx} \qquad \text{and} \qquad D_x \cos u = -\sin u \frac{du}{dx}$$

we obtain the more general integral formulas below.

Integral Formulas

For $u = u(x)$,

$$\int \sin u\, du = -\cos u + C$$

$$\int \cos u\, du = \sin u + C$$

Example 9 Integrate $\int x \sin x^2\, dx$.

Solution

$$\begin{aligned}\int x \sin x^2\, dx &= \frac{1}{2}\int 2x \sin x^2\, dx \\ &= \frac{1}{2}\int (\sin x^2)\, 2x\, dx \\ &= \frac{1}{2}\int \sin u\, du \qquad \text{Let } u = x^2\text{; then } du = 2x\, dx. \\ &= -\frac{1}{2}\cos x^2 + C\end{aligned}$$

Check To check, we differentiate the result to obtain the original integrand:

$$\begin{aligned}D_x\left(-\frac{1}{2}\cos x^2\right) &= -\frac{1}{2} D_x \cos x^2 \\ &= -\frac{1}{2}(-\sin x^2)\, D_x x^2 \\ &= -\frac{1}{2}(-\sin x^2)(2x) \\ &= x \sin x^2\end{aligned}$$

Problem 9 Integrate $\int \cos 20\pi t\, dt$.

Example 10 Integrate $\int (\sin x)^5 \cos x\, dx$.

Solution This is of the form $\int u^p\, du$, where $u = \sin x$ and $du = \cos x\, dx$. Thus,

$$\int (\sin x)^5 \cos x\, dx = \frac{(\sin x)^6}{6} + C$$

Problem 10 Integrate $\int \sqrt{\sin x} \cos x\, dx$.

Example 11 Evaluate $\int_2^{3.5} \cos x\, dx$ using a hand calculator.

Solution

$$\begin{aligned}\int_2^{3.5} \cos x\, dx &= \sin x \Big|_2^{3.5} \\ &= \sin 3.5 - \sin 2 \\ &= -0.3508 - 0.9093 \\ &= -1.2601\end{aligned}$$

Problem 11 Evaluate $\int_1^{1.5} \sin x\, dx$ using a hand calculator.

▪ Application

Example 12
Total Revenue

In Example 7 (Section 7-2), we were given the following revenue equation from the sale of ski jackets:

$$R(t) = 1.55 + 1.45 \cos \frac{\pi t}{26} \qquad 0 \leq t \leq 104$$

where $R(t)$ is revenue in thousands of dollars for a week of sales t weeks after January 1.

(A) Find the total revenue taken in over the 2 year period; that is, from $t = 0$ to $t = 104$.

(B) Find the total revenue taken in from $t = 39$ to $t = 65$.

Solutions (A) The area under the graph of the revenue equation for the 2 year period approximates the total revenue taken in for that period:

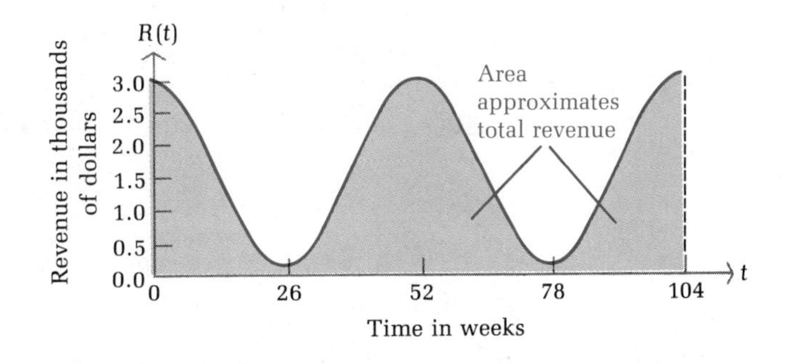

This area is given by the following definite integral:

$$\begin{aligned}\text{Total revenue} &\approx \int_0^{104} \left(1.55 + 1.45 \cos \frac{\pi t}{26}\right) dt \\ &= \left[1.55t + 1.45\left(\frac{26}{\pi}\right) \sin \frac{\pi t}{26}\right]\Big|_0^{104} \\ &= \$161.200 \text{ thousand or } \$161{,}200\end{aligned}$$

(B) The total revenue from $t = 39$ to $t = 65$ is approximated by the area under the curve from $t = 39$ to $t = 65$:

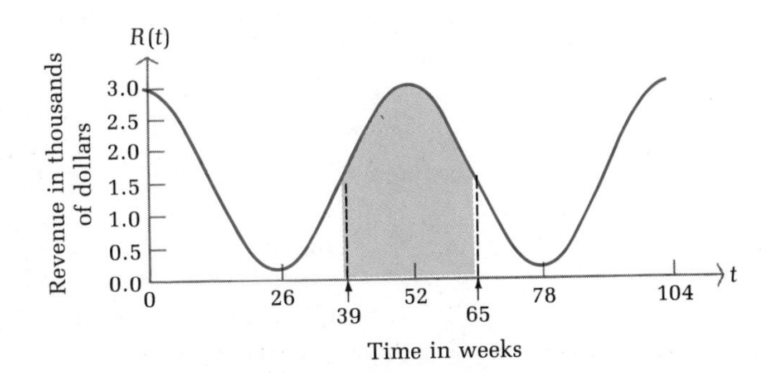

$$\begin{aligned}\text{Total revenue} &\approx \int_{39}^{65}\left(1.55 + 1.45\cos\frac{\pi t}{26}\right)dt\\ &= \left[1.55t + 1.45\left(\frac{26}{\pi}\right)\sin\frac{\pi t}{26}\right]\Bigg|_{39}^{65}\\ &= \$64.301 \text{ thousand or } \$64{,}301\end{aligned}$$

Problem 12 Suppose that in Example 12 revenue from the sale of ski jackets is given approximately by

$$R(t) = 6.2 + 5.8\cos\frac{\pi t}{6} \qquad 0 \le t \le 24$$

where $R(t)$ is revenue in thousands of dollars for a month of sales t months after January 1.

(A) Find the total revenue taken in over the 2 year period; that is, from $t = 0$ to $t = 24$.

(B) Find the total revenue taken in from $t = 4$ to $t = 8$.

Answers to Matched Problems

8. 1 9. $\frac{1}{20\pi}\sin 20\pi t + C$ 10. $\frac{2}{3}(\sin x)^{3/2} + C$ 11. 0.4696

12. (A) \$148.8 thousand or \$148,800 (B) \$5.614 thousand or \$5,614

Exercise 7-3

Find each of the following indefinite integrals:

A

1. $\int \sin t\, dt$
2. $\int \cos w\, dw$
3. $\int \cos 3x\, dx$
4. $\int \sin 2x\, dx$
5. $\int (\sin x)^{12}\cos x\, dx$
6. $\int \sin x \cos x\, dx$

B

7. $\int \sqrt[3]{\cos x}\,\sin x\, dx$
8. $\int \frac{\cos x}{\sqrt{\sin x}}\, dx$
9. $\int x^2 \cos x^3\, dx$
10. $\int (x + 1)\sin(x^2 + 2x)\, dx$

Evaluate each of the following definite integrals:

11. $\int_0^{\pi/2} \cos x\, dx$
12. $\int_0^{\pi/4} \cos x\, dx$
13. $\int_{\pi/2}^{\pi} \sin x\, dx$
14. $\int_{\pi/6}^{\pi/3} \sin x\, dx$

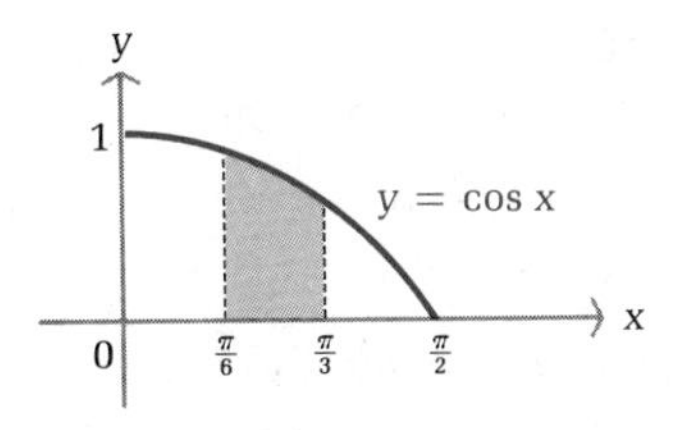

15. Find the shaded area under the cosine curve in the figure in the margin.
16. Find the shaded area under the sine curve in the figure below.

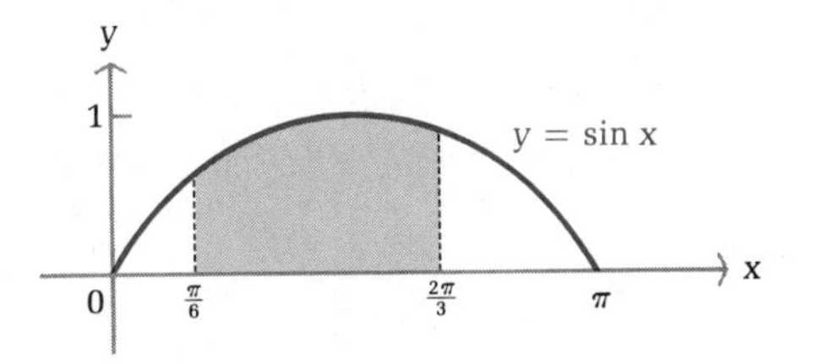

Use a hand calculator to evaluate the definite integrals below after performing the indefinite integration. (Remember that the limits are real numbers, hence the radian mode must be used on the calculator.)

17. $\int_0^2 \sin x \, dx$ 18. $\int_0^{0.5} \cos x \, dx$

19. $\int_1^2 \cos x \, dx$ 20. $\int_1^3 \sin x \, dx$

C *Find each of the following indefinite integrals:*

21. $\int e^{\sin x} \cos x \, dx$ 22. $\int e^{\cos x} \sin x \, dx$

23. $\int \frac{\cos x}{\sin x} \, dx$ 24. $\int \frac{\sin x}{\cos x} \, dx$

25. $\int \tan x \, dx$ 26. $\int \cot x \, dx$

Applications

Business & Economics

27. *Seasonal business cycle.* In Problem 49, Exercise 7-1, we were given the following profit equation for the sale of swimming suits:

$$P(t) = 5 - 5 \cos \frac{\pi t}{26} \qquad 0 \leq t \leq 104$$

where $P(t)$ is profit in hundreds of dollars for a week of sales t weeks after January 1. Use definite integrals to approximate:

(A) The total profit earned during the 2 year period.
(B) The total profit earned from $t = 13$ to $t = 26$.

28. *Seasonal business cycle.* In Problem 50, Exercise 7-1, we were given the following revenue equation from the sales of soft drinks:

$$R(t) = 4 - 3 \cos \frac{\pi t}{6} \qquad 0 \leq t \leq 24$$

where $R(t)$ is revenue in millions of dollars for a month of sales t months after February 1. Use definite integrals to approximate:

(A) Total revenues taken in over the 2 year period.
(B) Total revenues taken in from $t = 8$ to $t = 14$.

Life Sciences

29. *Pollution.* In Problem 52, Exercise 7-1, we were given the following equation for the amount of sulfur dioxide released into the atmosphere due to the burning of coal and oil for heating purposes:

$$P(n) = 1 + \cos \frac{\pi n}{26} \qquad 0 \leq n \leq 104$$

where $P(n)$ is the amount of sulfur dioxide in tons released during the nth week after January 1.

(A) How many tons of pollutants were emitted into the atmosphere over the 2 year period?
(B) How many tons of pollutants were emitted into the atmosphere from $n = 13$ to $n = 52$?

7-4 Chapter Review

Important Terms and Symbols

7-1 *Trigonometric functions review.* Angle, initial side, terminal side, vertex, degree measure, radian measure, standard position, generalized angle, $\sin x$, $\cos x$, $\tan x$, $\sec x$, $\csc x$, $\cot x$

7-2 *Derivatives of trigonometric functions.* $D_x \sin u = \cos u \, D_x u$, $D_x \cos u = -\sin u \, D_x u$

7-3 *Integration of trigonometric functions.* $\int \cos u \, du = \sin u + C$, $\int \sin u \, du = -\cos u + C$

Exercise 7-4 Chapter Review

Work through all the problems in this chapter review and check your answers in the back of the book. (Answers to all review problems are there.) Where weaknesses show up, review appropriate sections in the text.

A

1. Convert to radian measure in terms of π:

(A) $30°$ (B) $45°$ (C) $60°$ (D) $90°$

2. Evaluate without a table or calculator:

(A) $\cos \pi$ (B) $\sin 0$ (C) $\sin \frac{\pi}{2}$

Find:

3. $D_m \cos m$

4. $D_u \sin u$

5. $D_x \sin(x^2 - 2x + 1)$

6. $\int \sin 3t\, dt$

B

7. Convert to degree measure:

(A) $\frac{\pi}{6}$ (B) $\frac{\pi}{4}$ (C) $\frac{\pi}{3}$ (D) $\frac{\pi}{2}$

8. Evaluate without a table or calculator:

(A) $\sin \frac{\pi}{6}$ (B) $\cos \frac{\pi}{4}$ (C) $\sin \frac{\pi}{3}$

9. Evaluate using a hand calculator:

(A) $\cos 33.7$ (B) $\sin(-118.4)$

Find:

10. $D_x(x^2 - 1) \sin x$

11. $D_x(\sin x)^6$

12. $D_x \sqrt[3]{\sin x}$

13. $\int t \cos(t^2 - 1)\, dt$

14. $\int_0^{\pi} \sin u\, du$

15. $\int_0^{\pi/3} \cos x\, dx$

16. $\int_1^{2.5} \cos x\, dx$

17. Find the slope of the cosine curve $y = \cos x$ at $x = \pi/4$.

18. Find the area under the sine curve $y = \sin x$ from $x = \pi/4$ to $x = 3\pi/4$.

C

19. Convert 15° to radian measure.

20. Evaluate without a table or calculator:

(A) $\sin \frac{3\pi}{2}$ (B) $\cos \frac{5\pi}{6}$ (C) $\sin\left(\frac{-\pi}{6}\right)$

Find:

21. $D_u \tan u$

22. $D_x e^{\cos x^2}$

23. $\int e^{\sin x} \cos x\, dx$

24. $\int \tan x\, dx$

25. $\int_2^5 (5 + 2 \cos 2x)\, dx$

Applications

Business & Economics

Problems 26–28 refer to the following: Revenue from sweater sales in a sportswear chain is given approximately by

$$R(t) = 3 + 2\cos\frac{\pi t}{6} \qquad 0 \le t \le 24$$

where $R(t)$ is the revenue in thousands of dollars brought in for a month of sales t months after January 1.

26. (A) Find the exact values of $R(0)$, $R(2)$, $R(3)$, and $R(6)$ without using tables or a calculator.
(B) Use a calculator to find $R(1)$ and $R(22)$.

27. (A) What is the rate of change of revenue t months after January 1?
(B) What is the rate of change of revenue 3 months after January 1? 10 months after January 1? 18 months after January 1?
(C) Find all local maxima and minima for $0 < t < 24$.
(D) Find the absolute maximum and minimum for $0 \le t \le 24$.

28. (A) Find the total revenues taken in over the 2 year period.
(B) Find the total revenues taken in from $t = 5$ to $t = 9$.

Special Topics

A

APPENDIX A Contents

A-1 Integer Exponents and Square Root Radicals

- Integer Exponents
- Scientific Notation
- Square Root Radicals

Integer Exponents

Table 1 lists definitions for **integer exponents.**

Table 1 Definition of a^n

Given: n is an integer and a is a real number

1. For n a positive integer,

   $a^n = a \cdot a \cdot \cdots \cdot a$ $\quad$ n factors of a $\qquad$ $5^4 = 5 \cdot 5 \cdot 5 \cdot 5$

2. For $n = 0$,

   $a^0 = 1 \quad a \neq 0$ $\qquad$ $12^0 = 1$

   0^0 is not defined.

3. For n a negative integer,

   $a^n = \dfrac{1}{a^{-n}} \quad a \neq 0$ $\qquad$ $a^{-3} = \dfrac{1}{a^{-(-3)}} = \dfrac{1}{a^3}$

   [If n is negative, then $(-n)$ is positive.]
   Note: It can be shown that for *all* integers n,

   $a^{-n} = \dfrac{1}{a^n}$ and $a^n = \dfrac{1}{a^{-n}} \quad a \neq 0$ $\qquad$ $a^5 = \dfrac{1}{a^{-5}}, \quad a^{-5} = \dfrac{1}{a^5}$

Table 2 lists integer exponent properties that are very useful in manipulating integer exponent forms.

Table 2 Exponent Properties

Given: n and m are integers and a and b are real numbers		
1. $a^m a^n = a^{m+n}$		$a^8 a^{-3} = a^{8+(-3)} = a^5$
2. $(a^n)^m = a^{mn}$		$(a^{-2})^3 = a^{3(-2)} = a^{-6}$
3. $(ab)^m = a^m b^m$		$(ab)^{-2} = a^{-2}b^{-2}$
4. $\left(\frac{a}{b}\right)^m = \frac{a^m}{b^m}$	$b \neq 0$	$\left(\frac{a}{b}\right)^5 = \frac{a^5}{b^5}$
5. $\frac{a^m}{a^n} = a^{m-n} = \frac{1}{a^{n-m}}$	$a \neq 0$	$\frac{a^{-3}}{a^7} = \frac{1}{a^{7-(-3)}} = \frac{1}{a^{10}}$

Exponent and radical forms are frequently encountered in algebraic applications. You should sharpen your skills in using these forms by reviewing the basic definitions, properties, and exercises that follow.

Example 1 Simplify and express the answers using positive exponents only.

(A) $(2x^3)(3x^5) = 2 \cdot 3x^{3+5} = 6x^8$

(B) $x^5x^{-9} = x^{-4} = \frac{1}{x^4}$

(C) $\frac{x^5}{x^7} = x^{5-7} = x^{-2} = \frac{1}{x^2}$

or $\frac{x^5}{x^7} = \frac{1}{x^{7-5}} = \frac{1}{x^2}$

(D) $\frac{x^{-3}}{y^{-4}} = \frac{y^4}{x^3}$

(E) $(u^{-3}v^2)^{-2} = (u^{-3})^{-2}(v^2)^{-2} = u^6v^{-4} = \frac{u^6}{v^4}$

(F) $\left(\frac{y^{-5}}{y^{-2}}\right)^{-2} = \frac{(y^{-5})^{-2}}{(y^{-2})^{-2}} = \frac{y^{10}}{y^4} = y^6$

(G) $\frac{4m^{-3}n^{-5}}{6m^{-4}n^3} = \frac{2m^{-3-(-4)}}{3n^{3-(-5)}} = \frac{2m}{3n^8}$

(H) $\left(\frac{2x^{-3}x^3}{n^{-2}}\right)^{-3} = \left(\frac{2x^0}{n^{-2}}\right)^{-3} = \left(\frac{2}{n^{-2}}\right)^{-3} = \frac{2^{-3}}{n^6} = \frac{1}{2^3n^6} = \frac{1}{8n^6}$

Problem 1 Simplify and express the answers using positive exponents only.

(A) $(3y^4)(2y^3)$ (B) m^2m^{-6} (C) $(u^3v^{-2})^{-2}$

(D) $\left(\frac{y^{-6}}{y^{-2}}\right)^{-1}$ (E) $\frac{8x^{-2}y^{-4}}{6x^{-5}y^2}$ (F) $\left(\frac{3m^{-3}}{2x^2x^{-2}}\right)^{-2}$

▪ Scientific Notation

Writing and working with very large or very small numbers in standard decimal notation is often awkward, even with electronic hand calculators. It is often convenient to represent numbers of this type in **scientific notation;** that is, as the product of a number between 1 and 10 and a power of 10.

Example 2

Decimal Fractions and Scientific Notation

$7 = 7 \times 10^0$	$0.5 = 5 \times 10^{-1}$
$67 = 6.7 \times 10$	$0.45 = 4.5 \times 10^{-1}$
$580 = 5.8 \times 10^2$	$0.003\ 2 = 3.2 \times 10^{-3}$
$43{,}000 = 4.3 \times 10^4$	$0.000\ 045 = 4.5 \times 10^{-5}$
$73{,}400{,}000 = 7.34 \times 10^7$	$0.000\ 000\ 391 = 3.91 \times 10^{-7}$

Note that the power of 10 used corresponds to the number of places we move the decimal to form a number between 1 and 10. The power is positive if the decimal is moved to the left and negative if it is moved to the right. Positive exponents are associated with numbers greater than or equal to 10; negative exponents are associated with positive numbers less than 1.

Problem 2 Write each number in scientific notation.

(A) 370 (B) 47,300,000,000 (C) 0.047 (D) 0.000 000 089

▪ Square Root Radicals

To start, we define a **square root** of a number:

Definition of Square Root

x is a square root of y if $x^2 = y$.

2 is a square root of 4 since $2^2 = 4$.

-2 is a square root of 4 since $(-2)^2 = 4$.

How many square roots of a real number are there? The following theorem, which we state without proof, answers the question.

Theorem 1

(A) Every positive real number has exactly two real square roots, each the negative of the other.

(B) Negative real numbers have no real number square roots (since no real number squared can be negative—think about this).

(C) The square root of 0 is 0.

Square Root Notation

For a a positive number.

$\sqrt{a}$ is the positive square root of a.

$-\sqrt{a}$ is the negative square root of a.

[*Note:* $\sqrt{-a}$ is not a real number.]

Example 3 (A) $\sqrt{4} = 2$ (B) $-\sqrt{4} = -2$

(C) $\sqrt{-4}$ is not a real number. (D) $\sqrt{0} = 0$

Problem 3 Evaluate, if possible.

(A) $\sqrt{9}$ (B) $-\sqrt{9}$ (C) $\sqrt{-9}$ (D) $\sqrt{0}$

It can be shown that if a is a positive integer that is not the square of an integer, then

$$-\sqrt{a} \quad \text{and} \quad \sqrt{a}$$

are irrational numbers. Thus,

$$-\sqrt{7} \quad \text{and} \quad \sqrt{7}$$

name irrational numbers that are, respectively, the negative and positive square roots of 7.

Properties of Radicals

For a and b nonnegative real numbers.

1. $\sqrt{a^2} = a$
2. $\sqrt{a}\sqrt{b} = \sqrt{ab}$
3. $\dfrac{\sqrt{a}}{\sqrt{b}} = \sqrt{\dfrac{a}{b}}$

To see that property 2 holds, let $N = \sqrt{a}$ and $M = \sqrt{b}$, then $N^2 = a$ and $M^2 = b$. Hence,

$$\sqrt{a}\,\sqrt{b} = NM = \sqrt{(NM)^2} = \sqrt{N^2M^2} = \sqrt{ab}$$

Note how properties of exponents are used. The proof of the quotient part is left as an exercise.

Example 4

(A) $\sqrt{5}\sqrt{10} = \sqrt{5 \cdot 10} = \sqrt{50} = \sqrt{25 \cdot 2} = \sqrt{25}\sqrt{2} = 5\sqrt{2}$

(B) $\dfrac{\sqrt{32}}{\sqrt{8}} = \sqrt{\dfrac{32}{8}} = \sqrt{4} = 2$

(C) $\sqrt{\dfrac{7}{4}} = \dfrac{\sqrt{7}}{\sqrt{4}} = \dfrac{\sqrt{7}}{2}$ or $\dfrac{1}{2}\sqrt{7}$

Problem 4 Simplify as in Example 4.

(A) $\sqrt{3}\sqrt{6}$ (B) $\dfrac{\sqrt{18}}{\sqrt{2}}$ (C) $\sqrt{\dfrac{11}{9}}$

The foregoing definitions and theorems allow us to change algebraic expressions containing radicals to a variety of equivalent forms. One form that is often useful is called the *simplest radical form.*

Definition of the Simplest Radical Form

An algebraic expression that contains square root radicals is in **simplest radical form** if all three of the following conditions are satisfied:

1. No radicand (the expression within the radical sign) when expressed in completely factored form contains a factor raised to a power greater than 1. ($\sqrt{x^3}$ violates this condition.)
2. No radical appears in a denominator.
   $\left(\dfrac{3}{\sqrt{5}} \text{ violates this condition.}\right)$
3. No fraction appears within a radical.
   $\left(\sqrt{\dfrac{2}{3}} \text{ violates this condition.}\right)$

It should be understood that forms other than the simplest radical form may be more useful on occasion. The situation dictates what form to choose.

Example 5 Change to simplest radical form—all variables represent nonnegative real numbers.

(A) $\sqrt{8x^3}$ (B) $\dfrac{3x}{\sqrt{3}}$ (C) $\sqrt{\dfrac{3x}{8}}$

Solutions

(A) $\sqrt{8x^3}$ violates condition 1. Separate $8x^3$ into a perfect square part, (2^2x^2), and what is left over, $(2x)$, then use multiplication property 2.

$$\begin{aligned}\sqrt{8x^3} &= \sqrt{(2^2x^2)(2x)} \\ &= \sqrt{2^2x^2}\,\sqrt{2x} \\ &= 2x\sqrt{2x}\end{aligned}$$

(B) $3x/\sqrt{3}$ has a radical in the denominator; hence, it violates condition 2. To remove the radical from the denominator, we multiply the numerator and denominator by $\sqrt{3}$ to obtain $\sqrt{3^2}$ in the denominator (this is called **rationalizing a denominator**):

$$\begin{aligned}\frac{3x}{\sqrt{3}} &= \frac{3x}{\sqrt{3}} \cdot \frac{\sqrt{3}}{\sqrt{3}} \\ &= \frac{3x\sqrt{3}}{\sqrt{3^2}} \\ &= \frac{3x\sqrt{3}}{3} = x\sqrt{3}\end{aligned}$$

(C) $\sqrt{3x/8}$ has a fraction within the radical; hence it violates condition 3. To remove the fraction from the radical, we multiply the numerator and denominator of $3x/8$ by 2 to make the denominator a perfect square:

$$\begin{aligned}\sqrt{\frac{3x}{8}} &= \sqrt{\frac{3x \cdot 2}{8 \cdot 2}} \\ &= \sqrt{\frac{6x}{16}} \\ &= \frac{\sqrt{6x}}{\sqrt{16}} = \frac{\sqrt{6x}}{4}\end{aligned}$$

Problem 5 Change to simplest radical form. All variables represent positive real numbers.

(A) $\sqrt{18y^3}$ (B) $\dfrac{4xy}{\sqrt{2x}}$ (C) $\sqrt{\dfrac{5y}{18x}}$

Answers to Matched Problems

1. (A) $6y^7$ (B) $1/m^4$ (C) v^4/u^6
   (D) y^4 (E) $4x^3/3y^6$ (F) $4m^6/9$
2. (A) 3.7×10^2 (B) 4.73×10^{10}
   (C) 4.7×10^{-2} (D) 8.9×10^{-8}
3. (A) 3 (B) -3
   (C) Not a real number (D) 0
4. (A) $3\sqrt{2}$ (B) 3 (C) $\sqrt{11}/3$ or $\frac{1}{3}\sqrt{11}$
5. (A) $3y\sqrt{2y}$ (B) $2y\sqrt{2x}$ (C) $\dfrac{\sqrt{10xy}}{6x}$ or $\dfrac{1}{6x}\sqrt{10xy}$

Exercise A-1

A *Simplify and express answers using positive exponents only. Variables are restricted to avoid division by zero.*

1. $2x^{-9}$
2. $3y^{-5}$
3. $\frac{3}{2w^{-7}}$
4. $\frac{5}{4x^{-9}}$
5. $2x^{-8}x^{5}$
6. $3c^{-9}c^{4}$
7. $\frac{w^{-8}}{w^{-3}}$
8. $\frac{m^{-11}}{m^{-5}}$
9. $5v^{8}v^{-8}$
10. $7d^{-4}d^{4}$
11. $(a^{-3})^{2}$
12. $(b^{4})^{-3}$
13. $(x^{6}y^{-3})^{-2}$
14. $(a^{-3}b^{4})^{-3}$

Express in simplest radical form. All variables represent positive real numbers.

15. $\sqrt{x^2}$
16. $\sqrt{m^2}$
17. $\sqrt{a^5}$
18. $\sqrt{m^7}$
19. $\sqrt{18x^4}$
20. $\sqrt{8x^3}$
21. $\frac{1}{\sqrt{m}}$
22. $\frac{1}{\sqrt{A}}$
23. $\sqrt{\frac{2}{3}}$
24. $\sqrt{\frac{3}{5}}$
25. $\sqrt{\frac{2}{x}}$
26. $\sqrt{\frac{3}{y}}$

Write in scientific notation.

27. 82,300,000,000
28. 5,380,000
29. 0.783
30. 0.019
31. 0.000 034
32. 0.000 000 007 832

B *Simplify and express answers using positive exponents only.*

33. $(22 + 31)^{0}$
34. $(2x^{3}y^{4})^{0}$
35. $\frac{10^{-3} \cdot 10^{4}}{10^{-11} \cdot 10^{-2}}$
36. $\frac{10^{-17} \cdot 10^{-5}}{10^{-3} \cdot 10^{-14}}$
37. $(5x^{2}y^{-3})^{-2}$
38. $(2m^{-3}n^{2})^{-3}$
39. $\frac{8 \times 10^{-3}}{2 \times 10^{-5}}$
40. $\frac{18 \times 10^{12}}{6 \times 10^{-4}}$
41. $\frac{8x^{-3}y^{-1}}{6x^{2}y^{-4}}$
42. $\frac{9m^{-4}n^{3}}{12m^{-1}n^{-1}}$
43. $\left(\frac{6xy^{-2}}{3x^{-1}y^{2}}\right)^{-3}$
44. $\left(\frac{2x^{-3}y^{2}}{4xy^{-1}}\right)^{-2}$

Simplify and express answers in simplest radical form. All variables represent positive real numbers.

45. $\sqrt{18x^{8}y^{5}z^{2}}$
46. $\sqrt{8p^{3}q^{6}r^{5}}$
47. $\frac{12}{\sqrt{3x}}$

48. $\dfrac{10}{\sqrt{2y}}$

49. $\sqrt{\dfrac{6x}{7y}}$

50. $\sqrt{\dfrac{3m}{2n}}$

51. $\sqrt{\dfrac{4a^3}{3b}}$

52. $\sqrt{\dfrac{9m^5}{2n}}$

53. $\sqrt{18m^3n^4}\sqrt{2m^3n^2}$

54. $\sqrt{10x^3y}\sqrt{5xy}$

55. $\dfrac{\sqrt{4a^3}}{\sqrt{3b}}$

56. $\dfrac{\sqrt{9m^5}}{\sqrt{2n}}$

Convert each numeral to scientific notation and simplify. Express the answer in scientific notation and in standard decimal form. (Answers cannot have more significant digits than the number with the least number of significant digits in the problem.)

57. $\dfrac{9{,}600{,}000{,}000}{(1{,}600{,}000)(0.000\ 000\ 25)}$

58. $\dfrac{(60{,}000)(0.000\ 003)}{(0.000\ 4)(1{,}500{,}000)}$

59. $\dfrac{(1{,}250{,}000)(0.000\ 38)}{0.042\ 3}$

60. $\dfrac{(0.000\ 000\ 82)(230{,}000)}{(430{,}000)(0.008\ 2)}$

C *Simplify and write answers using positive exponents only.*

61. $\left[\left(\dfrac{x^{-2}y^3t}{x^{-3}y^{-2}t^2}\right)^2\right]^{-1}$

62. $\left[\left(\dfrac{u^3v^{-1}w^{-2}}{u^{-2}v^{-2}w}\right)^{-2}\right]^2$

63. $\left(\dfrac{2^2x^2y^0}{8x^{-1}}\right)^{-2}\left(\dfrac{x^{-3}}{x^{-5}}\right)^3$

64. $\left(\dfrac{3^3x^0y^{-2}}{2^3x^3y^{-5}}\right)^{-1}\left(\dfrac{3^3x^{-1}y}{2^2x^2y^{-2}}\right)^2$

Express in simplest radical form. All variables are restricted to avoid division by zero and square roots of negative numbers.

65. $\dfrac{\sqrt{2x}\sqrt{5}}{\sqrt{20x}}$

66. $\dfrac{\sqrt{x}\sqrt{8y}}{\sqrt{12y}}$

67. $\dfrac{2}{\sqrt{x-2}}$

68. $\sqrt{\dfrac{1}{x-5}}$

A-2 Rational Exponents and Radicals

- nth Roots of Real Numbers
- Rational Exponents and Radicals
- Properties of Radicals

A brief review of the material on integer exponents and square root radicals (Section A-1) might prove helpful before beginning this section.

▪ nth Roots of Real Numbers

Recall from Section A-1 that r is a **square root** of b if $r^2 = b$. There is no reason to stop there. We may also say that r is a **cube root** of b if $r^3 = b$.

In general:

> For any natural number n:
>
> r is an **nth root** of b if $r^n = b$

How many real square roots of 16 exist? Of 7? Of -4? How many real fourth roots of 7 exist? Of -7? How many real cube roots of -8 exist? Of 11? Theorem 2 (which we state without proof) answers these questions completely.

Theorem 2 Number of Real nth Roots of a Real Number b

	***n* even**	***n* odd**
***b* positive**	Two real nth roots	One real nth root
	-2 and 2 are both 4th roots of 16	2 is the only real cube root of 8
***b* negative**	No real nth root	One real nth root
	-4 has no real square roots	-2 is the only real cube root of -8
***b* zero**	One real nth root	One real nth root
	The nth root of 0 is 0 for any natural number n	

On the basis of Theorem 2, we conclude that

7 has two real square roots, two real 4th roots, and so on.

10 has one real cube root, one real 5th root, and so on.

-13 has one real cube root, one real 5th root, and so on.

-8 has no real square roots, no real 4th roots, and so on.

▪ Rational Exponents and Radicals

We now turn to the question of what symbols to use to represent the various kinds of real nth roots. For a natural number n greater than 1 we use

> $b^{1/n}$ or $\sqrt[n]{b}$

to represent one of the **real *n*th roots of *b*.** Which one? The symbols represent the real nth root of b if n is odd and the positive real nth root of b if b is positive and n is even. The symbol $\sqrt[n]{b}$ is called an ***n*th root radical.** The number n is the **index** of the radical and the number b is called the **radicand.** Note that we write $\sqrt{b}$ to indicate $\sqrt[2]{b}$.

Example 6 (A) $4^{1/2} = \sqrt{4} = 2$ $(\sqrt{4} \neq \pm 2)$ (B) $-4^{1/2} = -\sqrt{4} = -2$
(C) $(-4)^{1/2}$ and $\sqrt{-4}$ are not real numbers
(D) $8^{1/3} = \sqrt[3]{8} = 2$ (E) $(-8)^{1/3} = \sqrt[3]{-8} = -2$

Problem 6 Evaluate each of the following:

(A) $16^{1/2}$ (B) $-\sqrt{16}$ (C) $\sqrt[3]{-27}$
(D) $(-9)^{1/2}$ (E) $(\sqrt[4]{81})^3$

For m and n natural numbers without common prime factors,* b a real number, and b nonnegative when n is even, we define $b^{m/n}$ as follows (both definitions are equivalent under the indicated restrictions):

$$b^{m/n} = \begin{cases} (b^{1/n})^m = (\sqrt[n]{b})^m \\ (b^m)^{1/n} = (\sqrt[n]{b^m}) \end{cases}$$

Thus,

$$8^{2/3} = (8^{1/3})^2 = 2^2 = 4 \quad \text{or} \quad 8^{2/3} = (8^2)^{1/3} = 64^{1/3} = 4$$

We complete the definition of rational exponents with

$$b^{-m/n} = \frac{1}{b^{m/n}} \quad (b \neq 0)$$

All the properties listed for integer exponents in Section A-1 also hold for rational exponents, provided b is nonnegative when n is even. Unless stated to the contrary, all variables in the rest of the discussion represent positive real numbers.

Example 7 Change rational exponent form to radical form.

(A) $x^{1/7} = \sqrt[7]{x}$

(B) $(3u^2v^3)^{3/5} = \sqrt[5]{(3u^2v^3)^3}$ or $(\sqrt[5]{3u^2v^3})^3$ (The first is usually preferred.)

(C) $y^{-2/3} = \dfrac{1}{y^{2/3}} = \dfrac{1}{\sqrt[3]{y^2}}$ or $\sqrt[3]{y^{-2}}$ or $\sqrt[3]{\dfrac{1}{y^2}}$

* Prime factors will be discussed in detail in Section A-3.

Change radical form to rational exponent form.

(D) $\sqrt[5]{6} = 6^{1/5}$ (E) $-\sqrt[3]{x^2} = -x^{2/3}$

(F) $\sqrt{x^2 + y^2} = (x^2 + y^2)^{1/2}$ Note that $(x^2 + y^2)^{1/2} \neq x + y$. Why?

Problem 7 Convert to radical form.

(A) $u^{1/5}$ (B) $(6x^2y^5)^{2/9}$ (C) $(3xy)^{-3/5}$

Convert to rational exponent form.

(D) $\sqrt[4]{9u}$ (E) $-\sqrt[7]{(2x)^4}$ (F) $\sqrt[3]{x^3 + y^3}$

Example 8 Simplify each and express answers using positive exponents only. If rational exponents appear in final answers, convert to radical form.

(A) $(3x^{1/3})(2x^{1/2}) = 6x^{1/3+1/2} = 6x^{5/6} = 6\sqrt[6]{x^5}$

(B) $(-8)^{5/3} = [(-8)^{1/3}]^5 = (-2)^5 = -32$

(C) $(2x^{1/3}y^{-2/3})^3 = 8xy^{-2} = \dfrac{8x}{y^2}$

(D) $\left(\dfrac{4x^{1/3}}{x^{1/2}}\right)^{1/2} = \dfrac{4^{1/2}x^{1/6}}{x^{1/4}} = \dfrac{2}{x^{1/4-1/6}} = \dfrac{2}{x^{1/12}} = \dfrac{2}{\sqrt[12]{x}}$

Problem 8 Simplify each and express answers using positive exponents only. If rational exponents appear in final answers, convert to radical form.

(A) $9^{3/2}$ (B) $(-27)^{4/3}$ (C) $(5y^{1/4})(2y^{1/3})$

(D) $(2x^{-3/4}y^{1/4})^4$ (E) $\left(\dfrac{8x^{1/2}}{x^{2/3}}\right)^{1/3}$

Properties of Radicals

Changing or simplifying radical expressions is aided by the introduction of several properties of radicals that follow directly from properties of exponents considered earlier.

Properties of Radicals

If c, n, and m are natural numbers greater than or equal to 2, and if x and y are positive real numbers, then

1. $\sqrt[n]{x^n} = x$ $\qquad \sqrt[3]{x^3} = x$
2. $\sqrt[n]{xy} = \sqrt[n]{x}\sqrt[n]{y}$ $\qquad \sqrt[5]{xy} = \sqrt[5]{x}\sqrt[5]{y}$
3. $\sqrt[n]{\dfrac{x}{y}} = \dfrac{\sqrt[n]{x}}{\sqrt[n]{y}}$ $\qquad \sqrt[4]{\dfrac{x}{y}} = \dfrac{\sqrt[4]{x}}{\sqrt[4]{y}}$
4. $\sqrt[cn]{x^{cm}} = \sqrt[n]{x^m}$ $\qquad \sqrt[12]{x^8} = \sqrt[4\cdot 3]{x^{4\cdot 2}} = \sqrt[3]{x^2}$

The properties of radicals provide us with the means of changing algebraic expressions containing radicals into a variety of equivalent forms. One particularly useful form is the simplest radical form. An algebraic expression that contains radicals is said to be in the **simplest radical form** if all four of the following conditions are satisfied:

Simplest Radical Form

1. A radicand contains no factor to a power greater than or equal to the index of the radical.

   $\sqrt[3]{x^5}$ violates this condition.

2. The power of the radicand and the index of the radical have no common factor other than 1.

   $\sqrt[6]{x^4}$ violates this condition.

3. No radical appears in a denominator.

   $y/\sqrt[3]{x}$ violates this condition.

4. No fraction appears within a radical.

   $\sqrt[4]{\frac{3}{5}}$ violates this condition.

Example 9 Write in simplest radical form.

(A) $\sqrt[3]{x^3y^6} = \sqrt[3]{(xy^2)^3} = xy^2$

or $\sqrt[3]{x^3y^6} = (x^3y^6)^{1/3} = x^{3/3}y^{6/3} = xy^2$

(B) $\sqrt[3]{32x^8y^3} = \sqrt[3]{(2^3x^6y^3)(4x^2)} = \sqrt[3]{2^3x^6y^3}\,\sqrt[3]{4x^2}$

$= 2x^2y\sqrt[3]{4x^2}$

(C) $$\frac{6x^2}{\sqrt[3]{9x}} = \frac{6x^2}{\sqrt[3]{9x}} \cdot \frac{\sqrt[3]{3x^2}}{\sqrt[3]{3x^2}} = \frac{6x^2\sqrt[3]{3x^2}}{\sqrt[3]{3^3x^3}}$$

$$= \frac{6x^2\sqrt[3]{3x^2}}{3x} = 2x\sqrt[3]{3x^2}$$

(D) $$6\sqrt[4]{\frac{3}{4x^3}} = 6\sqrt[4]{\frac{3}{2^2x^3} \cdot \frac{2^2x}{2^2x}} = 6\sqrt[4]{\frac{12x}{2^4x^4}}$$

$$= 6\frac{\sqrt[4]{12x}}{\sqrt[4]{2^4x^4}} = 6\frac{\sqrt[4]{12x}}{2x} = \frac{3\sqrt[4]{12x}}{x}$$

(E) $\sqrt[6]{16x^4y^2} = \sqrt[6]{(4x^2y)^2}$

$= \sqrt[2\cdot 3]{(4x^2y)^{2\cdot 1}}$

$= \sqrt[3]{4x^2y}$

Note that in Examples 9C and 9D, we **rationalized the denominators;** that is, we performed operations to remove radicals from the denominators. This is a useful operation in some problems.

Problem 9 Write in simplest radical form.

(A) $\sqrt{12x^5y^6}$ (B) $\sqrt[3]{-27x^7y^5}$ (C) $\dfrac{8y^3}{\sqrt[4]{2y}}$

(D) $4x^2\sqrt[5]{\dfrac{y^2}{2x^3}}$ (E) $\sqrt[9]{8x^6y^3}$

Answers to Matched Problems

6. (A) 4 (B) -4 (C) -3 (D) Not a real number
   (E) 27
7. (A) $\sqrt[5]{u}$ (B) $\sqrt[9]{(6x^2y^5)^2}$ or $(\sqrt[9]{6x^2y^5})^2$ (C) $1/\sqrt[5]{(3xy)^3}$
   (D) $(9u)^{1/4}$ (E) $-(2x)^{4/7}$ (F) $(x^3 + y^3)^{1/3}$ (not $x + y$)
8. (A) 27 (B) 81 (C) $10y^{7/12} = 10\sqrt[12]{y^7}$ (D) $16y/x^3$
   (E) $2/x^{1/18} = 2/\sqrt[18]{x}$
9. (A) $2x^2y^3\sqrt{3x}$ (B) $-3x^2y\sqrt[3]{xy^2}$ (C) $4y^2\sqrt[4]{8y^3}$
   (D) $2x\sqrt[5]{16x^2y^2}$ (E) $\sqrt[3]{2x^2y}$

Exercise A-2

A *Change to radical form; do not simplify.*

1. $6x^{3/5}$
2. $7y^{2/5}$
3. $(4xy^3)^{2/5}$
4. $(7x^2y)^{5/7}$
5. $(x^2 + y^2)^{1/2}$
6. $x^{1/2} + y^{1/2}$

Change to rational exponent form; do not simplify.

7. $5\sqrt[4]{x^3}$
8. $7m\sqrt[5]{n^2}$
9. $\sqrt[5]{(2x^2y)^3}$
10. $\sqrt[9]{(3m^4n)^2}$
11. $\sqrt[3]{x} + \sqrt[3]{y}$
12. $\sqrt[3]{x^2 + y^3}$

Find rational number representations for each if they exist.

13. $25^{1/2}$
14. $64^{1/3}$
15. $16^{3/2}$
16. $16^{3/4}$
17. $-36^{1/2}$
18. $-32^{3/5}$
19. $(-36)^{1/2}$
20. $(-32)^{3/5}$
21. $(\frac{4}{25})^{3/2}$
22. $(\frac{8}{27})^{2/3}$
23. $9^{-3/2}$
24. $8^{-2/3}$

Simplify each expression and write answer using positive exponents only. All variables represent positive real numbers.

25. $x^{4/5}x^{-2/5}$
26. $y^{-3/7}y^{4/7}$
27. $\dfrac{m^{2/3}}{m^{-1/3}}$
28. $\dfrac{x^{1/4}}{x^{3/4}}$
29. $(8x^3y^{-6})^{1/3}$
30. $(4u^{-2}v^4)^{1/2}$

B

31. $\left(\dfrac{4x^{-2}}{y^4}\right)^{-1/2}$
32. $\left(\dfrac{w^4}{9x^{-2}}\right)^{-1/2}$
33. $\dfrac{8x^{-1/3}}{12x^{1/4}}$
34. $\dfrac{6a^{3/4}}{15a^{-1/3}}$
35. $\left(\dfrac{8x^{-4}y^3}{27x^2y^{-3}}\right)^{1/3}$
36. $\left(\dfrac{25x^5y^{-1}}{16x^{-3}y^{-5}}\right)^{1/2}$

Write in simplest radical form.

37. $\sqrt[3]{16m^4n^6}$
38. $\sqrt[3]{27x^7y^3}$
39. $\sqrt[4]{32m^9n^7}$
40. $\sqrt[5]{64u^{17}v^9}$
41. $\dfrac{x}{\sqrt[3]{x}}$
42. $\dfrac{u^2}{\sqrt[3]{u^2}}$
43. $\dfrac{4a^3b^2}{\sqrt[3]{2ab^2}}$
44. $\dfrac{8x^3y^5}{\sqrt[3]{4x^2y}}$
45. $\sqrt[4]{\dfrac{3x^3}{4}}$
46. $\sqrt[5]{\dfrac{3x^2}{2}}$
47. $\sqrt[12]{(x-3)^9}$
48. $\sqrt[8]{(t+1)^6}$

C

49. $\sqrt{x}\,\sqrt[3]{x^2}$
50. $\sqrt[3]{x}\,\sqrt{x}$
51. $\dfrac{\sqrt{x}}{\sqrt[3]{x^2}}$
52. $\dfrac{\sqrt{x}}{\sqrt[3]{x}}$

A-3 Algebraic Expressions: Basic Operations

- Algebraic Expressions and Polynomials
- Addition and Subtraction
- Multiplication
- Factoring

▪ Algebraic Expressions and Polynomials

Algebraic expressions are formed by using constants and variables* with the algebraic operations of addition, subtraction, multiplication, division,

* A **constant** is any symbol that is used to represent exactly one real number. For example, 4, $\sqrt{2}$, and π are all constants. A **variable** is a symbol used as a placeholder for any number in a given set of real numbers. This set is called the **replacement set** for the variable.

and the taking of roots. The following are examples of algebraic expressions:

$$\sqrt[3]{x^3 - 2x + 1} \qquad \frac{x - 5}{x^2 + 2x - 5} \qquad (3x^{-5} - 2x^{-3})^{2/3}$$

An algebraic expression that involves only the operations of addition, subtraction, and multiplication on variables and constants (such as $x^3 - 2x^2 + 5x - 1$) is called a **polynomial.** In general,

Polynomial in x

A **polynomial in x** is an algebraic expression of the form

$$a_n x^n + a_{n-1} x^{n-1} + \cdots + a_1 x + a_0$$

where the coefficients $a_0, a_1, \ldots, a_n$ are real numbers and n is a nonnegative integer.

Of course, we may consider polynomials in more than one variable. A polynomial in the two variables x and y is an algebraic expression formed by adding terms of the form $ax^m y^n$, where a is a real number and m and n are nonnegative integers. For example,

$$3x^3 - \sqrt{2}x^2y + xy - \frac{1}{2}xy^2 + y^3 + 2x - 3$$

is a polynomial in two variables. Polynomials in three or more variables are defined in a similar way.

Polynomial forms are encountered frequently in mathematics, and it is useful to classify them according to their degree. If a term in a polynomial has only one variable as a factor, then the **degree of that term** is the power of the variable. If a term has two or more variables as factors, then the **degree of the term** is the sum of the powers of the variables. The **degree of a polynomial** is the degree of the nonzero term with the highest degree in the polynomial. Any nonzero constant is defined to be a **polynomial of degree 0.** The number 0 is also a polynomial, but is not assigned a degree.

Example 10 (A) Polynomials in one variable:

$$x^2 - 3x + 2 \qquad 6x^3 - \sqrt{2}x - \frac{1}{3}$$

(B) Polynomials in several variables:

$$3x^2 - 2xy + y^2 \qquad 4x^3y^2 - \sqrt{3}xy^2z^5$$

(C) Nonpolynomials:

$$\sqrt{2x} - \frac{3}{x} + 5 \qquad \frac{x^2 - 3x + 2}{x - 3} \qquad \sqrt{x^2 - 3x + 1}$$

(D) The degree of the first term in $6x^3 - \sqrt{2}x - \frac{1}{3}$ is 3; the second term, 1; the third term, 0; and the whole polynomial, 3.

(E) The degree of the first term in $4x^3y^2 - \sqrt{3}xy^2$ is 5; the second, 3; and the whole polynomial, 5.

Problem 10 (A) Which of the following are polynomials?

$$3x^2 - 2x + 1 \qquad \sqrt{x - 3} \qquad x^2 - 2xy + y^2 \qquad \frac{x - 1}{x^2 + 2}$$

(B) Given the polynomial $3x^5 - 6x^3 + 5$, what is the degree of the first term? The second term? The whole polynomial?

(C) Given the polynomial $6x^4y^2 - 3xy^3$, what is the degree of the first term? The second term? The whole polynomial?

The rules behind manipulating algebraic expressions have their basis in the operational properties of real numbers. We list a few of these properties here for ready reference. Let R be the set of real numbers and let x, y, z be arbitrary elements of R; then

Commutative Properties	*Distributive Properties*
$x + y = y + x$	$x(y + z) = xy + xz$
$xy = yx$	$(y + z)x = yx + zx$
Associative Properties	*Subtraction*
$(x + y) + z = x + (y + z)$	$x - y = x + (-y)$
$(xy)z = x(yz)$	

These properties are either used or assumed almost any time you work with algebraic expressions.

▪ Addition and Subtraction

Example 11 (A) Add $5x^3 - 2x^2 + x - 3$ and $7x^3 + 5x^2 + 9$.

(B) Subtract $5x^3 - 2x^2 + x - 3$ from $7x^3 + 5x^2 + 9$.

Solutions (A) $(5x^3 - 2x^2 + x - 3) + (7x^3 + 5x^2 + 9)$

$$= 1(5x^3 - 2x^2 + x - 3) + 1(7x^3 + 5x^2 + 9)$$

$$= 5x^3 - 2x^2 + x - 3 + 7x^3 + 5x^2 + 9$$

$$= 5x^3 + 7x^3 + 5x^2 - 2x^2 + x + 6$$

$$= (5 + 7)x^3 + (5 - 2)x^2 + x + 6$$

$$= 12x^3 + 3x^2 + x + 6$$

(B) $(7x^3 + 5x^2 + 9) - (5x^3 - 2x^2 + x - 3)$

$$= 1(7x^3 + 5x^2 + 9) + (-1)(5x^3 - 2x^2 + x - 3)$$

$$= 7x^3 + 5x^2 + 9 - 5x^3 + 2x^2 - x + 3$$

$$= 7x^3 - 5x^3 + 5x^2 + 2x^2 - x + 12$$

$$= (7 - 5)x^3 + (5 + 2)x^2 - x + 12$$

$$= 2x^3 + 7x^2 - x + 12$$

Problem 11 Given the polynomials $x^3 - 7x + 2$ and $4x^3 - x^2 + x - 1$.

(A) Add the two polynomials.
(B) Subtract the first polynomial from the second.

Example 12 (A) Add $3\sqrt{x} + 5\sqrt{y} + 2$ and $\sqrt{x} + \sqrt[3]{y} - 4$.
(B) Subtract $4x^{2/3} - x^{1/3} + 2$ from $3x^{2/3} + 2x^{1/3} - 8$.

Solutions (A) $(3\sqrt{x} + 5\sqrt{y} + 2) + (\sqrt{x} + \sqrt[3]{y} - 4)$

$$= 3\sqrt{x} + 5\sqrt{y} + 2 + \sqrt{x} + \sqrt[3]{y} - 4$$

These are not polynomials.

$$= 3\sqrt{x} + \sqrt{x} + 5\sqrt{y} + \sqrt[3]{y} - 2$$

$$= (3 + 1)\sqrt{x} + 5\sqrt{y} + \sqrt[3]{y} - 2$$

$$= 4\sqrt{x} + 5\sqrt{y} + \sqrt[3]{y} - 2$$

(B) $(3x^{2/3} + 2x^{1/3} - 8) - (4x^{2/3} - x^{1/3} + 2)$

$$= 3x^{2/3} + 2x^{1/3} - 8 - 4x^{2/3} + x^{1/3} - 2$$

These are not polynomials.

$$= 3x^{2/3} - 4x^{2/3} + 2x^{1/3} + x^{1/3} - 10$$

$$= (3 - 4)x^{2/3} + (2 + 1)x^{1/3} - 10$$

$$= -x^{2/3} + 3x^{1/3} - 10$$

Problem 12 (A) Add $(2\sqrt[4]{m} + 3\sqrt{pq} - 3)$ and $(\sqrt[3]{m} - \sqrt{pq} + 5)$.
(B) Subtract the first algebraic expression from the second in part A.

▪ Multiplication

Multiplication of algebraic expressions requires extensive use of the properties for real numbers, especially the distributive properties.

Example 13 Multiply $(2x - 3)(3x^2 - 2x + 3)$.

Solution

$$(2x - 3)(3x^2 - 2x + 3)$$
$$= 2x(3x^2 - 2x + 3) - 3(3x^2 - 2x + 3)$$
$$= 6x^3 - 4x^2 + 6x - 9x^2 + 6x - 9$$
$$= 6x^3 - 13x^2 + 12x - 9$$

or

$$\begin{array}{r} 3x^2 - 2x + 3 \\ 2x - 3 \\ \hline 6x^3 - 4x^2 + 6x \phantom{{}-9} \\ - 9x^2 + 6x - 9 \\ \hline 6x^3 - 13x^2 + 12x - 9 \end{array}$$

Problem 13 Multiply $(2x - 3)(2x^2 + 3x - 2)$.

Certain types of products occur so frequently it is useful to note the following formulas for them:

Special Products

$$(ax + b)(cx + d) = acx^2 + (ad + bc)x + bd$$
$$(a - b)(a + b) = a^2 - b^2$$
$$(a + b)^2 = a^2 + 2ab + b^2$$
$$(a - b)^2 = a^2 - 2ab + b^2$$

Example 14

(A) $(2x - 3y)(5x + 2y) = 10x^2 - 11xy - 6y^2$

(B) $(\sqrt{2} - \sqrt{3})(\sqrt{2} + \sqrt{3}) = (\sqrt{2})^2 - (\sqrt{3})^2$
$= 2 - 3 = -1$

(C) $(3x^{1/2} - 2y^{1/2})^2 = 9x - 12x^{1/2}y^{1/2} + 4y$

(D) $(3\sqrt{x} + 2\sqrt{y})(2\sqrt{x} - \sqrt{y}) = 6x + \sqrt{xy} - 2y$

Problem 14 Multiply and simplify.

(A) $(5u + 3v)(4u - v)$ (B) $(\sqrt{x} - \sqrt{y})(\sqrt{x} + \sqrt{y})$

(C) $(2^{1/2} - 3^{1/2})^2$ (D) $(4\sqrt{a} - \sqrt{b})(3\sqrt{a} + 2\sqrt{b})$

■ Factoring

If a number is written as the product of other numbers, then each number in the product is called a **factor** of the original number. Similarly, if an algebraic expression is written as the product of other algebraic expressions, then each algebraic expression in the product is called a **factor** of the original algebraic expression. For example,

$30 = 2 \cdot 3 \cdot 5$ 2, 3, and 5 are factors of 30

$x^2 - 4 = (x - 2)(x + 2)$ $(x - 2)$ and $(x + 2)$ are factors of $x^2 - 4$

The process of writing a number or algebraic expression as the product of other numbers or algebraic expressions is called **factoring.** We start our discussion of factoring with the positive integers.

An integer such as 30 can be represented in a factored form in many ways. The products

$$6 \cdot 5 \qquad (\tfrac{1}{2})(10)(6) \qquad 15 \cdot 2 \qquad \sqrt{15} \cdot \sqrt{60} \qquad 2 \cdot 3 \cdot 5$$

all yield 30. A particularly useful way of factoring positive integers greater than 1 is in terms of prime numbers.

Prime and Composite Numbers

A positive integer greater than 1 is **prime** if its only positive integer factors are itself and 1. A positive integer greater than 1 that is not prime is called a **composite** number. The integer 1 is neither prime nor composite.

Prime numbers: 2, 3, 5, 7, 11, 13, . . .
Composite numbers: 4, 6, 8, 9, 10, 12, . . .

A composite integer greater than 1 is said to be **factored completely** if it is represented as a product of prime factors. The only factoring of 30 that meets this condition is $30 = 2 \cdot 3 \cdot 5$.

Example 15 Write 60 in a completely factored form.

Solution

$$60 = 6 \cdot 10 = 2 \cdot 3 \cdot 2 \cdot 5 = 2^2 \cdot 3 \cdot 5$$

or $$60 = 5 \cdot 12 = 5 \cdot 4 \cdot 3 = 2^2 \cdot 3 \cdot 5$$

or $$60 = 2 \cdot 30 = 2 \cdot 2 \cdot 15 = 2^2 \cdot 3 \cdot 5$$

Notice in Example 15 that we obtain the same prime factors for 60, regardless of how we progress through the factoring process. This illustrates the following basic property of integers:

Fundamental Theorem of Arithmetic

Each positive integer greater than 1 is either prime or has, except for the order of factors, a unique set of prime factors.

Problem 15 Write 180 in a completely factored form.

We can also talk about writing polynomials in a completely factored form. The following polynomials are written in a factored form:

$$x^2 - 9 = (x - 3)(x + 3)$$
$$2x^3 - 4x = 2x(x - \sqrt{2})(x + \sqrt{2})$$
$$2x^4 - 15x^2 - 27 = (x^2 - 9)(2x^2 + 3)$$
$$x^2 + 3x + \frac{9}{4} = \left(x + \frac{3}{2}\right)^2$$

But which are in a completely factored form? Paralleling our discussion with prime numbers, we define a **prime polynomial** as follows:

Prime Polynomials

A polynomial is said to be **prime** relative to a given set of numbers if:

1. It has coefficients from that set.
2. It cannot be written as a product of two polynomials of positive degree having coefficients from that set.

For example, $x^2 - 2$ is a prime polynomial relative to the integers, but is not prime relative to the real numbers [since $x^2 - 2 = (x - \sqrt{2})(x + \sqrt{2})$]. A nonprime polynomial is said to be **factored completely** relative to a given set of numbers if it is represented as a product of prime polynomials relative to that set of numbers.

Writing polynomials in a completely factored form is often a difficult task. But accomplishing it can lead to the simplification of certain algebraic expressions and to solutions of certain types of equations.

Example 16 Take out all factors common to all terms.

(A) $2x^3y - 8x^2y^2 - 6xy^3$ (B) $2x(3x - 2) - 7(3x - 2)$

Solutions

(A) $2x^3y - 8x^2y^2 - 6xy^3 = (2xy)x^2 - (2xy)4xy - (2xy)3y^2$

$= (2xy)(x^2 - 4xy - 3y^2)$

(B) $2x(3x - 2) - 7(3x - 2) = 2x(3x - 2) - 7(3x - 2)$

$= (2x - 7)(3x - 2)$

Problem 16 Take out all factors common to all terms.

(A) $3x^3y - 6x^2y^2 - 3xy^3$ (B) $3y(2y + 5) + 2(2y + 5)$

The factoring formulas given in the box show us how to factor certain polynomial forms that occur often.

Special Factoring Formulas

1. $u^2 + (a + b)u + ab = (u + a)(u + b)$
2. $acu^2 + (ad + bc)u + bd = (au + b)(cu + d)$
3. $a^2u^2 + 2abuv + b^2v^2 = (au + bv)^2$ Perfect square
4. $u^2 - v^2 = (u - v)(u + v)$ Difference of two squares
5. $u^3 - v^3 = (u - v)(u^2 + uv + v^2)$ Difference of two cubes
6. $u^3 + v^3 = (u + v)(u^2 - uv + v^2)$ Sum of two cubes

These formulas can be established by multiplying the factors on the right.

Example 17 Factor completely in the integers.

(A) $x^2 - 5x - 6$ (B) $6x^2 - 5x - 4$ (C) $x^2 + 6xy + 9y^2$
(D) $9x^2 - 4y^2$ (E) $8m^3 - 1$ (F) $x^3 + y^3z^3$

Solutions

(A) $x^2 - 5x - 6 = (x - 6)(x + 1)$
(B) $6x^2 - 5x - 4 = (3x - 4)(2x + 1)$
(C) $x^2 + 6xy + 9y^2 = (x + 3y)^2$
(D) $9x^2 - 4y^2 = (3x - 2y)(3x + 2y)$

(E)
$$8m^3 - 1 = (2m)^3 - 1^3 = (2m - 1)[(2m)^2 + (2m)(1) + 1^2] = (2m - 1)(4m^2 + 2m + 1)$$

(F)
$$x^3 + y^3z^3 = x^3 + (yz)^3 = (x + yz)(x^2 - xyz + y^2z^2)$$

Problem 17 Factor completely in the integers.

(A) $x^2 + 7x - 8$ (B) $4m^2 - 4mn - 3n^2$ (C) $4m^2 - 12mn + 9n^2$
(D) $x^2 - 16y^2$ (E) $z^3 - 1$ (F) $m^3 + n^3$

We now complete this section by considering factoring that involves combinations of the techniques discussed previously. Generally speaking, *when factoring a polynomial, we first take out all factors common to all terms (if any are present), then apply the special factoring formulas until all factors are prime.*

Example 18 Factor completely relative to the integers.

(A) $18x^3 - 8x$ (B) $3x^4y^2 + 12x^2y^4$ (C) $4m^3n - 2m^2n^2 + 2mn^3$

(D) $2t^4 - 16t$ (E) $2y^4 - 5y^2 - 12$

Solutions

(A) $18x^3 - 8x = 2x(9x^2 - 4)$
$= 2x(3x - 2)(3x + 2)$

(B) $3x^4y^2 + 12x^2y^4$
$= 3x^2y^2(x^2 + 4y^2)$ $(x^2 + 4y^2)$ is prime relative to the integers.

(C) $4m^3n - 2m^2n^2 + 2mn^3 = 2mn(2m^2 - mn + n^2)$

(D) $2t^4 - 16t = 2t(t^3 - 8)$
$= 2t(t - 2)(t^2 + 2t + 4)$

(E) $2y^4 - 5y^2 - 12 = (2y^2 + 3)(y^2 - 4)$
$= (2y^2 + 3)(y - 2)(y + 2)$

Problem 18 Factor completely relative to the integers.

(A) $3x^3 - 48x$ (B) $18m^2n + 2mn^3$ (C) $3u^4 - 3u^3v - 9u^2v^2$

(D) $3m^4 - 24mn^3$ (E) $3x^4 - 5x^2 + 2$

Remark: It should be noted that if one writes down a polynomial with random integer coefficients, then the resulting polynomial is more likely to be prime than not prime; that is, it most likely will not have polynomial factors of positive degree relative to the integers. But if it does, the results may be very useful, as we pointed out earlier.

Answers to Matched Problems

10. (A) $3x^2 - 2x + 1$, $x^2 - 2xy + y^2$ (B) 5, 3, 5 (C) 6, 4, 6
11. (A) $5x^3 - x^2 - 6x + 1$ (B) $3x^3 - x^2 + 8x - 3$
12. (A) $2\sqrt[4]{m} + \sqrt[3]{m} + 2\sqrt{pq} + 2$ (B) $-2\sqrt[4]{m} + \sqrt[3]{m} - 4\sqrt{pq} + 8$
13. $4x^3 - 13x + 6$
14. (A) $20u^2 + 7uv - 3v^2$ (B) $x - y$
    (C) $5 - 2 \cdot 2^{1/2} \cdot 3^{1/2}$ or $5 - 2(6)^{1/2}$
    (D) $12a + 5\sqrt{ab} - 2b$
15. $2^2 \cdot 3^2 \cdot 5$
16. (A) $3xy(x^2 - 2xy - y^2)$ (B) $(3y + 2)(2y + 5)$
17. (A) $(x + 8)(x - 1)$ (B) $(2m - 3n)(2m + n)$ (C) $(2m - 3n)^2$
    (D) $(x - 4y)(x + 4y)$ (E) $(z - 1)(z^2 + z + 1)$
    (F) $(m + n)(m^2 - mn + n^2)$
18. (A) $3x(x - 4)(x + 4)$ (B) $2mn(9m^2 + n^2)$
    (C) $3u^2(u^2 - uv - 3v^2)$ (D) $3m(m - 2n)(m^2 + 2mn + 4n^2)$
    (E) $(3x^2 - 2)(x - 1)(x + 1)$

Exercise A-3

Unless stated to the contrary, express all answers with positive exponents.

A *Consider the polynomials* $2x^3 - 3x^2 + x + 5$, $2x^2 + x - 1$, *and* $3x - 2$.

1. What is the degree of the first polynomial?
2. What is the degree of the second?
3. Add the first and second polynomials.
4. Add the second and third.
5. Subtract the second polynomial from the first.
6. Subtract the third from the second.
7. Multiply the first and third polynomials.
8. Multiply the second and third.

In Problems 9–18, perform the indicated operations and simplify.

9. $2(x - 3) - (4x + 5)$
10. $4(w + 1) - (2w - 3)$
11. $4m - 3[4 - 2(m - 1)]$
12. $3y - 2[6 - 3(y + 2)]$
13. $(4a - b)(2a + b)$
14. $(3m + 2n)(2m - 3n)$
15. $(3x - 2y)(3x + 2y)$
16. $(4m + 3n)(4m - 3n)$
17. $(4x - y)^2$
18. $(3u + 4v)^2$

Factor completely relative to the integers. Specify which polynomials are already prime (relative to the integers).

19. $x^2 - 9x + 14$
20. $y^2 + 7y + 12$
21. $w^2 + 3w - 40$
22. $x^2 - 4x - 21$
23. $2x^2 + 5x - 3$
24. $3y^2 - y - 2$
25. $x^2 - 4xy - 12y^2$
26. $u^2 - 2uv - 15v^2$
27. $x^2 + x - 4$
28. $m^2 - 6m - 3$
29. $A^2 - 36B^2$
30. $9m^2 - 1$
31. $25m^2 - 16n^2$
32. $w^2x^2 - y^2$
33. $x^2 + 10xy + 25y^2$
34. $9m^2 - 6mn + n^2$
35. $u^2 + 81$
36. $y^2 + 16$
37. $6x^2 + 48x + 72$
38. $4z^2 - 28z + 48$

B *In Problems 39–62, perform the indicated operations and simplify. All variables involved with radicals or fractional exponents represent positive real numbers.*

39. $2x(x - 3y) - y(x + 2y)$
40. $3u(2u + v) - v(u - 3v)$
41. $2\{x + 2[x - (x + 5)] + 1\}$
42. $u - \{u - [u - (u - 1)]\}$
43. $2(x - 2y)(x + y)$
44. $3(u + 3v)(u - v)$
45. $(a + b)(a^2 - ab + b^2)$
46. $(a - b)(a^2 + ab + b^2)$
47. $(2x^2 + x - 2)(x^2 - 3x + 5)$
48. $(x^2 - 2xy + y^2)(x^2 + 2xy + y^2)$
49. $(2x - 1)^2 - (3x + 2)(3x - 2)$
50. $(3a - b)(3a + b) - (2a - 3b)^2$

51. $(2m - n)^3$
52. $(x - 2y)^3$
53. $\sqrt{x}(\sqrt{x} - 3)$
54. $\sqrt{w}(5 - \sqrt{w})$
55. $(\sqrt{m} + 2)(\sqrt{m} - 2)$
56. $(4 - \sqrt{y})(4 + \sqrt{y})$
57. $(\sqrt{c} - \sqrt{d})(\sqrt{c} + \sqrt{d})$
58. $(x^{1/2} + y^{1/2})(x^{1/2} - y^{1/2})$
59. $(x^{1/2} + y^{1/2})^2$
60. $(\sqrt{y} + \sqrt{z})^2$
61. $(2x^{1/2} + 3)(x^{1/2} - 5)$
62. $(3u^{1/2} - 2)(2u^{1/2} + 4)$

In Problems 63–78, factor completely relative to the integers. Specify which polynomials are already prime (relative to the integers).

63. $2x^2 - 7xy + 6y^2$
64. $3x^2 - 11xy + 6y^2$
65. $3m^2 + 17m - 6$
66. $5z^2 - 18z - 8$
67. $2y^3 - 22y^2 + 48y$
68. $2x^4 - 24x^3 + 40x^2$
69. $6s^2 + 7st - 3t^2$
70. $6m^2 - mn - 12n^2$
71. $x^3y - 9xy^3$
72. $4u^3v - uv^3$
73. $m^3 + n^3$
74. $r^3 - t^3$
75. $3x^2 - 2xy - 4y^2$
76. $5u^2 + 4uv - 2v^2$
77. $m^4 - n^4$
78. $y^4 - 3y^2 - 4$

C

In Problems 79–84, perform the indicated operations and simplify. All variables represent positive real numbers.

79. $2\sqrt[3]{a} + 3\sqrt[3]{a} - \sqrt[4]{a}$
80. $4\sqrt[3]{y} - \sqrt[3]{y} + 2\sqrt{y}$
81. $2x^{1/2}(3x^{2/3} - x^6)$
82. $3m^{3/4}(4m^{1/4} - 2m^8)$
83. $(\sqrt{x+h} - \sqrt{x})(\sqrt{x+h} + \sqrt{x})$
84. $[(u + k)^{1/2} - u^{1/2}][(u + k)^{1/2} + u^{1/2}]$

A-4 Algebraic Fractions

- Fundamental Principle of Fractions
- Multiplication and Division
- Addition and Subtraction

Algebraic fractions represent quotients, and for those replacements of the variables by real numbers that result in the quotient of real numbers (division by 0 excluded), the properties of real fractions apply. We will review the use of these properties in this section. Note the following:

1. $\frac{1}{x}$, $\frac{1}{\sqrt{x}}$, and $\frac{x - 1}{x^{2/3}}$ are undefined for $x = 0$

2. $\frac{1}{x - 3}$, $\frac{1}{\sqrt{x - 3}}$, and $\frac{x - 1}{(x - 3)^{2/3}}$ are undefined for $x = 3$

3. $\frac{u - 1}{u^2 + u - 2} = \frac{u - 1}{(u - 1)(u + 2)}$ is undefined for $u = -2, 1$

We will not always explicitly state restrictions of the type just indicated, but it is important to remember that operations on algebraic fractions are valid only for values of the variables for which *all* fractions are defined.

Fundamental Principle of Fractions

A property of real fractions that is used frequently when working with algebraic fractions is the **fundamental principle of fractions.**

Fundamental Principle of Fractions

For real numbers a, b, and k:

$$\frac{ak}{bk} = \frac{a}{b} \qquad (b, k \neq 0)$$

Using the principle from left to right to eliminate all common factors from a numerator and denominator of a given fraction is referred to as **reducing a fraction to lowest terms.** We are actually dividing the numerator and denominator by the same nonzero common factor. Using the principle from right to left—that is, multiplying a numerator and denominator by the same nonzero factor—is referred to as **raising a fraction to higher terms.** We will use the principle in both directions in the material that follows.

A particular type of algebraic fraction, the quotient of two polynomials, is called a **rational expression.** We say that a rational expression is **reduced to lowest terms** if the numerator and denominator do not have any prime factors in common. (Unless stated to the contrary, "prime" will mean "prime relative to the integers.")

Example 19 Reduce to lowest terms:

(A) $$\frac{x^2 - 6x + 9}{x^2 - 9} = \frac{(x-3)^2}{(x-3)(x+3)} = \frac{x-3}{x+3}$$

Factor numerator and denominator completely. Divide numerator and denominator by $(x - 3)$, a valid operation as long as $x \neq 3$. Of course, $x \neq -3$, as well.

(B) $$\frac{x^3 - 1}{x^2 - 1} = \frac{(x-1)(x^2+x+1)}{(x-1)(x+1)} = \frac{x^2+x+1}{x+1}$$

(C) $$\frac{x^{2/3} - 1}{x^{1/3} + 1} = \frac{(x^{1/3} - 1)\overset{1}{\cancel{(x^{1/3}+1)}}}{\underset{1}{\cancel{(x^{1/3}+1)}}} = x^{1/3} - 1$$

Note: Throughout our work with fractions, we will always assume without specific statement for each case that variables are restricted in order to avoid division by 0.

Problem 19 Reduce to lowest terms:

(A) $\dfrac{6x^2 + x - 2}{2x^2 + x - 1}$ (B) $\dfrac{x^4 - 8x}{3x^3 - 2x^2 - 8x}$ (C) $\dfrac{u^{2/5} - 3}{u^{4/5} - 9}$

■ Multiplication and Division

Since in algebraic fractions we will restrict variable replacements to real numbers that produce real fractions, multiplication and division of algebraic fractions follow the rules for multiplication and division of fractions in real numbers. That is (excluding division by 0):

Multiplication and Division

$$\frac{a}{b} \cdot \frac{c}{d} = \frac{ac}{bd}$$

$$\frac{a}{b} \div \frac{c}{d} = \frac{a}{b} \cdot \frac{d}{c}$$

Example 20 (A) $\dfrac{10x^3y}{3xy + 9y} \cdot \dfrac{x^2 - 9}{4x^2 - 12x}$

$$= \frac{\overset{5x^2}{\cancel{10x^3y}}}{\underset{3 \cdot 1}{\cancel{3y(x+3)}}} \cdot \frac{\overset{1 \cdot 1}{\cancel{(x-3)(x+3)}}}{\underset{2 \cdot 1}{\cancel{4x(x-3)}}}$$

$$= \frac{5x^2}{6}$$

(B) $\dfrac{4 - 2x}{4} \div (x - 2)$ $x - 2$ is the same as $\dfrac{x - 2}{1}$

$$= \frac{\overset{1}{\cancel{2}}(2 - x)}{\underset{2}{\cancel{4}}} \cdot \frac{1}{x - 2}$$

$$= \frac{2 - x}{2(x - 2)} = \frac{\overset{1}{-\cancel{(x-2)}}}{\underset{1}{2\cancel{(x-2)}}}$$ $b - a = -(a - b)$, a useful change in some problems

$$= -\frac{1}{2}$$

(C) $\dfrac{2x^3 - 2x^2y + 2xy^2}{x^3y - xy^3} \div \dfrac{x^3 + y^3}{x^2 + 2xy + y^2}$

$$= \frac{\overset{2}{\cancel{2x}}\overset{1}{\cancel{(x^2 - xy + y^2)}}}{\underset{y}{\cancel{xy}}\underset{1}{\cancel{(x + y)}}(x - y)} \cdot \frac{\overset{1}{\cancel{(x + y)^2}}}{\underset{1}{\cancel{(x + y)}}\underset{1}{\cancel{(x^2 - xy + y^2)}}}$$

$$= \frac{2}{y(x - y)}$$

Problem 20 Perform the indicated operations and reduce to lowest terms.

(A) $\dfrac{12x^2y^3}{2xy^2 + 6xy} \cdot \dfrac{y^2 + 6y + 9}{3y^3 + 9y^2}$ (B) $(4 - x) \div \dfrac{x^2 - 16}{5}$

(C) $\dfrac{m^3 + n^3}{2m^2 + mn - n^2} \div \dfrac{m^3n - m^2n^2 + mn^3}{2m^3n^2 - m^2n^3}$

We will now use the fundamental principle of fractions to rationalize denominators and numerators in fractional expressions involving radicals.

Example 21 (A) Rationalize the denominator: $\dfrac{\sqrt{x} - \sqrt{y}}{\sqrt{x} + \sqrt{y}}$

(B) Rationalize the numerator: $\dfrac{\sqrt{x + h} - \sqrt{x}}{h}$

Solutions (A) $\dfrac{\sqrt{x} - \sqrt{y}}{\sqrt{x} + \sqrt{y}} = \dfrac{(\sqrt{x} - \sqrt{y})}{(\sqrt{x} + \sqrt{y})} \cdot \dfrac{(\sqrt{x} - \sqrt{y})}{(\sqrt{x} - \sqrt{y})}$

$$= \frac{x - 2\sqrt{xy} + y}{x - y}$$

Multiplying numerator and denominator by $(\sqrt{x} - \sqrt{y})$ eliminates radicals from the denominator since $(a + b)(a - b) = a^2 - b^2$.

(B) $\dfrac{\sqrt{x + h} - \sqrt{x}}{h} = \dfrac{(\sqrt{x + h} - \sqrt{x})}{h} \cdot \dfrac{(\sqrt{x + h} + \sqrt{x})}{(\sqrt{x + h} + \sqrt{x})}$

$$= \frac{x + h - x}{h(\sqrt{x + h} + \sqrt{x})}$$

$$= \frac{h}{h(\sqrt{x + h} + \sqrt{x})} = \frac{1}{\sqrt{x + h} + \sqrt{x}}$$

Problem 21 (A) Rationalize the denominator: $\dfrac{\sqrt{m} + \sqrt{n}}{\sqrt{m} - \sqrt{n}}$

(B) Rationalize the numerator: $\dfrac{\sqrt{3 + h} - \sqrt{3}}{h}$

Addition and Subtraction

Since in algebraic fractions we will restrict variable replacements to real numbers that produce real fractions, addition and subtraction of algebraic fractions follow the rules for addition and subtraction of fractions in real numbers. That is (excluding division by 0):

Addition and Subtraction

$$\frac{a}{b} + \frac{c}{b} = \frac{a + c}{b}$$

$$\frac{a}{b} - \frac{c}{b} = \frac{a - c}{b}$$

Thus, we add algebraic fractions, if their denominators are the same, by adding or subtracting their numerators and placing the result over the common denominator. If the denominators are not the same, we raise the fractions to higher terms (using the fundamental principle of fractions to obtain common denominators) and then proceed as indicated in the box.

Even though any common denominator will do, the problem is generally less involved if the **least common denominator (LCD)** is used. Often the LCD is obvious, but if it is not, use the following procedure to find it:

The Least Common Denominator

The LCD of two or more rational expressions is found as follows:

1. Factor each denominator completely.
2. Form a product containing each different factor from all denominators to the highest power it occurs in any one denominator. This product is the LCD.

Example 22 Combine into single fractions and simplify.

(A) $$\frac{x - 4}{x + 2} - \frac{x - 2}{x + 2} = \frac{x - 4 - (x - 2)}{x + 2} = \frac{x - 4 - x + 2}{x + 2} = \frac{-2}{x + 2}$$

(B) $$\frac{1}{3y^2} - \frac{1}{6y} + 1 = \frac{2(1)}{2(3y^2)} - \frac{y(1)}{y(6y)} + \frac{6y^2}{6y^2}$$

$$= \frac{2 - y + 6y^2}{6y^2} \qquad \text{LCD} = 6y^2$$

(C) $2 - \frac{x-2}{x+1} = \frac{2(x+1)}{x+1} - \frac{x-2}{x+1} = \frac{2(x+1)-(x-2)}{x+1}$

$= \frac{2x+2-x+2}{x+1} = \frac{x+4}{x+1}$ LCD $= x+1$

(D) $\frac{3}{x^2-1} - \frac{2}{x^2+2x+1} = \frac{3}{(x-1)(x+1)} - \frac{2}{(x+1)^2}$

$= \frac{3(x+1)}{(x-1)(x+1)^2} - \frac{2(x-1)}{(x-1)(x+1)^2}$

$= \frac{3(x+1)-2(x-1)}{(x-1)(x+1)^2} = \frac{3x+3-2x+2}{(x-1)(x+1)^2}$

$= \frac{x+5}{(x-1)(x+1)^2}$ LCD $= (x-1)(x+1)^2$

Problem 22 Combine into single fractions and simplify.

(A) $\frac{x-2}{x-3} - \frac{x+2}{x-3}$ (B) $\frac{1}{y} + \frac{1}{4y^2} - 1$

(C) $u - \frac{u-1}{u-2}$ (D) $\frac{4}{x^2-4} - \frac{3}{x^2-x-2}$

Example 23 Combine into single fractions and simplify. Write answers using positive exponents only.

(A) $-6y^2(y-1)^{-3} + 6y(y-1)^{-2} = \frac{-6y^2}{(y-1)^3} + \frac{6y}{(y-1)^2}$

$= \frac{-6y^2+6y(y-1)}{(y-1)^3}$

$= \frac{-6y}{(y-1)^3}$

(B) $x^{1/2}2x + \frac{1}{2}x^{-1/2}(x^2-1) = 2x^{3/2} + \frac{x^2-1}{2x^{1/2}}$

$= \frac{2x^{1/2}2x^{3/2}}{2x^{1/2}} + \frac{x^2-1}{2x^{1/2}}$

$= \frac{4x^2+x^2-1}{2x^{1/2}} = \frac{5x^2-1}{2x^{1/2}}$

(C) $\frac{x^{1/3} - \frac{1}{3}x^{-2/3}(x-1)}{x^{2/3}} = \frac{\frac{x^{1/3}}{1} - \frac{x-1}{3x^{2/3}}}{x^{2/3}}$

$= \frac{\frac{3x-(x-1)}{3x^{2/3}}}{x^{2/3}} = \frac{3x-x+1}{3x^{2/3}} \cdot \frac{1}{x^{2/3}}$

$= \frac{2x+1}{3x^{4/3}}$

Problem 23 Combine into single fractions and simplify. Write answers using positive exponents only.

(A) $-3n(n+3)^{-2} + 3(n+3)^{-1}$ (B) $2(x-1)^{1/2}x^2 + \frac{1}{2}(x-1)^{-1/2}x^3$

(C) $\dfrac{x^{1/2} - \frac{1}{2}x^{-1/2}(x+2)}{x}$

Answers to Matched Problems

19. (A) $\dfrac{3x+2}{x+1}$ (B) $\dfrac{x^2+2x+4}{3x+4}$ (C) $\dfrac{1}{u^{2/5}+3}$

20. (A) $2x$ (B) $\dfrac{-5}{x+4}$ (C) mn

21. (A) $\dfrac{m+2\sqrt{mn}+n}{m-n}$ (B) $\dfrac{1}{\sqrt{3+h}+\sqrt{3}}$

22. (A) $\dfrac{-4}{x-3}$ or $\dfrac{4}{3-x}$ (B) $\dfrac{1+4y-4y^2}{4y^2}$

(C) $\dfrac{u^2-3u+1}{u-2}$ (D) $\dfrac{1}{(x+2)(x+1)}$

23. (A) $\dfrac{9}{(n+3)^2}$ (B) $\dfrac{5x^3-4x^2}{2(x-1)^{1/2}}$ (C) $\dfrac{x-2}{2x^{3/2}}$

Exercise A-4

A *Perform the indicated operations and reduce to lowest terms.*

1. $\dfrac{3x^2y}{x-y} \div \dfrac{6xy}{x-y}$

2. $\dfrac{x+3}{2x^2} \div \dfrac{x+3}{4x}$

3. $\dfrac{v-1}{v^2} - \dfrac{1}{v} + \dfrac{3}{v^3}$

4. $\dfrac{2}{x} - \dfrac{x+1}{x^2} + \dfrac{5}{x^3}$

5. $\left(\dfrac{d^5}{3a} \div \dfrac{d^2}{6a^2}\right) \cdot \dfrac{a}{4d^3}$

6. $\dfrac{d^5}{3a} \div \left(\dfrac{d^2}{6a^2} \cdot \dfrac{a}{4d^3}\right)$

7. $1 - \dfrac{1}{x-3}$

8. $2 + \dfrac{3}{u+1}$

9. $\dfrac{2x^2+7x+3}{4x^2-1} \div (x+3)$

10. $\dfrac{x^2-9}{x^2-3x} \div (x^2-x-12)$

B

11. $\dfrac{x^2-6x+9}{x^2-x-6} \div \dfrac{x^2+2x-15}{x^2+2x}$

12. $\dfrac{m+n}{m^2-n^2} \div \dfrac{m^2-mn}{m^2-2mn+n^2}$

13. $\dfrac{3x+8}{4x^2} - \dfrac{2x-1}{x^3} - \dfrac{5}{8x}$

14. $\dfrac{4m-3}{18m^3} + \dfrac{3}{4m} - \dfrac{2m-1}{6m^2}$

15. $\dfrac{1}{a^2 - b^2} + \dfrac{1}{a^2 + 2ab + b^2}$

16. $\dfrac{3}{x^2 - 1} + \dfrac{2}{x^2 - 2x + 1}$

17. $m - 3 - \dfrac{m - 1}{m - 2}$

18. $\dfrac{x + 1}{x - 1} + x + 1$

19. $\dfrac{2}{y + 3} - \dfrac{1}{y - 3} + \dfrac{2y}{y^2 - 9}$

20. $\dfrac{2x}{x^2 - y^2} + \dfrac{1}{x + y} - \dfrac{1}{x - y}$

21. $\dfrac{x^2 - 16}{2x^2 + 10x + 8} \div \dfrac{x^2 - 13x + 36}{x^3 + 1}$

22. $\dfrac{x^3y - y^4}{xy^3 - y^4} \div \dfrac{x^2 + xy + y^2}{y^2}$

23. $\left(\dfrac{-b + \sqrt{b^2 - 4ac}}{2a}\right) \cdot \left(\dfrac{-b - \sqrt{b^2 - 4ac}}{2a}\right)$

24. $\dfrac{-b + \sqrt{b^2 - 4ac}}{2a} + \dfrac{-b - \sqrt{b^2 - 4ac}}{2a}$

Rationalize the denominators.

25. $\dfrac{3 - \sqrt{a}}{\sqrt{a} - 2}$

26. $\dfrac{2 + \sqrt{x}}{\sqrt{x} - 3}$

27. $\dfrac{x^2}{\sqrt{x^2 + 9} - 3}$

28. $\dfrac{-y^2}{2 - \sqrt{y^2 + 4}}$

Rationalize the numerators.

29. $\dfrac{\sqrt{t} - \sqrt{x}}{t - x}$

30. $\dfrac{\sqrt{x} - \sqrt{y}}{\sqrt{x} + \sqrt{y}}$

31. $\dfrac{\sqrt{x + h} - \sqrt{x}}{h}$

32. $\dfrac{\sqrt{2 + h} + \sqrt{2}}{h}$

Write as single fractions and simplify. Write answers using positive exponents only.

33. $-2x(x - 1)^{-2} + 2(x - 1)^{-1}$

34. $-4m(m + 3)^{-2} + 4(m + 3)^{-1}$

35. $15x^3(2 - 3x)^{-6} + 3x^2(2 - 3x)^{-5}$

36. $-10u^3(u + 2)^{-3} + 15u^2(u + 2)^{-2}$

C *Write as single fractions and simplify. Write answers using positive exponents only.*

37. $2x^{-1/3} + 1$

38. $1 - 3u^{-1/2}$

39. $(x + 1)^{-3/4} - (x + 1)^{1/4}$

40. $(u - 1)^{2/5} - u(u - 1)^{-3/5}$

41. $(x - 1)\frac{1}{4}x^{-1/4} + x^{3/4}$

42. $(x - 2)^{1/4} + \frac{1}{4}(x - 2)^{-3/4}x$

43. $\dfrac{u^{1/2}2u - (u^2 - 1)\frac{1}{2}u^{-1/2}}{u}$

44. $\dfrac{x^2\,\frac{1}{3}(x - 1)^{-2/3} - 2x(x - 1)^{1/3}}{x^4}$

A-5 Arithmetic Progressions

- Arithmetic Progressions—Definitions
- Special Formulas
- Application

Arithmetic Progressions – Definitions

Consider the sequence of numbers

1, 4, 7, 10, 13, . . .

Assuming the pattern continues, can you guess what the next two numbers are? If you guessed 16 and 19, you have observed that each number after the first can be obtained from the preceding one by adding 3 to it. This is an example of an *arithmetic progression.* In general,

Arithmetic Progression

A sequence of numbers

$$a_1, a_2, a_3, \ldots, a_n, \ldots$$

is called an **arithmetic progression** if there is a constant d, called the **common difference,** such that

$$a_n - a_{n-1} = d$$

That is,

$$a_n = a_{n-1} + d \qquad \text{for every} \quad n > 1 \tag{1}$$

Example 24 Which sequence of numbers is an arithmetic progression and what is its common difference?

(A) 2, 4, 8, 10, . . . (B) 3, 8, 13, 18, . . .

Solution Sequence A does not have a common difference, since $4 - 2 = 2$ and $8 - 4 = 4$; hence, it is not an arithmetic progression. Sequence B is an arithmetic progression, since the difference between any two successive terms is 5, the common difference, and each number after the first one can be obtained by adding 5 to the preceding number.

Problem 24 Which sequence of numbers is an arithmetic progression, and what is its common difference?

(A) 15, 13, 11, 9, . . . (B) 3, 9, 27, 81, . . .

Special Formulas

Arithmetic progressions have a number of convenient properties. For example, it is easy to derive formulas for the nth term in terms of n and the sum of any number of consecutive terms. To obtain a formula for the nth term of an arithmetic progression, we note that if a_1 is the first term and d is the common difference, then

$$a_2 = a_1 + d$$
$$a_3 = a_2 + d = (a_1 + d) + d = a_1 + 2d$$
$$a_4 = a_3 + d = (a_1 + 2d) + d = a_1 + 3d$$

This suggests that

$$a_n = a_1 + (n-1)d \qquad \text{for all} \quad n > 1 \qquad (2)$$

Example 25 Find the twenty-first term in the arithmetic progression 3, 8, 13, 18, . . .

Solution Find the common difference d and use formula (2):

$$d = 5, \qquad n = 21, \qquad a_1 = 3$$

Thus

$$a_{21} = 3 + (21 - 1)5$$
$$= 103$$

Problem 25 Find the fifty-first term in the arithmetic progression 15, 13, 11, 9, . . .

We now derive two simple and very useful formulas for the sum of n consecutive terms of an arithmetic progression. Let

$$S_n = a_1 + a_2 + \cdots + a_{n-1} + a_n$$

be the sum of n terms of an arithmetic progression with common difference d. Then,

$$S_n = a_1 + (a_1 + d) + \cdots + [a_1 + (n-2)d] + [a_1 + (n-1)d]$$

Reversing the order of the sum, we obtain

$$S_n = [a_1 + (n-1)d] + [a_1 + (n-2)d] + \cdots + (a_1 + d) + a_1$$

Something interesting happens if we combine these last two equations by addition (adding corresponding terms on the right sides):

$$2S_n = [2a_1 + (n-1)d] + [2a_1 + (n-1)d] + \cdots + [2a_1 + (n-1)d] + [2a_1 + (n-1)d]$$

All the terms on the right side are the same, and there are n of them. Thus,

$$2S_n = n[2a_1 + (n-1)d]$$

and

$$S_n = \frac{n}{2}[2a_1 + (n-1)d] \tag{3}$$

Replacing

$$[a_1 + (n-1)d] \quad \text{in} \quad \frac{n}{2}[a_1 + a_1 + (n-1)d]$$

by a_n from equation (2), we can obtain a second useful formula for the sum:

$$S_n = \frac{n}{2}(a_1 + a_n) \tag{4}$$

Example 26 Find the sum of the first 30 terms in the arithmetic progression 3, 8, 13, 18, . . .

Solution Use (3) with $n = 30$, $a_1 = 3$, and $d = 5$:

$$S_{30} = \frac{30}{2}[2 \cdot 3 + (30-1)5] = 2{,}265$$

Problem 26 Find the sum of the first 40 terms in the arithmetic progression 15, 13, 11, 9, . . .

Example 27 Find the sum of all the even numbers between 31 and 87.

Solution First, find n using (2):

$$a_n = a_1 + (n-1)d$$
$$86 = 32 + (n-1)2$$
$$n = 28$$

Now find S_{28} using (4):

$$S_n = \frac{n}{2}(a_1 + a_n)$$

$$S_{28} = \frac{28}{2}(32 + 86) = 1{,}652$$

Problem 27 Find the sum of all the odd numbers between 24 and 208.

▪ Application

Example 28 A person borrows \$3,600 and agrees to repay the loan in monthly installments over a 3 year period. The agreement is to pay 1% of the unpaid balance each month for using the money and \$100 each month to reduce the loan. What is the total cost of the loan over the 3 year period?

Solution Let us look at the problem relative to a time line:

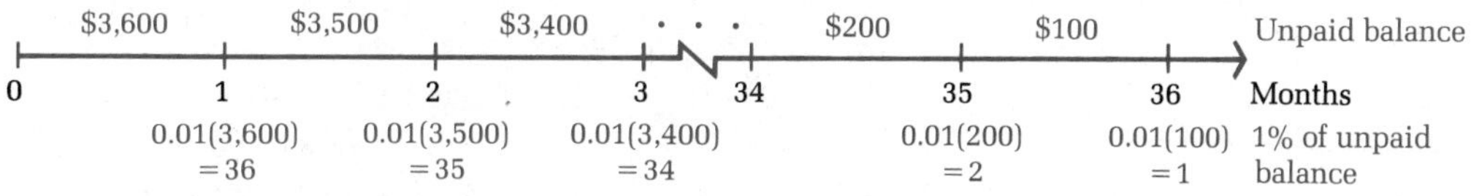

	\$3,600	\$3,500	\$3,400	. . .	\$200	\$100	Unpaid balance
0	1	2	3	34	35	36	Months
	0.01(3,600) = 36	0.01(3,500) = 35	0.01(3,400) = 34		0.01(200) = 2	0.01(100) = 1	1% of unpaid balance

The total cost of the loan is

$$1 + 2 + \cdots + 34 + 35 + 36$$

The terms form an arithmetic progression with $n = 36$, $a_1 = 1$, and $a_{36} = 36$, so we can use (4):

$$S_n = \frac{n}{2}(a_1 + a_n)$$

$$S_{36} = \frac{36}{2}(1 + 36) = \$666$$

And we conclude that the total cost of the loan over the 3 year period is \$666.

Problem 28 Repeat Example 28 with a loan of \$6,000 over a 5 year period.

Answers to Matched Problems
24. Sequence A; $d = -2$ 25. -85 26. -960
27. 10,672 28. \$1,830

Exercise A-5

A

1. Determine which of the following are arithmetic progressions. Find the common difference d and the next two terms for those progressions.

(A) 5, 8, 11, . . . (B) 4, 8, 16, . . .
(C) $-2, -4, -8, \ldots$ (D) $8, -2, -12, \ldots$

2. Repeat Problem 1 for:

   (A) 11, 16, 21, . . . (B) 16, 8, 4, . . .
   (C) 2, −3, −8, . . . (D) −1, −2, −4, . . .

Let $a_1, a_2, a_3, \ldots, a_n, \ldots$ be an arithmetic progression and S_n be the sum of the first n terms. In Problems 3–8 find the indicated quantities.

3. $a_1 = 7$, $d = 4$, $a_2 = ?$, $a_3 = ?$
4. $a_1 = -2$, $d = -3$, $a_2 = ?$, $a_3 = ?$

B

5. $a_1 = 2$, $d = 4$, $a_{21} = ?$, $S_{31} = ?$
6. $a_1 = 8$, $d = -10$, $a_{15} = ?$, $S_{23} = ?$
7. $a_1 = 18$, $a_{20} = 75$, $S_{20} = ?$
8. $a_1 = 203$, $a_{30} = 261$, $S_{30} = ?$
9. Find $f(1) + f(2) + f(3) + \cdots + f(50)$ if $f(x) = 2x - 3$.
10. Find $g(1) + g(2) + g(3) + \cdots + g(100)$ if $g(t) = 18 - 3t$.
11. Find the sum of all the odd integers between 12 and 68.
12. Find the sum of all the even integers between 23 and 97.

C

13. Show that the sum of the first n odd positive integers is n^2, using appropriate formulas from this section.
14. Show that the sum of the first n positive even integers is $n + n^2$, using formulas in this section.

Applications

Business & Economics

15. You are confronted with two job offers. Firm *A* will start you at $24,000 per year and guarantees you a $900 raise each year for 10 years. Firm *B* will start you at $22,000 per year but guarantees you a $1,300 raise each year for 10 years. Over the 10 year period, what is the total amount each firm will pay you?
16. In Problem 15, what would be your annual salary in each firm for the tenth year?
17. *Loan repayment.* If you borrow $4,800 and repay the loan by paying $200 per month to reduce the loan and 1% of the unpaid balance each month for the use of the money, what is the total cost of the loan over 24 months?
18. *Loan repayment.* Repeat Problem 17 replacing 1% with 1.5%.

A-6 Geometric Progressions

- Geometric Progressions—Definition
- Special Formulas
- Infinite Geometric Progressions

Geometric Progressions—Definition

Consider the sequence of numbers

2, 6, 18, 54, . . .

Assuming the pattern continues, can you guess what the next two numbers are? If you guessed 162 and 486, you have observed that each number after the first can be obtained from the preceding one by multiplying it by 3. This is an example of a *geometric progression*. In general,

Geometric Progression

A sequence of numbers

$a_1, a_2, a_3, \ldots, a_n, \ldots$

is called a **geometric progression** if there exists a nonzero constant r, called a **common ratio,** such that

$$\frac{a_n}{a_{n-1}} = r$$

That is,

$$a_n = ra_{n-1} \qquad \text{for every} \quad n \geqslant 1 \tag{1}$$

Example 29 Which sequence of numbers is a geometric progression and what is its common ratio?

(A) 5, 3, 1, −1, . . . (B) 1, 2, 4, 8, . . .

Solution Sequence A does not have a common ratio, since $3 \div 5 \neq 1 \div 3$; hence, it is not a geometric progression. Sequence B is a geometric progression, since the ratio of any two successive terms (the second divided by the first) is the constant 2, the common ratio, and each number after the first can be obtained by multiplying the preceding number by 2.

Problem 29 Which sequence of numbers is a geometric progression and what is its common ratio?

(A) $4, -2, 1, -\frac{1}{2}, \ldots$ (B) 2, 4, 6, 8, . . .

Special Formulas

Like arithmetic progressions, geometric progressions have several useful properties. It is easy to derive formulas for the nth term in terms of n and for the sum of any number of consecutive terms. To obtain a formula for the

nth term of a geometric progression, we note that if a_1 is the first term and r is the common ratio, then

$$a_2 = ra_1$$
$$a_3 = ra_2 = r(ra_1) = r^2a_1 = a_1r^2$$
$$a_4 = ra_3 = r(r^2a_1) = r^3a_1 = a_1r^3$$

This suggests that

$$a_n = a_1r^{n-1} \quad \text{for all} \quad n > 1 \tag{2}$$

Example 30 Find the eighth term in the geometric progression $\frac{1}{2}, \frac{1}{4}, \frac{1}{8}, \ldots$

Solution Find the common ratio r and use formula (2):

$$r = \tfrac{1}{2}, \qquad n = 8, \qquad a_1 = \tfrac{1}{2}$$

Thus,

$$a_8 = (\tfrac{1}{2})(\tfrac{1}{2})^{8-1}$$
$$= \tfrac{1}{256}$$

Problem 30 Find the seventh term in the geometric progression $\frac{1}{32}, -\frac{1}{16}, \frac{1}{8}, \ldots$

Example 31 If the first and tenth terms of a geometric progression are 2 and 4, respectively, find the common ratio r.

Solution

$$a_n = a_1r^{n-1}$$
$$4 = 2 \cdot r^{10-1}$$
$$2 = r^9$$
$$r = 2^{1/9} \approx 1.08 \qquad \text{Use a calculator or logarithms}$$

Problem 31 If the first and eighth terms of a geometric progression are 1,000 and 2,000, respectively, find the common ratio r.

We now derive two very useful formulas for the sum of n consecutive terms of a geometric progression. Let

$$a_1, a_1r, a_1r^2, \ldots, a_1r^{n-2}, a_1r^{n-1}$$

be n terms of a geometric progression. Their sum is

$$S_n = a_1 + a_1r + a_1r^2 + \cdots + a_1r^{n-2} + a_1r^{n-1}$$

If we multiply both sides by r, we obtain

$$rS_n = a_1r + a_1r^2 + a_1r^3 + \cdots + a_1r^{n-1} + a_1r^n$$

Now combine these last two equations by subtraction to obtain

$$\begin{aligned} rS_n - S_n &= (a_1r + a_1r^2 + a_1r^3 + \cdots + a_1r^{n-1} + a_1r^n) \\ &\qquad - (a_1 + a_1r + a_1r^2 + \cdots + a_1r^{n-2} + a_1r^{n-1}) \\ (r-1)S_n &= a_1r^n - a_1 \end{aligned}$$

Notice how many terms drop out on the right side. Hence,

$$S_n = \frac{a_1(r^n - 1)}{r - 1} \qquad r \neq 1 \qquad (3)$$

Since $a_n = a_1r^{n-1}$, or $ra_n = a_1r^n$, formula (3) can also be written in the form

$$S_n = \frac{ra_n - a_1}{r - 1} \qquad r \neq 1 \qquad (4)$$

Example 32 Find the sum of the first ten terms of the geometric progression $1, 1.05, 1.05^2, \ldots$

Solution Use formula (3) with $a_1 = 1$, $r = 1.05$, and $n = 10$:

$$S_n = \frac{a_1(r^n - 1)}{r - 1}$$

$$S_{10} = \frac{1(1.05^{10} - 1)}{1.05 - 1}$$

$$\approx \frac{0.6289}{0.05} \approx 12.58$$

Problem 32 Find the sum of the first eight terms of the geometric progression $100, 100(1.08), 100(1.08)^2, \ldots$

▪ Infinite Geometric Progressions

Given a geometric progression, what happens to the sum S_n of the first n terms as n increases without stopping? To answer this question, let us write formula (3) in the form

$$S_n = \frac{a_1 r^n}{r-1} - \frac{a_1}{r-1}$$

It is possible to show that if $|r| < 1$ (that is, $-1 < r < 1$), then r^n will tend to zero as n increases. (See what happens, for example, if you let $r = \frac{1}{2}$ and then increase n.) Thus, the first term above will tend to zero and S_n can be made as close as we please to the second term, $-a_1/(r-1)$ [which can be written as $a_1/(1-r)$], by taking n sufficiently large. Thus, if the common ratio r is between -1 and 1, we define the sum of an infinite geometric progression to be

$$S_\infty = \frac{a_1}{1-r} \qquad |r| < 1 \qquad (5)$$

If $r \leq -1$ or $r \geq 1$, then an infinite geometric progression has no sum.

Example 33 The government has decided on a tax rebate program to stimulate the economy. Suppose you receive \$600 and that you spend 80% of this, and that each of the people who receive what you spend also spend 80% of what they receive, and this process continues without end. According to the **multiplier doctrine** in economics, the effect of your \$600 tax rebate on the economy is multiplied many times. What is the total amount spent if the process continues as indicated?

Solution We need to find the sum of an infinite geometric progression with the first amount spent being $a_1 = (.08)(\$600) = \$480$ and $r = 0.8$. Using formula (5), we obtain

$$\begin{aligned} S_\infty &= \frac{a_1}{1-r} \\ &= \frac{\$480}{1-0.8} \\ &= \$2{,}400 \end{aligned}$$

Thus, assuming the process continues as indicated, we would expect the \$600 tax rebate to result in about \$2,400 of spending.

Problem 33 Repeat Example 33 with a tax rebate of \$1,000.

Answers to Matched Problems

29. Sequence A; $R = -\frac{1}{2}$
30. 2
31. Approximately 1.104
32. 1,063.66
33. \$4,000

Exercise A-6

A

1. Determine which of the following are geometric progressions. Find the common ratio r and the next two terms for those that are:

(A) $1, -2, 4, \ldots$ (B) $7, 6, 5, \ldots$ (C) $2, 1, \frac{1}{2}, \ldots$
(D) $2, -4, 6, \ldots$

2. Repeat Problem 1 for:

(A) $4, -1, -6, \ldots$ (B) $15, 5, \frac{5}{3}, \ldots$ (C) $\frac{1}{4}, -\frac{1}{2}, 1, \ldots$
(D) $\frac{1}{2}, \frac{2}{3}, \frac{3}{4}, \ldots$

Let $a_1, a_2, a_3, \ldots, a_n, \ldots$ be a geometric progression and S_n be the sum of the first n terms. In Problems 3–12 find the indicated quantities. Use logarithms or a calculator as needed.

3. $a_1 = 3, \quad r = -2, \quad a_2 = ?, \quad a_3 = ?, \quad a_4 = ?$
4. $a_1 = 32, \quad r = -\frac{1}{2}, \quad a_2 = ?, \quad a_3 = ?, \quad a_4 = ?$
5. $a_1 = 1, \quad a_7 = 729, \quad r = -3, \quad S_7 = ?$
6. $a_1 = 3, \quad a_7 = 2{,}187, \quad r = 3, \quad S_7 = ?$

B

7. $a_1 = 100, \quad r = 1.08, \quad a_{10} = ?$
8. $a_1 = 240, \quad r = 1.06, \quad a_{12} = ?$
9. $a_1 = 100, \quad a_9 = 200, \quad r = ?$
10. $a_1 = 100, \quad a_{10} = 300, \quad r = ?$
11. $a_1 = 500, \quad r = 0.6, \quad S_{10} = ?, \quad S_\infty = ?$
12. $a_1 = 8{,}000, \quad r = 0.4, \quad S_{10} = ?, \quad S_\infty = ?$
13. Find the sum of each infinite geometric progression (if it exists).

(A) $2, 4, 8, \ldots$ (B) $2, -\frac{1}{2}, \frac{1}{8}, \ldots$

14. Repeat Problem 13 for:

(A) $16, 4, 1, \ldots$ (B) $1, -3, 9, \ldots$

C

15. Find $f(1) + f(2) + \cdots + f(10)$ if $f(x) = (\frac{1}{2})^x$.
16. Find $g(1) + g(2) + \cdots + g(10)$ if $g(x) = 2^x$.

Applications

Business & Economics

17. *Economy stimulation.* The government, through a subsidy program, distributes \$5,000,000. If we assume each individual or agency spends 70% of what is received, and 70% of this is spent, and so on, how much total increase in spending results from this government action? (Let $a_1 = \$3{,}500{,}000$.)
18. *Economy stimulation.* Repeat Problem 17 using \$10,000,000 as the amount distributed and 80%.

19. *Cost-of-living adjustment.* If the cost-of-living index increased 5% for each of the past 10 years and you had a salary agreement that increased your salary by the same percentage each year, what would your present salary be if you had a $20,000 per year salary 10 years ago? What would be your total earnings in the past 10 years? [*Hint:* $r = 1.05$.]

20. *Depreciation.* In *straight-line depreciation*, an asset less its salvage value at the end of its useful life is depreciated (for tax purposes) in equal annual amounts over its useful life. Thus, a $100,000 company airplane with a salvage value of $20,000 at the end of 10 years would be depreciated at $8,000 per year for each of the 10 years.

    Since certain assets, such as airplanes, cars, and so on, depreciate more rapidly during the early years of their useful life, several methods of depreciation that take this into consideration are available to the taxpayer. One such method is called the *method of declining balance.* The rate used cannot exceed double that used for straight-line depreciation (ignoring salvage value) and is applied to the remaining value of an asset after the previous year's depreciation has been deducted. In our airplane example, the annual rate of straight-line depreciation over the 10 year period is 10%. Let us assume we can double this rate for the method of declining balance. At some point before the salvage value is reached (taxpayer's choice), we must switch over to the straight-line method to depreciate the final amount of the asset.

    The table below illustrates the two methods of depreciation for the company airplane.

	Straight-Line		**Declining Balance**	
Year end	Amount depreciated	Asset value	Amount depreciated	Asset value
0	$ 0	$100,000	$ 0	$100,000
1	0.1(80,000) = 8,000	92,000	0.2(100,000) = 20,000	80,000
2	0.1(80,000) = 8,000	84,000	0.2(80,000) = 16,000	64,000
3	0.1(80,000) = 8,000	76,000	0.2(64,000) = 12,800	51,200
.	.	.	.	.
.	.	.	.	.
.	.	.	.	.
7	0.1(80,000) = 8,000	44,000	0.2(26,214) = 5,243	20,972
8	0.1(80,000) = 8,000	36,000	$\frac{972}{3}$ = 324	20,648
9	0.1(80,000) = 8,000	28,000	$\frac{972}{3}$ = 324	20,324
10	0.1(80,000) = 8,000	20,000	$\frac{972}{3}$ = 324	20,000

Shift to straight-line, otherwise next entry will drop below salvage value

Arithmetic progression

Geometric progressions above dashed line

(A) For the declining balance, find the sum of the depreciation amounts above the dashed line using formula (4) and then add the entries below the line to this result.
(B) Repeat part A using formula (3).
(C) Find the asset value under declining balance at the end of the fifth year using formula (2).
(D) Find the asset value under straight-line at the end of the fifth year using formula (2) in the preceding section.

A-7 The Binomial Formula

- Factorial
- Binomial Theorem—Development

The binomial form

$$(a + b)^n$$

where n is a natural number, appears more frequently than you might expect. The coefficients in the expansion play an important role in probability studies. The binomial formula, which we will informally derive, enables us to expand $(a + b)^n$ directly for n any natural number. Since the formula involves **factorials,** we digress for a moment here to introduce this important concept.

Factorial

For n a natural number, ***n* factorial**—denoted by ***n*!**—is the product of the first n natural numbers. **Zero factorial** is defined to be one. Symbolically,

$$n! = n(n-1) \cdot \cdots \cdot 2 \cdot 1$$

$$1! = 1$$

$$0! = 1$$

It is also useful to note that

$$n! = n \cdot (n-1)!$$

Example 34 Evaluate each.

(A) $5! = 5 \cdot 4 \cdot 3 \cdot 2 \cdot 1 = 120$ (B) $\dfrac{8!}{7!} = \dfrac{8 \cdot \not{7!}}{\not{7!}} = 8$

(C) $\dfrac{10!}{7!} = \dfrac{10 \cdot 9 \cdot 8 \cdot \not{7!}}{\not{7!}} = 720$

Problem 34 Evaluate each: (A) $4!$ (B) $\dfrac{7!}{6!}$ (C) $\dfrac{8!}{5!}$

A special formula involving factorials is

$$C_{n,r} = \frac{n!}{r!(n-r)!} \qquad n \geq r \geq 0$$

Example 35 (A) $C_{9,2} = \dfrac{9!}{2!(9-2)!} = \dfrac{9!}{2!7!} = \dfrac{9 \cdot 8 \cdot 7!}{2 \cdot 7!} = 36$

(B) $C_{5,5} = \dfrac{5!}{5!(5-5)!} = \dfrac{5!}{5!0!} = \dfrac{5!}{5!} = 1$

Problem 35 Find: (A) $C_{5,2}$ (B) $C_{6,0}$

▪ Binomial Theorem—Development

Let us expand $(a+b)^n$ for several values of n to see if we can observe a pattern that leads to a general formula for the expansion for any natural number n:

$$(a+b)^1 = a + b$$
$$(a+b)^2 = a^2 + 2ab + b^2$$
$$(a+b)^3 = a^3 + 3a^2b + 3ab^2 + b^3$$
$$(a+b)^4 = a^4 + 4a^3b + 6a^2b^2 + 4ab^3 + b^4$$
$$(a+b)^5 = a^5 + 5a^4b + 10a^3b^2 + 10a^2b^3 + 5ab^4 + b^5$$

Observations

1. The expansion of $(a + b)^n$ has $(n + 1)$ terms.
2. The power of a decreases by 1 for each term as we move from left to right.
3. The power of b increases by 1 for each term as we move from left to right.
4. In each term the sum of the powers of a and b always equals n.
5. Starting with a given term, we can get the coefficient of the next term by multiplying the coefficient of the given term by the exponent of a and dividing by the number that represents the position of the term in the series of terms. For example, in the expansion of $(a + b)^4$, above, the coefficient of the third term is found from the second term by multiplying 4 and 3, and then dividing by 2 [that is, the coefficient of the third term = $(4 \cdot 3)/2 = 6$].

We now postulate these same properties for the general case:

$$(a + b)^n = a^n + \frac{n}{1} a^{n-1}b + \frac{n(n-1)}{1 \cdot 2} a^{n-2}b^2 + \frac{n(n-1)(n-2)}{1 \cdot 2 \cdot 3} a^{n-3}b^3 + \cdots + b^n$$

$$= \frac{n!}{0!(n-0)!} a^n + \frac{n!}{1!(n-1)!} a^{n-1}b + \frac{n!}{2!(n-2)!} a^{n-2}b^2 + \frac{n!}{3!(n-3)!} a^{n-3}b^3 + \cdots + \frac{n!}{n!(n-n)!} b^n$$

$$= C_{n,0}a_n + C_{n,1}a^{n-1}b + C_{n,2}a^{n-2}b^2 + C_{n,3}a^{n-3}b^3 + \cdots + C_{n,n}b^n$$

And we are led to the formula in the binomial theorem (a formal proof requires mathematical induction, which is beyond the scope of this book):

Binomial Theorem

For all natural numbers n,

$$(a + b)^n = C_{n,0}a_n + C_{n,1}a^{n-1}b + C_{n,2}a^{n-2}b^2 + C_{n,3}a^{n-3}b^3 + \cdots + C_{n,n}b^n$$

Example 36 Use the binomial formula to expand $(u + v)^6$.

Solution

$$(u + v)^6 = C_{6,0}u^6 + C_{6,1}u^5v + C_{6,2}u^4v^2 + C_{6,3}u^3v^3 + C_{6,4}u^2v^4 + C_{6,5}uv^5 + C_{6,6}v^6$$

$$= u^6 + 6u^5v + 15u^4v^2 + 20u^3v^3 + 15u^2v^4 + 6uv^5 + v^6$$

Problem 36 Use the binomial formula to expand $(x+2)^5$.

Example 37 Use the binomial formula to find the sixth term in the expansion of $(x-1)^{18}$.

Solution

$$\begin{aligned}\text{Sixth term} &= C_{18,5}x^{13}(-1)^5\\ &= \frac{18!}{5!(18-5)!}x^{13}(-1)\\ &= -8{,}568x^{13}\end{aligned}$$

Problem 37 Use the binomial formula to find the fourth term in the expansion of $(x-2)^{20}$.

Answers to Matched Problems

34. (A) 24 (B) 7 (C) 336 35. (A) 10 (B) 1

36. $x^5 + 5x^4 \cdot 2 + 10x^3 \cdot 2^2 + 10x^2 \cdot 2^3 + 5x \cdot 2^4 + 2^5$
$= x^5 + 10x^4 + 40x^3 + 80x^2 + 80x + 32$

37. $-9{,}120x^{17}$

Exercise A-7

A *Evaluate.*

1. $6!$ 2. $7!$ 3. $\frac{10!}{9!}$ 4. $\frac{20!}{19!}$

5. $\frac{12!}{9!}$ 6. $\frac{10!}{6!}$ 7. $\frac{5!}{2!3!}$ 8. $\frac{7!}{3!4!}$

9. $\frac{6!}{5!(6-5)!}$ 10. $\frac{7!}{4!(7-4)!}$ 11. $\frac{20!}{3!17!}$ 12. $\frac{52!}{50!2!}$

B *Evaluate.*

13. $C_{5,3}$ 14. $C_{7,3}$ 15. $C_{6,5}$ 16. $C_{7,4}$
17. $C_{5,0}$ 18. $C_{5,5}$ 19. $C_{18,15}$ 20. $C_{18,3}$

Expand each expression using the binomial formula.

21. $(a+b)^4$ 22. $(m+n)^5$ 23. $(x-1)^6$
24. $(u-2)^5$ 25. $(2a-b)^5$ 26. $(x-2y)^5$

Find the indicated term in each expansion.

27. $(x-1)^{18}$, fifth term 28. $(x-3)^{20}$, third term
29. $(p+q)^{15}$, seventh term 30. $(p+q)^{15}$, thirteenth term
31. $(2x+y)^{12}$, eleventh term 32. $(2x+y)^{12}$, third term

C

33. Show that: $C_{n,0} = C_{n,n}$
34. Show that: $C_{n,r} = C_{n,n-r}$
35. The triangle below is called **Pascal's triangle.** Can you guess what the next two rows at the bottom are? Compare these numbers with the coefficients of binomial expansions.

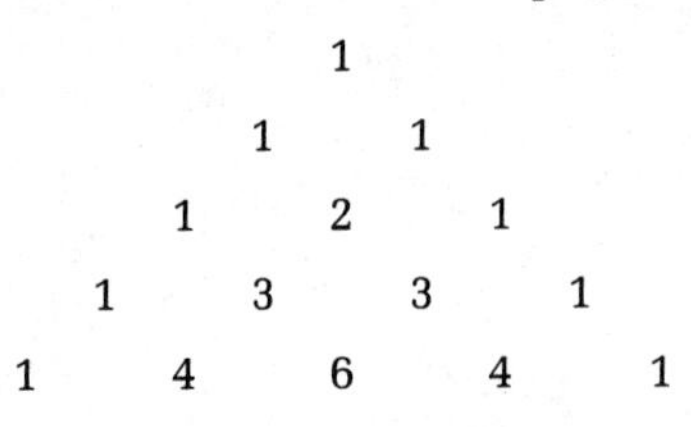

Tables

B

APPENDIX B Contents

Table I Exponential Functions (e^x and e^{-x})

x	e^x	e^{-x}	x	e^x	e^{-x}	x	e^x	e^{-x}
0.00	1.0000	1.00 000	0.50	1.6487	0.60 653	1.00	2.7183	0.36 788
0.01	1.0101	0.99 005	0.51	1.6653	0.60 050	1.01	2.7456	0.36 422
0.02	1.0202	0.98 020	0.52	1.6820	0.59 452	1.02	2.7732	0.36 059
0.03	1.0305	0.97 045	0.53	1.6989	0.58 860	1.03	2.8011	0.35 701
0.04	1.0408	0.96 079	0.54	1.7160	0.58 275	1.04	2.8292	0.35 345
0.05	1.0513	0.95 123	0.55	1.7333	0.57 695	1.05	2.8577	0.34 994
0.06	1.0618	0.94 176	0.56	1.7507	0.57 121	1.06	2.8864	0.34 646
0.07	1.0725	0.93 239	0.57	1.7683	0.56 553	1.07	2.9154	0.34 301
0.08	1.0833	0.92 312	0.58	1.7860	0.55 990	1.08	2.9447	0.33 960
0.09	1.0942	0.91 393	0.59	1.8040	0.55 433	1.09	2.9743	0.33 622
0.10	1.1052	0.90 484	0.60	1.8221	0.54 881	1.10	3.0042	0.33 287
0.11	1.1163	0.89 583	0.61	1.8404	0.54 335	1.11	3.0344	0.32 956
0.12	1.1275	0.88 692	0.62	1.8589	0.53 794	1.12	3.0649	0.32 628
0.13	1.1388	0.87 810	0.63	1.8776	0.53 259	1.13	3.0957	0.32 303
0.14	1.1503	0.86 936	0.64	1.8965	0.52 729	1.14	3.1268	0.31 982
0.15	1.1618	0.86 071	0.65	1.9155	0.52 205	1.15	3.1582	0.31 664
0.16	1.1735	0.85 214	0.66	1.9348	0.51 685	1.16	3.1899	0.31 349
0.17	1.1853	0.84 366	0.67	1.9542	0.51 171	1.17	3.2220	0.31 037
0.18	1.1972	0.83 527	0.68	1.9739	0.50 662	1.18	3.2544	0.30 728
0.19	1.2092	0.82 696	0.69	1.9937	0.50 158	1.19	3.2871	0.30 422
0.20	1.2214	0.81 873	0.70	2.0138	0.49 659	1.20	3.3201	0.30 119
0.21	1.2337	0.81 058	0.71	2.0340	0.49 164	1.21	3.3535	0.29 820
0.22	1.2461	0.80 252	0.72	2.0544	0.48 675	1.22	3.3872	0.29 523
0.23	1.2586	0.79 453	0.73	2.0751	0.48 191	1.23	3.4212	0.29 229
0.24	1.2712	0.78 663	0.74	2.0959	0.47 711	1.24	3.4556	0.28 938
0.25	1.2840	0.77 880	0.75	2.1170	0.47 237	1.25	3.4903	0.28 650
0.26	1.2969	0.77 105	0.76	2.1383	0.46 767	1.26	3.5254	0.28 365
0.27	1.3100	0.76 338	0.77	2.1598	0.46 301	1.27	3.5609	0.28 083
0.28	1.3231	0.75 578	0.78	2.1815	0.45 841	1.28	3.5966	0.27 804
0.29	1.3364	0.74 826	0.79	2.2034	0.45 384	1.29	3.6328	0.27 527
0.30	1.3499	0.74 082	0.80	2.2255	0.44 933	1.30	3.6693	0.27 253
0.31	1.3634	0.73 345	0.81	2.2479	0.44 486	1.31	3.7062	0.26 982
0.32	1.3771	0.72 615	0.82	2.2705	0.44 043	1.32	3.7434	0.26 714
0.33	1.3910	0.71 892	0.83	2.2933	0.43 605	1.33	3.7810	0.26 448
0.34	1.4049	0.71 177	0.84	2.3164	0.43 171	1.34	3.8190	0.26 185
0.35	1.4191	0.70 469	0.85	2.3396	0.42 741	1.35	3.8574	0.25 924
0.36	1.4333	0.69 768	0.86	2.3632	0.42 316	1.36	3.8962	0.25 666
0.37	1.4477	0.69 073	0.87	2.3869	0.41 895	1.37	3.9354	0.25 411
0.38	1.4623	0.68 386	0.88	2.4109	0.41 478	1.38	3.9749	0.25 158
0.39	1.4770	0.67 706	0.89	2.4351	0.41 066	1.39	4.0149	0.24 908
0.40	1.4918	0.67 032	0.90	2.4596	0.40 657	1.40	4.0552	0.24 660
0.41	1.5068	0.66 365	0.91	2.4843	0.40 252	1.41	4.0960	0.24 414
0.42	1.5220	0.65 705	0.92	2.5093	0.39 852	1.42	4.1371	0.24 171
0.43	1.5373	0.65 051	0.93	2.5345	0.39 455	1.43	4.1787	0.23 931
0.44	1.5527	0.64 404	0.94	2.5600	0.39 063	1.44	4.2207	0.23 693
0.45	1.5683	0.63 763	0.95	2.5857	0.38 674	1.45	4.2631	0.23 457
0.46	1.5841	0.63 128	0.96	2.6117	0.38 289	1.46	4.3060	0.23 224
0.47	1.6000	0.62 500	0.97	2.6379	0.37 908	1.47	4.3492	0.22 993
0.48	1.6161	0.61 878	0.98	2.6645	0.37 531	1.48	4.3939	0.22 764
0.49	1.6323	0.61 263	0.99	2.6912	0.37 158	1.49	4.4371	0.22 537
0.50	1.6487	0.60 653	1.00	2.7183	0.36 788	1.50	4.4817	0.22 313

x	e^x	e^{-x}	x	e^x	e^{-x}	x	e^x	e^{-x}
1.50	4.4817	0.22 313	2.00	7.3891	0.13 534	2.50	12.182	0.082 085
1.51	4.5267	0.22 091	2.01	7.4633	0.13 399	2.51	12.305	0.081 268
1.52	4.5722	0.21 871	2.02	7.5383	0.13 266	2.52	12.429	0.080 460
1.53	4.6182	0.21 654	2.03	7.6141	0.13 134	2.53	12.554	0.079 659
1.54	4.6646	0.21 438	2.04	7.6906	0.13 003	2.54	12.680	0.078 866
1.55	4.7115	0.21 225	2.05	7.7679	0.12 873	2.55	12.807	0.078 082
1.56	4.7588	0.21 014	2.06	7.8460	0.12 745	2.56	12.936	0.077 305
1.57	4.8066	0.20 805	2.07	7.9248	0.12 619	2.57	13.066	0.076 536
1.58	4.8550	0.20 598	2.08	8.0045	0.12 493	2.58	13.197	0.075 774
1.59	4.9037	0.20 393	2.09	8.0849	0.12 369	2.59	13.330	0.075 020
1.60	4.9530	0.20 190	2.10	8.1662	0.12 246	2.60	13.464	0.074 274
1.61	5.0028	0.19 989	2.11	8.2482	0.12 124	2.61	13.599	0.073 535
1.62	5.0531	0.19 790	2.12	8.3311	0.12 003	2.62	13.736	0.072 803
1.63	5.1039	0.19 593	2.13	8.4149	0.11 884	2.63	13.874	0.072 078
1.64	5.1552	0.19 398	2.14	8.4994	0.11 765	2.64	14.013	0.071 361
1.65	5.2070	0.19 205	2.15	8.5849	0.11 648	2.65	14.154	0.070 651
1.66	5.2593	0.19 014	2.16	8.6711	0.11 533	2.66	14.296	0.069 948
1.67	5.3122	0.18 825	2.17	8.7583	0.11 418	2.67	14.440	0.069 252
1.68	5.3656	0.18 637	2.18	8.8463	0.11 304	2.68	14.585	0.068 563
1.69	5.4195	0.18 452	2.19	8.9352	0.11 192	2.69	14.732	0.067 881
1.70	5.4739	0.18 268	2.20	9.0250	0.11 080	2.70	14.880	0.067 206
1.71	5.5290	0.18 087	2.21	9.1157	0.10 970	2.71	15.029	0.066 537
1.72	5.5845	0.17 907	2.22	9.2073	0.10 861	2.72	15.180	0.065 875
1.73	5.6407	0.17 728	2.23	9.2999	0.10 753	2.73	15.333	0.065 219
1.74	5.6973	0.17 552	2.24	9.3933	0.10 646	2.74	15.487	0.064 570
1.75	5.7546	0.17 377	2.25	9.4877	0.10 540	2.75	15.643	0.063 928
1.76	5.8124	0.17 204	2.26	9.5831	0.10 435	2.76	15.800	0.063 292
1.77	5.8709	0.17 033	2.27	9.6794	0.10 331	2.77	15.959	0.062 662
1.78	5.9299	0.16 864	2.28	9.7767	0.10 228	2.78	16.119	0.062 039
1.79	5.9895	0.16 696	2.29	9.8749	0.10 127	2.79	16.281	0.061 421
1.80	6.0496	0.16 530	2.30	9.9742	0.10 026	2.80	16.445	0.060 810
1.81	6.1104	0.16 365	2.31	10.074	0.099 261	2.81	16.610	0.060 205
1.82	6.1719	0.16 203	2.32	10.176	0.098 274	2.82	16.777	0.059 606
1.83	6.2339	0.16 041	2.33	10.278	0.097 296	2.83	16.945	0.059 013
1.84	6.2965	0.15 882	2.34	10.381	0.096 328	2.84	17.116	0.058 426
1.85	6.3598	0.15 724	2.35	10.486	0.095 369	2.85	17.288	0.057 844
1.86	6.4237	0.15 567	2.36	10.591	0.094 420	2.86	17.462	0.057 269
1.87	6.4883	0.15 412	2.37	10.697	0.093 481	2.87	17.637	0.056 699
1.88	6.5535	0.15 259	2.38	10.805	0.092 551	2.88	17.814	0.056 135
1.89	6.6194	0.15 107	2.39	10.913	0.091 630	2.89	17.993	0.055 576
1.90	6.6859	0.14 957	2.40	11.023	0.090 718	2.90	18.174	0.055 023
1.91	6.7531	0.14 808	2.41	11.134	0.089 815	2.91	18.357	0.054 476
1.92	6.8210	0.14 661	2.42	11.246	0.088 922	2.92	18.541	0.053 934
1.93	6.8895	0.14 515	2.43	11.359	0.088 037	2.93	18.728	0.053 397
1.94	6.9588	0.14 370	2.44	11.473	0.087 161	2.94	18.916	0.052 866
1.95	7.0287	0.14 227	2.45	11.588	0.086 294	2.95	19.106	0.052 340
1.96	7.0993	0.14 086	2.46	11.705	0.085 435	2.96	19.298	0.051 819
1.97	7.1707	0.13 946	2.47	11.822	0.084 585	2.97	19.492	0.051 303
1.98	7.2427	0.13 807	2.48	11.941	0.083 743	2.98	19.688	0.050 793
1.99	7.3155	0.13 670	2.49	12.061	0.082 910	2.99	19.886	0.050 287
2.00	7.3891	0.13 534	2.50	12.182	0.082 085	3.00	20.086	0.049 787

Table I **(Continued)**

x	e^x	e^{-x}	x	e^x	e^{-x}	x	e^x	e^{-x}
3.00	20.086	0.049 787	3.50	33.115	0.030 197	4.00	54.598	0.018 316
3.01	20.287	0.049 292	3.51	33.448	0.029 897	4.01	55.147	0.018 133
3.02	20.491	0.048 801	3.52	33.784	0.029 599	4.02	55.701	0.017 953
3.03	20.697	0.048 316	3.53	34.124	0.029 305	4.03	56.261	0.017 774
3.04	20.905	0.047 835	3.54	34.467	0.029 013	4.04	56.826	0.017 597
3.05	21.115	0.047 359	3.55	34.813	0.028 725	4.05	57.397	0.017 422
3.05	21.328	0.046 888	3.56	35.163	0.028 439	4.06	57.974	0.017 249
3.07	21.542	0.046 421	3.57	35.517	0.028 156	4.07	58.557	0.017 077
3.08	21.758	0.045 959	3.58	35.874	0.027 876	4.08	59.145	0.016 907
3.09	21.977	0.045 502	3.59	36.234	0.027 598	4.09	59.740	0.016 739
3.10	22.198	0.045 049	3.60	36.598	0.027 324	4.10	60.340	0.016 573
3.11	22.421	0.044 601	3.61	36.966	0.027 052	4.11	60.947	0.016 408
3.12	22.646	0.044 157	3.62	37.338	0.026 783	4.12	61.559	0.016 245
3.13	22.874	0.043 718	3.63	37.713	0.026 516	4.13	62.178	0.016 083
3.14	23.104	0.043 283	3.64	38.092	0.026 252	4.14	62.803	0.015 923
3.15	23.336	0.042 852	3.65	38.475	0.025 991	4.15	63.434	0.015 764
3.16	23.571	0.042 426	3.66	38.861	0.025 733	4.16	64.072	0.015 608
3.17	23.807	0.042 004	3.67	39.252	0.025 476	4.17	64.715	0.015 452
3.18	24.047	0.041 586	3.68	39.646	0.025 223	4.18	65.366	0.015 299
3.19	24.288	0.041 172	3.69	40.045	0.024 972	4.19	66.023	0.015 146
3.20	24.533	0.040 762	3.70	40.447	0.024 724	4.20	66.686	0.014 996
3.21	24.779	0.040 357	3.71	40.854	0.024 478	4.21	67.357	0.014 846
3.22	25.028	0.039 955	3.72	41.264	0.024 234	4.22	68.033	0.014 699
3.23	25.280	0.039 557	3.73	41.679	0.023 993	4.23	68.717	0.014 552
3.24	25.534	0.039 164	3.74	42.098	0.023 754	4.24	69.408	0.014 408
3.25	25.790	0.038 774	3.75	42.521	0.023 518	4.25	70.105	0.014 264
3.26	26.050	0.038 388	3.76	42.948	0.023 284	4.26	70.810	0.014 122
3.27	26.311	0.038 006	3.77	43.380	0.023 052	4.27	71.522	0.013 982
3.28	26.576	0.037 628	3.78	43.816	0.022 823	4.28	72.240	0.013 843
3.29	26.843	0.037 254	3.79	44.256	0.022 596	4.29	72.966	0.013 705
3.30	27.113	0.036 883	3.80	44.701	0.022 371	4.30	73.700	0.013 569
3.31	27.385	0.036 516	3.81	45.150	0.022 148	4.31	74.440	0.013 434
3.32	27.660	0.036 153	3.82	45.604	0.021 928	4.32	75.189	0.013 300
3.33	27.938	0.035 793	3.83	46.063	0.021 710	4.33	75.944	0.013 168
3.34	28.219	0.035 437	3.84	46.525	0.021 494	4.34	76.708	0.013 037
3.35	28.503	0.035 084	3.85	46.993	0.021 280	4.35	77.478	0.012 907
3.36	28.789	0.034 735	3.86	47.465	0.021 068	4.36	78.257	0.012 778
3.37	29.079	0.034 390	3.87	47.942	0.020 858	4.37	79.044	0.012 651
3.38	29.371	0.034 047	3.88	48.424	0.020 651	4.38	79.838	0.012 525
3.39	29.666	0.033 709	3.89	48.911	0.020 445	4.39	80.640	0.012 401
3.40	29.964	0.033 373	3.90	49.402	0.020 242	4.40	81.451	0.012 277
3.41	30.265	0.033 041	3.91	49.899	0.020 041	4.41	82.269	0.012 155
3.42	30.569	0.032 712	3.92	50.400	0.019 841	4.42	83.096	0.012 034
3.43	30.877	0.032 387	3.93	50.907	0.019 644	4.43	83.931	0.011 914
3.44	31.187	0.032 065	3.94	51.419	0.019 448	4.44	84.775	0.011 796
3.45	31.500	0.031 746	3.95	51.935	0.019 255	4.45	85.627	0.011 679
3.46	31.817	0.031 430	3.96	52.457	0.019 063	4.46	86.488	0.011 562
3.47	32.137	0.031 117	3.97	52.985	0.018 873	4.47	87.357	0.011 447
3.48	32.460	0.030 807	3.98	53.517	0.018 686	4.48	88.235	0.011 333
3.49	32.786	0.030 501	3.99	54.055	0.018 500	4.49	89.121	0.011 221
3.50	33.115	0.030 197	4.00	54.598	0.018 316	4.50	90.017	0.011 109

x	e^x	e^{-x}	x	e^x	e^{-x}	x	e^x	e^{-x}
4.50	90.017	0.011 109	5.00	148.41	0.006 7379	7.50	1,808.0	0.000 5531
4.51	90.922	0.010 998	5.05	156.02	0.006 4093	7.55	1,900.7	0.000 5261
4.52	91.836	0.010 889	5.10	164.02	0.006 0967	7.60	1,998.2	0.000 5005
4.53	92.759	0.010 781	5.15	172.43	0.005 7994	7.65	2,100.6	0.000 4760
4.54	93.691	0.010 673	5.20	181.27	0.005 5166	7.70	2,208.3	0.000 4528
4.55	94.632	0.010 567	5.25	190.57	0.005 2475	7.75	2,321.6	0.000 4307
4.56	95.583	0.010 462	5.30	200.34	0.004 9916	7.80	2,440.6	0.000 4097
4.57	96.544	0.010 358	5.35	210.61	0.004 7482	7.85	2,565.7	0.000 3898
4.58	97.514	0.010 255	5.40	221.41	0.004 5166	7.90	2,697.3	0.000 3707
4.59	98.494	0.010 153	5.45	232.76	0.004 2963	7.95	2,835.6	0.000 3527
4.60	99.484	0.010 052	5.50	244.69	0.004 0868	8.00	2,981.0	0.000 3355
4.61	100.48	0.009 9518	5.55	257.24	0.003 8875	8.05	3,133.8	0.000 3191
4.62	101.49	0.009 8528	5.60	270.43	0.003 6979	8.10	3,294.5	0.000 3035
4.63	102.51	0.009 7548	5.65	284.29	0.003 5175	8.15	3,463.4	0.000 2887
4.64	103.54	0.009 6577	5.70	298.87	0.003 3460	8.20	3,641.0	0.000 2747
4.65	104.58	0.009 5616	5.75	314.19	0.003 1828	8.25	3,827.6	0.000 2613
4.66	105.64	0.009 4665	5.80	330.30	0.003 0276	8.30	4,023.9	0.000 2485
4.67	106.70	0.009 3723	5.85	347.23	0.002 8799	8.35	4,230.2	0.000 2364
4.68	107.77	0.009 2790	5.90	365.04	0.002 7394	8.40	4,447.1	0.000 2249
4.69	108.85	0.009 1867	5.95	383.75	0.002 6058	8.45	4,675.1	0.000 2139
4.70	109.95	0.009 0953	6.00	403.43	0.002 4788	8.50	4,914.8	0.000 2035
4.71	111.05	0.009 0048	6.05	424.11	0.002 3579	8.55	5,166.8	0.000 1935
4.72	112.17	0.008 9152	6.10	445.86	0.002 2429	8.60	5,431.7	0.000 1841
4.73	113.30	0.008 8265	6.15	468.72	0.002 1335	8.65	5,710.1	0.000 1751
4.74	114.43	0.008 7386	6.20	492.75	0.002 2094	8.70	6,002.9	0.000 1666
4.75	115.58	0.008 6517	6.25	518.01	0.001 9305	8.75	6,310.7	0.000 1585
4.76	116.75	0.008 5656	6.30	544.57	0.001 8363	8.80	6,634.2	0.000 1507
4.77	117.92	0.008 4804	6.35	572.49	0.001 7467	8.85	6,974.4	0.000 1434
4.78	119.10	0.008 3960	6.40	601.85	0.001 6616	8.90	7,332.0	0.000 1364
4.79	120.30	0.008 3125	6.45	632.70	0.001 5805	8.95	7,707.9	0.000 1297
4.80	121.51	0.008 2297	6.50	665.14	0.001 5034	9.00	8,103.1	0.000 1234
4.81	122.73	0.008 1479	6.55	699.24	0.001 4301	9.05	8,518.5	0.000 1174
4.82	123.97	0.008 0668	6.60	735.10	0.001 3604	9.10	8,955.3	0.000 1117
4.83	125.21	0.007 9865	6.65	772.78	0.001 2940	9.15	9,414.4	0.000 1062
4.84	126.47	0.007 9071	6.70	812.41	0.001 2309	9.20	9,897.1	0.000 1010
4.85	127.74	0.007 8284	6.75	854.06	0.001 1709	9.25	10,405	0.000 0961
4.86	129.02	0.007 7505	6.80	897.85	0.001 1138	9.30	10,938	0.000 0914
4.87	130.32	0.007 6734	6.85	943.88	0.001 0595	9.35	11,499	0.000 0870
4.88	131.63	0.007 5970	6.90	992.27	0.001 0078	9.40	12,088	0.000 0827
4.89	132.95	0.007 5214	6.95	1,043.1	0.000 9586	9.45	12,708	0.000 0787
4.90	134.29	0.007 4466	7.00	1,096.6	0.000 9119	9.50	13,360	0.000 0749
4.91	135.64	0.007 3725	7.05	1,152.9	0.000 8674	9.55	14,045	0.000 0712
4.92	137.00	0.007 2991	7.10	1,212.0	0.000 8251	9.60	14,765	0.000 0677
4.93	138.38	0.007 2265	7.15	1,274.1	0.000 7849	9.65	15,522	0.000 0644
4.94	139.77	0.007 1546	7.20	1,339.4	0.000 7466	9.70	16,318	0.000 0613
4.95	141.17	0.007 0834	7.25	1,408.1	0.000 7102	9.75	17,154	0.000 0583
4.96	142.59	0.007 0129	7.30	1,480.3	0.000 6755	9.80	18,034	0.000 0555
4.97	144.03	0.006 9431	7.35	1,556.2	0.000 6426	9.85	18,958	0.000 0527
4.98	145.47	0.006 8741	7.40	1,636.0	0.000 6113	9.90	19,930	0.000 0502
4.99	146.94	0.006 8057	7.45	1,719.9	0.000 5814	9.95	20,952	0.000 0477
5.00	148.41	0.006 7379	7.50	1,808.0	0.000 5531	10.00	22,026	0.000 0454

N	0	1	2	3	4	5	6	7	8	9
1.0	0.0000	0.004321	0.008600	0.01284	0.01703	0.02119	0.02531	0.02938	0.03342	0.03743
1.1	0.04139	0.04532	0.04922	0.05308	0.05690	0.06070	0.06446	0.06819	0.07188	0.07555
1.2	0.07918	0.08279	0.08636	0.08991	0.09342	0.09691	0.1004	0.1038	0.1072	0.1106
1.3	0.1139	0.1173	0.1206	0.1239	0.1271	0.1303	0.1335	0.1367	0.1399	0.1430
1.4	0.1461	0.1492	0.1523	0.1553	0.1584	0.1614	0.1644	0.1673	0.1703	0.1732
1.5	0.1761	0.1790	0.1818	0.1847	0.1875	0.1903	0.1931	0.1959	0.1987	0.2014
1.6	0.2041	0.2068	0.2095	0.2122	0.2148	0.2175	0.2201	0.2227	0.2253	0.2279
1.7	0.2304	0.2330	0.2355	0.2380	0.2405	0.2430	0.2455	0.2480	0.2504	0.2529
1.8	0.2553	0.2577	0.2601	0.2625	0.2648	0.2673	0.2695	0.2718	0.2742	0.2765
1.9	0.2788	0.2810	0.2833	0.2856	0.2878	0.2900	0.2923	0.2945	0.2967	0.2989
2.0	0.3010	0.3032	0.3054	0.3075	0.3096	0.3118	0.3139	0.3160	0.3181	0.3201
2.1	0.3222	0.3243	0.3263	0.3284	0.3304	0.3324	0.3345	0.3365	0.3385	0.3404
2.2	0.3424	0.3444	0.3464	0.3483	0.3502	0.3522	0.3541	0.3560	0.3579	0.3598
2.3	0.3617	0.3636	0.3655	0.3674	0.3692	0.3711	0.3729	0.3747	0.3766	0.3784
2.4	0.3802	0.3820	0.3838	0.3856	0.3874	0.3892	0.3909	0.3927	0.3945	0.3962
2.5	0.3979	0.3997	0.4014	0.4031	0.4048	0.4065	0.4082	0.4099	0.4116	0.4133
2.6	0.4150	0.4166	0.4183	0.4200	0.4216	0.4232	0.4249	0.4265	0.4281	0.4298
2.7	0.4314	0.4330	0.4346	0.4362	0.4378	0.4393	0.4409	0.4425	0.4440	0.4456
2.8	0.4472	0.4487	0.4502	0.4518	0.4533	0.4548	0.4564	0.4579	0.4594	0.4609
2.9	0.4624	0.4639	0.4654	0.4669	0.4683	0.4698	0.4713	0.4728	0.4742	0.4757
3.0	0.4771	0.4786	0.4800	0.4814	0.4829	0.4843	0.4857	0.4871	0.4886	0.4900
3.1	0.4914	0.4928	0.4942	0.4955	0.4969	0.4983	0.4997	0.5011	0.5024	0.5038
3.2	0.5051	0.5065	0.5079	0.5092	0.5105	0.5119	0.5132	0.5145	0.5159	0.5172
3.3	0.5185	0.5198	0.5211	0.5224	0.5237	0.5250	0.5263	0.5276	0.5289	0.5302
3.4	0.5315	0.5328	0.5340	0.5353	0.5366	0.5378	0.5391	0.5403	0.5416	0.5428
3.5	0.5441	0.5453	0.5465	0.5478	0.5490	0.5502	0.5514	0.5527	0.5539	0.5551
3.6	0.5563	0.5575	0.5587	0.5599	0.5611	0.5623	0.5635	0.5647	0.5658	0.5670
3.7	0.5682	0.5694	0.5705	0.5717	0.5729	0.5740	0.5752	0.5763	0.5775	0.5786
3.8	0.5798	0.5809	0.5821	0.5832	0.5843	0.5855	0.5866	0.5877	0.5888	0.5899
3.9	0.5911	0.5922	0.5933	0.5944	0.5955	0.5966	0.5977	0.5988	0.5999	0.6010
4.0	0.6021	0.6031	0.6042	0.6053	0.6064	0.6075	0.6085	0.6096	0.6107	0.6117
4.1	0.6128	0.6138	0.6149	0.6160	0.6170	0.6180	0.6191	0.6201	0.6212	0.6222
4.2	0.6232	0.6243	0.6253	0.6263	0.6274	0.6284	0.6294	0.6304	0.6314	0.6325
4.3	0.6335	0.6345	0.6355	0.6365	0.6375	0.6385	0.6395	0.6405	0.6415	0.6425
4.4	0.6435	0.6444	0.6454	0.6464	0.6474	0.6484	0.6493	0.6503	0.6513	0.6522
4.5	0.6532	0.6542	0.6551	0.6561	0.6571	0.6580	0.6590	0.6599	0.6609	0.6618
4.6	0.6628	0.6637	0.6646	0.6656	0.6665	0.6675	0.6684	0.6693	0.6702	0.6712
4.7	0.6721	0.6730	0.6739	0.6749	0.6758	0.6767	0.6776	0.6785	0.6794	0.6803
4.8	0.6812	0.6821	0.6830	0.6839	0.6848	0.6857	0.6866	0.6875	0.6884	0.6893
4.9	0.6902	0.6911	0.6920	0.6928	0.6937	0.6946	0.6955	0.6964	0.6972	0.6981
5.0	0.6990	0.6998	0.7007	0.7016	0.7024	0.7033	0.7042	0.7050	0.7059	0.7067
5.1	0.7076	0.7084	0.7093	0.7101	0.7110	0.7118	0.7126	0.7135	0.7143	0.7152
5.2	0.7160	0.7168	0.7177	0.7185	0.7193	0.7202	0.7210	0.7218	0.7226	0.7235
5.3	0.7243	0.7251	0.7259	0.7267	0.7275	0.7284	0.7292	0.7300	0.7308	0.7316
5.4	0.7324	0.7332	0.7340	0.7348	0.7356	0.7364	0.7372	0.7380	0.7388	0.7396

N	0	1	2	3	4	5	6	7	8	9
5.5	0.7404	0.7412	0.7419	0.7427	0.7435	0.7443	0.7451	0.7459	0.7466	0.7474
5.6	0.7482	0.7490	0.7497	0.7505	0.7513	0.7520	0.7528	0.7536	0.7543	0.7551
5.7	0.7559	0.7566	0.7574	0.7582	0.7589	0.7597	0.7604	0.7612	0.7619	0.7627
5.8	0.7634	0.7642	0.7649	0.7657	0.7664	0.7672	0.7679	0.7686	0.7694	0.7701
5.9	0.7709	0.7716	0.7723	0.7731	0.7738	0.7745	0.7752	0.7760	0.7767	0.7774
6.0	0.7782	0.7789	0.7796	0.7803	0.7810	0.7818	0.7825	0.7832	0.7839	0.7846
6.1	0.7853	0.7860	0.7868	0.7875	0.7882	0.7889	0.7896	0.7903	0.7910	0.7917
6.2	0.7924	0.7931	0.7938	0.7945	0.7952	0.7959	0.7966	0.7973	0.7980	0.7987
6.3	0.7993	0.8000	0.8007	0.8014	0.8021	0.8028	0.8035	0.8041	0.8048	0.8055
6.4	0.8062	0.8069	0.8075	0.8082	0.8089	0.8096	0.8102	0.8109	0.8116	0.8122
6.5	0.8129	0.8136	0.8142	0.8149	0.8156	0.8162	0.8169	0.8176	0.8182	0.8189
6.6	0.8195	0.8202	0.8209	0.8215	0.8222	0.8228	0.8235	0.8241	0.8248	0.8254
6.7	0.8261	0.8267	0.8274	0.8280	0.8287	0.8293	0.8299	0.8306	0.8312	0.8319
6.8	0.8325	0.8331	0.8338	0.8344	0.8351	0.8357	0.8363	0.8370	0.8376	0.8382
6.9	0.8388	0.8395	0.8401	0.8407	0.8414	0.8420	0.8426	0.8432	0.8439	0.8445
7.0	0.8451	0.8457	0.8463	0.8470	0.8476	0.8482	0.8488	0.8494	0.8500	0.8506
7.1	0.8513	0.8519	0.8525	0.8531	0.8537	0.8543	0.8549	0.8555	0.8561	0.8567
7.2	0.8573	0.8579	0.8585	0.8591	0.8597	0.8603	0.8609	0.8615	0.8621	0.8627
7.3	0.8633	0.8639	0.8645	0.8651	0.8657	0.8663	0.8669	0.8675	0.8681	0.8686
7.4	0.8692	0.8698	0.8704	0.8710	0.8716	0.8722	0.8727	0.8733	0.8739	0.8745
7.5	0.8751	0.8756	0.8762	0.8768	0.8774	0.8779	0.8785	0.8791	0.8797	0.8802
7.6	0.8808	0.8814	0.8820	0.8825	0.8831	0.8837	0.8842	0.8848	0.8854	0.8859
7.7	0.8865	0.8871	0.8876	0.8882	0.8887	0.8893	0.8899	0.8904	0.8910	0.8915
7.8	0.8921	0.8927	0.8932	0.8938	0.8943	0.8949	0.8954	0.8960	0.8965	0.8971
7.9	0.8976	0.8982	0.8987	0.8993	0.8998	0.9004	0.9009	0.9015	0.9020	0.9025
8.0	0.9031	0.9036	0.9042	0.9047	0.9053	0.9058	0.9063	0.9069	0.9074	0.9079
8.1	0.9085	0.9090	0.9096	0.9101	0.9106	0.9112	0.9117	0.9122	0.9128	0.9133
8.2	0.9138	0.9143	0.9149	0.9154	0.9159	0.9165	0.9170	0.9175	0.9180	0.9186
8.3	0.9191	0.9196	0.9201	0.9206	0.9212	0.9217	0.9222	0.9227	0.9232	0.9238
8.4	0.9243	0.9248	0.9253	0.9258	0.9263	0.9269	0.9274	0.9279	0.9284	0.9289
8.5	0.9294	0.9299	0.9304	0.9309	0.9315	0.9320	0.9325	0.9330	0.9335	0.9340
8.6	0.9345	0.9350	0.9355	0.9360	0.9365	0.9370	0.9375	0.9380	0.9385	0.9390
8.7	0.9395	0.9400	0.9405	0.9410	0.9415	0.9420	0.9425	0.9430	0.9435	0.9440
8.8	0.9445	0.9450	0.9455	0.9460	0.9465	0.9469	0.9474	0.9479	0.9484	0.9489
8.9	0.9494	0.9499	0.9504	0.9509	0.9513	0.9518	0.9523	0.9528	0.9533	0.9538
9.0	0.9542	0.9547	0.9552	0.9557	0.9562	0.9566	0.9571	0.9576	0.9581	0.9586
9.1	0.9590	0.9595	0.9600	0.9605	0.9609	0.9614	0.9619	0.9624	0.9628	0.9633
9.2	0.9638	0.9643	0.9647	0.9652	0.9657	0.9661	0.9666	0.9671	0.9675	0.9680
9.3	0.9685	0.9689	0.9694	0.9699	0.9703	0.9708	0.9713	0.9717	0.9722	0.9727
9.4	0.9731	0.9736	0.9741	0.9745	0.9750	0.9754	0.9759	0.9763	0.9768	0.9773
9.5	0.9777	0.9782	0.9786	0.9791	0.9795	0.9800	0.9805	0.9809	0.9814	0.9818
9.6	0.9823	0.9827	0.9832	0.9836	0.9841	0.9845	0.9850	0.9854	0.9859	0.9863
9.7	0.9868	0.9872	0.9877	0.9881	0.9886	0.9890	0.9894	0.9899	0.9903	0.9908
9.8	0.9912	0.9917	0.9921	0.9926	0.9930	0.9934	0.9939	0.9943	0.9948	0.9952
9.9	0.9956	0.9961	0.9965	0.9969	0.9974	0.9978	0.9983	0.9987	0.9991	0.9996

Table III Natural Logarithms ($\ln N = \log_e N$)

$\ln 10 = 2.3026$ | $5 \ln 10 = 11.5130$ | $9 \ln 10 = 20.7233$
$2 \ln 10 = 4.6052$ | $6 \ln 10 = 13.8155$ | $10 \ln 10 = 23.0259$
$3 \ln 10 = 6.9078$ | $7 \ln 10 = 16.1181$
$4 \ln 10 = 9.2103$ | $8 \ln 10 = 18.4207$

N	.00	.01	.02	.03	.04	.05	.06	.07	.08	.09
1.0	0.0000	0.0100	0.0198	0.0296	0.0392	0.0488	0.0583	0.0677	0.0770	0.0862
1.1	0.0953	0.1044	0.1133	0.1222	0.1310	0.1398	0.1484	0.1570	0.1655	0.1740
1.2	0.1823	0.1906	0.1989	0.2070	0.2151	0.2231	0.2311	0.2390	0.2469	0.2546
1.3	0.2624	0.2700	0.2776	0.2852	0.2927	0.3001	0.3075	0.3148	0.3221	0.3293
1.4	0.3365	0.3436	0.3507	0.3577	0.3646	0.3716	0.3784	0.3853	0.3920	0.3988
1.5	0.4055	0.4121	0.4187	0.4253	0.4318	0.4383	0.4447	0.4511	0.4574	0.4637
1.6	0.4700	0.4762	0.4824	0.4886	0.4947	0.5008	0.5068	0.5128	0.5188	0.5247
1.7	0.5306	0.5365	0.5423	0.5481	0.5539	0.5596	0.5653	0.5710	0.5766	0.5822
1.8	0.5878	0.5933	0.5988	0.6043	0.6098	0.6152	0.6206	0.6259	0.6313	0.6366
1.9	0.6419	0.6471	0.6523	0.6575	0.6627	0.6678	0.6729	0.6780	0.6831	0.6881
2.0	0.6931	0.6981	0.7031	0.7080	0.7129	0.7178	0.7227	0.7275	0.7324	0.7372
2.1	0.7419	0.7467	0.7514	0.7561	0.7608	0.7655	0.7701	0.7747	0.7793	0.7839
2.2	0.7885	0.7930	0.7975	0.8020	0.8065	0.8109	0.8154	0.8198	0.8242	0.8286
2.3	0.8329	0.8372	0.8416	0.8459	0.8502	0.8544	0.8587	0.8629	0.8671	0.8713
2.4	0.8755	0.8796	0.8838	0.8879	0.8920	0.8961	0.9002	0.9042	0.9083	0.9123
2.5	0.9163	0.9203	0.9243	0.9282	0.9322	0.9361	0.9400	0.9439	0.9478	0.9517
2.6	0.9555	0.9594	0.9632	0.9670	0.9708	0.9746	0.9783	0.9821	0.9858	0.9895
2.7	0.9933	0.9969	1.0006	1.0043	1.0080	1.0116	1.0152	1.0188	1.0225	1.0260
2.8	1.0296	1.0332	1.0367	1.0403	1.0438	1.0473	1.0508	1.0543	1.0578	1.0613
2.9	1.0647	1.0682	1.0716	1.0750	1.0784	1.0818	1.0852	1.0886	1.0919	1.0953
3.0	1.0986	1.1019	1.1053	1.1086	1.1119	1.1151	1.1184	1.1217	1.1249	1.1282
3.1	1.1314	1.1346	1.1378	1.1410	1.1442	1.1474	1.1506	1.1537	1.1569	1.1600
3.2	1.1632	1.1663	1.1694	1.1725	1.1756	1.1787	1.1817	1.1848	1.1878	1.1909
3.3	1.1939	1.1969	1.2000	1.2030	1.2060	1.2090	1.2119	1.2149	1.2179	1.2208
3.4	1.2238	1.2267	1.2296	1.2326	1.2355	1.2384	1.2413	1.2442	1.2470	1.2499
3.5	1.2528	1.2556	1.2585	1.2613	1.2641	1.2669	1.2698	1.2726	1.2754	1.2782
3.6	1.2809	1.2837	1.2865	1.2892	1.2920	1.2947	1.2975	1.3002	1.3029	1.3056
3.7	1.3083	1.3110	1.3137	1.3164	1.3191	1.3218	1.3244	1.3271	1.3297	1.3324
3.8	1.3350	1.3376	1.3403	1.3429	1.3455	1.3481	1.3507	1.3533	1.3558	1.3584
3.9	1.3610	1.3635	1.3661	1.3686	1.3712	1.3737	1.3762	1.3788	1.3813	1.3838
4.0	1.3863	1.3888	1.3913	1.3938	1.3962	1.3987	1.4012	1.4036	1.4061	1.4085
4.1	1.4110	1.4134	1.4159	1.4183	1.4207	1.4231	1.4255	1.4279	1.4303	1.4327
4.2	1.4351	1.4375	1.4398	1.4422	1.4446	1.4469	1.4493	1.4516	1.4540	1.4563
4.3	1.4586	1.4609	1.4633	1.4656	1.4679	1.4702	1.4725	1.4748	1.4770	1.4793
4.4	1.4816	1.4839	1.4861	1.4884	1.4907	1.4929	1.4951	1.4974	1.4996	1.5019
4.5	1.5041	1.5063	1.5085	1.5107	1.5129	1.5151	1.5173	1.5195	1.5217	1.5239
4.6	1.5261	1.5282	1.5304	1.5326	1.5347	1.5369	1.5390	1.5412	1.5433	1.5454
4.7	1.5476	1.5497	1.5518	1.5539	1.5560	1.5581	1.5602	1.5623	1.5644	1.5665
4.8	1.5686	1.5707	1.5728	1.5748	1.5769	1.5790	1.5810	1.5831	1.5851	1.5872
4.9	1.5892	1.5913	1.5933	1.5953	1.5974	1.5994	1.6014	1.6034	1.6054	1.6074
5.0	1.6094	1.6114	1.6134	1.6154	1.6174	1.6194	1.6214	1.6233	1.6253	1.6273
5.1	1.6292	1.6312	1.6332	1.6351	1.6371	1.6390	1.6409	1.6429	1.6448	1.6467
5.2	1.6487	1.6506	1.6525	1.6544	1.6563	1.6582	1.6601	1.6620	1.6639	1.6658
5.3	1.6677	1.6696	1.6715	1.6734	1.6752	1.6771	1.6790	1.6808	1.6827	1.6845
5.4	1.6864	1.6882	1.6901	1.6919	1.6938	1.6956	1.6974	1.6993	1.7011	1.7029

Note: $\ln 35,200 = \ln(3.52 \times 10^4) = \ln 3.52 + 4 \ln 10$
$\ln 0.00864 = \ln(8.64 \times 10^{-3}) = \ln 8.64 - 3 \ln 10$

N	.00	.01	.02	.03	.04	.05	.06	.07	.08	.09
5.5	1.7047	1.7066	1.7084	1.7102	1.7120	1.7138	1.7156	1.7174	1.7192	1.7210
5.6	1.7228	1.7246	1.7263	1.7281	1.7299	1.7317	1.7334	1.7352	1.7370	1.7387
5.7	1.7405	1.7422	1.7440	1.7457	1.7475	1.7492	1.7509	1.7527	1.7544	1.7561
5.8	1.7579	1.7596	1.7613	1.7630	1.7647	1.7664	1.7681	1.7699	1.7716	1.7733
5.9	1.7750	1.7766	1.7783	1.7800	1.7817	1.7834	1.7851	1.7867	1.7884	1.7901
6.0	1.7918	1.7934	1.7951	1.7967	1.7984	1.8001	1.8017	1.8034	1.8050	1.8066
6.1	1.8083	1.8099	1.8116	1.8132	1.8148	1.8165	1.8181	1.8197	1.8213	1.8229
6.2	1.8245	1.8262	1.8278	1.8294	1.8310	1.8326	1.8342	1.8358	1.8374	1.8390
6.3	1.8405	1.8421	1.8437	1.8453	1.8469	1.8485	1.8500	1.8516	1.8532	1.8547
6.4	1.8563	1.8579	1.8594	1.8610	1.8625	1.8641	1.8656	1.8672	1.8687	1.8703
6.5	1.8718	1.8733	1.8749	1.8764	1.8779	1.8795	1.8810	1.8825	1.8840	1.8856
6.6	1.8871	1.8886	1.8901	1.8916	1.8931	1.8946	1.8961	1.8976	1.8991	1.9006
6.7	1.9021	1.9036	1.9051	1.9066	1.9081	1.9095	1.9110	1.9125	1.9140	1.9155
6.8	1.9169	1.9184	1.9199	1.9213	1.9228	1.9242	1.9257	1.9272	1.9286	1.9301
6.9	1.9315	1.9330	1.9344	1.9359	1.9373	1.9387	1.9402	1.9416	1.9430	1.9445
7.0	1.9459	1.9473	1.9488	1.9502	1.9516	1.9530	1.9544	1.9559	1.9573	1.9587
7.1	1.9601	1.9615	1.9629	1.9643	1.9657	1.9671	1.9685	1.9699	1.9713	1.9727
7.2	1.9741	1.9755	1.9769	1.9782	1.9796	1.9810	1.9824	1.9838	1.9851	1.9865
7.3	1.9879	1.9892	1.9906	1.9920	1.9933	1.9947	1.9961	1.9974	1.9988	2.0001
7.4	2.0015	2.0028	2.0042	2.0055	2.0069	2.0082	2.0096	2.0109	2.0122	2.0136
7.5	2.0149	2.0162	2.0176	2.0189	2.0202	2.0215	2.0229	2.0242	2.0255	2.0268
7.6	2.0281	2.0295	2.0308	2.0321	2.0334	2.0347	2.0360	2.0373	2.0386	2.0399
7.7	2.0412	2.0425	2.0438	2.0451	2.0464	2.0477	2.0490	2.0503	2.0516	2.0528
7.8	2.0541	2.0554	2.0567	2.0580	2.0592	2.0605	2.0618	2.0631	2.0643	2.0656
7.9	2.0669	2.0681	2.0694	2.0707	2.0719	2.0732	2.0744	2.0757	2.0769	2.0782
8.0	2.0794	2.0807	2.0819	2.0832	2.0844	2.0857	2.0869	2.0882	2.0894	2.0906
8.1	2.0919	2.0931	2.0943	2.0956	2.0968	2.0980	2.0992	2.1005	2.1017	2.1029
8.2	2.1041	2.1054	2.1066	2.1078	2.1090	2.1102	2.1114	2.1126	2.1138	2.1150
8.3	2.1163	2.1175	2.1187	2.1199	2.1211	2.1223	2.1235	2.1247	2.1258	2.1270
8.4	2.1282	2.1294	2.1306	2.1318	2.1330	2.1342	2.1353	2.1365	2.1377	2.1389
8.5	2.1401	2.1412	2.1424	2.1436	2.1448	2.1459	2.1471	2.1483	2.1494	2.1506
8.6	2.1518	2.1529	2.1541	2.1552	2.1564	2.1576	2.1587	2.1599	2.1610	2.1622
8.7	2.1633	2.1645	2.1656	2.1668	2.1679	2.1691	2.1702	2.1713	2.1725	2.1736
8.8	2.1748	2.1759	2.1770	2.1782	2.1793	2.1804	2.1815	2.1827	2.1838	2.1849
8.9	2.1861	2.1872	2.1883	2.1894	2.1905	2.1917	2.1928	2.1939	2.1950	2.1961
9.0	2.1972	2.1983	2.1994	2.2006	2.2017	2.2028	2.2039	2.2050	2.2061	2.2072
9.1	2.2083	2.2094	2.2105	2.2116	2.2127	2.2138	2.2148	2.2159	2.2170	2.2181
9.2	2.2192	2.2203	2.2214	2.2225	2.2235	2.2246	2.2257	2.2268	2.2279	2.2289
9.3	2.2300	2.2311	2.2322	2.2332	2.2343	2.2354	2.2364	2.2375	2.2386	2.2396
9.4	2.2407	2.2418	2.2428	2.2439	2.2450	2.2460	2.2471	2.2481	2.2492	2.2502
9.5	2.2513	2.2523	2.2534	2.2544	2.2555	2.2565	2.2576	2.2586	2.2597	2.2607
9.6	2.2618	2.2628	2.2638	2.2649	2.2659	2.2670	2.2680	2.2690	2.2701	2.2711
9.7	2.2721	2.2732	2.2742	2.2752	2.2762	2.2773	2.2783	2.2793	2.2803	2.2814
9.8	2.2824	2.2834	2.2844	2.2854	2.2865	2.2875	2.2885	2.2895	2.2905	2.2915
9.9	2.2925	2.2935	2.2946	2.2956	2.2966	2.2976	2.2986	2.2996	2.3006	2.3016

Answers

Chapter 0

Exercise 0-1

1. T **3.** T **5.** T **7.** T **9.** {1, 2, 3, 4, 5} **11.** {3, 4} **13.** ∅ **15.** {2} **17.** {−7, 7} **19.** {1, 3, 5, 7, 9} **21.** $A' = \{1, 5\}$ **23.** 40 **25.** 60 **27.** 60 **29.** 20 **31.** 95 **33.** 40 **35.** (A) {1, 2, 3, 4, 6} (B) {1, 2, 3, 4, 6} **37.** {1, 2, 3, 4, 6} **39.** Yes **41.** Yes **43.** Yes **45.** (A) 2 (B) 4 (C) 8; 2^n **47.** 800 **49.** 200 **51.** 200 **53.** 800 **55.** 200 **57.** 200 **59.** 6 **61.** A+, AB+ **63.** A−, A+, B+, AB−, AB+, O+ **65.** O+, O− **67.** B−, B+ **69.** Everybody in the clique relates to each other.

Exercise 0-2

1. $m = 5$ **3.** $x < -9$ **5.** $x \le 4$ **7.** $x < -3$ or $(-\infty, -3)$ 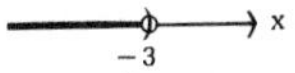

9. $-1 \le x \le 2$ or $[-1, 2]$ 

11. $y = 8$ **13.** $x > -6$ **15.** $y = 8$ **17.** $x = 10$ **19.** $y \ge 3$

21. $x = 36$ **23.** $m < 3$ **25.** $x = 10$ **27.** $3 \le x < 7$ or $[3, 7)$ 

29. $-20 \le C \le 20$ or $[-20, 20]$ 

31. $y = \frac{3}{4}x - 3$ **33.** $y = -(A/B)x + (C/B) = (-Ax + C)/B$ **35.** $C = \frac{5}{9}(F - 32)$ **37.** $B = A/(m - n)$

39. $-2 < x \le 1$ or $(-2, 1]$ 

41. 3,000 \$10 tickets, 5,000 \$6 tickets

43. \$7,200 at 10%, \$4,800 at 15% **45.** \$7,800 **47.** 5,000 **49.** 12.6 yr

Exercise 0-3

1. ± 2 **3.** $\pm\sqrt{11}$ **5.** $-2, 6$ **7.** $0, 2$ **9.** $3 \pm 2\sqrt{3}$ **11.** $-2 \pm \sqrt{2}$ **13.** $0, 2$ **15.** $\pm\frac{3}{2}$ **17.** $\frac{1}{2}, -3$ **19.** $(-1 \pm \sqrt{5})/2$ **21.** $(3 \pm \sqrt{3})/2$ **23.** No real solutions **25.** $-4 \pm \sqrt{11}$ **27.** $r = \sqrt{A/P} - 1$ **29.** \$2 **31.** 8 ft/sec; $4\sqrt{2}$ or 5.66 ft/sec

Exercise 0-4

1.

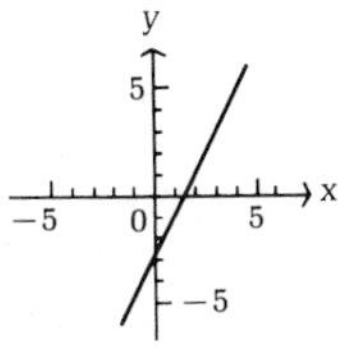

3.

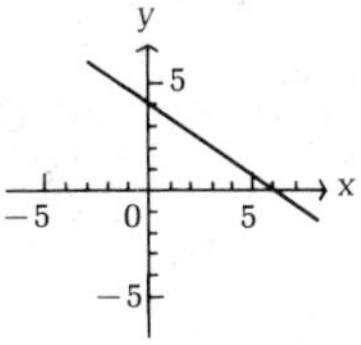

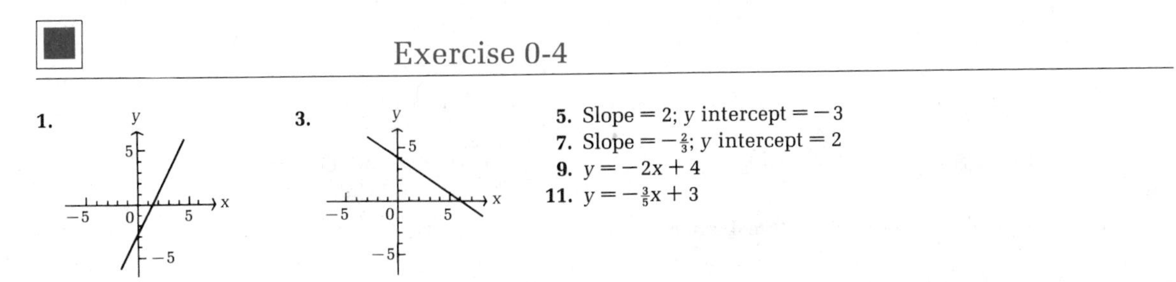

5. Slope = 2; y intercept = −3
7. Slope = $-\frac{2}{3}$; y intercept = 2
9. $y = -2x + 4$
11. $y = -\frac{3}{5}x + 3$

13. **15.** **17.**

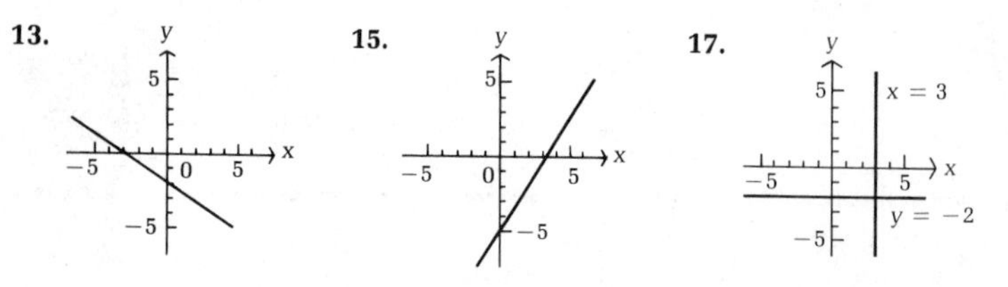

19. $y = -3x + 5$, $m = -3$
21. $y = -\frac{2}{3}x + 4$, $m = -\frac{2}{3}$
23. $y + 1 = -3(x - 4)$, $y = -3x + 11$
25. $y + 5 = \frac{2}{3}(x + 6)$, $y = \frac{2}{3}x - 1$
27. $\frac{1}{3}$ **29.** $-\frac{1}{5}$

31. $(y - 3) = \frac{1}{3}(x - 1)$, $x - 3y = -8$ **33.** $(y + 2) = -\frac{1}{5}(x + 5)$, $x + 5y = -15$ **35.** $x = 3$, $y = -5$ **37.** $x = -1$, $y = -3$ **39.** $y = -\frac{1}{2}x + 4$ **41.** (A) $y = -\frac{1}{2}x + 1$ (B) $y = 2x + 6$ **43.** (A) $y = (\frac{1}{2})x$ (B) $y = -2x - 5$

45.

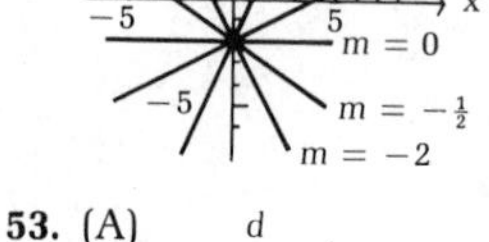

47. $x = 2$ **49.** $y = 3$ **51.** (A) \$130; \$220 (B) (C) 6

53. (A) (B) $d = -60p + 12{,}000$ **55.** $0.2x + 0.1y = 20$

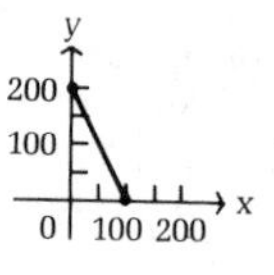

57. (A) 64 grams; 35 grams (B) (C) $-\frac{1}{5}$

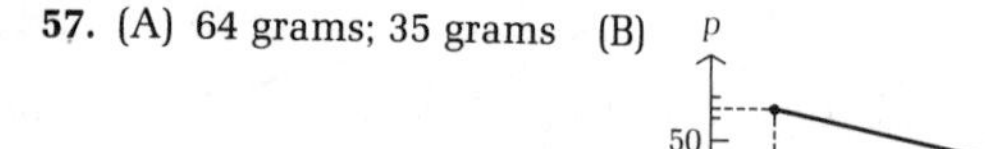

Exercise 0-5

1. Function **3.** Not a function **5.** Function **7.** Function **9.** Not a function **11.** Function **13.** 4 **15.** −5 **17.** −6 **19.** −2 **21.** −12 **23.** −1 **25.** −6 **27.** 12 **29.** $\frac{3}{4}$

31. **33.** **35.**

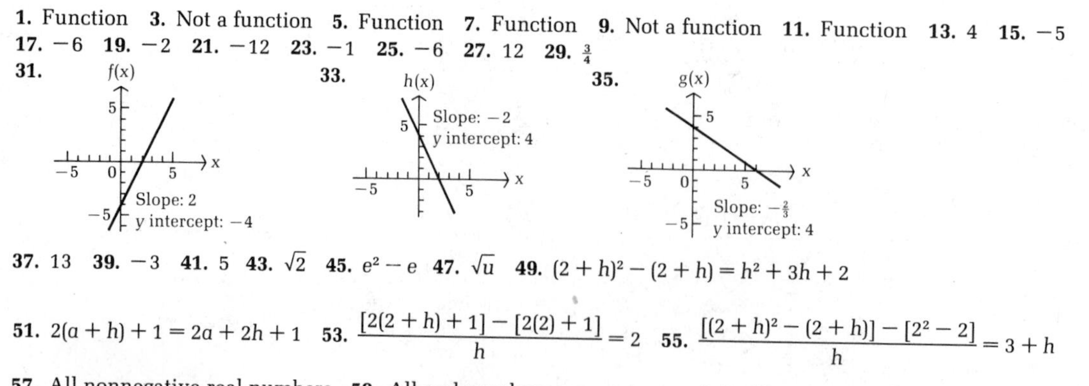

37. 13 **39.** −3 **41.** 5 **43.** $\sqrt{2}$ **45.** $e^2 - e$ **47.** $\sqrt{u}$ **49.** $(2 + h)^2 - (2 + h) = h^2 + 3h + 2$

51. $2(a + h) + 1 = 2a + 2h + 1$ **53.** $\frac{[2(2 + h) + 1] - [2(2) + 1]}{h} = 2$ **55.** $\frac{[(2 + h)^2 - (2 + h)] - [2^2 - 2]}{h} = 3 + h$

57. All nonnegative real numbers **59.** All real numbers x except $x = -3, 5$ **61.** $x \geq -5$ or $[-5, \infty)$

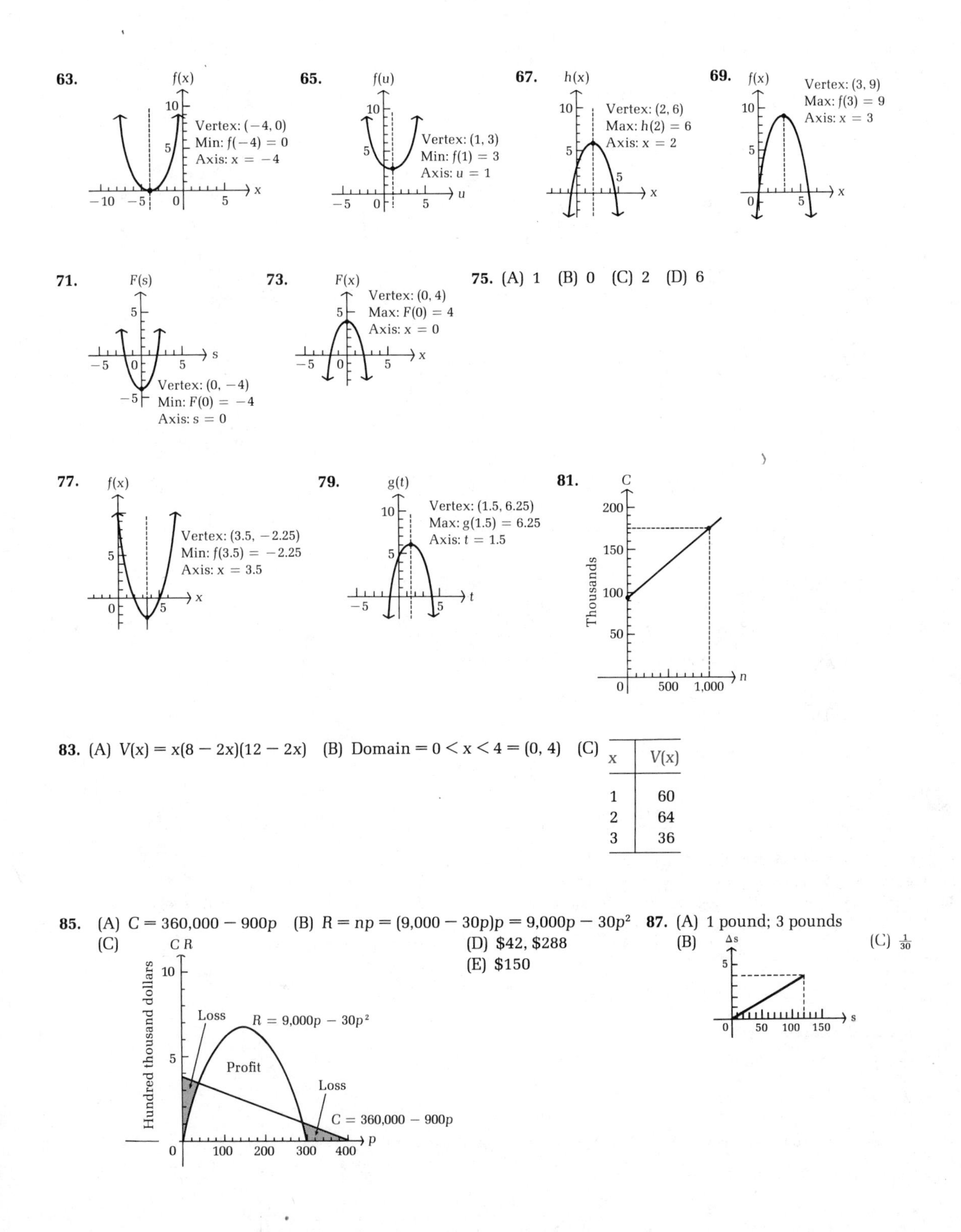

75. (A) 1 (B) 0 (C) 2 (D) 6

83. (A) $V(x) = x(8 - 2x)(12 - 2x)$ (B) Domain $= 0 < x < 4 = (0, 4)$ (C)

x	V(x)
1	60
2	64
3	36

85. (A) $C = 360{,}000 - 900p$ (B) $R = np = (9{,}000 - 30p)p = 9{,}000p - 30p^2$
(C)
(D) \$42, \$288
(E) \$150

87. (A) 1 pound; 3 pounds
(B)
(C) $\frac{1}{30}$

Exercise 0-6

1.
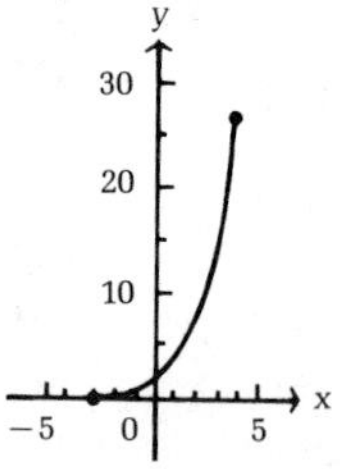

3.
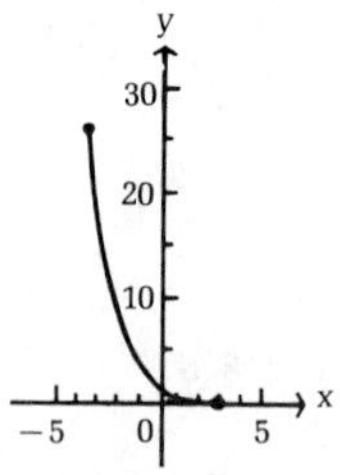

5.
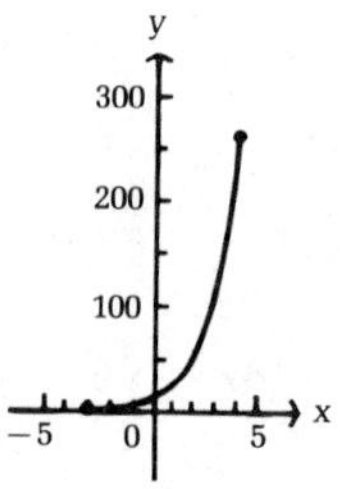

7.
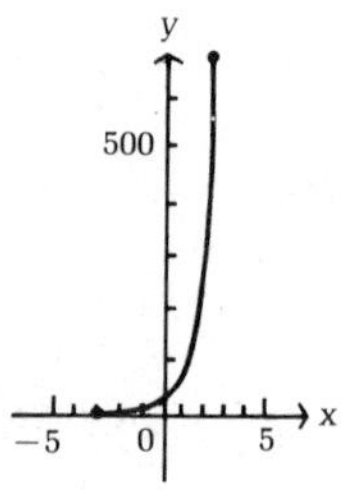

9.
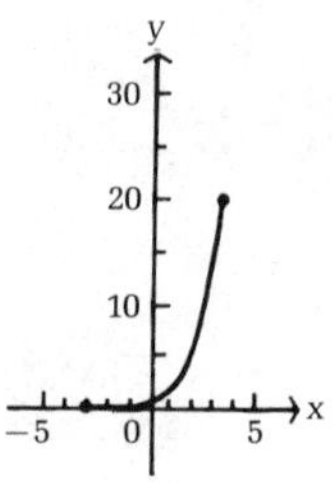

11.
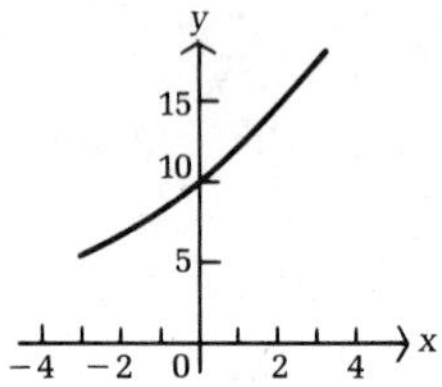

13.
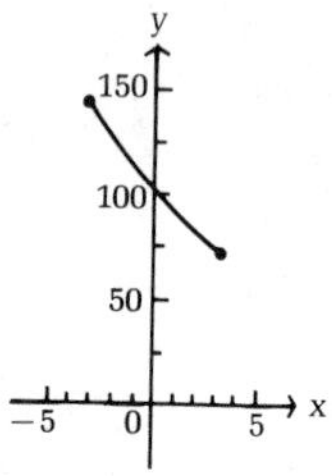

15.
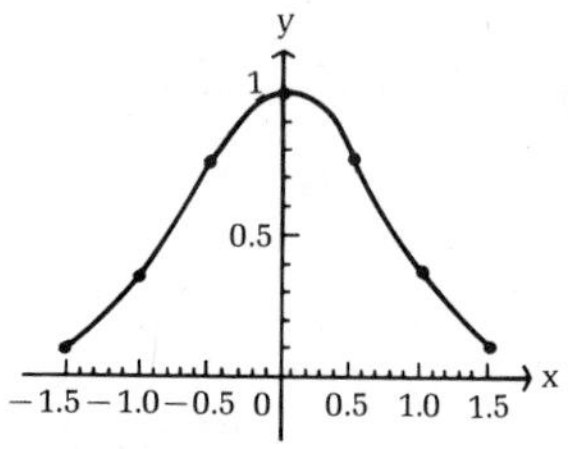

17.
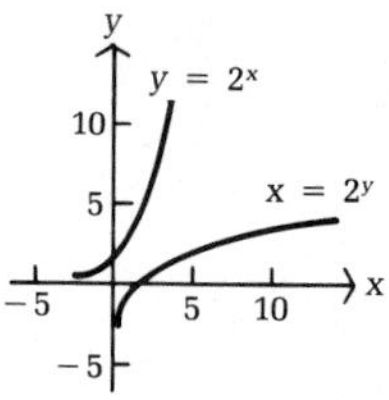

19.
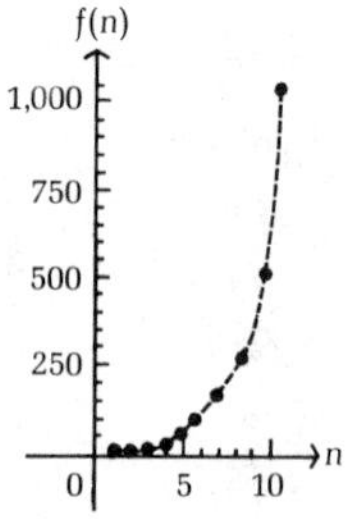

21.
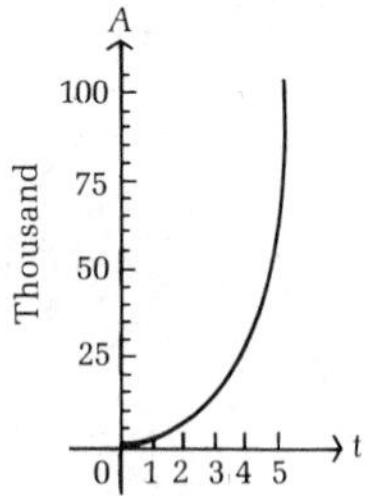

23.
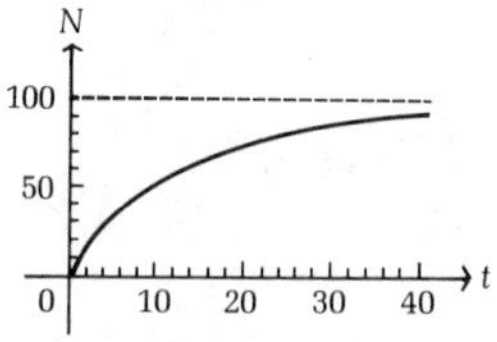

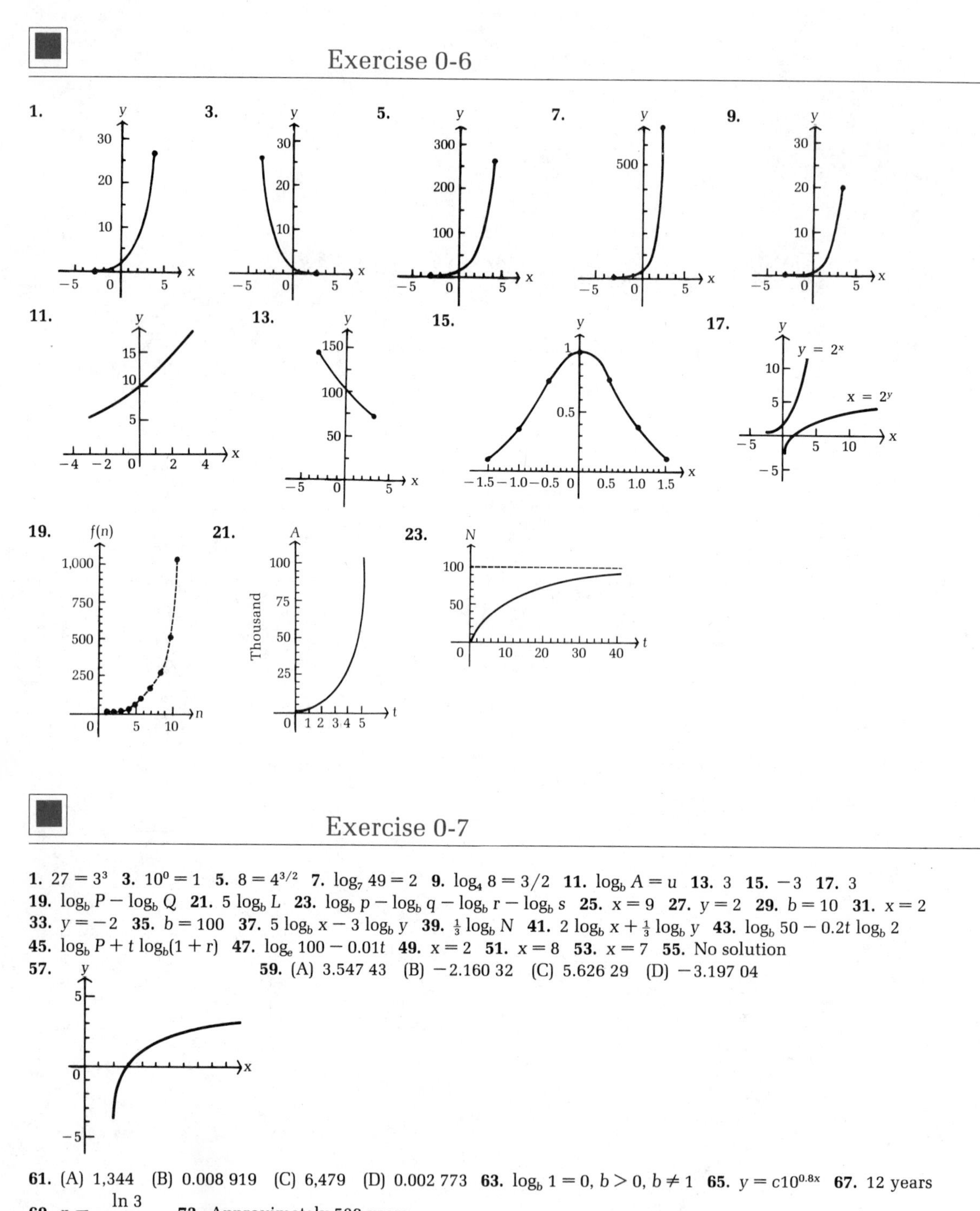

Exercise 0-7

1. $27 = 3^3$ **3.** $10^0 = 1$ **5.** $8 = 4^{3/2}$ **7.** $\log_7 49 = 2$ **9.** $\log_4 8 = 3/2$ **11.** $\log_b A = u$ **13.** 3 **15.** -3 **17.** 3
19. $\log_b P - \log_b Q$ **21.** $5 \log_b L$ **23.** $\log_b p - \log_b q - \log_b r - \log_b s$ **25.** $x = 9$ **27.** $y = 2$ **29.** $b = 10$ **31.** $x = 2$
33. $y = -2$ **35.** $b = 100$ **37.** $5 \log_b x - 3 \log_b y$ **39.** $\frac{1}{3} \log_b N$ **41.** $2 \log_b x + \frac{1}{3} \log_b y$ **43.** $\log_b 50 - 0.2t \log_b 2$
45. $\log_b P + t \log_b(1 + r)$ **47.** $\log_e 100 - 0.01t$ **49.** $x = 2$ **51.** $x = 8$ **53.** $x = 7$ **55.** No solution

57.
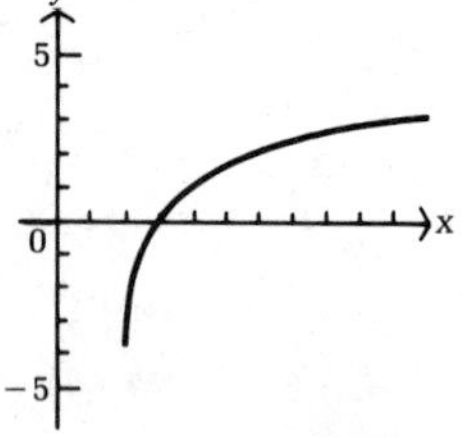

59. (A) 3.547 43 (B) −2.160 32 (C) 5.626 29 (D) −3.197 04

61. (A) 1,344 (B) 0.008 919 (C) 6,479 (D) 0.002 773 **63.** $\log_b 1 = 0$, $b > 0$, $b \neq 1$ **65.** $y = c10^{0.8x}$ **67.** 12 years
69. $n = \dfrac{\ln 3}{\ln(1 + i)}$ **73.** Approximately 538 years

Exercise 0-8 Chapter Review

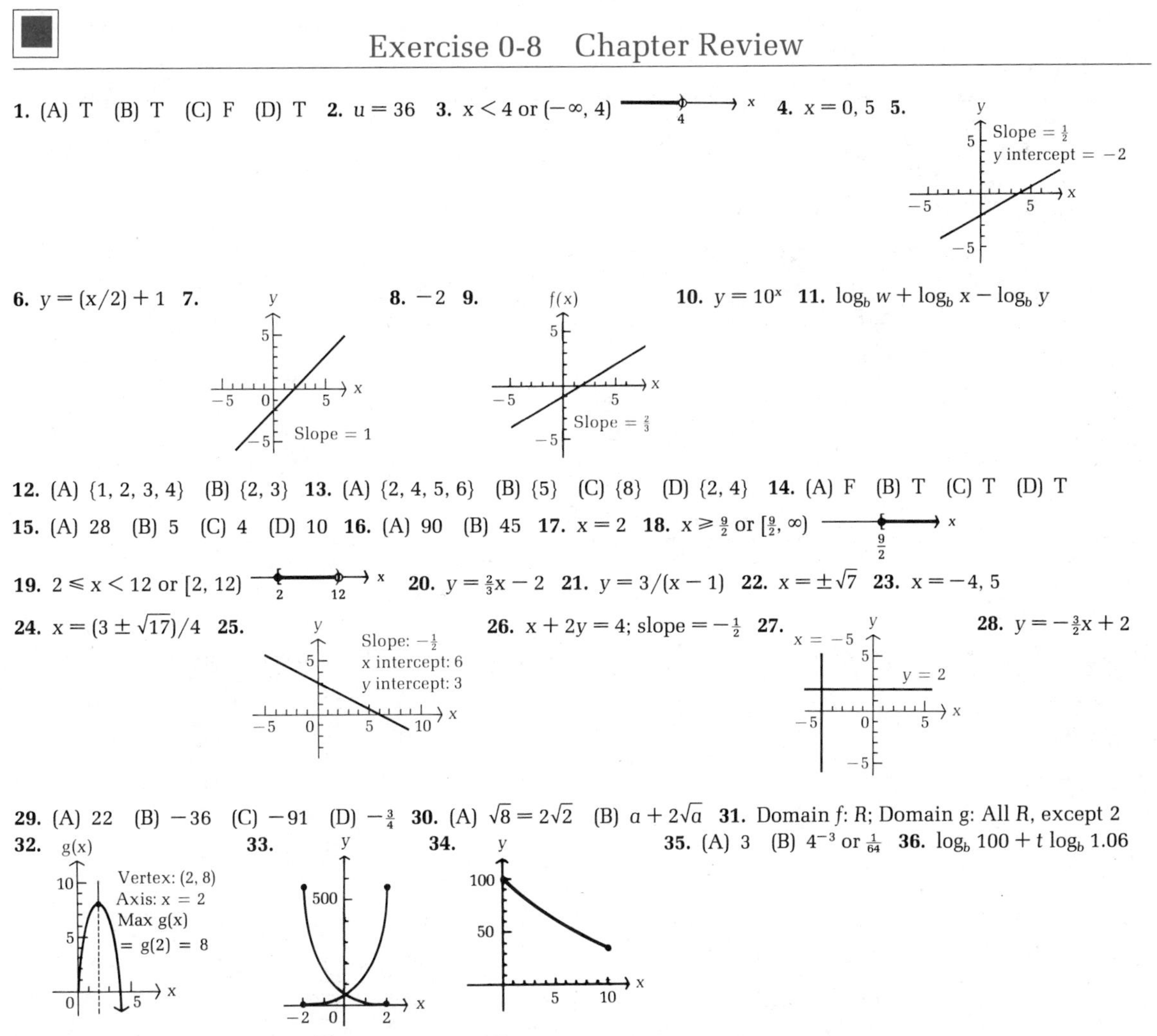

1. (A) T (B) T (C) F (D) T **2.** $u = 36$ **3.** $x < 4$ or $(-\infty, 4)$ **4.** $x = 0, 5$ **5.** [graph] **6.** $y = (x/2) + 1$ **7.** [graph] **8.** -2 **9.** [graph] **10.** $y = 10^x$ **11.** $\log_b w + \log_b x - \log_b y$

12. (A) {1, 2, 3, 4} (B) {2, 3} **13.** (A) {2, 4, 5, 6} (B) {5} (C) {8} (D) {2, 4} **14.** (A) F (B) T (C) T (D) T **15.** (A) 28 (B) 5 (C) 4 (D) 10 **16.** (A) 90 (B) 45 **17.** $x = 2$ **18.** $x \geq \frac{9}{2}$ or $[\frac{9}{2}, \infty)$

19. $2 \leq x < 12$ or $[2, 12)$ **20.** $y = \frac{2}{3}x - 2$ **21.** $y = 3/(x - 1)$ **22.** $x = \pm\sqrt{7}$ **23.** $x = -4, 5$

24. $x = (3 \pm \sqrt{17})/4$ **25.** [graph] **26.** $x + 2y = 4$; slope $= -\frac{1}{2}$ **27.** [graph] **28.** $y = -\frac{3}{2}x + 2$

29. (A) 22 (B) -36 (C) -91 (D) $-\frac{3}{4}$ **30.** (A) $\sqrt{8} = 2\sqrt{2}$ (B) $a + 2\sqrt{a}$ **31.** Domain f: R; Domain g: All R, except 2 **32.** [graph] **33.** [graph] **34.** [graph] **35.** (A) 3 (B) 4^{-3} or $\frac{1}{64}$ **36.** $\log_b 100 + t \log_b 1.06$

37. $x = \frac{1}{6}$ **38.** (A) $-2.040\ 55$ (B) $9.194\ 55$ **39.** (A) $0.000\ 156\ 5$ (B) 367,400 **40.** $x = 7$ **41.** Yes

42. $x = (-j \pm \sqrt{j^2 - 4k})/2$ **43.** $x = 4$ **44.** (A) $x - 2y = 8$ (B) $2x + y = 1$

45. (A) All real numbers except 3 (B) All real numbers ≥ 1 **46.** 2 **47.** $t = 9.15$ **48.** $y = ce^{-0.2x}$

49. (A) 900 (B) 350 **50.** \$20,000 at 8%, \$40,000 at 14% **51.** $x/800 = \frac{247}{89}$; $x = \$2,220.22$ **52.** 10%

53. (A) $V(t) = -1,250t + 12,000$ (B) \$5,750 **54.** (A) $R = \frac{8}{5}C$ (B) \$168

55. (A) $A(x) = x(20 - 2x)$ (B) Domain = (0, 10) (C)

x	A(x)
2	32
4	48
5	50
6	48
8	32

56. 8 years **57.** 6.93 years

Chapter 1

Exercise 1-1

1. (A) -5 (B) 3 (C) 4 (D) 0 **3.** 47 **5.** -4 **7.** $\frac{5}{3}$ **9.** 243

11.

x	0.9	0.99	0.999	$\rightarrow$	1	$\leftarrow$	1.001	1.01	1.1
$f(x)$	-19	-199	$-1{,}999$	$\rightarrow$	?	$\leftarrow$	2,001	201	21

$\lim_{x\to 1} f(x)$ does not exist

13.

x	0.9	0.99	0.999	$\rightarrow$	1	$\leftarrow$	1.001	1.01	1.1
$f(x)$	-1	-1	-1	$\rightarrow$	?	$\leftarrow$	1	1	1

$\lim_{x\to 1} f(x)$ does not exist

15.

x	10	100	1,000	10,000	$\rightarrow$	∞
$f(x)$	0.091	0.009 9	0.000 999	0.000 099 9	$\rightarrow$	?

$\lim_{x\to\infty} f(x) = 0$

17.

x	10	100	1,000	10,000	$\rightarrow$	∞
$f(x)$	9.09	99.01	999.001	9,999.000 1	$\rightarrow$	?

$\lim_{x\to\infty} f(x)$ does not exist

19. (A) 0 (B) Does not exist (C) 1 (D) 0 **21.** -3 **23.** 10 **25.** 0 **27.** -5 **29.** $\frac{5}{6}$ **31.** $\frac{1}{2}$ **33.** $\frac{2}{3}$ **35.** 4 **37.** 2 **39.** 4 **41.** $\frac{2}{3}$ **43.** 0 **45.** Does not exist

47. (A) Does not exist
(B) 2
(C) Does not exist

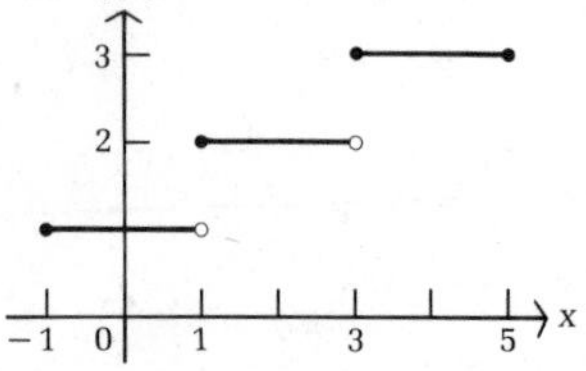

49. (A) Does not exist
(B) 0
(C) 1

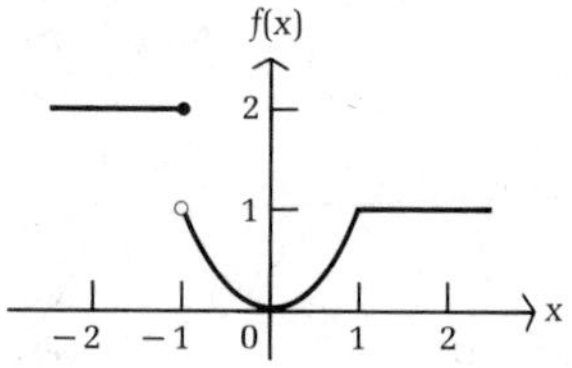

51. $\sqrt[3]{4}$ **53.** 0 **55.** $\frac{1}{4}$ **57.** Does not exist **59.** 8 **61.** 12 **63.** $\frac{1}{12}$ **65.** $2a$ **67.** $1/(2\sqrt{a})$
69. (A) \$23 (B) \$3.20 (C) \$5 (D) \$3

71. (A)

Compounded	n	$A(n)$
Annually	1	\$108.00
Semiannually	2	\$108.16
Quarterly	4	\$108.24
Monthly	12	\$108.30
Weekly	52	\$108.32
Daily	365	\$108.33
Hourly	8,760	\$108.33

(B) \$108.33

73. (A) 0.056 (B) 0.07 (C) 0.07 (D) 0 **75.** (A) 30 (B) 44 (C) 44 (D) 60

Exercise 1-2

1. (A) 1 (B) 1 (C) Yes **3.** (A) Does not exist (B) 1 (C) No **5.** (A) 1 (B) 3 (C) No **7.** $(-\infty, \infty)$
9. $(-\infty, 5), (5, \infty)$ **11.** $(-\infty, -2), (-2, 3), (3, \infty)$ **13.** $-3 < x < 4$ **15.** $x < 3$ or $x > 7$ **17.** $0 \leq x \leq 8$ **19.** Continuous
21. Discontinuous: $\lim_{x\to -1} f(x) \neq f(-1)$ **23.** Discontinuous: f is not defined at $x = 2$
25. Discontinuous at $x = 1$ **27.** Continuous for all x **29.** Discontinuous at $x = 0$

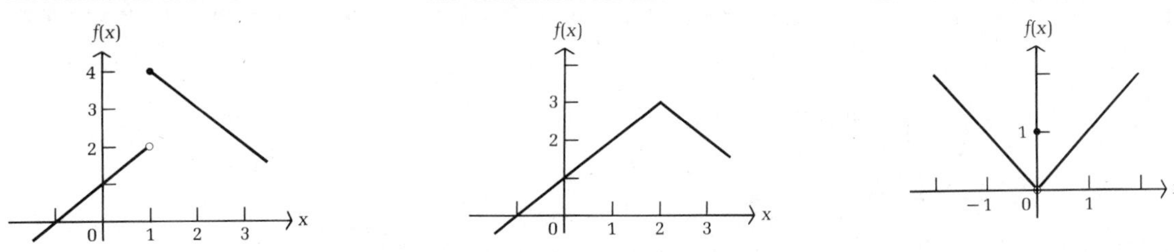

31. $(-\infty, \infty)$ **33.** $(5, \infty)$ **35.** $(-\infty, \infty)$ **37.** $(-\infty, 1), (1, 2), (2, \infty)$ **39.** $1 < x < 3$ or $x > 5$ **41.** $x < 4$ **43.** $-4 < x \leq 2$
45. $-5 \leq x \leq 0$ or $x > 3$ **47.** $x < -4$ **49.** $x \leq -3, 0 \leq x < \frac{3}{2}$, or $x > \frac{3}{2}$ **51.** $(-2, 2)$ **53.** $(-3, 0), (2, \infty)$ **55.** $(-1, 1)$
57. Not possible; $\lim_{x\to 0} f(x)$ does not exist **59.** Define $f(1) = 2$

61. (A) **63.** (A)

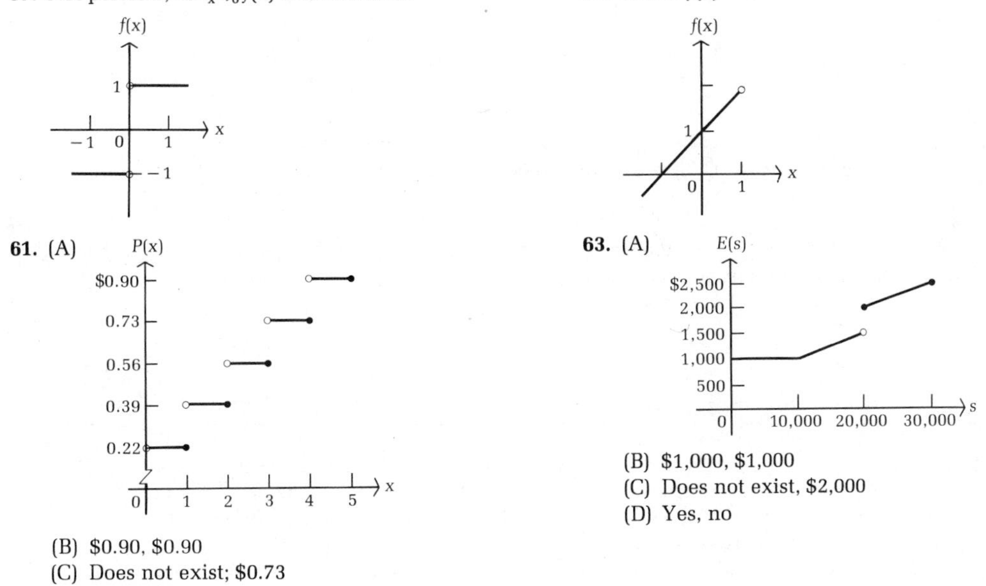

63. (B) \$1,000, \$1,000
(C) Does not exist, \$2,000
(D) Yes, no

61. (B) \$0.90, \$0.90
(C) Does not exist; \$0.73
(D) Yes, no

65. Loss: $\$0 \leq p < \$4$ or $p > \$7$; Profit: $\$4 < p < \$7$ **67.** (A) t_2, t_3, t_4, t_6, t_7 (B) 7, 7 (C) Does not exist; 4

Exercise 1-3

1. $\Delta x = 3$; $\Delta y = 45$; $\Delta y/\Delta x = 15$ **3.** 12 **5.** 12 **7.** 12 **9.** 15 **11.** (A) $12 + 3\Delta x$ (B) 12 **13.** (A) $24 + 3\Delta x$ (B) 24
15. (A) 5 meters per second (B) $3 + \Delta x$ meters per second (C) 3 meters per second
17. (A) 5 (B) $3 + \Delta x$ (C) 3 (D) $y = 3x - 1$ **19.** 3 meters per second **21.** (A) \$200 per year (B) \$450 per year
23. (A) 25 dryers per ad (B) $30 - \Delta x$ (C) 30 dryers per ad
25. (A) -110 square millimeters per day (B) -15 square millimeters per day
27. (A) 0.6 birth per year (B) 8 births per year

Exercise 1-4

1. $f'(a)$ exists **3.** $f'(c)$ does not exist **5.** $f'(e)$ does not exist **7.** $f'(g)$ exists **9.** $f'(x) = 2; f'(1) = f'(2) = f'(3) = 2$
11. $f'(x) = 6 - 2x; f'(1) = 4, f'(2) = 2, f'(3) = 0$ **13.** $f'(x) = -1/(x+1)^2; f'(1) = -\frac{1}{4}, f'(2) = -\frac{1}{9}, f'(3) = -\frac{1}{16}$
15. $f'(x) = 1/(2\sqrt{x}); f'(1) = \frac{1}{2}, f'(2) = 1/(2\sqrt{2}), f'(3) = 1/(2\sqrt{3})$ **17.** $f'(x) = -2/x^3; f'(1) = -2, f'(2) = -\frac{1}{4}, f'(3) = -\frac{2}{27}$
19. $v = f'(x) = 8x - 2$; $f'(1) = 6$ feet per second; $f'(3) = 22$ feet per second; $f'(5) = 38$ feet per second
21. (A) $m = f'(x) = 2x$ (B) $m_1 = f'(-2) = -4$; $m_2 = f'(0) = 0$; $m_3 = f'(2) = 4$ (C) $y = -4x - 4$; $y = 0$; $y = 4x - 4$
(D)

23. (A) $f'(x) = 3x^2 + 2$ (B) $f'(1) = 5; f'(3) = 29$ **25.** f is nondifferentiable at $x = 1$

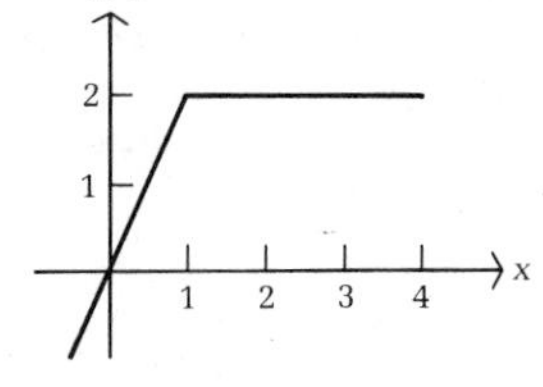

27. No **29.** Yes **31.** No
33. (A) $C'(x) = 10 - 2x$
(B) $C'(1) = \$8$ hundred per unit increase; $C'(3) = \$4$ hundred per unit increase; $C'(4) = \$2$ hundred per unit increase
35. (A) $N'(t) = 2t - 8$
(B) $N'(1) = -6$ thousand per hour; $N'(2) = -4$ thousand per hour; $N'(3) = -2$ thousand per hour [*Note:* A negative rate indicates the population is decreasing.]

Exercise 1-5

1. 0 **3.** 0 **5.** $12x^{11}$ **7.** 1 **9.** $-7x^{-8}$ **11.** $\frac{5}{2}x^{3/2}$ **13.** $-5x^{-6}$ **15.** $8x^3$ **17.** $2x^5$ **19.** $x^4/3$ **21.** $-10x^{-6}$ **23.** $-16x^{-5}$
25. x^{-3} **27.** $-x^{-2/3}$ **29.** $4x - 3$ **31.** $15x^4 - 6x^2$ **33.** $-12x^{-5} - 4x^{-3}$ **35.** $-\frac{1}{2}x^{-2} + 2x^{-4}$ **37.** $2x^{-1/3} - \frac{5}{3}x^{-2/3}$
39. $-\frac{9}{5}x^{-8/5} + 3x^{-3/2}$ **41.** $-\frac{1}{3}x^{-4/3}$ **43.** $-6x^{-3/2} + 6x^{-3} + 1$ **45.** $f'(x) = 6 - 2x$ (B) $y = 2x + 4$; $y = -2x + 16$
47. (A) $f'(x) = 3x^2 - 6x$ (B) $y = -2$; $y = 24x - 78$
49. (A) $v = 176 - 32x$ (B) 176 feet per second; 80 feet per second; -16 feet per second (C) $x = 5.5$ seconds
51. (A) $v = 3x^2 - 18x + 15$
(B) 15 feet per second; -12 feet per second; 15 feet per second
(C) $x = 1$ second; $x = 5$ seconds
53. (A) $f'(x) = 3x^2 + 12x - 15$ (B) $x = 1$; $x = -5$ **55.** (A) $f'(x) = 12x^3 - 12x^2$ (B) $x = 0$; $x = 1$ **57.** $-20x^{-2}$
59. $2x - 3 - 10x^{-3}$
65. (A) $N'(x) = 60 - 2x$
(B) $N'(10) = 40$ (at the \$10,000 level of advertising, there would be an approximate increase of 40 units of sales per \$1,000 increase in advertising); $N'(20) = 20$ (at the \$20,000 level of advertising, there would be an approximate increase of only 20 units of sales per \$1,000 increase in advertising); the effect of advertising levels off as the amount spent increases.
67. (A) -1.37 beats per minute (B) -0.58 beat per minute **69.** (A) 25 items per hour (B) 8.33 items per hour

Exercise 1-6

1. $2x^3(2x) + (x^2 - 2)(6x^2) = 10x^4 - 12x^2$ **3.** $(x - 3)(2) + (2x - 1)(1) = 4x - 7$ **5.** $\frac{(x - 3)(1) - x(1)}{(x - 3)^2} = \frac{-3}{(x - 3)^2}$

7. $\frac{(x - 2)(2) - (2x + 3)(1)}{(x - 2)^2} = \frac{-7}{(x - 2)^2}$ **9.** $(x^2 + 1)(2) + (2x - 3)(2x) = 6x^2 - 6x + 2$

11. $\frac{(2x - 3)(2x) - (x^2 + 1)(2)}{(2x - 3)^2} = \frac{2x^2 - 6x - 2}{(2x - 3)^2}$ **13.** $2x(x^2 - 3) + 2x(x^2 + 2) = 4x^3 - 2x$

15. $\frac{(x^2 - 3)2x - (x^2 + 2)2x}{(x^2 - 3)^2} = \frac{-10x}{(x^2 - 3)^2}$ **17.** $(2x + 1)(2x - 3) + (x^2 - 3x)(2) = 6x^2 - 10x - 3$

19. $(2x - x^2)(5) + (5x + 2)(2 - 2x) = -15x^2 + 16x + 4$ **21.** $\frac{(x^2 + 2x)(5) - (5x - 3)(2x + 2)}{(x^2 + 2x)^2} = \frac{-5x^2 + 6x + 6}{(x^2 + 2x)^2}$

23. $\frac{(x^2 - 1)(2x - 3) - (x^2 - 3x + 1)(2x)}{(x^2 - 1)^2} = \frac{3x^2 - 4x + 3}{(x^2 - 1)^2}$ **25.** $f'(x) = (1 + 3x)(-2) + (5 - 2x)(3)$; $y = -11x + 29$

27. $f'(x) = \frac{(3x - 4)(1) - (x - 8)(3)}{(3x - 4)^2}$; $y = 5x - 13$ **29.** $f'(x) = (2x - 15)(2x) + (x^2 + 18)(2) = 6(x - 2)(x - 3)$; $x = 2$, $x = 3$

31. $f'(x) = \frac{(x^2 + 1)(1) - x(2x)}{(x^2 + 1)^2} = \frac{1 - x^2}{(x^2 + 1)^2}$; $x = -1$, $x = 1$ **33.** $7x^6 - 3x^2$ **35.** $-27x^{-4}$

37. $(2x^4 - 3x^3 + x)(2x - 1) + (x^2 - x + 5)(8x^3 - 9x^2 + 1)$ **39.** $\frac{(4x^2 + 5x - 1)(6x - 2) - (3x^2 - 2x + 3)(8x + 5)}{(4x^2 + 5x - 1)^2}$

41. $9x^{1/3}(3x^2) + (x^3 + 5)(3x^{-2/3})$ **43.** $\frac{(x^2 - 3)(2x^{-2/3}) - 6x^{1/3}(2x)}{(x^2 - 3)^2}$ **45.** $x^{-2/3}(3x^2 - 4x) + (x^3 - 2x^2)(-\frac{2}{3}x^{-5/3})$

47. $\frac{(x^2 + 1)[(2x^2 - 1)(2x) + (x^2 + 3)(4x)] - (2x^2 - 1)(x^2 + 3)(2x)}{(x^2 + 1)^2}$

49. (A) $S'(t) = \frac{7{,}200 - 200t^2}{(t^2 + 36)^2}$

(B) $S(2) = 10$; $S'(2) = 4$; at $t = 2$ months, monthly sales are 10,000 and increasing at 4,000 albums per month
(C) $S(8) = 16$; $S'(8) = -0.56$; at $t = 8$ months, monthly sales are 16,000 and decreasing at 560 albums per month

51. (A) $d'(x) = \frac{-50{,}000(2x + 10)}{(x^2 + 10x + 25)^2} = \frac{-100{,}000}{(x + 5)^3}$

(B) $d'(5) = -100$ radios per \$1 increase in price; $d'(10) = -30$ radios per \$1 increase in price

53. (A) $C'(t) = \frac{0.14 - 0.14t^2}{(t^2 + 1)^2}$

(B) $C'(0.5) = 0.0672$, concentration is increasing at 0.0672 unit per hour; $C'(3) = -0.0112$, concentration is decreasing at 0.0112 unit per hour

55. (A) $N'(x) = \frac{(x + 32)(100) - (100x + 200)}{(x + 32)^2} = \frac{3{,}000}{(x + 32)^2}$ (B) $N'(4) = 2.31$; $N'(68) = 0.30$

Exercise 1-7

1. $6(2x + 5)^2$ **3.** $-8(5 - 2x)^3$ **5.** $30x(3x^2 + 5)^4$ **7.** $8(x^3 - 2x^2 + 2)^7(3x^2 - 4x)$ **9.** $(2x - 5)^{-1/2}$ **11.** $-8x^3(x^4 + 1)^{-3}$
13. $24x(x^2 - 2)^3$ **15.** $-6(x^2 + 3x)^{-4}(2x + 3)$ **17.** $x(x^2 + 8)^{-1/2}$ **19.** $(3x + 4)^{-2/3}$

21. $\frac{1}{2}(x^2-4x+2)^{-1/2}(2x-4)=(x-2)/(x^2-4x+2)^{1/2}$ **23.** $(-1)(2x+4)^{-2}(2)=-2/(2x+4)^2$ **25.** $-15x^2(x^3+4)^{-6}$

27. $(-1)(4x^2-4x+1)^{-2}(8x-4)=-4/(2x-1)^3$ **29.** $-2(x^2-3x)^{-3/2}(2x-3)=\dfrac{-2(2x-3)}{(x^2-3x)^{3/2}}$

31. $f'(x)=(4-x)^3-3x(4-x)^2$; $y=-16x+48$ **33.** $f'(x)=\dfrac{(2x-5)^3-6x(2x-5)^2}{(2x-5)^6}$; $y=-17x+54$

35. $f'(x)=(2x+2)^{1/2}+x(2x+2)^{-1/2}$; $y=\frac{5}{2}x-\frac{1}{2}$

37. $f'(x)=2x(x-5)^3+3x^2(x-5)^2=5x(x-5)^2(x-2)$; $x=0$, $x=2$, $x=5$

39. $f'(x)=\dfrac{(2x+5)^2-4x(2x+5)}{(2x+5)^4}=\dfrac{5-2x}{(2x+5)^3}$; $x=\frac{5}{2}$ **41.** $f'(x)=\dfrac{x-4}{\sqrt{x^2-8x+20}}$; $x=4$

43. $18x^2(x^2+1)^2+3(x^2+1)^3=3(x^2+1)^2(7x^2+1)$ **45.** $\dfrac{2x^3 4(x^3-7)^3 3x^2-(x^3-7)^4 6x^2}{4x^6}=\dfrac{3(x^3-7)^3(3x^3+7)}{2x^4}$

47. $(2x-3)^2[3(2x^2+1)^2(4x)]+(2x^2+1)^3[2(2x-3)(2)]=4(2x^2+1)^2(2x-3)(8x^2-9x+1)$

49. $4x^2[\frac{1}{2}(x^2-1)^{-1/2}(2x)]+(x^2-1)^{1/2}(8x)=\dfrac{12x^3-8x}{\sqrt{x^2-1}}$ **51.** $\dfrac{(x-3)^{1/2}(2)-2x[\frac{1}{2}(x-3)^{-1/2}]}{x-3}=\dfrac{x-6}{(x-3)^{3/2}}$

53. $(2x-1)^{1/2}(x^2+3)(11x^2-4x+9)$

55. (A) $\overline{C}'(x)=2(2x-8)2=8x-32$
(B) $\overline{C}'(2)=-16$; $\overline{C}'(4)=0$; $\overline{C}'(6)=16$. An increase in production at the 2,000 level will reduce costs; at the 4,000 level, no increase or decrease will occur; and at the 6,000 level, an increase in production will increase the costs.

57. $\dfrac{(4\times 10^6)x}{(x^2-1)^{5/3}}$

59. (A) $f'(n)=n(n-2)^{-1/2}+2(n-2)^{1/2}=\dfrac{3n-4}{(n-2)^{1/2}}$
(B) $f'(11)=\frac{29}{3}$ (rate of learning is $\frac{29}{3}$ units per minute at the $n=11$ level); $f'(27)=\frac{77}{5}$ (rate of learning is $\frac{77}{5}$ units per minute at the $n=27$ level)

Exercise 1-8

1. (A) \$29.50 (B) \$30

3. (A) \$420
(B) $\overline{C}'(500)=-0.24$; at a production level of 500 units, a unit increase in production will decrease average cost by approximately 24¢.

5. (A) $R'(1{,}600)=20$; at a production level of 1,600 units, a unit increase in production will increase revenue by approximately \$20.
(B) $R'(2{,}500)=-25$; at a production level of 2,500 units, a unit increase in production will decrease revenue by approximately \$25.

7. (A) \$4.50 (B) \$5

9. (A) $P'(450)=0.5$; at a production level of 450 units, a unit increase in production will increase profit by approximately 50¢.
(B) $P'(750)=-2.5$; at a production level of 750 units, a unit increase in production will decrease profit by approximately \$2.50.

11. (A) \$1.25
(B) $\overline{P}'(150)=0.015$; at a production level of 150 units, a unit increase in production will increase average profit by approximately 1.5¢.

13. (A) $C'(x)=60$ (B) $R(x)=200x-x^2/30$ (C) $R'(x)=200-x/15$
(D) $R'(1{,}500)=100$; at a production level of 1,500 units, a unit increase in production will increase revenue by approximately \$100. $R'(4{,}500)=-100$; at a production level of 4,500 units, a unit increase in production will decrease revenue by approximately \$100.

(E) Break-even points: (600, 108,000) and (3,600, 288,000)

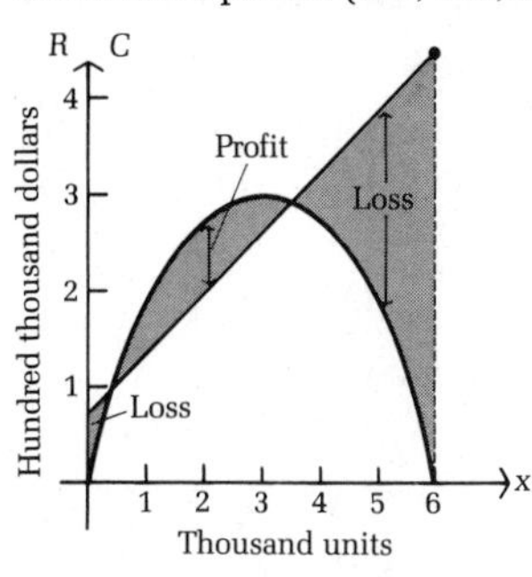

(F) $P(x) = -x^2/30 + 140x - 72{,}000$
(G) $P'(x) = -x/15 + 140$
(H) $P'(1{,}500) = 40$; at a production level of 1,500 units, a unit increase in production will increase profit by approximately \$40. $P'(3{,}000) = -60$; at a production level of 3,000 units, a unit increase in production will decrease profit by approximately \$60.

15. (A) $p = 20 - x/50$ (B) $R(x) = 20x - x^2/50$ (C) $C(x) = 4x + 1{,}400$
(D) Break-even points: (100, 1,800) and (700, 4,200)

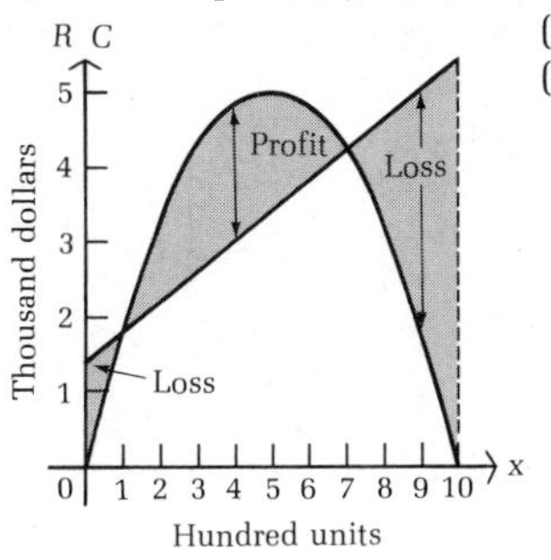

(E) $P(x) = 16x - x^2/50 - 1{,}400$
(F) $P'(250) = 6$; at a production level of 250 units, a unit increase in production will increase profit by approximately \$6. $P'(475) = -3$; at a production level of 475 units, a unit increase in production will decrease profit by approximately \$3.

Exercise 1-9 Chapter Review

1. $12x^3 - 4x$ **2.** $x^{-1/2} - 3 = \frac{1}{x^{1/2}} - 3$ **3.** 0 **4.** $-x^{-3} + x$ **5.** $(2x-1)(3) + (3x+2)(2) = 12x + 1$

6. $(x^2-1)(3x^2) + (x^3-3)(2x) = 5x^4 - 3x^2 - 6x$ **7.** $\frac{(x^2+2)2 - 2x(2x)}{(x^2+2)^2} = \frac{4-2x^2}{(x^2+2)^2}$ **8.** $(-1)(3x+2)^{-2}3 = \frac{-3}{(3x+2)^2}$

9. $3(2x-3)^2 2 = 6(2x-3)^2$ **10.** $-2(x^2+2)^{-3}2x = \frac{-4x}{(x^2+2)^3}$ **11.** $12x^3 + 6x^{-4}$

12. $(2x^2-3x+2)(2x+2) + (x^2+2x-1)(4x-3) = 8x^3 + 3x^2 - 12x + 7$ **13.** $\frac{(x-1)^2 2 - (2x-3)2(x-1)}{(x-1)^4} = \frac{4-2x}{(x-1)^3}$

14. $x^{-1/2} - 2x^{-3/2} = \frac{1}{\sqrt{x}} - \frac{2}{\sqrt{x^3}}$ **15.** $(x^2-1)[2(2x+1)2] + (2x+1)^2(2x) = 2(2x+1)(4x^2+x-2)$

16. $\frac{1}{3}(x^3-5)^{-2/3}3x^2 = \frac{x^2}{\sqrt[3]{(x^3-5)^2}}$ **17.** $-8x^{-3}$ **18.** $\frac{(2x-3)4(x^2+2)^3 2x - (x^2+2)^4 2}{(2x-3)^2} = \frac{2(x^2+2)^3(7x^2-12x-2)}{(2x-3)^2}$

19. (A) $m = f'(1) = 2$ (B) $y = 2x + 3$ **20.** (A) $m = f'(1) = 16$ (B) $y = 16x - 12$ **21.** $x = 5$ **22.** $x = -5, x = 3$
23. $x = -2, x = 2$ **24.** $x = 0, x = 3, x = \frac{15}{2}$ **25.** (A) $v = f'(x) = 32x - 4$ (B) $f'(3) = 92$ feet per second
26. (A) $v = f'(x) = 96 - 32x$ (B) $x = 3$ seconds **27.** $-3 < x < 4$ **28.** $x \leq -2$ or $x \geq 4$ **29.** $-\infty < x < \infty$
30. $x < -3$ or $0 < x \leq 5$ **31.** $0 < x < 2$ or $2 < x < 5$ **32.** $x < -2$ or $x > 3$ **33.** (A) Does not exist (B) 6 (C) No
34. (A) 3 (B) 3 (C) Yes **35.** $(-\infty, \infty)$ **36.** $(-\infty, -5), (-5, \infty)$ **37.** $(-\infty, -2), (-2, 3), (3, \infty)$ **38.** $(3, \infty)$ **39.** $(-\infty, \infty)$

40. $(-4, 5)$ **41.** $\frac{2(3)-3}{3+5} = \frac{3}{8}$ **42.** $2(3^2) - 3 + 1 = 16$ **43.** -1 **44.** 4 **45.** $\frac{1}{6}$ **46.** Does not exist **47.** $\frac{1}{2\sqrt{7}}$ **48.** $\sqrt{2}$

49. 3 **50.** $\frac{2}{3}$ **51.** 0 **52.** Does not exist **53.** $2x - 1$ **54.** $\frac{1}{2\sqrt{x}}$ **55.** No **56.** No **57.** No **58.** Yes

59. Discontinuous at $x = 0$ **60.** Continuous for all x

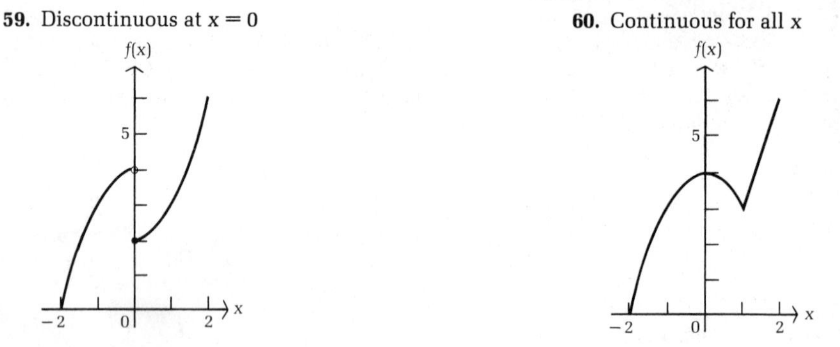

61. (A) $\frac{5}{8}$ (B) Does not exist (C) No **62.** (A) $\frac{1}{2}$ (B) $\frac{1}{2}$ (C) Yes **63.** $x = -\frac{2}{3}, 2$ **64.** $7x(x-4)^3(x+3)^2$ **65.** $\frac{x^4(2x+5)}{(2x+1)^5}$

66. $\frac{1}{x^2\sqrt{x^2-1}}$ **67.** $\frac{4}{(x^2+4)^{3/2}}$ **68.** (A) $C = 48 - 2p$ (B) $R = 14p - p^2$ (C) $\$4 < p < \$12$

69. (A) $C'(x) = 2$; $\overline{C}(x) = 2 + 56x^{-1}$; $\overline{C}'(x) = -56x^{-2}$
(B) $R(x) = xp = 20x - x^2$; $R'(x) = 20 - 2x$; $\overline{R}(x) = 20 - x$; $\overline{R}'(x) = -1$
(C) $P(x) = R(x) - C(x) = 18x - x^2 - 56$; $P'(x) = 18 - 2x$; $\overline{P}(x) = 18 - x - 56x^{-1}$; $\overline{P}'(x) = -1 + 56x^{-2}$
(D) Solving $R(x) = C(x)$, we find break-even points at $x = 4, 14$.
(E) $P'(7) = 4$ (increasing production increases profit); $P'(9) = 0$ (stable); $P'(11) = -4$ (increasing production decreases profit)
(F)

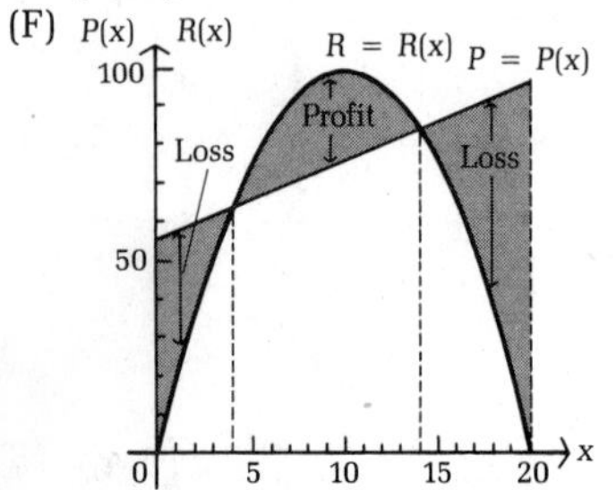

70. (A) 2 components per day (B) 3.2 components per day (C) 40 components per day
71. $C'(9) = -1$ part per million per meter; $C'(99) = -0.001$ part per million per meter
72. (A) 10 items per hour (B) 5 items per hour

Chapter 2

Exercise 2-1

1. (a, b); (d, f); (g, h) **3.** c, d, f **5.** b, f **7.** Decreasing on $(-\infty, 8)$; increasing on $(8, \infty)$; local minimum at $x = 8$
9. Increasing on $(-\infty, 5)$; decreasing on $(5, \infty)$; local maximum at $x = 5$ **11.** Increasing for all x; no local extrema
13. Decreasing for all x; no local extrema
15. Increasing on $(-\infty, -2)$ and $(2, \infty)$; decreasing on $(-2, 2)$; local maximum at $x = -2$; local minimum at $x = 2$

17. Increasing on $(-\infty, -2)$ and $(4, \infty)$; decreasing on $(-2, 4)$; local maximum at $x = -2$; local minimum at $x = 4$
19. Increasing on $(-\infty, -1)$ and $(0, 1)$; decreasing on $(-1, 0)$ and $(1, \infty)$; local maxima at $x = -1$ and $x = 1$; local minimum at $x = 0$

21. Increasing on $(-\infty, 4)$
Decreasing on $(4, \infty)$
Horizontal tangent at $x = 4$

23. Increasing on $(-\infty, -1)$, $(1, \infty)$
Decreasing on $(-1, 1)$
Horizontal tangents at $x = -1, 1$

25. Decreasing for all x
Horizontal tangent at $x = 2$

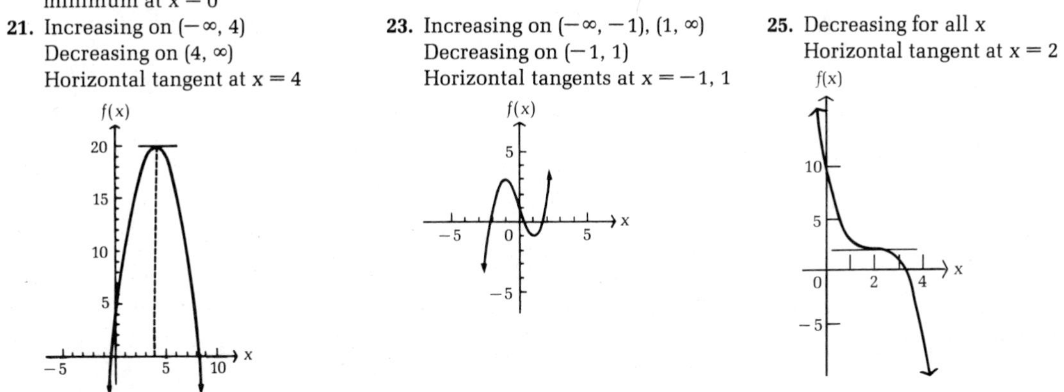

27. No critical values; increasing on $(-\infty, -2)$ and $(-2, \infty)$; no local extrema
29. Critical values: $x = -2$, $x = 2$; increasing on $(-\infty, -2)$ and $(2, \infty)$; decreasing on $(-2, 0)$ and $(0, 2)$; local maximum at $x = -2$; local minimum at $x = 2$
31. Critical value: $x = -2$; increasing on $(-2, 0)$; decreasing on $(-\infty, -2)$ and $(0, \infty)$; local minimum at $x = -2$
33. Critical values: $x = 0$, $x = 4$; increasing on $(-\infty, 0)$ and $(4, \infty)$; decreasing on $(0, 2)$ and $(2, 4)$; local maximum at $x = 0$; local minimum at $x = 4$
35. Critical values: $x = 0$, $x = 4$, $x = 6$; increasing on $(0, 4)$ and $(6, \infty)$; decreasing on $(-\infty, 0)$ and $(4, 6)$; local maximum at $x = 4$; local minima at $x = 0$ and $x = 6$
37. Critical value: $x = 2$; increasing on $(2, \infty)$; decreasing on $(-\infty, 2)$; local minimum at $x = 2$
39. Critical value: $x = 1$; increasing on $(0, 1)$; decreasing on $(1, \infty)$; local maximum at $x = 1$
41. Critical value: $x = 80$; decreasing for $0 < x < 80$; increasing for $80 < x < 150$; local minimum at $x = 80$
43. $P(x)$ is increasing over (a, b) if $P'(x) = R'(x) - C'(x) > 0$ over (a, b); that is, if $R'(x) > C'(x)$ over (a, b)
45. Critical value: $t = 1$; increasing for $0 < t < 1$; decreasing for $1 < t < 24$; local maximum at $t = 1$
47. Critical value: $t = 7$; increasing for $0 < t < 7$; decreasing for $7 < t < 24$; local maximum at $t = 7$

Exercise 2-2

1. (a, c), (c, d), (e, g) **3.** d, e, g **5.** $6x - 4$ **7.** $40x^3$ **9.** $6x$ **11.** $24x^2(x^2 - 1) + 6(x^2 - 1)^2 = 6(x^2 - 1)(5x^2 - 1)$
13. $6x^{-3} + 12x^{-4}$ **15.** $f(2) = -2$ is a local minimum **17.** $f(-1) = 2$ is a local maximum; $f(2) = -25$ is a local minimum
19. No local extrema **21.** $f(-2) = -6$ is a local minimum; $f(0) = 10$ is a local maximum; $f(2) = -6$ is a local minimum
23. $f(0) = 2$ is a local minimum **25.** $f(-4) = -8$ is a local maximum; $f(4) = 8$ is a local minimum
27. Concave upward for all x; no inflection points
29. Concave upward on $(6, \infty)$; concave downward on $(-\infty, 6)$; inflection point at $x = 6$
31. Concave upward on $(-\infty, -2)$ and $(2, \infty)$; concave downward on $(-2, 2)$; inflection points at $x = -2$ and $x = 2$
33. Concave upward on $(0, 2)$; concave downward on $(-\infty, 0)$ and $(2, \infty)$; inflection points at $x = 0$ and $x = 2$

35. Local maximum at $x = 0$
Local minimum at $x = 4$
Inflection point at $x = 2$

37. Inflection point at $x = 0$

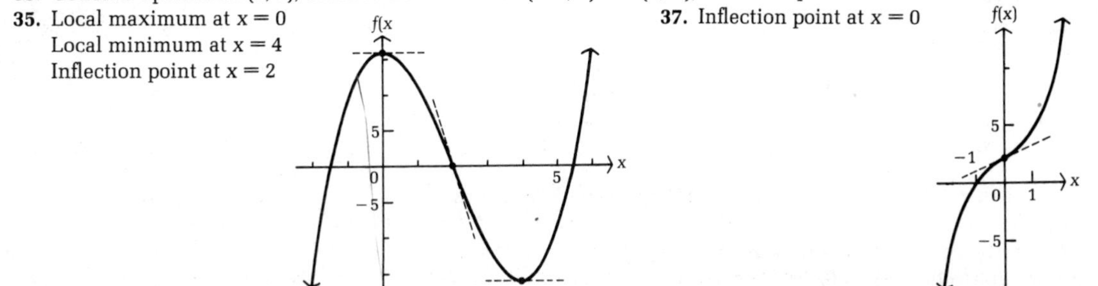

39. Inflection point at $x = 0$

41. Local maximum at $x = -2$
Local minimum at $x = 2$
Inflection point at $x = 0$

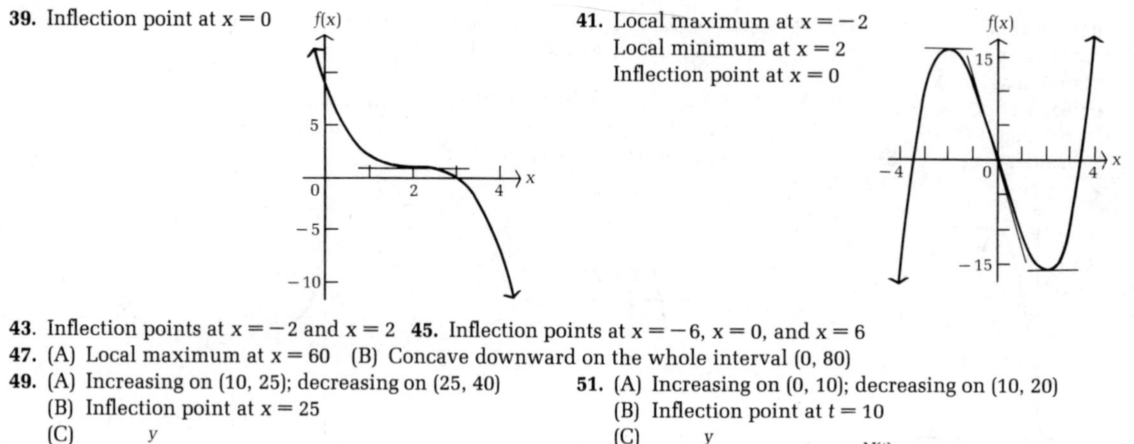

43. Inflection points at $x = -2$ and $x = 2$ **45.** Inflection points at $x = -6$, $x = 0$, and $x = 6$

47. (A) Local maximum at $x = 60$ (B) Concave downward on the whole interval $(0, 80)$

49. (A) Increasing on $(10, 25)$; decreasing on $(25, 40)$
(B) Inflection point at $x = 25$
(C)

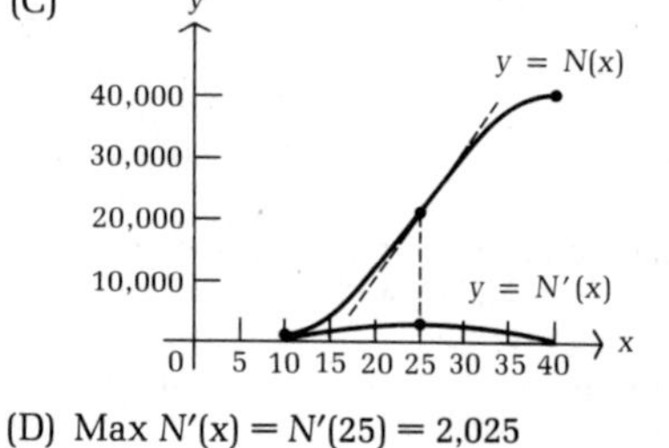

(D) Max $N'(x) = N'(25) = 2{,}025$

51. (A) Increasing on $(0, 10)$; decreasing on $(10, 20)$
(B) Inflection point at $t = 10$
(C)

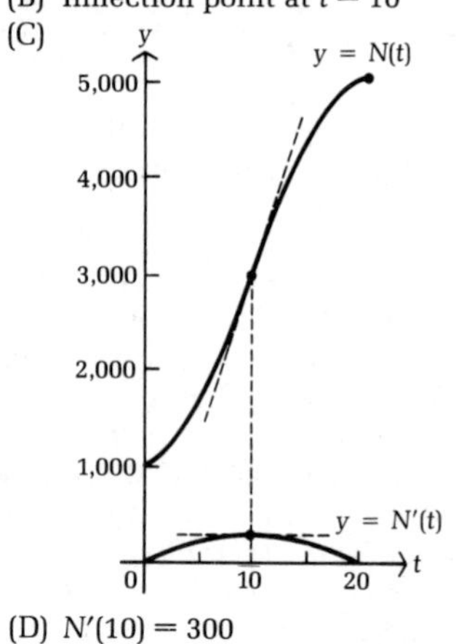

(D) $N'(10) = 300$

53. (A) Increasing on $(5, \infty)$; decreasing on $(0, 5)$
(B) Inflection point at $n = 5$

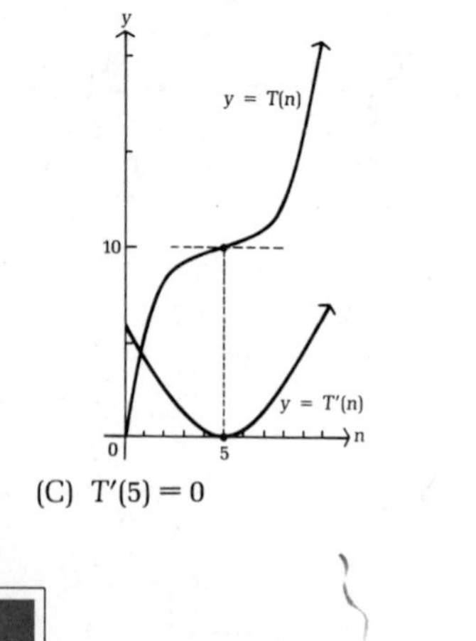

(C) $T'(5) = 0$

Exercise 2-3

1. Min $f(x) = f(2) = 1$; no maximum **3.** Max $f(x) = f(4) = 26$; no minimum **5.** No absolute extrema exist
7. Max $f(x) = f(2) = 16$ **9.** Min $f(x) = f(2) = 14$

11. (A) Max $f(x) = f(5) = 14$; min $f(x) = f(-1) = -22$ (B) Max $f(x) = f(1) = -2$; min $f(x) = f(-1) = -22$
(C) Max $f(x) = f(5) = 14$; min $f(x) = f(3) = -6$
13. (A) Max $f(x) = f(0) = 126$; min $f(x) = f(2) = -26$ (B) Max $f(x) = f(7) = 49$; min $f(x) = f(2) = -26$
(C) Max $f(x) = f(6) = 6$; min $f(x) = f(3) = -15$
15. Exactly in half **17.** 15 and -15 **19.** A square of side 25 cm; maximum area = 625 cm^2
21. 3,000 pairs; \$3.00 per pair **23.** \$35; \$6,125 **25.** 40 trees; 1,600 lb **27.** $(10 - 2\sqrt{7})/3 = 1.57$ in. squares
29. 20 ft by 40 ft (with the expensive side being one of the short sides) **31.** 10,000 books in 5 printings
33. (A) $x = 5.1$ mi (B) $x = 10$ mi **35.** 4 days; 20 bacteria per cm^3 **37.** 50 mice per order **39.** 1 month; 2 ft
41. 4 years from now

Exercise 2-4

1. (b, d), $(d, 0)$, (g, ∞) **3.** $x = 0$ **5.** (a, d), (e, h) **7.** $x = a$, $x = h$ **9.** $x = d$, $x = e$
11. Horizontal asymptote: $y = 2$; vertical asymptote: $x = -2$
13. Horizontal asymptote: $y = 1$; vertical asymptotes: $x = -1$ and $x = 1$ **15.** No horizontal or vertical asymptotes
17. Horizontal asymptote: $y = 0$; no vertical asymptotes **19.** No horizontal asymptote; vertical asymptote: $x = 3$

21. Decreasing on $(-\infty, 3)$
Increasing on $(3, \infty)$
Local minimum at $x = 3$
Concave upward on $(-\infty, \infty)$
$f(1) = 0$; $f(5) = 0$; $f(0) = 5$

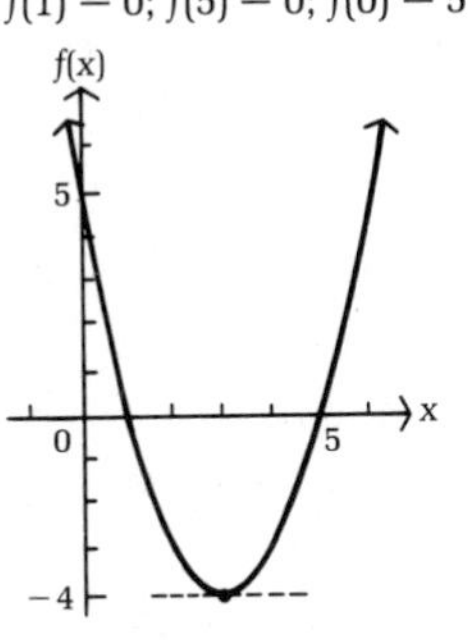

23. Increasing on $(-\infty, 0)$ and $(4, \infty)$
Decreasing on $(0, 4)$
Local maximum at $x = 0$
Local minimum at $x = 4$
Concave upward on $(2, \infty)$
Concave downward on $(-\infty, 2)$
Inflection point at $x = 2$
$f(0) = 0$; $f(6) = 0$

f(x)
10
-2
2
4
x
-30

25. Increasing on $(-\infty, -2)$ and $(2, \infty)$
Decreasing on $(-2, 2)$
Local maximum at $x = -2$
Local minimum at $x = 2$
Concave upward on $(0, \infty)$
Concave downward on $(-\infty, 0)$
Inflection point at $x = 0$
$f(0) = 16$; $f(-4) = 0$; $f(2) = 0$

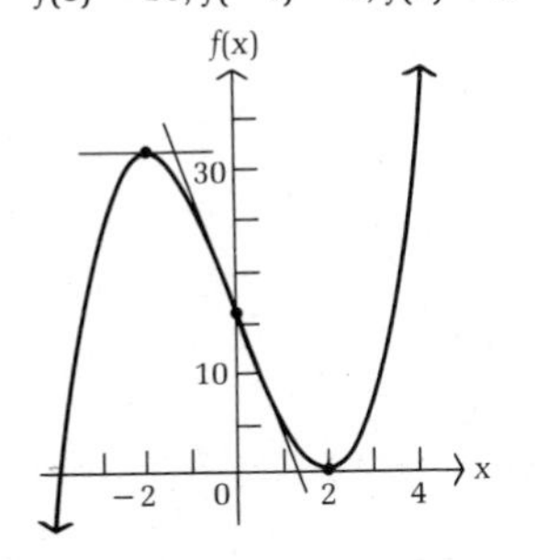

27. Increasing on $(-\infty, 3)$
Decreasing on $(3, \infty)$
Local maximum at $x = 3$
Concave upward on $(0, 2)$
Concave downward on $(-\infty, 0)$ and $(2, \infty)$
Inflection points at $x = 0$ and $x = 2$
$f(0) = 0$; $f(4) = 0$

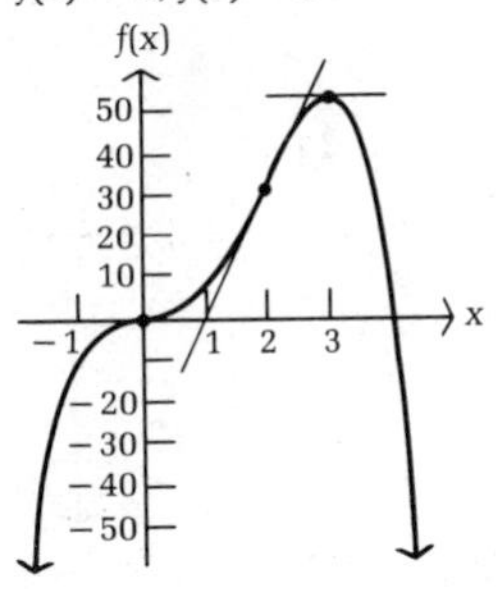

29. Decreasing on $(-\infty, 3)$ and $(3, \infty)$
Concave upward on $(3, \infty)$
Concave downward on $(-\infty, 3)$
Horizontal asymptote: $y = 1$
Vertical asymptote: $x = 3$
$f(0) = -1$; $f(-3) = 0$

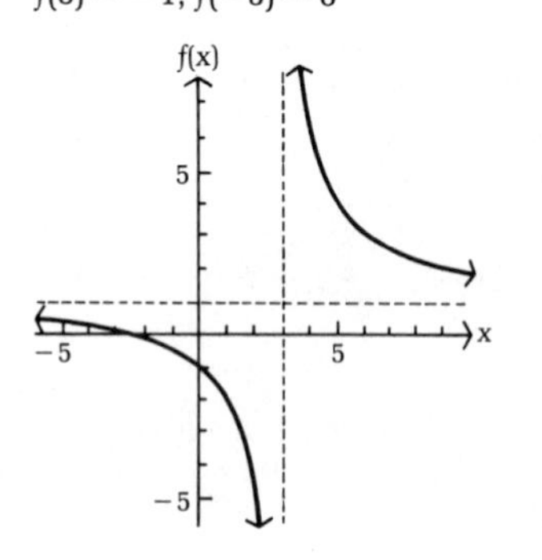

31. Decreasing on $(-\infty, 2)$ and $(2, \infty)$
Concave downward on $(-\infty, 2)$
Concave upward on $(2, \infty)$
Horizontal asymptote at $y = 1$
Vertical asymptote at $x = 2$
$f(0) = 0$

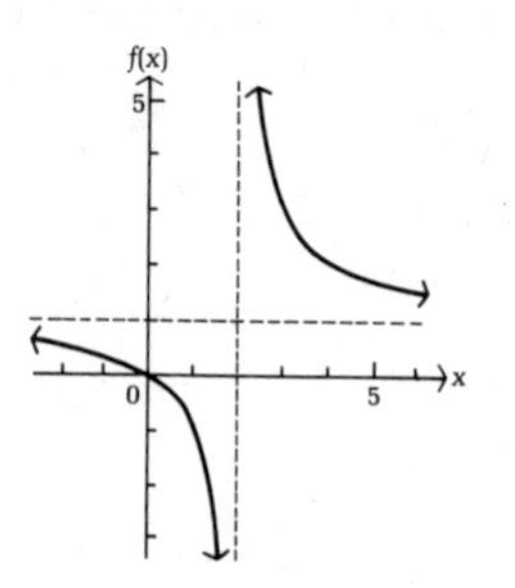

33. Increasing on $(-\infty, -1)$ and $(1, \infty)$
Decreasing on $(-1, 0)$ and $(0, 1)$
Local maximum at $x = -1$
Local minimum at $x = 1$
Concave upward on $(0, \infty)$
Concave downward on $(-\infty, 0)$
Vertical asymptote: $x = 0$

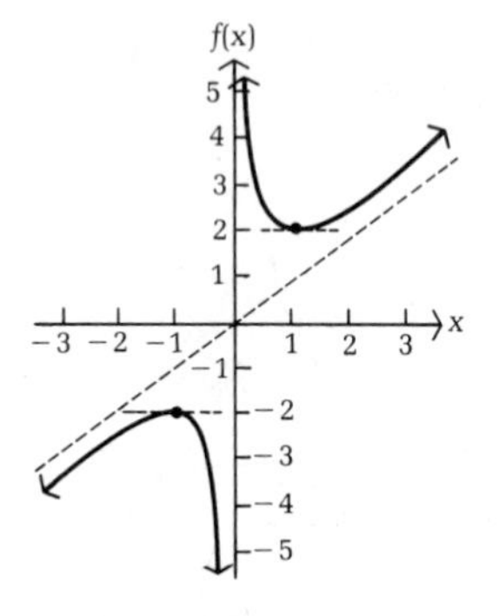

35. Increasing on $(-\infty, -\sqrt{3}/3)$ and $(\sqrt{3}/3, \infty)$
Decreasing on $(-\sqrt{3}/3, \sqrt{3}/3)$
Local maximum at $x = -\sqrt{3}/3$
Local minimum at $x = \sqrt{3}/3$
Concave downward on $(-\infty, 0)$
Concave upward on $(0, \infty)$
Inflection point at $x = 0$
$f(0) = 0$; $f(1) = 0$; $f(-1) = 0$

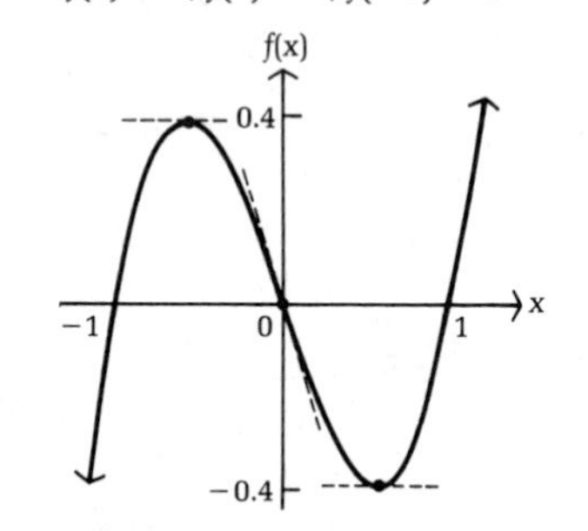

37. Increasing on $(-\infty, -\sqrt{3})$ and $(0, \sqrt{3})$
Decreasing on $(-\sqrt{3}, 0)$ and $(\sqrt{3}, \infty)$
Local maxima at $x = -\sqrt{3}$ and $x = \sqrt{3}$
Local minimum at $x = 0$
Concave upward on $(-1, 1)$
Concave downward on $(-\infty, -1)$ and $(1, \infty)$
$f(0) = 27$; $f(-3) = 0$; $f(3) = 0$

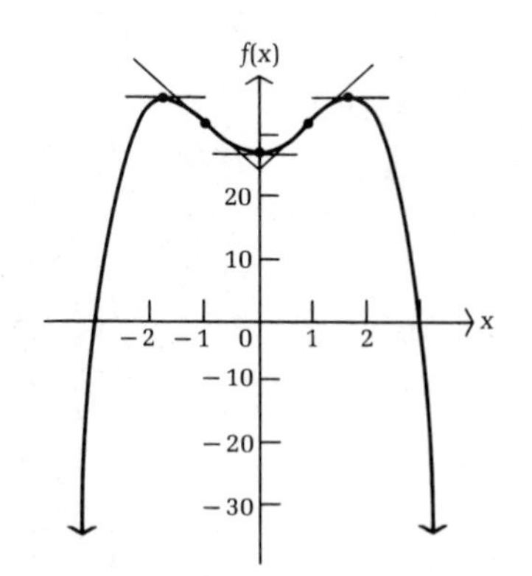

39. Decreasing on $(-\infty, -2)$ and $(0, 2)$
Increasing on $(-2, 0)$ and $(2, \infty)$
Local minima at $x = -2$ and $x = 2$
Local maximum at $x = 0$
Concave upward on $(-\infty, -2\sqrt{3}/3)$ and $(2\sqrt{3}/3, \infty)$
Concave downward on $(-2\sqrt{3}/3, 2\sqrt{3}/3)$
Inflection points at $x = -2\sqrt{3}/3$ and $x = 2\sqrt{3}/3$
$f(-2) = 0$; $f(2) = 0$; $f(0) = 16$

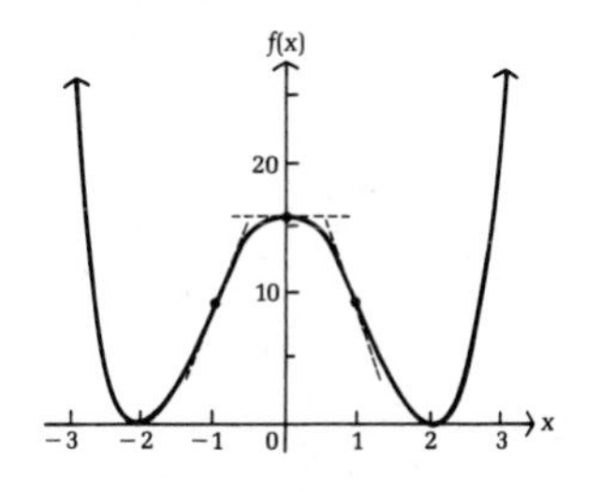

41. Decreasing on $(-\infty, 0)$ and $(0, 1.25)$
Increasing on $(1.25, \infty)$
Local minimum at $x = 1.25$
Concave upward on $(-\infty, 0)$ and $(1, \infty)$
Concave downward on $(0, 1)$
Inflection points at $x = 0$ and $x = 1$
$f(0) = 0$; $f(1.5) = 0$

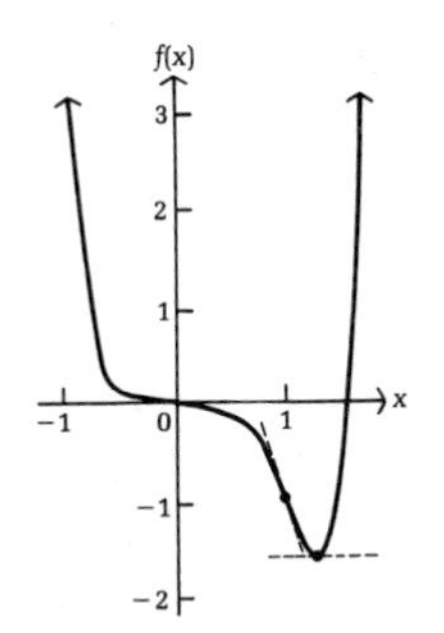

43. Decreasing on $(-\infty, -2)$, $(-2, 2)$, and $(2, \infty)$
Concave upward on $(-2, 0)$ and $(2, \infty)$
Concave downward on $(-\infty, -2)$ and $(0, 2)$
Inflection point at $x = 0$
Horizontal asymptote: $y = 0$
Vertical asymptotes: $x = -2$ and $x = 2$
$f(0) = 0$

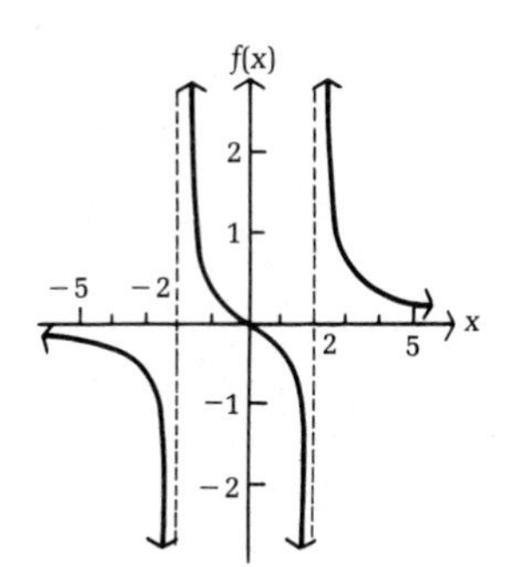

45. Increasing on $(-\infty, 0)$
Decreasing on $(0, \infty)$
Local maximum at $x = 0$
Concave upward on $(-\infty, -\sqrt{3}/3)$ and $(\sqrt{3}/3, \infty)$
Concave downward on $(-\sqrt{3}/3, \sqrt{3}/3)$
Inflection points at $x = -\sqrt{3}/3$ and $x = \sqrt{3}/3$
Horizontal asymptote: $y = 0$
$f(0) = 1$

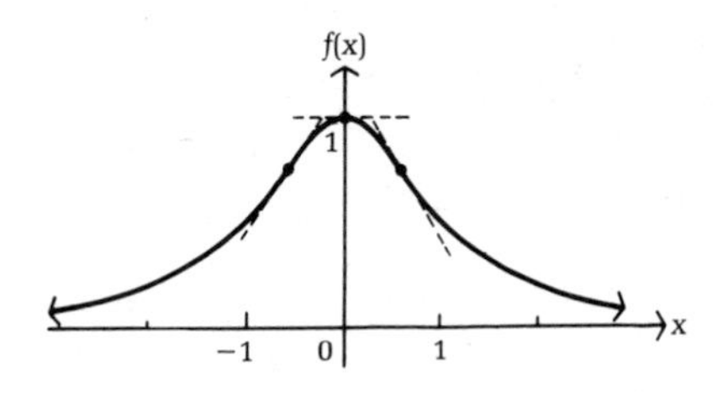

47.

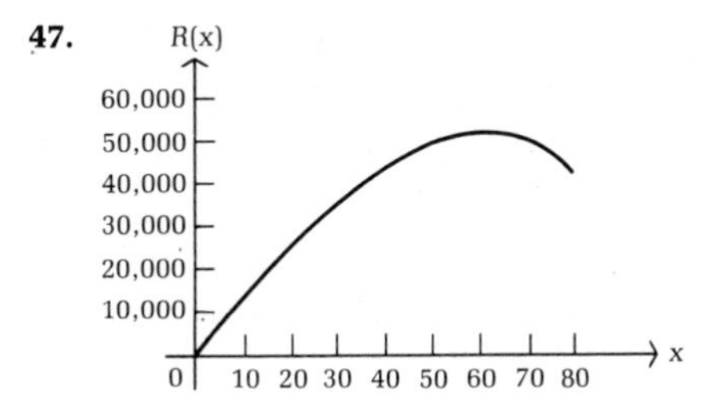

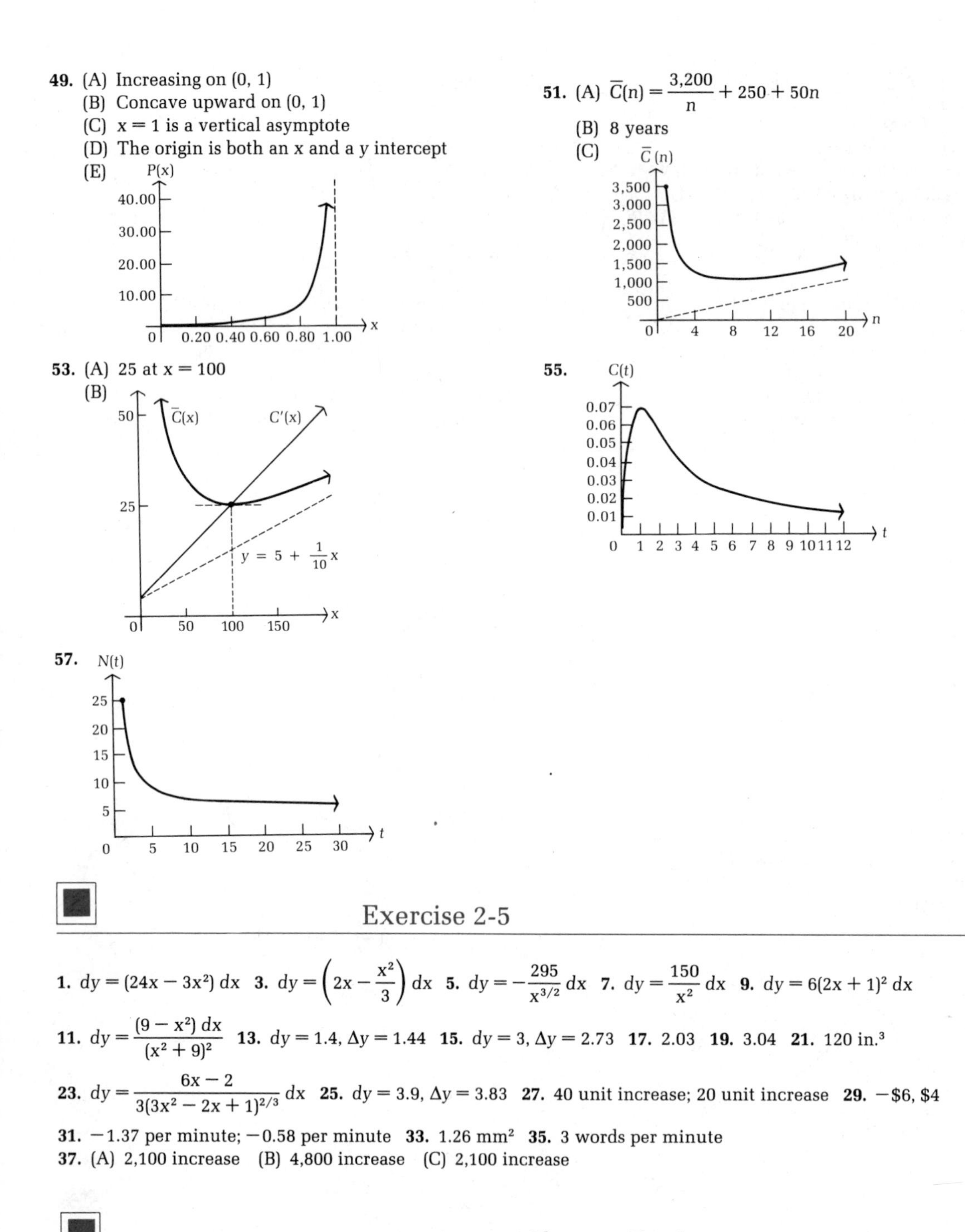

49. (A) Increasing on (0, 1)
(B) Concave upward on (0, 1)
(C) $x = 1$ is a vertical asymptote
(D) The origin is both an x and a y intercept
(E)

51. (A) $\bar{C}(n) = \dfrac{3{,}200}{n} + 250 + 50n$
(B) 8 years
(C)

53. (A) 25 at $x = 100$
(B)

55.

57.

Exercise 2-5

1. $dy = (24x - 3x^2)\,dx$ **3.** $dy = \left(2x - \dfrac{x^2}{3}\right) dx$ **5.** $dy = -\dfrac{295}{x^{3/2}}\,dx$ **7.** $dy = \dfrac{150}{x^2}\,dx$ **9.** $dy = 6(2x+1)^2\,dx$
11. $dy = \dfrac{(9 - x^2)\,dx}{(x^2+9)^2}$ **13.** $dy = 1.4, \Delta y = 1.44$ **15.** $dy = 3, \Delta y = 2.73$ **17.** 2.03 **19.** 3.04 **21.** 120 in.^{3}
23. $dy = \dfrac{6x - 2}{3(3x^2 - 2x + 1)^{2/3}}\,dx$ **25.** $dy = 3.9, \Delta y = 3.83$ **27.** 40 unit increase; 20 unit increase **29.** $-\$6$, $4
31. -1.37 per minute; -0.58 per minute **33.** 1.26 mm^2 **35.** 3 words per minute
37. (A) 2,100 increase (B) 4,800 increase (C) 2,100 increase

Exercise 2-6 Chapter Review

1. $(a, c_1), (c_3, c_5), (c_5, c_6)$ **2.** $(c_1, c_3), (c_6, b)$ **3.** $(a, c_2), (c_4, c_5), (c_7, b)$ **4.** c_3 **5.** c_6 **6.** c_1, c_3, c_5 **7.** c_6 **8.** c_2, c_4, c_5, c_7
9. $f''(x) = 12x^2 + 30x$ **10.** $y'' = 8/x^3$ **11.** $dy = (3x^2 + 4)\,dx$ **12.** $dy = 18x(3x^2 - 7)^2\,dx$ **13.** $-4, 2$

14. Increasing on $(-\infty, -4)$ and $(2, \infty)$; decreasing on $(-4, 2)$
15. Local maximum at $x = -4$; local minimum at $x = 2$
16. Concave upward on $(-1, \infty)$; concave downward on $(-\infty, -1)$
17. Inflection point at $x = -1$
18.

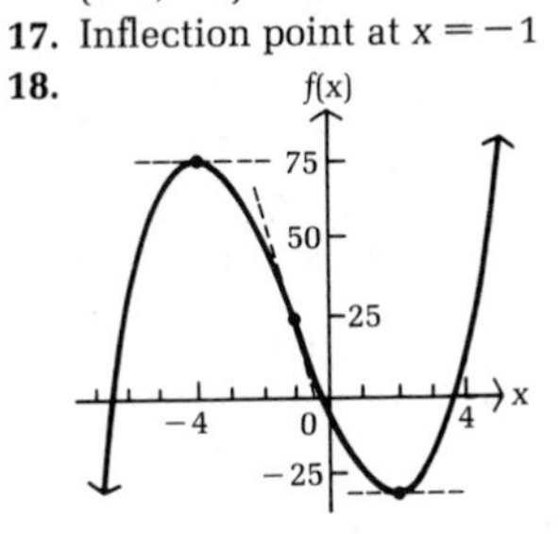

19. Horizontal asymptote at $y = 3$
20. Vertical asymptote at $x = -2$
21. Increasing on $(-\infty, -2)$ and $(-2, \infty)$
22. Concave upward on $(-\infty, -2)$; concave downward on $(-2, \infty)$
23.

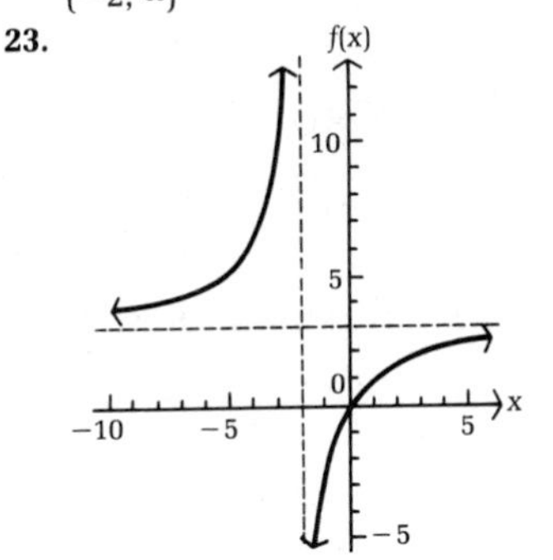

24. Min $f(x) = f(2) = -4$; max $f(x) = f(5) = 77$ **25.** Min $f(x) = f(2) = 8$
26. Horizontal asymptote: $y = 0$; no vertical asymptotes
27. No horizontal asymptotes; vertical asymptotes: $x = -3$ and $x = 3$ **28.** $dy = 7.3$, $\Delta y = 7.45$ **29.** 4.13
30. Max $f'(x) = f'(2) = 12$

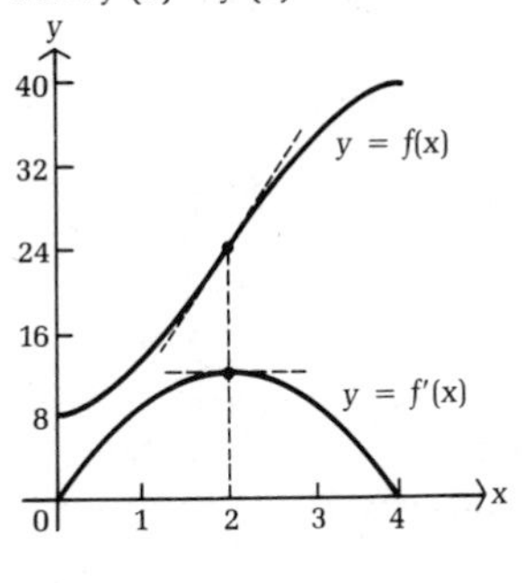

31. Each number is 20; minimum sum is 40
32. Increasing on $(-2, \infty)$
Decreasing on $(-\infty, -2)$
Local minimum at $x = -2$
Concave upward on $(-\infty, -1)$ and $(1, \infty)$
Concave downward on $(-1, 1)$
Inflection points at $x = -1$ and $x = 1$
$f(0) = -3$; $f(-3) = 0$; $f(1) = 0$

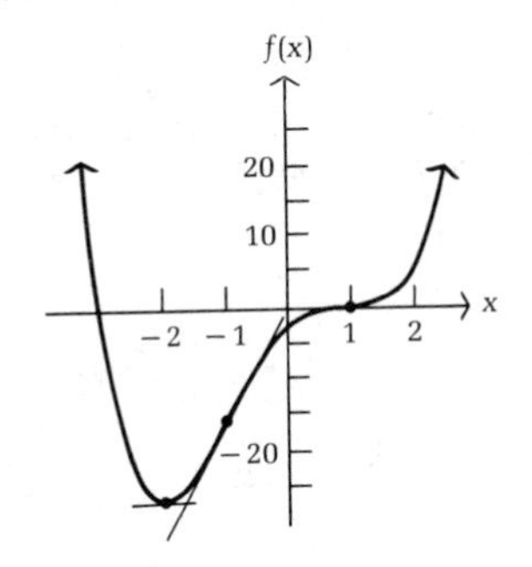

33. $dy = -0.0031$, $\Delta y = -0.0031$
34. Max $P(x) = P(3{,}000) = \$175{,}000$
35. Min $\overline{C}(x) = \overline{C}(200) = 50$

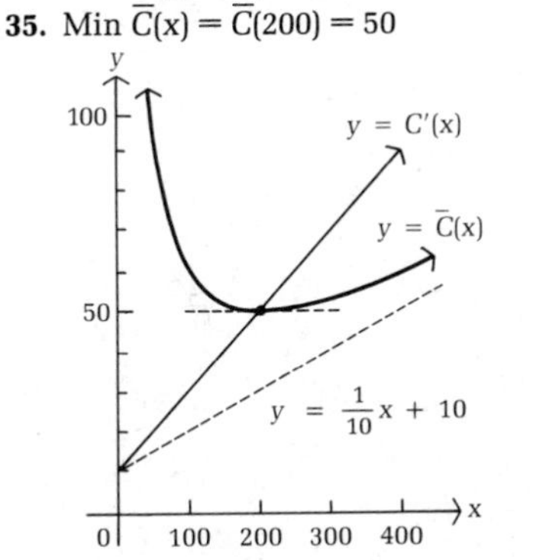

36. \$49; \$6,724
37. 12 orders per year
38. \$110
39. 3 days
40. In 2 years

Chapter 3

Exercise 3-1

1. \$1,221.40; \$1,648.72; \$2,225.54 **3.** 11.55 **5.** 10.99 **7.** 0.14 **9.**

n	$(1 + 1/n)^n$
10	2.593 74
100	2.704 81
1,000	2.716 92
10,000	2.718 15
100,000	2.718 27
1,000,000	2.718 28
10,000,000	2.718 28
↓	↓
∞	e = 2.718 281 8 . . .

11. \$55,463.90

13. \$9,931.71 **15.** $r = \frac{1}{4} \ln 1.5 \approx 0.1014$ or 10.14%

17. (A) P

10,000
8,000
6,000
4,000
2,000
0 10 20 30 40 50 t

(B) $\lim_{x\to\infty} 10{,}000e^{-0.08t} = 0$ **19.** 2.77 years **21.** 13.86%

23. $A = Pe^{rt}$; $2P = Pe^{rt}$; $e^{rt} = 2$; $\ln e^{rt} = \ln 2$; $rt = \ln 2$; $t = (\ln 2)/r$ **25.** 34.66 years **27.** 3.47%
29. $t = -(\ln 0.5)/0.000\ 433\ 2 \approx 1{,}600$ years **31.** $r = (\ln 0.5)/30 \approx -0.0231$ **33.** Approximately 538 years

Exercise 3-2

1. $6e^x - \frac{7}{x}$ **3.** $2ex^{e-1} + 3e^x$ **5.** $\frac{5}{x}$ **7.** $\frac{2 \ln x}{x}$ **9.** $x^3 + 4x^3 \ln x = x^3(1 + 4 \ln x)$ **11.** $x^3e^x + 3x^2e^x = x^2e^x(x + 3)$

13. $\frac{(x^2+9)e^x - 2xe^x}{(x^2+9)^2} = \frac{e^x(x^2 - 2x + 9)}{(x^2+9)^2}$ **15.** $\frac{x^3 - 4x^3 \ln x}{x^8} = \frac{1 - 4 \ln x}{x^5}$

17. $3(x+2)^2 \ln x + \frac{(x+2)^3}{x} = (x+2)^2\left(3 \ln x + \frac{x+2}{x}\right)$ **19.** $(x+1)^3e^x + 3(x+1)^2e^x = (x+1)^2e^x(x+4)$

21. $\frac{2xe^x - (x^2+1)e^x}{(e^x)^2} = \frac{2x - x^2 - 1}{e^x}$ **23.** $(\ln x)^3 + 3(\ln x)^2 = (\ln x)^2(\ln x + 3)$ **25.** $-15e^x(4 - 5e^x)^2$ **27.** $\frac{1}{2x\sqrt{1 + \ln x}}$

29. xe^x **31.** $4x \ln x$ **33.** $y = ex$ **35.** $y = \frac{1}{e}x$ **37.** Max $f(x) = f(e^3) = e^3 \approx 20.086$ **39.** Min $f(x) = f(1) = e \approx 2.718$

41. Max $f(x) = f(e^{1/2}) = 2e^{-1/2} \approx 1.213$

43. Decreasing on $(-\infty, \infty)$
Concave downward on $(-\infty, \infty)$
Horizontal asymptote: $y = 1$
$f(0) = 0$

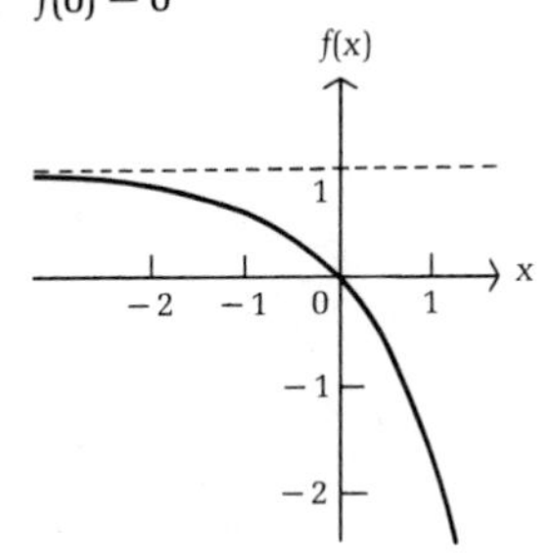

45. Increasing on $(1, \infty)$
Decreasing on $(0, 1)$
Local minimum at $x = 1$
Concave upward on $(0, \infty)$
Vertical asymptote: $x = 0$

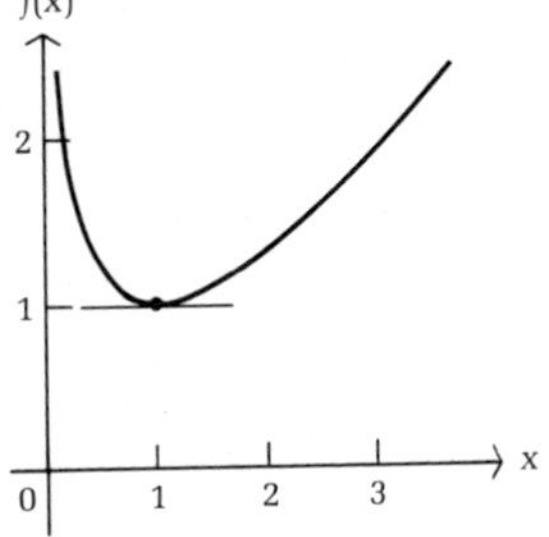

47. Increasing on $(-\infty, 2)$
Decreasing on $(2, \infty)$
Local maximum at $x = 2$
Concave upward on $(-\infty, 1)$
Concave downward on $(1, \infty)$
Inflection point at $x = 1$
Horizontal asymptote: $y = 0$
$f(0) = 3$; $f(3) = 0$

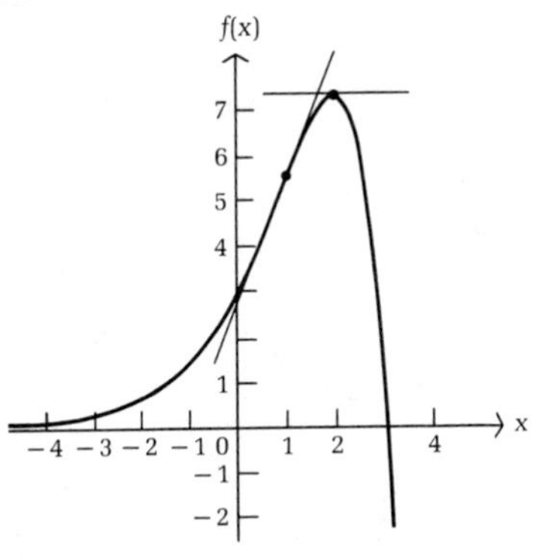

49. Increasing on $(e^{-1/2}, \infty)$
Decreasing on $(0, e^{-1/2})$
Local minimum at $x = e^{-1/2}$
Concave upward on $(e^{-3/2}, \infty)$
Concave downward on $(0, e^{-3/2})$
Inflection point at $x = e^{-3/2}$
$f(1) = 0$

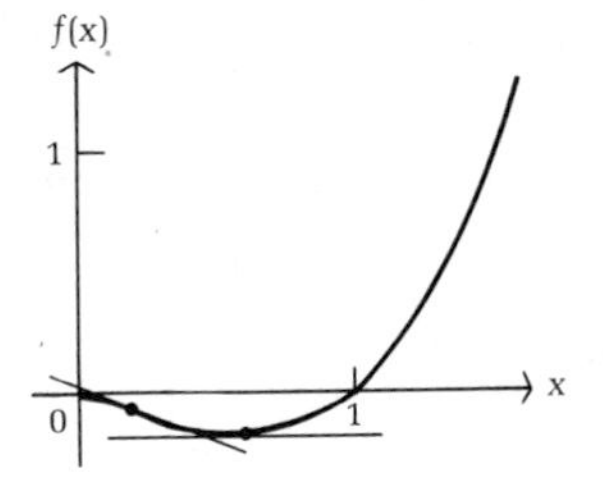

53. $p = \$2$ 55. Min $\overline{C}(x) = \overline{C}(e^7) \approx \$99.91$

57. (A) At \$3.68 each, the maximum revenue will be \$3,678.79 per week (in the test city).
(B)

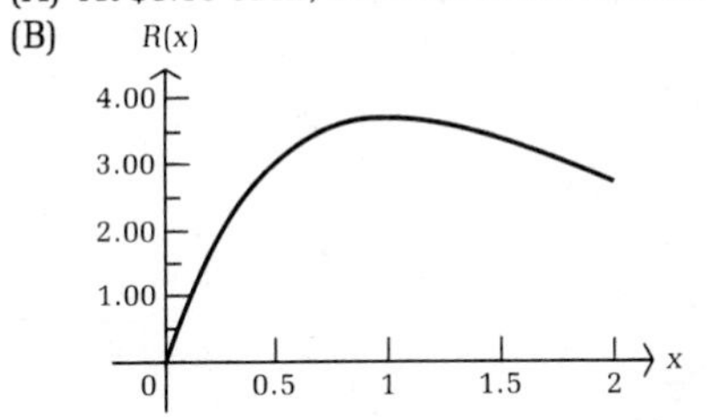

59. At the 40 lb weight level, blood pressure would increase at the rate of 0.44 in. of mercury per pound of weight gain. At the 90 lb weight level, blood pressure would increase at the rate of 0.19 in. of mercury per pound of weight gain.

61. (A) After 1 hour, the concentration is decreasing at the rate of 1.60 mg/ml per hour; after 4 hours, the concentration is decreasing at the rate of 0.08 mg/ml per hour.

(B)

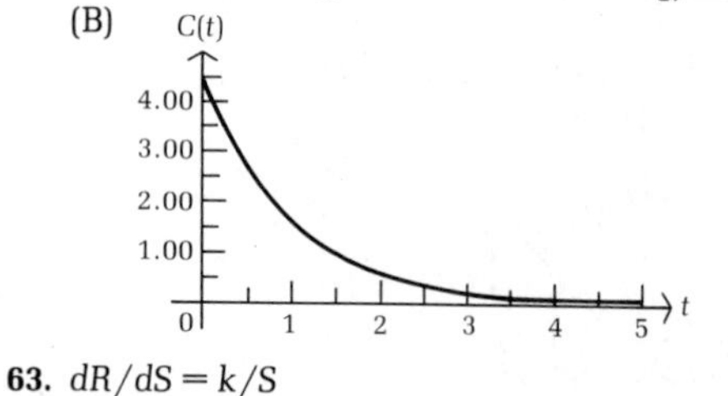

63. $dR/dS = k/S$

Exercise 3-3

1. $y = u^3$, $u = 2x + 5$ **3.** $y = \ln u$, $u = 2x^2 + 7$ **5.** $y = e^u$, $u = x^2 - 2$ **7.** $y = (2 + e^x)^2$, $dy/dx = 2e^x(2 + e^x)$

9. $y = e^{2-x^4}$, $dy/dx = -4x^3e^{2-x^4}$ **11.** $y = \ln(4x^5 - 7)$, $\dfrac{dy}{dx} = \dfrac{20x^4}{4x^5 - 7}$ **13.** $\dfrac{1}{x-3}$ **15.** $\dfrac{-2}{3-2t}$ **17.** $6e^{2x}$ **19.** $-8e^{-4t}$

21. $-3e^{-0.03x}$ **23.** $\dfrac{4}{x+1}$ **25.** $4e^{2x} - 3e^x$ **27.** $(6x - 2)e^{3x^2-2x}$ **29.** $\dfrac{2t+3}{t^2+3t}$ **31.** $\dfrac{x}{x^2+1}$

33. $\dfrac{4[\ln(t^2+1)]^3(2t)}{t^2+1} = \dfrac{8t[\ln(t^2+1)]^3}{t^2+1}$ **35.** $4(e^{2x} - 1)^3(2e^{2x}) = 8e^{2x}(e^{2x} - 1)^3$ **37.** $\dfrac{(x^2+1)(2e^{2x}) - e^{2x}(2x)}{(x^2+1)^2} = \dfrac{2e^{2x}(x^2 - x + 1)}{(x^2+1)^2}$

39. $(x^2 + 1)(-e^{-x}) + e^{-x}(2x) = e^{-x}(2x - x^2 - 1)$ **41.** $\dfrac{e^{-x}}{x} - e^{-x}\ln x$ **43.** $\dfrac{-2x}{(1+x^2)[\ln(1+x^2)]^2}$ **45.** $\dfrac{-2x}{3(1-x^2)[\ln(1-x^2)]^{2/3}}$

47. Increasing on $(-\infty, \infty)$
Concave downward on $(-\infty, \infty)$
Horizontal asymptote: $y = 1$
$f(0) = 0$

49. Decreasing on $(-\infty, 1)$
Concave downward on $(-\infty, 1)$
Vertical asymptote: $x = 1$
$f(0) = 0$

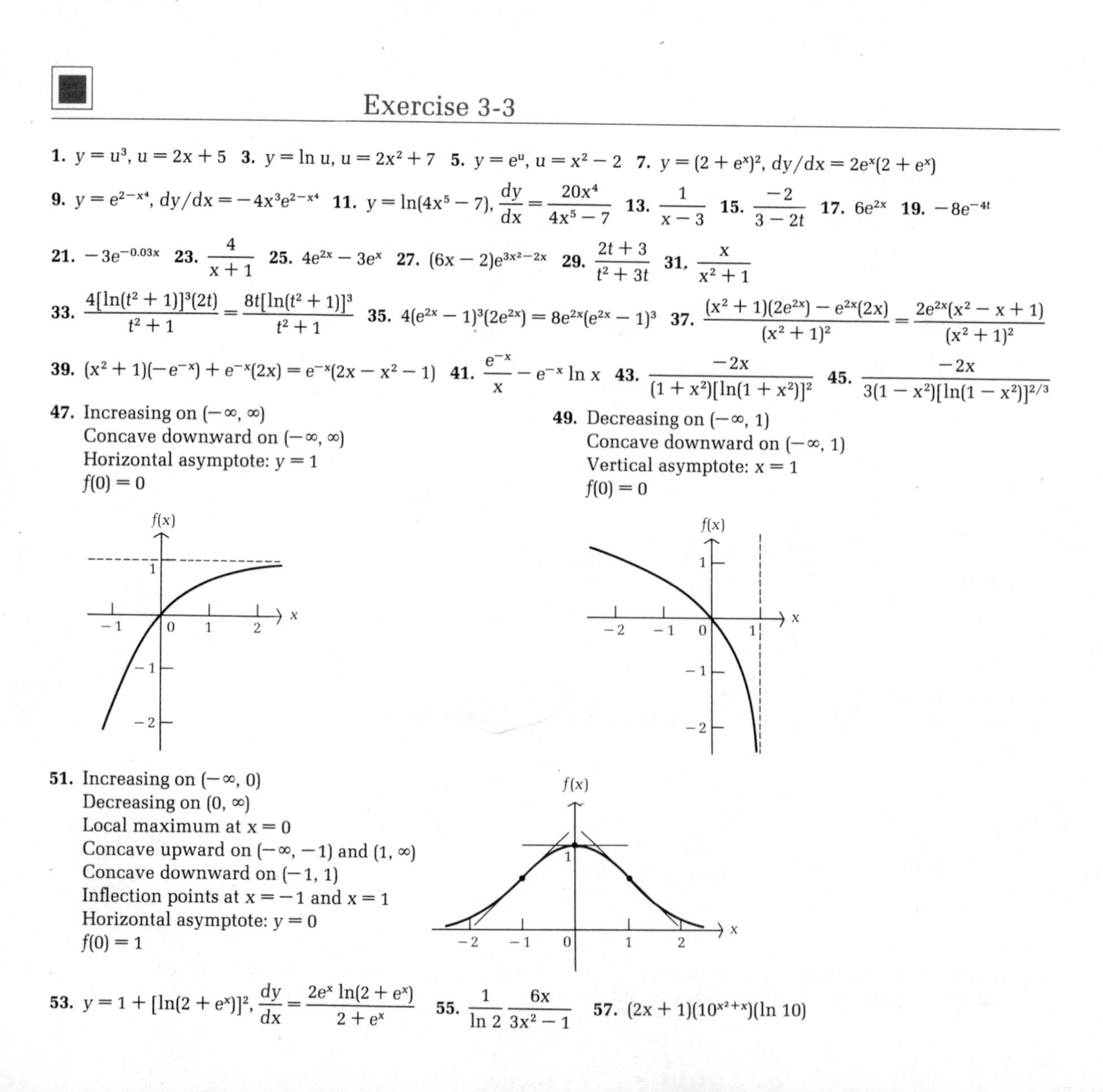

51. Increasing on $(-\infty, 0)$
Decreasing on $(0, \infty)$
Local maximum at $x = 0$
Concave upward on $(-\infty, -1)$ and $(1, \infty)$
Concave downward on $(-1, 1)$
Inflection points at $x = -1$ and $x = 1$
Horizontal asymptote: $y = 0$
$f(0) = 1$

53. $y = 1 + [\ln(2 + e^x)]^2$, $\dfrac{dy}{dx} = \dfrac{2e^x \ln(2 + e^x)}{2 + e^x}$ **55.** $\dfrac{1}{\ln 2}\dfrac{6x}{3x^2 - 1}$ **57.** $(2x + 1)(10^{x^2+x})(\ln 10)$

61. A maximum revenue of \$735.76 is realized at a production level of 20 units at \$36.79 each.
63. −\$27,145 per year; −\$18,196 per year; −\$11,036 per year **65.** (A) 23 days; \$26,685; about 50%
(B)

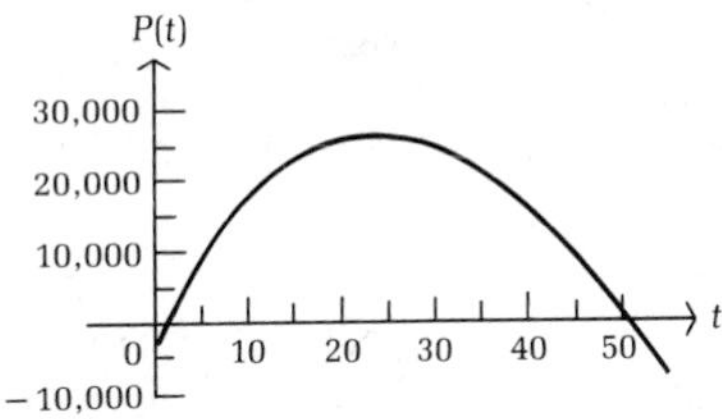

67. 2.27 mm of mercury per year; 0.81 mm of mercury per year; 0.41 mm of mercury per year
69. $A'(t) = 2(\ln 2)5{,}000e^{2t\ln 2} = 10{,}000(\ln 2)2^{2t}$; $A'(1) = 27{,}726$ bacteria per hour (rate of change at the end of the first hour); $A'(5) = 7{,}097{,}827$ bacteria per hour (rate of change at the end of the fifth hour)
71.

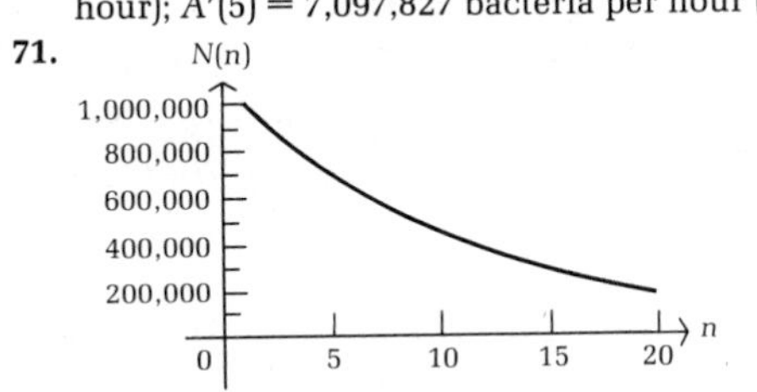

Exercise 3-4 Chapter Review

1. \$3,136.62; \$4,919.21; \$12,099.29 **2.** $\frac{2}{x} + 3e^x$ **3.** $2e^{2x-3}$ **4.** $\frac{2}{2x+7}$ **5.** (A) $\ln(3 + e^x)$ (B) $\frac{e^x}{3 + e^x}$

6. Decreasing on $(-\infty, \infty)$
Concave upward on $(-\infty, \infty)$
Horizontal asymptote: $y = 0$
$y = 100$ when $x = 0$

7. $\frac{7[(\ln z)^6 + 1]}{z}$ **8.** $x^5(1 + 6 \ln x)$

9. $\frac{e^x(x-6)}{x^7}$ **10.** $\frac{6x^2 - 3}{2x^3 - 3x}$

11. $(3x^2 - 2x)e^{x^3 - x^2}$ **12.** $\frac{e^{-2x}}{x - 2e^{-2x} \ln 5x}$

13. $y = -x + 2$; $y = -ex + 1$
14. Max $f(x) = f(e^{4.5}) = 2e^{4.5} \approx 180.03$
15. Max $f(x) = f(0.5) = 5e^{-1} \approx 1.84$

16. Increasing on $(-\infty, \infty)$
Concave downward on $(-\infty, \infty)$
Horizontal asymptote: $y = 5$
$f(0) = 0$

17. Increasing on $(e^{-1/3}, \infty)$
Decreasing on $(0, e^{-1/3})$
Local minimum at $x = e^{-1/3}$
Concave upward on $(e^{-5/6}, \infty)$
Concave downward on $(0, e^{-5/6})$
Inflection point at $x = e^{-5/6}$
$f(1) = 0$

18. (A) $[\ln(4 - e^x)]^3$ (B) $\dfrac{-3e^x[\ln(4 - e^x)]^2}{4 - e^x}$ **19.** $2x(5^{x^2-1})(\ln 5)$ **20.** $\left(\dfrac{1}{\ln 5}\right)\dfrac{2x - 1}{x^2 - x}$ **21.** $\dfrac{2x + 1}{2(x^2 + x)\sqrt{\ln(x^2 + x)}}$
22. (A) 14.2 years (B) 13.9 years **23.** $A'(t) = 10e^{0.1t}$; $A'(1) = \$11.05$ per year; $A'(10) = \$27.18$ per year
24. $R'(x) = (1{,}000 - 20x)e^{-0.02x}$
25. A maximum revenue of \$18,394 is realized at a production level of 50 units at \$367.88 each.
26.

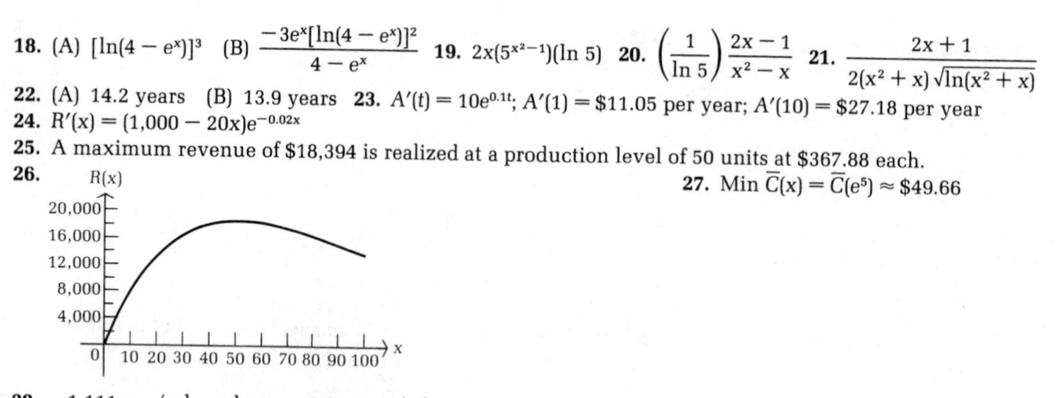

27. Min $\overline{C}(x) = \overline{C}(e^5) \approx \$49.66$

28. −1.111 mg/ml per hour; −0.335 mg/ml per hour
29. (A) Increasing at the rate of 2.68 units per day at the end of 1 day of training; increasing at the rate of 0.54 unit per day after 5 days of training
(B)

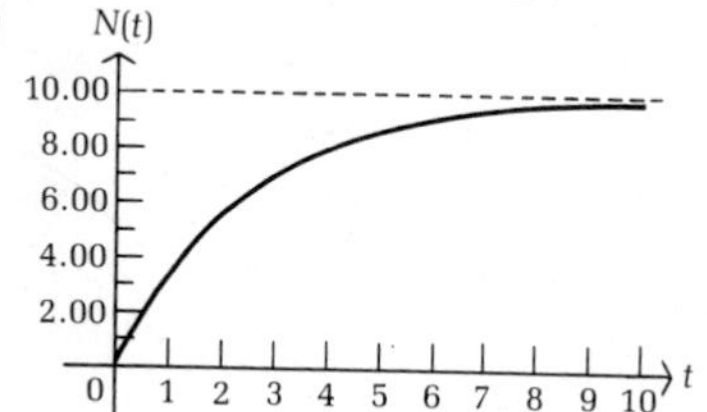

Chapter 4

Exercise 4-1

1. $7x + C$ **3.** $(x^7/7) + C$ **5.** $2t^4 + C$ **7.** $u^2 + u + C$ **9.** $x^3 + x^2 - 5x + C$ **11.** $(s^5/5) - \frac{4}{3}s^6 + C$ **13.** $3e^t + C$
15. $2\ln|z| + C$ **17.** $y = 40x^5 + C$ **19.** $P = 24x - 3x^2 + C$ **21.** $y = \frac{1}{3}u^6 - u^3 - u + C$ **23.** $y = e^x + 3x + C$
25. $x = 5\ln|t| + t + C$ **27.** $4x^{3/2} + C$ **29.** $-4x^{-2} + C$ **31.** $2\sqrt{u} + C$ **33.** $-(x^{-2}/8) + C$ **35.** $-(u^{-4}/8) + C$
37. $x^3 + 2x^{-1} + C$ **39.** $2x^5 + 2x^{-4} - 2x + C$ **41.** $2x^{3/2} + 4x^{1/2} + C$ **43.** $\frac{3}{5}x^{5/3} + 2x^{-2} + C$ **45.** $(e^x/4) - (3x^2/8) + C$
47. $-z^{-2} - z^{-1} + \ln|z| + C$ **49.** $y = x^2 - 3x + 5$ **51.** $C(x) = 2x^3 - 2x^2 + 3{,}000$ **53.** $x = 40\sqrt{t}$
55. $y = -2x^{-1} + 3\ln|x| - x + 3$ **57.** $x = 4e^t - 2t - 3$ **59.** $y = 2x^2 - 3x + 1$ **61.** $x^2 + x^{-1} + C$ **63.** $\frac{1}{2}x^2 + x^{-2} + C$
65. $e^x - 2\ln|x| + C$ **67.** $M = t + t^{-1} + \frac{3}{4}$ **69.** $y = 3x^{5/3} + 3x^{2/3} - 6$ **71.** $p(x) = 10x^{-1} + 10$
73. $P(x) = 50x - 0.02x^2$; $P(100) = \$4{,}800$ **75.** $R(x) = 100x - (x^2/10)$; $p = 100 - (x/10)$; $p = \$30$
77. $L(x) = 4{,}800x^{1/2}$; $L(25) = 24{,}000$ labor-hours **79.** $W(h) = 0.0005h^3$; $W(70) = 171.5$ lb **81.** 19,400

Exercise 4-2

1. $\frac{1}{6}(x^2 - 4)^6 + C$ **3.** $e^{4x} + C$ **5.** $\ln|2t + 3| + C$ **7.** $\frac{1}{24}(3x - 2)^8 + C$ **9.** $\frac{1}{16}(x^2 + 3)^8 + C$ **11.** $-20e^{-0.5t} + C$
13. $\frac{1}{10}\ln|10x + 7| + C$ **15.** $\frac{1}{4}e^{2x^2} + C$ **17.** $\frac{1}{3}\ln|x^3 + 4| + C$ **19.** $-\frac{1}{18}(3t^2 + 1)^{-3} + C$ **21.** $\frac{1}{3}(4 - x^3)^{-1} + C$

23. $\frac{1}{8}(1+e^{2x})^4+C$ **25.** $\frac{1}{2}\ln|4+2x+x^2|+C$ **27.** $e^{x^2+x+1}+C$ **29.** $\frac{1}{4}(e^x-2x)^4+C$ **31.** $-\frac{1}{12}(x^4+2x^2+1)^{-3}+C$
33. $\frac{1}{9}(3x^2+7)^{3/2}+C$ **35.** $\frac{1}{4}(2x^4+3)^{1/2}+C$ **37.** $\frac{1}{4}(\ln x)^4+C$ **39.** $e^{-1/x}+C$ **41.** $x=\frac{1}{3}(t^3+5)^7+C$
43. $y=3(t^2-4)^{1/2}+C$ **45.** $p=-(e^x-e^{-x})^{-1}+C$ **47.** $p(x)=2{,}000/(3x+50)$; 250 bottles
49. $C(x)=12x+500\ln(x+1)+2{,}000$; $\overline{C}(1{,}000)=\$17.45$
51. $S(t)=10t+100e^{-0.1t}-100$, $0\leq t\leq 24$; $S(12)\approx\$50$ million
53. $Q(t)=100\ln(t+1)+5t$, $0\leq t\leq 20$; $Q(9)\approx 275$ thousand barrels **55.** $W(t)=2e^{0.1t}$; $W(8)\approx 4.45$ g
57. $N(t)=5{,}000-1{,}000\ln(1+t^2)$; $N(10)\approx 385$ bacteria per milliliter
59. $N(t)=100-60e^{-0.1t}$, $0\leq t\leq 15$; $N(15)\approx 87$ words per minute
61. $E(t)=12{,}000-10{,}000(t+1)^{-1/2}$; $E(15)=9{,}500$ students

Exercise 4-3

1. $A=1{,}000e^{0.08t}$ **3.** $A=8{,}000e^{0.06t}$ **5.** $p(x)=100e^{-0.05x}$ **7.** $N=L(1-e^{-0.051t})$ **9.** $I=I_0e^{-0.00942x}$; $x\approx 74$ feet
11. $Q=3e^{-0.04t}$; $Q(10)=2.01$ milliliters **13.** 24,200 years (approximately) **15.** 104 times; 67 times
17. (A) 7 people; 353 people (B) 400

Exercise 4-4

1. 5 **3.** 5 **5.** 2 **7.** 48 **9.** $-\frac{7}{3}$ **11.** 2 **13.** $\frac{1}{2}(e^2-1)$ **15.** $2\ln 3.5$ **17.** -2 **19.** 14 **21.** $5^6=15{,}625$ **23.** $\ln 4$
25. $\frac{1}{6}[(e^2-2)^3]-1$ **27.** $-3-\ln 2$ **29.** $\frac{1}{6}(15^{3/2}-5^{3/2})$ **31.** $\frac{1}{2}(\ln 2-\ln 3)$ **33.** 0
35. $\int_0^5 500(t-12)\,dt=-\$23{,}750$; $\int_5^{10} 500(t-12)\,dt=-\$11{,}250$
37. Useful life $=\sqrt{\ln 55}\approx 2$ years; Total profit $=\frac{51}{22}-\frac{5}{2}e^{-4}\approx 2.272$ or \$2,272 **39.** 4,800 labor-hours
41. $100\ln 11+50\approx 290$ thousand barrels; $100\ln 21-100\ln 11+50\approx 115$ thousand barrels
43. $20+100e^{-1.2}\approx\$50$ million; $120+100e^{-2.4}-100e^{-1.2}\approx\$99$ million **45.** 134 billion ft^3
47. $2e^{0.8}-2\approx 2.45$ g; $2e^{1.6}-2e^{0.8}\approx 5.45$ g
49. An increase of $60-60e^{-0.5}\approx 24$ words per minute; an increase of $60e^{-0.5}-60e^{-1}\approx 14$ words per minute; an increase of $60e^{-1}-60e^{-1.5}\approx 9$ words per minute

Exercise 4-5

1. 16 **3.** 7 **5.** $\frac{7}{3}$ **7.** 9 **9.** e^2-e^{-1} **11.** $-\ln 0.5$ **13.** 15 **15.** 32 **17.** 36 **19.** 9 **21.** $\frac{5}{2}$ **23.** $2e+\ln 2-2e^{0.5}$ **25.** $\frac{23}{3}$
27. $\frac{4}{3}$ **29.** $\frac{343}{3}$ **31.** 8
33. Total production from the end of the fifth year to the end of the tenth year is $50+100\ln 20-100\ln 15\approx 79$ thousand barrels.
35. Total profit over the 5 year useful life of the game is $20-30e^{-1.5}\approx 13.306$ or \$13,306.
37. (A) 0.1825, 0.41 (B) $\frac{3}{25}$ **39.** $\frac{4.482}{5.482}\approx 0.82$ **41.** Total weight gain during the 10 hr is $3e-3\approx 5.15$ g.
43. Average number of words learned during the second 2 hr is $15\ln 4-15\ln 2\approx 10$.

Exercise 4-6 Chapter Review

1. t^3-t^2+C **2.** 12 **3.** $-3t^{-1}-3t+C$ **4.** $\frac{15}{2}$ **5.** $-2e^{-0.5x}+C$ **6.** $2\ln 5$ **7.** $y=f(x)=x^3-2x+4$ **8.** 12
9. $\frac{1}{8}(6x-5)^{4/3}+C$ **10.** 2 **11.** $-2x^{-1}-e^{x^2}+C$ **12.** $(20^{3/2}-8)/3$ **13.** $-\frac{1}{2}e^{-2x}+\ln|x|+C$ **14.** $-500(e^{-0.2}-1)\approx 90.63$
15. $y=f(x)=3\ln|x|+x^{-1}+4$ **16.** $y=3x^2+x-4$ **17.** $\frac{64}{3}$ **18.** $\frac{1}{2}\ln 10$ **19.** 0.45 **20.** $\frac{1}{48}(2x^4+5)^6+C$
21. $-\ln(e^{-x}+3)+C$ **22.** $-(e^x+2)^{-1}+C$ **23.** $\frac{1}{3}(\ln x)^3+C$ **24.** $y=3e^{x^3}-1$ **25.** $N=800e^{0.06t}$ **26.** $\frac{46}{3}$
27. $P(x)=100x-0.01x^2$; $P(10)=\$999$ **28.** $\int_0^{15}(60-4t)\,dt=450$ thousand barrels **29.** $\int_{10}^{40}\left(150-\frac{x}{10}\right)dx=\$4{,}425$
30. Useful life $=10\ln\frac{20}{3}\approx 19$ years; total profit $=143-200e^{-1.9}\approx 113.086$ or \$113,086

31. $500e^{-0.36} - 320 \approx \$29$ million **32.** $\frac{1}{6}$ **33.** 1 cm^2 **34.** 800 gal **35.** (A) $226e^{0.11} \approx 252$ million (B) $\dfrac{\ln 2}{0.011} \approx 63$ years
36. $\dfrac{-\ln 0.04}{0.000\ 123\ 8} \approx 26{,}000$ years **37.** $N(t) = 95 - 70e^{-0.1t}$; $N(15) \approx 79$ words per minute

Chapter 5

Exercise 5-1

1. (A) 120 (B) 124 **3.** (A) 123 (B) 124 **5.** (A) -4 (B) -5.33 **7.** (A) -5 (B) -5.33 **9.** 23.4 **11.** 55.3 **13.** 250
15. 2 **17.** $\frac{45}{28} \approx 1.61$ **19.** $2(1 - e^{-2}) \approx 1.73$ **21.** 0.791 **23.** 1.650 **25.** 0.748 **27.** 0.747 **29.** $[f(b) - f(a)]/(b - a)$
31. (A) $I = -200t + 600$ (B) $\frac{1}{3}\int_0^3 (-200t + 600)\,dt = 300$ **33.** \$16,000 **35.** (A) \$420 (B) \$135,000
37. \$149.18; \$122.96 **39.** $50e^{0.6} - 50e^{0.4} - 10 \approx \$6.51$
41. $(40 + 100e^{-0.7})/7 \approx 12.8$ or 12,800 hamburgers; $(140 + 100e^{-1.4} - 100e^{-0.7})/7 \approx 16.4$ or 16,400 hamburgers
43. 3,120,000 ft^2 **45.** 10°C **47.** $0.6 \ln 2 + 0.1 \approx 0.516$; $(4.2 \ln 625 + 2.4 - 4.2 \ln 49)/24 \approx 0.546$

Exercise 5-2

1. \$12,500 **3.** $8{,}000(e^{0.15} - 1) \approx \$1{,}295$ **5.** $10{,}000(1 - e^{-0.8}) \approx \$5{,}507$ **7.** $12{,}500(1 - e^{-0.48}) \approx \$4{,}765$ **9.** \$625,000
11. \$9,500 **13.** $\bar{p} = 24$; $CS = \$3{,}380$; $PS = \$1{,}690$
15. $\bar{p} \approx 49$; $CS = 55{,}990 - 80{,}000e^{-0.49} \approx \$6{,}980$; $PS = 54{,}010 - 30{,}000e^{0.49} \approx \$5{,}041$ **17.** \$2,308 **19.** \$1,310
21. Clothing store: $PV = 120{,}000(1 - e^{-0.5}) \approx \$47{,}216$; computer store: $PV = 200{,}000(1 - e^{-0.25}) \approx \$44{,}240$; the clothing store is the better choice.
23. $k(1 - e^{-rT})/r$

Exercise 5-3

1. $\frac{1}{3}xe^{3x} - \frac{1}{9}e^{3x} + C$ **3.** $\dfrac{x^3}{3}\ln x - \dfrac{x^3}{9} + C$ **5.** $\frac{1}{8}x(x - 6)^8 - \frac{1}{72}(x - 6)^9 + C$ **7.** $-xe^{-x} - e^{-x} + C$ **9.** $\frac{1}{2}e^{x^2} + C$
11. $\frac{1}{12}x(2x - 1)^6 - \frac{1}{168}(2x - 1)^7 + C$ **13.** $\frac{1}{21}(3x + 2)^7 + C$ **15.** $(xe^x - 4e^x)\Big|_0^1 = -3e + 4 \approx -4.1548$
17. $(x \ln 2x - x)\Big|_1^3 = (3 \ln 6 - 3) - (\ln 2 - 1) \approx 2.6821$ **19.** $\ln(x^2 + 1) + C$ **21.** $(\ln x)^2/2 + C$ **23.** $\frac{2}{3}x^{3/2}\ln x - \frac{4}{9}x^{3/2} + C$
25. $(x^2 - 2x + 2)e^x + C$ **27.** $\dfrac{xe^{ax}}{a} - \dfrac{e^{ax}}{a^2} + C$ **29.** $\left(-\dfrac{\ln x}{x} - \dfrac{1}{x}\right)\Big|_1^e = -\dfrac{2}{e} + 1 \approx 0.2642$
31. $\frac{1}{8}x^2(x + 1)^8 - \frac{1}{36}x(x + 1)^9 + \frac{1}{360}(x + 1)^{10} + C$ **33.** $-\frac{1}{2}x(x + 6)^{-2} - \frac{1}{2}(x + 6)^{-1} + C$ **35.** $x(\ln x)^2 - 2x \ln x + 2x + C$
37. $x(\ln x)^3 - 3x(\ln x)^2 + 6x \ln x - 6x + C$ **39.** $\frac{2}{3}x(x + 4)^{3/2} - \frac{4}{15}(x + 4)^{5/2} + C$ **41.** $(e^2 - 3)/2$ **43.** $2e^2 - 3$
45. $P(t) = t^2 + te^{-t} + e^{-t} - 1$ **47.** $31{,}250e^{-0.4} - 18{,}750 \approx \$2{,}198$ **49.** $\frac{37}{80} = 0.4625$
51. $(10 - 2 \ln 6)/3 \approx 2.1388$ parts per million **53.** 20,980

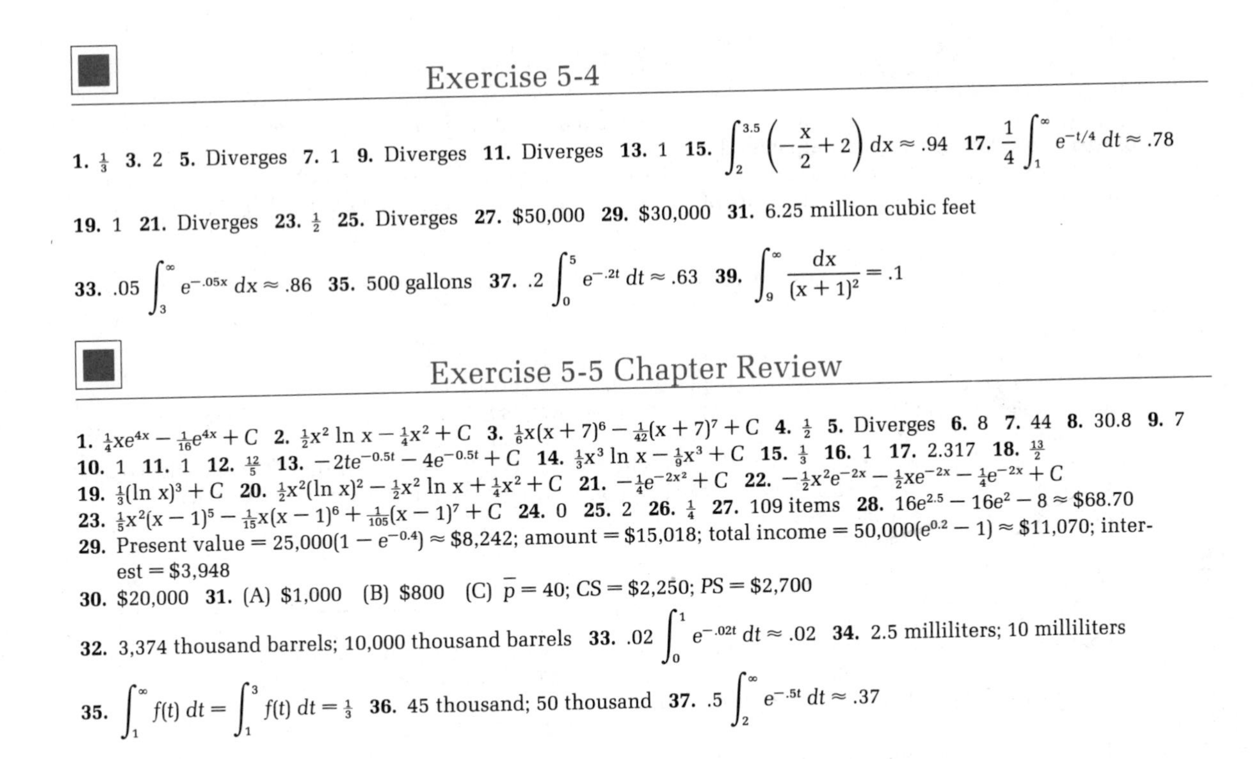

Exercise 5-4

1. $\frac{1}{3}$ **3.** 2 **5.** Diverges **7.** 1 **9.** Diverges **11.** Diverges **13.** 1 **15.** $\int_2^{3.5}\left(-\frac{x}{2}+2\right)dx \approx .94$ **17.** $\frac{1}{4}\int_1^{\infty} e^{-t/4}\,dt \approx .78$
19. 1 **21.** Diverges **23.** $\frac{1}{2}$ **25.** Diverges **27.** \$50,000 **29.** \$30,000 **31.** 6.25 million cubic feet
33. $.05\int_3^{\infty} e^{-.05x}\,dx \approx .86$ **35.** 500 gallons **37.** $.2\int_0^5 e^{-.2t}\,dt \approx .63$ **39.** $\int_9^{\infty}\frac{dx}{(x+1)^2} = .1$

Exercise 5-5 Chapter Review

1. $\frac{1}{4}xe^{4x} - \frac{1}{16}e^{4x} + C$ **2.** $\frac{1}{2}x^2 \ln x - \frac{1}{4}x^2 + C$ **3.** $\frac{1}{6}x(x+7)^6 - \frac{1}{42}(x+7)^7 + C$ **4.** $\frac{1}{2}$ **5.** Diverges **6.** 8 **7.** 44 **8.** 30.8 **9.** 7
10. 1 **11.** 1 **12.** $\frac{12}{5}$ **13.** $-2te^{-0.5t} - 4e^{-0.5t} + C$ **14.** $\frac{1}{3}x^3 \ln x - \frac{1}{9}x^3 + C$ **15.** $\frac{1}{3}$ **16.** 1 **17.** 2.317 **18.** $\frac{13}{2}$
19. $\frac{1}{3}(\ln x)^3 + C$ **20.** $\frac{1}{2}x^2(\ln x)^2 - \frac{1}{2}x^2 \ln x + \frac{1}{4}x^2 + C$ **21.** $-\frac{1}{4}e^{-2x^2} + C$ **22.** $-\frac{1}{2}x^2e^{-2x} - \frac{1}{2}xe^{-2x} - \frac{1}{4}e^{-2x} + C$
23. $\frac{1}{5}x^2(x-1)^5 - \frac{1}{15}x(x-1)^6 + \frac{1}{105}(x-1)^7 + C$ **24.** 0 **25.** 2 **26.** $\frac{1}{4}$ **27.** 109 items **28.** $16e^{2.5} - 16e^2 - 8 \approx \$68.70$
29. Present value $= 25{,}000(1 - e^{-0.4}) \approx \$8{,}242$; amount = \$15,018; total income $= 50{,}000(e^{0.2} - 1) \approx \$11{,}070$; interest = \$3,948
30. \$20,000 **31.** (A) \$1,000 (B) \$800 (C) $\bar{p} = 40$; CS = \$2,250; PS = \$2,700
32. 3,374 thousand barrels; 10,000 thousand barrels **33.** $.02\int_0^1 e^{-.02t}\,dt \approx .02$ **34.** 2.5 milliliters; 10 milliliters
35. $\int_1^{\infty} f(t)\,dt = \int_1^3 f(t)\,dt = \frac{1}{3}$ **36.** 45 thousand; 50 thousand **37.** $.5\int_2^{\infty} e^{-.5t}\,dt \approx .37$

Chapter 6

Exercise 6-1

1. 10 **3.** 1 **5.** 0 **7.** 1 **9.** 6 **11.** 150 **13.** 16π **15.** 791 **17.** 0.192 **19.** 118 **21.** $100e^{0.8} \approx 222.55$ **23.** $2x + \Delta x$
25. $2y^2$ **27.** E(0, 0, 3); F(2, 0, 3) **29.** \$4,400; \$6,000; \$7,100
31. $R(p, q) = -5p^2 + 6pq - 4q^2 + 200p + 300q$; $R(2, 3) = \$1{,}280$; $R(3, 2) = \$1{,}175$ **33.** \$272,615.08
35. $T(70, 47) \approx 29$ minutes; $T(60, 27) = 33$ minutes **37.** $C(6, 8) = 75$; $C(8.1, 9) = 90$ **39.** $Q(12, 10) = 120$; $Q(10, 12) \approx 83$

Exercise 6-2

1. 3 **3.** 2 **5.** $-4xy$ **7.** -6 **9.** $10xy^3$ **11.** 60 **13.** $2x - 2y + 6$ **15.** 6 **17.** -2 **19.** 2 **21.** $2e^{2x+3y}$ **23.** $6e^{2x+3y}$
25. $6e^2$ **27.** $4e^3$ **29.** $f_x(x, y) = 6x(x^2 - y^3)^2$; $f_y(x, y) = -9y^2(x^2 - y^3)^2$
31. $f_x(x, y) = 24xy(3x^2y - 1)^3$; $f_y(x, y) = 12x^2(3x^2y - 1)^3$ **33.** $f_x(x, y) = 2x/(x^2 + y^2)$; $f_y(x, y) = 2y/(x^2 + y^2)$
35. $f_x(x, y) = y^4e^{xy^2}$; $f_y(x, y) = 2xy^3e^{xy^2} + 2ye^{xy^2}$ **37.** $f_x(x, y) = 4xy^2/(x^2 + y^2)^2$; $f_y(x, y) = -4x^2y/(x^2 + y^2)^2$
39. $f_{xx}(x, y) = 2y^2 + 6x$; $f_{xy}(x, y) = 4xy = f_{yx}(x, y)$; $f_{yy}(x, y) = 2x^2$
41. $f_{xx}(x, y) = -2y/x^3$; $f_{xy}(x, y) = (-1/y^2) + (1/x^2) = f_{yx}(x, y)$; $f_{yy}(x, y) = 2x/y^3$

43. $f_{xx}(x, y) = (2y + xy^2)e^{xy}$; $f_{xy}(x, y) = (2x + x^2y)e^{xy} = f_{yx}(x, y)$; $f_{yy}(x, y) = x^3e^{xy}$ **45.** $x = 2$ and $y = 4$
47. $f_{xx}(x, y) + f_{yy}(x, y) = (2y^2 - 2x^2)/(x^2 + y^2)^2 + (2x^2 - 2y^2)/(x^2 + y^2)^2 = 0$ **49.** (A) $2x$ (B) $4y$
51. $P_x(1, 2) = 4$: Profit will increase approximately $4 thousand per 1,000 increase in production of type A calculator at the (1, 2) output level; $P_y(1, 2) = -2$: Profit will decrease approximately $2 thousand per 1,000 increase in production of type B calculator at the (1, 2) output level
53. $\partial x/\partial p = -5$: A $1 increase in the price of brand A will decrease the demand for brand A by 5 pounds at any price level (p, q); $\partial y/\partial p = 2$: A $1 increase in the price of brand A will increase the demand for brand B by 2 pounds at any price level (p, q)
55. (A) $f_x(x, y) = 7.5x^{-0.25}y^{0.25}$; $f_y(x, y) = 2.5x^{0.75}y^{-0.75}$
(B) Marginal productivity of labor $= f_x(625, 81) = 4.50$, Marginal productivity of capital $= f_y(625, 81) = 11.57$
(C) Capital
57. Competitive **59.** Complementary
61. (A) $f_w(w, h) = 6.65w^{-0.575}h^{0.725}$; $f_h(w, h) = 11.34w^{0.425}h^{-0.275}$
(B) $f_w(65, 57) = 11.31$: For a 65 lb child 57 in. tall, the rate of change in surface area is 11.31 in.^{2} for each pound gained in weight (height is held fixed); $f_h(65, 57) = 21.99$: For a 65 lb child 57 in. tall, the rate of change in surface area is 21.99 in.^{2} for each inch gained in height (weight is held fixed).
63. $C_W(6, 8) = 12.5$: Index increases approximately 12.5 units for 1 in. increase in width of the head (length held fixed) when $W = 6$ and $L = 8$; $C_L(6, 8) = -9.38$: Index decreases approximately 9.38 units for 1 in. increase in length (width held fixed) when $W = 6$ and $L = 8$

Exercise 6-3

1. $f(-2, 0) = 10$ is a local maximum **3.** $f(-1, 3) = 4$ is a local minimum **5.** f has a saddle point at $(3, -2)$
7. $f(3, 2) = 33$ is a local maximum **9.** $f(2, 2) = 8$ is a local minimum **11.** f has a saddle point at (0, 0)
13. f has a saddle point at (0, 0); $f(1, 1) = -1$ is a local minimum
15. f has a saddle point at (0, 0); $f(3, 18) = -162$ and $f(-3, -18) = -162$ are local minima
17. The test fails at (0, 0); f has saddle points at (2, 2) and $(2, -2)$
19. 2,000 type A and 4,000 type B; Max $P = P(2, 4) = \$15$ million

21. (A)

p	q	x	y
$10	$12	56	16
$11	$11	6	56

(B) A maximum weekly profit of $288 is realized for p = $10 and q = $12.

23. $P(x, y) = P(4, 2)$ **25.** 8 by 4 by 2 in. **27.** 20 by 20 by 40 in.

Exercise 6-4

1. Max $f(x, y) = f(3, 3) = 18$ **3.** Min $f(x, y) = f(3, 4) = 25$
5. Max $f(x, y) = f(3, 3) = f(-3, -3) = 18$; min $f(x, y) = f(3, -3) = f(-3, 3) = -18$
7. Maximum product is 25 when each number is 5 **9.** Min $f(x, y, z) = f(-4, 2, -6) = 56$
11. Max $f(x, y, z) = f(2, 2, 2) = 6$; min $f(x, y, z) = f(-2, -2, -2) = -6$
13. 60 of model A and 30 of model B will yield a minimum cost of $32,400 per week
15. 8,000 labor units and 1,000 capital units **17.** 8 by 8 by $\frac{8}{3}$ in. **19.** $x = 50$ ft and $y = 200$ ft; maximum area is 10,000 ft^2

Exercise 6-5

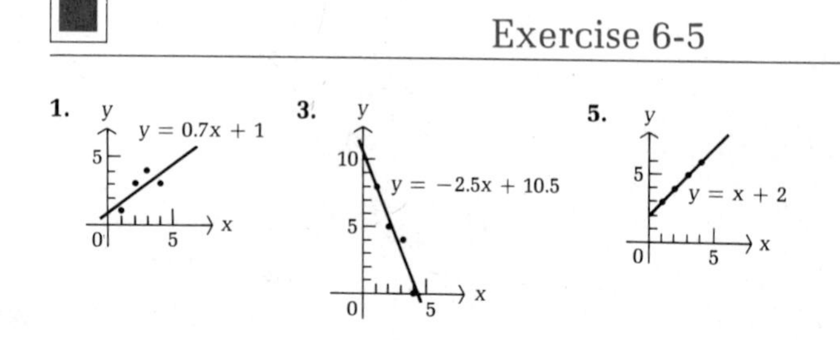

7. $y = 2.12x + 10.8$; $y = 63.8$ when $x = 25$ **9.** $y = -1.2x + 12.6$; $y = 10.2$ when $x = 2$
11. $y = -1.53x + 26.67$; $y = 14.4$ when $x = 8$
13. $y = 0.75x^2 - 3.45x + 4.75$ **15.** (A) $y = 0.382x + 1.265$ (B) \$10,815
17. (A) $y = -0.48x + 4.38$ (B) \$6.56 per bottle
19. (A) $P = -0.66T + 48.8$ (B) 11.18 beats per minute
21. (A) $D = -3.1A + 54.6$ (B) 45%

Exercise 6-6

1. (A) $3x^2y^4 + C(x)$ (B) $3x^2$ **3.** (A) $2x^2 + 6xy + 5x + E(y)$ (B) $35 + 30y$ **5.** (A) $\sqrt{y + x^2} + E(y)$ (B) $\sqrt{y+4} - \sqrt{y}$
7. 9 **9.** 330 **11.** $(56 - 20\sqrt{5})/3$ **13.** 16 **15.** 49 **17.** $\frac{1}{8}\int_1^5 \int_{-1}^1 (x+y)^2\, dy\, dx = \frac{32}{3}$
19. $\frac{1}{15}\int_1^4 \int_2^7 \frac{x}{y}\, dy\, dx = \frac{1}{2}\ln\frac{7}{2} \approx 0.626\ 4$ **21.** $\frac{4}{3}$ **23.** $\frac{32}{3}$ **25.** $\int_0^1 \int_1^2 xe^{xy}\, dy\, dx = \frac{1}{2} + \frac{1}{2}e^2 - e$
27. $\int_0^1 \int_{-1}^1 \frac{2y + 3xy^2}{1 + x^2}\, dy\, dx = \ln 2$ **29.** $\frac{1}{0.4}\int_{0.6}^{0.8} \int_5^7 \frac{y}{1 - x}\, dy\, dx = 30 \ln 2 \approx \$20.8$ billion
31. $\frac{1}{10}\int_{10}^{20} \int_1^2 x^{0.75}y^{0.25}\, dy\, dx = \frac{8}{175}(2^{1.25} - 1)(20^{1.75} - 10^{1.75}) \approx 8.375$ or 8,375 items
33. $\frac{1}{192}\int_{-8}^{8} \int_{-6}^{6} \left[10 - \frac{1}{10}(x^2 + y^2)\right] dy\, dx = \frac{20}{3}$ insects per square foot
35. $\frac{1}{8}\int_{-2}^{2} \int_{-1}^{1} [100 - 15(x^2 + y^2)]\, dy\, dx = 75$ parts per million **37.** $\frac{1}{10{,}000}\int_{2{,}000}^{3{,}000} \int_{50}^{60} 0.000\ 013\ 3xy^2\, dy\, dx \approx 100.86$ feet
39. $\frac{1}{16}\int_8^{16} \int_{10}^{12} 100\frac{x}{y}\, dy\, dx = 600 \ln 1.2 \approx 109.4$

Exercise 6-7 Chapter Review

1. $f(5, 10) = 2{,}900$; $f_x(x, y) = 40$; $f_y(x, y) = 70$ **2.** $\partial^2 z/\partial x^2 = 6xy^2$; $\partial^2 z/\partial x\, \partial y = 6x^2y$ **3.** $2xy^3 + 2y^2 + C(x)$
4. $3x^2y^2 + 4xy + E(y)$ **5.** 1 **6.** $f(2, 3) = 7$; $f_y(x, y) = -2x + 2y + 3$; $f_y(2, 3) = 5$ **7.** $(-8)(-6) - 4^2 = 32$
8. $y = -1.5x + 15.5$; $y = 0.5$ when $x = 10$ **9.** 18 **10.** $f_x(x, y) = 2xe^{x^2+2y}$; $f_y(x, y) = 2e^{x^2+2y}$; $f_{xy}(x, y) = 4xe^{x^2+2y}$
11. $f_x(x, y) = 10x(x^2 + y^2)^4$; $f_{xy}(x, y) = 80xy(x^2 + y^2)^3$
12. $f(2, 3) = -25$ is a local minimum; f has a saddle point at $(-2, 3)$ **13.** $y = \frac{116}{165}x + \frac{100}{3}$ **14.** $\frac{27}{5}$ **15.** 4 cubic units
16. (A) $P_x(1, 3) = 8$; profit will increase \$8,000 for 100 units increase in product A if production of product B is held fixed at an output level of (1, 3)
(B) For 200 units of A and 300 units of B, $P(2, 3) = \$100$ thousand is a local maximum
17. 8 by 6 by 2 in. **18.** $y = 0.63x + 1.33$; profit in sixth year is \$5.11 million
19. $\frac{1}{4}\int_{10}^{12} \int_1^3 x^{0.8}y^{0.2}\, dy\, dx \approx 7.764$ or 7,764 units
20. $T_x(70, 17) = -0.924$ minute per foot increase in depth when $V = 70$ cubic feet and $x = 17$ feet **21.** 36 ppm
22. 50,000 **23.** $y = \frac{1}{2}x + 48$; $y = 68$ when $x = 40$

Chapter 7

Exercise 7-1

1. $\pi/3$ radians **3.** $\pi/4$ radian **5.** $\pi/6$ radian **7.** II **9.** IV **11.** I **13.** 1 **15.** 0 **17.** -1 **19.** 60° **21.** 45° **23.** 30° **25.** $\sqrt{3}/2$ **27.** $\frac{1}{2}$ **29.** $-\sqrt{3}/2$ **31.** 0.1411 **33.** 0.6840 **35.** 0.7970 **37.** $3\pi/20$ radian **39.** 15° **41.** 1 **43.** 2 **45.** $1/\sqrt{3}$ or $\sqrt{3}/3$ **49.** (A) $P(13) = 5$, $P(26) = 10$, $P(39) = 5$, $P(52) = 0$ (B) $P(30) \approx 9.43$, $P(100) \approx 0.57$ **51.** (A) $V(0) = 0.10$, $V(1) = 0.45$, $V(2) = 0.80$, $V(3) = 0.45$, $V(7) = 0.45$ (B) $V(3.4) \approx 0.20$, $V(5.7) \approx 0.76$ **53.** (A) $-5.6°$ (B) $-4.7°$

Exercise 7-2

1. $-\sin t$ **3.** $3x^2 \cos x^3$ **5.** $t \cos t + \sin t$ **7.** $(\cos x)^2 - (\sin x)^2$ **9.** $5(\sin x)^4 \cos x$ **11.** $\dfrac{\cos x}{2\sqrt{\sin x}}$

13. $-\dfrac{x^{-1/2}}{2} \sin \sqrt{x} = \dfrac{-\sin \sqrt{x}}{2\sqrt{x}}$ **15.** $f'\left(\dfrac{\pi}{6}\right) = \cos \dfrac{\pi}{6} = \dfrac{\sqrt{3}}{2}$ **17.** $\dfrac{(\cos x)^2 + (\sin x)^2}{(\cos x)^2} = \dfrac{1}{(\cos x)^2} = (\sec x)^2$ **19.** $\dfrac{x \cos \sqrt{x^2-1}}{\sqrt{x^2-1}}$

21. (A) $P'(t) = \dfrac{5\pi}{26} \sin \dfrac{\pi t}{26}$, $0 \le t \le 104$

(B) $P'(8) = \$0.50$ hundred ($50 per week); $P'(26) = \$0$ per week; $P'(50) = -\$0.14$ hundred (−$14 per week)

(C)

t	$P(t)$	
26	$1,000	Local maximum
52	$0	Local minimum
78	$1,000	Local maximum

(D)

t	$P(t)$	
0	$0	Absolute minimum
26	$1,000	Absolute maximum
52	$0	Absolute minimum
78	$1,000	Absolute maximum
104	$0	Absolute minimum

23. (A) $V'(t) = \dfrac{0.35\pi}{2} \sin \dfrac{\pi t}{2}$, $0 \le t \le 8$

(B) $V'(3) = -0.55$ liter per second; $V'(4) = 0.00$ liter per second; $V'(5) = 0.55$ liter per second

(C)

t	$V(t)$	
2	0.80	Local maximum
4	0.10	Local minimum
6	0.80	Local maximum

(D)

t	$V(t)$	
0	0.10	Absolute minimum
2	0.80	Absolute maximum
4	0.10	Absolute minimum
6	0.80	Absolute maximum
8	0.10	Absolute minimum

Exercise 7-3

1. $-\cos t + C$ **3.** $\frac{1}{3}\sin 3x + C$ **5.** $\frac{1}{13}(\sin x)^{13} + C$ **7.** $-\frac{3}{4}(\cos x)^{4/3} + C$ **9.** $\frac{1}{3}\sin x^3 + C$ **11.** 1 **13.** 1
15. $\sqrt{3}/2 - \frac{1}{2} \approx 0.366$ **17.** 1.4161 **19.** 0.0678 **21.** $e^{\sin x} + C$ **23.** $\ln|\sin x| + C$ **25.** $-\ln|\cos x| + C$
27. (A) \$520 hundred or \$52,000 (B) \$106.38 hundred or \$10,638 **29.** (A) 104 tons (B) 31 tons

Exercise 7-4 Chapter Review

1. (A) $\pi/6$ (B) $\pi/4$ (C) $\pi/3$ (D) $\pi/2$ **2.** (A) -1 (B) 0 (C) 1 **3.** $-\sin m$ **4.** $\cos u$ **5.** $(2x - 2)\cos(x^2 - 2x + 1)$
6. $-\frac{1}{3}\cos 3t + C$ **7.** (A) 30° (B) 45° (C) 60° (D) 90° **8.** (A) $\frac{1}{2}$ (B) $\sqrt{2}/2$ (C) $\sqrt{3}/2$ **9.** (A) -0.6543 (B) 0.8308
10. $(x^2 - 1)\cos x + 2x \sin x$ **11.** $6(\sin x)^5 \cos x$ **12.** $(\cos x)/[3(\sin x)^{2/3}]$ **13.** $\frac{1}{2}\sin(t^2 - 1) + C$ **14.** 2 **15.** $\sqrt{3}/2$
16. -0.243 **17.** $-\sqrt{2}/2$ **18.** $\sqrt{2}$ **19.** $\pi/12$ **20.** (A) -1 (B) $-\sqrt{3}/2$ (C) $-\frac{1}{2}$ **21.** $1/(\cos u)^2 = (\sec u)^2$
22. $-2x \sin x^2 e^{\cos x^2}$ **23.** $e^{\sin x} + C$ **24.** $-\ln|\cos x| + C$ **25.** 15.2128
26. (A) $R(0) = 5$, $R(2) = 4$, $R(3) = 3$, $R(6) = 1$ (B) $R(1) = 4{,}732$, $R(22) = 4$
27. (A) $R'(t) = -\dfrac{\pi}{3}\sin\dfrac{\pi t}{6}$, $0 \le t \le 24$
(B) $R'(3) = -\$1.047$ thousand ($-\$1{,}047$ per month); $R'(10) = \$0.907$ thousand (\$907 per month); $R'(18) = \$0.000$ thousand

(C)

t	$P(t)$	
6	\$1,000	Local minimum
12	\$5,000	Local maximum
18	\$1,000	Local minimum

(D)

t	$P(t)$	
0	\$5,000	Absolute maximum
6	\$1,000	Absolute minimum
12	\$5,000	Absolute maximum
18	\$1,000	Absolute minimum
24	\$5,000	Absolute maximum

28. (A) \$72 thousand or \$72,000 (B) \$6.270 thousand or \$6,270

Appendix A

Exercise A-1

1. $2/x^9$ **3.** $3w^7/2$ **5.** $2/x^3$ **7.** $1/w^5$ **9.** 5 **11.** $1/a^6$ **13.** y^6/x^{12} **15.** x **17.** $a^2\sqrt{a}$ **19.** $3x^2\sqrt{2}$ **21.** $\sqrt{m}/m$
23. $\sqrt{6}/3$ **25.** $\sqrt{2x}/x$ **27.** 8.23×10^{10} **29.** 7.83×10^{-1} **31.** 3.4×10^{-5} **33.** 1 **35.** 10^{14} **37.** $y^6/(25x)^4$ **39.** 4×10^2
41. $4y^3/(3x)^5$ **43.** $y^{12}/(8x)^6$ **45.** $3x^4y^2z\sqrt{2y}$ **47.** $4\sqrt{3x}/x$ **49.** $\sqrt{42xy}/(7y)$ **51.** $2a\sqrt{3ab}/(3b)$ **53.** $6m^3n^3$
55. $2a\sqrt{3ab}/(3b)$ **57.** 2.4×10^{10}; 24,000,000,000 **59.** 1.1×10^4; 11,000 **61.** $t^2/(x^2y^{10})$ **63.** 4 **65.** $\sqrt{2}/2$
67. $2\sqrt{x-2}/(x-2)$

Exercise A-2

1. $6\sqrt[5]{x^3}$ **3.** $\sqrt[5]{(4xy^3)^2}$ **5.** $\sqrt{x^2+y^2}$; not $x+y$ **7.** $5x^{3/4}$ **9.** $(2x^2y)^{3/5}$ **11.** $x^{1/3}+y^{1/3}$ **13.** 5 **15.** 64 **17.** -6 **19.** Not a rational number (not even a real number) **21.** $\frac{8}{125}$ **23.** $\frac{1}{27}$ **25.** $x^{2/5}$ **27.** m **29.** $2x/y^2$ **31.** $xy^2/2$ **33.** $2/(3x^{7/12})$ **35.** $2y^2/(3x^2)$ **37.** $2mn^2\sqrt[3]{2m}$ **39.** $2m^2n\sqrt[4]{2mn^3}$ **41.** $\sqrt[3]{x^2}$ **43.** $2a^2b\sqrt[3]{4a^2b}$ **45.** $\sqrt[4]{12x^3}/2$ **47.** $\sqrt[4]{(x-3)^3}$ **49.** $x\sqrt[6]{x}$ **51.** $\sqrt[6]{x^5}/x$

Exercise A-3

1. 3 **3.** $2x^3-x^2+2x+4$ **5.** $2x^3-5x^2+6$ **7.** $6x^4-13x^3+9x^2+13x-10$ **9.** $-2x+1$ **11.** $10m-18$ **13.** $8a^2+2ab-b^2$ **15.** $9x^2-4y^2$ **17.** $16x^2-8xy+y^2$ **19.** $(x-7)(x-2)$ **21.** $(w+8)(w-5)$ **23.** $(2x-1)(x+3)$ **25.** $(x-6y)(x+2y)$ **27.** Prime **29.** $(A-6B)(A+6B)$ **31.** $(5m-4n)(5m+4n)$ **33.** $(x+5y)^2$ **35.** Prime **37.** $6(x+2)(x+6)$ **39.** $2x^2-7xy-2y^2$ **41.** $2x-18$ **43.** $2x^2-2xy-8y^2$ **45.** a^3+b^3 **47.** $2x^4-5x^3+5x^2+11x-10$ **49.** $-5x^2-4x+5$ **51.** $8m^3-12m^2n+6mn^2-n^3$ **53.** $x-3\sqrt{x}$ **55.** $m-4$ **57.** $c-d$ **59.** $x+2x^{1/2}y^{1/2}+y$ **61.** $2x-7x^{1/2}-15$ **63.** $(2x-3y)(x-2y)$ **65.** $(3m-1)(m+6)$ **67.** $2y(y-3)(y-8)$ **69.** $(3s-t)(2s+3t)$ **71.** $xy(x-3y)(x+3y)$ **73.** $(m+n)(m^2-mn+n^2)$ **75.** Prime **77.** $(m-n)(m+n)(m^2+n^2)$ **79.** $5\sqrt[3]{a}-\sqrt[4]{a}$ **81.** $6x-2x^{19/3}$ **83.** h

Exercise A-4

1. $x/2$ **3.** $(3-v)/v^2$ **5.** $a^2/2$ **7.** $(x-4)/(x-3)$ **9.** $1/(2x-1)$ **11.** $x/(x+5)$ **13.** $(x^2+8)/(8x^3)$ **15.** $2a/[(a+b)^2(a-b)]$ **17.** $(m^2-6m+7)/(m-2)$ **19.** $3/(y+3)$ **21.** $(x^2-x+1)/[2(x-9)]$ **23.** c/a **25.** $(6+\sqrt{a}-a)/(a-4)$ **27.** $\sqrt{x^2+9}+3$ **29.** $1/(\sqrt{t}+\sqrt{x})$ **31.** $1/(\sqrt{x+h}+\sqrt{x})$ **33.** $-2/(x-1)^2$ **35.** $(6x^2+6x^3)/(2-3x)^6$ or $6x^2(1+x)/(2-3x)^6$ **37.** $(2+x^{1/3})/x^{1/3}$ **39.** $-x/(x+1)^{3/4}$ **41.** $(5x-1)/(4x^{1/4})$ **43.** $(3u^2+1)/(2u^{3/2})$

Exercise A-5

1. (A) $d=3$; 14, 17 (B) Not an arithmetic progression (C) Not an arithmetic progression (D) $d=-10$; -22, -32 **3.** $a_2=11$, $a_3=15$ **5.** $a_{21}=82$, $S_{31}=1{,}922$ **7.** $S_{20}=930$ **9.** 2,400 **11.** 1,120 **13.** Use $a_1=1$ and $d=2$ in $S_n=(n/2)[2a_1+(n-1)d]$ **15.** Firm A: \$280,500; firm B: \$278,500 **17.** \$48 + \$46 + · · · + \$4 + \$2 = \$600

Exercise A-6

1. (A) $r=-2$; $a_4=-8$, $a_5=16$ (B) Not a geometric progression (C) $r=\frac{1}{2}$, $a_4=\frac{1}{4}$, $a_5=\frac{1}{8}$ (D) Not a geometric progression **3.** $a_2=-6$, $a_3=12$, $a_4=-24$ **5.** $S_7=547$ **7.** $a_{10}=199.90$ **9.** $r=1.09$ **11.** $S_{10}=1{,}242$, $S_\infty=1{,}250$ **13.** (B) $S_\infty=\frac{8}{5}=1.6$ **15.** 0.999 **17.** About \$11,670,000 **19.** \$31,027; \$251,600

Exercise A-7

1. 720 **3.** 10 **5.** 1,320 **7.** 10 **9.** 6 **11.** 1,140 **13.** 10 **15.** 6 **17.** 1 **19.** 816 **21.** $C_{4,0}a^4+C_{4,1}a^3b+C_{4,2}a^2b^2+C_{4,3}ab^3+C_{4,4}b^4=a^4+4a^3b+6a^2b^2+4ab^3+b^4$ **23.** $x^6-6x^5+15x^4-20x^3+15x^2-6x+1$ **25.** $32a^5-80a^4b+80a^3b^2-40a^2b^3+10ab^4-b^5$ **27.** $3{,}060x^{14}$ **29.** $5{,}005p^9q^6$ **31.** $264x^2y^{10}$ **33.** $C_{n,0}=\dfrac{n!}{0!n!}=1$; $C_{n,n}=\dfrac{n!}{n!0!}=1$ **35.** 1 5 10 10 5 1; 1 6 15 20 15 6 1

Index

Applications Index

■ Business & Economics

Life Sciences

Social Sciences

Designer: Janet Bollow
Cover designer: John Williams
Cover photographer: John Jensen
Interior photographer: John Drooyan
Technical artists: Vantage Art, Art by AYXA
Production coordinators: Phyllis Niklas, Janet Bollow
Typesetter: Progressive Typographers
Printer and binder: R. R. Donnelley & Sons